U0386896

主编简介

付祥胜 成都医学院第一附属医院消化内科主任医师，成都医学院临床医学院教授，医学博士，研究生导师。美国 University of Cincinnati 访问学者，四川省卫健委学术技术带头人后备人选。

在消化系疾病的诊治及内镜下微创治疗方面具有丰富的经验。主要从事消化道微生态与消化系统疾病领域研究，主持国家自然科学基金面上项目、四川省科技厅项目等十余项。主译《肠道菌群 - 对营养和健康的交互影响》专著。以第一、通讯作者在国际期刊 OncoImmunology，International Journal of Cancer 等发表 SCI 收录论文 40 余篇，担任 BioMed Research International 编委，Clinical Cancer Research, Gut Microbes 等二十余个 SCI 杂志审稿人。

主编简介

郑淑梅　中国人民解放军西部战区总医院消化内科主任医师，医学博士，硕士研究生导师。美国约翰霍普金斯大学胃肠与肝病科访问学者，现任四川省医促会消化专委会副主任委员，四川省医学会消化专委会炎症性肠病学组委员，四川省医学会消化专委会委员，四川省医学会消化内镜专委会委员及四川省医师协会消化医师分会委员。

致力于胃肠、肝胆胰等疾病的基础与临床研究二十余年。擅长炎症性肠病及相关肠道疾病、慢性肝病等消化系疾病的诊治。主持四川省卫健委、四川省中医药管理局、西南交通大学医工结合课题 3 项，获军队和四川省科技进步奖 7 项，发表论文 60 余篇，其中 SCI 论文 10 余篇。

肠道菌群与机体免疫

主　编◎付祥胜　郑淑梅

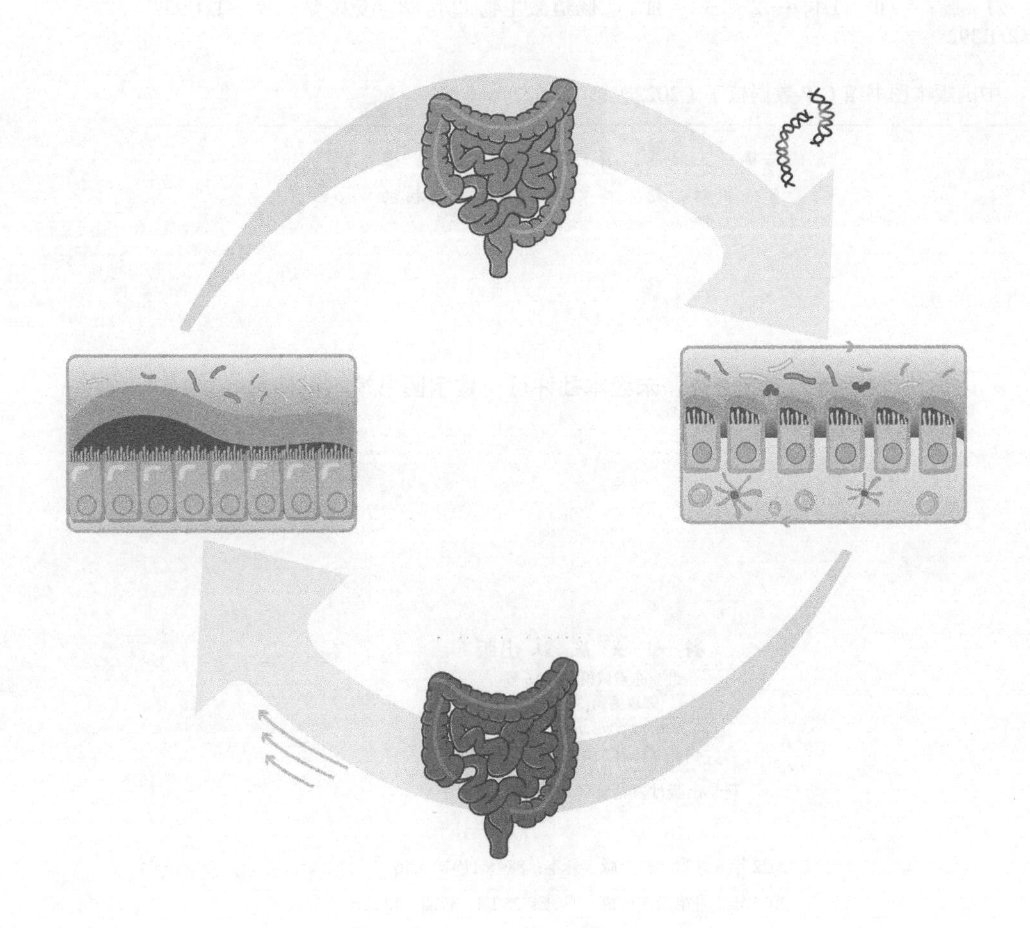

科学出版社

北　京

内 容 简 介

本书共10章，介绍了肠道微生物群与机体免疫的相关基础知识、与人类健康的密切关系、在疾病发病机制中的作用，以及在疾病诊断和治疗中的应用。重点介绍肠道微生物群与机体免疫领域国内外最新的研究进展，以及本团队的研究成果。本书绝大部分内容都有参考文献来源，并简明扼要地列举了相关实验数据，方便读者查阅。

本书适合从事消化、心血管、内分泌与代谢、神经、儿科、产科、肿瘤科等临床医师参考，也可供对此领域感兴趣的本科生、科研人员、营养专家及大众参考。

图书在版编目（CIP）数据

肠道菌群与机体免疫 / 付祥胜，郑淑梅主编 . — 北京 : 科学出版社 , 2022.8
ISBN 978-7-03-072696-4

Ⅰ. 肠… Ⅱ. ①付… ②郑… Ⅲ. ①肠道微生物 ②医学－免疫学 Ⅳ. ① Q939
② R392

中国版本图书馆 CIP 数据核字（2022）第 114622 号

责任编辑：王海燕　肖　芳 / 责任校对：张　娟
责任印制：赵　博 / 封面设计：吴朝洪

科学出版社 出版
北京东黄城根北街 16 号
邮政编码：100717
http://www.sciencep.com
三河市春园印刷有限公司印刷
科学出版社发行　各地新华书店经销
*
2022 年 8 月第　一　版　开本：889 × 1194　1/16
2025 年 1 月第四次印刷　印张：25 1/4　彩插：12
字数：633 000

定价：198.00 元

（如有印装质量问题，我社负责调换）

编著者名单

主　编　付祥胜　郑淑梅

副主编　汤小伟　陈　烨　张发明　牟　东

编　委　（以姓氏汉语拼音为序）

蔡微尘　川北医学院

陈　霞　成都医学院第一附属医院

陈　烨　南方医科大学深圳医院

陈柏君　成都医学院第一附属医院

陈桂美　川北医学院

陈科全　广州医科大学

陈兰芳　成都医学院第一附属医院

杜　超　西部战区总医院

杜鑫浩　成都医学院第一附属医院

付祥胜　成都医学院第一附属医院

侯娟妮　西部战区总医院

胡　青　川北医学院

金治安　西部战区总医院

江　丰　中国科学院大学重庆医院

黎　军　成都医学院第一附属医院

李雪梅　成都医学院第一附属医院

吕一品　西部战区总医院

牟　东　西部战区总医院

倪　阵　西部战区总医院

彭　娟　成都医学院第一附属医院

彭　惟　绵阳市中心医院
宋聆梦　成都医学院第一附属医院
覃　刚　遂宁市第一人民医院
谭晓华　深圳市第三人民医院
汤善宏　西部战区总医院
汤小伟　西南医科大学附属医院
唐　斌　中江县人民医院
唐桢桢　川北医学院
田　倩　成都医学院
王铭铭　南方医科大学深圳医院
汪岩巍　成都医学院第一附属医院
温　彪　成都医学院第一附属医院
吴维宇　成都医学院第一附属医院
伍芳丽　成都医学院第一附属医院
闫　丽　川北医学院
岳华文　川北医学院
曾祥浩　川北医学院
张　玲　川北医学院
张发明　南京医科大学
张同琴　成都医学院第一附属医院
张雪莲　成都医学院第一附属医院
张兆红　广汉市人民医院
郑　巧　川北医学院
郑淑梅　西部战区总医院
钟　瑞　成都医学院第一附属医院
钟　月　川北医学院
朱季香　遂宁市第一人民医院

前 言

人类消化道是数万亿微生物的栖息地。近年来大量重磅研究成果发现，这些微生物与人类健康和疾病息息相关。消化道不再是传统观念认为的仅参与营养的消化和吸收，其中的微生物群在代谢、内分泌、神经、免疫功能中也发挥了重要作用。胃肠道是人体内免疫细胞最集中的地方，是机体最大的免疫器官。其中数量庞大的微生物群与众多黏膜免疫细胞共生，进化出相互制衡、互惠互利的奇妙关系。对肠道菌群与机体免疫之间这种微妙关系的深入理解，正在革新人类对许多疾病发病机制的认识，并将对许多顽固性疾病的治疗策略带来革命性变化。

随着消化道微生物与机体免疫领域大量重要研究成果的不断涌现，本领域和相关学科的研究人员迫切需要了解这些重要进展。因为这些成果正在或即将改变很多基础、临床、交叉学科的理论构架，并对大众的营养、生活方式，以及疾病的预防、治疗及预后策略产生实质性的影响。

例如，饮食通过介导肠道微生物群的组成和功能变化，以及直接通过特定食物成分的抗炎特性来影响机体免疫。发酵食品可能是人类微生物 - 免疫系统轴的强大调节器，食用发酵食品使机体许多炎症指标水平下降。补充微量营养素（例如锌）和 ω-3 脂肪酸，可能有助于减轻全球性传染病（包括新冠肺炎）的负担。这些新发现可能改变人们的膳食结构，从而保持机体的最佳健康状态。

有趣的是，生命早期肠道菌群与不良环境的相互作用可能会“规划”终身健康和疾病风险。如果儿童早期缺乏寄生虫感染，则成年后对免疫性疾病明显易感。大量研究认为，过分清洁的环境对幼儿免疫系统的发育和稳态可能是有害的。这些有趣的发现为免疫系统疾病提供了新的视角，并将影响婴幼儿家庭保健和儿科临床实践。

此外，现在认为肥胖是一种低度炎症性疾病，肠道菌群已经成为导致体重增加和免疫 / 炎症系统失衡的新因素。二甲双胍在糖尿病中的治疗作用与其对肠道菌群的影响有关，这是药物说明书之外一种独特的治疗作用。我们预测，未来糖尿病的新治疗方式将直接针对肠道菌群。肠道菌群还通过一系列神经、免疫、内分泌和代谢信号通路，影响中枢神经系统的发育和功能。分析肠道菌群在神经系统疾病中的作用已成为一个有希望的研究领域，将有可能对神经系统难治性疾病如阿尔茨海默病、帕金森病提供新的有效治疗选择。

正是惊叹于肠道菌群的神奇世界及其在机体免疫发育、成熟及疾病中的重大作用，我们研究团队近年来一直专注于肠道菌群与机体免疫领域，并取得了初步的研究成果。一些前期结果已经发表在国际重要期刊上，例如 *OncoImmunology*，*International Journal of Cancer* 等。本书相关章节的部分内容来自于我们团队的研究成果，供国内同行探讨交流。

另外，我们采用了更加学术的写作方式组织本书内容。我们查阅了近年来肠道菌群与机体免疫方面的大量国内外专著、学术期刊，因此内容包含许多来自世界顶级期刊的重要新发现，例如 *Science*、*Nature* 等，因而本书内容新颖、权威。书中绝大部分内容都有参考文献来源，并简明扼要地列举了相关实验数据，以方便本领域专家、学者查阅、甄别与借鉴。同时，各章节也对相关基础理论知识做了介绍，适合对此领域感兴趣的本科生、研究生作为入门参考。

本书还重点介绍了肠道菌群与免疫在人类疾病发病机制和治疗干预中的作用，很多发现都具有革命性的意义，适合消化、心血管、内分泌与代谢、神经、儿科、产科、肿瘤科等临床医师参考。此外，本书介绍了肠道菌群与营养、健康之间有趣的研究成果，可供营养专家及大众参考。

需要指出的是，消化道微生物群构成非常复杂，包括了从口腔到直肠整个消化管道，甚至消化道附属腺等多个部位的上千种微生物，这些微生物包括细菌、古细菌、病毒及真菌。目前研究较为充分的是大肠细菌，因此本书名称及相关内容着重描述了“肠道菌群”。随着研究的进一步深入，特别是近年来对肠道病毒组的初步研究，以及对肠道真菌作用的逐步认识，相信人类将最终揭开肠道微生物群的神秘面纱，为维护人类健康开辟一个新纪元。

最后，感谢国家自然科学基金（项目编号：81972315）对本书出版的资助。书中若有不妥之处，敬请各位同仁批评指正。

成都医学院第一附属医院消化内科主任医师、教授　付祥胜

目录

第1章
肠道微生物群概述

第一节　对肠道微生物群的再认识

一、消化道不仅发挥“消化”功能

消化道包括口腔、咽、食管、胃、肠道、肛门，本书以肠道作为主体，探讨消化道的复杂功能。作为人体最大的器官，肠道主要包括小肠和大肠两个部分，其主要器官功能是进行食物的消化和吸收。此外，随着人们对肠道功能的进一步发掘，肠道在免疫、代谢、神经调节等方面的作用也逐渐得到了广泛的认同。其中，肠道微生物在肠道诸多功能的正常运转中发挥了至关重要的作用。

二、对肠道微生物群复杂性的再认识

“人类微生物群”是生活在人体内和体表的所有微生物的总称[1]。人类微生物存在于皮肤、口腔和胃肠道、呼吸道和泌尿生殖道，占人体总体重的1%～3%。其中，胃肠道是人体微生物最丰富的部位，人体肠道生存着超过100万亿个微生物，形成了一个独特的基因组，这些微生物随年龄、饮食、地理位置及营养状况变化。肠道微生物包括几种不同的生物群，包括细菌、病毒、真菌、古细菌等，迄今研究最为广泛的生物群是细菌，因此本书主要阐述肠道菌群[2, 3]。肠道内的大部分细菌定植于人体结肠内，每克肠内容物的细菌含量高达10^{12}CFU。在同一个体中，由于肠道各部位pH、氧化还原电位和食物转运时间上的差异，导致微生物在肠道不同部位的分布也有所差异。此外，由于局部、个人、遗传和随机因素的差异，肠道菌群的组成在个人和种群之间也存在很大的不同。

探讨肠道微生物群落组成的研究最早是采用培养法，通过这种方法，人们发现了可以在成年人类肠道中定居的多种不同的微生物物种。但是这种方法对于微肠道微生物群分布的研究具有明显的局限性，因为它只能用于研究个别可培养的细菌[2]。此后，随着16S rRNA测序法的快速发展，可满足一次对几十万至几百万DNA分子的测序需求，极大地促进了对复杂多样的肠道菌群的深入研究。高通量测序技术的发展使得人们对复杂生态系统的认识得到了进一步提高。

三、对肠道微生物群功能的再认识

人类肠道微生物与人类共同进化，两者密不可分，肠道微生物在人体的健康和疾病方面都发挥着重要作用。当肠道微生物群与人体和谐相处时，宿主和微生物可以彼此协调以维持体内平衡，促进人体健康。肠道微生物对人

体的正面作用，主要包括代谢、营养和保护三个方面。这种作用可以直接通过微生物基因的表达，为人类宿主提供其自身基因组所缺乏的一些代谢能力，也可以间接地通过与人体生理的相互作用，特别是与免疫系统的相互作用来实现。

利用微生物来调节和改善人体健康状况具有极大的可行性，目前在临床上，益生菌和粪菌移植（fecal microbiota transplantation，FMT）对肠道菌群的调节作用和部分疾病的防治作用已得到了一定的肯定[4, 5]。反之，若肠道微生物群之间或宿主和常驻微生物群之间的平衡被打破，则会触发遗传易感个体的免疫炎症反应，可能引发肥胖症、糖尿病、心血管动脉粥样硬化、结直肠癌等多种疾病。

人体代谢是由宿主自身基因组调节的各种代谢途径及微生物基因组调节的代谢过程共同完成的，这种宿主与微生物之间的共代谢过程最终调节着宿主的整体代谢。因此研究肠道菌群与宿主整体代谢及其与慢性代谢性疾病的相关性，寻找影响宿主代谢变化的重要功能成员，能够更好地理解肠道菌群对人体健康和疾病的重要作用。然而肠道菌群的复杂性使得确定每种细菌详细的代谢功能十分困难，因此研究者们选择运用气相色谱 - 质谱法来分析生物样本中微生物代谢产物，并且使用微生物的代谢模型来模拟特定微生物群中物种之间的代谢情况。

下面将从肠道菌群的定义、分类、与宿主的关系、研究方法、组成、其代谢产物以及可行的研究方法等几个方面介绍人体肠道菌群。

第二节　什么是人体肠道菌群

一、人体肠道简述

肠道是指从胃幽门至肛门的消化管，在解剖学上主要包括小肠（包括十二指肠、空肠、回肠）、大肠（包括盲肠、升结肠、横结肠、降结肠、乙状结肠、直肠）两个部分。小肠长 5 ～ 6m，是消化道中最长的一段，位于腹中，上端接幽门与胃相通，下端通过回盲瓣在右下腹与盲肠连接，是食物消化吸收的主要场所。食物经过小肠内胰液、胆汁和小肠液的化学性消化及小肠运动的机械性消化后，基本完成了消化过程，同时营养物质（特别是脂质和简单碳水化合物）大部分都被小肠黏膜吸收[6]。小肠液的成分比较复杂，主要含有多种消化酶、脱落的肠上皮细胞及微生物等。所含有的各种消化酶中，有肠激活酶、淀粉酶、肽酶、脂肪酶及蔗糖酶、麦芽糖酶和乳糖酶等，这些酶对于将各种营养成分进一步分解为最终可吸收的产物具有重要作用。由回肠末端突入盲肠而形成的上、下两个半月形的瓣称回盲瓣，位于回肠和结肠的交界处，有阻止小肠内容物过快流入大肠和防止盲肠内容物逆流到回肠的作用[7]。结肠在右髂窝内续于盲肠，分为升结肠、横结肠、降结肠和乙状结肠 4 个部分。结肠是水分持续再吸收、摄取微生物衍生的维生素和粪便成形的场所。直肠是结肠的肌肉末端部分，肛门是消化道的末端，也是粪便排出的部位[8]。

作为人体最大的器官，肠道具有诸多重要的生理功能：①人体肠道的主要器官功能是进行食物的消化、营养吸收、分泌和运动，此外肠道还在这种消化环境中建立了一个保护性的上皮屏障，可以在一定程度上抵御病原微生物的入侵[9, 10]。②肠道可以通过代谢药物调节全身生理[11]。③通过门静脉血流与肝脏和胰腺等其他器官进行交流[12, 13]。④人体肠道含有一个独立的肠神经系统。结肠由肠系膜上、下神经丛支配，它们所含的交感神经纤维来自腰交感神经节，分布于全部结肠。迷走神经纤维仅分布于结肠脾曲以上的结肠，降结肠和乙状结肠则由骶 2 ～ 4 脊髓节的副交感神经分布。肠神经系统可以与中枢神经系统关联，构成肠 - 脑轴的一部分[14, 15]。⑤同时，肠道也是肠道微生物群中的共生微生物与肠道淋巴组织和宿主免疫系统生活及相互作用的主要场所，这对肠内稳态有重要作用[16, 17]。

这些功能需要极为复杂的生物系统的交织。以前的研究主要采用药理学、病理生理学、生化和遗传学方法来描述肠道器官中的细胞群体和类型。最近在单细胞转录分析方面的技术进步使人们能够更精确、全面地描述许多器官的细胞类型和亚群[18, 19]。

二、肠道菌群定义

如前文所述，肠道微生物包括细菌、病毒、真菌、古细菌等不同的微生物群，而我们主要对目前研究最为广泛的肠道细菌进行阐述。在人类漫长的生存过程中，细菌是从何时开始定植于肠道中的呢？一些学者认为，婴儿在子宫内就开始接触母亲的粪便细菌或细菌成分，有许多研究都在胎粪样本中检测出了细菌 DNA[20]。另有学者对此持有不同观点，他们认为在母亲子宫内的胎儿肠道处于无菌状态，随着婴儿出生时与母体产道的接触及外界环境因子的影响，细菌才开始在肠道内迅速定植。Katherine 等认为此前大多数研究中用于分析的胎粪均取自婴儿出生后，他们在 2021 年的最新研究中利用直肠试纸进行出生前胎粪的收集，通过 16S rRNA 基因测序证实出生前的胎粪中无细菌定植[21]。因此，关于人类肠道细菌的初始定植，学者们众说纷纭，目前尚存在争议。在婴儿刚出生的一段时间内，兼性厌氧菌在肠道中占据优势地位，约在 3 岁，人类厌氧肠道环境开始建立，严格厌氧细菌，如梭状杆菌（*Clostridium*）、拟杆菌（*Bacteroides*）、双歧杆菌（*Bifidobacteria*）等开始在肠道定植，导致肠道微生物的日益多样化和复杂化，此后微生物多样性逐渐增加；5 岁儿童的肠道菌群就基本能达到成人水平，但仍存在一些重要差异，如甲烷短杆菌（*Methanobrevibacter*）、克里斯滕森菌(*Christensenellaceae*)等细菌丰度低于成人，而活泼瘤胃球菌（*Ruminococcusgnavus*）等细菌丰度则高于成人；在青春期和成年时期人体肠道菌群达到相对稳定，成人肠道内的严格厌氧菌占主要地位，其数量远超兼性厌氧菌和需氧菌[22, 23]。此外，有研究指出，母乳喂养在人体早期微生物群的形成中起着重要作用，与配方奶粉喂养相比，母乳喂养的婴儿肠道中的双歧杆菌（*Bifdobacterium*）和拟杆菌（*Bacteroides*）占主导地位[24, 25]。肠道菌群的多样性很大程度上是在生命的最初几年里建立起来的，此后则受环境因素，如年龄、饮食、生活方式、卫生条件和疾病状态的影响。通过研究不同人体肠道组织和粪便中菌群的组成和结构，可以发现不同人体肠道菌群的多样性存在明显差异[26, 27]。最新公布的人类肠道菌群基因组图谱通过从 11 850 个人类肠道微生物群落中重建基因组，确认了 1952 种人体肠道细菌。这一研究扩大了人类肠道微生物群的已知物种储备，使得系统发育多样性增加了 281%，并且使非洲和南美洲人群的肠道菌群分类提高了 200% 以上[28]。

人类胃肠道是一个复杂的系统，始于食管，止于肛门，但出于标本采集的实际考虑，迄今获得的绝大多数微生物群数据均来自于远端结肠。重要生理条件，如 pH、胆汁含量等，会随着消化道的不同部位而发生变化，这就造成了上、下消化道的微生物群落组成的差异。在本节中，我们将对口腔、食管、胃以及小肠和结肠中健康细菌群落的组成进行描述（图 1-1）。

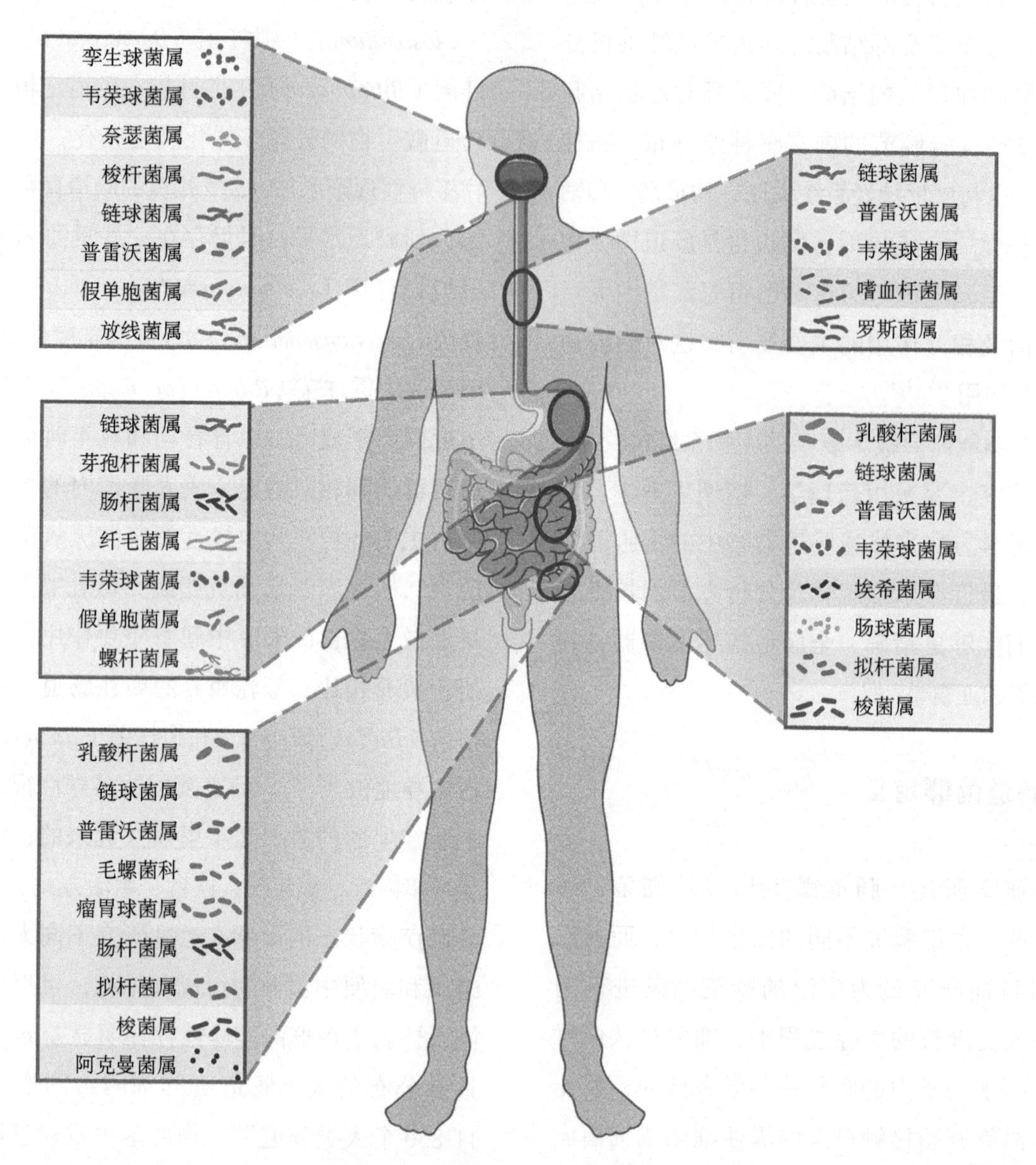

图 1-1　人类胃肠道微生物群分布

人类胃肠道不同部位具有不同的微生物群组成。口腔中的优势菌群为孪生球菌属、韦荣球菌属、奈瑟菌属、梭杆菌属、链球菌属、普雷沃菌属、假单胞菌属、放线菌属；食管的优势菌群为链球菌属、普雷沃菌属、韦荣球菌属、嗜血杆菌属、罗斯菌属；胃内优势菌群为链球菌属、芽胞杆菌属、肠杆菌属、纤毛菌属、韦荣球菌属、假单胞菌属、螺杆菌属；小肠中的优势菌群为乳酸杆菌属、链球菌属、普雷沃菌属、韦荣球菌属、埃希菌属、肠球菌属、拟杆菌属、梭菌属；大肠中优势菌群为乳酸杆菌属、链球菌属、普雷沃菌属、毛螺菌科、瘤胃球菌属、肠杆菌属、拟杆菌属、梭菌属、阿克曼菌属

三、肠道微生物群的分类

（一）消化道不同部位的微生物群组成

1. 口腔　口腔是消化道的起始部分，由两唇、两颊、硬腭、软腭等构成，腔内有牙齿、舌、唾液腺等器官，其微生物丰富度仅次于肠道。这些丰富的口腔微生物与人体健康息息相关，它们与2型糖尿病、类风湿关节炎、神经系统及心血管系统的多种疾病具有密切联系[29, 30]。2008年美国波士顿福赛斯研究所和伦敦国王学院的科学家联手创立了人类口腔微生物组数据库（Human Oral Microbiome Database，HOMD），对口腔

细菌的类型、新陈代谢、致病能力等进行了详细记录[31]。口腔中的优势菌属为孪生球菌属（*Gemella*）、韦荣球菌属（*Veillonella*）、奈瑟菌属（*Neisseria*）、梭杆菌属（*Fusobacterium*）、链球菌属（*Streptococcus*）、普雷沃菌属（*Prevotella*）、假单胞菌属（*Pseudomonas*）、放线菌属（*Actinomyces*）[32]。

2. 食管　食管是咽和胃之间的消化管，过去常被认为没有独特的细菌群。究其原因，一是人们认为任何存在于食管的细菌大多都是从口咽或胃移位而来的，二是食管微生物组在取样和培养方面具有挑战性，在一定程度上限制了人们对食管菌群的研究。近年来，随着食管组织活检等技术的普及，人们对于食管菌群有了全新的认识，有研究揭示了健康人群中 3 种不同的细菌群落类型：链球菌属（包括轻型链球菌、口腔链球菌、肺炎链球菌）为主型、普雷沃菌属（产黑色素普雷沃菌和苍白普雷沃菌）和韦荣球菌属为主型，以及嗜血杆菌属（*Haemophilus*）和罗氏菌属（*Rothia*）为主型[32]。同时，食管菌群与人体健康与疾病有密切联系，包括胃食管反流病、Barrett 食管、食管癌、嗜酸性食管炎等疾病都被证明与食管菌群紊乱相关[33]。

3. 胃　胃是人体非常重要的一个消化器官，它含有蛋白水解酶和胃酸，能够对摄入体内的食物进行消化。由于其特殊的酸性环境，许多细菌的生长都会受到抑制。幽门螺杆菌（*Helicobacter pylori*，*H.pylori*）是人类最知名的致病性胃菌群成员，在没有幽门螺杆菌感染的情况下，胃微生物群与口腔菌群有很高的相似性；有幽门螺杆菌感染的个体其胃内微生物的 α 多样性（Alpha 多样性：指一个特定区域或生态系统内的多样性，是反映丰富度和均匀度的综合指标）会显著降低[34]。正常成人胃内的优势菌属包括链球菌属、杆菌属、肠杆菌属、纤毛菌属、韦荣球菌属、假单胞菌属和螺杆菌属[32]。

4. 小肠　小肠位于腹中，上衔幽门，下接盲肠，其功能主要包括消化、吸收、分泌和运动。由于肠道蠕动速度快、肠分泌杀菌物质导致小肠菌群数量较少（10^3 ～ 10^5CFU/ml），而小肠中细菌异常增多的现象称为小肠细菌过度生长，这一现象往往会导致一系列的胃肠道症状[35]。小肠菌群易受环境、饮食、时间影响，随着小肠部位的变化会出现独特的微生物组成：十二指肠中以需氧菌和兼性厌氧菌为主，如链球菌属（*Streptococcus*）、普雷沃菌属（*Prevotella*）、韦荣球菌属（*Veillonella*）、梭形杆菌属（*Fusobacterium*）、埃希菌属（*Escherichia*）、克雷伯杆菌属（*Klebsiella*）和柠檬酸杆菌属（Citrobacter）；空肠和回肠中最常见的细菌群落包括链球菌、乳酸菌、伽马变形菌（*Gamma proteobacteria*）、肠球菌（*Enterococcus*）和拟杆菌（*Bacteroides*）；十二指肠含有与胃相似的菌属，如链球菌、肠球菌、韦荣球菌、假单胞菌等，而回肠末端的微生物成分更接近于结肠，如梭菌科（Clostridiaceae）、毛螺菌科（Lachnospiraceae）、消化链球菌科（Peptostreptococcaceae）、瘤胃菌科（Ruminococcaceae）、肠杆菌科（Enterobacteriaceae）、拟杆菌科（Bacteroidaceae）[32]。

5. 大肠　大肠由盲肠、升结肠、横结肠、降结肠、乙状结肠及直肠组成，肠壁由内向外依次为黏膜、黏膜下层，肌层及浆膜层。黏膜和部分黏膜下层向肠腔内突起成半环形皱襞，肌层局部的膨大为结肠袋。它是人体微生物丰度最高的部位，共含有约 38 万亿个细菌[36]。拟杆菌门（Bacteroidetes）、厚壁菌门（Firmicutes）是健康人体大肠中最主要的两大门类，其比例占 90% 以上，其次是疣微菌门（Verrucomicrobia）、变形菌门（Proteobacteria）和放线菌门（Actinobacteria）[32]。结肠细菌通常是沿着结肠横轴定植的，可吸收淀粉和营养成分的微生物在结肠腔内处于优势地位，如阿克曼菌（*Akkermansia*）、瘤胃球菌属（*Ruminococcus*）

和一些拟杆菌类等[37-39]。此外，氧梯度随着肠腔到黏膜及结肠纵轴的位置不同而存在差异，从而影响结肠的微生物组成[40]。

结肠的主要功能是吸收水分和电解质，形成、储存和排泄粪便。结肠细菌可对食物残渣中的碳水化合物和肠道上皮细胞分泌的糖蛋白进行发酵分解，产生可被人体利用的乙酸、丙酸、丁酸等短链脂肪酸（short-chain fatty acid，SCFA），促使肠道 pH 下降，有利于钙、铁、维生素 D 的吸收。其中丁酸为结肠细胞的主要能量来源，并且具有预防和抑制结肠癌、抗炎、抗氧化、修复肠黏膜防御功能、调节内脏敏感度和肠道运动功能的作用[41]。肠道菌群还可以促进肠道发育，比如促进肠道血管新生及肠道免疫系统成熟等[42]。此外，肠黏膜及其定植的微生物还是机体抵御病原体入侵最重要的屏障之一，肠道内定植的共生菌能够通过与肠源性病原菌竞争肠上皮的黏附位点而抵御病原菌侵入肠上皮[43]。

近来的微生物全基因组研究表明，结肠微生物群改变导致的机体代谢和免疫紊乱与多种人类疾病密切相关[44]。其中较为典型的疾病是炎性肠病，包括溃疡性结肠炎和克罗恩病。在世界范围内，其发病率均呈明显增加的趋势，其特点是免疫激活和肠上皮黏膜屏障功能紊乱，导致对共生细菌的异常免疫激活。最新研究表明，胆型螺杆菌（Helicobacter_bilis，H. bilis）可能通过诱导宿主免疫紊乱，继而参与炎性肠病的发生[45]。全基因组关联研究已经鉴定了超过 200 个人类炎性肠病相关敏感基因，其中一些基因参与介导宿主对肠道微生物的反应（如 NOD2、ATG16L1、IRGM 等易感位点），这就为肠道微生物参与炎性肠病的发病提供了依据[46]。此外，在一些全身性疾病的发病过程中，也常能发现肠道细菌的踪影，如 2 型糖尿病患者的肠道菌群与血糖正常者相比，其双歧杆菌等有益菌减少，乳酸杆菌、肠球菌、肠杆菌等增多。因此，结肠微生物群在维持肠道上皮稳态及人类健康方面起着关键作用[47]。

（二）肠道微生物群的不同分类

广义的肠道微生物群包括细菌、古细菌、真菌和病毒等微生物集群。已经在肠道内检测到的真菌包括念珠菌属、酵母菌属、曲霉菌属、青霉菌属、红酵母属、栓菌属、格孢腔菌属、核盘菌属、布勒掷孢酵母属、耐碱酵母属等[48, 49]。古细菌如广古生菌门、泉古菌门等构成另一大类肠道菌群，它们在发酵细菌产物的代谢中发挥重要作用。此外，肠道细菌因其种群多样性和高丰度使其成为肠道核心微生物，根据不同的分类方法，可分为不同的类别。

1. 依据细菌的天然属性分类　已经鉴定的 400 余种肠道细菌分属 11 个门，包括厚壁菌门、拟杆菌门、变形菌门、放线菌门、疣微菌门、梭杆菌门、蓝藻菌门、螺旋菌门、VadinBE97 门、异常球菌 - 栖热菌门、脱铁杆菌门[50, 51]。绝大多数肠道细菌集中归属于前四类，根据菌群数量或种群密集度又分为主要（优势）菌群和次要菌群。

（1）主要（优势）菌群：指菌群数量或种群密度大的细菌，通常是原籍菌。16sRNA 测序研究表明，人体肠道中数量上占优势的微生物有两个分支，即拟杆菌门和厚壁菌门（Firmicutes）。Eckburg 等对 11 831 条细菌近全长非嵌合的 16S rRNA 序列进行系统性分析，得到了 395 种细菌类型。其中 93% 是厚壁菌或 CFB 菌群[51]。有趣的是，该研究获得的 1524 个古细菌序列都属于史氏甲烷短杆菌（methanobrevibacter smithii）。这种古生菌菌株细胞形态为短杆状，偶见链状细胞，是从人体分离出来的一类数量上占主导地位的产甲烷菌落集群[52]。

胃肠道中发现的大多数厚壁菌门可分为两大类，即拟球梭状芽孢杆菌和柔嫩梭菌群。拟球梭状芽孢杆菌群（又称梭菌属 rRNA 亚群 XIVa）由梭状芽孢杆菌属、真细菌属、瘤胃球菌属、粪球菌属、毛螺菌属、罗氏菌属和丁酸弧菌属的细菌组成[53]。一些分析胃肠道菌群多样性的研究指出，拟球梭菌属通常占肠道细菌总量的 11% ～ 43%，其中许多细菌至今仍不能单独培养[54, 55]。类球梭菌属含有大量的产丁酸盐细菌，如罗斯拜瑞弧菌，该菌被认为有助于肠道健康，因为丁酸盐是结肠上皮细胞的首选能量来源[56]。从人类粪便中随机分离的产丁酸盐的厌氧菌中，80% 属于这一组[57]。厚壁菌门的第二类优势菌群是柔嫩梭菌群。该类群包括梭菌属、优杆菌属、瘤胃球菌属和厌氧细杆菌属，又称为梭菌属 rRNA 亚群Ⅳ，由高度氧敏感的厌氧菌组成，含有大量产丁酸盐的纤维降解菌[57]。Lay 等采用荧光原位杂交（fluorescence in situ hybridization，FISH）技术分析了人类粪便微生物群中柔嫩梭菌的组成，发现 22% 的粪菌都归属于柔嫩梭菌[58]。无独有偶，Saunier 等对粪菌的组成分析获得了类似的结果，柔嫩梭菌占全部粪菌总量的 19% ± 7%[59]。Lay 等还指出，在柔嫩梭菌群中，普拉梭菌占 64%，其次是布氏瘤胃球菌（12%）、生黄瘤胃球菌（1.8%）和灵巧瘤胃球菌（1.4%）。尽管普拉梭菌（旧称普氏栖粪杆菌）最初的命名与梭菌属没有密切的关系，但其序列中 G+C 含量达到 47% ～ 57%[60]。寡核苷酸探针分析表明，普拉梭菌相关菌株是人类粪便中最丰富的菌株之一[61]。采用纯培养微生物群落解析技术的其他研究也发现，普拉梭菌在人类粪菌群落中很常见，这种系统发育型序列在 16sRNA 克隆文库占 3.8% ～ 10%[62]。在迄今为止报道的最全面的研究中，普拉梭菌的种系型在 11 831 条 16sRNA 序列占 1556 条[51]。在地理上相互独立的 5 个不同的国家中，普拉梭菌被认为是柔嫩梭菌属中含量最丰富的成员，占该菌群总数的 13% ～ 17.6%[58]。总之，这些研究结果表明，这种菌群可能对胃肠道生态做出重要贡献，尤其是在厌氧增殖过程中，普拉梭菌样细菌能够产生＞ 10 mM 的丁酸盐[63]。

拟杆菌属是革兰阴性专性厌氧菌，G+C 组成为 40% ～ 48%。由 Eckburg 等鉴定的普通类杆菌占 31%，多形类杆菌占 12%，吉氏类杆菌占 0.8%，它们与脆弱拟杆菌一起是该群落最常见的菌种[51]。Matsuki 等通过聚合酶链反应（polymerase chain reaction，PCR）检测粪便中的主要可培养菌，发现大多数分离菌株属于脆弱拟杆菌（117/300）[64]。这一高比例可能反映了脆弱拟杆菌群体相对较高的耐氧性和易培养性。普雷沃菌属通常存在于口腔，其在胃肠道中的存在是可变的。此后，Matsuki 等在 46% 的志愿者肠道粪便中发现了普雷沃菌(21/46)[65]。此外，Eckburg 等还发现，CFB 菌群中拟杆菌属分支在宿主之间的分布变异最大，99.8% 的普氏菌序列是从受检者 B 分离出来的，75% 的普通拟杆菌与受检者 A 相关。

无论采用哪种培养方法，胃肠道生态学研究均表明，厚壁菌门（又分为拟球梭菌属、柔嫩梭菌属）和 CFB 菌群在人类粪菌群落中占主导地位，这三类菌群的分布情况见表 1-1。

表 1-1　不同分子技术测定的三类主要菌群在人类粪便中的分布

方法	拟球梭菌属[a]	柔嫩梭菌属[b]	拟杆菌属	参考文献
rRNA library	23.7% ～ 58.8%	11.0% ～ 22.7%	5.0% ～ 16.3%	Hayashi, et al [62]
rRNA library	44%	20%	30%	Suau, et al [61]
rRNA library[c]	43.3% ～ 48.7%	10.8% ～ 17.9%	20.5% ～ 35.1%	Hold, et al [55]
Dot Blot	14% ± 6%	16% ± 7%	37% ± 16%	Sghir, et al [66]
Dot Blot	22.8% ± 2.2%	13.0% ± 0.78%	8.0% ± 0.32%	Marteau, et al [67]

续表

方法	拟球梭菌属[a]	柔嫩梭菌属[b]	拟杆菌属	参考文献
TRAC[d]	42%～43%	9%～12%	NT[e]	Maukonen, et al [68]
FISH	16.9%	NT	NT	Zoetendal, et al [54]
FISH	29%	NT	20%[f]	Franks, et al [69]
FISH/ 流式细胞术	28% ± 11.3%	25.2% ± 7.6%	8.5% ± 7.1%	Lay, et al [70]
FISH/ 流式细胞术	12.7%～29.7%	NT	3.2%～16.8%	Mueller ,et al [71]

注：a. 指类球梭菌群（梭菌属 rRNA 亚群 XIVa）；b. 指柔嫩梭菌群（梭菌属 rRNA 亚群 IV）；c. 结果来自人结肠组织样本；d. 亲和捕获辅助转录分析，一种通过与寡核苷酸探针杂交，然后采用亲和捕获来定量分析细菌 16S rRNA 的技术 86；e. 未检测；f. 采用种特异性探针，这个数值仅代表脆弱类杆菌和吉氏类杆菌

基于公共数据集—人类微生物组计划，欧洲肠道宏基因组计划（MetaHIT）和众多其他的肠道菌群研究数据汇集而成的 gutMEGA（http://gutmega.omicsbio.info/）、GMrepo（https://gmrepo.humangut.info）、人类虚拟代谢数据库[VMH（http://www.vmh.life）]，都证明肠道菌群具有相似的局部子结构，即群落结构中总是存在上述拟杆菌属（*Bacteroides*）、普雷沃菌属（*Prevotella*）和厚壁菌门（Firmicutes）为主导的集群，且前两者是个体间最主要的差异类群[72]，因此有研究者将个体间不同的优势菌群群落组成的结构模式定义为“肠型”（enterotype），作为一种有效区分人体肠道微生物的方法[73]。Holmes 等采用多项式混合模型对主要用于肠型分类的 3 种肠道菌群核心物种进行分析，发现可能存在 4 种肠型，其中两个集群类似 ET B（拟杆菌是最好的指示类群）和 ET P（由普雷沃菌驱动，其丰度常与拟杆菌的丰度成反比），第三个集群 ET F（通过厚壁菌占比高低来区分，其优势类群为瘤胃球菌），最后一个集群则含有大量未鉴定的类群[74]。但由于聚类分析的本质特征，以及对分类级别、距离指标、聚类算法和集群优化得分等方面的多种选择，这一概念从产生之初就倍受争议。尽管肠型可能不是对数据的最佳解释，但它可为我们研究肠道菌群提供直接框架，用于功能和生态学研究及潜在临床应用。

（2）次要菌群：主要为需氧菌或兼性厌氧菌，大部分属于外来菌群或过路菌群。最初采用培养的方法估计双歧杆菌属约占微生物群落的 10%，但由于引用了纯培养方法，现在估计这一数值要低得多。双歧杆菌是革兰阳性杆菌，1899 年首次从母乳喂养的婴儿粪便中分离得到[75]。此后发现它们的存在与肠道微生态的健康密切相关，且被添加在许多保健食品制剂中。Langendijk 等检测了人类粪便中双歧杆菌的阳性率，并与 DAPI 总细胞计数进行比较，发现双歧杆菌占总人群的 0.8% ± 0.4%。使用纯培养和混合培养技术发现的双歧杆菌数量大致相同，表明粪便中的大多数双歧杆菌是可培养的[76]。基于这些新发现，有学者认为双歧杆菌的丰度被高估了 10 倍，其他研究者也证实了该结论[77]。Matsuki 等通过定量 PCR 发现，每克粪便中的双歧杆菌总数平均为 $\log_{10} 9.4 \pm 0.7$，最常见的双歧杆菌分离株是青春双歧杆菌、链状双歧杆菌和长双歧杆菌[78]。

胃肠道中的第二类次要菌群是乳酸菌（LAB）。顾名思义，这类细菌产生乳酸作为发酵的终产物。典型 LAB 如干酪乳杆菌、乳球菌、肠球菌、链球菌和明串珠菌属于低 G+C 含量的革兰阳性菌。乳杆菌 - 肠球菌群落以前曾使用纯培养技术，估计占成人粪菌的 2%。Harmsen 等采用一种肠球菌乳杆菌 FISH 探针进行检测，将这个数值精确到总微生物群的

0.01%，这表明由于这些细菌的可培养性，早期的估值被夸大了[79]。胃肠道中发现的其他细菌亚分支包括变形菌、疣微菌、梭杆菌、放线菌、蓝藻菌、螺旋菌、异常球菌 - 栖热菌、脱铁杆菌和 VadinBE97。

2. 按照与宿主的关系分类　肠道菌群可分为共生菌、条件致病菌和致病菌。共生菌长期寄居于肠道内组成相对稳定的微生态，占肠道细菌数量的 99%，在进化过程中承受个体适应和自然选择压力，与宿主相互依存、制约，是机体不可分割的一部分。条件致病菌是一定条件下可致病的细菌。这类细菌在肠道内由于大量共生菌的存在，通常不容易大量繁殖造成危害，常见的条件致病菌有肠球菌和肠杆菌。致病菌可导致疾病，一般不常驻于肠道内，多从外界摄入后在肠道内大量增殖导致疾病的发生。常见的致病菌有沙门菌和致病性大肠埃希菌等。

3. 根据对氧气的需求分类　肠道菌群可以分为专性厌氧菌、兼性厌氧菌和需氧菌。以厌氧菌居多，共生菌一般都是专性厌氧菌。

4. 根据肠道的不同部位分类　即肠道微生物群的空间异质性，是肠道微生物群生物地理学研究的范畴。由于人体胃肠道内的生理状态不同，不同的微生物群落定植在不同的人体部位，因此菌群的种类、丰度分布存在胃肠道横向延续部位上的差异。如上消化道内细菌数量为 10^2 ～ 10^4CFU/ml；回肠内细菌数量为 10^6 ～ 10^{12}CFU/ml，而远端结肠内细菌数量最多，为 10^{10} ～ 10^{12}CFU/ml。这是由于小肠转运速度更快，单糖和氨基酸的代谢更有利，群落主要为快速分裂的兼性厌氧菌，如变形菌门和乳酸杆菌属；相比之下大肠的代谢更慢，代谢更利于从未消化的植物纤维或宿主黏液中提取复杂多糖发酵，导致该处物种丰富度远多于小肠，形成分解糖的拟杆菌属和梭状芽孢杆菌属的优势群落。从肠腔内到肠黏膜上皮纵向来看，肠道微生物群形成 3 个生物层：底层为生物膜菌群（膜菌群又称为原籍菌，在出生后数天就能贴近黏膜表面定植，是长期居住的细菌），菌群紧贴黏膜表面并黏附于黏膜上皮细胞，主要为双歧杆菌和乳酸杆菌，为肠道共生菌；中层为链球菌、粪杆菌、韦荣球菌和优杆菌等厌氧菌；表层为腔菌群，可在肠腔内游动，主要为大肠埃希菌、肠球菌等需氧和兼性需氧菌。化学梯度（如 pH）、氧水平、营养物质可获得性和免疫效应（如分泌抗菌肽）被认为是促进这种横向和纵向空间异质性的假设因素，然而，特定微生物类群定植和局限于特定解剖部位的机制仍在很大程度上是未知的，需要系统研究来确定异质空间的稳定定植对宿主生物学的影响。肠道微生物群也存在时间异质性，随着时间推移，个体的肠道微生物群落变化还表现出不同的演替阶段，可分为婴幼儿快速变化期、成人稳定期和老年退化期[80, 81]。这种演替如同黏液在整个结肠中的变化：在向直肠方向蠕动推进的过程中，由于含水量减少、组织结构和其他代谢转运等方面的原因，黏液变得更密集、更连续，并且黏液层的厚度也会随昼夜节律变化而波动，从而影响黏膜相关微生态系统的区系组成。疾病期间微生物群落也存在空间区系重编程现象，可能是慢性炎症疾病的共同特征和功能相关性特征。

四、肠道菌群与宿主的关系

人类微生物组计划和人类肠道宏基因组学是最早研究宿主与微生物关系的大规模微生物群研究计划[82]。哺乳动物的免疫系统在维持宿主与微生物群落的动态平衡方面起着至关重要的作用，从而确保了宿主 - 微生物互动关系的稳定性。另一方面微生物也会对其寄生宿主的免疫系统带来巨大影响[83, 84]。肠道菌群能促进宿主免疫系统成熟，它们在宿主免疫的激活、训练和调节中起关键作用，从而促进宿主黏膜免疫的建立[85, 86]。有研究表明，新生儿黏膜免疫系统的成熟与微生物对肠道的初始定植是密切相关的，两者之间的相互作用是最终确定微生物群落组成的关键，这个过程主要依赖于分泌型免疫球蛋白 A 的黏膜保护作用[42]。此外，肠道微生物可通过分泌丁酸盐、L- 色氨酸、吲哚、胆汁酸和视黄酸等代谢产物，以及 Toll 样受体和 Nod 样受体等信号通路和小非编码 RNA 对宿主免疫系统产生影响[87, 88]。肠道黏膜表面屏障允许宿主 - 微生物共生。然而，这些障碍很容易受到持续不断的环境刺激，必须快速修复才能重建稳态[89]。值得强调的是，一旦肠道菌群紊乱，超过了宿主的调节范围，肠道表面黏膜障碍就会被突破，哺乳动物的免疫系统将会被激活，继而出现一系列的适应性免疫反应，例如白细胞介素 -23、白细胞介素 -17、TNF-α、IL-6 轴、Sirtuin-3 和 STAT 3 的激活，这些不同的分子途径最终可能促成肿瘤的发生[90]。此外，最新研究表明，肠道菌群还可以通过调节肠道黏膜细胞中 DNA 甲基化，对肠道的稳态和炎症反应产生很大的影响，由此证明了微生物在表观遗传调控中的关键作用[91]。

为了更好地了解肠道细菌和人类宿主的关系，Kylie 等首次创建了人类结肠免疫细胞和肠道细菌的详细细胞图谱，该图谱完整显示了人体结肠中免疫细胞和细菌的变化。通过对人体肠道中 41 000 个免疫细胞及其相应部位的细菌的活性基因测序，他们发现，不仅结肠不同部位的免疫细胞之间存在差异，而且微生物组也随着部位的不同发生了微妙的变化，结肠下方的细菌种类更多。作为“人类细胞图谱”计划的一部分，这些结果使得人们认识到结肠特定区域的细菌对疾病 [例如炎性肠病和结直肠癌（colorectal cancer，CRC）] 的重要影响。更重要的是，这项研究首次表明抑制免疫反应的调节性免疫细胞从淋巴结转移到结肠，这可能是肠道耐受甚至是“欢迎”微生物组的一种方式[92]。

肠道菌群与人体健康息息相关，部分细菌会增加机体对于剩余脂肪、蛋白以及如钙、镁、铁等矿物的吸收，还可以提供如维生素 B_1、维生素 B_2、烟酸、生物素、泛酸、叶酸及维生素 K 等营养物质[93]。此外，肠道中微生物的存在与否，与肠道的蠕动速度之间有着明显的关系，微生物可以通过帮助肠道内的神经细胞调节结肠肌肉壁的收缩和松弛来支持健康的消化[94]。许多人体疾病也与结肠黏膜或结肠肠腔 / 粪便中的菌群紊乱有着至关重要的关系。目前所证实的与 CRC 相关的细菌主要包括拟杆菌门（Bacteroidetes）、厚壁菌门（Firmicutes）和梭杆菌门（Fusobacteria）等[95]。另外，有研究发现 CRC 患者的肠道微生物与健康人的肠道微生物结构存在明显的不同，因此，人们认为促进 CRC 发生发展可能是多种细菌共同作用的结果，而非归咎于某种单一的细菌[96, 97]。

目前，人们已经开始将细菌运用到人体生理功能的维持和疾病的预防和治疗中，其中研究最为广泛的是益生菌。益生菌可定义为“足够的数量能给宿主带来健康益处的活性微生物”，是靶向调节微生物群的主要手段之一，现已经被用于临床上某些疾病的治疗中。但这种调节作用是微弱的，Kristensen 等对 7 项随机临床试验的系统性回顾文章中指出，不同的益生菌对微生物组成没有影响，也没有证据表明益生菌植入的可持续性。这可能是因为肠道微生物群对益生菌具有定植抗性[98]。有研究指出，益生菌能否植入成功，可能取决于益生菌对受体基线微生物群组成的互补程度。例如，裂解梭菌（*Clostridium scindens*）的定植可以补充宿主微生物群的代谢缺陷，并且能够增强宿主对艰难梭菌（*Clostridium difficile*）的定植抗性[99]。因此，在二代益生菌的运用上，我们不仅需要选择有益于宿主功能的细菌，而且需要提供必要的生态环境来维持它们。有趣的是，最近的一项研究发现，小鼠肠道中的双歧杆菌可以迁移到肿瘤内部，并激活干扰素基因刺激因子（stimulator of interferon genes，STING）通路，激活免疫细胞的表达，当与“抗 CD47 疗法”（CD47 是许多癌细胞表面表达的蛋白质，抑制该蛋白能够促进患者机体的免疫系统，靶向作用于 CD47 的抗体目前正在多项癌症临床试验中使用）相结合时，这些被激活的免疫细胞就会攻击并破坏周围的肿瘤组织。也就是说，肠道菌群或能通过在肿瘤内部定植来增强 CD47 疗法的抗肿瘤效应，给予特定的细菌或其工程菌就能作为一种有效的策略来调节多种抗肿瘤免疫疗法的功效 [100]。

除益生菌外，粪菌移植也是目前利用细菌改善人体健康的一个研究方向。FMT 是非靶向微生物群调节的主要例子：将（健康）供体粪便中的功能菌群，移植到患者胃肠道内，重建新的肠道菌群，其目的是改善受体的健康或不理想的微生物群状态，实现肠道及肠道外疾病的治疗。在医学史中，FMTs 的运用由来已久，至少有 1700 年的历史，甚至在中国传统医学中也有用人粪治疗疾病的记载 [101]。FMTs 已被证明在治疗复发性艰难梭菌感染（recurrent Clostridium difficile infection，rCDI）中具有很高的有效性，并且比抗生素更适合这种疾病的治疗 [102]。在 FMTs 治疗后，受体的微生物多样性增加，rCDI 患者的微生物组成出现了向供体微生物组成改变的趋势，并且在炎症性肠病患者中也出现这种趋势 [103，104]。截至目前，已有近 200 项注册临床试验将 FMTs 作为疾病的治疗方案，相信未来会发现越来越多的 FMTs 使用适应证。此外，有研究证明，供体菌株和受体肠道微生物可以在受者体内实现共存，并且至少在 FMTs 后可以持续共存几个月 [105]。在临床实践中，粪菌移植前通常先对受体进行预防性抗生素治疗或肠道清洁。南京医科大学第二附属医院的张发明教授是第一位在中国开展现代标准化粪菌移植的医师。所谓标准化粪菌移植，是在实验室内借助现代仪器设备人性化地分离获得高度纯化的菌群，再经内镜或引流管将量化的菌液输注到患者肠内 [101]。从长远发展来看，使用配方的、受者定制的脱脂微生物菌株混合物来替换供体粪便样本是更加稳妥的方法。虽然目前的结果令人鼓舞，但我们对受体微生物群和 FMT 反应的理解到目前为止仍然不够深刻。

第三节　肠道微生物群的研究方法

一、肠道微生物群研究方法概述

肠道菌群分析是进行复杂微生态分析的基础，如何对肠道菌群的丰度和数量变化进行全面分析是开展微生态研究的瓶颈问题。随着时间的推移，用于研究肠道微生物的方法发生了重大变化，每种方法都有其独特的优点和局限性。这些方法大体而言分为 3 类：培养法、分子法、基于小分子代谢物检测和定量的方法（表 1-2）[2]。其中，培养法是最为经典的一种微生物研究方法，其发源最早且沿用至今，具有独特的优势。细菌培养是一种用人工方法使细菌生长繁殖的技术，该方法根据细菌种类和筛选目的等选择培养方法、培养基，利用不同的培养条件将细菌进行分离，最终对所得单个菌落进行形态、生化及血清学反应鉴定。因此，培养法有助于研究细菌的功能和不同生长条件下的生理活性，时至今日仍是肠道细菌分析中必不可少的一种研究方法。然而，随着人们对肠道菌群研究的深入，培养法的不足之处也日益显现。如前文所述，人体肠道中以厌氧菌为主，在采样、运输和储存的过程中易出现损耗，使得专性厌氧菌的相对丰度低于实际值。此外，培养法无法用于研究肠道微生物群的分布，只能用于研究个别可培养的细菌群。因此，这种单纯基于细菌表型特征的研究方法在很大程度上限制了人们对人类肠道微生物群组成的分析[106]。总的来说，该方法只能对部分能够在体外培养的细菌进行分析，而且耗时，对于种类和数量庞大的肠道微生态系统而言就显得不够全面，也不能反映整个微生态系统与疾病发生发展的关系，分析的结果与结论也具有一定局限性。

表 1-2 用于研究肠道和其他身体部位微生物群的技术

A. 培养法

B. 分子（核酸）法

a. 非测序法

i. 荧光原位杂交流式细胞术

ii. 脉冲场凝胶电泳

iii. 变性梯度凝胶电泳

iv. 温度梯度凝胶电泳

v. 单链构象多态性法

b. 测序法

i. 16S rRNA 基因或其高变区的测序（靶向基因测序）

ii. 全细菌基因组 DNA（宏基因组）测序

iii. 全细菌 mRNA（meta-trascriptome）测序

C. 基于小代谢物检测和定量的方法

i. 气相色谱 - 质谱法

ii. 毛细管电泳与质谱联用法

iii. 傅里叶变换红外光谱

iv. 核子和质子磁共振波谱

为了弥补传统培养技术的不足，分子生物学检测方法应运而生，rRNA 序列数据的高特异性激发了生物多样性的发现热潮，大量的新型肠道细菌被发掘及鉴定。同时基于小代谢物检测和定量的方法打开了细菌代谢产物研究的大门，极大地拓展了微生物的研究领域。接下来，我们将对目前运用最为广泛的两种分子生物学及菌群代谢产物检测方法——16S rRNA 测序法和气相色谱 - 质谱法进行详细介绍。

二、16S rRNA 测序法

除哺乳动物成熟的红细胞、植物筛管细胞外，细胞中都有核糖体存在，核糖体没有膜包被、由大小两个亚基组成。在原核细胞（包括细菌）中小核糖体亚基包含的 RNA 为 16S rRNA，在真核细胞中为 18S rRNA（“S”是一个沉降系数，亦即反映生物大分子在离心场中向下沉降速度的一个指标，值越高，说明分子越大）[107, 108]。16S rRNA 基因是细菌上编码 rRNA 相对应的 DNA 序列，存在于所有细菌的基因组中，具有高度保守性和特异性。细菌 16S rRNA 约包含 1500 核苷酸，不同细菌间存在一定的变异[109]。通常认为，当 16S rRNA 基因序列同源性小于 97% 时属于不同的种，当同源性小于 93% ～ 95% 时属于不同的属。16S rRNA 基因测序法大体上分为以下几个步骤。

1. 标本采集　用于分析人体肠道微生物的标本主要包括粪便、肠道组织及肠腔灌洗液（该标本主要用于动物模型研究），标本的适当选择、采集在一定程度上决定了肠道微生物分析结果的准确性[110]。粪便标本主要用于远端肠道细菌的检测，其获取途径最为简单。肠道组织活检可以同时对宿主和微生物群进行评估，因此可用于评估宿主肠道与微生物的相互作用关系。然而，在实际操作中一些胃肠道部位（如小肠）的组织活检标本难以获得，并且在胃肠道活检前，通常需要进行胃肠道的灌洗，这样就会在一定程度上改变原本的微生物群。值得注意的是，对于标本的选择需指定严格的纳入和排除标准，在标本采集的前 2 周需停止服用抗生素、非甾体抗炎药、微生态制剂、促胃肠动力药、抑酸药等影响肠道菌群的药物。在同一实验项目中，使用的所有标本都应以相同的方式收集、储存和处理 [111]。

2. 宏基因组 DNA 提取及扩增　制备基因组 DNA 是进行基因结构和功能研究的重要步骤，通常要求得到的片段长度不小于 100 ～ 200kb，提取方法因所用的样本类型而异 [112, 113]。此外，所获得的实验结果可能随所使用的方法而变化。DNA 提取有以下注意事项：①在提取过程中，应保证核酸一级结构的完整性；②核酸样品中不应存在对酶有抑制作用的有机溶剂和过高浓度的金属离子；③其他生物大分子如蛋白质、多糖和脂类分子的污染应降低到最低程度；④应尽量去除其他核酸分子，如 RNA。目前用于多样性分析的高变区（hypervariable regions，HVRs）主要分为两类，一类是单独的 V 区，如 V3 区、V4 区、V6 区，另一类是连续的 V 区。研究者们认为，在连续的 V 区中，V4 ～ V5 区尤其适用于微生物群的研究，因为它提供了跨平台最具可比性的结果及高分类分辨率 [114, 115]。选择 16S rRNA 基因的扩增和测序区域基于以下几点：①该区域能否准确地分类为尽可能多的属或种，这就需要研究者在实验开始前大量查阅相关文献；②整个微生物物种的侧翼区域的保护水平（越高越好）；③选择的测序平台是否能以成本效益高的方式对所选择区域的长度进行排序。一旦选择了要研究的 16S rRNA 基因的高变区（或者研究者感兴趣的区域），就用聚合酶链反应放大该区域。现有的排序方法可以在一次运行中生成大量数据，得到的这些数据量远远超过一个样本需要研究的序列数。因此我们常使用不同的反向引物，使得每一个序列中都包含一个独特的六核苷酸“索引”序列（图 1-2）。每个样本的扩增产物中都包含这个“索引”的不同序列，这些带有不同“索引”标记的产品可以汇集在一起，并在相同的测序实验中运行。这种汇集不同样本的过程称为“复用”。一旦获得序列数据，通过读取每个序列的“索引”区域来识别其起源，这些数据就可以被分隔开（这一过程称为“解复用”）[2]。

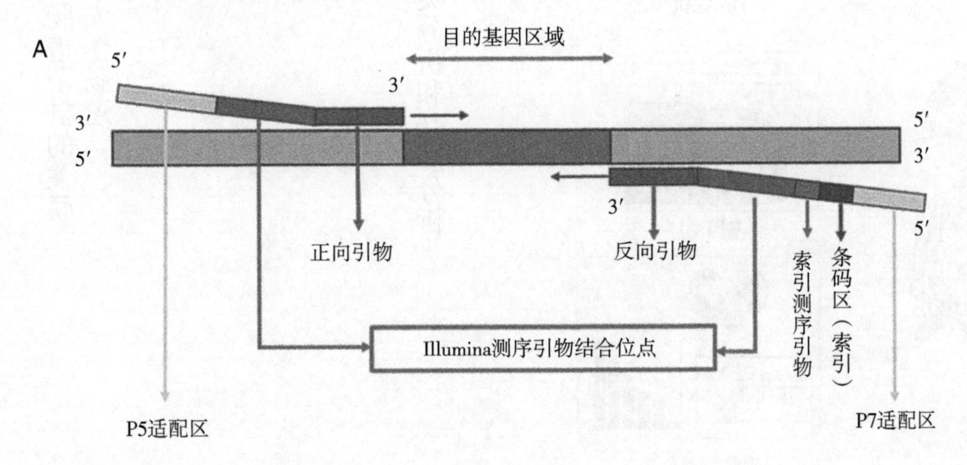

B

引物名称	引物核苷酸序列
V3F 5′	–aatgatacggcgaccaccgagatct acactctttccctacacgacgctcttccgatct NNNNCCTACGGGAGGCAGCAG–3′
V3R 5′	–caagcagaagacggcatacgagat xxxxx gtgactggagttcagacgtgtgctcttccgatct ATTACCGCGGCTGCTGG–3′

图 1-2　用于扩增 16S rRNA 基因区域，以生成 Illumina DNA 测序文库的引物中包含的各种功能元素

3. 16S rRNA 基因测序及生物信息学数据分析　目前常用的两个测序平台为 Illumina MiSeq 和 Illumina HiSeq，两者的测序原理相同，但 Illumina HiSeq 的测序通量大于 Illumina

MiSeq，而Illumina MiSeq的测序耗时短于Illumina HiSeq。16S rRNA基因克隆文库是目前微生物群落多样性研究中最常用的方法。使用通用引物扩增群落中所有细菌的16S rRNA基因，然后用克隆建库的方法，把每一个16S rRNA基因插入文库中的每一个克隆里，再通过测序与已知公共的16S rRNA基因序列数据库比对，得到每一个16S rRNA基因的序列信息和系统发育地位。常用的生物信息分析程序主要是微生物生态学定量研究（QIIME），它是Quantitative Insights Into Microbial Ecology的首字母缩写，目前已经被用于分析和解释来自真菌、病毒、细菌和古细菌群落的核酸测序数据，它包含了操作分类单元（operational taxonomic unit，OTU）聚类及多样性等多种分析。通过相似性对个体进行分组，根据序列彼此的相似性进行聚类，然后基于设置的相似性阈值来定义OTU。简单来说，我们通常将序列相同或相近的克隆划分为一个OTU，通过OTU分析就可以知道样品中的微生物多样性和丰度[116]。为了便于读者理解，下面我们以粪便标本为例对16S rRNA基因测序的主要步骤做一图示（图1-3）。

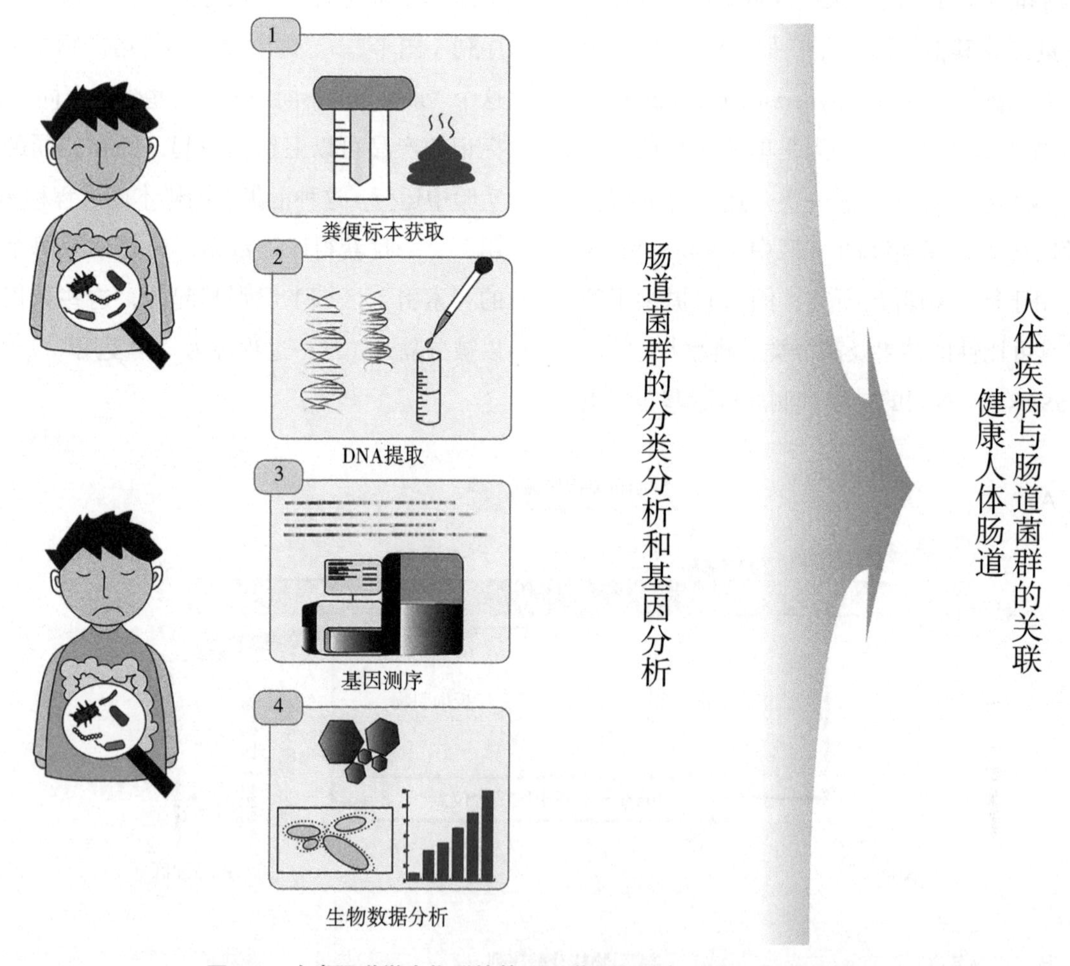

图1-3 人类肠道微生物群的基因组学研究通常包括4个实验步骤

在过去的10～15年中，16S rRNA基因测序的迅速发展使得高通量测序技术（high-throughput sequencing，HTS），又称“下一代”测序技术，能一次对几十万到几百万条DNA分子进行序列测定）得到广泛运用，并且其价格较为合理，这些技术实际上成为目前研究肠道微生物的金标准[117]。基于16S rRNA基因测序的细菌鉴定方法的优点有：①这种方法无须进行细菌培养，很大程度上减少了科研人员的时间成本，并且为体外培养不佳的细菌检测提供了可能；②传统的核酸测序技术需要相对纯净的 DNA 作为起始材料，且每个实验只能检测一个序列，这就意味着这种技术不可能对含有相关核酸混合物的样品进行测序[118]，而 16S rRNA 基因测序则可以弥补这些缺陷，更适用于含有相关核酸混合物的样本检测；③此外，当这些方法应用于细菌混合物时，可以为培养基中各种生长能力和生长速率不同的细菌群的相对丰度提供无偏移评估。

三、气相色谱 - 质谱法

气相色谱 - 质谱（gas chromatography-mass spectrometry，GC-MS）技术分析的主要任务是分析复杂样品中的众多未知物，以检测时间短、耗费溶剂少等特点成为人们普遍采用的检测手段，该技术是一种快速、准确、灵敏度高的定性定量分析技术[119]。肠道菌群的代谢产物组成极为复杂，因此我们常借助 GC-MS 技术对复杂样本（如粪便、组织等）中的细菌代谢产物进行分析。肠道菌群的常见代谢产物有叶酸、吲哚、次级胆汁酸、三甲胺 -N- 氧化物、神经递质（如 5- 羟色胺，γ 氨基丁酸）及 SCFAs[120]。其中，研究最为广泛的代谢产物是 SCFAs，目前常用的检测脂肪酸的方法有 GC-MS、气相色谱 - 氢火焰离子化检测法（gas chromatography with flame ionization detection，GC-FID）及液相色谱 - 质谱法（liquid chromatography-mass spectrometry，LC-MS）。这 3 种方法各有其优缺点：LC-MS 具有分离效能高、灵敏度高、应用范围广等优点，但其溶剂用量大、选择性低[121]；GC-MS 与 GC-FID 同属于气相色谱法，并且两者都适用于挥发性成分的分析，但 GC-MS 是气相色谱和质谱的联用，定量能力和定性能力都更强，且具有更好的选择性[122, 123]。下面以 SCFAs 为例，对 GC-MS 技术的步骤进行详细讲解。

1. 脂肪酸的萃取　目前常用的 SCFAs 萃取方法主要包括液 - 液萃取法和固相萃取法。

（1）液 - 液萃取法：又称溶剂萃取或抽提法，这种方法由有机相和水相相互混合，是通过在液体混合物中加入与其不相混溶（或稍相混溶）的溶剂，利用其组分在溶剂中的不同溶解度而达到分离或提取目的。有机相一般由 3 种物质组成，即萃取剂、稀释剂、溶剂。有时还要在萃取剂中加入一些调节剂，以使萃取剂的性能更好。目前，常用的萃取溶剂有两种，一种采用比例为 2 ∶ 1 的氯仿和甲醇的混合物作为萃取溶剂，萃取溶剂体积为组织样品体积的 20 倍（将 1g 组织加入 20ml 萃取溶剂中）。然后，加入 0.2 倍体积的水或 0.9%NaCl 溶液清洗，离心后收集下相进行脂肪酸分析。这种提取方法是目前脂质提取最常用的方法[124]。另一种采用氯仿 / 甲醇 / 水的混合物（比例通常为 2 ∶ 2 ∶ 1.8）作为萃取溶剂，通常用于含水量 80% 以上的生物样品（如组织和血液）中脂肪酸的检测，具有溶剂消耗率低的优势，同时还能保证高回收率[125]。然而，这两种方法都采用氯仿作为萃取溶剂的一部分，由于氯仿的高毒性，使得这两种方法具有一定的安全隐患[126]。因此，目前已有研究采用其他萃取溶剂来取代氯仿。例如，甲基叔丁基醚、BUME 混合物（丁醇：甲醇 = 3 ∶ 1）等，这些方法可以提供更高效和安全的脂质提取[127, 128]。

（2）固相微萃取法：由液固萃取和柱液相色谱技术相结合发展而来，对于挥发性和半挥发性物质，我们推荐使用固相微萃取法（solid-phase microextraction，SPME）[129]。它是在固相萃取技术上发展起来的一种微萃取分离技术，是一种集采样、萃取、浓缩和进样于一体的无溶剂样品微萃取新技术。它具有操作简单、携带方便、无需溶剂等优点[130]。其装置类似于一支气相色谱的微量进样器，萃取头是在一根石英纤维上涂上固相微萃取涂层，用其直接萃取含水样本，目前可粗略地分为静态批量平衡微萃取和动态流动平衡微萃取的方法[131]。氯化钠和硫酸钠是SPME体系中常用的盐析试剂。而Fiorini等则使用（NH_4）$_2SO_4$/NaH_2PO_4作为盐析试剂，他们使用生物样本和食物样本证明了该方法的适用性，并证明了与传统SPME法相比，该方法具有更高的灵敏度，从而提高SPME的回收率。这种新型盐析试剂提高了顶空萃取的有效性，并且可用于测定自由形式的短链脂肪酸和中链脂肪酸[132]。

2. 脂肪酸的衍生化　使用色谱分离原理检测化学成分，当被检测目标成分因理化性质不便分离检测时，经常采用衍生化技术，使其达到可以使用色谱仪检测的目的。一般来说，一个特定功能的化合物参与衍生反应，溶解度、沸点、熔点，聚集态或化学成分会产生偏离，由此产生的新的化学性质可用于量化或分离。气相色谱中常用的衍生化方法包括硅烷化衍生化方法、酯化衍生化方法、酰化衍生化方法和卤化衍生化方法。其中，硅烷化是一种经典的衍生化方法，也是脂肪酸衍生化中最常用的衍生化方法。它是利用质子性化合物（如醇、酚、酸、胺，硫醇等）与硅烷化试剂反应，形成挥发性的硅烷衍生物。可以通过取代活性氢原子（例如羟基、羧基和氨基）将硅基引入代谢物。活性氢被硅烷基取代后降低了化合物的极性，减少了氢键束缚，因此所形成的硅烷化衍生物更容易挥发。同时，由于含活性氢的反应位点数目减少，化合物的稳定性也得以加强，最终形成更为稳定的、更具挥发性和更低极性的代谢产物[133]。在利用GC-MS法进行脂肪酸分析时，通常需要将脂肪酸衍生化，形成脂肪酸甲酯[134]。

此外，还有研究提出了一些新型衍生化方法，例如采用三甲基氢氧化硫（trimethylsulfonium hydroxide，TMSH）作为衍生化试剂，可以在一步内快速衍生，具有减少伪影的优势。因此，该方法可用于大批量样本的分析，但TMSH法的局限性是多不饱和脂肪酸的衍生效率不高[135]。有研究建议使用N，O-双（三甲基甲硅烷基）三氟乙酰胺-三甲基氯硅烷作为游离脂肪酸的衍生化试剂，研究者认为该方法简单、特异性好，且完全不受已酯化脂肪酸（例如三酰甘油）的干扰，可用于临床实验室或药理研究中游离脂肪酸的色谱分析[136]。然而，有研究认为五氟苄基溴能将脂肪酸转化为卤化衍生物，因此可以作为游离脂肪酸研究的首选衍生化试剂[137]。由此可以看出，衍生化试剂的选择具有多样性，因此我们需要更好地理解每种方法的利弊，并考虑方法的局限性。此外，还可以优化衍生条件，以满足特定应用的需要。

3. GC-MS分析　气相色谱-质谱联合技术结合了气相色谱的高效分离方法、质谱的精确检测方法，可以准确定性、精确定量样品的结果。因此在进行GC-MS分析前，我们应选择合适的GC条件和MS条件，然后再进行数据处理，最后与标准质谱库进行对比[134]。首先是色谱柱的选择，一个合适的色谱柱，是对混合物中的脂肪酸进行良好分离的必备条件。不同的色谱柱适用于不同链长、饱和程度、双键位置、顺式或反式异构体的脂肪酸的分离。高极性柱，如HP-88柱（88%-氰丙基芳基聚硅氧烷）、DB-FFAP柱（硝基苯二甲酸改性聚乙二醇）和SLB-IL系列柱（离子液体）常用于生物样品中的脂肪酸分析[138, 139]。有研究表明，IL柱对FAME混合物的选择性优于HP-88柱，此外，IL柱还可以分离出不同的脂肪酸异构体[140, 141]。例如，几种FAME几何异构体，如C18：2n6t和C18：2n6c，不能用非极性柱（SLB-5ms）分离，但使用IL柱就可以进行很好的分离。值得注意的是，不同的IL柱具有不同的分离优势。例如，与SLB-IL59、SLB-IL60、SLB-IL61和SLB-IL76柱相比，SLB-IL82、SLB-IL100和SLB-IL111柱可以提供更好的顺式和反式异构体分辨率[142-144]。总之，高极性柱，如 HP-88 柱和 DB-FFAP 柱，能够分离不同碳链长度的脂肪酸，而离子液体系列柱（特别是 SLB-IL82、SLB-IL 110、SLB-IL111）对分离脂肪酸异构体具有特别的优势。

GC 条件设置包括色谱柱的选择（上面已详细介绍）、升温程序设置、进样口温度设置、载气设置（一般为高纯氦，纯度＞ 99.999%）、流速设置、分流比设置及进样量设置（一般为 1μl）。MS 条件设置包括电离方式设置、电子能量设置、电子源温度设置、传输线温度设置、电子倍增器电压设置及扫描模式设置[145]。设置条件根据不同的生物学样本及不同的脂肪酸类型而定，这里不再赘述。最后，通过检索标准质谱库获得标准化合物信息，并结合对照品比对、化合物质谱裂解规律解析生物样本中的代谢产物。

第四节　内外环境对肠道微生物群组成的影响

一、外界因素：环境、饮食、生活方式、药物等

目前已有大量研究证实了成人肠道微生物群与宿主外在因素的关联，肠道微生物群的组成在一定程度上取决于种群的地理起源。有研究对高海拔和低海拔人群的肠道菌群进行了基因测序，发现高海拔和低海拔人群的肠道菌群存在很大的差异，即不同生存环境会对肠道菌群的组成产生重要影响[146]。环境因素极为复杂，迄今仍未得到充分研究，因为居住在不同地理位置人群的生活方式在很大程度上也会有差异[147, 148]。因此，有学者在研究生存环境对肠道菌群的影响时将生活方式（如饮食习惯、家庭成员、婴幼儿喂养方式等）也归在其中[149]。

众所周知，肠道微生物群暴露于由数百万种饮食来源化合物组成的复杂组合中。食物是肠道菌群生长繁殖的主要营养物，肠道菌群的组成与数量是否处于平衡状态与个体的膳食结构和食物成分有着密切的联系，因此从不同水平探索饮食与微生物群之间的联系一直是研究的热点[150, 151]。目前的研究大致分为两种模式：①大范围的研究以植物和动物为基础的饮食，以及这种长期饮食模式对肠道菌群的影响[152, 153]；②单个食物或营养物质（脂肪、碳水化合物等）与肠道内特定微生物的关系[154]。近来有研究表明，短期摄入高糖会显著改变小鼠肠道微生物的组成，尤其是使得黏液降解细菌的丰度增加，细菌衍生的黏解酶富集，导致了细菌对小鼠结肠黏液层的侵蚀，最终加重葡聚糖硫酸钠诱导的小鼠结肠炎[155]。而大量摄入超级加工食物（包括已经包装好的烘焙食物和零食、发泡饮料、含糖谷物、含有添加剂的即食食品、重组的肉类和鱼类产品）可能会增加机体患炎性肠病的风险[156]。然而，尽管在横断面研究证实了饮食 - 微生物群的关联，饮食因素只能解释在调整协变量后微生物群的个位数百分比变化，即饮食因素并非微生物群变异的主要因素[157-159]。

时至今日，人们已经充分认识到一些生活方式因素，如吸烟、饮酒和体育锻炼等对于调节体内微生物的重要性，并且阐明了它们对人体健康的复杂协同作用。研究表明，吸烟可以使得肠道菌群的组成丰度发生变化，其中变形菌门、拟杆菌门、梭状芽孢杆菌属及普雷沃菌属（*Prevotella*）丰度增加，而放线菌门、厚壁菌门、双歧杆菌和乳球菌减少；同时吸烟还会降低肠道微生物群的多样性[160]。酒精性脂肪肝

是长期酗酒所致酒精中毒性肝脏疾病，而这种酒精依赖的肝功能障碍对肠道微生物有很强的负面影响。有研究表明，酒精性肝硬化患者的肠道菌群的组成结构和代谢都发生了明显变化，与酒精依赖有关的机会病原体（如促炎性肠杆菌科）增加，肠道微生物合成有毒乙醛的倾向增加，提示酗酒者患结直肠癌和其他疾病的风险较高[161]。最新研究证明，生存在同一海拔高度的汉族和藏族人群，由于其饮食习惯的不同，他们的肠道菌群组成存在着明显差异[162]。有趣的是，不仅仅是食物的种类，摄入食物的方式也会影响肠道菌群的组成。近期，有研究人员对78名脑卒中患者进行了经口进食训练，通过纳入排除标准，最终选择了8名能够完全经口进食的脑卒中患者，并比较其口服食物摄入前后口腔和肠道微生物组的变化。他们发现经口摄入食物会极大地改变口腔和肠道微生物组，口腔和肠道微生物组中胡萝卜菌科和颗粒菌属的丰度都有所增加，并且其多样化进一步提升。尽管口腔微生物群改变比肠道微生物改变更为显著，但宏基因组预测显示，肠道中的差异信号通路更加丰富，特别是与脂肪酸代谢相关的信号通路[163]。这就证明了营养和饮食对于肠道菌群的组成和平衡具有深远影响。微生物群、体重指数（body mass index，BMI）及肥胖三者之间的关系受到了极大关注，尤其是BMI与肠道菌群亚种优势具有很大的相关性[164]。与宿主BMI具有最强联系的细菌亚种是直肠真杆菌（*Eubacterium rectale*，可从人粪便中分离得到，主要起到降解产物的转运与加工作用），它又可以进一步分为3个亚种，其中一个亚种——MGSS3的样本几乎全部来自于中国[165]。值得注意的是，不仅是成人，婴幼儿的肠道菌群组成也与其生活方式有关。有研究表明，与配方奶粉喂养的婴儿相比，母乳喂养婴儿的微生物多样性和丰富度更高，并且具有更多的蛋白菌、更少的拟杆菌和厚壁菌门[166]。

在佛兰德肠道菌群计划中，研究员发现药物（包括抗生素、抗组胺药和激素在内）是肠道微生物群改变的主要协变量之一[157]。人们普遍认为，不管是长期还是短期，使用广谱抗生素对肠道微生物群的稳定具有消极影响[167, 168]。有研究者对662名丹麦儿童进行了抗生素耐药性基因的检测，他们惊讶地发现其中一名婴儿的肠道菌群中携带着数百种抗生素抗性基因，这有力地佐证了抗生素对肠道菌群的巨大影响，也为抗生素的合理、规范使用敲响了警钟[169]。近年来，越来越多的报道开始将非抗生素药物与微生物群调节联系起来[170, 171]。有研究表明，2型糖尿病药物二甲双胍对微生物组成的影响比疾病本身对微生物组成的影响更大[172]。再比如，长期使用质子泵抑制剂（一种目前治疗消化性溃疡最常用的一类药物，可以通过高效快速抑制胃酸分泌，与阿莫西林、克拉霉素等药物联用治疗幽门螺杆菌感染，而达到快速治愈溃疡的目的）则可能会导致小肠细菌过生长[173]。此外，一些非典型抗精神病药物（如齐拉西酮和喹硫平）及非甾体抗炎药物等也能够在一定程度上影响肠道微生物群的组成[174, 175]。有趣的是，Han等证明，一种临床上广泛使用的止泻药——蒙脱石散，可以促进双歧杆菌、乳酸杆菌等益生菌形成生物膜，从而延长益生菌在小鼠肠道内的停留时间，最终抑制小鼠肠道肿瘤的生长。其机制与蒙脱石表面电荷分布不均匀、具有阳离子交换特性和高比表面积有关[176]。

然而，这些外界因素并不是相互独立的（例如，饮食、宿主BMI和药物使用情况等都有一定的相关性），而且可能与宿主内在或外在环境因素有关。因此，我们必须意识到，许多微生物群特征可能是由混合效应驱动的。

二、内在因素：宿主和微生物的基因与遗传

上面谈到的一些外界因素（例如 BMI）可以部分归因于遗传学。而对于其他因素，宿主的遗传成分则更为明显：例如，微生物群与先天免疫系统和适应性免疫系统有着错综复杂的联系，尽管量化免疫系统对肠道微生物的影响仍然具有挑战性[86, 177]。同样，越来越多的证据表明了“微生物 - 肠 - 脑”（microbiota–gut–brain，MGB）轴之间的相互作用[178]。MGB 轴是一个整合了肠道和大脑之间的神经、激素和免疫信号的通信系统，并为肠道微生物及其代谢产物提供了一个进入大脑的潜在途径（图 1-4）。肠道微生物群对神经系统的影响是双向的。一方面，微生物可以影响神经系统的发育、认知和行为；另一方面，行为的变化也可以改变肠道微生物群的组成。在某种程度而言，我们可以认为肠道微生物是精神病理学的直接介质[179, 180]。

目前已有研究探讨了微生物与宿主遗传位点之间更为直接的关联。Goodrich 等利用 16S rRNA 基因测序技术对 126 对英国双胞胎肠道的微生物群进行了横断面研究，发现部分细菌的相对丰度是可遗传的，这一观点后来在物种水平上得到证实[181, 182]。此外，Turpin 等对 1561 名北美健康人的粪便细菌进行了分析，发现几乎 1/3 的粪便细菌分类群是可以遗传的。更令人振奋的是，他们证明了 58 个人类单核苷酸多态性与 33 个细菌类群的相对丰度有关，其中 rs62171178（nearest gene UBR3）与理研菌科相关，rs1394174（CNTN6）与大肠埃希菌相关，rs59846192（DMRTB1）与毛螺菌属相关，rs28473221（SALL3）与真菌属相关[183]。近期，Grieneisen 等对野生狒狒的肠道微生物群进行了为期 14 年的纵向研究，通过对收集到的粪便样本进行 16s rRNA 测序分析，他们发现 97% 的单个分类单元和群落表型都具有显著遗传性[184]。这些研究结果有力地佐证了特定遗传变异与肠道微生物群之间的关联性。

值得注意的是，微生物之间的直接相互作用也被认为是微生物群组成的重要驱动因素[185]。微生物与微生物及微生物与宿主之间的交流主要通过群体感应（quorum sensing，QS）系统来实现[186]。QS 是通过感知密度变化来调节基因表达的细菌行为，是物种内部和物种之间交流的重要机制，使细菌能够作为一个团队而不是单个细胞发挥作用[187]。它可以控制特定的过程，如生物膜的形成、毒力因子的表达、次级代谢物的产生以及包括分泌系统（secretion systems，SS）在内的细菌竞争系统等应激适应机制[188, 189]。而参与这一效应的分子，被称为自体诱导物（autoinducers，AIs），AIs 以极低的阈浓度调控基因的表达，它们可以促进细菌的稳态、生长、孢子形成、程序性细胞死亡、毒力因子分泌及生物膜的形成[190]。常见的自诱导因子包括酰化高丝氨酸内酯（Acylated homoserine lactones，AHLs，又称 AI-1）、AI-2、AI-3、霍乱弧菌 I 类自诱导分子（cholerae autoinducer-1，CAI-1）、PQS（Pseudomonas quinolone signal）和自诱导肽（auto-inducing peptides，AIPs）[191]。AI-2 是 QS 的主要信号分子，是一种非特异性自诱导剂，参与调节种间的通信。AI-2 可介导肠道细菌定植，从而改变肠道细菌的组成，调节宿主免疫反应，导致肠道疾病[192]。这部分内容将在本书第 4 章详细阐述。

分泌系统在细菌交流过程中扮演了重要角色，截至目前，分泌系统共有 8 种类型（分别为 T1SS、T2SS、T3SS、T4SS、T5SS、T6SS、T7SS 和 T9SS），它们广泛存在于革兰阴性菌、革兰阳性菌及分枝杆菌中，不仅参与蛋白酶、脂肪酶、黏附素、血红素结合蛋白和酰胺酶的转运，还能在宿主细胞中合成蛋白质、分泌效应物以建立感染性生态位、转移、吸收和释放 DNA、转移效

应蛋白或DNA，以及分泌自体转运体等[189，193]。

有趣的是，Coyte等认为，相比肠道细菌之间的合作关系，细菌间的拮抗作用才是维持肠道菌群稳定的关键因素，当细菌之间的竞争削弱了其合作网络时，宿主则可以从细菌的竞争中受益[194]。肠道细菌的基因组中包含了多种介导接触依赖性细菌间拮抗作用的途径，如LXG蛋白等介导的革兰阳性菌之间的拮抗作用[195]，以及由T6SS介导的革兰阴性菌之间的相互拮抗作用[196-198]。

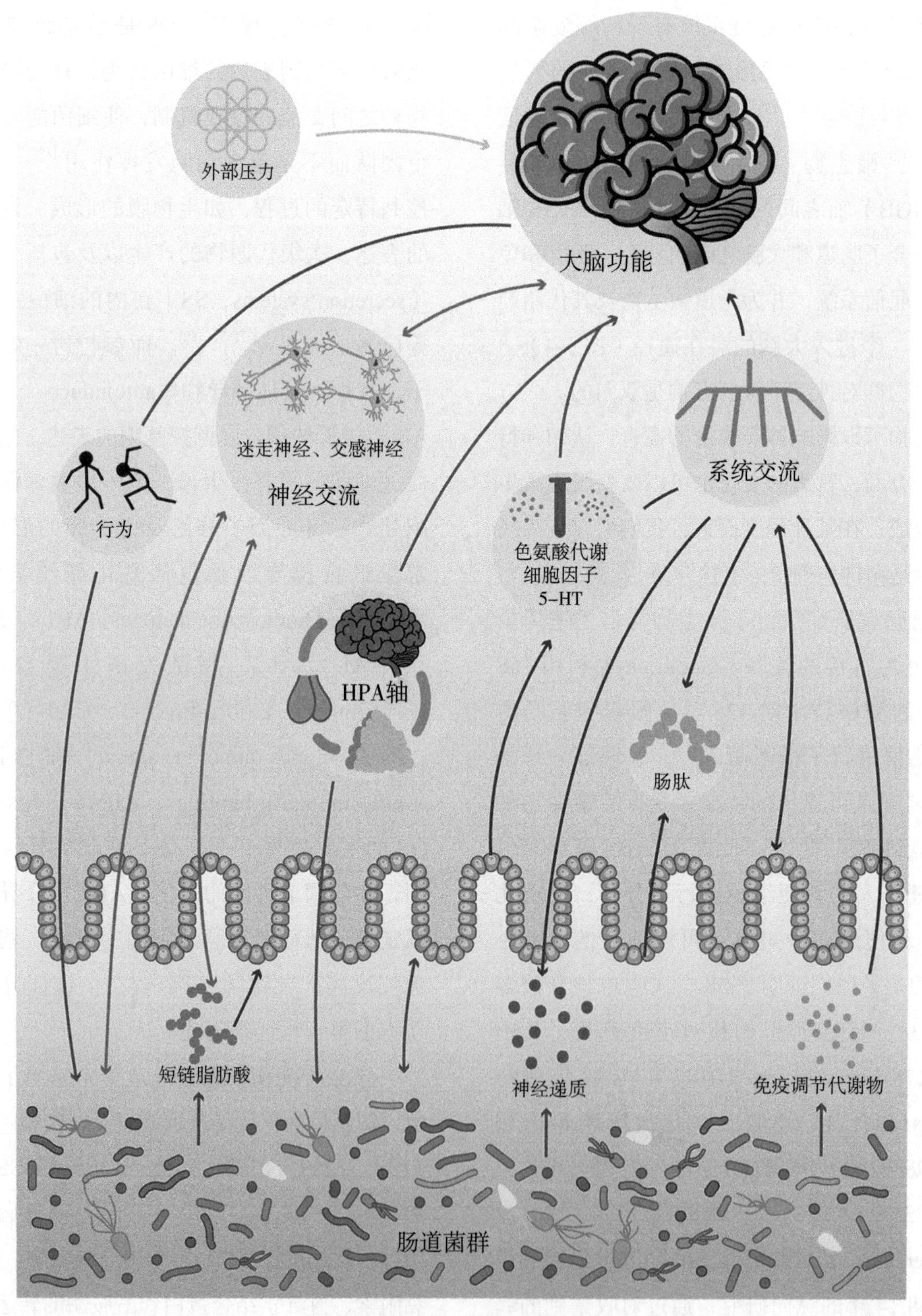

图1-4　微生物-肠-脑轴示意图

HPA轴．即the hypothalamic–pituitary–adrenal axis，下丘脑-垂体-肾上腺轴，是神经内分泌系统的重要部分，参与控制应激反应，并调节许多躯体活动；5-HT.5-羟色胺

第五节　肠道菌群的代谢产物

一、肠道菌群的代谢产物及其作用

随着微生物代谢组学、元基因组学及元蛋白质组学的发展，使得人们能够识别和量化微生物代谢产物，并探索代谢物的生化途径，迄今为止人们已经发现肠道微生物中携带着编码数千种微生物酶和代谢物的基因[199, 200]。肠道菌群代谢的底物为食物中未消化的碳水化合物、蛋白质和肽类，其主要代谢终产物为单糖和寡糖分子，或乙酸、乳酸、琥珀酸等有机酸。人体肠道菌群在大量膳食营养素的刺激下，产生胆汁酸、SCFAs、氨、酚类、内毒素等生物活性化合物，这些代谢物是细菌与宿主沟通的媒介，是维持宿主正常生理状态必不可少的[201]。其中，某些微生物在代谢过程中可以产生许多生物活性化合物，包括γ-氨基丁酸、色氨酸代谢物、蛋白质等，这些代谢产物是脑肠轴的重要组成部分[202]。肠道细菌可以与宿主发生共代谢关系，如胆固醇和胆酸代谢、激素代谢等都是由肠道菌群和宿主共同协作完成的。有研究表明，一些代谢产物可以通过与特定的宿主膜或核受体结合来影响宿主代谢[203, 204]。微生物的代谢产物具有重要的生理功能，可以作用于微生物与微生物之间及微生物与宿主之间（图 1-5）。

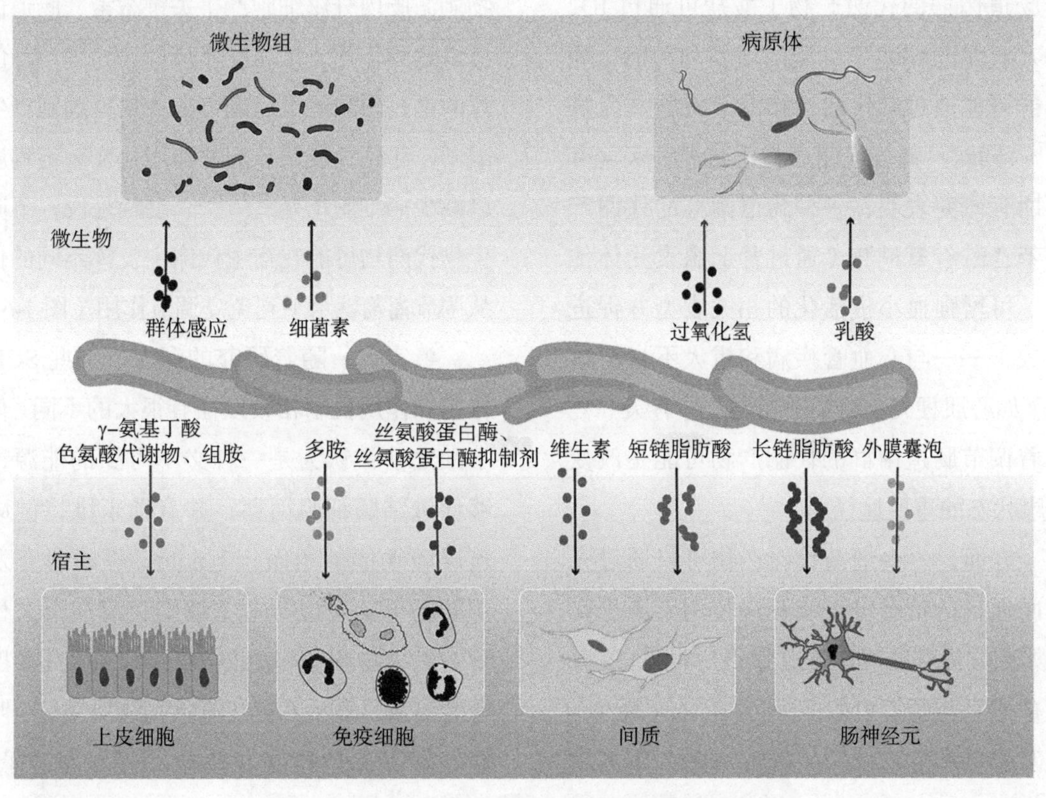

图 1-5　胃肠道中产生的微生物代谢物具有多种功能

微生物代谢产物可以调节微生物内部（微生物 - 微生物）和微生物 - 宿主之间的关系，从而影响人类健康。细菌参与群体感应，并能释放细菌素、过氧化氢和乳酸，从而对肠道微生物群和病原体产生影响。同时，细菌还可以分泌γ-氨基丁酸、色氨酸代谢物、组胺、多胺、丝氨酸蛋白酶、丝氨酸蛋白酶抑制剂、维生素、短链脂肪酸、长链脂肪酸和外膜囊泡等，这些代谢产物对宿主上皮细胞、免疫细胞、间质和肠神经元产生影响一方面，细菌的代谢产物可以对宿主的生理功能产生重要影响。短链脂肪酸是一种重要的肠道细菌代谢产物，它是组蛋白去乙酰化酶的抑制剂和G蛋白偶联受体（G protein-coupled receptors，GPCRs）的配体，因此可以作为影响造血和非造血细胞系增殖和功能的信号分子[205]。此外，次级胆汁酸、吲哚衍生物和单链脂肪酸等代谢产物，可以通过肠内分泌细胞和肠嗜铬细胞发出信号，调节神经肽（如食欲调节激素GLP1）和神经调节剂（如激素和神经递质5-羟色胺）的分泌。此外，肠道细菌的亚群可以直接合成和释放神经递质和神经调节剂。如果这些物质被吸收并释放到门静脉循环中，它们直接影响中枢神经元活动的潜力将取决于其穿越血脑屏障的能力[206]。有意思的是，He等提出肠道菌群的代谢产物丁酸盐可通过ID2依赖的方式促进$CD8^{+}$T细胞的增殖和功能，即肠道菌群可能通过其代谢产物调节抗肿瘤免疫反应[207]。同时，部分代谢产物还与多种疾病有关。近期有研究表明，一种肠道菌群的代谢产物——苯乙酰谷氨酰胺（通过肾上腺素受体传递信号，可增强血小板活化的相关表型并促进血栓形成）——与心血管疾病和重大不良心血管事件，如心肌梗死、脑卒中和死亡等有关[208]。这意味着调节肠道菌群的代谢产物可能是改变宿主健康状态的重要途径。

另一方面，一些细菌代谢产物可以通过充当细菌间通信的信号分子来靶向作用于微生物群，例如上文提到的群体感应分子。常见QS分子AI-2的水平与肠道中拟杆菌门（Bacteroidetes）丰度相对降低和厚壁菌门（Firmicutes）丰度增加有关[192]。值得注意的是，细菌不仅可以分泌促进其他细菌生长的QS化合物，还可以分泌抑制竞争对手生长的化合物。这些具有抑制作用的化合物可以是蛋白质（细菌素）或小分子（乳酸、过氧化氢和反应醛），通过定植抗性来增强宿主健康[209, 210]。有趣的是，宿主细胞已进化出干扰细菌QS信号的能力，称为群体感应淬灭，以抵御细菌入侵[211]。近来发现，哺乳动物肿瘤上皮细胞能分泌AI-2模拟物，影响细菌QS系统的活性[212]。

接下来，我们对肠道菌群的主要代谢产物进行详细介绍。

短链脂肪酸：主要包括乙酸、丙酸、异丁酸、丁酸、异戊酸、戊酸、己酸，是所有细菌代谢产物中研究最广泛的，在维持肠道菌群稳态、促进结肠上皮细胞代谢、减少炎症反应等方面都发挥着重要作用。现已证实，SCFAs可以被宿主的G蛋白偶联受体（如GPR-41和GPR-43）识别[213, 214]。刺激这些受体可触发参与葡萄糖代谢或食物摄入的肠道肽（如胰高血糖素样肽-1或多肽YY）的分泌[215, 216]。因此，通过代谢产物刺激肠内分泌细胞产生关键激素，肠道微生物可以远程作用于宿主的不同器官[217, 218]。值得注意的是，丙酸还可以调节宿主免疫细胞产生抗菌因子，起到免疫调节剂的作用，在一定程度上可以减少癌细胞增殖[219, 220]。也就是说，不同的微生物代谢物可能在宿主代谢中发挥不同的作用，从调节葡萄糖水平到免疫调节作用（图1-6）。

事实上，随着研究的深入，一些SCFAs的作用和作用机制相较以往有很大的不同。例如，丁酸之前被认为是一种必不可少的能源，它能够促进结肠细胞增殖，并有助于维持健康的肠道屏障功能。然而，Byndloss等的研究发现，丁酸还可能通过与宿主细胞交流在很大程度上影响微生物环境和生态[221]。尚有研究表明，丁酸能够通过激活β氧化来指示结肠细胞“呼吸”氧气，以保护宿主免受肠道潜在致病菌的侵袭[222]。肠腔内氧含量极低（即厌氧状态）是防止沙门菌属、埃希菌属等兼性厌氧病原菌扩张所必需的条件[223]。因此宿主细胞消耗线粒体中β氧化的氧有助于限制氧从结肠细胞扩散到肠腔内，最终维持肠腔中的厌氧条件。这一部分内容将在第6章详细阐述。

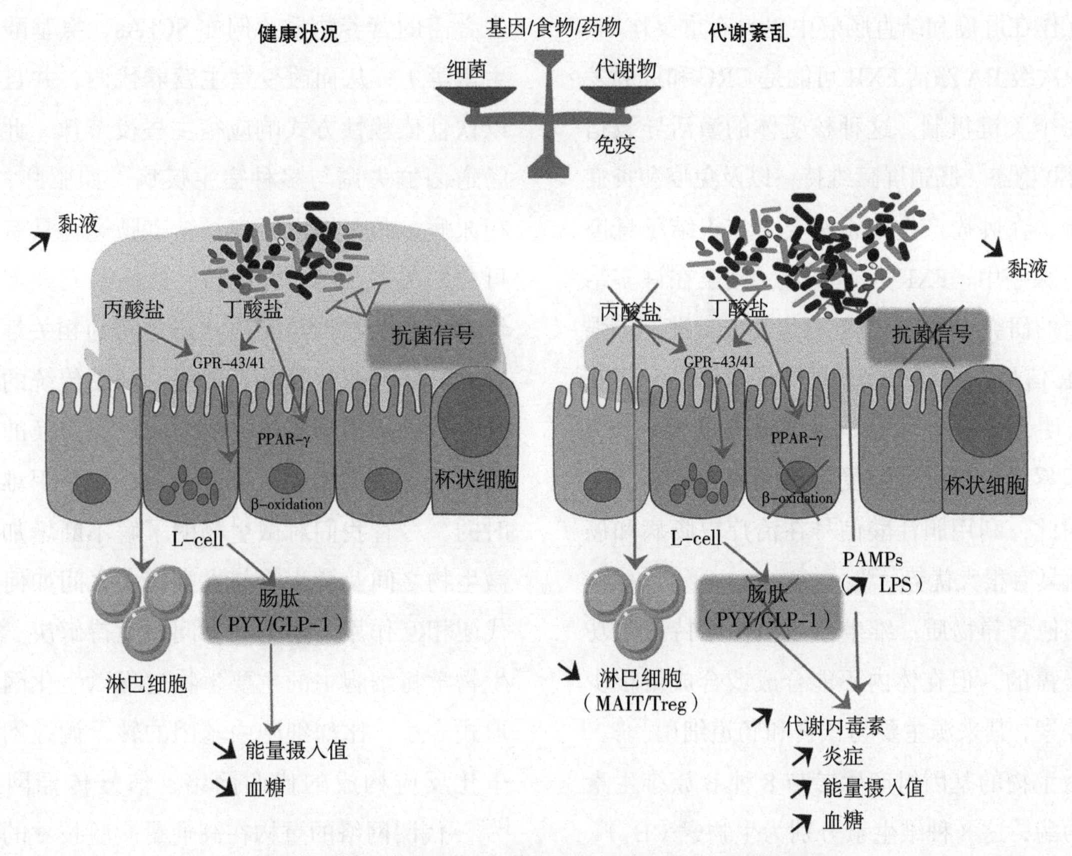

图 1-6　微生物与宿主之间交流的主要机制：代谢的影响

图的左边部分显示，在健康情况下，肠道微生物群的组成与较高的黏液层厚度、抗菌信号的产生以及丁酸和丙酸等不同短链脂肪酸的产生有关。丁酸和丙酸均与肠内分泌 L 细胞上表达的 GPR-43 和 GPR-41 结合，从而刺激胰高血糖素样肽 -1（GLP-1）或多肽 YY（PYY）等肠道肽的分泌。这种作用有助于减少食物摄入和改善葡萄糖代谢。丙酸还可以与淋巴细胞上表达的 GPR-43 结合，以维持宿主的免疫防御。丁酸激活 PPA R-γ 促进 β 氧化和肠道细胞耗氧的增加，该作用有助于维持肠腔厌氧状态。图的右边部分显示，在代谢紊乱过程中，肠道微生物群的变化与较低的黏液层厚度、抗菌防御能力的降低以及丁酸和丙酸的产生减少有关。在这种情况下，L 细胞分泌的肠肽减少。缺乏 PPA R-γ 激活将导致肠道细胞耗氧的减少，肠腔内可供微生物生存的氧气增加，进而促进肠杆菌科的增殖。丙酸的减少也导致包括 MAIT 及 Treg 细胞等在内的特定 T 细胞的丰度降低。总之，微生物环境和代谢物的这种变化将导致 PAMPs 的泄漏，如脂多糖在血液中增加，以及轻度炎症的发生。PPAR-γ. 氧化物酶体增殖物激活受体 -γ；GPR. G 蛋白偶联受体；MAIT. 黏膜相关不变 T 细胞；Treg 细胞 . 调节性 T 细胞；PAMPs. 病原体相关分子模式；LPS. 脂多糖；β-oxidation. β 氧化

胆汁酸：初级胆汁酸（bile acid，BA）从肝脏中的胆固醇、牛磺酸和甘氨酸结合物中转化，分泌到肠道，然后通过肠道微生物体中的胆盐水解酶将它们转化为次级胆汁酸[224]。胆汁酸在宿主体内主要通过激活特定的受体而发挥改变代谢的功能，这些受体包括法尼醇 X 受体（farnesoid X receptor，FXR）和 G 蛋白偶联受体，如 TGR5[224]。TGR5 是一种广泛表达的跨膜受体，通过促进棕色脂肪组织中的细胞内甲状腺激素活性，增加棕色脂肪组织和肌肉的能量消耗，并诱导肠 L 细胞释放胰岛素分泌 GLP-1 来调节能量平衡。FXR 是一种胞质配体激活的转录因子，可转运到细胞核以诱导靶基因转录。除了作为 BAs 合成和转运的调节因子外，FXR 还在调节炎症和免疫以及肝再生中起着重要作用，并在肝细胞和胃肠道中诱导保护性细胞反应。肠道菌群是胆汁酸代谢的重要调节因子。宿主 - 微生物 - 胆汁酸网络在影响宿主免疫反应中也起着至关重要的作用，哈佛大学的最新研究发现，胆汁酸是调节结肠 RORγ+T 细胞的一个重要群体，通过调节胆汁酸核受体从而影响宿主对炎症性肠病的易感性[225]。胆汁酸 - 肠道

菌群互作在肝癌和结直肠癌中也具有重要作用。肠道中次级 BA 激活 FXR 可能是 CRC 和肝癌发病的一个关键机制，这种核受体的激活导致宿主 - 菌群稳态、肠道屏障维持，以及免疫和炎症的控制，在肿瘤产生中很重要。在人结肠息肉和结肠腺癌中，FXR 表达下降，甚至在许多结肠肿瘤的研究中发现 FXR 表达丢失，进一步强调 FXR 信号在胃肠道癌发展中的重要性。还有研究表明，梭菌产生次级 BA 引起的自然杀伤 T 细胞积累可保护宿主免受肝癌和肝转移瘤的侵袭。因此，利用胆汁酸信号在治疗胃肠病和癌症方面具有很大优势[225]。

其他营养物质：维生素是机体维持正常功能所必需的，但在体内不能合成或合成量很少的营养素，其来源主要为饮食和肠道细菌[226]。肠道微生物的基因组产生参与 8 种 B 族维生素合成的酶，这 8 种维生素分别为生物素（B_7）、钴胺（B_{12}）、叶酸（B_9）、烟酸（B_3）、泛酸（B_5）、吡多辛（B_6）和核黄素（B_2）[226]。肠道微生物代谢产物的维生素与饮食中的维生素不同，微生物维生素如单、聚谷氨酰叶酸的关键修饰能影响其在宿主体内的吸收和功能，证明了微生物维生素对宿主健康的独特贡献[227-229]。此外，肠道微生物还能利用蛋白质残渣合成非必需氨基酸，如天冬氨酸、丙氨酸、缬氨酸和苏氨酸等，并参与糖类和蛋白质的代谢，同时还能促进铁、镁、锌等矿物元素的吸收。这些营养物质对人类的健康起着至关重要的作用，一旦缺少会引起多种疾病[230]。除上述代谢产物外，还有许多其他肠道微生物的代谢物在宿主代谢中发挥着不可忽视的作用，这里不一一介绍。

二、常见代谢模型

如前文所述，肠道微生物可以从饮食中分解出大量人体自身无法消化的化合物，同时产生必需的营养物质（例如 SCFAs、氨基酸和维生素等），从而改变宿主营养代谢，并且可以以饮食依赖性方式响应宿主昼夜节律。此外，肠道菌群失调与多种宿主疾病，如肥胖、2 型糖尿病、肠道慢性炎症、结直肠癌等具有密不可分的关系[230, 231]。然而，这些研究大多数仅针对于微生物群与疾病状态之间的相关性，其中的具体代谢途径并不清楚。利用传统的研究方法，要阐明肠道微生物之间代谢交换的详细机制及这些交换如何影响人类代谢是困难且耗时的。尽管我们对微生物的了解不断增加，但微生物之间及微生物与人类宿主之间如何通过代谢相互作用，还有许多问题亟待解决。系统生物学将细胞中的主要生物过程以生化网络的形式表达，比如细胞中基因的转录调控网络、生化反应构成的代谢网络、信号传导网络等[232]。代谢网络的重构在高通量实验技术的支持下，已经取得了很大成功，其中，基于约束的重建和分析法（constraint-based reconstruction and analysis，COBRA）为代谢网络的重构提供了标准化方式。这种方法已成功应用于代谢途径、个体物种代谢和物种间代谢相互作用的研究[233]。2007 年，Stolyar 等利用普通脱硫弧菌（*Desulfovibrio vulgaris*）和海藻甲烷球菌（*Methanococcus maripaludis*）的全序列基因组建立了全球第一个多种化学计量代谢模型，这也是第一个微生物群落 COBRA 模型。他们的研究表明，重建的代谢网络和化学计量模型不仅可以预测单个生物体的代谢通量和生长表型，而且还可以捕获细菌群落的参数和组成[234]。

肠道微生物群的 COBRA 研究多使用基因组水平代谢网络重构（genome-scale metabolic reconstructions，GENREs）模型，它以基因组序列为基础，通过构建基因 - 蛋白质反应的关系重构生物体内代谢过程，整合了给定的微生物内部发生的绝大部分生化反应[235]。GENREs 模型是一种包含了某种特定生物或细胞基因组范围的代谢反应及其酶学和基因关联的数学模型。近 10 年以来，基因组规模的代谢网络分析领域迅速扩大，截至目前已陆续经发表了多个基因组规模的代谢重建报告。大致而言，GENREs 模型的构建包括 4 个步骤：①首先，根据基因注释数据和在线数据库（如 KEGG 和 EXPASY）的信息构建初始重构模型，将已知基因与功能类别联系起来，并缩小基因型 - 表型差距；②查阅主要文献优化重构模型，然后将知识库转化为数学模型；③通过模型预测与表型数据的比较，验证重建过程；④代谢重构模型进行持续的实验室循环，以提高模型准确性 [236]。目前，GENRES 已经可以在短短几小时内通过几个不同的平台（比如 Model SEED、KBase、Pathway Tools）实现自动创建 [237-239]。然而，这些自动模型的重建需要进一步手工改进，以解决包括化学计量一致性、反应方向性，以及基因注释在内的诸多问题 [240-242]。由于不同研究采用的算法不同，因此需要生化和表型数据来手动评估计算结果的生物学相关性并验证其差异 [243]。

由于微生物处于动态变化中，因此研究者们多采用 Logistic、Gompertz、Schunte、Lopz 等微生物生长动力学模型来描述微生物数量随生长时间的对应关系。同样，微生物代谢也存在时间和空间的动态变化而传统的 COBRA 方法不能复制微生物群落的这种动态。因此，为了弥补不足，一些新型能够反映微生物代谢动态变化的模型应运而生。其中较为著名的几种模型如下：①微生物生态系统的时空计算（computation of microbial ecosystems in time and space，COMETS）模型，这种模型能够动态模拟多物种细菌菌落之间的代谢交换 [244]；② d-OptCom 模型，该方法是对 OptCom 框架的扩展，可以模拟细胞外代谢物和单个微生物生物量浓度的动态变化 [245]；③ BacArena 模型，该方法可以模拟微生物群落代谢相互作用的时空动力学，此外，由于 BacArena 模型可以同时模拟多个菌株，因此可以利用 BacArena 模型来探索单个菌株在微生物群落中不同时间或地点的不同代谢表型 [246, 247]。2017 年，Hoek 等提出了空间动态通量平衡分析建模框架，以研究微生物种群动态和人类肠道微生物群的进化，为肠道微生物建模提供了新的方向 [248]。

除此之外，微生物与宿主的代谢相互影响。第一项关于哺乳动物宿主与微生物之间代谢模型的研究由 Bordbar 等于 2010 年发表 [249]。此后 Heinken 等进行了一项关于不同微生物群落对人小肠细胞代谢影响的研究。该研究证明，群落中病原体的存在会导致宿主细胞重要代谢功能的丧失，同时微生物群能够合成许多激素代谢的前体，也就是说在一定程度上肠道微生物群可以具有内分泌器官的功能 [250]。2017 年，Stefanía 等收集了 773 种肠道菌株的已知代谢数据，由此研发出了 AGORA 模型。利用这一模型，能够模拟肠道微生物代谢过程，分析它们对其他微生物和宿主代谢的影响，可以有针对性地检索具有重要功能的代谢途径，分析这些代谢进程如何出错，从而明确宿主 - 微生物组相互作用对特殊疾病的影响，最终有望实现个体精准治疗 [235]。

在基因重建过程中，组学数据是至关重要的，各种数据类型均可以与模型集成，以模拟个性化微生物、营养、体液中检测到的代谢物或基因表达数据（图 1-7）。值得注意的是，将组学数据应用于细胞外和（或）细胞内反应，就可以将通用的 GENRE 转化为条件特异性代谢模型。宏基因组学基于二代测序技术，又称微生物环境基因组学、元基因组学，它通过基因测序、序列拼接实现对研究样品中所包含的全部微生物的遗传组成及其群落功能的研究 [251]。SAMtools 是目前常用的一种高通量测序数据序列对比工具，其运行基于 SAM 格式（sequence alignment/map，是一种用于存储针对参考序列的通用对齐格式），可用于对比不同测序平台产生的序列（其可识别的基因片段大小高达 128Mbp），能够实现二进制查看、格式转换、排序及合并等功能，可以完成比对结果的统计汇总[252]。利用这类序列对比工具可以将粪便样本的元基因组读数映射至肠道微生物群落结构[253，254]。然而，微生物样品中微生物的基因编码功能并非都是活跃的，还存在大量非编码 RNA（non-coding RNA），因此需要额外的数据，如宏转录组学、宏蛋白组学和宏代谢物组学数据[255]。尽管组学数据和集成平台的可用性越来越高，截至目前，只有少数研究使用组学数据来描述肠道微生物群落的 GENRES，这就为微生物群代谢模型的研究留下了许多机会。

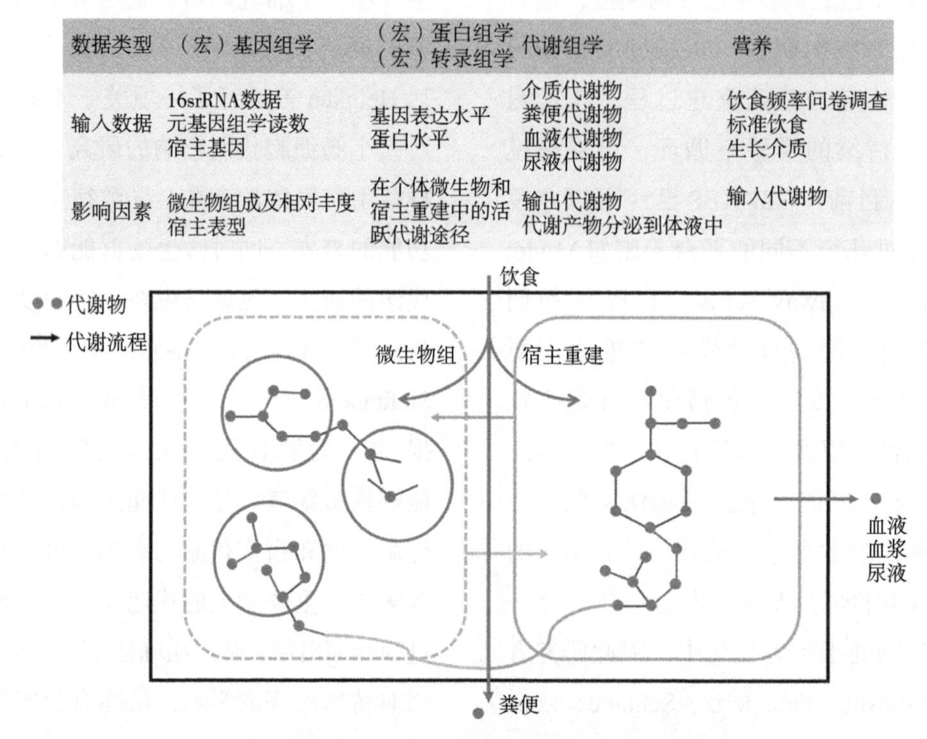

数据类型	（宏）基因组学	（宏）蛋白组学 （宏）转录组学	代谢组学	营养
输入数据	16srRNA数据 元基因组学读数 宿主基因	基因表达水平 蛋白水平	介质代谢物 粪便代谢物 血液代谢物 尿液代谢物	饮食频率问卷调查 标准饮食 生长介质
影响因素	微生物组成及相对丰度 宿主表型	在个体微生物和宿主重建中的活跃代谢途径	输出代谢物 代谢产物分泌到体液中	输入代谢物

图 1-7　宿主 - 微生物代谢模型设置的原理图概述

各种数据类型均可以与模型集成，以模拟个性化微生物、营养、体液中检测到的代谢物或基因表达数据。图中不同的颜色代表了模型中可以与不同数据类型关联的部分

第六节　小结与展望

我们的身体是人类细胞和微生物（包括细菌、古细菌、病毒和真菌）的共生体，其中肠道是微生物聚集最为密切的场所。肠道微生物可以弥补部分人体自身功能的缺乏，与人体的健康息息相关，被认为是众多生理功能的重要调节剂，例如消化、情绪、睡眠和对药物的反应，以及对糖尿病、自闭症、肥胖症和癌症等的易感性。总的来说，这些微生物群落具有以下 3 个方面的功能：①代谢功能，肠道细菌能够代谢食物中宿主自身不能消化、分解的物质，并将其转化为具有一定生理功能的代谢终产物（如 SCFA、乳酸、甘油、胆碱等），为宿主提供能量及细菌生长繁殖所需的营养物质；②屏障效应，肠道微生物可以增强对外源性或机会性病原体定植的抵抗力；③促进肠上皮和免疫系统的成熟。同时，人类宿主也为微生物的生存和繁殖提供了一个营养丰富的环境，并且随着人体肠道各部分的变化，肠道各处定植的微生物

群有着明显的差异。

大量的DNA测序研究扩大了我们对肠道微生物群的生态和功能特征的理解。先进的测序技术使我们能够了解肠道微生物群与人类健康和危重疾病的密切联系。通过调节肠道微生物群可以干预人体健康和疾病，这一观点逐渐获得世界范围的认可。目前，一种基于微生物群信息的个性化饮食建议已经被用于作为糖尿病等代谢性疾病管理的指南[256]。因此，我们认为来自微生物组的数据将为今后各种疾病的治疗方向提供重要信息。随着对粪便取样、DNA提取方法、大量DNA测序、生物信息学分析和优化方法的知识积累，我们可以从人类肠道微生物群中获得更重要的信息，从而促进未来个性化保健制度的发展。目前，许多分子生物学分析工具已被迅速开发并应用于肠道微生物群的研究中，然而，研究人员之间的方法差异可能导致数据不一致，并且限制了数据的广泛共享。因此，规范目前的方法，并为人类肠道微生物群的研究建立适当的交流平台是非常重要的。在本章节中，我们回顾了目前用于研究人类肠道微生物群的方法（培养法、分子基法、基于小代谢物检测和定量的方法），并对16S rRNA测序法和气相色谱 - 质谱法的标准步骤、适用范围、数据分析，以及优缺点做了详尽描述。我们相信，这一部分的内容将有助于规范目前检测肠道微生物群的方法，从而有助于拓展肠道微生物群在人类健康的临床和实践方面的研究进展。

目前大量研究已经在分析微生物的组成、关键代谢物，以及新细菌的发现与分离等方面取得了巨大进展。微生物群落根据年龄、饮食和胃肠道位置的不同而发生相应改变。健康肠道微生物群的特征包括相对较大的生物多样性和特定门和属的相对丰度。在生命的早期阶段，肠道菌群的组成和结构相对稳定。此外，妊娠期间母体肠道菌群可能对刺激胎儿免疫系统发育具有重要的作用，并且可能塑造后代机体的代谢状况。随着时间的推移，外界因素的干扰（如饮食、营养、生活环境的差异，以及抗生素的使用）增加，细菌的组成和多样性也随之发生变化。人类肠道微生物的组成和多样性变化往往与肠道功能障碍有着密切的关系，肠道菌群失调将导致人体代谢、免疫等生理功能出现紊乱，最终导致多种疾病的发生。

细菌含有能够分解食物残渣的酶，它们对于糖和脂肪的分解称为发酵，发酵产物主要包括SCFA、乳酸、甘油、胆碱、二氧化碳、甲烷等，其中二氧化碳和甲烷是肠道气体的主要来源。细菌对蛋白质的分解称为腐败，产物有氨基酸、氨气、硫化氢、组胺、吲哚等。这些微生物代谢物具有非常重要的生理功能，能够提供关键信号，可以作用于微生物与微生物之间和微生物与宿主之间，有助于维持健康的生理状态。虽然目前关于肠道微生物的研究成果颇丰，但大多数研究仍然停留于简单地探索细菌组成和人体生理、病理特征变化的相关性这一层面。

当我们发现特定的微生物与人体疾病或健康情况之间存在相关性时，很难判定该细菌对疾病发生或健康情况具有确切作用。换句话说，相关性仅能反映特定细菌对人体的潜在有益或有害作用。因此，作为研究人员，如果我们想超越简单的相关性研究而探讨条件变化与肠道菌群改变的因果关系并了解其机制的话，就需要尽可能多地采用更复杂的分析方法（例如，多组学和时间序列测量分析）。但是，通过16s rRNA测序法分离特定细菌、鉴定、培养，并在复杂的体外模型中进行研究，这一过程是非常耗时耗力的。尽管研究员们最近在细菌培养方面已经取得了很大进展，但厌氧菌的分离和鉴定仍然是困难的[120, 257]。此外，当特定的细菌被成功分离和培养后，还需要经历漫长的体内试验才能证实细菌对人体某种生理或病理状态的确切作用。由于上述原因以及现有知识水平的缺乏，时至今日，我们仍然很难完全破译肠道微生物等复杂群落在人体健康与疾病中的确切作用和机制。

代谢模型是研究代谢途径、个体物种代谢和物种间代谢相互作用的重要工具。在上一小节中，我们回顾性分析了肠道微生物群代谢功能研究中运用到的代谢模型，并且重点介绍了研究人类肠道微生物代谢相互作用的 COBRA 法。微生物群落代谢模型与代表性体外研究的代谢组学数据相匹配，表明 COBRA 方法适用于微生物群落的研究[258]。此外，COBRA 方法也可用于来构建和模拟大规模的微生物群落模型[259]。我们设想未来对人类肠道微生物群的 COBRA 研究将探索个性化肠道微生物群代谢模型的使用，使之应用于个性化医学，例如，用于制订代谢疾病的个性化治疗方案。

综上，肠道菌群这一陪伴我们一生的极为复杂的微生物群落，对于人体而言具有非常重要的生理功能，与我们的健康与疾病休戚相关。同时，随着人们对肠道菌群研究的深入和研究方法的完善，肠道菌群与一些非感染性慢性疾病的关系被逐步发现，成为研究新的治疗方法的切入点。

参考文献

[1] Marchesi JR, Ravel J. The vocabulary of microbiome research: a proposal. Microbiome,2015,3:31. dio: 10.1186/S40168-015-0094-5.

[2] Sarangi AN, Goel A, Aggarwal R. Methods for studying gut microbiota: a primer for physicians. Journal of Clinical and Experimental Hepatology,2019,9(1):62-73.

[3] Colov EP, Degett TH, Raskov H, et al. The impact of the gut microbiota on prognosis after surgery for colorectal cancer - a systematic review and meta-analysis. Acta Pathologica, Microbiologica, et Immunologica Scandinavica,2020，128(2):162-176.

[4] Wang JW, Kuo CH, Kuo FC, et al. Fecal microbiota transplantation: review and update. Journal of the Formosan Medical Association, 2019,118 Suppl 1:S23-S31.

[5] Islam SU. Clinical uses of probiotics. Medicine (Baltimore), 2016,95(5):e2658.

[6] Sokolis DP. Variation of passive biomechanical properties of the small intestine along its length: microstructure-based characterization. Bioengineering (Basel),2021,8(3):32.

[7] Lew D, Kashani A, Lo SK, et al. Efficacy and safety of cap-assisted endoscopic mucosal resection of ileocecal valve polyps. Endoscopy International Open,2020,8(3):E241-E246.

[8] Liu Q, Chen Z, Chen Y, et al. Microplastics and nanoplastics: emerging contaminants in food. Journal of Agricultural and Food Chemistry, 2021,69(36):10450-10468.

[9] Duerkop BA, Vaishnava S, Hooper LV. Immune responses to the microbiota at the intestinal mucosal surface. Immunity, 2009,31(3):368-376.

[10] Bein A, Shin W, Jalili-Firoozinezhad S, et al. Microfluidic organ-on-a-chip models of human intestine. Cellular and Molecular Gastroenterology and Hepatology,2018,5(4):659-668.

[11] Fanous J, Swed A, Joubran S, et al. Superiority of the S,S conformation in diverse pharmacological processes: Intestinal transport and entry inhibition activity of novel anti-HIV drug lead. International Journal of Pharmaceutics,2015,495(2):660-663.

[12] Bloemen JG, Venema K, van de Poll MC, et al. Short chain fatty acids exchange across the gut and liver in humans measured at surgery. Clinical Nutrition (Edinburgh, Scotland),2009,28(6):657-661.

[13] Ahuja M, Schwartz DM, Tandon M, et al. Orai1-mediated antimicrobial secretion from pancreatic acini shapes the gut microbiome and regulates gut innate immunity. Cell Metabolism, 2017,25(3):635-646.

[14] Mayer EA. Gut feelings: the emerging biology of gut-brain communication. Nature Reviews Neuroscience,2011,12(8):453-466.

[15] Cryan JF, Dinan TG. Mind-altering microorganisms: the impact of the gut microbiota on brain and behaviour. Nature Reviews Neuroscience,2012, 13(10):701-712.

[16] Ahern PP, Maloy KJ. Understanding immune-microbiota interactions in the intestine. Immunology, 2020,159(1):4-14.

[17] Peterson LW, Artis D. Intestinal epithelial cells: regulators of barrier function and immune homeostasis. Nature Reviews Immunology,2014,14(3):141-153.

[18] Tabula Muris C, Overall C, Logistical C, et al. Single-cell transcriptomics of 20 mouse organs creates a

Tabula Muris. Nature,2018,562(7727):367-372.

[19] Han X, Wang R, Zhou Y, et al. Mapping the mouse cell atlas by microwell-seq. Cell,2018,172(5):1091-107 e17.

[20] Rackaityte E, Halkias J, Fukui EM, et al. Viable bacterial colonization is highly limited in the human intestine in utero. Nature Medicine,2020,26(4):599-607.

[21] Kennedy KM, Gerlach MJ, Adam T, et al. Fetal meconium does not have a detectable microbiota before birth. Nature Microbiology,2021,6(7):865-873.

[22] Mondot S, de Wouters T, Dore J, et al. The human gut microbiome and its dysfunctions. Digestive Diseases,2013,31(3-4):278-285.

[23] Roswall J, Olsson LM, Kovatcheva-Datchary P, et al. Developmental trajectory of the healthy human gut microbiota during the first 5 years of life. Cell Host & Microbe,2021,29(5):765-776, e3.

[24] Marcobal A, Sonnenburg JL. Human milk oligosaccharide consumption by intestinal microbiota. Clinical Microbiology and Infection,2012,18 Suppl 4:12-15.

[25] Matamoros S, Gras-Leguen C, Le Vacon F, et al. Development of intestinal microbiota in infants and its impact on health. Trends in Microbiology, 2013,21(4):167-173.

[26] Carr VR, Witherden EA, Lee S, et al. Abundance and diversity of resistomes differ between healthy human oral cavities and gut. Nature Communications, 2020,11(1):693.

[27] Martinez I, Stegen JC, Maldonado-Gomez MX, et al. The gut microbiota of rural papua new guineans: composition, diversity patterns, and ecological processes. Cell Reports,2015,11(4):527-538.

[28] Almeida A, Mitchell AL, Boland M, et al. A new genomic blueprint of the human gut microbiota. Nature, 2019,568(7753):499-504.

[29] Kumar PS, Monteiro MF, Dabdoub SM, et al. Subgingival host-microbial interactions in hyperglycemic individuals. Journal of Dental Research, 2020,99(6):650-657.

[30] Watanabe K, Katagiri S, Takahashi H, et al. Porphyromonas gingivalis impairs glucose uptake in skeletal muscle associated with altering gut microbiota. FASEB Journal,2021,35(2):e21171.

[31] Escapa IF, Chen T, Huang Y, et al. New insights into human nostril microbiome from the expanded human oral microbiome database (eHOMD): a resource for the microbiome of the human aerodigestive tract. mSystems,2018,3(6) :e00187-18.

[32] Ruan W, Engevik MA, Spinler JK, et al. Healthy human gastrointestinal microbiome: composition and function after a decade of exploration. Digestive Diseases and Sciences,2020,65(3):695-705.

[33] Corning B, Copland AP, Frye JW. The esophageal microbiome in health and disease. Current Gastroenterology Reports,2018,20(8):39.

[34] Pereira-Marques J, Ferreira RM, Pinto-Ribeiro I, et al. Helicobacter pylori infection, the gastric microbiome and gastric cancer. Advances in Experimental Medicine and Biology,2019,1149:195-210.

[35] Rao SSC, Bhagatwala J. Small intestinal bacterial overgrowth: clinical features and therapeutic management. Clinical and Translational Gastroenterology, 2019, 10(10):e00078.

[36] Sender R, Fuchs S, Milo R. Revised estimates for the number of human and bacteria cells in the body. PLoS Biology,2016,14(8):e1002533.

[37] Tropini C, Earle KA, Huang KC, et al. The gut microbiome: connecting spatial organization to function. Cell Host & Microbe,2017,21(4):433-442.

[38] Crost EH, Tailford LE, Monestier M, et al. The mucin-degradation strategy of Ruminococcus gnavus: the importance of intramolecular trans-sialidases. Gut Microbes,2016,7(4):302-312.

[39] Yasuda K, Oh K, Ren B, et al. Biogeography of the intestinal mucosal and lumenal microbiome in the rhesus macaque. Cell Host & Microbe, 2015,17(3):385-391.

[40] Albenberg L, Esipova TV, Judge CP, et al. Correlation between intraluminal oxygen gradient and radial partitioning of intestinal microbiota. Gastroenterology, 2014,147(5):1055-1063.

[41] Silva YP, Bernardi A, Frozza RL. The role of short-chain fatty acids from gut microbiota in gut-brain communication. Front Endocrinol (Lausanne), 2020, 11:25.

[42] Pabst O, Cerovic V, Hornef M. Secretory IgA in the coordination of establishment and maintenance of the microbiota. Trends in Immunology,2016,37(5):287-296.

[43] Bernet MF, Brassart D, Neeser JR, et al. Lactobacillus acidophilus LA 1 binds to cultured human intestinal cell lines and inhibits cell attachment and cell invasion by enterovirulent bacteria. Gut,1994,35(4):483-489.

[44] Hall AB, Tolonen AC, Xavier RJ. Human genetic variation and the gut microbiome in disease. Nature Reviews Genetics,2017,18(11):690-699.

[45] Peng W, Li H, Xu Y, et al. Association of helicobacter bilis infection with the development of colorectal cancer. Nutrition and Cancer,2020:1-11.

[46] Liu JZ, van Sommeren S, Huang H, et al. Association analyses identify 38 susceptibility loci for inflammatory bowel disease and highlight shared genetic risk across populations. Nature Genetics,2015,47(9):979-986.

[47] Ma Q, Li Y, Li P, et al. Research progress in the relationship between type 2 diabetes mellitus and intestinal flora. Biomed Pharmacother, 2019, 117: 109138.

[48] Cui L, Morris A, Ghedin E. The human mycobiome in health and disease. Genome Medicine,2013,5(7):63.

[49] Erdogan A, Rao SS. Small intestinal fungal overgrowth. Current Gastroenterology Reports. 2015,17(4):16.

[50] Backhed F, Ley RE, Sonnenburg JL, et al. Host-bacterial mutualism in the human intestine. Science, 2005, 307(5717):1915-1920.

[51] Eckburg PB, Bik EM, Bernstein CN, et al. Diversity of the human intestinal microbial flora. Science, 2005, 308(5728):1635-1638.

[52] Knobbe TJ, Douwes RM, Kremer D, et al. Altered gut microbial fermentation and colonization with methanobrevibacter smithii in renal transplant recipients. Journal of Clinical Medicine,2020, 9(2):518.

[53] Collins MD, Lawson PA, Willems A, et al. The phylogeny of the genus clostridium: proposal of five new genera and eleven new species combinations. International Journal of Systematic Bacteriology, 1994, 44(4):812-826.

[54] Zoetendal EG, Ben-Amor K, Harmsen HJ, et al. Quantification of uncultured ruminococcus obeum-like bacteria in human fecal samples by fluorescent in situ hybridization and flow cytometry using 16S rRNA-targeted probes. Applied and Environmental Microbiology, 2002,68(9):4225-4232.

[55] Hold GL, Schwiertz A, Aminov RI, et al. Oligonucleotide probes that detect quantitatively significant groups of butyrate-producing bacteria in human feces. Applied and Environmental Microbiology, 2003, 69(7):4320-4324.

[56] Debedat J, Clement K, Aron-Wisnewsky J. Gut microbiota dysbiosis in human obesity:impact of bariatric surgery. Current Obesity Reports,2019,8(3):229-242.

[57] Barcenilla A, Pryde SE, Martin JC, et al. Phylogenetic relationships of butyrate-producing bacteria from the human gut. Applied and Environmental Microbiology, 2000, 66(4):1654-1661.

[58] Lay C, Sutren M, Rochet V, et al. Design and validation of 16S rRNA probes to enumerate members of the Clostridium leptum subgroup in human faecal microbiota. Environmental Microbiology, 2005, 7(7):933-946.

[59] Saunier K, Rouge C, Lay C, et al. Enumeration of bacteria from the Clostridium leptum subgroup in human faecal microbiota using Clep1156 16S rRNA probe in combination with helper and competitor oligonucleotides. Systematic and Applied Microbiology, 2005,28(5):454-464.

[60] Duncan SH, Hold GL, Harmsen HJM, et al. Growth requirements and fermentation products of Fusobacterium prausnitzii, and a proposal to reclassify it as Faecalibacterium prausnitzii gen. nov., comb. nov. International Journal of Systematic and Evolutionary Microbiology, 2002,52(Pt 6):2141-2146.

[61] Suau A, Rochet V, Sghir A, et al. Fusobacterium prausnitzii and related species represent a dominant group within the human fecal flora. Systematic and Applied Microbiology,2001,24(1):139-145.

[62] Hayashi H, Sakamoto M, Benno Y. Phylogenetic analysis of the human gut microbiota using 16S rDNA clone libraries and strictly anaerobic culture-based methods. Microbiology and Immunology, 2002, 46(8):535-548.

[63] Pryde SE, Duncan SH, Hold GL, et al. The microbiology of butyrate formation in the human colon. FEMS Microbiology Letters,2002,217(2):133-139.

[64] Matsuki T, Watanabe K, Fujimoto J, et al. Development of 16S rRNA-gene-targeted group-specific primers for the detection and identification of predominant bacteria in human feces. Applied and Environmental Microbiology, 2002,68(11):5445-5451.

[65] Matsuki T, Watanabe K, Fujimoto J, et al. Use of 16S rRNA gene-targeted group-specific primers for real-time PCR analysis of predominant bacteria in human feces. Applied and Environmental Microbiology, 2004,70(12):7220-7228.

[66] Sghir A, Gramet G, Suau A, et al. Quantification of bacterial groups within human fecal flora by oligonucleotide probe hybridization. Applied and Environmental Microbiology,2000,66(5):2263-2266.

[67] Marteau P, Pochart P, Dore J, et al. Comparative study of bacterial groups within the human cecal and fecal microbiota. Applied and Environmental Microbiology, 2001,67(10):4939-4942.

[68] Maukonen J, Satokari R, Matto J, et al. Prevalence and temporal stability of selected clostridial groups in irritable bowel syndrome in relation to predominant faecal bacteria. Journal of Medical Microbiology,2006,55(Pt 5):625-633.

[69] Franks AH, Harmsen HJ, Raangs GC, et al. Variations of bacterial populations in human feces measured by fluorescent in situ hybridization with group-specific 16S rRNA-targeted oligonucleotide probes. Applied and Environmental Microbiology,1998,64(9):3336-3345.
[70] Lay C, Rigottier-Gois L, Holmstrom K, et al. Colonic microbiota signatures across five northern European countries. Applied and Environmental Microbiology, 2005,71(7):4153-4155.
[71] Mueller S, Saunier K, Hanisch C, et al. Differences in fecal microbiota in different European study populations in relation to age, gender, and country: a cross-sectional study. Applied and Environmental Microbiology, 2006,72(2):1027-1033.
[72] Costea PI, Hildebrand F, Arumugam M, et al. Enterotypes in the landscape of gut microbial community composition. Nature Microbiology, 2018, 3(1): 8-16.
[73] Arumugam M, Raes J, Pelletier E, et al. Enterotypes of the human gut microbiome. Nature, 2011, 473(7346):174-180.
[74] Holmes I, Harris K, Quince C. Dirichlet multinomial mixtures: generative models for microbial metagenomics. PLoS One,2012,7(2):e30126.
[75] Klijn A, Mercenier A, Arigoni F. Lessons from the genomes of bifidobacteria. FEMS Microbiology Reviews,2005,29(3):491-509.
[76] Langendijk PS, Schut F, Jansen GJ, et al. Quantitative fluorescence in situ hybridization of Bifidobacterium spp. with genus-specific 16S rRNA-targeted probes and its application in fecal samples. Applied and Environmental Microbiology,1995,61(8):3069-3075.
[77] Welling GW, Elfferich P, Raangs GC, et al. 16S ribosomal RNA-targeted oligonucleotide probes for monitoring of intestinal tract bacteria. Scandinavian Journal of Gastroenterology. Supplement, 1997, 222: 17-19.
[78] Matsuki T, Watanabe K, Fujimoto J, et al. Quantitative PCR with 16S rRNA-gene-targeted species-specific primers for analysis of human intestinal bifidobacteria. Applied and Environmental Microbiology, 2004,70(1):167-173.
[79] Harmsen HJ, Raangs GC, He T, et al. Extensive set of 16S rRNA-based probes for detection of bacteria in human feces. Applied and Environmental Microbiology, 2002,68(6):2982-2990.
[80] Knight DJ, Girling KJ. Gut flora in health and disease. Lancet,2003,361(9371):1831.
[81] O'Hara AM, Shanahan F. The gut flora as a forgotten organ. EMBO Reports,2006,7(7):688-693.
[82] Turnbaugh PJ, Ley RE, Hamady M, et al. The human microbiome project. Nature,2007,449(7164):804-810.
[83] Zhang CX, Wang HY, Chen TX. Interactions between intestinal microflora/probiotics and the immune system. BioMed Research International ,2019,2019:6764919.
[84] Hooper LV, Littman DR, Macpherson AJ. Interactions between the microbiota and the immune system. Science, 2012, 336(6086):1268-1273.
[85] Kurilshikov A, Wijmenga C, Fu J, et al. Host genetics and gut microbiome: challenges and perspectives. Trends in Immunology ,2017,38(9):633-647.
[86] Belkaid Y, Hand TW. Role of the microbiota in immunity and inflammation. Cell,2014,157(1):121-141.
[87] Wang X, Yang Y, Huycke MM. Microbiome-driven carcinogenesis in colorectal cancer: Models and mechanisms. Free Radical Biology & Medicine, 2017, 105:3-15.
[88] Gao J, Xu K, Liu H, et al. Impact of the gut microbiota on intestinal immunity mediated by tryptophan metabolism. Frontiers in Cellular and Infection Microbiology,2018,8:13.
[89] Vallianou NG, Tzortzatou-Stathopoulou F. Microbiota and cancer: an update. Journal of Chemotherapy, 2019, 31(2):59-63.
[90] Reis SAD, da Conceicao LL, Peluzio M. Intestinal microbiota and colorectal cancer: changes in the intestinal microenvironment and their relation to the disease. Journal of Medical Microbiologyl, 2019, 68(10):1391-1407.
[91] Ansari I, Raddatz G, Gutekunst J, et al. The microbiota programs DNA methylation to control intestinal homeostasis and inflammation. Nature Microbiology, 2020, 5(4):610-619.
[92] James KR, Gomes T, Elmentaite R, et al. Distinct microbial and immune niches of the human colon. Nature Immunology,2020,21(3):343-353.
[93] Biesalski HK. Nutrition meets the microbiome: micronutrients and the microbiota. Annals of the New York Academy of Sciences,2016,1372(1):53-64.
[94] Obata Y, Castano A, Boeing S, et al. Neuronal programming by microbiota regulates intestinal physiology. Nature,2020, 578(7794):284-289.
[95] Pan HW, Du LT, Li W, et al. Biodiversity and richness shifts of mucosa-associated gut microbiota with progression of colorectal cancer. Research in Microbiology, 2020, 171(3-4):107-114.
[96] Norouzi-Beirami MH, Marashi SA, Banaei-Moghaddam AM, et al. Beyond taxonomic analysis

of microbiomes: a functional approach for revisiting microbiome changes in colorectal cancer. Frontiers in Microbiology, 2019,10:3117.

[97] Han YW, Wang X. Mobile microbiome: oral bacteria in extra-oral infections and inflammation. Journal of Dental Research,2013,92(6):485-491.

[98] Kristensen NB, Bryrup T, Allin KH, et al. Alterations in fecal microbiota composition by probiotic supplementation in healthy adults: a systematic review of randomized controlled trials. Genome Medicine,2016,8(1):52.

[99] Buffie CG, Bucci V, Stein RR, et al. Precision microbiome reconstitution restores bile acid mediated resistance to clostridium difficile. Nature, 2015, 517(7533):205-208.

[100] Shi Y, Zheng W, Yang K, et al. Intratumoral accumulation of gut microbiota facilitates CD47-based immunotherapy via STING signaling. The Journal of Experimental Medicine,2020,217(5): e20192282.

[101] Zhang F, Luo W, Shi Y, et al. Should we standardize the 1,700-year-old fecal microbiota transplantation? The American Journal of Gastroenterology, 2012, 107(11): 1755-1756.

[102] Van Nood E, Vrieze A, Nieuwdorp M, et al. Duodenal infusion of donor feces for recurrent clostridium difficile. The New England Journal of Medicine, 2013,368(5):407-415.

[103] Fuentes S, van Nood E, Tims S, et al. Reset of a critically disturbed microbial ecosystem: faecal transplant in recurrent clostridium difficile infection. The ISME Journal,2014,8(8):1621-1633.

[104] Vermeire S, Joossens M, Verbeke K, et al. Donor species richness determines faecal microbiota transplantation success in inflammatory bowel disease. Journal of Crohn's & Colitis, 2016, 10(4): 387-394.

[105] Li SS, Zhu A, Benes V, et al. Durable coexistence of donor and recipient strains after fecal microbiota transplantation. Science,2016,352(6285):586-589.

[106] Blaut M, Collins MD, Welling GW, et al. Molecular biological methods for studying the gut microbiota: the EU human gut flora project. The British Journal of Nutrition,2002,87 Suppl 2:S203-S211.

[107] Gillespie JJ, Johnston JS, Cannone JJ, et al. Characteristics of the nuclear (18S, 5.8S, 28S and 5S) and mitochondrial (12S and 16S) rRNA genes of Apis mellifera (Insecta: Hymenoptera): structure, organization, and retrotransposable elements. Insect Molecular Biology,2006,15(5):657-686.

[108] Chen YZ, Deng WA, Wang JM, et al. Phylogenetic relationships of Scelimeninae genera (Orthoptera: Tetrigoidea) based on COI, 16S rRNA and 18S rRNA gene sequences. Zootaxa,2018,4482(2):392-400.

[109] Bouchet V, Huot H, Goldstein R. Molecular genetic basis of ribotyping. Clinical Microbiology Rreviews,2008,21(2):262-273.

[110] Tong M, Jacobs JP, McHardy IH, et al. Sampling of intestinal microbiota and targeted amplification of bacterial 16S rRNA genes for microbial ecologic analysis. Current Protocols in Immunology, 2014, 107: 7411-74111.

[111] Flemer B, Lynch DB, Brown JM, et al. Tumour-associated and non-tumour-associated microbiota in colorectal cancer. Gut,2017,66(4):633-643.

[112] Maleki A, Zamirnasta M, Taherikalani M, et al. The characterization of bacterial communities of oropharynx microbiota in healthy children by combining culture techniques and sequencing of the 16S rRNA gene. Microbial Pathogenesis,2020, 143: 104115.

[113] Kim H, Kim S, Jung S. Instruction of microbiome taxonomic profiling based on 16S rRNA sequencing. Journal of Microbiology (Seoul, Korea),2020, 58(3): 193-205.

[114] Fouhy F, Clooney AG, Stanton C, et al. 16S rRNA gene sequencing of mock microbial populations-impact of DNA extraction method, primer choice and sequencing platform. BMC Microbiology, 2016, 16(1):123.

[115] Clooney AG, Fouhy F, Sleator RD, et al. Comparing apples and oranges?: next generation sequencing and its impact on microbiome analysis. PLoS One, 2016, 11(2):e0148028.

[116] Schloss PD. Application of a database-independent approach to assess the quality of operational taxonomic unit picking methods. mSystems,2016,1(2) : e00027-16.

[117] Perez-Losada M, Arenas M, Galan JC, et al. High-throughput sequencing (HTS) for the analysis of viral populations. Infection, Genetics and Evolution, 2020, 80:104208.

[118] Podnar JW, Kolling FW, Zeller MJ, et al. Cross site evaluation of sanger sequencing dye chemistries. Journal of Biomolecular Techniques: JBT, 2020, 31(3): 88-93.

[119] Zhang X, Liu W, Lu Y, et al. Recent advances in the application of headspace gas chromatography-mass

spectrometry. Se Pu,2018,36(10):962-971.

[120] Cani PD. Human gut microbiome: hopes, threats and promises. Gut,2018,67(9):1716-1725.

[121] Davies SL, Davison AS. Liquid chromatography tandem mass spectrometry for plasma metadrenalines. Clinica Chimica Acta; International Journal of Clinical Chemistry, 2019,495:512-521.

[122] Ecker J, Scherer M, Schmitz G, et al. A rapid GC-MS method for quantification of positional and geometric isomers of fatty acid methyl esters. Journal of Chromatography. B, Analytical Technologies in the Biomedical and Life Sciences,2012,897:98-104.

[123] Aparicio-Ruiz R, Garcia-Gonzalez DL, Morales MT, et al. Comparison of two analytical methods validated for the determination of volatile compounds in virgin olive oil: GC-FID vs GC-MS. Talanta, 2018,187:133-141.

[124] Folch J, Lees M, Sloane Stanley GH. A simple method for the isolation and purification of total lipides from animal tissues. The Journal of Biological Chemistry, 1957, 226(1):497-509.

[125] Bligh EG, Dyer WJ. A rapid method of total lipid extraction and purification. Canadian Journal of Biochemistry and Physiology, 1959,37(8):911-917.

[126] Breil C, Abert Vian M, Zemb T, et al. "Bligh and dyer" and folch methods for solid-liquid-liquid extraction of lipids from microorganisms. Comprehension of Solvatation Mechanisms and towards Substitution with Alternative Solvents. International Journal of Molecular Sciences, 2017, 18(4): 708.

[127] Lofgren L, Forsberg GB, Stahlman M. The BUME method: a new rapid and simple chloroform-free method for total lipid extraction of animal tissue. Scientific Reports,2016,6:27688.

[128] Julien M, Gori D, Hohener P, et al. Intramolecular isotope effects during permanganate oxidation and acid hydrolysis of methyl tert-butyl ether. Chemosphere, 2020,248:125975.

[129] Zacharis CK, Tzanavaras PD. Solid-phase microextraction. Molecules,2020,25(2):379.

[130] Huang S, Chen G, Ye N, et al. Solid-phase microextraction: An appealing alternative for the determination of endogenous substances - A review. Analytica Chimica Acta,2019,1077:67-86.

[131] 郭慧，徐卓．固相微萃取技术在制药和生物医学分析中的当前发展和未来趋势．中国医药指南，2019, 17(13):292-293.

[132] Fiorini D, Pacetti D, Gabbianelli R, et al. A salting out system for improving the efficiency of the headspace solid-phase microextraction of short and medium chain free fatty acids. Journal of Chromatography. A,2015,1409:282-287.

[133] Shahwar D, Young LW, Shim YY, et al. Extractive silylation method for high throughput GC analysis of flaxseed cyanogenic glycosides. Journal of Chromatography. B, Analytical Technologies in the Biomedical and Life Sciences,2019,1132:121816.

[134] Chiu HH, Kuo CH. Gas chromatography-mass spectrometry-based analytical strategies for fatty acid analysis in biological samples. Journal of Food and Drug Analysis,2020,28(1):60-73.

[135] Pflaster EL, Schwabe MJ, Becker J, et al. A high-throughput fatty acid profiling screen reveals novel variations in fatty acid biosynthesis in Chlamydomonas reinhardtii and related algae. Eukaryot Cell,2014,13(11):1431-1438.

[136] 杜嘉琳，杨凉九，卫克昭，等．三甲基硅烷衍生化 GC-MS 测定兔血浆中的游离脂肪酸．Journal of Chinese Pharmaceutical Sciences, 2020,29(06):411-421.

[137] Kulkarni BV, Wood KV, Mattes RD. Quantitative and qualitative analyses of human salivary NEFA with gas-chromatography and mass spectrometry. Front Physiol,2012,3:328.

[138] Turner TD, Karlsson L, Mapiye C, et al. Dietary influence on the m. longissimus dorsi fatty acid composition of lambs in relation to protein source. Meat Science,2012,91(4):472-477.

[139] Jiang W, Zhang N, Zhang F, et al. Determination of olive oil content in olive blend oil by headspace gas chromatography-mass spectrometry. Se Pu,2017,35(7):760-765.

[140] Blasius J, Elfgen R, Holloczki O, et al. Glucose in dry and moist ionic liquid: vibrational circular dichroism, IR, and possible mechanisms. Physical Chemistry Chemical Physics: PCCP,2020, 22(19):10726-10737.

[141] Delmonte P, Fardin Kia AR, Kramer JK, et al. Separation characteristics of fatty acid methyl esters using SLB-IL111, a new ionic liquid coated capillary gas chromatographic column. Journal of Chromatography. A, 2011,1218(3):545-554.

[142] Zeng AX, Chin ST, Nolvachai Y, et al. Characterisation of capillary ionic liquid columns for gas chromatography-mass spectrometry analysis of fatty acid methyl esters. Analytica Chimica Acta,2013,803:166-173.

[143] Ragonese C, Sciarrone D, Tranchida PQ, et al. Use

of ionic liquids as stationary phases in hyphenated gas chromatography techniques. Journal of Chromatography. A,2012,1255:130-144.

[144] Weatherly CA, Zhang Y, Smuts JP, et al. Analysis of long-chain unsaturated fatty acids by ionic liquid gas chromatography. Journal of Agricultural and Food Chemistry, 2016,64(6):1422-1432.

[145] 吴丹，杨君君，杨帆，等．气相色谱 - 质谱联用法分析粪便中碱性和中性挥发性代谢产物．分析化学,2017,45(06):837-843 .

[146] Li K, Dan Z, Gesang L, et al. Comparative analysis of gut microbiota of native tibetan and han populations living at different altitudes. PLoS One,2016,11(5):e0155863.

[147] Suzuki TA, Worobey M. Geographical variation of human gut microbial composition. Biology Letters,2014,10(2):20131037.

[148] Jha AR, Davenport ER, Gautam Y, et al. Gut microbiome transition across a lifestyle gradient in Himalaya. PLoS Biology,2018,16(11):e2005396.

[149] Chan CW, Wong RS, Law PT, et al. Environmental factors associated with altered gut microbiota in children with eczema: a systematic review. International Journal of Molecular Sciences, 2016, 17(7): 1147.

[150] Sonnenburg JL, Backhed F. Diet-microbiota interactions as moderators of human metabolism. Nature, 2016,535(7610):56-64.

[151] Fulling C, Lach G, Bastiaanssen TFS, et al. Adolescent dietary manipulations differentially affect gut microbiota composition and amygdala neuroimmune gene expression in male mice in adulthood. Brain, Behavior, and Immunity,2020 ,87:666-678.

[152] David LA, Maurice CF, Carmody RN, et al. Diet rapidly and reproducibly alters the human gut microbiome. Nature,2014,505(7484):559-563.

[153] Chen Z, Zuurmond MG, van der Schaft N, et al. Plant versus animal based diets and insulin resistance, prediabetes and type 2 diabetes: the Rotterdam Study. European Journal of Epidemiology, 2018, 33(9):883-893.

[154] Jiang L, Xie M, Chen G, et al. Phenolics and carbohydrates in buckwheat honey regulate the human intestinal microbiota. Evid Based Complement AlterNature Medicine,2020,2020:6432942.

[155] Khan S, Waliullah S, Godfrey V, et al. Dietary simple sugars alter microbial ecology in the gut and promote colitis in mice. Science Translational Medicine,2020,12(567):e6218.

[156] Narula N, Wong ECL, Dehghan M, et al. Association of ultra-processed food intake with risk of inflammatory bowel disease: prospective cohort study. BMJ, 2021,374:n1554.

[157] Falony G, Joossens M, Vieira-Silva S, et al. Population-level analysis of gut microbiome variation. Science,2016,352(6285):560-564.

[158] Zeevi D, Korem T, Zmora N, et al. Personalized nutrition by prediction of glycemic responses. Cell,2015,163(5):1079-1094.

[159] Zhernakova A, Kurilshikov A, Bonder MJ, et al. Population-based metagenomics analysis reveals markers for gut microbiome composition and diversity. Science,2016,352(6285):565-569.

[160] Savin Z, Kivity S, Yonath H, et al. Smoking and the intestinal microbiome. Archives of Microbiology, 2018,200(5):677-684.

[161] Dubinkina VB, Tyakht AV, Odintsova VY, et al. Links of gut microbiota composition with alcohol dependence syndrome and alcoholic liver disease. Microbiome,2017,5(1):141.

[162] Li K, Peng W, Zhou Y, et al. Host genetic and environmental factors shape the composition and function of gut microbiota in populations living at high altitude. BioMed Research International,2020,2020:1482109.

[163] Katagiri S, Shiba T, Tohara H, et al. Re-initiation of oral food intake following enteral nutrition alters oral and gut microbiota communities. Frontiers in Cellular and Infection Microbiology,2019,9:434.

[164] Turnbaugh PJ, Hamady M, Yatsunenko T, et al. A core gut microbiome in obese and lean twins. Nature,2009,457(7228):480-484.

[165] Costea PI, Coelho LP, Sunagawa S, et al. Subspecies in the global human gut microbiome. Molecular Systems Biology,2017,13(12):960.

[166] Davis EC, Dinsmoor AM, Wang M, et al. Microbiome composition in pediatric populations from birth to adolescence: impact of diet and prebiotic and probiotic interventions. Digestive Diseases Science,2020,65(3):706-722.

[167] Becattini S, Taur Y, Pamer EG. Antibiotic-induced changes in the intestinal microbiota and disease. Trends in Molecular Medicine,2016,22(6):458-478.

[168] Langdon A, Crook N, Dantas G. The effects of antibiotics on the microbiome throughout development and alternative approaches for therapeutic modulation. Genome Medicine, 2016, 8(1):39.

[169] Li X, Stokholm J, Brejnrod A, et al. The infant gut resistome associates with E. coli, environmental exposures, gut microbiome maturity, and asthma-associated bacterial composition. Cell Host & Microbe,2021,29(6):975-987, e4.

[170] Le Bastard Q, Al-Ghalith GA, Gregoire M, et al. Systematic review: human gut dysbiosis induced by non-antibiotic prescription medications. Alimentary Pharmacology & Therapeutics,2018,47(3):332-345.

[171] Maier L, Typas A. Systematically investigating the impact of medication on the gut microbiome. Current Opinion in Microbiology, 2017, 39:128-135.

[172] Forslund K, Hildebrand F, Nielsen T, et al. Disentangling type 2 diabetes and metformin treatment signatures in the human gut microbiota. Nature, 2015, 528(7581):262-266.

[173] Freedberg DE, Toussaint NC, Chen SP, et al. Proton pump inhibitors alter specific taxa in the human gastrointestinal microbiome: a crossover trial. Gastroenterology, 2015,149(4):883-885, e9.

[174] Flowers SA, Evans SJ, Ward KM, et al. Interaction between atypical antipsychotics and the gut microbiome in a bipolar disease cohort. Pharmacotherapy, 2017, 37(3):261-267.

[175] Rogers MAM, Aronoff DM. The influence of non-steroidal anti-inflammatory drugs on the gut microbiome. Clinical Microbiology and Infection,2016,22(2):178, e1- e9.

[176] Han C, Song J, Hu J, et al. Smectite promotes probiotic biofilm formation in the gut for cancer immunotherapy. Cell Reports,2021,34(6):108706.

[177] Lei-Leston AC, Murphy AG, Maloy KJ. Epithelial cell inflammasomes in intestinal immunity and inflammation. Frontiers in Immunology,2017,8:1168.

[178] Carabotti M, Scirocco A, Maselli MA, et al. The gut-brain axis: interactions between enteric microbiota, central and enteric nervous systems. Annals of Gastro enterology,2015,28(2):203-209.

[179] Rogers GB, Keating DJ, Young RL, et al. From gut dysbiosis to altered brain function and mental illness: mechanisms and pathways. Molecular Psychiatry,2016,21(6):738-748.

[180] Srikantha P, Mohajeri MH. The possible role of the microbiota-gut-brain-axis in autism spectrum disorder. International Journal of Molecular Sciences,2019,20(9) :2115.

[181] Goodrich JK, Davenport ER, Beaumont M, et al. Genetic determinants of the gut microbiome in UK twins. Cell Host & Microbe,2016,19(5):731-743.

[182] Xie H, Guo R, Zhong H, et al. Shotgun metagenomics of 250 adult twins reveals genetic and environmental impacts on the gut microbiome. Cell Systems, 2016, 3(6):572-584,e3.

[183] Turpin W, Espin-Garcia O, Xu W, et al. Association of host genome with intestinal microbial composition in a large healthy cohort. Nature Genetics, 2016, 48(11):1413-1417.

[184] Grieneisen L, Dasari M, Gould TJ, et al. Gut microbiome heritability is nearly universal but environmentally contingent. Science, 2021, 373(6551): 181-186.

[185] Ross BD, Verster AJ, Radey MC, et al. Human gut bacteria contain acquired interbacterial defence systems. Nature, 2019,575(7781):224-228.

[186] Kendall MM, Sperandio V. What a Dinner Party! Mechanisms and functions of interkingdom signaling in host-pathogen associations. mBio,2016,7(2):e01748.

[187] Kaper JB, Sperandio V. Bacterial cell-to-cell signaling in the gastrointestinal tract. Infection and Immunity,2005,73(6):3197-3209.

[188] Moreno-Gamez S, Sorg RA , Domenech A, et al. Quorum sensing integrates environmental cues, cell density and cell history to control bacterial competence. Nature Communications,2017,8(1):854.

[189] Pena RT, Blasco L, Ambroa A, et al. Relationship between quorum sensing and secretion systems. Frontiers in Microbiology,2019,10:1100.

[190] Li Q, Ren Y, Fu X. Inter-kingdom signaling between gut microbiota and their host. Cellular and Molecular Life Sciences ,2019,76(12):2383-2389.

[191] Li YH, Tian X. Quorum sensing and bacterial social interactions in biofilms. Sensors (Basel), 2012, 12(3): 2519-2538.

[192] Thompson JA, Oliveira RA, Djukovic A, et al. Manipulation of the quorum sensing signal AI-2 affects the antibiotic-treated gut microbiota. Cell Reports,2015,10(11):1861-1871.

[193] Boudaher E, Shaffer CL. Inhibiting bacterial secretion systems in the fight against antibiotic resistance. Medchemcomm,2019,10(5):682-692.

[194] Coyte KZ, Schluter J, Foster KR. The ecology of the microbiome: Networks, competition, and stability. Science,2015,350(6261):663-666.

[195] Whitney JC, Peterson SB, Kim J, et al. A broadly distributed toxin family mediates contact-dependent antagonism between gram-positive bacteria. Elife, 2017, 6:e26938.

[196] Russell AB, Wexler AG, Harding BN, et al. A type Ⅵ secretion-related pathway in Bacteroidetes mediates interbacterial antagonism. Cell Host & Microbe, 2014, 16(2):227-236.

[197] Verster AJ, Ross BD, Radey MC, et al. The landscape of type Ⅵ secretion across human gut microbiomes reveals its role in community composition. Cell Host & Microbe,2017,22(3):411-419, e4.

[198] Coyne MJ, Roelofs KG, Comstock LE. Type Ⅵ secretion systems of human gut Bacteroidales segregate into three genetic architectures, two of which are contained on mobile genetic elements. BMC Genomics, 2016,17:58.

[199] Rath CM, Dorrestein PC. The bacterial chemical repertoire mediates metabolic exchange within gut microbiomes. Current Opinion in Microbiology, 2012, 15(2):147-154.

[200] Wu J, Wang K, Wang X, et al. The role of the gut microbiome and its metabolites in metabolic diseases. Protein Cell,2021,12(5):360-373.

[201] Schroeder BO, Backhed F. Signals from the gut microbiota to distant organs in physiology and disease. Nature Medicine,2016,22(10):1079-1089.

[202] Doifode T, Giridharan VV, Generoso JS, et al. The impact of the microbiota-gut-brain axis on Alzheimer's disease pathophysiology. Pharmacological Research,2021,164:105314.

[203] Rastelli M, Knauf C, Cani PD. Gut microbes and health: a focus on the mechanisms linking microbes, obesity, and related disorders. Obesity (Silver Spring),2018,26(5):792-800.

[204] James SC, Fraser K, Young W, et al. Gut microbial metabolites and biochemical pathways involved in irritable bowel syndrome: effects of diet and nutrition on the microbiome. The Journal of Nutrition, 2020 ,150(5):1012-1021.

[205] Rooks MG, Garrett WS. Gut microbiota, metabolites and host immunity. Nature Reviews Immunology, 2016,16(6):341-352.

[206] Agirman G, Hsiao EY. SnapShot: The microbiota-gut-brain axis. Cell,2021,184(9):2524, e1.

[207] He Y, Fu L, Li Y, et al. Gut microbial metabolites facilitate anticancer therapy efficacy by modulating cytotoxic CD8(+) T cell immunity. Cell Metabolism,2021,33(5):988-1000, e7.

[208] Nemet I, Saha PP, Gupta N, et al. A cardiovascular disease-linked gut microbial metabolite acts via adrenergic receptors. Cell,2020,180(5):862-877, e22.

[209] Hegarty JW, Guinane CM, Ross RP, et al. Bacteriocin production: a relatively unharnessed probiotic trait? F1000 Research,2016,5:2587.

[210] Ventura M, Turroni F, Motherway MO, et al. Host-microbe interactions that facilitate gut colonization by commensal bifidobacteria. Trends in Microbiology, 2012, 20(10):467-476.

[211] Pacheco AR, Sperandio V. Inter-kingdom signaling: chemical language between bacteria and host. Current Opinion in Microbiology,2009,12(2):192-198.

[212] Ismail AS, Valastyan JS, Bassler BL. A host-produced autoinducer-2 mimic activates bacterial quorum sensing. Cell Host & Microbe,2016,19(4):470-480.

[213] Kobayashi M, Mikami D, Kimura H, et al. Short-chain fatty acids, GPR41 and GPR43 ligands, inhibit TNF-alpha-induced MCP-1 expression by modulating p38 and JNK signaling pathways in human renal cortical epithelial cells. Biochemical and Biophysical Research Communications,2017,486(2):499-505.

[214] Kimura I, Inoue D, Hirano K, et al. The SCFA receptor GPR43 and energy metabolism. Front Endocrinol (Lausanne),2014,5:85.

[215] McKenzie C, Tan J, Macia L, et al. The nutrition-gut microbiome-physiology axis and allergic diseases. Immunological Reviews,2017,278(1):277-295.

[216] Brooks L, Viardot A, Tsakmaki A, et al. Fermentable carbohydrate stimulates FFAR2-dependent colonic PYY cell expansion to increase satiety. Molecular Metabolism,2017,6(1):48-60.

[217] Postler TS, Ghosh S. Understanding the Holobiont: How microbial metabolites affect human health and shape the immune system. Cell Metabolism, 2017, 26(1):110-130.

[218] Cani PD, Everard A, Duparc T. Gut microbiota, enteroendocrine functions and metabolism. Current Opinion in Pharmacology,2013,13(6):935-940.

[219] Maslowski KM, Vieira AT, Ng A, et al. Regulation of inflammatory responses by gut microbiota and chemoattractant receptor GPR43. Nature, 2009, 461(7268):1282-1286.

[220] Sivaprakasam S, Gurav A, Paschall AV, et al. An essential role of Ffar2 (Gpr43) in dietary fibre-mediated promotion of healthy composition of gut microbiota and suppression of intestinal carcinogenesis. Oncogenesis,2016,5(6):e238.

[221] Byndloss MX, Olsan EE, Rivera-Chavez F, et al. Microbiota-activated PPAR-gamma signaling inhibits dysbiotic Enterobacteriaceae expansion. Science, 2017, 357(6351):570-575.

[222] Cani PD. Gut Cell metabolismolism shapes the

microbiome. Science,2017,357(6351):548-549.

[223] Rivera-Chavez F, Lopez CA, Baumler AJ. Oxygen as a driver of gut dysbiosis. Free Radical Biology & Medicine,2017,105:93-101.

[224] Matsubara T, Li F, Gonzalez FJ. FXR signaling in the enterohepatic system. Molecular and Cellular Endocrinology, 2013,368(1-2):17-29.

[225] Song X, Sun X, Oh SF, et al. Microbial bile acid metabolites modulate gut RORgamma(+) regulatory T cell homeostasis. Nature,2020,577(7790):410-415.

[226] Said HM. Recent advances in transport of water-soluble vitamins in organs of the digestive system: a focus on the colon and the pancreas. American journal of physiology. Gastrointestinal and Liver Physiology, 2013, 305(9):G601-G610.

[227] Engevik MA, Morra CN, Roth D, et al. Microbial metabolic capacity for intestinal folate production and modulation of host folate receptors. Frontiers in Microbiology, 2019,10:2305.

[228] Thomas CM, Saulnier DM, Spinler JK, et al. FolC2-mediated folate metabolism contributes to suppression of inflammation by probiotic Lactobacillus reuteri. Microbiologyopen,2016,5(5):802-818.

[229] Spinler JK, Sontakke A, Hollister EB, et al. From prediction to function using evolutionary genomics: human-specific ecotypes of Lactobacillus reuteri have diverse probiotic functions. Genome Biology and Evolution,2014,6(7):1772-1789.

[230] Lynch SV, Pedersen O. The human intestinal microbiome in health and disease. The New England Journal of Medicine,2016,375(24):2369-2379.

[231] Kim SK, Guevarra RB, Kim YT, et al. Role of probiotics in human gut microbiome-associated diseases. Journal of Microbiology (Seoul, Korea) Biotechnol, 2019, 29(9):1335-1340.

[232] 谢东强 . COBRA 模型上代谢调控约束的定量化研究 . 复旦大学 ,2012.

[233] Heirendt L, Arreckx S, Pfau T, et al. Creation and analysis of biochemical constraint-based models using the COBRA Toolbox v.3.0. Nature Protocols,2019,14(3):639-702.

[234] Stolyar S, Van Dien S, Hillesland KL, et al. Metabolic modeling of a mutualistic microbial community. Molecular Systems Biology,2007,3:92.

[235] Magnusdottir S, Heinken A, Kutt L, et al. Generation of genome-scale metabolic reconstructions for 773 members of the human gut microbiota. Nature Biotechnology,2017,35(1):81-89.

[236] Oberhardt MA, Palsson BO, Papin JA. Applications of genome-scale metabolic reconstructions. Molecular Systems Biology,2009,5:320.

[237] Henry CS, DeJongh M, Best AA, et al. High-throughput generation, optimization and analysis of genome-scale metabolic models. Nature Biotechnology, 2010,28(9):977-982.

[238] Allen B, Drake M, Harris N, et al. Using KBase to assemble and annotate prokaryotic genomes. Current Protocols in Microbiology,2017,46: 1E.13.1-1E.13.18.

[239] Karp PD, Latendresse M, Paley SM, et al. Pathway tools version 19.0 update: software for pathway/genome informatics and systems biology. Brief Bioinform,2016,17(5):877-890.

[240] Fleming RMT, Vlassis N, Thiele I, et al. Conditions for duality between fluxes and concentrations in biochemical networks.Journal of Theoretical Biology,2016,409:1-10.

[241] Haraldsdottir HS, Thiele I, Fleming RM. Quantitative assignment of reaction directionality in a multicompartmental human metabolic reconstruction. Biophysical Journal,2012,102(8):1703-1711.

[242] Green ML, Karp PD. Genome annotation errors in pathway databases due to semantic ambiguity in partial EC numbers. Nucleic Acids Research, 2005, 33(13):4035-4039.

[243] Rolfsson O, Palsson BO, Thiele I. The human metabolic reconstruction Recon 1 directs hypotheses of novel human metabolic functions. BMC Systems Biology,2011,5:155.

[244] Harcombe WR, Riehl WJ, Dukovski I, et al. Metabolic resource allocation in individual microbes determines ecosystem interactions and spatial dynamics. Cell Reports,2014,7(4):1104-1115.

[245] Zomorrodi AR, Islam MM, Maranas CD. d-OptCom: Dynamic multi-level and multi-objective metabolic modeling of microbial communities. ACS Synthetic Biology,2014,3(4):247-257.

[246] Bauer E, Zimmermann J, Baldini F, et al. BacArena: Individual-based metabolic modeling of heterogeneous microbes in complex communities. PLoS Computational Biology,2017,13(5):e1005544.

[247] Bauer E, Thiele I. From metagenomic data to personalized in silico microbiotas: predicting dietary supplements for Crohn's disease. NPJ Systems Biology and Applications,2018,4:27.

[248] Hoek M, Merks RMH. Emergence of microbial diversity due to cross-feeding interactions in a spatial model of gut microbial metabolism. BMC Systems

Biology,2017,11(1):56.

[249] Bordbar A, Lewis NE, Schellenberger J, et al. Insight into human alveolar macrophage and Mtuberculosis interactions viametabolic reconstructime Mole cular systems biology, 2010, 6:422.

[250] Heinken A, Thiele I. Systematic prediction of health-relevant human-microbial co-metabolism through a computational framework. Gut Microbes, 2015, 6(2):120-130.

[251] 敬晓棋，李娟，郝婷婷，等．宏基因组学技术在动物传染病原挖掘中的应用与进展．基因组学与应用生物学，2021: 1-7.

[252] Li H, Handsaker B, Wysoker A, et al. The Sequence Alignment/Map format and SAMtools. Bioinformatics, 2009,25(16):2078-2079.

[253] Etherington GJ, Ramirez-Gonzalez RH, MacLean D. Bio-samtools 2: a package for analysis and visualization of sequence and alignment data with SAMtools in Ruby. Bioinformatics, 2015, 31(15): 2565-2567.

[254] Human Microbiome Project C. Structure, function and diversity of the healthy human microbiome. Nature,2012,486(7402):207-214.

[255] Heintz-Buschart A, May P, Laczny CC, et al. Integrated multi-omics of the human gut microbiome in a case study of familial type 1 diabetes. Nature Microbiology,2016,2:16180.

[256] Song EJ, Lee ES, Nam YD. Progress of analytical tools and techniques for human gut microbiome research. Journal of Microbiology, 2018,56(10):693-705.

[257] Guilhot E, Khelaifia S, La Scola B, et al. Methods for culturing anaerobes from human specimen. Future Microbiol,2018,13:369-381.

[258] Shoaie S, Ghaffari P, Kovatcheva-Datchary P, et al. Quantifying diet-induced metabolic changes of the human gut microbiome. Cell Metabolism, 2015,22(2):320-331.

[259] Heirendt L, Thiele I, Fleming MT. DistributedFBA.jl: high-level, high-performance flux balance analysis in Julia. Bioinformatics, 2017,33(9):1421-1423.

第 2 章

肠道菌群与机体健康

第一节　概　述

肠道菌群对人体健康具有重要作用，包括代谢碳水化合物、合成维生素、预防病原菌感染和调节宿主免疫反应等。此外，通过建立、维持和补充肠道菌群多样性，有利于人体维持健康状态。然而，决定个体最佳细菌群落的因素多种多样，包括年龄、宿主遗传、饮食和环境等。因此，并不存在人类共有的核心“健康”菌群。并且，共生菌和病原菌之间可以相互转换，即同一种细菌既可以是共生菌，也可以是病原菌，也就是说细菌具有促进机体健康和导致机体致病的两面性，这取决于特定的菌株及它们在体内的位置（图 2-1）。例如，脆弱类杆菌可以产生具有免疫调节作用的荚膜多糖，可刺激抗炎细胞因子产生；如果这种细菌转移到腹膜，那么荚膜多糖将会导致腹膜脓肿。大肠埃希菌可以是肠道菌群的正常共生体，也可以是肠道感染的病原体，这取决于它不同的菌株亚型及其产生的毒力因子；同样，胃内幽门螺杆菌的感染，可增加消化性溃疡和胃癌的发病率，但相反，高丰度的幽门螺杆菌与较好的食管癌预后有关。

一般说来，我们认为健康状况良好的个体，需要富含多种有益物种的菌群，而不能有以变形杆菌门为代表的病原菌。自人类菌群组计划启动以来，研究发现人体肠道内存在 1000 多种共生菌，主要包括拟杆菌门、厚壁菌门、变形菌门等，它们寄居在肠道的不同部位，通过特有的菌群结构、菌群活动、代谢产物等来影响宿主的新陈代谢，维持宿主内环境稳态。肠道菌群相对稳定，却又受到宿主年龄、遗传背景及生活环境的影响，在数量、结构、菌群丰度及生理状态方面表现出明显的差异。近年来，肠道菌群与人类疾病之间的关系受到国内外研究者的广泛关注。大量研究发现，肠道菌群与人类疾病息息相关，它不仅影响机体的生理状态，还会影响机体的精神状态；这也提示了在人类疾病的治疗、康复方面，需要制订个体化的营养支持方案和药物治疗策略[1]。

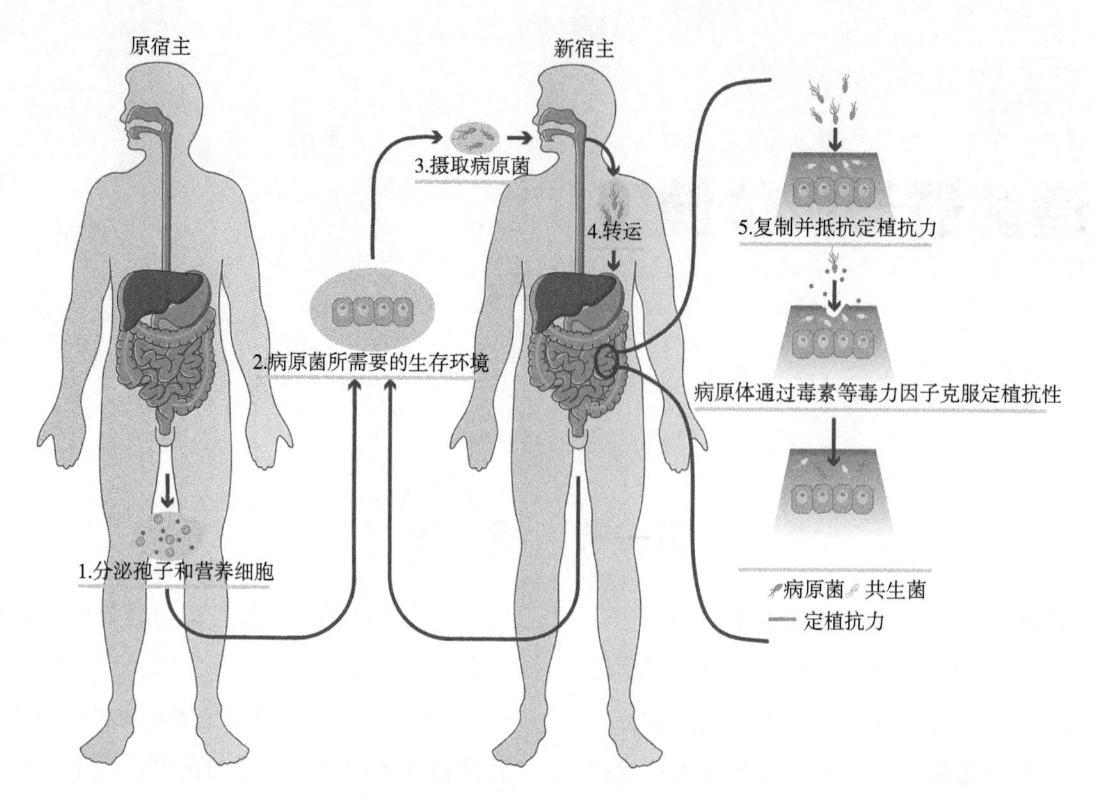

图 2-1　致病性和共生性肠道细菌的传播

肠道病原体和共生菌使用相似的机制在宿主之间传播。细菌从宿主的粪便中排出是传播的第一阶段（步骤 1）。为了促进新宿主的传播和随后的摄取，病原体可能会导致原宿主腹泻。一旦进入外部环境，这些以厌氧为主的细菌就会利用耐氧性、休眠和产生孢子等生存机制来生存和传播。如人、食物、动物和建筑环境，将作为传播的源头发挥作用（步骤 2）。一旦被新宿主摄取（步骤 3），细菌就转移到肠道（步骤 4）。来自新宿主原有菌群的竞争可以阻止定植（步骤 5）；但是，如阻力不足，病原菌就可以定植（步骤 5）。细菌的恢复性具有维持定植抗性和维持肠道健康相关的菌群多样性的功能。病原体可以通过诱导毒力因子（如毒素）的表达来克服定植抗力，这种毒力因子可能导致炎症并扰乱原有的菌群（步骤 5）

第二节　影响肠道菌群的具体因素

一、食物

与人类基因组不同的是，肠道菌群表现出了极大的可塑性，很容易受各种环境和宿主因素的影响。在这些因素中，饮食是影响肠道菌群的关键性因素。饮食对个人和全人类的健康都有着深远的影响。因此，食物营养研究是医学、经济、文化和社会的焦点 [2]。2000 年前，希波克拉底就提出了“让食物成为你的药”这一概念，世界各地的卫生组织也一直在努力制订“健康饮食”的标准，以推荐可促进健康的微量营养素、宏观营养素和总卡路里的摄入量。通过研究饮食和肠道菌群之间的相互关系，以及它们对宿主的共同影响，从而制订人类健康饮食标准，对健康饮食的标准予以相应的解释，并阐明一些无法理解的个体间差异 [3]。本节内容旨在讲述饮食因素对肠道菌群平衡状态的影响，并阐述影响肠道细菌群落结构和功能的关键概念。

食物中蕴含的营养素可以直接与菌群相互作用，促进或抑制它们的生长，部分细菌可以在特定饮食成分中获取能量而拥有竞争优势，淘汰获取能力弱的细菌成员 [4]。例如源自小米的牡荆素，可使以嗜黏蛋白阿克曼菌为代表的特定菌群获得竞争优势，从而控制炎症、保护神经 [5]。

除上述直接作用外，饮食还可通过其衍生

的抗原和化合物影响宿主新陈代谢和免疫系统，以间接的方式塑造肠道菌群[6]。细菌发酵膳食纤维会产生短链脂肪酸，短链脂肪酸不仅在肠道动态平衡中发挥作用，还可以通过其在血液中的循环对肠道以外的组织和器官产生影响。例如短链脂肪酸可通过与内分泌和免疫细胞上同源的 G 蛋白偶联受体（G protein-coupled receptors，GPCRs）结合而发出信号，通过影响组蛋白乙酰化酶和组蛋白脱乙酰基酶的活性，诱导基因组的表观遗传学变化[7]。

饮食成分还可能破坏肠道屏障的保护功能，从而影响宿主 - 菌群平衡，导致生物失调，促进炎症的发生。例如，在日常生活中摄入大量红肉，其富含的血红素铁可催化 N- 亚硝基化合物形成，导致肠道上皮细胞过度增殖，引发肠道慢性炎症和持续性菌群失调，可能驱动结直肠癌的发生[8]。在加工食品中使用乳化剂可以侵蚀肠道保护性上皮黏液层，并导致肠道低度炎症和代谢综合征[9]。

食物所包含的宏观营养素及微量营养素也会影响肠道菌群。在现代社会中，世界各地农村地区的肠道菌群丰度更高，这说明从膳食纤维中摄取能量需要更多的菌群帮助[10]。而城市人口的菌群组成则更加多样化，说明细菌对饮食的干预反应灵敏。人们食用完全由动物产品组成的饮食会导致耐胆汁细菌（双歧杆菌和拟杆菌）的富集，以及代谢植物多糖的细菌（罗斯氏菌、直肠真杆菌和布氏瘤胃球菌）减少[9]。此外，在超重或肥胖的人群中，细菌丰富度与水果、蔬菜和鱼类的摄入呈正相关[11]。

营养物质的缺乏对细菌和人体都有深远的影响。对发展中国家人口的研究表明，营养不良通常是一个“双重打击的过程”，即指匮乏的菌群和营养不良的双重打击。通过计算肠道菌群成熟的程度，可以预测孟加拉儿童的严重急性营养不良，并可以评估对其营养干预的效果[12]。发达国家的人群倾向于食用低纤维的饮食。纤维摄入量过低会导致厚壁菌增加，拟杆菌减少。而食用大量植物多糖的非洲儿童的菌群表现为厚壁菌丰度降低和拟杆菌丰度增高（主要是普雷沃特氏菌）[13]。此外，当饮食在富含纤维和无纤维之间转换，会导致肠道菌群组成发生显著变化。在缺乏膳食纤维的情况下，黏液降解菌的数量增加，纤维降解菌的数量减少。此外，饮食中缺乏纤维可以选择性地使多形拟杆菌等菌群增加，从而将这种营养缺乏的后果从菌群扩展到全身[14]。

食物的摄入量亦会影响肠道菌群。摄入过量的蔗糖会导致小鼠拟杆菌门 / 厚壁菌门的比例增加、α 多样性降低、短链脂肪酸水平降低，促进高脂血症和脂肪肝发生[15]。在人体试验中，短期的碳水化合物限制（每天 24 ～ 164g，持续 4 周）将导致产生丁酸的细菌减少，从而减少丁酸的产生，而限制卡路里的方法（连续 10 周减少 10% ～ 40% 的能量摄入）将会导致产乙酸菌的减少和拟杆菌的增加。而延长至 1 年的饮食干预将会导致粪便拟杆菌增加和放线菌相对丰度降低[16]。

目前，我们很难区分饮食是直接影响人体还是通过改变菌群组成而间接影响人体。因为人体基因构成差异很大，暴露在众多外源因素中，他们的饮食往往由大量不同的营养成分组成。这众多的变量可能会与肠道菌群产生协同或相反的结果，因此很难预测饮食干预对肠道菌群的净影响。部分微量营养素的缺乏在人类、小鼠、大鼠和小猪身上引发了不同模式的菌群结构改变，包括铁、镁、锌、硒、亚硝酸盐或硝酸盐、维生素 A、维生素 D 和类黄酮。这些微量营养素成为预防、诊断和治疗饮食诱导的肥胖和代谢综合征的潜在药物[2]。例如，蔓越莓提取物增加了高脂饮食小鼠中嗜黏蛋白阿克曼菌的丰度，并改善了代谢综合征的症状。在美国城市环境中采样的健康人类、纯素食动物和杂食动物在菌群结构和宿主代谢方面没有明显的差异，而非洲裔美国人的饮食特点是动物脂肪和蛋白质含量高，纤维含量低。这在一定程度上说明草食和肉食动物之间的菌群差异是来自全球环境的影响。这些地理位置不同的人群进行饮食转换，会导致菌群组成、功能、分泌的代谢物，以及增殖和炎症标志物的变化[17]。2018 年的一项样本分析表明，与纯素食等饮食结构单一的人群相比，保持多样性食物摄入的人群维持肠道菌群稳定的能力更强。同时，该研究发现，那些被认为有益于发达国家人群的饮食建议，有时可能对发展中国家人群有害。其中一个突出的例子是补充铁和叶酸，由于补充铁和叶酸将导致肠道病原体丰富（如大肠埃希菌、志贺菌和梭状芽孢杆菌）和炎症加剧，这可能与桑给巴尔的儿童感染疟疾后病死率增加有关[18]。此外，饮食的细菌环境也可能对肠道菌群的形成起作用，因为生存在同一环境中的细菌可以通过物种间基因重排、基因复制和横向基因转移动态进化。这些基因修饰增强了肠道细菌代谢能力，扩大了可消化底物的范围[18]。例如，日本人食用海藻有助于将基因从海洋菌群转移到肠道菌群，使后者能够消化藻类，这一特征可用于基于饮食的生态位调节，以植入有益细菌[19]。最后，宿主的基因构成会影响消化。例如，食用富含淀粉食物的人群比食用低淀粉食物的人群拥有更多的唾液淀粉酶基因拷贝数。此外，携带信号转导途径或类固醇生成突变的小鼠可表现出生物失调，导致肥胖、脂肪组织炎症和胰岛素抵抗[19]。这些结果都说明，饮食不仅在影响肠道菌群组成方面优于遗传，在预测人体生理个体差异方面也更具优势，例如血糖水平和肥胖指标。

二、遗传

人类肠道菌群的组成是由多种因素决定的，但宿主遗传学的相对贡献仍然不可忽视。宿主和共栖菌群构成“共生体”，宿主的各种生物学行为受到自身基因和肠道菌群基因组的共同影响。近年来，生物信息分析法和宏基因组学广泛应用于肠道菌群的研究，揭示了肠道菌群的多样性及其结构和功能。但目前尚不清楚肠道菌群中不同遗传元件的总数。具体来说，在人类菌群组遗传分析中，其宏基因组分析和方法都集中在核心基因家族上[20]，功能基因组的内容则通过制作“基因目录”来解决，“基因目录”是指通过组装大量样本而确定所有基因的集合。在人类肠道菌群组中，主要的测序联合体已经使用新方法确定了多达 1000 万个基因[21]。此外，如果将宏基因组学与传统菌群遗传学联系起来，菌群遗传元素可以归类到“泛基因组”中，它描述了在特定物种的所有菌株中发现的所有基因的集合。泛基因组的大小受其有效种群大小和迁移到新生态位能力的影响[22]。

双胞胎样本是塑造肠道菌群从而解析人类基因组的良好模型。早期基于 DNA 指纹技术和细胞培养发现，与异卵双胞胎和无亲缘关系个体相比，同卵双胞胎的粪便菌群结构更为相似。近年来，在大规模的双胞胎人群中观察到这一类似结果，但未达到显著水平，这些研究更多的是强调环境因素在塑造肠道菌群中发挥关键性作用。肠道经历了由少到多、由简单到复杂、由不稳定直至相对稳定的菌群定植过程，但是尚不能确定肠道菌群结构的相似性是由于遗传背景的相似还是由于生活环境的相似所造成的。对 250 对双胞胎的粪便样品进行宏基因组测序，结果表明人体基因组对肠道内诸多菌群及其潜在功能具有显著贡献。双胞胎之间，尤其是同卵双胞胎之间的肠道菌群组成、单核苷酸多态性（single nucleotide polymorphism，SNP）和功能分类高度相似，但是这种相似性又会随着异地居住而逐渐减弱[23]。双胞胎和母女的菌群组成，比不相关的个体有更多相似的菌群，这表明基因可能对菌群有影响。一项针对双胞胎的研究证实，一种可以防止机体发胖的细菌克里斯滕森菌受遗传影响最大；但是存在不同观点，Rothschild 等对 1046 名以色列健康个体的基因分析后发现，个体的肠道菌群与遗传没有任何显著性关联 [24]。

亲代的健康状况或饮食习惯可以通过表观遗传影响其后代的疾病易感性，这种饮食诱导的跨代疾病易感性主要是由肠道菌群介导的。将肠道菌群的遗传力理解为肠道菌群组成和相对丰度能够稳定遗传的能力是一种误解，更为合适的解释是其反映了宿主遗传因素对肠道菌群的影响程度 [25]。例如，对高炎性肠病遗传风险的个体肠道菌群整体遗传性进行分析，发现与无风险的健康人群相比，具有高炎性肠病遗传风险的健康个体的肠道菌群发生显著改变，即说明与炎性肠病相关的遗传风险变异影响健康个体的肠道菌群 [26]。而另一项对有潜在心脏代谢风险因子的 441 名哥伦比亚人进行基因分型及肠道菌群测定，发现多个心脏代谢风险等位基因与有益微生物群的丰度增加相关，表明肠道菌群处于宿主遗传控制之下并且呈现宿主 - 微生物相互作用 [27]。通过对宿主自身基因组和肠道菌群分别采用基因芯片分型和 16S rRNA 测序的方法，Goodrich 等分析了来自英国的 416 对双胞胎的粪便样本，结果显示同卵双胞胎相较异卵双胞胎和无亲缘个体具有更加相似的菌群结构，且三组样本间的相似性达到显著差异水平；进一步评估肠道菌群的遗传力，发现 33 个菌群的相对丰度受到宿主遗传因素影响，尤其是克里斯滕森菌科，该结果在加拿大和韩国人群粪便样本中也得到了验证。

2016 年，Goodrich 等将双胞胎数量提高到 1126 对，发现克里斯滕森菌科的遗传力升高，同时观察到更多的可遗传性肠道菌群，这些菌群的相对丰度在相当长的时间内保持稳定，其中厚壁菌门和放线菌门更易遗传，而拟杆菌门则遗传性较小；除了估计分类群的遗传力外，α 多样性也具有一定的遗传力 [28]。同时，Turpin 等在加拿大招募了 1561 名志愿者，评估了从门至属水平共计 249 种粪便菌群的遗传力，结果发现 94 种细菌类群受宿主遗传背景的影响；Lim 等基于 655 名韩国人体的粪便样本，估计了 85 种不同分类水平下的菌群遗传力，其中 50 种菌群具有显著的遗传力，放线菌门的遗传力尤为明显 [29]。

孕妇产前体重指数可能会影响婴儿最初的肠道菌群，孕妇肥胖可能导致分娩时某些特定的肠道菌群存在于婴儿肠道中，提示与肥胖有关菌群的代际转移 [30]。受干扰的菌群遗传增加了后代群落组成的不稳定性，并降低了丰富性。抗生素扰乱的肠道菌群遗传力不仅在下一代忠实地重演，而且赋予下一代更大的疾病风险。例如早期抗生素暴露与儿童炎性肠病风险之间呈正相关。Larsen 将受扰乱的肠道菌群移植到小鼠体内，发现后代小鼠患炎性肠病的概率明显增加，提示接种物和基因型都塑造了子代的肠道菌群 [31]。抗生素暴露重新塑造了母体肠道菌群，并会对后代的肠道微生态和健康产生长期影响 [32]。

三、生命不同时期

在人类生命不同阶段，肠道菌群的组成各不相同。从出生开始，随着接触内外界环境，肠道菌群不断定植和发育，从相对简单的菌群，不断发展变化，其数量及种群多样性逐步增加，直到形成稳态，之后又随着年龄的增长，肠道菌群同样出现衰老，菌群丰度及多样性随之降低 [33]。

新生儿肠道菌群结构较为简单，但有多种因素可影响新生儿肠道微生物组成，例如，母亲在妊娠和哺乳期间额外补充营养剂可增加婴儿肠道菌群多样性 [34]；生产时接受了抗生素治疗的孕妇，其婴儿肠道菌群在 3 个月内存在明显差异，主要表现为双歧杆菌减少、梭菌和葡萄球菌增加，这与早产儿的肠道菌群表现一致，如未接受母乳喂养，这种差异将持续至12个月[35]；新生儿的出生时间和方式不同也将影响肠道菌群。顺产新生儿的肠道菌群以乳酸杆菌、普雷沃菌属或纤毛菌属为主，与母体直肠菌群相似，而剖宫产新生儿肠道菌群中以葡萄球菌属、棒状杆菌属和丙酸杆菌属为主，与母亲皮肤菌群相似[36]；母乳喂养婴儿肠道内双歧杆菌占主导地位，配方奶喂养婴儿肠道内以肠球菌、肠杆菌、拟杆菌、梭状芽孢杆菌及其他厌氧链球菌为主[37]。

幼儿断奶后，放线菌门的丰度发生了显著变化，当幼儿进食固体食物，其肠道菌群会形成以拟杆菌门和厚壁菌门为主导的肠道菌群。但5岁以前，幼儿的细菌多样性并没有增加；5岁以后，随着拟杆菌门的持续性增加，儿童的肠道菌群组成将逐步向着成人肠道菌群转变[38]。在这个发育阶段，肠道菌群可能会影响多个器官和系统的发育。如双歧杆菌和乳酸杆菌可促进婴幼儿的神经系统发育和精细运动发育[39]；瘤胃球菌参与控制细胞膜通透性，维持肠道屏障，在防御感染及营养、内分泌和免疫功能方面发挥着重要作用[40]；幼儿期的菌群变化会直接影响中枢神经系统和肠神经系统的发育[41]。

当从幼儿期进入儿童期，菌群的复杂程度及生物多样性进一步增加，肠道菌群种类基本与成年人持平，其中放线菌、杆菌、梭状芽孢杆菌Ⅳ和拟杆菌的丰度明显高于成年人[42]。有研究对281名学龄儿童（6～9岁）的肠道菌群进行了大规模的综合横断面分析，发现儿童的肠道菌群富含拟杆菌和双歧杆菌，其功能组成与健康成年人相似，尤其是β多样性指标[43]。越来越多的证据表明，儿童时期的肠道菌群组成对宿主的生长发育起到了重要作用，肠道菌群可在营养、生长激素信号传导、定植抗性和免疫耐受等多方面参与儿童的生长发育[44]。肠道菌群在介导宿主代谢和免疫稳态之间起着至关重要的作用，生命早期阶段菌群定植的改变可以影响肠道相关淋巴组织的发育，从而可能导致免疫发育异常[45]。

青春期肠道菌群开始出现性别特异性，与此同时，随着年龄的增长，需氧菌和兼性厌氧菌的数量逐渐减少，同时厌氧菌数量激增。与此同时，肠道菌群可以改变性激素水平，并调节高遗传风险个体的自身免疫疾病发病率[46]。青少年有着与成人相似的核心肠道菌群组成，但青少年的梭菌和双歧杆菌属的水平显著升高，且在功能上也有所不同。例如青少年肠道菌群主要表达与发育和生长相关的基因，并可通过影响神经系统，以减轻青春期的焦虑、抑郁等情绪障碍[47]，通过益生菌治疗亦可明显改善压力引起的情绪脆弱[48]。

成人的肠道菌群具有高度个体特异性，随着基因组学的不断发展，已经确定了健康成人所共有的核心肠道菌群，包括厚壁菌门、拟杆菌门、变形菌门和放线菌门。健康成年人的核心肠道菌群基因编码了宿主生存所需的绝大部分蛋白质，并可参与包括多糖消化、免疫系统发育、抗感染、维生素合成、脂肪储存、血管生成调节和行为发育等多种生理功能[49]。随着研究的不断深入，了解肠道菌群与疾病的关系，肠道菌群将很有可能成为诊断和治疗疾病的新靶点。

肠道内的细菌本身不会老化，但随着人体不断衰老，免疫力下降，食物摄取减少，以及抗生素使用的增加，老年人肠道菌群会出现稳态失衡，多样性降低，肠道菌群的主要构成部分也由厚壁菌门向拟杆菌门转变[50]。老年人粪便中的拟杆菌种类多样性略有增加，双歧杆菌种类多样性有所减少，而梭状芽孢杆菌较丰富[50]。而百岁老人与普通老年人的肠道菌群组成差别也较大，这种差别主要是由饮食成分的改变引起的。肠道本身的衰老过程也会影响人体肠道菌群的构成，这主要是由于老年性便秘引起的，便秘不利于肠道发酵，进而影响肠道细菌稳态[51]。此外，长期接受护理的老年人，肠道菌群更是显示出了明显差异，并出现与身体衰弱程度相关的特殊菌种的定居。这些研究将确定微生物群是否可以代表衰老状况的诊断指标，亦或是否可以通过调节肠道菌群直接干预衰老[52]。

四、运动

目前，肠道菌群功能与运动相关性研究还处于初级阶段。越来越多的证据指出，运动对肠道菌群具有积极的调控作用，可有效调节肠道菌群的成分和结构，提高肠道菌群的丰富度，增加有益菌群数量，保持菌群平衡从而促进机体健康。运动有可能逆转肥胖、代谢性疾病、不良饮食习惯以及神经和行为障碍等相关疾病[53]。心肺运动可改变人体的初始肠道菌群，而抗阻力运动则对人体肠道菌群没有明显影响[54]，通俗来说，即有氧运动更能有效地改变肠道菌群。

运动有助于增加肠道菌群的生物多样性，并可通过影响肠道菌群来预防身体虚弱或认知功能障碍，脑卒中、抑郁症、精神分裂症患者均在有氧运动训练后得到了改善[55]，有规律的有氧运动可以防止脑萎缩，并增加额叶和左上颞叶的脑容量，这对认知、注意力和记忆的控制非常重要。中等强度的有氧运动还促进了 60 ～ 79 岁老年人大脑功能的激活，并调节行为和情绪[56]。目前已有针对性的运动治疗来应对肠易激综合征，并已经取得了一定疗效[57]。

在人类运动研究中，Allen 等比较了 6 周有氧运动计划前后苗条与肥胖人类之间的菌群变化，并进一步评估了训练后的变化。运动引起了苗条和肥胖受试者肠道菌群多样性和短链脂肪酸产量的变化，厚壁菌门的菌群明显升高，这些菌群的数量随着运动训练的增加而增加，并在停止运动后恢复到原始水平。值得注意的是，肠道菌群的变化与肌肉数量、脂肪质量的变化有关，这证明有氧运动影响了人类肠道菌群的内容和多样性，并提出了肠道菌群如何影响运动适应的问题。这项研究证实了有氧运动训练后增加的菌群都是产生短链脂肪酸的菌群，更具体地说是锻炼能产生丁酸盐。丁酸盐和其他短链脂肪酸是结肠细胞的能量来源[58]。这可能解释了为什么运动可以改善肠道症状并降低肠易激综合征和结肠癌的风险。此外，已经证明丁酸盐可以刺激小鼠齿状脑回区的神经增殖，也可诱导成年啮齿动物缺血性脑损伤后的神经再生[59]。这些结果说明，运动诱导的短链脂肪酸的产生与神经增殖再生有一定的联系。然而，尚不能说明锻炼对肠道和大脑的积极益处是否通过调节菌群来实现。对运动影响菌群的研究表明，来自厚壁菌门的菌属似乎对运动诱导的变化最敏感。目前的证据支持这样一种假设，即：运动可能通过改变菌群来调节肠道和大脑之间的双向关系。这种关系可以解释为什么锻炼可以成为心理和胃肠道疾病的治疗策略。未来的工作应该研究运动如何具体影响菌群的潜在机制，以及运动参与“脑 - 肠轴”的介质。这可能催生出运动 + 益生菌的治疗组合，以改善特定的疾病状态[60]。

五、抗生素

抗生素改变了大多数传染病的自然病史，拯救了数百万人的生命。它们的开发不仅改变了医学，也改变了药理学、保健学和整个人类社会的进程。但抗生素也有潜在的危害，因为越来越多的证据表明，过度使用抗生素会导致肠道菌群改变从而导致相关疾病发生[61]。此外，抗生素过度暴露也可能导致肠道菌群耐药性的发展，并有可能将其转移到病原菌中，抗生素还可以丰富噬菌体编码的基因，使其对药物和无关抗生素产生耐药性，促进噬菌体和细菌之间的相互作用。由于市场上大多数抗生素的作用范围很广，它们不仅对有害细菌有影响，而且对健康细菌也有影响[62]。抗生素对肠道菌群的影响既有药物相关因素，也有宿主相关因素。药物相关因素中，影响这种关系的抗生素特性包括它们的类别、药动学、药效学、作用范围，以及它们的剂量、持续时间和给药途径[63]。例如，林可霉素类抗生素（克林霉素）可减少产丁酸盐菌株丰度，诱导肠道屏障功能障碍，造成抗生素相关性腹泻和营养不良[64]；大环内酯类化合物能引起总细菌多样性的减少，双歧杆菌、乳杆菌丰度减少，拟杆菌和变形杆菌相对丰度的增加；而青霉素类药物如阿莫西林，对拟杆菌、厚壁菌门和双歧杆菌产生不利影响，并持续减少产短链脂肪酸的菌株[65]。另一方面，包括氨苄西林和头孢菌素在内的β-内酰胺类药物，导致厚壁菌门减少，拟杆菌和变形杆菌（特别是肠杆菌）增加。此外，一些氟喹诺酮类药物，包括环丙沙星和莫西沙星，增加了革兰阳性需氧菌的丰度，并降低细菌多样性[62]。氟喹诺酮类抗生素影响菌群的能力也取决于它们的给药途径。口服给药途径比静脉给药更能刺激健康的菌群成员产生抗生素耐药性[66]。当然，抗生素也可以对肠道菌群起积极作用，提供所谓的“共生”效应。例如，当呋喃妥因被用于尿路感染的患者时，可增加患者肠道双歧杆菌和粪杆菌的丰度[67]。在一项对15名肠易激综合征患者进行的小型试验中，患者使用利福昔明后增加了乳酸菌/类杆菌的比率，增加了普拉梭菌（具有很强的抗炎特性）的丰度，而对细菌多样性没有影响[68]。

此外，宿主的年龄、生活方式和菌群的原有组成等因素也会影响抗生素对健康菌群的损害程度。并且，抗生素从婴幼儿时期就开始改变人类肠道菌群。有关数据表明，婴儿菌群极不稳定，因此对抗生素的反应更敏感。几项研究表明，广谱抗生素对新生儿和婴儿的菌群产生长期影响，包括变形杆菌，特别是肠杆菌科的数量增加，放线杆菌的丰度和多样性减少，特别是双歧杆菌科。加拿大的一项研究表明，母亲在分娩期间接受抗生素治疗的婴儿表现出长期的菌群失调，特别是肠球菌属和梭菌属的增加，拟杆菌的比例降低；总而言之，抗生素治疗后婴儿菌群均会发生变化[69]。相比之下，成人肠道菌群的初始组成可能会影响其对抗生素的反应；成人的生活方式也可能会影响其菌群对抗生素的反应。而对于肠道菌群本身就不稳定的个体，如短期内接受多个抗生素治疗的群体或老年人，抗生素对肠道菌群的影响将会呈现出更为复杂的变化，具体变化趋势还需要进一步研究证实[70]。

第三节　肠道菌群对机体健康的影响

一、肠道菌群与代谢

肠道共生菌能够执行多种人体自身不能进行的代谢活动，可通过分解食物中宿主不能完全消化的多糖、蛋白质和脂肪类物质获取能量，并产生一系列代谢产物影响宿主健康。在这个过程中，肠道微生态与宿主代谢能力息息相关。已有大量研究证实，肠道微生态系统失衡与多种疾病的发生关系密切，特别是肥胖、2型糖尿病等慢性代谢性疾病。根据世界卫生组织的数据，2014年，超过19亿成年人超重，其中超过6亿人肥胖。肥胖是正能量平衡的结果，当摄入的能量超过消耗的能量时，就会发生肥胖，它是其他代谢并发症（如2型糖尿病、高脂血症）

的强烈危险因素。随着卫生组织努力寻找预防肥胖的解决方案，关于消化道菌群与人体代谢的研究领域正在不断扩大，这为我们提供了新的见解和干预途径。现有研究表明，比较肥胖人群和苗条人群之间的肠道菌群，发现肥胖人群的厚壁菌门水平增加，拟杆菌门水平降低。在小鼠中也观察到了类似的变化，这表明肠道菌群可能在肥胖的发病机制中起作用[71]。肥胖人群不仅与苗条人群的肠道菌群组成存在差异，其肝脏细菌 DNA 负荷也明显高于苗条人群，这说明肠道菌群的改变可能是非酒精性脂肪肝的早期危险因素[72]。

越来越多的证据表明，菌群的早期变化可能会对宿主的新陈代谢产生长期影响。例如，婴儿在生命的前 6 个月内使用抗生素，会增加 7 岁时超重的风险[73]。同样，小鼠断奶前的抗生素治疗会导致成年后的肥胖和代谢性疾病。研究推测，人类的某个时期对菌群改变特别敏感，如肠道菌群遭受巨大改变，则有很大概率导致后期的代谢性疾病。值得注意的是，与肥胖诊断相关的特定微生物特征尚未确定。目前的共识是肥胖人群出现肠道产丁酸盐的菌群减少并伴随机会性病原体增加。这些发现证明了肠道菌群和代谢之间相互作用的重要性[74]。

肥胖会增加患 2 型糖尿病等多因素疾病的风险。最近有研究发现，人类 2 型糖尿病与丁酸产生菌的丰度降低和乳杆菌属丰度的增加有关。此外，基于肠道宏基因组的计算模型能够预测糖耐量受损患者的 2 型糖尿病相关表型，提示肠道菌群组可能成为预测 2 型糖尿病新的生物标志物。代谢综合征患者接受万古霉素治疗会减少产生丁酸的细菌数量，并会导致胰岛素敏感性降低，这表明产丁酸的细菌水平降低可能导致 2 型糖尿病[75]。在一项里程碑式的研究中，Vrieze 等使用苗条身材捐赠者的粪便菌群移植给代谢综合征的胰岛素抵抗患者，提升了丁酸盐产生菌的数量，并改善了胰岛素敏感性[76]。这说明菌群可以直接调节胰岛素敏感性。然而，菌群如何调节宿主的葡萄糖代谢仍不清楚。

肠道菌群能够产生一些特殊代谢物从而影响宿主新陈代谢。其中一些代谢物是细菌原有的结构成分，如革兰阴性细菌膜的主要成分脂多糖，已知它会导致 2 型糖尿病并激活天然免疫系统（图 2-2）。有研究通过给小鼠喂食 4 周脂多糖，发现小鼠体内脂肪细胞可通过多糖受体 CD14 激活天然免疫信号，并促进趋化因子 CCL2 分泌，随后刺激巨噬细胞在脂肪组织中的积累，并增加肝脏促炎脂质介质的水平导致肝脏损伤，这项研究发现了脂多糖和代谢损伤之间的直接联系。该研究同时发现通过注射脂多糖可触发炎症反应诱导人胰岛素抵抗，导致患有 2 型糖尿病的受试者脂蛋白水平升高。脂多糖是由肠道菌群中的革兰阴性细菌产生的，通过使用膳食纤维或抗生素改变菌群组成，可以减少脂多糖相关的代谢变化。另一方面，肠源性脂多糖虽然促进了巨噬细胞向脂肪组织的聚集，但对葡萄糖代谢没有影响[77]。菌群与饮食的相互作用也会产生生物活性成分，如口服燕麦 β- 葡聚糖（一种聚合益生元）可以增加肠道菌群的丰度，特别是产丁酸的瘤胃球菌科和毛螺菌科，从而增强了肠道屏障功能、减轻代谢综合征[78]。

在小鼠中补充丁酸可以通过促进产热和脂肪酸氧化来预防饮食诱导的肥胖和胰岛素抵抗。部分原因是肌肉中组蛋白脱乙酰化酶活性降低，这可以调节过氧化物酶体增殖物激活受体 -γ 共激活因子 1α 的表达[79]。肠道菌群也是胆汁酸代谢的重要调节因子；它是合成次级胆汁酸所必需的，但也通过自然产生的法尼醇 X 受体拮抗剂的代谢来调节胆汁酸的合成（图 2-2）。值得注意的是，在无菌条件下抑制法尼醇 X 受体导致胆汁酸显著增加，有助于保护无菌小鼠因饮食诱导的肥胖。次级胆汁酸如胆石酸和脱氧胆酸是 GPCR5 的激动剂。GPCR5 被认为是改善代谢性疾病的潜在靶点。GPCR5 在组织中的广泛表达提示它可能通过多条途径调节代谢。GPCR5 的激活可以刺激 L 细胞分泌胰高血糖素样肽 -1，胰高血糖素样肽 -1 不仅促进棕色脂肪产热，还可以增加能量消耗，减少饮食诱导的肥胖。GPCR5 在肥胖受试者皮下脂肪组织中的表达增强，但在减肥后表达下降，并与静息代谢率呈正相关，表明 GPCR5 与人类的新陈代谢有关[80]。

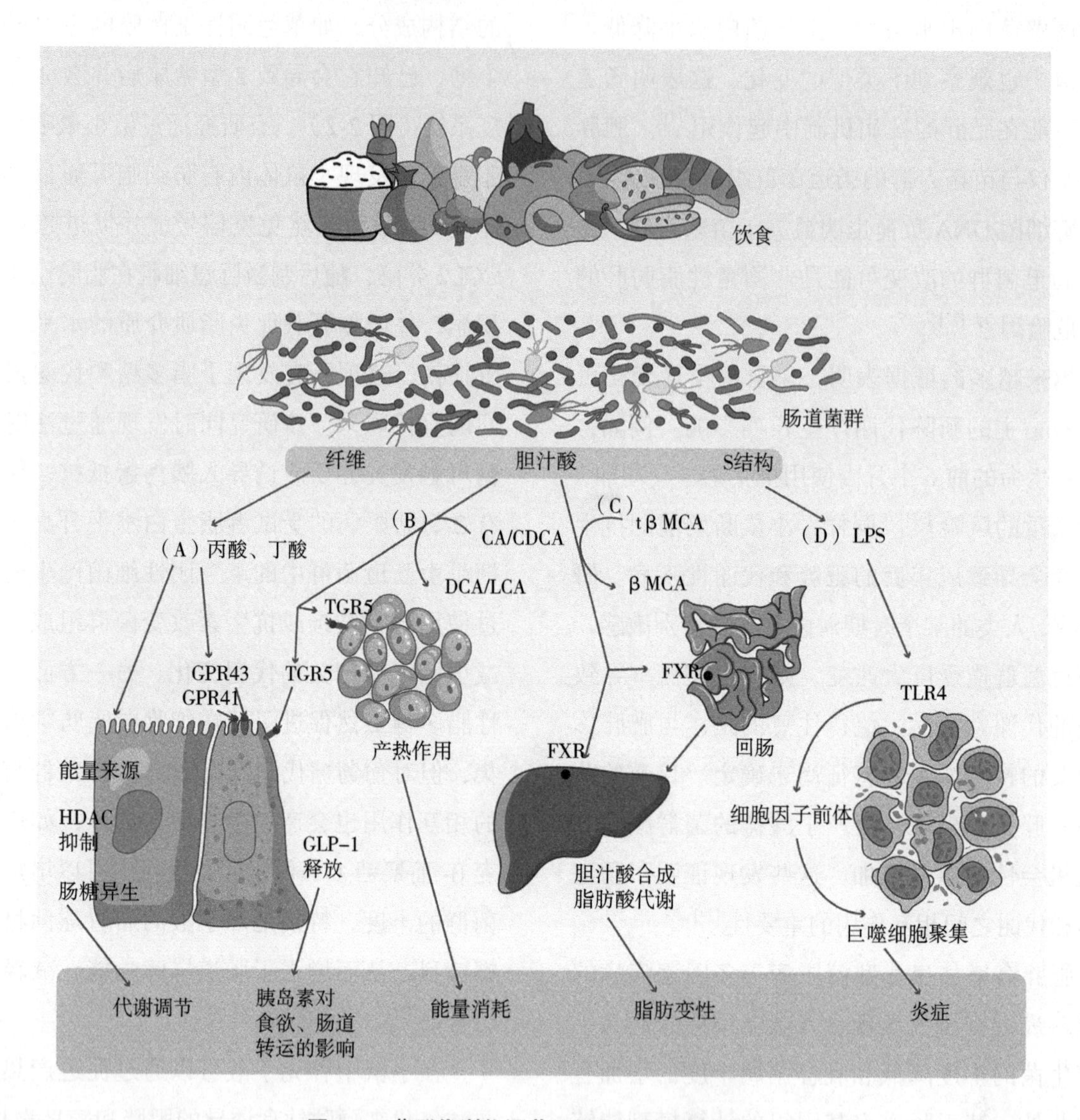

图 2-2　菌群代谢物调节不同组织的新陈代谢

A. 肠道菌群发酵膳食纤维产生短链脂肪酸。短链脂肪酸中的丁酸可作为能量来源并促进肠细胞中 HDAC 的抑制；丙酸和丁酸可刺激肠道糖异生，促进代谢调节。短链脂肪酸还能通过 G 蛋白偶联受体 43 和 41，促进胰高血糖素样肽 -1 的释放，菌群衍生的次级胆汁酸胆石酸和脱氧胆酸也可以通过胆汁酸受体 5 同样促进胰高血糖素样肽 -1 的释放，达到抑制食欲、减少肠道转运的效果。B. 胆石酸和脱氧胆酸可以促进棕色脂肪产热，增加能量消耗。C. 肠道菌群减轻肠道和肝脏的法尼醇 X 受体抑制，从而减少胆汁酸的合成并改变脂肪酸的代谢。D. 脂多糖是一种来源于革兰阴性细菌膜的促炎分子，通过 Toll 样受体 4 促进白色脂肪组织中巨噬细胞的募集和极化，诱导炎症。HDAC. 组蛋白去乙酰化酶；GPR. G 蛋白偶联受体；TGR. 胆汁酸受体；GLP-1. 胰高血糖素样肽 -1；CA. 胆酸；CDCA. 鹅脱氧胆酸；DCA. 脱氧胆酸；LCA. 石胆酸；MCA. 鼠胆酸；FXR. 法尼醇 X 受体；LPS. 脂多糖；TLR4. Toll 样受体 4

近年来，探索菌群对代谢性疾病影响的研究数量急剧增加。然而，这些研究大多是横断面的，仅证明了改变菌群与特定代谢性疾病之间的联系。还需要进行前瞻性研究，调查菌群在疾病发生前是否发生了改变，这些结果可能为人类干预代谢性疾病提供依据。然而，考虑到人群中的遗传和菌群多样性及复杂的菌群 - 饮食相互作用，个性化的治疗策略是必要的，并需要进一步的研究证实。

二、肠道菌群与衰老

人们普遍认为，人类的寿命是由遗传和环境（主要是饮食和生活方式）因素之间的复杂相互作用决定的[81]。随着年龄的增长，患慢性病的风险也会逐年上升，包括肌肉骨骼和免疫系统受损及代谢紊乱，这会导致肠道菌群的生态破坏，而人类共栖菌群在调节健康和寿命方面起了关键作用。具体来说，健康的肠道菌群在控制新陈代谢、抵抗感染和炎症、预防自身免疫和癌症，以及调节脑 - 肠轴等方面发挥着关键作用。此外，肠道菌群能够影响一系列疾病，如结直肠癌、炎性肠病、肠易激综合征及一些肠外疾病（如支气管哮喘、过敏、囊性纤维化等）[82]。最近，菌群靶向治疗被认为是治疗年龄相关的代谢性疾病和神经退行性疾病的一种有前途的治疗模式。通过肠道菌群组靶向干预特定的年龄相关性疾病具有很大的治疗潜力，而且在减缓衰老过程和促进人类健康及长寿方面也具有治疗潜力[83]。

生命早期肠道菌群与不良环境的相互作用可能会“规划”终身健康和疾病风险。越来越多的证据表明，肠道菌群在成人健康和疾病的发生发展中扮演着重要角色。肠道早期的菌群定植可以影响随后的共生宿主 - 细菌相互作用，并影响此后生活中疾病的发生，例如包括肥胖和 2 型糖尿病在内的代谢紊乱、慢性肠炎、自身免疫性疾病、过敏、肠易激综合征、过敏性胃肠炎和坏死性小肠结肠炎。这一过程被称为“菌群编程”[84]。如果能够掌握“菌群编程”的奥妙，则可以通过控制早期的菌群定植，达到减少疾病发生和延缓衰老的目的。

肠道菌群的组成随着年龄和相关疾病的变化而发生实质性变化（图 2-3）。这些变化是由多种压力事件、胃肠系统的生理衰老以及生活方式和饮食习惯的积累所驱动的。肠道菌群中与年龄相关的变化包括生物多样性减少、机会性革兰阴性致病菌的数量增加，以及具有促进健康功能的物种数量减少，特别是产短链脂肪酸的细菌，而在患病的老年人中，这些改变更为明显，患病老年人的肠道菌群包含更多的病原细菌，整个肠道菌群均处于营养不良状态，从而导致碳水化合物消化能力减弱，聚糖和硫胺素代谢产物及脂肪酸合成减少[85]。百岁老人的菌群多样性进一步降低，包括双歧杆菌、拟杆菌和肠杆菌科的数量减少，梭菌属的数量增加[86]。这种与衰老相关的肠道菌群差异多伴随着营养不良和对药物（非甾体抗炎药和抗生素）的需求增加。

菌群多样性的丧失与其说与实际年龄本身有关，不如说与衰老相关的脆弱性有关。肠道菌群是与年龄相关的病理状态的重要决定因素，例如慢性炎症、神经变性、认知衰退、虚弱、1 型和 2 型糖尿病，以及非酒精性脂肪性肝病和心血管疾病。肠道菌群也被认为是调节癌症发展风险的一个重要因素，菌群可通过特定的毒素、代谢物及诱导的免疫反应，在致癌过程中发挥作用[86]。因此，肠道菌群可能成为人类长寿的预测因子。青壮年和 70 岁老年人的菌群组成非常相似，但与百岁老人的菌群组成有很大不同。在 100 岁以上的人群中，肠道菌群的特点是厚壁菌门的重排和变形杆菌的富集，包括机会性促炎细菌（“病原体”）。这种菌群 - 宿主动态平衡受损与百岁老人病理状态增加有关。百岁老人菌群的重塑也伴随着普氏粪杆菌及其近亲数量的大幅减少，这些细菌是具有显著抗炎活性的共生物种。此外，黏液真杆菌及其近亲种细菌被特别确定为长寿的标志性细菌，因为在百岁老人的肠道菌群中，这种细菌数量增加了10倍以上。在一项百岁老人的研究中，肠道菌群年龄相关的轨迹特征是菌群的短链脂肪酸合成能力降低和糖酵解潜力下降，而老年人肠道宏基因组的蛋白水解功能比年轻人更丰富[87]。这种功能变化与“致病细菌”（即成年人肠道生态系统中普遍存在的机会性促炎细菌）的富集有关。此外，116个与衰老显著相关的菌群基因被确定为长寿的标志性基因，包括粪杆菌属、柔嫩梭菌属，以及低水平的脂多糖。

图 2-3　人类肠道菌群组成随年龄增长而变化

①青少年核心肠道菌群组成与成年人相似，但青少年的梭菌和双歧杆菌属的水平显著高于成年人；②成人肠道菌群个体差异极大，但以厚壁菌门、拟杆菌门、变形菌门和放线菌门为核心的肠道菌群是健康所必需的；③老年人肠道菌群会出现稳态失衡，多样性降低，肠道菌群的主要构成部分也由厚壁菌门向拟杆菌门转变，拟杆菌种类多样性增加，双歧杆菌种类多样性减少，而梭状芽孢杆菌较丰富；④百岁老人肠道菌群的特点是厚壁菌门的重排和变形杆菌的富集，包括机会性促炎细菌，并出现黏液真杆菌的明显增加

菌群可能有利于宿主健康和延缓衰老过程，其分子机制还需要通过益生菌临床试验进一步研究，这些益生菌和益生元被证明能有效预防老年人病理情况的出现。其中最重要的机制是抑制慢性炎症。它包括使老年人促炎细胞因子的合成减少，如IL-6、IL-8、IL-10和肿瘤坏死因子，激活的淋巴细胞数量、自然杀伤细胞数量及其吞噬活性增加。肠道菌群的调节可以减少炎症反应，改善适应性免疫反应，从而对抗免疫衰老。介导这些效应的机制包括降解不可消化的碳水化合物、增强抗氧化活性、产生维生素B和共轭亚油酸、调节宿主脂肪沉积和代谢、预防胰岛素抵抗，以及改善黏膜屏障完整性和保持免疫动态平衡等[88]。此外，肠道菌群代谢物丁酸在衰老和年龄相关疾病中扮演了重要角色。丁酸是结肠中产生的一种基本代谢物，它是结肠上皮细胞的主要能源，有助于维持肠道屏障功能，并显示出免疫调节、抗炎和抗癌能力。它同样可以通过预防代谢性内毒素血症、增强线粒体活性和激活肠道糖异生

来发挥其有益的代谢作用；不仅如此，它还通过抑制组蛋白脱乙酰化酶活性来调节表观遗传过程[89]。最近有证据表明，丁酸盐对免疫性疾病、癌症、神经系统疾病、2型糖尿病均有治疗潜力。

近年来，研究人员在了解肠道菌群对人类健康、衰老和长寿方面的作用取得了很大的进展。事实上，越来越多的证据表明，包括抗生素在内的各种药物可能会显著影响菌群基因的表达水平，并且菌群可以影响药物的生物利用度。而饮食、益生菌补充剂、体力活动水平、心理应激源暴露及宿主遗传学同样也对肠道菌群有巨大影响[86]。然而，在大多数情况下，适当控制混杂因素和最小化偏差是一项非常困难的任务。一些研究结果表明，直接调节肠道菌群不仅可以用于治疗特定的年龄相关性疾病，而且可以成为对抗衰老过程的治疗选择。例如，给老年小鼠喂食益生元低聚半乳糖，可通过调节肠道菌群组成来调控衰老肠道的稳态，其作用包括改善小鼠的营养不良、减轻慢性炎症、修复肠道屏障等[90]；口服健康百岁老人肠道菌群中分离出的活双歧杆菌亚株，可改善细胞免疫、体液免疫和非特异性免疫功能的参数，以及增强肠道的免疫屏障功能。

因此，分离并培养出与人类长寿相关的肠道细菌，并利用这些培养物开发标准化的粪菌移植治疗程序，似乎有希望促进老年人的健康和寿命。此外，使用转基因菌群似乎非常有效，因为它们可以生产出具有潜在促进健康和抗衰老特性的物质。例如，添加经过治疗性修饰的大肠埃希菌用于合成N-酰基磷脂酰乙醇胺（N-酰基磷脂酰乙醇胺的前体，是在小肠中合成的一系列脂质相关信号分子，能减少食物摄入量），可防止高脂饮食小鼠的代谢异常，包括胰岛素抵抗、肥胖症和肝性骨病。通过生物工程，不仅有可能改善现有菌株的有益性质，而且可能创造出具有全新性质和功能的益生菌。这样的生物工程结构不仅可以由细菌成分组成，还可以包括人类来源的特定酶。由于工程菌可以比传统治疗剂更精确地输送特定的蛋白质、药物或基因治疗载体，因此使用益生菌似乎是一个有吸引力的治疗选择，特别是如果这些新的治疗策略经过全面测试和优化，能满足当前的临床标准，将来极有可能会给医疗带来革命性的变化[91]。

三、肠道菌群与精神健康

近几年来，“菌群-肠道-脑轴”的发现引起了人们的极大兴趣，它是指发生在肠道和中枢神经系统之间的双向通信。共生菌群能够通过一系列神经、免疫、内分泌和代谢信号通路影响中枢神经系统的发育和神经功能。这个网络中，肠道菌群的改变可能影响与健康和疾病相关的潜在神经生物学机制，包括下丘脑-垂体-肾上腺轴的激活、神经递质和免疫系统活性的改变。焦虑和抑郁是影响全球数百万人的两种最突出的神经精神疾病，在人类和动物研究中表明，肠道菌群在焦虑与抑郁中发挥着重要作用，因此可以推测，焦虑与抑郁可能受“菌群-肠道-脑轴”的影响[92]。

现有的研究成果已经让人们较好地认识到肠道和大脑之间的相互交流。包括神经内分泌、免疫、自主神经和肠神经系统在内的多条路径，允许跨不同身体区域双向交换传入和传出信息。这些信息的相互作用在胃肠道内同时发生，胃肠道是约5亿个神经末梢的所在地，也是人体内免疫细胞最集中的地方。这些神经约20%被归类为内源性初级传入神经元，通过迷走神经将胃肠道内的细微变化传递给大脑。免疫细胞释放细胞因子，在宿主对炎症和感染的反应中扮演着重要角色。而神经内分泌激素（如皮质醇）可改变肠道的通透性、屏障功能，并与细胞因子的分泌进行沟通。这种神经元和生化信号的传递过程发生在机体全身，包括胃肠道和中枢神经系统之间建立的通路。研究人员将这种动态信号通路称为“肠 - 脑轴”，肠道菌群积极参与该轴调节的各种过程。因此，“肠 - 脑轴”也被称为“肠道菌群 - 肠 - 脑轴”[93]。肠道菌群通过多种途径与中枢神经系统的不同区域进行交流，包括自主神经系统、肠神经系统、免疫系统（趋化因子和细胞因子）和内分泌系统（下丘脑 - 垂体 - 肾上腺轴）。若“肠 - 脑轴”出现功能障碍，则可能导致广泛的病理生理后果，如炎症、慢性腹痛、进食障碍、恶心等。阐明肠道和大脑相互作用的途径和涉及的生物学机制，可能有助于推进胃肠道和中枢神经系统疾病的综合治疗[94]。

迷走神经是肠 - 脑轴的重要组成部分，可将肠道菌群和细胞的反应传递到中枢神经系统。通俗来说，就是肠道菌群和中枢神经系统之间的通信是通过迷走神经的初级传入通路介导的。研究表明，肠道菌群与神经系统相互作用可能是通过神经元 c-FOS 的表达和 c-FOS mRNA 的上调实现的[95]，该研究对感染了鼠伤寒沙门菌的大鼠进行迷走神经切断术，证实了迷走神经通路能将肠道免疫信号传递到中枢神经系统；一旦切断迷走神经通路，这些神经元中 c-FOS 的表达就会减弱，免疫细胞的百分比就会下降[96]。因此，探索迷走神经传入通路在调节肠道菌群和大脑之间的“传话”作用，可能证明肠道菌群在行为障碍的治疗干预措施方面是有用的。

肽聚糖细胞壁作为细菌的特有结构，可激活宿主的先天性和适应性黏膜免疫系统。先天免疫反应主要是由对致炎细菌成分进行识别，这些菌群成分通常被称为致病相关分子模式，它们与防御细胞上的模式识别受体结合，触发炎性细胞因子的产生，这些细胞因子可以通过外周迷走神经通路间接影响大脑，也可以通过血脑屏障的通透区直接影响大脑。例如，促炎细胞因子 IL-6 和趋化因子配体 2 可以通过两条途径作用于大脑：①致病相关分子模式作用于特定脑区 Toll 样受体的体液途径；②通过传入神经的神经途径。与革兰阴性菌细胞壁相关的致病相关分子包括肽聚糖单体、脂多糖、孔蛋白和富含甘露糖的糖链。与革兰阳性菌相关的致病相关分子包括肽聚糖单体和脂磷壁酸。这些细胞壁成分可以刺激肠上皮细胞产生与神经信号有关的其他分子[97]。这说明肠道菌群对黏膜免疫系统的强大免疫调节作用可能是肠道菌群影响大脑功能和行为的潜在机制。

饮食和营养成分的消化和菌群发酵产生的代谢物可能会对大脑和免疫反应过程产生重大影响。例如，控制肠道菌群的组成会影响脂肪酸和色氨酸的可用性和调节性。反过来，脂肪酸和色氨酸可以与免疫系统相互作用，从而调节细胞免疫反应。大脑富含脂肪酸，有助于调节几个过程，如神经传递、细胞存活和神经炎症[98]。此外，膳食脂肪酸还参与类二十烷酸的产生，这是一类化学信使，通过基因调节、细胞因子生物合成，以及膜成分和功能改变，参与免疫调节和炎症反应。脂肪酸还可以通过结合和激活一系列过氧化物酶体增殖剂激活受体的核受体来调节基因转录，这些受体影响细胞分化和功能特性。而且菌群衍生的短链脂肪酸已被证明能对人体产生诸多益处，如改善肠道屏障功能、减轻炎症等；对于神经系统也不例外，如吲哚丙酸具有抗炎、免疫调节和淀粉样蛋白抗聚集特性，可改善神经变性，并有效改善迟发性阿尔茨海默病的症状[99]。此外，短链脂肪酸通过与同源受体（如 GPR43 或 GPR41）结合来刺激神经肽的释放，如 YY 肽和胰高血糖素样肽 -1，在肠道内分泌信号中具有特殊作用。这些肽一旦释放，就可以通过激活肠道和初级传入迷走神经通路，影响能量稳态的调节[100]。另外，丁酸和丙酸也具有神经活性。鉴于菌群衍生的短链脂肪酸对能量代谢和次生肽产生的影响，说明肠道菌群是影响人类行为的另一种机制[101]。

通过控制 5- 羟色胺的产生，肠道菌群可以直接影响中枢神经系统的功能。大肠埃希菌和肠球菌能直接产生 5- 羟色胺，肠道菌群代谢生产短链脂肪酸也能促进 5- 羟色胺的生成。喂食婴儿双歧杆菌益生菌的大鼠在强迫游泳实验（一项评价抗抑郁作用的动物模型）后，几种促炎细胞因子减少，血浆色氨酸浓度增加，类抑郁行为减少；重要的是，益生菌治疗还降低了额叶皮质和杏仁核中 5- 羟色胺和多巴胺代谢物的浓度[102]。色氨酸代谢失调与许多大脑和胃肠道疾病有关，色氨酸代谢既可以生成神经毒性代谢物（如喹啉酸），也可以生成神经保护性代谢物犬尿酸和吲哚。增加犬尿酸的转化可加强神经保护作用，减轻应激诱导的抑郁。在婴儿芽孢杆菌治疗的动物中，发现色氨酸和犬尿酸增加[103]。吲哚可通过芳烃受体介导的信号通路，提高促神经成熟基因（如 β- 连环蛋白、Neurog2 和 VEGF-α）的表达，诱导小鼠神经突触成熟，促进海马齿状回神经发生[104]。这些发现表明，肠道菌群能影响色氨酸代谢，发挥保护神经、减轻精神情绪障碍、促进神经发育等作用。

研究表明，乳杆菌、双歧杆菌、大肠埃希菌、肠球菌等细菌能产生神经递质和神经肽。γ- 氨基丁酸、5- 羟色胺和脑源性神经营养因子等神经元信使失衡，会对大脑和行为产生显著影响。γ- 氨基丁酸是中枢神经系统的主要抑制性神经递质，是调节神经元兴奋性的主要化学物质。乳杆菌和双歧杆菌，能够在培养中产生 γ- 氨基丁酸。此外，鼠李糖乳杆菌能够调节小鼠中枢神经系统关键区域 γ- 氨基丁酸受体的表达，因此可能对抑郁症和焦虑症的治疗有益，从而进一步证明了细菌、肠道和大脑之间的功能交流途径。脑源性神经营养因子是在中枢神经系统中广泛表达的神经营养因子，大量证据支持脑源性神经营养因子的神经保护功能及其在突触生长和可塑性、神经元存活和分化中的作用，脑源性神经营养因子水平的降低与慢性抑郁症有关。接受抗菌剂和粪菌移植治疗改善肠道菌群后，无菌小鼠脑源性神经营养因子的水平上升，说明脑源性神经营养因子的水平与“肠 - 脑轴”相关[92]。

焦虑，是最常见的精神障碍之一，越来越多的证据支持肠道内的共生生物和应激系统的程序性和响应性之间的双向关系[92]。肠道菌群可能通过应激相关的神经内分泌、自主神经和免疫途径在治疗和预防焦虑和抑郁方面发挥重要作用。成年后的慢性压力也会影响肠道菌群的组成。暴露于应激源的小鼠出现拟杆菌属和肉毒杆菌属丰度下降[105]，循环血中 IL-6 水平的增加，脾脏巨噬细胞对菌群刺激的反应性增强，表明菌群是应激源诱导免疫激活的潜在必要环节。目前已在动物水平乃至人体水平上证明了调节肠道菌群在精神、神经系统疾病中的治疗潜力，例如给食物成瘾的小鼠模型喂食单形拟杆菌，可增加健康代谢菌种的丰度，影响大脑奖励反应，改善暴饮暴食并减少焦虑样行为[106]。以上研究代表了一种预防和治疗各种精神疾病的创新方法，未来这一领域的工作有助于阐明人类生理学中关键的感觉神经间的联系，从而改进中枢神经系统疾病的治疗方式。

四、肠道菌群与病原菌的相互作用

肠道菌群在病原菌感染中扮演了完全相反的两种角色，它既可增强对病原菌感染的抵抗力，也可能促进感染。一方面，病原菌可利用菌群的碳和氮源作为营养物质和调节信号来促进自己的生长，从而诱发炎症；另一方面，稳定状态下的肠道菌群可提高对病原菌定植的抵抗力[107]。例如，经抗生素治疗或在无菌环境中饲养的小鼠更容易受到福氏志贺菌、柠檬酸杆菌、李斯特菌和鼠伤寒沙门菌等肠道致病菌的影响；而拥有丰富多尔菌属和粪球菌属的个体对空肠弯曲菌感染具有显著的抵抗力。

食物中摄取的营养素同样是肠道病原菌成长的重要一环。因单糖被小肠吸收，肠道菌群仅能利用未消化的植物多糖和宿主多糖。糖类经过菌群分解后，可形成含岩藻糖、半乳糖、唾液酸、N- 乙酰半乳糖胺、N- 乙酰氨基葡萄糖和甘露糖的黏液，从而保护肠上皮[108]。然而，肠道中的病原菌也可以利用这些糖来促进自己的生长。如唾液酸是某些黏膜多糖的末端糖，多形拟杆菌具有唾液酸酶活性，可释放唾液酸，以获得底层多糖的碳来源。而艰难梭菌和鼠伤寒沙门菌同样能分解多形拟杆菌释放的唾液酸，为它们的生长提供优势。多形拟杆菌也可以将岩藻糖从宿主多糖中分离出来，使岩藻糖在肠道的管腔中有很高的利用率。这种游离岩藻糖也可以作为小鼠伤寒沙门菌的碳源[109]。而肠道菌群同样通过自身独有的机制对抗病原菌入侵，例如属于脱铁杆菌门的 *Mucispirillum schaedleri* 菌，一种存在于肠黏膜，在人类肠道菌群中普遍存在但丰度很低的细菌，它可以通过干扰病原体入侵，保护小鼠免受肠伤寒沙门菌的感染[110]。

菌群多位于肠腔内黏液层，然而，肠出血性大肠埃希菌则是通过紧密黏附于肠上皮细胞来实现独特的生存环境。为了成功定植，肠出血性大肠埃希菌必须与大肠埃希菌争夺营养，大肠埃希菌可优先利用岩藻糖作为碳源。肠出血性大肠埃希菌可利用大肠埃希菌分解的糖，如半乳糖、己糖醛酸酯、甘露糖和核糖。肠出血性大肠埃希菌同时可利用岩藻糖作为信号分子来调节其新陈代谢，并调节其毒力谱系在肠腔和黏液层的表达，并促进编码岩藻糖信号转导系统的致病基因的表达。肠出血性大肠埃希菌通过菌群利用宿主来源的信号系统，感知肠腔内环境，并调节自己的新陈代谢和毒力[111]。多形拟杆菌的代谢产物可促使肠出血性大肠埃希菌产生黏蛋白酶，从而促使其到达上皮内层，破坏黏液层并创造一个不适宜肠道原生菌群生长的环境。肠出血性大肠埃希菌利用来自多形拟杆菌的代谢产物，精确地规划其新陈代谢和毒力。

菌群产生的几种短链脂肪酸是菌群与肠道致病菌相互作用的重要决定因素。短链脂肪酸的浓度在肠道的每个部位都是不同的，病原菌可通过感知其浓度从而识别生态位。肠道中含量最丰富的短链脂肪酸是乙酸、丙酸和丁酸。鼠伤寒沙门菌在回肠定植，因为回肠含有浓度为 30mmol/L 的乙酸盐，这种浓度的乙酸盐增强了鼠伤寒沙门菌致病基因 SPI-1 编码的Ⅲ型分泌系统（type Ⅲ secretion systems，T3SS）中 T3SS-1 的表达，T3SS-1 参与细菌对宿主的侵袭行为；而结肠存在 70mmol/L 浓度的丙酸和 20mmol/L 浓度的丁酸，可抑制 T3SS-1 的表达，故鼠伤寒沙门菌通常不定植于结肠。

肠道病原菌还使用其他营养物质成功地克服菌群对其定植的抵抗力[112]。乙醇胺可作为细菌的碳源和氮源，它在哺乳动物肠道中含量丰富，但不能被大多数共生的菌群代谢。而鼠伤寒沙门菌、肠出血性大肠埃希菌和单核细胞增多性李斯特菌可使用这种化合物在肠道中获得生长优势。乙醇胺也被肠出血性大肠埃希菌和鼠伤寒沙门菌用作信号分子来激活其毒力基因的表达[113]。病原菌感染宿主往往是利用菌群衍生的分子作为营养物质和信号。如肠出血性大肠埃希菌在肠道的定植能力源于其和大肠埃希菌使用的糖源的差异[114]。

五、肠道菌群、免疫与炎症

肠道菌群与人类宿主免疫系统之间的相互

作用同样不可忽视，在数百万年的共同进化中，宿主免疫系统和共生菌群不断地相互适应。免疫系统在出生前并未完全发育，生命早期适当的菌群刺激对免疫系统的成熟具有不可替代的作用。与自然分娩和母乳喂养的婴儿相比，剖宫产和配方奶喂养的婴儿在早期生活中更有可能出现免疫紊乱，这种易感性可能与肠道菌群有关。同样的情况也发生在接受抗生素治疗的婴儿上。剖宫产婴儿的肠道菌群种类比自然出生的婴儿少。另外，如前所述，剖宫产婴儿的肠道菌群与其母亲的皮肤菌群相似，主要由葡萄球菌和棒状杆菌组成；自然分娩婴儿与母亲阴道菌群相似，主要由乳杆菌和普雷沃菌组成。而从喂养角度来看，配方奶喂养婴儿的肠道菌群与母乳喂养婴儿的肠道菌群有显著差别，这种差异更多地体现在配方奶喂养婴儿的双歧杆菌种类的多样性，而不是丰富性。

肠道菌群可促使肠腔形成黏液层，而黏液层致密而坚固的结构可阻止细菌与肠上皮细胞的接触，使细菌局限于肠腔[115]。肠道菌群产生的代谢产物短链脂肪酸能为免疫系统的 B 细胞反应提供动力。一般来说，免疫球蛋白 A 反应可避免细菌与抗原呈递细胞接触，从而阻止 T 细胞的激活；然而，在健康状态下，可通过特定的免疫识别来启动免疫球蛋白 A 反应，例如识别拟杆菌的荚膜多糖 C。肠道菌群还可产生以梭状芽孢杆菌相关的节段性丝状细菌（segmented filamentous bacteria，*SFB*）为代表的促炎作用。在机体内，*SFB* 与肠内皮细胞膜紧密黏附，这种黏附特性是依赖 *SFB* 诱导 17 型辅助性 T 细胞（type 17 T helper，Th17）产生促炎反应的先决条件。除了 Th 17 外，*SFB* 还可能驱动 Th1 和 Th2 细胞应答，从而导致 IFN-γ、IL-4 和 IL-13 增加，促进炎症发生。

从抗炎方面来说，肠道菌群的代谢产物短链脂肪酸可作为特异性 GPCR 的配体和组蛋白脱乙酰化酶的抑制剂，促进幼稚 T 细胞向调节性 T 细胞和 $Foxp3^{+}$ 调节性 T 细胞分化，进而通过控制炎症反应和促进低亲和力免疫球蛋白 A 的产生来维持肠道菌群的动态平衡。此外，短链脂肪酸可通过抑制组蛋白脱乙酰化酶的活性，抑制促炎效应分子表达[116]。而普雷沃菌的抗炎作用与几种未知的代谢产物有关，这些代谢产物可以增加 IL-10 的分泌，阻断核因子 κB（Nuclear factor-κ B，NF-κB）的活化和促炎细胞因子 IL-8 的释放[117]。此外，特定的乳酸菌菌株具有抗炎作用，副乳杆菌基因编码的蛋白酶乳霉素也具有抗炎作用，能够在 IL-6 和 IL-23 的参与下，以类似于 Th17 细胞的方式驱动 $RoR\gamma t^{+}$ 调节性 T 细胞的分化，参与抗炎作用。

第四节　肠道菌群与人类疾病

细菌和宿主间的共生关系与宿主的健康息息相关，肠道菌群为宿主提供了许多益处，包括脂质和碳水化合物代谢、激素调节、抵抗病原体、上皮细胞增殖，以及维生素和氨基酸的合成；另一方面，肠道菌群在多种疾病中同样扮演了重要角色，包括心血管疾病、营养不良、炎性肠病、神经系统疾病、癌症，以及糖尿病和肥胖症等[118]。

一、肠道菌群与心血管疾病

心血管疾病仍然是发达国家人群死亡和致残的主要原因，因此，制订预防和控制心血管疾病的新策略至关重要。肠道菌群的组成改变和功能改变被称为菌群失调。肠道菌群及其相关代谢物的多样性和成分的变化，可能会影响疾病的发病机制和进展。肠道菌群失调正成为心血管疾病强有力的决定因素之一。越来越多的证据表明，肠道菌群失调与心血管疾病的进展有关，包括动脉粥样硬化、高血压和心力衰竭[119]。

（一）动脉粥样硬化

动脉粥样硬化是一种慢性炎症状态，初始出现动脉血管内皮的损伤，随后出现动脉脂质积聚和斑块形成。其危险因素包括高胆固醇血症、高血压、糖尿病、肥胖和静坐少动生活方式。动脉粥样硬化斑块含有人体肠道细菌 DNA，这说明粥样硬化斑块中的细菌来源于肠道[119]。

氧化三甲胺（trimethylamine-N-oxide，TMAO）与动脉粥样硬化之间的联系，展现了肠道菌群作用于心血管疾病的新途径[120]。代谢组学将 TMAO 确定为人血浆中与心血管疾病风险相关的一种小分子代谢物，这种分子来源于磷脂酰胆碱的代谢。磷脂酰胆碱和其他含三甲胺的化合物如左旋肉碱被肠道菌群三甲胺裂解酶代谢产生气态三甲胺（trimethylamine，TMA），TMA 再由黄素单加氧酶家族的肝酶代谢氧化，从而生成 TMAO[121]。

研究发现，给小鼠补充 TMAO 饮食，促进了小鼠的动脉粥样硬化[122]。人体中 TMAO 浓度的升高与心血管疾病的风险增加相关，并可增加动脉粥样硬化的发生[122]。给小鼠补充左旋肉碱时，盲肠菌群组成发生变化，TMA 和 TMAO 的合成增加，动脉粥样硬化增加；然而，当肠道菌群同时受到抑制时，则不会出现这种情况[123]。有证据表明，脑卒中和心血管疾病风险较高的个体，其 TMAO 水平都有所升高[119]。使用 3，3- 二甲基 -1- 丁醇（一种三甲胺抑制剂）可预防由高糖高脂饮食引起的心脏功能障碍[119]。

TMAO 水平和心血管疾病不良临床后果的联系已经在许多独立队列研究中显示出来[119]。一项有 4000 多人参加的临床研究证实，血清 TMAO 水平可剂量依赖性地预测心血管风险的发生率[124]。在一项对 608 例冠心病患者和对应健康人群的前瞻性研究中发现，人们患冠心病的风险与饮食习惯密切相关，健康人群食物普遍以谷物和植物纤维为主，其血浆 TMAO、三酰甘油和 C 反应蛋白的浓度更低[125]。最近证实，菌群影响血浆 TMAO 水平，同时发现 TMAO 水平与血小板活化有关[126]。动物模型研究也发现，TMAO 能改变血小板钙信号，并在体内引发促血栓形成效应[127]。

（二）高血压

高血压是心血管疾病最普遍的可变危险因素，世界卫生组织将高血压列为全球人类过早死亡的主要原因之一[127]。与正常血压的动物相比，在几种啮齿动物高血压模型中都观察到菌群的丰富性和多样性降低。在门水平上，厚壁菌门 / 拟杆菌门比率的增加似乎是实验性高血压模型的一个恒定特征，并且在人类高血压患者中也观察到此现象[128]。给大鼠输注 TMAO 可增强血管紧张素Ⅱ的高血压作用[129]。使用抗生素米诺环素消耗肠道菌群，可显著降低注入了血管紧张素Ⅱ动物的平均血压，同时降低了厚壁菌门 / 拟杆菌门的比率[128]。需要进一步研究以确认这些联系的相关性及其内在机制。

短链脂肪酸由菌群在肠道中发酵不可消化的膳食纤维产生[128]，粪便中短链脂肪酸水平的升高与肠道菌群多样性降低、肠道通透性增加、全身炎症、血糖水平、血脂异常、肥胖和高血压有关，即使在调整饮食、总热量和体力活动等混杂因素后也是如此[128]。其中丁酸盐可能在调节血压方面起着重要作用。一项研究发现，自发性高血压大鼠的血浆丁酸水平降低，并且直接在下丘脑给予丁酸盐可以降低高血压大鼠的血压[130]。另外，丙酸和乙酸盐在近年来的各项研究中已被认为是血压调节剂。

（三）心力衰竭

越来越多的文献支持肠道菌群在心力衰竭发病机制中的作用，即所谓的“心力衰竭的肠道假说”。肠道假说表明，心排血量减少和全身性充血增加可导致肠道黏膜缺血和(或)水肿，细菌易位增加，循环内毒素增加，从而导致心力衰竭患者出现潜在炎症[131]。与健康对照组相比，心力衰竭患者体内的菌群发生了改变[132]。具体而言，在属水平上，梭状芽孢杆菌减少，而在种水平上，直肠真杆菌减少，这突出了肠道菌群在心力衰竭进展中的潜在临床意义[133]。其他研究表明，肠道菌群失调涉及保护性代谢物如丁酸盐，以及有害代谢物如心力衰竭患者的 TMAO 水平发生变化[132]。

TMAO 除了和动脉粥样硬化的风险有明确联系外，还与心力衰竭的发展和不良预后相关[134]。研究发现，与年龄和性别匹配的无心力衰竭受试者相比，心力衰竭患者循环中的 TMAO 水平更高[132]，这提示 TMAO 与心力衰竭的进程密切相关。

（四）治疗干预

肠道菌群失调与心血管疾病之间的许多联系，使得肠道菌群成为潜在的治疗靶点。营养干预的饮食方法已被证明是降低心血管风险的有效策略[135]。据报道，高纤维饮食可增加产生乙酸的菌群，降低血压，减少心肌肥大和纤维化。目前，饮食调节是临床实践中用于治疗慢性代谢性疾病的主要治疗工具[135]。例如多酚是一大类芳香化合物，广泛存在于植物饮料和食品中，肠道菌群将摄入的多酚分解、代谢和吸收，并产生心脏保护和抗糖尿病作用[136]。

粪菌移植是指将健康人粪便中的功能菌群通过一定方式移植到患者肠道内，重建新的肠道菌群，从而实现治疗肠道及肠道外疾病的目的[119]。目前，粪菌移植已在艰难梭菌感染、炎性肠病、肿瘤合并的肠道疾病、自身免疫性疾病、肝病、糖尿病及脑 - 肠轴疾病（如癫痫、抽动症、自闭症）等多种肠道菌群相关性疾病中显示出重要的治疗价值[137]。

益生菌是活的“有益”细菌，可以维持或促进肠道菌群平衡。大多数益生菌产品可能含有有益的双歧杆菌、乳杆菌、乳球菌和链球菌[119]。益生菌的性质决定了它们可能会抑制致病菌的过度生长，刺激免疫系统，调节 pH 并预防炎症[138]。益生菌可以通过刺激消化酶的活性，抑制细菌酶的活性和氨的产生来改善宿主的新陈代谢。益生元是一种不可消化的食品成分，可以促进肠道菌群的组成和活性，对宿主有积极的影响[119]。据报道，高纤维饮食会改变肠道菌群，从而增加产生乙酸盐的细菌，有助于减少肠道菌群的营养不良并提供心脏保护作用[139]。用抗生素消除致病菌群是一个显而易见的方法，但普遍的共识是，这种非特异性抗菌方法可能弊大于利[119]。

TMAO 与心血管疾病不良预后联系密切，并对宿主心脏代谢带来不利影响，因此选择性作用于菌群 TMAO 合成酶就成为一种潜在的治疗策略。有研究开发了一种抑制菌群胆碱三甲胺裂解酶活性的小分子药物，证明了这一策略的可能性[119]。最近的一项研究也成功开发了一种以此机制为基础的抑制剂药物（3,3- 二甲基 -1- 丁醇），主要以菌群三甲胺生成酶为靶点，其中单次口服剂量可使血浆 TMAO 水平显著降低。最多 3 天，可以减少饮食引起的血小板反应性增强和血栓形成[140]。

将经典疗法与调节微生物群及其代谢物相结合的治疗策略，用于心血管疾病的防治，仍需要更多设计严谨的临床对照试验验证。

二、肠道菌群与肥胖和糖尿病

在世界各地，肥胖及其相关并发症是一个主要的公共卫生问题[141]。肥胖被定义为身体脂肪过多导致体重指数升高，现行标准将肥胖定义为体重指数（body mass index，BMI）≥ 30kg/m^2。其患病率在过去 40 年中显著增加，2016 年全世界超过 5 亿成年人被诊断为肥胖[141]。

较高的体重指数会增加患多种疾病的风险，包括心血管疾病、糖尿病、慢性肾病、视网膜病变和癌症[142]。遗传因素和环境因素之间的相互作用与体重增加有关[141]，但这些环境因素和遗传因素的结合并不能完全解释肥胖的发病机制。因此，其他因素如肠道菌群，它将基因、环境和免疫系统联系在一起，就可以解释肥胖及其代谢并发症的发展[141]。越来越多的证据表明，肠道菌群，特别是其组成和多样性的变化，在肥胖及其相关并发症的发展中发挥作用[143]。

2 型糖尿病是一种以高血糖为特征的慢性代谢性疾病，由胰岛素分泌缺陷、胰岛素抵抗或两者兼而有之引起[141]。2017 年，国际糖尿病联合会估计，全球有 4.25 亿人被诊断为糖尿病，估计到 2030 年，全球将有 3.66 亿人死于糖尿病[142]。糖尿病的慢性高血糖与各种器官的长期损害有关，特别是眼睛、肾脏、神经、心脏和血管。最近的大量研究表明，2 型糖尿病也与肠道菌群失调有关[141]。肠道菌群失调可以影响代谢活动，导致肥胖和代谢紊乱，如代谢综合征、血脂异常、糖尿病和非酒精性脂肪性肝病[144]。因此，保持肠道菌群的多样性和平衡是促进人类健康的关键点[141]。

（一）肠道菌群与肥胖

肥胖和肠道菌群的联系一直是许多研究的主题。动物研究表明，改变啮齿动物的肠道菌群可以减少其脂肪量和体重，并改善胰岛素敏感性[141]。将肥胖小鼠的肠道菌群转移到无菌小鼠（无任何细菌的小鼠）的肠道中，会导致无菌小鼠的脂肪量和体重增加[145]。当给予小鼠影响大多数肠道细菌的广谱抗生素时，部分小鼠能免受饮食诱导的肥胖和相关代谢紊乱的影响[146]。对遗传性肥胖小鼠的盲肠内容物进行分析后发现，与野生型和杂合瘦削型小鼠相比，遗传性肥胖小鼠的肠道菌群表现出厚壁菌门增加，拟杆菌门减少[142]。后者与乳酸盐、乙酸盐和丁酸盐水平升高有关，而乳酸盐、乙酸盐和丁酸盐是脂肪酸合成的底物[141]。总的来说，目前基于动物研究的数据强调了肠道菌群和肥胖的因果关系。

肠道菌群的改变与肥胖有关，但是，这些变化是否包括厚壁菌门 / 拟杆菌门比率的增加，存在相互矛盾的数据。有研究检测了肥胖人群受试者减肥前后的菌群，报道了与上述小鼠模型相似的发现，无论患者是否在 1 年内保持限制性脂肪或限制性碳水化合物饮食，体重下降与厚壁菌门减少和拟杆菌门增加相关[147]。然而，据 Turnbaugh 等报道，肥胖者与苗条者相比，菌群多样性降低，拟杆菌门水平降低，但厚壁菌门丰度没有显著差异。研究还发现，肥胖者与苗条者厚壁菌门 / 拟杆菌门的比率没有显著差异[142]。因此，这一比例与肥胖之间的联系很弱，说明这一特定的菌群比例特征尚不能区分肥胖和非肥胖受试者的肠道菌群[148]。

肥胖是一种低度炎症性疾病，部分原因是脂肪组织的炎症损伤最终导致全身的慢性炎症[149]。肠道菌群已经成为体重增加和免疫 / 炎症系统失衡的新因素。动物实验发现，喂食高脂饮食的无菌小鼠与喂食相同饮食的常规饲养小鼠相比，肠道炎症增加、脂肪量减少[143]。然而，无菌小鼠移植了肥胖小鼠肠道菌群后，在接下来的 2 周内变得肥胖[143]。同时人类研究发现，部分肥胖受试者的肠道菌群表现出基因失调，其特征是菌群基因丰度降低，具有促炎和抗炎特性的物种相应增加或减少[150]；在高脂饮食诱导的肥胖小鼠中，肠道通透性增加，随后细菌和细菌成分转移到全身，导致循环脂多糖增加和巨噬细胞积聚。因此，肠道通透性的改变可能导致炎症和相关的代谢改变[150]。

肠黏膜层起着屏障作用，可防止肠道上皮细胞与病毒、毒素和致病菌之间产生不良的相互作用[143]。黏蛋白是肠屏障功能的重要组成部分。嗜黏蛋白阿克曼菌是人类肠道菌群中最丰富的单一物种之一，这是一种能够降解黏蛋白的菌株，由于该菌数量的减少与啮齿动物和人类的肥胖、胰岛素抵抗、糖尿病和其他心脏代谢紊乱有关，因此受到了广泛关注。在肥胖小鼠中发现嗜黏蛋白阿克曼菌显著减少，当其水平恢复后，肥胖小鼠的代谢性内毒素血症、肥胖和胰岛素抵抗逐渐恢复正常 [143]。

短链脂肪酸是一种含有 2 ～ 6 个碳原子的有机脂肪酸，由肠道菌群在宿主的结肠和盲肠中合成，发酵不可消化的食物蛋白、糖蛋白和纤维。短链脂肪酸中乙酸、丙酸和丁酸最为丰富。与非肥胖受试者相比，肥胖和胰岛素抵抗受试者的特征是短链脂肪酸产量减少 [141]。不同短链脂肪酸有不同的致饱机制：丁酸作用于肠细胞，增加胰高血糖素样肽 -1 的产生，丙酸影响肠糖异生，这两种途径都导致葡萄糖稳态的改善和饱腹感的增强 [151]。在饮食中添加丁酸和丙酸已被证明能防止小鼠肥胖和减轻胰岛素抵抗 [143]。丁酸还可以促进棕色脂肪组织的产热，这是防止肥胖的一个重要因素 [152]。总的来说，短链脂肪酸能通过防止脂肪堆积来减少体重增加，但短链脂肪酸的有益效果是否能应用于人类是一个需要深入研究的问题。

肠道菌群参与胆汁酸的代谢，突出了其对调节糖、脂代谢的作用。迄今为止，在动物及人类中，关于胆汁酸在体重调节和肥胖中作用的研究还很少。大多数人类研究表明，体重指数与血清总胆汁酸和某些特定胆汁酸之间存在正相关。肥胖与胆汁酸合成增加和转运受损有关 [153]；然而，还需要更多的研究来阐明肠道菌群在调节胆汁酸中的作用，这是肥胖和代谢紊乱的潜在机制。

（二）肠道菌群与糖尿病

糖耐量异常和 2 型糖尿病都与肠道菌群的变化有关。研究发现，与健康对照组相比，糖耐量异常受试者的肠道菌群存在显著差异，包括菌群多样性降低，这说明健康的肠道依赖于丰富多样的菌群 [154]。

糖尿病患者的肠道菌群可能出现机会致病菌增多，产丁酸细菌的减少 [155]。值得注意的是，肠道菌群产生的代谢物变化可能与 2 型糖尿病和胰岛素抵抗的发生有关。特别是丁酸、乙酸和丙酸等短链脂肪酸 [156]。已知丁酸梭状芽孢杆菌能上调丁酸，用丁酸梭状芽孢杆菌治疗糖尿病小鼠后，发现小鼠葡萄糖水平改善和胰岛素抵抗降低、炎性标志物降低、线粒体代谢增加等有益改变 [156]。普拉梭菌是一种重要的丁酸盐产生菌，一项宏基因组的分析显示，相比于肥胖、糖尿病患者，在瘦削受试者的粪便样本中，普拉梭菌丰度更高 [157]。事实上，普拉梭菌的丰度在患有和不患有糖尿病的肥胖受试者之间也存在显著差异 [157]，两组之间的丁酸盐水平也存在差异。

在糖尿病和糖耐量异常患者中发现，嗜黏蛋白阿克曼菌减少。有趣的是，最近研究表明，嗜黏蛋白阿克曼菌对免疫系统具有调节作用，从而延缓糖尿病易感动物 1 型糖尿病的发生 [158]。

最近许多研究认为，二甲双胍能促进胰高血糖素样肽 -1 的分泌和降低血清胆汁酸水平，这与厚壁菌门 / 拟杆菌门比值的变化有关。二甲双胍处理的供体向高脂饮食喂养的无菌小鼠转移肠道菌群后，改善了无菌小鼠的葡萄糖耐量，证明二甲双胍改变的菌群可以产生降糖效应，同时还发现了嗜黏蛋白阿克曼菌丰度的增加。在体外实验证明，二甲双胍能直接促进这种细菌的生长 [159]。肠道菌群似乎通过与二甲双胍的协同作用来调节高血糖状态下的代谢。

二甲双胍作用的另一个靶点是胆汁酸，胆汁酸不仅参与了肝脏调节的胆固醇代谢，而且还参与了葡萄糖稳态的调节。自 20 世纪 70 年代以来，双胍类药物对胆汁酸吸收的抑制作用已得到公认，由此引起的粪便胆汁酸排泄增加，再加上胆固醇合成胆汁酸的增加，可能是二甲双胍降低血清胆固醇水平的原因 [143]。

近两年来，对小鼠和人类的初步研究表明，

二甲双胍可以同时针对血糖、血脂和血管通路起作用，降低糖尿病视网膜病变的发病率，但其具体机制尚不清楚。二甲双胍的治疗作用及其在糖尿病并发症中的保护作用，与其对肠道菌群的影响有关，这是药物说明书之外一种独特的治疗作用。我们预测，未来糖尿病及其并发症的新治疗方式将直接针对肠道菌群，在获得最大疗效的同时改善不良反应[160]。

（三）预防及治疗

目前，很难制订旨在调节肠道菌群数量和组成的策略，以预防或治疗与菌群失调相关的代谢紊乱。然而，肠道菌群生态系统与健康的生活方式密切相关，因此，调节肠道菌群组分最有效的方法是通过均衡健康的饮食[141]。

一般来说，饮食导致的体重减轻与肠道菌群的丰富性增加和全身慢性炎症的减少有关。当肥胖患者食用限制碳水化合物或限制脂肪的低热量饮食时，可以观察到厚壁菌门减少，拟杆菌门增加[141]。体育锻炼也可以调节肠道菌群的组成。益生菌、益生元或合生元有助于超重个体减重，以及非酒精性脂肪肝和2型糖尿病患者的代谢改善[161]。尽管口服抗生素治疗对根除细菌病原体有好处，但抗生素治疗会导致肠道菌群代谢途径的改变，因此需要更谨慎地使用抗生素。减肥手术是迄今为止治疗肥胖症最成功的方法，因为它既能减轻体重，又能改善代谢，60%的患者甚至能达到糖尿病缓解[143]。粪菌移植作为治疗糖尿病的最新研究，已经在欧洲和中国进行。

三、肠道菌群与神经系统疾病

神经系统疾病如阿尔茨海默病、帕金森病的发病率在全世界范围内都在迅速增加，但是它们的发病机制仍然不清楚，并且治疗效果往往很不理想[162]。如前节所述，中枢神经系统和肠道菌群之间的双向通信可能在上述疾病的发病机制中起着一定作用[162]。肠道菌群的产物例如代谢物作为信号分子，对神经系统有直接或间接的影响。越来越多的研究表明，肠道菌群可以调节神经系统发育过程中的重要过程，包括神经形成、髓鞘形成、胶质细胞功能和血脑屏障通透性[163]。大量研究表明，在生命早期存在一个发育窗口，在此期间肠道菌群失调会对神经系统的发育产生长期影响。此外，在成年动物中有些功能也很容易被肠道菌群调节，包括小胶质细胞的激活和神经炎症[163]。

肠道菌群通过不同途径和大脑之间相互作用，包括代谢产物、激素、免疫系统和传入神经[163]。例如在肠道中，菌群代谢了其他不可消化的复杂碳水化合物，产生短链脂肪酸。短链脂肪酸可被宿主吸收并作为能量[164]。然而，短链脂肪酸也可作为信号分子，对中枢神经系统小胶质细胞的成熟很重要[165]。此外，大肠埃希菌蛋白ClpB是α-黑素细胞刺激激素的模拟物[166]，它已被证明能直接作用于下丘脑前阿片样蛋白神经元并增加这些神经元的放电，从而可能诱导饱腹感。

肠道菌群的功能障碍在某些神经系统疾病中发挥作用，恢复其健康和肠道屏障完整性可以积极影响临床进程和症状[167]。

（一）阿尔茨海默病

阿尔茨海默病是一种神经退行性疾病，其各种病理生理学机制仍在研究中。这是一种以认知功能逐渐衰退和特定类型神经元和突触丧失为特征的疾病。阿尔茨海默病中最常见的病理学表现是淀粉样β肽的聚集和过度磷酸化的Tau蛋白缠结[168]。目前认为，环境和遗传因素都参与了阿尔茨海默病的发病。同时，越来越多的研究表明，阿尔茨海默病可能与肠道菌群的异常有关。

阿尔茨海默病患者中，肠道菌群最显著的变化是抗炎细菌种类丰度减少、促炎细菌种类丰度增加。这可以导致血浆中的炎症因子水平上升，中枢神经系统的炎症增加[169]。肠道菌群损伤会加剧肠道炎症，从而减少结肠上皮细胞中紧密连接蛋白的表达。肠上皮连接受损导致菌群分泌物进入循环系统，引发炎症反应。血浆中炎性前细胞因子可通过血脑屏障到达大脑，并通过改变小胶质细胞成熟和星形胶质细胞激活触发炎症反应[170]。

高脂肪和（或）高热量的饮食似乎是患阿尔茨海默病的一个危险因素，而饮食可以通过多种方式影响肠道菌群。例如，深海鱼类、坚果和植物油等食物富含 ω-3 多不饱和脂肪酸，该脂肪酸可以降低患阿尔茨海默病的风险，并且阿尔茨海默病患者通常出现二十二碳六烯酸水平降低，二十二碳六烯酸是一种 ω-3 多不饱和脂肪酸。重要的是，肠道菌群参与 ω-3 多不饱和脂肪酸的吸收和代谢[171]。

阿尔茨海默病导致的认知障碍可能与各种生长因子的活性改变有关，包括神经营养因子，如脑源性神经营养因子，而脑源性神经营养因子的表达可能受到肠道菌群的调节。据报道，与无菌动物相比，无特定病原体的小鼠在海马和杏仁核中具有更高的脑源性神经营养因子基因水平[172]。对阿尔茨海默病患者大脑的尸检研究显示，相比于正常人，脑源性神经营养因子减少了 30%，而在晚期患者中，这一减少高达 40%。阿尔茨海默病患者的脑源性神经营养因子血清水平低于患有其他类型痴呆的受试者或健康对照者[173]。然而，尽管脑源性神经营养因子的血清水平易于测量，但它们既不能预测阿尔茨海默病的发展，也与功能评估、分期无关。

益生菌是对宿主健康有益的细菌，而益生元是作为这些细菌食物的物质（主要是纤维）。关于益生菌和益生元在阿尔茨海默病中作用的数据尚不丰富。研究发现，以益生菌和益生元的高摄入量为特征的健康饮食模式，结合其他营养素，可延缓神经认知能力下降，并降低患病的风险。此外，益生菌饮食补充不仅对正常的大脑活动有影响[174]，而且还显著改善了阿尔茨海默病患者的认知障碍[175]。最近研究发现，与未治疗的阿尔茨海默病小鼠相比，益生菌治疗的阿尔茨海默病小鼠具有更好的认知能力，并且在海马中减少了淀粉样 β 肽斑块的数量[176]。

（二）帕金森病

帕金森病是老年人第二常见的神经退行性疾病，从诊断到死亡的平均时间为 15 年。帕金森病发病的中位年龄为 60 岁，随后的残疾以运动和非运动症状为特征，包括威胁生命的痴呆。非运动症状，包括情绪和认知障碍、睡眠障碍、感觉障碍和胃肠功能障碍[177]。

帕金森病的关键特征是存在由聚集和磷酸化的 α- 突触核蛋白组成的神经元内含物（Lewy 体或 Lewy 神经突）[178]。虽然帕金森病的病因尚不清楚，但胃肠道是影响中枢神经系统的许多毒素的来源。

已在帕金森病患者中发现，胃肠道功能和结构变化会引起肠道神经系统中的 α- 突触核蛋白积聚[162]。肠道菌群异常在帕金森病患者中很常见，主要是普雷沃菌丰度的降低和乳杆菌丰度的增加，这可能与肠道生长激素释放肽水平的降低相关[162]。肠道生长激素释放肽可调节黑质纹状体多巴胺功能并限制帕金森病的神经变性，其分泌在帕金森病患者中减少[162]。研究表明，帕金森病患者中普雷沃菌减少了 77.6%，因此，普雷沃菌的相对丰度低于 6.5% 被认为是帕金森病的诊断指标之一，敏感度为 86.1%，特异度为 38.9%[179]。另有报道，肠杆菌科的高水平与姿势不稳定和步态困难的严重程度呈正相关，这支持了肠道菌群组成和帕金森病运动障碍之间的联系。此外，有报道称肠易激综合征患者患帕金森病的风险增加[162]。

越来越多的研究表明，肠道菌群在神经系统的发育和维持中发挥着重要作用。分析肠道菌群组成与神经系统疾病之间的关系已成为一个新的、有希望的研究领域，它可以提供对疾病发病机制的理解，并提供新的和有效的治疗选择。益生元、益生菌或粪菌移植对肠道菌群的调节作用已被用于某些神经系统疾病的治疗，包括自闭症和抑郁症及其相关的胃肠道症状，越来越多的临床试验表明，这种治疗是有益的。然而，还需要更多的研究来确定不同菌群在宿主生理和发病机制中的作用，以及如何调节肠道菌群以达到有益的效果。

四、肠道菌群与慢性肾病

慢性肾病是一个世界性的主要健康问题，给全世界的医疗保健系统带来了沉重的经济负担[180]。其定义为各种原因引起的慢性肾脏结构或功能异常≥ 3个月，包括出现肾脏损伤标志（白蛋白尿、尿沉渣异常、肾小管相关病变、组织学检查异常及影像学检查异常）或有肾移植病史，伴或不伴肾小球滤过率下降；或者不明原因的肾小球滤过率下降（<60ml/min）≥ 3个月[180]。慢性肾脏病的特征是慢性肾损害和（或）功能障碍，并且在其早期到晚期的发展过程中，经常伴随着高比率的病态肥胖症和死亡率，此时需要肾脏替代治疗（即透析或移植）。越来越多的证据表明，肠道菌群与肾脏相互作用，一方面，慢性肾脏病能导致肠道菌群的组成和功能变化，这种变化通过破坏肠上皮屏障和产生毒性代谢物而导致全身炎症及相关并发症。另一方面，肠道菌群通过引发炎症、蛋白尿、高血压和糖尿病，在慢性肾脏病的发展中发挥作用[181]。

在晚期慢性肾病中，尿毒症改变了肠内生化环境，促进了肠道菌群和肠道屏障的紊乱[182]。除了尿毒症之外，一些代谢物的积累、纤维摄入不足（为了防止高钾血症，对所需的水果和蔬菜限制）以及多种药物治疗方案改变了尿毒症肠内的生化环境，从而导致了生物代谢障碍[183]。代谢障碍的存在有利于某些特定细菌的生长，这些细菌具有能够产生尿毒症毒素的酶，如硫酸吲哚酚（indoxyl sulfate，IS）、硫酸对甲酚（p-cresyl sulfate，PCS）、吲哚-3-乙酸（indole-3-acetic acid，IAA）和TMAO，它们在慢性肾脏病中积累[182]。此外，肠道菌群失调导致上皮紧密连接的破坏，导致脂多糖的移位、免疫调节障碍和炎症[182]。

目前的证据表明，在患有慢性肾病的人和动物中，宿主与肠道菌群的关系是双向的[184]。肠道菌群组成和功能的变化导致尿毒症毒素的过量和保护性代谢物的减少，这些变化与氧化应激、尿毒症、炎症和肾功能恶化有关[182]。

（一）慢性肾病的肠道菌群变化

正常的肠道菌群可以保护肾脏，而肠道菌群失调则可以促进慢性肾病的发展[182]。与健康对照组相比，5/6肾切除的大鼠及血液透析的晚期肾病患者显示出不同的菌群[183]。在动物研究中发现，慢性肾病动物的肠道菌群中，乳杆菌科和普雷沃菌的丰度显著降低[185]。在人类研究中发现，在终末期肾病患者中，放线菌门、厚壁菌门和变形菌门的丰度显著增加[183]。

此外，慢性肾病患者肠道菌群中产丁酸细菌减少，其体内高浓度的尿素会导致含有脲酶的细菌家族过度生长[183]。脲酶在肠道中水解尿素形成氨，氨水解成氢氧化铵，大量的氨和氢氧化铵会破坏肠道的上皮屏障，改变菌群的组成和腔内生化环境，并导致局部和全身炎症[181]。

炎症在慢性肾脏病的发病机制中起着核心作用，它直接造成肾损伤，并间接促进引起慢性肾脏病的疾病的进展，包括糖尿病、高血压和蛋白尿。值得注意的是，慢性肾病中增加的菌株通常能够直接和间接诱导局部和全身炎症。例如，变形菌门可以通过以下方式诱导炎症反应：损害肠黏膜屏障功能（增加肠黏膜渗透性），增加Th17细胞与调节性T细胞的比率，以及促使脂多糖和肠细菌成分向体循环转移[181]。此外，慢性肾病中增加的菌株能产生促炎物质及尿毒症毒素。同时，慢性肾病中的保护性菌株减少，包括增强肠道屏障功能的细菌、产生抗炎物质和细胞保护物质的细菌，以及能够刺激抗炎性迷走神经系统的菌株，这类菌株能通过减轻疼痛和焦虑来抑制促炎性交感神经的活性[186]。

肠道菌群和肾脏系统之间的关系不是单方面的。随着肾功能恶化，肾脏对水、矿物质、代谢废物和毒素的处理以及肾脏内分泌功能会发生变化。肾衰竭导致体液超载、尿毒症毒素和废物积累、电解质紊乱等，这些因素通过多种机制协同作用促进肠道菌群的变化。肠道菌群失调通过促进慢性炎症导致慢性肾病，从而形成恶性循环[187]。

（二）慢性肾病的尿毒症毒素

尿毒症毒素分为 3 组：内源性、外源性和菌群来源[182]。硫酸吲哚酚、硫酸对甲酚、吲哚 -3- 乙酸和 TMAO，是主要的细菌来源毒素[188]。上述肠道菌群的变化表明，在慢性肾病患者中，能够产生尿毒症毒素酶的菌种增加，导致尿毒症毒素产生增加，从而促进全身炎症[181]。

硫酸吲哚酚、硫酸对甲酚、吲哚 -3- 乙酸不能通过常规血液透析有效去除，因为它们与白蛋白高度结合，而 TMAO 是水溶性的且可透析去除。硫酸吲哚酚、硫酸对甲酚和 TMAO 与慢性肾病患者心血管发病率和死亡率增加相关[181]。在动物模型中，口服 TMAO 已被证明能导致肾小管间质纤维化和进行性肾功能障碍[181]。硫酸对甲酚能促进心肌纤维化并诱导内皮细胞氧化应激[189]。除了已知的主要肠源性毒素外，终末期肾病患者体内许多尚未确定的毒素可能来自胃肠道菌群。

（三）预防及治疗

许多尿毒症毒素来源于肠道菌群对营养物质的吸收，因为这些营养物质会促进细菌生长，进而产生更多的尿毒症毒素。因此，可能的治疗方法包括改变饮食、改变菌群落、减少菌群产生的尿毒症毒素、增加毒素的排泄或靶向清除某些尿毒症毒素[190]。其中，饮食是肠道菌群调节中最强大和最容易控制的因素。然而，到目前为止，还没有关于慢性肾脏病对肠道菌群影响的深入研究，大多数相关问题仍未得到解答。

前文提到的多酚是我们日常饮食中必不可少的抗氧化剂，并显示出潜在的抗菌效果。近年的研究表明，多酚类物质如花青素、儿茶素、绿原酸和白藜芦醇对肾脏有保护作用，如白藜芦醇可通过抑制血管紧张素Ⅱ对衰老的肾脏产生保护作用，从而减少氧化应激、炎症和纤维化。白藜芦醇显著降低了慢性肾病小鼠模型中肌酐介导的间质损伤，其中白藜芦醇治疗的小鼠显示出更好的肾功能[191]。

益生元可以促进肠道菌群的组成和活性，对宿主产生有益的影响。已经证明了高抗性淀粉饮食能改变肠环境，减轻了氧化应激和炎症，并改善了慢性肾病大鼠的肾功能[192]。在腺嘌呤诱导的慢性肾病大鼠模型中，高抗性淀粉饮食减少了肾功能下降、间质纤维化、肾小管损伤和促炎因子的活化[193]。

合生元含有益生菌和益生元，在慢性肾病患者中发挥有益作用。合生元治疗可显著降低血清硫酸对甲酚的水平，改善粪便中双歧杆菌的数量，血清中的硫酸吲哚酚水平也有显著下降[192]。

盐酸司维拉姆是一种非金属磷酸盐黏合剂，除了结合磷酸盐外，它还在体外结合尿滞留溶质，包括吲哚和对甲酚。然而，在慢性肾病小鼠模型或血液透析患者中，它并不降低硫酸吲哚酚或硫酸对甲酚的浓度[194]。同样，α- 葡萄糖苷酶抑制剂阿卡波糖被认为是增加结肠多糖含量的一种方法。阿卡波糖的作用是抑制 α- 葡萄糖苷酶，一种存在于小肠刷状缘的酶。使用阿卡波糖后，增加了到达胃肠道远端肠腔的寡糖和多糖的数量。在最近的一项研究中，用阿卡波糖治疗后，硫酸对甲酚和硫酸吲哚酚的血清浓度降低，这两种尿毒症毒素的排泄量增加[181]。

五、肠道菌群与结直肠癌

结直肠癌仍然是全世界癌症相关死亡的第二大原因。尽管过去10年的研究进展在癌症治疗方面带来了一些激动人心的突破，但这些新疗法对大多数结直肠癌患者并没有产生良好效果[195]。因此迫切需要开发结直肠癌早期检测、预防和治疗的新策略。

在自发和化学诱导的结肠肿瘤小鼠的肠道中可以检测到菌群多样性的减少。例如，由于Apc肿瘤抑制基因密码子850的点突变，$Apc^{Min/+}$小鼠自发形成肠道肿瘤。与对照组相比，$Apc^{Min/+}$小鼠在肉眼可见肿瘤形成之前，菌群多样性就已经减少[196]，提示菌群多样性的降低可能导致小鼠模型中结肠肿瘤发病率的增加。

慢性炎症产生相当多的炎症介质，如TNF-α、IL-6、IL-1b和其他细胞因子，它们NF-κ B，导致结肠癌发生[197]。炎性肠病与结直肠癌的高风险有关，炎性肠病和结直肠癌有一个共同的过程，即TGF-β、TNF-α、NF-κ B、活性氧和其他信号分子水平的增加，导致肠道内的菌群失调。已经证明，伴有炎性肠病的结直肠癌患者的预后较差[197]。

肠道细菌通过对肠道的碳水化合物、脂质和氨基酸的代谢，从而影响结肠黏膜微环境中的pH、氧化环境、能量利用率和致癌物的产生[198]。这些代谢产物中大多数可促进或抑制结直肠癌的发生。例如，大肠埃希菌产生的基因毒素通过诱导DNA损伤促进肿瘤的发生。丁酸盐具有抗炎特性，与其他化学预防性膳食成分合用，能显著降低炎症的严重程度、DNA损伤的累积、肿瘤形成风险和肿瘤细胞的生长速度[195]。

特定细菌和菌群失调可能在结直肠肿瘤的发生发展中发挥作用。与健康人相比，具核梭杆菌（*fusobacterium nucleatum*，*Fn*）在结肠腺瘤和结肠癌患者的结肠组织中丰度增高。*Fn*通过在结直肠癌微环境中诱导炎症和抑制宿主免疫反应来促进肿瘤的发生发展[199]。细胞表面蛋白FadA是*Fn*的一个重要毒力因子，调控细菌的黏附和侵袭。FadA基因在人结直肠癌组织中的表达明显高于癌旁正常组织。*Fn*可表达细胞表面蛋白FadA、Fap2和RadD，这些蛋白与肠上皮细胞黏附后，可导致宿主产生炎性因子并募集炎性细胞，从而创造有利于肿瘤生长的环境。

最近有报道称，FadA通过E-钙黏蛋白上调Wnt/β-catenin调节因子膜联蛋白A1的表达。FadA还可以与内皮细胞血管内皮钙黏蛋白结合，血管内皮钙黏蛋白是内皮细胞上的连接分子[199]。这种结合改变了内皮细胞的完整性，增加了内皮细胞的通透性，使细菌能够克服血脑屏障、胎盘屏障，并定植于身体的不同部位[199]。*Fn*的Fap2蛋白能与T细胞和ITIM结构域受体相互作用，抑制自然杀伤细胞和T细胞的活性[200]。最近，我们的研究还发现，侵袭性*Fn*能通过TLR4/p-PAK1信号通路，促进结直肠癌中β-catenin转运进入细胞核，从而促进肿瘤细胞的增殖[201]。

此外，侵袭性*Fn*能与结肠黏膜多种免疫细胞相互作用，导致T细胞密度降低，M2巨噬细胞极化增强，NK细胞活性抑制，树突状细胞和肿瘤相关中性粒细胞增多，从而降低抗肿瘤免疫，有利于肿瘤进展[199]（图2-4）。

脆弱拟杆菌（*bacteroides fragilis*，*BF*）是另一种常见的共生菌，也与人类和小鼠结肠肿瘤的发生发展有关[195]。产肠毒素脆弱拟杆菌通过诱导整个结肠黏膜的促炎性IL-17和NF-κ B信号转导，诱导$Apc^{Min/+}$小鼠结肠癌的发生[195]。产肠毒素脆弱拟杆菌表达脆弱拟杆菌毒素，这是一种可以激活结肠上皮细胞中WNT和NF-κB信号通路的毒力因子[202]。

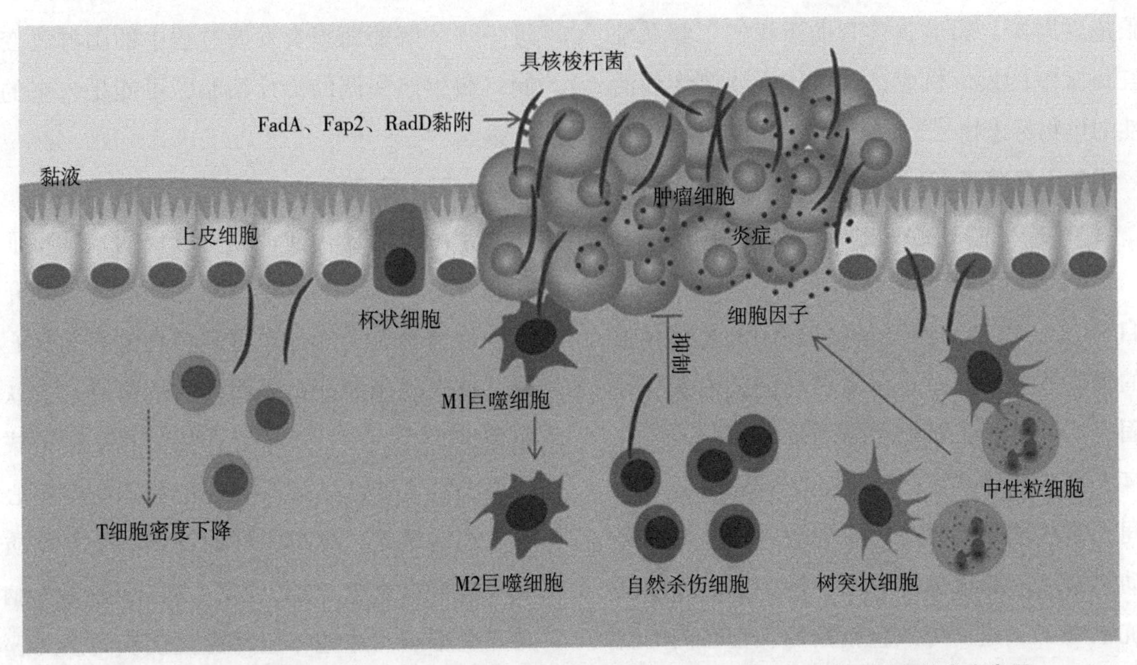

图 2-4 具核梭杆菌诱导促炎性肿瘤微环境和抑制宿主免疫，有利于结直肠肿瘤的发生发展

具核梭杆菌通过黏附分子 FadA、Fap2 和 RadD 与结肠上皮结合，侵入黏膜。具核梭杆菌的入侵增加了炎症细胞的浸润，刺激细胞因子的释放，从而促进细胞增殖。此外，具核梭杆菌还与多种免疫细胞相互作用，导致 T 细胞密度降低，巨噬细胞 M2 极化增强，自然杀伤细胞活性抑制，树突状细胞和肿瘤相关中性粒细胞增多，从而抑制抗肿瘤免疫反应，促进肿瘤进展。

［Wu J，Li Q，Fu XS. Fusobacterium nucleatum Contributes to the Carcinogenesis of Colorectal Cancer by Inducing Inflammation and Suppressing Host Immunity. Translational oncology，2019，12（6）:846-851.］

肠道菌群调节促炎淋巴细胞的募集和激活，包括 Th17 和 γδT 细胞在结肠内的分布 [195]。例如，与接种健康受试者粪便的小鼠相比，接种来自结直肠癌患者粪便的无菌小鼠 Th17 在结肠中的募集增加 [195]。这些淋巴细胞活化后又产生促炎细胞因子，包括 IL-23 和 IL-17。肿瘤浸润 Th17 和 γδT 调节性 T 细胞产生的 IL-17 促进了 $Apc^{Min/+}$ 小鼠结肠肿瘤的发展 [203]。

对肠道菌群与癌症发生机制的研究仍处于早期阶段，目前主要揭示的是相关性而不是因果关系。然而，人们已经认识到肠道菌群与人类有着密切的联系，在人类健康和疾病中发挥着重要而独特的作用。人们已经开始采取大胆措施，试图调控肠道菌群。这些目标是多方面的，从调节人体新陈代谢、免疫和炎症反应，到预防癌症发生、抑制癌症发展和提高癌症个体化治疗的疗效。基于肠道菌群的研究，人们也在探索癌症新的治疗靶点及预后标志物。

六、肠道菌群和其他疾病

（一）系统性红斑狼疮

系统性红斑狼疮是一种自身免疫性疾病，其特征在于Ⅰ型 IFN 的表达和自身抗体的产生增多，从而引发多器官炎症和损伤。越来越多的研究发现，系统性红斑狼疮与肠道菌群密切相关，但肠道菌群如何引发系统性自身免疫，从而导致系统性红斑狼疮的机制仍然不明确。通过高通量测序研究了系统性红斑狼疮患者的肠道菌群，结果发现，患者厚壁菌门与拟杆菌门的比率持续降低。

如果肠上皮屏障受损，鸡肠球菌和罗伊乳杆菌都是可能的致病微生物，会通过受损的肠道屏障转移到全身组织中。然后，这些菌群与宿主免疫系统相互作用，激活Ⅰ型 IFN 途径并诱导自身抗体的产生。此外，鸡肠球菌和罗伊乳杆菌的丰度升高表明疾病活动性增加，并预示狼疮性肾炎。肠道菌群有助于我们了解系统性红斑狼疮的发病机制，并可作为预测疾病活动性的生物标志物[204]。

（二）葡萄膜炎

自身免疫性葡萄膜炎是人类失明的主要原因。已有研究证明肠道菌群可启动表达IL-17的自体反应性T细胞，进而直接触发视网膜中的葡萄膜炎。而对小鼠注射广谱抗生素可导致固有层调节性T细胞数量增加，同时效应性$CD4^{+}$T细胞数量减少，这可证明肠道菌群和眼之间的直接联系。另一项研究表明，如果人类或动物模型的肠道菌群失调，将产生人类白细胞抗原-B27，进而导致包括葡萄膜炎在内的自身免疫性疾病。据报道，葡萄膜炎患者的肠道菌群组成与正常对照组相比存在差异，这也间接说明肠道菌群在自身免疫性葡萄膜炎中的作用，抗原模仿和微生态失调导致的免疫系统调节失常和肠道屏障破坏可能是发病机制之一。进一步了解葡萄膜炎发展过程中的菌群改变并制订微生态失调的治疗策略，可能是未来的研究重点[205]。

（三）痤疮

寻常痤疮是一种常见皮肤疾病，其特征是皮脂分泌过多，毛囊过度角化，以及炎症介质化学物质释放增加，严重影响青少年的身心健康。越来越多的证据表明，肠道菌群在皮肤炎症和情绪行为中发挥中介作用，且饮食模式和肠道菌群可能会共同影响痤疮和随后发生心理后遗症的风险，肠道菌群能够影响全身脂质和组织脂肪酸分布，通过菌群-脑-肠轴对大脑功能产生影响，这些作用都能直接或间接对痤疮产生影响，随着我们对痤疮病理生理学认识的进一步加深，随着与菌群介导的神经、免疫和营养通路相关的重要发现，我们对痤疮的治疗方案和目标有了新的要求，从益生菌到影响菌群的饮食策略，都需要进一步深入研究[206]。

第五节　小结与展望

目前大量关于肠道菌群的基础研究都是为了探索改善健康的新策略，这一举措可能有助于改善社会经济条件，维护人体健康，降低医疗成本。人类胃肠道中含有大量丰富的菌群，人体健康与肠道菌群有着密切关系。一方面，人类饮食、遗传、生命周期、运动、使用抗生素等因素影响调节肠道菌群组成和功能；另一方面，肠道菌群的不同组成及其代谢物可通过各种机制参与人体的生理、病理进程，例如营养、代谢、神经、内分泌、炎症与免疫等，并在多种特定疾病中扮演独特角色。简单来说，人体会影响肠道细菌的组成和代谢功能，而且这种改变可能与人体中的病理状态或相关疾病有关。因此，在肠道中保持有益的菌群可能有助于预防或延缓与年龄相关的疾病。

随着人们对肠道菌群与人类健康认识的加深，人们试图通过使用益生元/益生菌来调节肠道菌群组成，或者通过粪菌移植引入新的肠道菌种。此外，维持并增加植物性饮食比例，可有效维持有益的肠道菌群比例，从而维持人体健康。基因组代谢模型可以用来模拟饮食对肠道微菌群组成的影响及其与人体代谢的相互作用，从而为筛选适合的健康饮食成分提供了一种有效而快捷的途径；此外，肠道菌群的代谢能力不同，且菌群间存在相互作用，而通用效能模型驱动的建模则通过对此进行计算机分析，从而确定可作为候选物种的益生菌。这些数学建模方法的应用有可能成为开发个性化益生菌或营养饮食策略的有效方法，以维持人体的健康状态。目前的主要障碍是饮食、微生物和宿主之间鲜为人知的代谢联系，因为菌群代谢物和细菌产物如何影响宿主健康的研究还很少。

因此，还需要更多研究来确定不同菌群在宿主生理和发病机制中的作用，以及如何调节肠道菌群以达到有益的效果。

参考文献

[1] Lloyd-Price J, Abu-Ali G, Huttenhower C. The healthy human microbiome. Genome Medicine, 2016, 8(1): 51.

[2] Sonnenburg JL, Bäckhed F. Diet-microbiota interactions as moderators of human metabolism. Nature, 2016, 535(7610): 56-64.

[3] Trends in adult body-mass index in 200 countries from 1975 to 2014: a pooled analysis of 1698 population-based measurement studies with 19·2 million participants. The Lancet, 2016, 387(10026): 1377-1396.

[4] Marsh A, Eslick EM, Eslick GD. Does a diet low in FODMAPs reduce symptoms associated with functional gastrointestinal disorders? A comprehensive systematic review and meta-analysis. European Journal of Nutrition, 2016, 55(3): 897-906.

[5] Eor J-Y, Son Y-J, Kim J-Y, et al. Neuroprotective effect of both synbiotics and ketogenic diet in a pentylenetetrazol-induced acute seizure murine model. Epilepsy Research, 2021, 174:106668.

[6] Song X, Sun X, Oh S-F, et al. Microbial bile acid metabolites modulate gut RORY regulatory T cell homeostasis. Nature, 2020, 577(7790): 410-415.

[7] van der Hee B, Wells J-M. Microbial Regulation of Host Physiology by Short-chain Fatty Acids. Trends in Microbiology, 2021, 29(8): 700-712.

[8] Seiwert N, Adam J, Steinberg P, et al. Chronic intestinal inflammation drives colorectal tumor formation triggered by dietary heme iron in vivo. Archives of Toxicology, 2021, 95(7): 2507-2522.

[9] David LA, Maurice CF, RN Carmody, et al. Diet rapidly and reproducibly alters the human gut microbiome. Nature, 2014, 505(7484): 559-563.

[10] Smits SA, Leach J, ED Sonnenburg, et al. Seasonal cycling in the gut microbiome of the Hadza hunter-gatherers of Tanzania. Science (New York, N.Y.), 2017, 357(6353): 802-806.

[11] Dao MC, Everard A, Aron-Wisnewsky J, et al. Akkermansia muciniphila and improved metabolic health during a dietary intervention in obesity: relationship with gut microbiome richness and ecology. Gut, 2016, 65(3): 426-436.

[12] Gehrig JL, Venkatesh S, Chang HW, et al. Effects of microbiota-directed foods in gnotobiotic animals and undernourished children. Science (New York, N.Y.), 2019, 365(6449): eaau4732.

[13] De Filippo C, Cavalieri D, Di Paola M, et al. Impact of diet in shaping gut microbiota revealed by a comparative study in children from Europe and rural Africa. Proceedings of The National Academy of Sciences of The United States of America, 2010, 107(33): 14691-14696.

[14] Desai MS, Seekatz AM, Koropatkin NM, et al. A Dietary Fiber-Deprived Gut Microbiota Degrades the Colonic Mucus Barrier and Enhances Pathogen Susceptibility. Cell, 2016, 167(5): 1339-1353.

[15] Sun S, Araki Y, Hanzawa F, et al. High sucrose diet-induced dysbiosis of gut microbiota promotes fatty liver and hyperlipidemia in rats. The Journal of Nutritional Biochemistry, 2021, 93:108621.

[16] Zhang L, Xue X, Zhai R, et al. Timing of Calorie Restriction in Mice Impacts Host Metabolic Phenotype with Correlative Changes in Gut Microbiota. mSystems, 2019, 4(6):e00348-19.

[17] O'Keefe SJ, Li JV, Lahti L, et al. Fat, fibre and cancer risk in African Americans and rural Africans. Nature Communications, 2015, 6:6342.

[18] Jaeggi T, Kortman GA, Moretti D, et al. Iron fortification adversely affects the gut microbiome, increases pathogen abundance and induces intestinal inflammation in Kenyan infants. Gut, 2015, 64(5): 731-742.

[19] Shepherd ES, DeLoache WC, Pruss KM, et al. An exclusive metabolic niche enables strain engraftment in the gut microbiota. Nature, 2018, 557(7705): 434-438.

[20] Lloyd-Price J, Mahurkar A, Rahnavard G, et al. Erratum: Strains, functions and dynamics in the expanded Human Microbiome Project. Nature, 2017, 551(7679): 256.

[21] Forster SC, Browne HP, Kumar N, et al. HPMCD: the database of human microbial communities from metagenomic datasets and microbial reference genomes. Nucleic Acids Research, 2016, 44: D604-D609.

[22] McInerney JO, McNally A, O'Connell MJ. Why prokaryotes have pangenomes. Nature Microbiology, 2017, 2:17040.

[23] Xie H, Guo R, Zhong H, et al. Shotgun Metagenomics of 250 Adult Twins Reveals Genetic and Environmental Impacts on the Gut Microbiome. Cell Systems, 2016, 3(6): 572-584.

[24] Rothschild D, Weissbrod O, Barkan E, et al. Environment dominates over host genetics in shaping

human gut microbiota. Nature, 2018, 555(7695): 210-215.

[25] Pannaraj PS, Li F, Cerini C, et al. Association Between Breast Milk Bacterial Communities and Establishment and Development of the Infant Gut Microbiome. JAMA Pediatrics, 2017, 171(7): 647-654.

[26] Imhann F, Vich Vila A, Bonder MJ, et al. Interplay of host genetics and gut microbiota underlying the onset and clinical presentation of inflammatory bowel disease. Gut, 2018, 67(1): 108-119.

[27] E-L Ortega-Vega, Guzmán-Castañeda SJ, Campo O, et al. Variants in genes of innate immunity, appetite control and energy metabolism are associated with host cardiometabolic health and gut microbiota composition. Gut Microbes, 2020, 11(3): 556-568.

[28] Goodrich JK, Davenport ER, Beaumont M, et al. Genetic Determinants of the Gut Microbiome in UK Twins. Cell Host & Microbe, 2016, 19(5): 731-743.

[29] Lim MY, You HJ, Yoon HS, et al. The effect of heritability and host genetics on the gut microbiota and metabolic syndrome. Gut, 2017, 66(6): 1031-1038.

[30] Stanislawski MA, Dabelea D, Wagner BD, et al. Pre-pregnancy weight, gestational weight gain, and the gut microbiota of mothers and their infants. Microbiome, 2017, 5(1): 113.

[31] Larsen PE, Dai Y. Metabolome of human gut microbiome is predictive of host dysbiosis. Giga Science, 2015,4:42.

[32] Schulfer AF, Battaglia T, Alvarez Y, et al. Intergenerational transfer of antibiotic-perturbed microbiota enhances colitis in susceptible mice. Nature Microbiology, 2018, 3(2): 234-242.

[33] Bayer F, Dremova O, Khuu MP, et al. The Interplay between Nutrition, Innate Immunity, and the Commensal Microbiota in Adaptive Intestinal Morphogenesis. Nutrients, 2021, 13(7):2198.

[34] Zaidi AZ, Moore SE, Okala SG. Impact of Maternal Nutritional Supplementation during Pregnancy and Lactation on the Infant Gut or Breastmilk Microbiota: A Systematic Review. Nutrients, 2021, 13(4):1137.

[35] Azad MB, Konya T, Persaud RR, et al. Impact of maternal intrapartum antibiotics, method of birth and breastfeeding on gut microbiota during the first year of life: a prospective cohort study. An International Journal of Obstetrics and Gynaecology, 2016, 123(6): 983-993.

[36] Sakwinska O, Foata F, Berger B, et al. Does the maternal vaginal microbiota play a role in seeding the microbiota of neonatal gut and nose? Beneficial Microbes, 2017, 8(5): 763-778.

[37] Conlon MA, Bird AR. The impact of diet and lifestyle on gut microbiota and human health. Nutrients, 2014, 7(1): 17-44.

[38] Cheng J, Ringel-Kulka T, Heikamp-de Jong I, et al. Discordant temporal development of bacterial phyla and the emergence of core in the fecal microbiota of young children. The ISME Journal, 2016, 10(4): 1002-1014.

[39] Acuña I, Cerdó T, Ruiz A, et al. Infant Gut Microbiota Associated with Fine Motor Skills. Nutrients, 2021, 13(5):1673.

[40] Kaczmarczyk M, Löber U, Adamek K, et al. The gut microbiota is associated with the small intestinal paracellular permeability and the development of the immune system in healthy children during the first two years of life. Journal of Translational Medicine, 2021, 19(1): 177.

[41] Collins J, Borojevic R, Verdu EF, et al. Intestinal microbiota influence the early postnatal development of the enteric nervous system. Neurogastroenterology and Motility: The Official Journal of The European Gastrointestinal Motility Society, 2014, 26(1): 98-107.

[42] Hollister EB, Riehle K, RA Luna, et al. Structure and function of the healthy pre-adolescent pediatric gut microbiome. Microbiome, 2015, 3:36.

[43] Zhong H, Penders J, Shi Z, et al. Impact of early events and lifestyle on the gut microbiota and metabolic phenotypes in young school-age children. Microbiome, 2019, 7(1): 2.

[44] Robertson RC. The Gut Microbiome in Child Malnutrition. Nestle Nutrition Institute Workshop Series, 2020, 93: 133-144.

[45] West CE, Jenmalm MC, Prescott SL. The gut microbiota and its role in the development of allergic disease: a wider perspective. Clinical and Experimental Allergy: Journal of The British Society for Allergy and Clinical Immunology, 2015, 45(1): 43-53.

[46] Markle JG, Frank DN, Mortin-Toth S, et al. Sex differences in the gut microbiome drive hormone-dependent regulation of autoimmunity. Science (New York, N.Y.), 2013, 339(6123): 1084-1088.

[47] Ojeda J, Ávila A, Vidal PM. Gut Microbiota Interaction with The Central Nervous System throughout Life. Journal of Clinical Medicine, 2021, 10(6):1299.

[48] Murray E, Sharma R, Smith KB, et al. Probiotic consumption during puberty mitigates LPS-induced immune responses and protects against stress-induced depression- and anxiety-like behaviors in adulthood in

a sex-specific manner. Brain, Behavior, and Immunity, 2019, 81:198-212.

[49] Lee JG, Han DS, Jo SV, et al. Characteristics and pathogenic role of adherent-invasive Escherichia coli in inflammatory bowel disease: Potential impact on clinical outcomes. Plos One, 2019, 14(4): e216165.

[50] VTE Aho, Pereira PAB, Voutilainen S, et al. Gut microbiota in Parkinson's disease: Temporal stability and relations to disease progression. EBio Medicine, 2019, 44:691-707.

[51] BW Ji, Sheth RU, Dixit PD, et al. Macroecological dynamics of gut microbiota. Nature Microbiology, 2020,5(5):768-775.

[52] Ling Z, Liu X, Cheng Y, et al. Gut microbiota and aging. Critical Reviews in Food Science and Nutrition, 2020: 1-56,doi:10.1080/10408398.2020.1867054.

[53] Clauss M, Gérard P, Mosca A, et al. Interplay Between Exercise and Gut Microbiome in the Context of Human Health and Performance. Frontiers in Nutrition, 2021, 8:637010.

[54] Bycura D, Santos AC, Shiffer A, et al. Impact of Different Exercise Modalities on the Human Gut Microbiome. Sports (Basel, Switzerland), 2021, 9(2):14.

[55] Shin HE, Kwak SE, Lee JH, et al. Exercise, the Gut Microbiome, and Frailty. Annals of Geriatric Medicine and Research, 2019, 23(3): 105-114.

[56] Gomez-Pinilla F, Hillman C. The influence of exercise on cognitive abilities. Comprehensive Physiology, 2013, 3(1): 403-428.

[57] Wu MH, Lee CP, Hsu SC, et al. Effectiveness of high-intensity interval training on the mental and physical health of people with chronic schizophrenia. Neuropsychiatric Disease and Treatment, 2015, 11:1255-1263.

[58] Allen JM, Mailing LJ, Niemiro GM, et al. Exercise Alters Gut Microbiota Composition and Function in Lean and Obese Humans. Medicine and Science in Sports and Exercise, 2018, 50(4): 747-757.

[59] Astbury SM, Corfe BM. Uptake and metabolism of the short-chain fatty acid butyrate, a critical review of the literature. Current Drug Metabolism, 2012, 13(6): 815-821.

[60] Lamoureux EV, Grandy SA, Langille MGI. Moderate Exercise Has Limited but Distinguishable Effects on the Mouse Microbiome. mSystems, 2017, 2(4):e00006.

[61] Dubinsky V, Reshef L, Bar N, et al. Predominantly Antibiotic-resistant Intestinal Microbiome Persists in Patients With Pouchitis Who Respond to Antibiotic Therapy. Gastroenterology, 2020, 158(3): 610-624.

[62] Blaser MJ. Antibiotic use and its consequences for the normal microbiome. Science (New York, N.Y.), 2016, 352(6285): 544-545.

[63] Li B, Qiu H, Zheng N, et al. Integrated Metagenomic and Transcriptomic Analyses Reveal the Dietary Dependent Recovery of Host Metabolism From Antibiotic Exposure. Frontiers in Cell and Developmental Biology, 2021, 9:680174.

[64] Mao J, Yan Y, Li H, et al. Glutamine deficiency links clindamycin-induced dysbiosis and intestinal barrier dysfunction in mice. The British Journal of Nutrition, 2021, 126(3): 366-374.

[65] Duysburgh C, Van den Abbeele P, Morera M, et al. Lacticaseibacillus rhamnosus GG and supplementation exert protective effects on human gut microbiome following antibiotic administration. Beneficial Microbes, 2021, 12(4):59-73.

[66] Stewardson AJ, Gaïa N, François P, et al. Collateral damage from oral ciprofloxacin versus nitrofurantoin in outpatients with urinary tract infections: a culture-free analysis of gut microbiota.Clinical microbiology and infection : the official publication of the European Society of Clinical Microbiology and Infectious Diseases, 2015, 21(4): 341-344.

[67] Al-Mitwalli A, Kyriazis G, El-Taji O, et al. Selective transperineal prostate biopsy for fluoroquinolone-resistance patients reduces sepsis and cost. Current urology, 2021,15(2):115-118.

[68] Ponziani FR, Scaldaferri F, Petito V, et al. The Role of Antibiotics in Gut Microbiota Modulation: The Eubiotic Effects of Rifaximin. Digestive diseases (Basel, Switzerland), 2016, 34(3): 269-278.

[69] Azad MB, Konya T, Persaud RR, et al. Impact of maternal intrapartum antibiotics, method of birth and breastfeeding on gut microbiota during the first year of life: a prospective cohort study. BJOG : An International Journal of Obstetrics and Gynaecology, 2016, 123(6): 983-993.

[70] Raymond F, Ouameur AA, Déraspe M, et al. The initial state of the human gut microbiome determines its reshaping by antibiotics. The ISME Journal, 2016, 10(3): 707-720.

[71] Swain-Ewald HA, Ewald PW. Natural Selection, The Microbiome, and Public Health. The Yale Journal of Biology and Medicine, 2018, 91(4): 445-455.

[72] Suppli MP, Bagger JI, Lelouvier B, et al. Hepatic microbiome in healthy lean and obese humans. JHEP Reports : Innovation in Hepatology, 2021, 3(4):

100299.

[73] Ajslev TA, Andersen CS, Gamborg M, et al. Childhood overweight after establishment of the gut microbiota: the role of delivery mode, pre-pregnancy weight and early administration of antibiotics. International Journal of Obesity (2005), 2011, 35(4): 522-529.

[74] Cunningham AL, Stephens JW, Harris DA. A review on gut microbiota: a central factor in the pathophysiology of obesity. Lipids in Health and Disease, 2021, 20(1): 65.

[75] Vangipurapu J, Fernandes-Silva L, Kuulasmaa T, et al. Microbiota-Related Metabolites and the Risk of Type 2 Diabetes. Diabetes Care, 2020;43(6):1319-1325.

[76] Vrieze A, Van Nood E, Holleman F, et al. Transfer of intestinal microbiota from lean donors increases insulin sensitivity in individuals with metabolic syndrome. Gastroenterology, 2012, 143(4): 913-916.

[77] WMT Kuwabara, Yokota CNF, Curi R, et al. Obesity and Type 2 Diabetes mellitus induce lipopolysaccharide tolerance in rat neutrophils. Scientific Reports, 2018, 8(1): 17534.

[78] ChengWY, Lam KL, Li X, et al. Circadian disruption-induced metabolic syndrome in mice is ameliorated by oat β-glucan mediated by gut microbiota. Carbohydrate Polymers, 2021， 267:118216.

[79] De Vadder F, Kovatcheva-Datchary P, Goncalves D, et al. Microbiota-generated metabolites promote metabolic benefits via gut-brain neural circuits. Cell, 2014,156:84-96.

[80] Garibay D, Zaborska KE, Shanahan M, et al. TGR5 Protects Against Colitis in Mice, but Vertical Sleeve Gastrectomy Increases Colitis Severity. Obesity Surgery, 2019, 29(5): 1593-1601.

[81] Moncrieft AE, Llabre MM, McCalla JR, et al. Effects of a Multicomponent Life-Style Intervention on Weight, Glycemic Control, Depressive Symptoms, and Renal Function in Low-Income, Minority Patients With Type 2 Diabetes: Results of the Community Approach to Lifestyle Modification for Diabetes Randomized Controlled Trial. Psychosomatic Medicine, 2016, 78(7): 851-860.

[82] Finlay BB, Pettersson S, Melby MK, et al. The Microbiome Mediates Environmental Effects on Aging. BioEssays : News and Reviews in Molecular, Cellular and Developmental Biology, 2019, 41(10): e1800257.

[83] Biagi E, Franceschi C, Rampelli S, et al. Gut Microbiota and Extreme Longevity. Current Biology : CB, 2016, 26(11): 1480-1485.

[84] Koleva PT, Kim JS, Scott JA, et al. Microbial programming of health and disease starts during fetal life. Birth Defects Research. Part C, Embryo today : reviews, 2015, 105(4): 265-277.

[85] Zhang S, Zeng B, Chen Y, et al. Gut microbiota in healthy and unhealthy long-living people. Gene, 2021, 779:145510.

[86] Bischoff SC. Microbiota and aging. Current Opinion in Clinical Nutrition and Metabolic Care, 2016, 19(1): 26-30.

[87] Rampelli S, Soverini M, D’Amico F, et al. Shotgun Metagenomics of Gut Microbiota in Humans with up to Extreme Longevity and the Increasing Role of Xenobiotic Degradation. mSystems, 2020, 5(2):e00124-120.

[88] Lowry CA, Smith DG, Siebler PH, et al. The Microbiota, Immunoregulation, and Mental Health: Implications for Public Health. Current Environmental Health Reports, 2016, 3(3): 270-286.

[89] Rivière A, Selak M, Lantin D, et al. Bifidobacteria and Butyrate-Producing Colon Bacteria: Importance and Strategies for Their Stimulation in the Human Gut. Frontiers in Microbiology, 2016,7:979.

[90] Arnold JW, Roach J, Fabela S, et al. The pleiotropic effects of prebiotic galacto-oligosaccharides on the aging gut. Microbiome, 2021, 9(1): 31.

[91] Kumar M, Yadav AK, Verma V, et al. Bioengineered probiotics as a new hope for health and diseases: an overview of potential and prospects. Future Microbiology, 2016, 11(4): 585-600.

[92] Järbrink-Sehgal E, Andreasson A. The gut microbiota and mental health in adults. Current Opinion in Neurobiology, 2020,62:102-114.

[93] Sherwin E, Sandhu KV, Dinan TG, et al. May the Force Be With You: The Light and Dark Sides of the Microbiota-Gut-Brain Axis in Neuropsychiatry. CNS Drugs, 2016, 30(11): 1019-1041.

[94] Saurman V, Margolis KG, Luna RA. Autism Spectrum Disorder as a Brain-Gut-Microbiome Axis Disorder. Digestive Diseases and Sciences, 2020, 65(3): 818-828.

[95] Molina-Torres G, Rodriguez-Arrastia M, Roman P, et al. Stress and the gut microbiota-brain axis. Behav Pharmacol, 2019, 30(2 and 3-Spec Issue): 187-200.

[96] Wang X, Wang BR, Zhang XJ, et al. Evidences for vagus nerve in maintenance of immune balance and transmission of immune information from gut to brain in STM-infected rats. World Journal of Gastroenterology, 2002, 8(3): 540-545.

[97] Ghezzi L, Cantoni C, Pinget GV, et al. Targeting the gut to treat multiple sclerosis. The Journal of Clinical

Investigation, 2021, 131(13):e143774.

[98] Bhatti UF, Karnovsky A, Dennahy IS, et al. Pharmacologic modulation of brain metabolism by valproic acid can induce a neuroprotective environment. The Journal of Trauma and Acute Care Surgery, 2021, 90(3): 507-514.

[99] Pappolla MA, Perry G, Fang X, et al. Indoles as essential mediators in the gut-brain axis. Their role in Alzheimer's disease. Neurobiology of Disease, 2021, 156:105403.

[100] Ziętek M, Celewicz Z, Szczuko M. Short-Chain Fatty Acids, Maternal Microbiota and Metabolism in Pregnancy. Nutrients, 2021, 13(4):1244.

[101] Pekmez CT, Dragsted LO, Brahe LK. Gut microbiota alterations and dietary modulation in childhood malnutrition - The role of short chain fatty acids. Clinical Nutrition (Edinburgh, Scotland), 2019, 38(2): 615-630.

[102] Reigstad CS, Salmonson CE , Rainey JF, et al. Gut microbes promote colonic serotonin production through an effect of short-chain fatty acids on enterochromaffin cells. FASEB Journal : Official Publication of The Federation of American Societies for Experimental Biology, 2015, 29(4): 1395-1403.

[103] Huang W, Cho KY, Meng D, et al. The impact of indole-3-lactic acid on immature intestinal innate immunity and development: a transcriptomic analysis. Scientific Reports, 2021, 11(1): 8088.

[104] Wei GZ, Martin KA, Xing PY, et al. Tryptophan-metabolizing gut microbes regulate adult neurogenesis via the aryl hydrocarbon receptor. Proceedings of The National Academy of Sciences of the United States of America, 2021, 118(27):91-118.

[105] Bailey MT, Dowd SE, Galley JD, et al. Exposure to a social stressor alters the structure of the intestinal microbiota: implications for stressor-induced immunomodulation. Brain, Behavior, and Immunity, 2011, 25(3): 397-407.

[106] Agustí A, Campillo I, Balzano T, et al. Bacteroides uniformis CECT 7771 Modulates the Brain Reward Response to Reduce Binge Eating and Anxiety-Like Behavior in Rat. Molecular Neurobiology, 2021, 58(10):4959-4979.

[107] Sassone-Corsi M, Raffatellu M. No vacancy: how beneficial microbes cooperate with immunity to provide colonization resistance to pathogens. Journal of Immunology, 2015, 194(9): 4081-4087.

[108] Rakoff-Nahoum S, Coyne MJ, Comstock LE. An ecological network of polysaccharide utilization among human intestinal symbionts. Current Biology, 2014, 24(1): 40-49.

[109] Ng KM, Ferreyra JA, Higginbottom SK, et al. Microbiota-liberated host sugars facilitate post-antibiotic expansion of enteric pathogens. Nature, 2013, 502(7469): 96-99.

[110] Herp S, Durai Raj AC, Salvado Silva M, et al. The human symbiont Mucispirillum schaedleri: causality in health and disease. Medical Microbiology and Immunology, 2021, 210(4): 173-179.

[111] Pacheco AR, Curtis MM, Ritchie JM, et al. Fucose sensing regulates bacterial intestinal colonization. Nature, 2012, 492(7427): 113-117.

[112] Takao M, Yen H, Tobe T. LeuO enhances butyrate-induced virulence expression through a positive regulatory loop in enterohaemorrhagic Escherichia coli. Molecular microbiology, 2014, 93(6): 1302-1313.

[113] Zafar M, Jahan H, Shafeeq S, et al. Clarithromycin Exerts an Antibiofilm Effect against Salmonella enterica Serovar Typhimurium rdar Biofilm Formation and Transforms the Physiology towards an Apparent Oxygen-Depleted Energy and Carbon Metabolism. Infection and Immunity, 2020, 88(11):e00510-520.

[114] Anderson CJ, Clark DE, Adli M, et al. Ethanolamine Signaling Promotes Salmonella Niche Recognition and Adaptation during Infection. PLoS Pathog, 2015, 11(11): e1005278.

[115] Kristensen K, Henriksen L. Cesarean section and disease associated with immune function. The Journal of Allergy and Clinical Immunology, 2016, 137(2): 587-590.

[116] Kim M, Qie Y, Park J, et al. Gut Microbial Metabolites Fuel Host Antibody Responses. Cell Host Microbe, 2016, 20(2): 202-214.

[117] Gang Wang, Huang Shuo, Wang Yuming, et al. Bridging intestinal immunity and gut microbiota by metabolites. Cellular and molecular life sciences: CMLS, 2019, 76(20): 3917-3937.

[118] Leystra AA, Clapper ML. Gut Microbiota Influences Experimental Outcomes in Mouse Models of Colorectal Cancer. Genes, 2019, 10(11):900.

[119] Ahmad AF, Dwivedi G, O'Gara F, et al. The gut microbiome and cardiovascular disease: current knowledge and clinical potential. American Journal of Physiology. Heart and Circulatory Physiology, 2019, 317(5): H923-H938.

[120] Nicole Farmer, Gutierrez-Huerta Cristhian-A, Turner Briana-S, et al. Neighborhood Environment

Associates with Trimethylamine-N-Oxide (TMAO) as a Cardiovascular Risk Marker. International Journal of Environmental Research and Public Health, 2021, 18(8): 4296.

[121] Manor O, Zubair N, Conomos MP, et al. A Multi-omic Association Study of Trimethylamine N-Oxide. Cell reports, 2018, 24(4): 935-946.

[122] Koopen AM, Groen AK, Nieuwdorp M. Human microbiome as therapeutic intervention target to reduce cardiovascular disease risk. Current Opinion in Lipidology, 2016, 27(6): 615-622.

[123] Samulak JJ, Sawicka AK, Hartmane D, et al. L-Carnitine Supplementation Increases Trimethylamine-N-Oxide but not Markers of Atherosclerosis in Healthy Aged Women. Annals of Nutrition & Metabolism, 2019, 74(1): 11-17.

[124] Tang WH, Wang Z, Levison BS, et al. Intestinal microbial metabolism of phosphatidylcholine and cardiovascular risk. The New England Journal of Medicine, 2013, 368(17): 1575-1584.

[125] Liu G, Li J, Li Y, et al. Gut microbiota-derived metabolites and risk of coronary artery disease: a prospective study among US men and women. The American Journal of Clinical Nutrition, 2021, 114(1): 238-247.

[126] Tilg H. A Gut Feeling about Thrombosis. The New England Journal of Medicine, 2016, 374(25): 2494-2496.

[127] Lau K, Srivatsav V, Rizwan A, et al. Bridging the Gap between Gut Microbial Dysbiosis and Cardiovascular Diseases. Nutrients, 2017, 9(8): 859.

[128] Adnan S, Nelson JW, Ajami NJ, et al. Alterations in the gut microbiota can elicit hypertension in rats. Physiological Genomics, 2017, 49(2): 96-104.

[129] Ufnal M, Jazwiec R, Dadlez M, et al. Trimethylamine-N-oxide: a carnitine-derived metabolite that prolongs the hypertensive effect of angiotensin II in rats. The Canadian Journal of Cardiology, 2014, 30(12): 1700-1705.

[130] Hsu CN, Chang-Chien GP, Lin S, et al. Targeting on Gut Microbial Metabolite Trimethylamine-N-Oxide and Short-Chain Fatty Acid to Prevent Maternal High-Fructose-Diet-Induced Developmental Programming of Hypertension in Adult Male Offspring. Molecular Nutrition & Food Research, 2019, 63(18): e1900073.

[131] Francesc Formiga, Ferreira Teles Cristiana-Isabel, Chivite David. Impact of intestinal microbiota in patients with heart failure: A systematic review. Medicina Clinica, 2019, 153(10): 402-409.

[132] Adilah-F Ahmad, Ward Natalie-C, Dwivedi Girish. The gut microbiome and heart failure. Current Opinion in Cardiology, 2019, 34(2): 225-232.

[133] Kamo T, Akazawa H, Suda W, et al. Dysbiosis and compositional alterations with aging in the gut microbiota of patients with heart failure. PloS one, 2017, 12(3): e174099.

[134] Zabell A, Tang WH. Targeting the Microbiome in Heart Failure. Current Treatment Options in Cardiovascular. Medicine, 2017, 19(4): 27.

[135] Estruch-Ramón-Ros-Emilio Fernando. Primary Prevention of Cardiovascular Disease with a Mediterranean Diet. The New England Journal of Medicine, 2013, 368(14): 1279-1290.

[136] Williamson G. The role of polyphenols in modern nutrition. Nutrition Bulletin, 2017, 42(3): 226-235.

[137] Wang JW, Kuo CH, Kuo FC, et al. Fecal microbiota transplantation: Review and update. Journal of The Formosan Medical Association, 2019,118 Suppl 1: S23-S31.

[138] Zabell A, Tang WH. Targeting the Microbiome in Heart Failure.Current treatment options in cardiovascular. Medicine, 2017, 19(4): 27.

[139] Marques FZ, Nelson E, Chu PY, et al. High-Fiber Diet and Acetate Supplementation Change the Gut Microbiota and Prevent the Development of Hypertension and Heart Failure in Hypertensive Mice. Circulation, 2017, 135(10): 964-977.

[140] Roberts AB, Gu X, Buffa JA, et al. Development of a gut microbe-targeted nonlethal therapeutic to inhibit thrombosis potential. Nature Medicine, 2018, 24(9): 1407-1417.

[141] Vallianou N, Stratigou T, Christodoulatos GS, et al. Understanding the Role of the Gut Microbiome and Microbial Metabolites in Obesity and Obesity-Associated Metabolic Disorders: Current Evidence and Perspectives. Current Obesity Reports, 2019, 8(3): 317-332.

[142] Singer-Englar T, Barlow G, Mathur R. Obesity, diabetes, and the gut microbiome: an updated review. Expert Review of Gastroenterology & Hepatology, 2019, 13(1): 3-15.

[143] Pascale A, Marchesi N, Govoni S, et al. The role of gut microbiota in obesity, diabetes mellitus, and effect of metformin: new insights into old diseases. Current Opinion in Pharmacology, 2019: 49: 1-5.

[144] Jeremiah-J Faith, Guruge Janaki-L, Charbonneau Mark, et al. The Long-Term Stability of The Human Gut Microbiota. Science, 2013, 341(6141): 1237439.

[145] Yufeng Qin, Roberts John-D, Grimm Sara-A, et al. An obesity-associated gut microbiome reprograms the intestinal epigenome and leads to altered colonic gene expression. Genome Biology, 2018, 19(1): 7.

[146] Carvalho BM, Guadagnini D, Tsukumo DML, et al. Expression of Concern: Modulation of gut microbiota by antibiotics improves insulin signalling in high-fat fed mice. Diabetologia, 2017, 10.1007/s00125-017-4293-4.

[147] Fahad Alhusain. Microbiome: Role and functionality in human nutrition. CycleSaudi Medical Journal, 2021, 42(2): 146-150.

[148] Sze MA, Schloss PD. Looking for a Signal in the Noise: Revisiting Obesity and the Microbiome. mBio, 2016, 7(4):e01018-16.

[149] Debédat J, Amouyal C, Aron-Wisnewsky J, et al. Impact of bariatric surgery on type 2 diabetes: contribution of inflammation and gut microbiome? Seminars in Immunopathology, 2019, 41(4): 461-475.

[150] Le Chatelier E, Nielsen T, Qin J, et al. Richness of human gut microbiome correlates with metabolic markers. Nature, 2013, 500(7464): 541-546.

[151] Gijs den Besten, Bleeker Aycha, Gerding Albert, et al. Short-Chain Fatty Acids Protect Against High-Fat Diet-Induced Obesity via a PPAR γ -Dependent Switch From Lipogenesis to Fat Oxidation. Diabetes, 2015, 64(7): 2398-2408.

[152] Esther Mezhibovsky, Knowles Kim-A, He Qiyue, et al. Grape Polyphenols Attenuate Diet-Induced Obesity and Hepatic Steatosis in Mice in Association With Reduced Butyrate and Increased Markers of Intestinal Carbohydrate Oxidation. Frontiers in Nutrition, 2021, 8:675267.

[153] Haeusler RA, Camastra S, Nannipieri M, et al. Increased Bile Acid Synthesis and Impaired Bile Acid Transport in Human Obesity. The Journal of Clinical Fndocrinology and Metabolism, 2016, 101(5): 1935-1944.

[154] Allin KH, Tremaroli V, Caesar R, et al. Aberrant intestinal microbiota in individuals with prediabetes. Diabetologia, 2018, 61(4): 810-820.

[155] Rodriguez J, Hiel S, Delzenne NM. Metformin: old friend, new ways of action-implication of the gut microbiome? Current Opinion in Clinical Nutrition and Metabolic Care, 2018, 21(4): 294-301.

[156] Chambers ES, Preston T, Frost G, et al. Role of Gut Microbiota-Generated Short-Chain Fatty Acids in Metabolic and Cardiovascular Health. Current Nutrition Reports, 2018, 7(4): 198-206.

[157] Hippe B, Remely M, Aumueller E, et al. Faecalibacterium prausnitzii phylotypes in type two diabetic, obese, and lean control subjects. Beneficial Microbes, 2016, 7(4): 511-517.

[158] Hänninen A, Toivonen R, Pöysti S, et al. Akkermansia muciniphila induces gut microbiota remodelling and controls islet autoimmunity in NOD mice. Gut, 2018, 67(8): 1445-1453.

[159] Wu H, Esteve E, Tremaroli V, et al. Metformin alters the gut microbiome of individuals with treatment-naive type 2 diabetes, contributing to the therapeutic effects of the drug. Nature Medicine, 2017, 23(7): 850-858.

[160] Bauer PV, Duca FA. Targeting the gastrointestinal tract to treat type 2 diabetes. The Journal of Endocrinology, 2016, 230(3): R95-R113.

[161] Crovesy L, Ostrowski M, Ferreira DMTP, et al. Effect of Lactobacillus on body weight and body fat in overweight subjects: a systematic review of randomized controlled clinical trials. International Journal of Obesity, 2017, 41(11): 1607-1614.

[162] Grochowska M, Laskus T, Radkowski M. Gut Microbiota in Neurological Disorders. Archivum Immunologiaeet Therapiae Experimentalis, 2019, 67(6): 375-383.

[163] Heiss CN, Olofsson LE. The role of the gut microbiota in development, function and disorders of the central nervous system and the enteric nervous system. Journal of Neuroendocrinology, 2019, 31(5): e12684.

[164] Heiss CN, Olofsson LE. Gut Microbiota-Dependent Modulation of Energy Metabolism. Journal of innate immunity, 2018, 10(3): 163-171.

[165] Erny D, de Angelis AL Hrab ě , Jaitin D, et al. Host microbiota constantly control maturation and function of microglia in the CNS. Nature Neuroscience, 2015, 18(7): 965-977.

[166] Breton J, Tennoune N, Lucas N, et al. Gut Commensal E. coli Proteins Activate Host Satiety Pathways following Nutrient-Induced Bacterial Growth. Cell Metabolism, 2016, 23(2): 324-334.

[167] Julio-Pieper M, Bravo JA, Aliaga E, et al. Review article: intestinal barrier dysfunction and central nervous system disorders-a controversial association. Alimentary Pharmacology & Therapeutics, 2014, 40(10): 1187-1201.

[168] Jagust W. Imaging the evolution and pathophysiology of Alzheimer disease. Nature Reviews. Neuroscience, 2018, 19(11): 687-700.

[169] Kobayashi Y, Sugahara H, Shimada K, et al.

Therapeutic potential of Bifidobacterium breve strain A1 for preventing cognitive impairment in Alzheimer's disease. Scientific Reports, 2017, 7(1): 13510.

[170] Rothhammer V, Mascanfroni ID, Bunse L, et al. Type I interferons and microbial metabolites of tryptophan modulate astrocyte activity and central nervous system inflammation via the aryl hydrocarbon receptor. Nature Medicine, 2016, 22(6): 586-597.

[171] Hu X, Wang T, Jin F. Alzheimer's disease and gut microbiota. Science China. Life Sciences, 2016, 59(10): 1006-1023.

[172] Welcome MO. Gut Microbiota Disorder, Gut Epithelial and Blood-Brain Barrier Dysfunctions in Etiopathogenesis of Dementia: Molecular Mechanisms and Signaling Pathways. Neuromolecular Medicine, 2019, 21(3): 205-226.

[173] Carlino D, De Vanna M, Tongiorgi E. Is altered BDNF biosynthesis a general feature in patients with cognitive dysfunctions? The Neuroscientist a review journal bringing neurobiology. Neurology and Psychiatry, 2013, 19(4): 345-353.

[174] Katerina-Cechova-Jana-Amlerova Francesco Angelucci. Antibiotics, gut microbiota, and Alzheimer's disease. Journal of Neuroinflammation, 2019, 16(1):108.

[175] Akbari E, Asemi Z, Daneshvar-Kakhaki R, et al. Effect of Probiotic Supplementation on Cognitive Function and Metabolic Status in Alzheimer's Disease: A Randomized, Double-Blind and Controlled Trial. Frontiers in Aging Neuroscience, 2016, 8:256.

[176] Abraham D, Feher J, Scuderi GL, et al. Exercise and probiotics attenuate the development of Alzheimer's disease in transgenic mice: Role of microbiome. Experimental gerontology, 2019, 115: 122-131.

[177] Vascellari S, Melis M, Palmas V, et al. Clinical Phenotypes of Parkinson's Disease Associate with Distinct Gut Microbiota and Metabolome Enterotypes. Biomolecules, 2021, 11(2):144.

[178] Ikuko Miyazaki, Asanuma Masato. Neuron-Astrocyte Interactions in Parkinson's Disease. Cells, 2020, 9(12): 2623.

[179] Scheperjans F, Aho V, Pereira PA, et al. Gut microbiota are related to Parkinson's disease and clinical phenotype. Movement Disorders : Official Journal of The Movement Disorder Society, 2015, 30(3): 350-358.

[180] Jazani NH, Savoj J, Lustgarten M, et al. Impact of Gut Dysbiosis on Neurohormonal Pathways in Chronic Kidney Disease. Diseases (Basel, Switzerland), 2019, 7(1):21.

[181] Kanbay M, Onal EM, Afsar B, et al. The crosstalk of gut microbiota and chronic kidney disease: role of inflammation, proteinuria, hypertension, and diabetes mellitus. International Urology and Nephrology, 2018, 50(8): 1453-1466.

[182] Mafra D, Borges N, Alvarenga L, et al. Dietary Components That May Influence the Disturbed Gut Microbiota in Chronic Kidney Disease. Nutrients, 2019, 11(3):406.

[183] Hobby GP, Karaduta O, Dusio GF, et al. Chronic kidney disease and the gut microbiome.American journal of physiology. Renal Physiology, 2019, 316(6): F1211-F1217.

[184] Mahmoodpoor F, Rahbar-Saadat Y, Barzegari A, et al. The impact of gut microbiota on kidney function and pathogenesis. Biomedicine & pharmacotherapy, 2017, 93:412-419.

[185] Vaziri ND, Wong J, Pahl M, et al. Chronic kidney disease alters intestinal microbial flora. Kidney International, 2013, 83(2): 308-315.

[186] Luis Vitetta, Llewellyn Hannah, Oldfield Debbie. Gut Dysbiosis and the Intestinal Microbiome: Streptococcus thermophilus a Key Probiotic for Reducing Uremia. Microorganisms, 2019, 7(8): 228.

[187] Vaziri ND, Zhao YY, Pahl MV. Altered intestinal microbial flora and impaired epithelial barrier structure and function in CKD: the nature, mechanisms, consequences and potential treatment. Nephrology, Dialysis, Transplantation : Official Publication of The European Dialysis and Transplant Association - European Renal Association, 2016, 31(5): 737-746.

[188] Lau WL, Savoj J, Nakata MB, et al. Altered microbiome in chronic kidney disease: systemic effects of gut-derived uremic toxins. Clinical Science (London, England : 1979), 2018, 132(5): 509-522.

[189] Wing MR, Patel SS, Ramezani A, et al. Gut microbiome in chronic kidney disease. Experimental Physiology, 2016, 101(4): 471-477.

[190] Castillo-Rodriguez E, Fernandez-Prado R, Esteras R, et al. Impact of Altered Intestinal Microbiota on Chronic Kidney Disease Progression. Toxins, 2018, 10(7):300.

[191] Jing Wang, Ghosh Siddhartha-S, Ghosh Shobha. Curcumin improves intestinal barrier function: modulation of intracellular signaling, and organization of tight junctionsAmerican journal of physiology.

Cell Physiology, 2017, 312(4): C438-C445.

[192] Natarajan R, Pechenyak B, Vyas U, et al. Randomized controlled trial of strain-specific probiotic formulation (Renadyl) in dialysis patients. BioMed Research International, 2014, 2014:568571.

[193] Al-Sadi R, Boivin M, Ma T. Mechanism of cytokine modulation of epithelial tight junction barrier. Frontiers in Bioscience (Landmark Edition), 2009, 14: 2765-2778.

[194] Poesen R, Meijers B, Evenepoel P. The colon: an overlooked site for therapeutics in dialysis patients. Seminars in Dialysis, 2013, 26(3): 323-332.

[195] Leystra AA, Clapper ML. Gut Microbiota Influences Experimental Outcomes in Mouse Models of Colorectal Cancer. Genes, 2019, 10(11):900.

[196] Son JS, Khair S, Pettet DW, et al. Altered Interactions between the Gut Microbiome and Colonic Mucosa Precede Polyposis in APCMin/+ Mice. PloS One, 2015, 10(6): e127985.

[197] Meng C, Bai C, Brown TD, et al. Human Gut Microbiota and Gastrointestinal Cancer. Genomics, Proteomics & Bioinformatics, 2018, 16(1): 33-49.

[198] Han S, Gao J, Zhou Q, et al. Role of intestinal flora in colorectal cancer from the metabolite perspective: a systematic review. Cancer Management and Research, 2018, 10:199-206.

[199] Wu J, Li Q, Fu X. Fusobacterium nucleatum Contributes to the Carcinogenesis of Colorectal Cancer by Inducing Inflammation and Suppressing Host Immunity. Translational Oncology, 2019, 12(6): 846-851.

[200] Mima K, Ogino S, Nakagawa S, et al. The role of intestinal bacteria in the development and progression of gastrointestinal tract neoplasms. Surgical Oncology, 2017, 26(4): 368-376.

[201] Chen Y, Peng Y, Yu J, et al. Invasive Fusobacterium nucleatum activates beta-catenin signaling in colorectal cancer via a TLR4/P-PAK1 cascade. Oncotarget, 2017, 8(19): 31802-31814.

[202] Aref Shariati, Razavi Shabnam, Ghaznavi-Rad Ehsanollah, et al. Association between colorectal cancer and Fusobacterium nucleatum and Bacteroides fragilis bacteria in Iranian patients: a preliminary study. Infectious Agents and Cancer, 2021, 16(1): 41.

[203] Housseau F, Wu S, Wick EC, et al. Redundant Innate and Adaptive Sources of IL17 Production Drive Colon Tumorigenesis. Cancer Research, 2016, 76(8): 2115-2124.

[204] Kim JW, Kwok SK, Choe JY, et al. Recent Advances in Our Understanding of the Link between the Intestinal Microbiota and Systemic Lupus Erythematosus. International Journal of Molecular Sciences, 2019, 20(19): 4871.

[205] Baim AD, Movahedan A, Farooq AV, et al. The microbiome and ophthalmic disease. Experimental Biology and Medicine, 2019, 244(6): 419-429.

[206] Bowe W, Patel NB, Logan AC. Acne vulgaris, probiotics and the gut-brain-skin axis: from anecdote to translational medicine. Beneficial Microbes, 2014, 5(2): 185-199.

第3章
肠道免疫功能概述

人体与外界环境的接触部位大致可分为两大部分：一部分是人体表面覆盖的皮肤，另一部分是覆盖消化、呼吸、泌尿生殖器官等的黏膜组织。人体黏膜系统表面积有 400 ～ 500m^2，是人体表面积的 200 倍以上。在黏膜组织中，消化系统的黏膜表面积约占 80%，主要用于吸收营养素。消化系统除每日摄取食物中的营养物质外，还接触许多微生物，这类微生物能在肠道内增殖，形成"肠道微生物群"。肠道微生物与机体共存，可抑制病原微生物及毒素的侵犯，进而特异性地清除入侵的有害物质。肠道屏障构成机体防御感染的重要防线，在维持正常生理功能方面，发挥极其重要的免疫功能。深入认识肠道免疫功能，将有助于肠道疾病及与之相关肠外疾病的预防与治疗。

第一节　肠道屏障

肠道是宿主与环境接触的重要部位，尤其适于营养物质的消化和吸收[1]。同时，肠道亦是受外界环境影响最大的部位。每天，肠道会接触大量微生物及饮食产生的化合物。因此，肠道屏障的作用至关重要[2]。一个大表面积的且具有可选择通透性的界面增加了外界微生物和物质侵入机体的可能性，然而很多定植肠道的微生物不会对宿主构成威胁，甚至可能发挥有益的作用，因此肠道屏障系统必须适时、适当地做出反应[1]。

实际上，肠道除了消化、吸收、分泌功能外，还具有重要的屏障功能。肠道要执行这些功能，必须形成复杂的平衡网络系统，其中至关重要的是有效的屏障功能[3]。肠道的屏障功能是指肠道上皮能够分隔肠腔内的物质，防止致病性物质的侵入及阻止肠道内细菌和内毒素的移位。肠道屏障并非一层结构，包含黏膜机械屏障、微生物屏障、化学屏障及免疫屏障等多层结构（图 3-1）。屏障之间并非孤立，而是相互协作。肠道屏障由肠道共生菌、肠道黏液层、肠道上皮及固有层中的多种免疫细胞等多种要素共同组成，保护机体免受外来病原体和潜在有害物质的伤害[4]。肠上皮形成一个半透膜，既允许营养物质、电解质、水和抗原的吸收，又保护宿主不受微生物群及肠腔中潜在有毒分子的影响[5]。

一、机械屏障

机械屏障又称物理屏障，是指完整的彼此紧密连接的肠黏膜上皮结构。肠黏膜屏障以机械屏障最为重要，其结构基础为完整的肠黏膜上皮细胞及上皮细胞间的紧密连接。机械屏障包括单层上皮细胞、细胞间紧密连接和覆盖上皮表面的黏液[6]。肠道黏液层覆盖于肠上皮，由肠黏膜上皮细胞分泌，成为菌群与肠道上皮间的物理屏障，同时为肠道菌群提供营养物质和生活环境。此外，机械屏障离不开化学屏障的支持，如酸度（低 pH）、去垢剂（胆盐）、蛋白水解酶（胰蛋白酶）、细胞壁降解酶（溶菌酶）和抗菌蛋白（防御素等），这些化学屏障使微生物种群处于受机体控制的状态。此外，肠道的单向蠕动也有助于防止远端肠道内容物及微生物进入小肠[6]。机械屏障表面可以防止细菌渗透和黏附，并调控通向宿主深部组织的细胞旁扩散[7]。

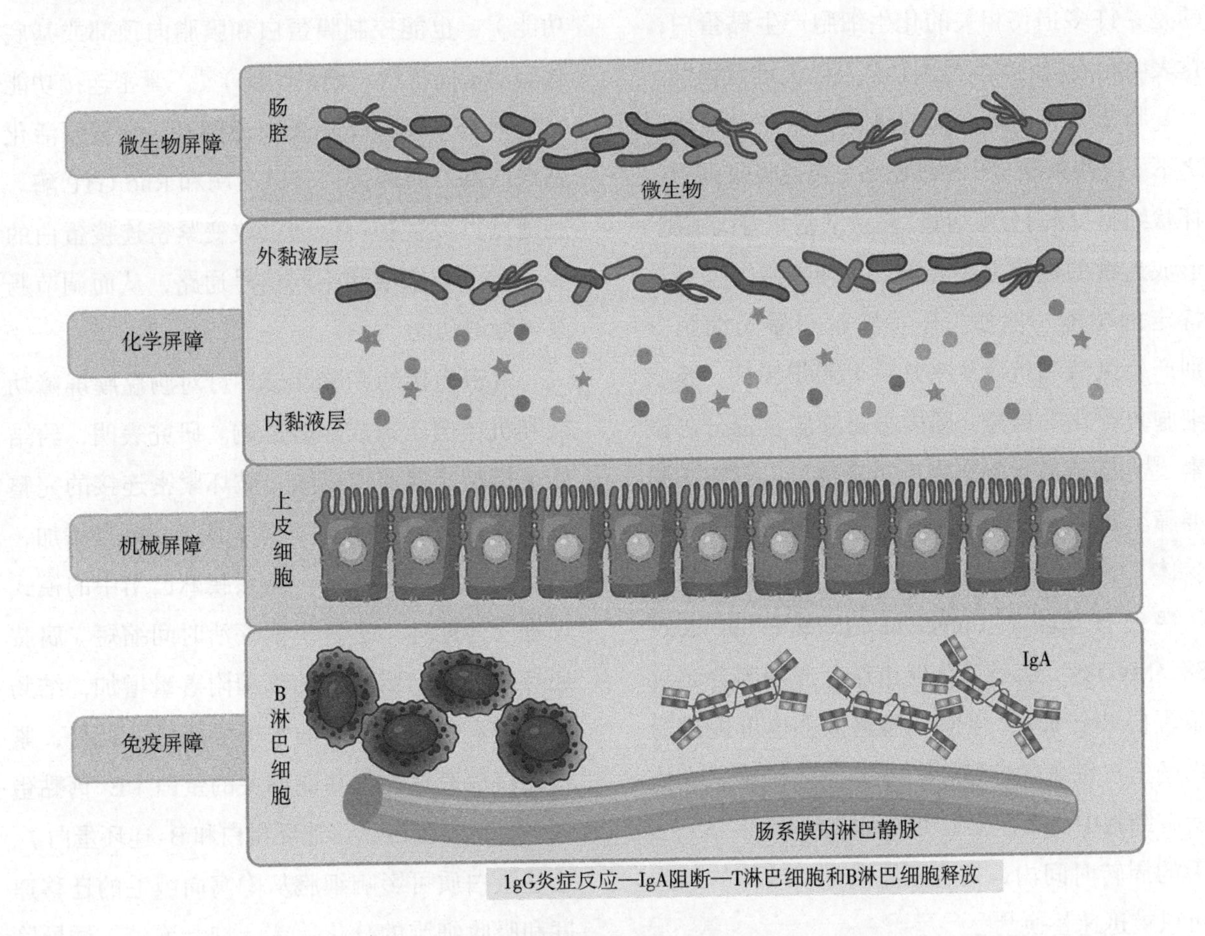

图 3-1　肠道屏障的 4 个组成部分

微生物屏障是指肠道正常菌群，各种菌株相互依赖，有益细菌通过生物拮抗和免疫功能形成宿主生物防御。化学屏障是指消化道分泌胃酸、黏液、黏蛋白、胆汁、糖蛋白、黏多糖、各种消化酶、溶菌酶等化学物质。机械屏障由肠腔和内环境之间的一层柱状上皮细胞组成。上皮细胞之间是细胞间连接复合体，包括紧密连接、黏附连接、桥粒和缝隙连接。免疫屏障由肠道上皮细胞、肠上皮内淋巴组织、固有层淋巴细胞、派尔集合淋巴结、肠系膜淋巴结、其他肠组织和 sIgA 组成

黏液层由杯状细胞分泌的各种糖蛋白组成[8]。黏液是微生物群在消化道内遇到的第一道机械屏障。它可以保护上皮细胞免受有害微生物和抗原的侵害，同时也是肠道运动的润滑剂[2]。在大肠中，黏液层分为内黏液层和外黏液层：内黏液层薄而致密（约 50μm），与上皮黏附牢固，富含抗菌肽；外黏液层是由退化的黏液、稀释的抗菌肽和一些细菌组成，厚而松散（约 100μm），是肠道共生菌的栖息地[2, 8, 9]。虽然外层为共生细菌提供了附着点和营养源，但内层通常不会定植细菌，它能够形成一道屏障，防止细菌侵入上皮[8]。肠上皮细

胞表达C型凝集素Reg Ⅲ γ，该凝集素在离肠上皮组织约50μm处形成无菌隔离区，以限制革兰阳性细菌与肠表面的接触。在Reg Ⅲ γ基因敲除小鼠的肠道中，肠上皮表面的细菌定植增加[10]。此外，胃肠道损伤与损伤相关细胞的出现及去分化有关，包括表面黏膜细胞、溃疡相关细胞、颈黏液细胞、假幽门腺化生。有趣的是，许多损伤相关的化生细胞产生黏蛋白，这表明黏液层的修复是损伤恢复的关键过程[11]。

肠黏膜表面最重要的机械屏障位于黏液层之下。这些细胞与几种细胞系（包括吸收细胞、杯状细胞、肠内分泌细胞、簇状细胞和潘氏细胞）形成连续的极化上皮屏障。每种细胞类型都有特定的功能。例如，杯状细胞和潘氏细胞分别产生黏液和抗菌肽来加固黏膜屏障[5]。簇状细胞负责化学传感，肠内分泌细胞负责分泌激素[12]。肠道微生物产物反过来也可以影响黏膜屏障。脂多糖或鞭毛蛋白，可通过激活Toll样受体（toll-like receptors，TLRs）——髓样分化初级应答基因88（myeloid differentiation factor 88，MyD88）信号通路促进肠上皮细胞分泌抗菌蛋白Reg Ⅲ γ，该信号通路似乎也可促进受损肠上皮细胞的修复[10]。肠道上皮细胞是人体所有细胞中更新速度最快的细胞之一。在3～5天的周转时间内，被病原体或毒素破坏的细胞可以被迅速替换[11]。

机械屏障除包含多种表型和功能不同的上皮细胞外，细胞间连接也能防止微生物侵入机体，它主要调控细胞旁路的通透性[12]。肠道上皮细胞为一层柱状上皮细胞，单个上皮细胞通过上皮间连接复合体整合为一道完整屏障，在肠腔和内环境之间形成第一道防线。上皮细胞间的连接复合体包括紧密连接、黏附连接、桥粒和缝隙连接。紧密连接是构成肠道机械屏障的关键因素[13]。紧密连接由位于内皮细胞顶端的不同跨膜多蛋白复合物组成。它作为一种动态的通透性屏障，具有双重功能：阻止潜在的有害物质或病原体进入人体，同时允许营养物质、离子和水进入机体[14]。由于紧密连接位于顶端和基底外侧质膜之间的连接处，它们还可以通过细胞旁路和跨细胞途径调节上皮极性，从而允许物质进出管腔，同时节省能量[15]。紧密连接的主要功能被形容为门和栅栏，紧密连接可选择性地控制管腔内成分进入上皮间隙（门功能），也能控制膜蛋白和膜脂向顶部或基底外侧的横向扩散（栅栏功能）[4]。紧密连接功能由信号分子调控，如蛋白激酶C、丝裂原活化蛋白激酶、肌球蛋白轻链激酶和Rho GTP酶。肠道菌群或食物可以通过改变紧密连接蛋白的表达和分布来激活不同信号通路，从而调节肠道屏障的功能[12]。

紧密连接的完整性破坏可对肠黏膜屏障功能和机体健康造成显著影响。研究表明，给结肠炎模型小鼠服用腐胺（破坏紧密连接的完整性）后，小鼠结肠炎症加重，肠道通透性增加，结肠跨上皮电阻降低，肠系膜淋巴结中的模式识别受体配体增多，小鼠存活时间缩短。腐胺还导致小鼠结肠上皮的细菌附着量增加，结肠组织中炎性细胞因子的水平增高[16]。此外，紧密连接也表达黏附功能相关的蛋白（E-钙黏蛋白、N-钙黏蛋白、A-连环蛋白和B-连环蛋白）。这些蛋白质可影响细胞从隐窝向绒毛的迁移速度和吸收细胞的分化速率，抑制增殖，诱导隐窝中的细胞凋亡[17]。另一方面，Toll样受体2（TLR2）可以通过紧密连接蛋白（如ZO-1和闭塞蛋白）的重新分布影响肠上皮屏障功能，从而改变与微生物和代谢物的相互作用[12]。

二、化学屏障

肠黏膜化学屏障是指由消化道分泌的胃酸、黏液、黏蛋白、胆汁、糖蛋白、黏多糖、消化酶、溶菌酶等化学物质构成的屏障[14]。这些物质可以通过改变病原菌的攻击部位，破坏食物中的抗原和抵御微生物入侵，从而起到化学屏障的作用[14]。外黏液层含数量较多的肠道微生物，而内黏液层极少有微生物。黏液层含有抗菌肽，有助于防止细菌与上皮细胞接触。抵抗素样分子 β（resistin like molecule beta，RELMβ）是一种限制革兰阴性菌与结肠上皮表面接触的抗菌蛋白。缺乏 RELMβ 的小鼠结肠内黏液层的变形杆菌数量增加[10]。

化学屏障可以保护肠黏膜免受酶和酸碱环境的侵蚀。胃酸和胆汁可以灭活细菌。胃肠黏液和消化液的 pH 不利于细菌的生长。胃酸是胃肠道最好的杀菌剂[14]。胆盐可以结合内毒素，胆酸可以降解内毒素，溶菌酶可以破坏细菌细胞壁，消灭细菌。肠道分泌的消化液可以稀释毒素，清洁肠腔，使潜在的致病菌很难黏附于肠上皮细胞[13, 14]。肠道分泌物不仅保护肠上皮免受物理损伤，而且能黏附抗原物质，使其更容易被蛋白酶降解[18]。此外，分泌物中还有一些补充性成分，可以帮助肠道免疫细胞清除病原体[19]。肝脏分泌的胆汁酸不仅在乳化脂肪、吸收脂溶性维生素和调节糖脂代谢平衡方面起重要作用，而且通过抑制小肠细菌过度生长和菌群移位来维持肠道屏障功能和动态平衡[20, 21]。胆盐在小肠中以结合物的形式分泌，具有促进脂肪分散和酶降解的去垢剂特性。这种特性对某些细菌的细胞膜也有影响，因此，胆盐被认为具有抗菌作用[22]。

三、微生物屏障

肠道微生物屏障是指正常肠道菌群中，各种菌株之间相互依赖，有益细菌通过生物拮抗和免疫功能形成宿主生物防御[14]。肠道常驻菌与宿主的微空间结构形成了一个相互依赖又相互作用的微生态系统。在通常情况下，肠道内微生物群构成一个对抗病原体的重要的保护屏障。当正常微生态的稳定性遭到破坏后，肠道定植抵抗力降低，导致肠道中潜在致病性病原体（包括条件致病菌）的定植和入侵。人体肠道菌群中的基因数量是人类基因组的 100 倍，在正常情况下，肠道菌群与内外环境保持平衡，在激活人体免疫系统、促进消化吸收方面具有不可替代的作用[23]。

肠道内的厌氧菌通过发酵碳水化合物产生短链脂肪酸。在人类和小鼠中，短链脂肪酸有助于肠上皮细胞以及其他细胞类型（如固有淋巴细胞 3）中缺氧诱导因子 -1（hypoxia-inducible factor-1，HIF-1）的激活。HIF-1 调节多种基因的表达，包括黏蛋白、抗菌肽和紧密连接蛋白，这些基因对上皮的屏障功能非常重要[24]。肠道内的短链脂肪酸（主要是丁酸）被肠上皮细胞摄取，有助于维持完整的肠道屏障，降低肠道通透性，从而防止细菌移位，避免内毒素血症和相关免疫反应的发生。短链脂肪酸对肠道免疫细胞也有重要的调节作用。例如，丁酸刺激调节性 T 细胞分化，增加 IL-10 水平，同时减少 IL-6 的产生，抑制促炎性 Th17 细胞的增殖[25]。

另一项研究证实，肌动蛋白结合蛋白突触足蛋白（synaptopodin，SYNPO）位于肠上皮紧密连接处和 F- 肌动蛋白应力纤维内，对屏障完整性和细胞运动至关重要。丁酸可能通过抑制组蛋白去乙酰化酶的机制在上皮细胞和小鼠结肠中诱导 SYNPO 表达。在肠道菌群耗竭的小鼠结肠中，SYNPO 的表达显著减少，而服用丁酸盐可诱导 SYNPO 恢复表达。Synpo 基因敲除小鼠肠道通透性增加，对右旋糖酐硫酸钠诱导的结肠炎易感性显著增加。这些研究发现了肠道菌群及其产物（特别是丁酸盐）在调节 SYNPO 表达中具有关键作用，并揭示了微生物群产生的丁酸盐有助于肠道屏障功能的恢复[26]。

四、免疫屏障

肠黏膜的免疫屏障主要位于肠上皮以下的

层面，其中一些成分位于肠黏膜细胞之间，称为黏膜免疫系统，是人体最大的免疫器官[5]。肠道免疫屏障由肠上皮细胞、上皮内淋巴细胞、固有层淋巴细胞、派尔集合淋巴结、肠系膜淋巴结及肠道浆细胞分泌型免疫球蛋白A（secretory immunoglobulin A，sIgA）等构成。

肠道免疫屏障主要由肠道相关淋巴组织（gut-associated lymphoid tissue，GALT）通过分泌sIgA发挥的体液免疫及细胞免疫组成。在胃肠黏膜中，25%为淋巴组织，它们通过细胞免疫和体液免疫作用，防止致病性抗原对机体的伤害。GALT是由派尔集合淋巴结、肠系膜淋巴结及分散在黏膜固有层和肠上皮中的大量淋巴细胞组成，它包含了整个人体70%的免疫细胞。GALT是一个能保护机体免受外部伤害的功能性深层屏障，它紧邻黏膜，通过特定的免疫细胞（如树突状细胞和派尔集合淋巴结内的微折叠细胞）与外部环境接触，这些细胞能捕获微生物和大分子物质，并向T淋巴细胞呈递抗原，T淋巴细胞产生细胞因子，激活对病原微生物的免疫反应[27]，并对共生细菌产生免疫耐受。

GALT产生的特异性sIgA进入肠道，选择性地包被革兰阴性菌，形成抗原抗体复合物，阻碍细菌与上皮细胞受体相结合，同时刺激肠道黏液分泌并加速黏液层的流动，可有效阻止细菌对肠黏膜的黏附，并且sIgA在穿胞过程中对已潜入细胞内的病毒同样具有中和作用。在创伤、感染、休克等应激状态下，GALT呈现选择性的抑制状态，sIgA分泌减少，增加了细菌黏附机会进而发生易位。GALT既能抵御致病因子入侵，又可耐受肠道非致病菌群。其作用机制就是在抗原刺激下产生局部的免疫反应，中和抗原物质（主要是产生大量的sIgA）。sIgA是免疫球蛋白A单体通过J链连结而成的二聚体，主要由肠黏膜固有层浆细胞产生。sIgA是黏膜表面含量最丰富的免疫球蛋白，它与肠腔内的细菌有一定程度的特异性结合，并通过包裹细菌、抑制与上皮细胞的黏附，以及中和细菌毒素来阻止微生物的入侵，从而在维持肠道黏液屏障功能方面发挥着重要作用[5]。sIgA阻抑黏附的可能机制是：使病原微生物发生凝集，丧失活动能力而不能黏附于黏膜上皮细胞。与病原微生物结合后，阻断了微生物表面的特异结合点，因而使其丧失了黏附能力。

概括来说，肠道免疫防御系统主要是由GALT及其分泌的IgA、IgE、IgM等构成。另一方面，树突状细胞识别病原菌表面的病原体相关分子模式（pathogen associated molecular pattern，PAMP），刺激来自派尔集合淋巴结的B细胞产生IgA[8]。因此，IgA在维持上皮屏障的完整性方面起着关键作用，可阻止因肠上皮受损而引起的全身免疫反应[8]。

第二节　肠道免疫相关细胞及其功能

哺乳动物的肠道是一个复杂的环境，经常暴露于来自食物、微生物群和代谢物的抗原[28]。哺乳动物的肠道中存在大量的微生物群，特别是在远端的肠道[6]，其代谢功能非常丰富，与肝脏的代谢功能相当。例如，微生物群可以影响视网膜和晶状体的脂肪酸组成，影响骨密度，促进肠道血管形成[6]。肠道微生物群竞争性地夺取食物、生存空间、黏附点，阻止入侵的肠道病原体定植[6]。但是，研究表明，微生物群也有助于致病性病毒的传播。此外，微生物群的组成是疾病严重程度的决定因素，这表明微生物群也可能对宿主构成重大威胁[6]。

要维持肠道复杂生态系统的动态平衡，就需要借助“黏膜免疫系统”（mucosal immune

system，MIS），它可以快速检测和清除暂居菌，同时使有益菌留存于肠道适当的位置。换言之，上皮细胞和微生物群之间存在持续的相互作用，即上皮细胞能够分泌免疫细胞趋化物质，为应对病原体的入侵做好充分的准备[6]。黏膜免疫系统具有与微生物群相关的复杂信号系统，主要由 Toll 样受体和核苷酸结合寡聚化结构域样受体（nucleotide-binding oligomerization domain-like receptor，NLR）这类模式识别受体（pattern-recognition receptors，PRRs）介导[6]，在识别并清除病原微生物、减少对有益微生物和宿主的损伤中发挥重要作用。

哺乳动物的肠道是机体最大的免疫器官，由非造血细胞（上皮细胞、潘氏细胞、杯状细胞）和造血细胞（巨噬细胞、树突状细胞、T 细胞）组成，也是数万亿微生物的栖息地[6]。肠道是潜在病原体的主要通道，包括来自食物的抗原和形成耐受的多种共生体。因此，肠道屏障在肠道生理过程中起着重要作用，如物理屏障、免疫耐受、病原体清除等。它的功能在很大程度上依赖于组织中的细胞和介质，如结构细胞（如上皮细胞、杯状细胞、潘氏细胞）和免疫细胞（如肥大细胞、树突状细胞、巨噬细胞和淋巴细胞）[7]（图 3-2）。

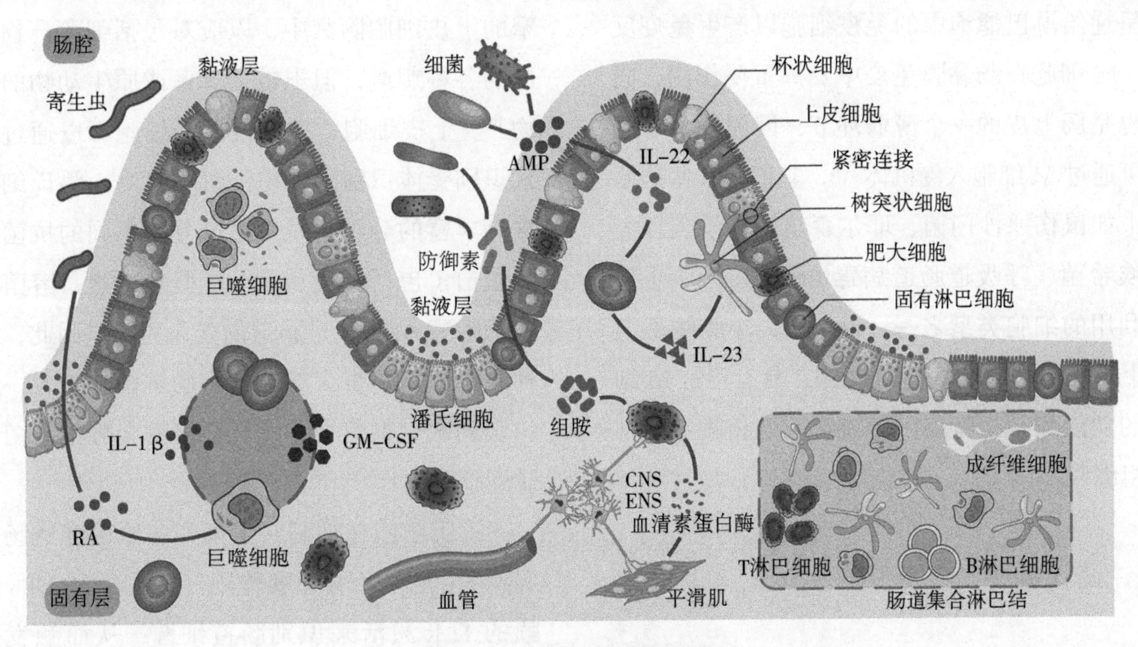

图 3-2　肠道屏障结构和功能

化学屏障中的抗菌肽，如两性调节素，可攻击包括细菌和蠕虫等病原体。机械屏障就像一堵墙，在空间上将入侵的微生物和宿主隔开。肠上皮中有许多类型的细胞可调节肠黏膜免疫功能。肠道屏障被破坏后，肠道细菌从肠腔渗漏到固有层，诱导宿主免疫细胞产生过度免疫反应。巨噬细胞或树突状细胞释放的维 A 酸有助于抵抗蠕虫感染。树突状细胞产生的 IL-23 可调控固有淋巴细胞释放 IL-22，后者可促进上皮细胞分泌 AMP 以应对细菌感染。此外，巨噬细胞来源的 IL-1β 可促进固有淋巴细胞产生粒细胞 - 巨噬细胞集落刺激因子，进一步刺激单核细胞向巨噬细胞分化。RA. 维甲酸；GM-CSF. 粒细胞 - 巨噬细胞集落刺激因子；AMP. 抗菌肽；CNS. 中枢神经系统；ENS. 肠道神经系统

一、M 细胞

微折叠细胞（microfold cells，也称 M 细胞），存在于滤泡相关上皮（follicle associated epithelial，FAE）中，形态不规则，基底质膜内陷形成可容纳免疫活性细胞的褶皱或囊袋[12]。与其他肠上皮细胞相比，FAE 更多地表达与膜转运、宿主防御和转录调节相关的多种基因，如泛素 D、肿瘤坏死因子受体超家族 12a、四次跨膜超家族 4（TM4SF4）。M 细胞约占 FAE

细胞的10%，专门摄取肠腔内抗原，主要分布在肠道淋巴组织表面的FAE中，如派尔集合淋巴结和孤立淋巴滤泡[29]。M细胞依赖核因子κB受体活化因子配体（receptor activator for nuclear Factor-κ B ligand，RANKL）进行分化，RANKL在派尔集合淋巴结穹窿的上皮下基质细胞中选择性表达[12]。M细胞有短而少的微绒毛，表面糖萼较其他肠道细胞少，有助于从管腔摄取颗粒并将其输送到基底外侧的淋巴组织中[6]。M细胞糖蛋白外壳上的岩藻糖残基可减少静电斥力，从而增强M细胞与肠腔内抗原的相互作用[30]。

M细胞可将肠内抗原和完整的微生物跨上皮呈递给淋巴滤泡中的免疫细胞以产生免疫反应。M细胞在肠黏膜免疫中发挥重要作用，同时也是肠上皮的一个薄弱环节，因为许多病原体可通过M细胞入侵机体[29]，它们是某些病原体［如鼠伤寒沙门菌、耶尔森菌属、侵袭性大肠埃希菌、呼吸道肠道病毒（reoviruses）］最易利用的细胞类型之一。近年来，小肠绒毛上皮上发现了“绒毛”型M细胞，具有正常M细胞的功能和结构特点，但缺乏与之相联系的淋巴组织[6]。

二、潘氏细胞

1872年，Gustav Schwalbe最先对潘氏细胞进行了形态学描述。1888年，奥地利科学家Joseph Paneth将潘氏细胞细致描述为“肠嗜酸颗粒细胞”，故以Joseph Paneth命名。后来人们认识到潘氏细胞具有高度分泌功能，能将颗粒释放至肠腔中。潘氏细胞存在于哺乳动物（猫犬除外）、鸟类、两栖动物和爬行动物体内。潘氏细胞产物在胎儿发育期间即可检测到，在出生后肠道微生物定植期间产物丰度会更显著。然而，微生物刺激（即微生物群定植）并非潘氏细胞发育成熟所必需的条件，实验证明无菌小鼠的潘氏细胞仍能发育成熟。

潘氏细胞的丰度是动态变化的，因为在生命早期潘氏细胞数量可能会增加，而在老年时期潘氏细胞数量则会下降[31]。潘氏细胞是黏膜免疫系统的重要组成部分[6]，是小肠特有的上皮细胞系[29]，主要存在于小肠上皮，位于李氏腺（Lieberkühn crypt，亦或称Lieberkühn隐窝）的底部（肠腺干细胞的下方），每个隐窝包含约15个干细胞和10个潘氏细胞。若潘氏细胞出现在胃和结肠中，则称为肠化生。肠上皮细胞的寿命为3～5天，潘氏细胞则不同，它的寿命相对较长（＞30天）。潘氏细胞顶端有大量富含防御素的分泌颗粒，通过胞吐释放到狭窄的上皮细胞隐窝中，以应对包括细菌产物在内的各种刺激，但不包括真菌或原生动物的产物[6]。上皮细胞，特别是潘氏细胞可以通过模式识别受体识别微生物代谢产物[32]。潘氏细胞含有丰富的分泌颗粒，通过分泌不同的抗菌肽和蛋白（包括α防御素、C型凝集素、溶菌酶和磷脂酶A2）促进肠道的先天免疫。因此，这种潘氏细胞负责大部分的小肠抗菌功能。一旦识别到微生物信号，潘氏细胞就会分泌抗菌物质到肠道中[7, 29]。

潘氏细胞在抑制共生菌和病原菌穿透肠道屏障的过程中发挥重要作用。它通过MyD88依赖的TLR激活来识别肠道细菌，从而触发多种抗菌因子的表达，如溶菌酶和防御素。防御素是潘氏细胞分泌的主要分子，含有一个疏水结构域，可与细菌细胞膜上的磷脂相互作用，从而引起细菌裂解。潘氏细胞是由肠道干细胞分化而来的子代细胞。有趣的是，潘氏细胞失去分泌表达特征后，可重新进入细胞周期，并能获得干细胞特性，促进组织对炎症的再生反应[7]。目前认为，潘氏细胞的生理功能包括：①为肠道隐窝内小肠干细胞提供生长环境；②向肠腔中分泌大量有抑菌活性的蛋白多肽，调控肠道菌群[33]。

潘氏细胞中包含大量内质网和高尔基复合体，具有强大的分泌蛋白质功能，其主要分泌物是具有杀菌能力的蛋白多肽（抗菌肽），如α-防御素、隐窝素相关序列肽、溶菌酶、Ⅱ A型分泌性磷脂酶A2、再生胰岛衍生蛋白3β、再生胰岛衍生蛋白3γ及血管生成素4等。抗菌肽的主要作用是对抗病原微生物感染，它是天然免疫反应的重要分子。α-防御素是最早被确认的抗菌肽家族之一，它是吞噬细胞中分泌颗粒的主要成分。除了吞噬细胞，多种黏膜上皮细胞也可分泌α-防御素[33]。人潘氏细胞表达两种α-防御素：人防御素5和人防御素6[6]。潘氏细胞是小肠中抗菌物质的主要来源，如α-防御素（在小鼠称为隐窝防御素）[31]。小鼠潘氏细胞分泌多达6个亚型的防御素。潘氏细胞特异地表达α-防御素，分泌到胞外的α-防御素具有重要的防御功能。

防御素是一类分子量很小（15～20个残基）的富含半胱氨酸的阳离子蛋白质，它是两性分子，通过与细菌细胞膜结合，在膜上形成跨膜的离子通道，破坏细胞膜的完整性，造成细胞内容物释放以杀死细菌。同时，防御素可通过与细菌毒素相结合使其发生变性，因此具有灭活多种细菌毒素的作用。迄今，肠道共生菌与大量具有杀菌能力的防御素共存的机制仍未阐明。最近的研究发现，肠道共生菌表达去磷酸酶LpxF以去除细菌表面的负电荷，从而抵御防御素等阳离子抗菌类多肽的杀伤[33]。

研究表明，潘氏细胞分泌的α-防御素在应对病原菌感染中起重要作用。革兰阳性菌、革兰阴性菌、脂多糖、胞壁酸、胞壁酰二肽及脂类A都能刺激小肠潘氏细胞分泌防御素，而真菌和原生动物并不能刺激潘氏细胞脱颗粒。小鼠小肠的潘氏细胞接触到病原菌或病原菌抗原后的几分钟内，就会分泌富含抗菌肽的颗粒，从而杀死病原微生物，这种杀菌作用是病原菌或病原菌抗原剂量依赖性的。α-防御素的抗菌活性占总抗菌肽活性的70%[33]。

小肠潘氏细胞是抗菌肽的重要来源。抗菌肽可以抑制革兰阳性菌与上皮细胞的黏附。此外，诸如α-防御素之类的防御素，通过直接破坏细菌细胞壁，参与微生物群落调控[8]。肠道缺乏α-防御素会增加小肠内厚壁菌门与拟杆菌门的比例，而人类α-防御素5在小鼠中的转基因表达会导致厚壁菌数量减少，最显著的是那些已知直接接触上皮的厚壁菌[34]。因此，抗菌肽除了阻断上皮细胞与微生物接触外，还有助于调节细菌群落的组成[8]。在微生物刺激下，潘氏细胞不仅会分泌囊泡中的抗菌肽，而且可以通过感受微生物而调控抗菌肽的产生。肠道菌群可大幅提高潘氏细胞中Reg Ⅲ -γ的表达，而Reg Ⅲ -γ的上调表达依赖于TLR的下游MyD88信号通路，Reg Ⅲ -γ的表达是阻止微生物入侵到宿主组织中所必需的。潘氏细胞MyD88信号通路的激活足以阻止微生物的入侵，并不仅仅依赖骨髓等其他来源细胞的MyD88信号通路。这项研究发现，去除潘氏细胞的小鼠不能阻止肠道共生菌和病原菌入侵到脾脏和黏膜相关淋巴结等组织内，提示潘氏细胞来源的抗菌肽在防御微生物的入侵和扩散中发挥重要作用[33]。

T细胞特异性转录因子4通过调控SRY-Box 9（SOX9）和EPH受体B2（EPHB2）基因转录而促进潘氏细胞分化和定位。此外，WNT/βcatenin、成纤维细胞生长因子受体3、Notch的下游靶点MATh1、锌指蛋白GFI1及Hippo/Yap信号通路均参与潘氏细胞的分化[31]。以往认为，潘氏细胞产物可参与营养物质的消化，目前发现，潘氏细胞分泌的溶菌酶和免疫球蛋白与先天免疫密切相关[31]。

在稳定状态下，小鼠潘氏细胞可分泌受钙释放调节的抗生素，这种抗生素可被特定的细菌及其表面蛋白进一步诱导。作为病原体相关分子模式传感器的TLR和NLR是潘氏细胞脱颗粒所必需的。一种机制可能是共生菌将核苷酸结合寡聚化结构域蛋白 2（nucleotide-binding oligomerization domain-containing protein 2，NOD2）募集到潘氏细胞溶酶体上。当这一机制失效时（由于 NOD2 基因缺失），溶菌酶就会降解溶酶体。受体相互作用的丝氨酸 / 苏氨酸激酶 2 被认为是 NOD2 介导的潘氏细胞溶菌酶分选的调节因子。然而，NOD2 在潘氏细胞中的作用存在争议。参与潘氏细胞脱颗粒的不仅有细菌抗原及其受体，还有内源性免疫反应（如干扰素信号）。此外，上皮溶菌酶的表达需要维生素 D 受体信号，类似于胰岛素受体信号 [31]。哺乳动物实验表明，潘氏细胞可调控肠道微生物群落，影响干细胞功能，并可能是小肠炎症的始作俑者 [31]。

三、杯状细胞

杯状细胞不仅是黏膜屏障的主要成分，也是先天防御系统的主要细胞成分 [35]。杯状细胞是一种特殊的分泌型上皮细胞，广泛存在于小肠和大肠黏膜上皮中 [29]。杯状细胞可分泌多种因子，其中，小肠杯状细胞分泌的黏蛋白 MUC2是第一个被人类鉴定出的分泌型黏蛋白，它是构成黏液层的主要成分。除了分泌型黏蛋白 MUC2 外，杯状细胞还合成多种生物活性分子，如上皮膜结合型黏蛋白（MUC1、MUC3 和 MUC17）、三叶因子、抵抗素样分子 β 和 Fc-γ 结合蛋白 [7]。这些因子有助于保护黏膜层，在维持组织内稳态方面发挥重要作用 [35]。

杯状细胞分泌黏蛋白，而黏蛋白构成覆盖在肠黏膜表面的水合凝胶。黏液层是天然防御的前线，可防止大颗粒分子和细菌直接接触黏液层下的上皮。杯状细胞的其他分泌产物包括调节 T 细胞免疫的抵抗素样分子 β 和促进黏膜损伤后上皮修复的三叶因子。最近的研究表明，杯状细胞可从肠腔中获得可溶性抗原并将其传递给上皮下的树突状细胞。因此，杯状细胞参与抗原的摄取和向底层免疫细胞的抗原呈递，这种功能在以前被认为是 M 细胞独有的 [29]。

黏蛋白由富含丝氨酸和苏氨酸的串联重复序列的大型糖蛋白组成，其羟基残基富含 O- 连接低聚糖 [36]。迄今为止，已检测到 21 种不同的黏蛋白基因，根据发现顺序将命名为 MUC1 ～ MUC21[37]。此外，根据其结构特征和生物学功能，黏蛋白分为两大类，即膜相关黏蛋白和分泌黏蛋白。肠道膜相关黏蛋白为 MUC1、MUC3A/B、MUC4、MUC12、MUC13、MUC15、MUC17、MUC20 和 MUC21。膜相关黏蛋白主要构成细胞表面糖萼，这些多聚糖参与细胞与外部环境之间的相互作用 [37]。分泌性黏蛋白可分为凝胶形成黏蛋白（MUC2、MUC5AC、MUC5B、MUC6 和 MUC19）和非凝胶形成黏蛋白（MUC7、MUC8 和 MUC9），凝胶形成黏蛋白主要参与黏膜表面黏液屏障的形成 [38]，起保护、运输、润滑和水合作用。关于非凝胶形成黏蛋白的功能，目前研究较少。

肠腔黏液在肠腔内容物和肠上皮之间起着润滑剂和物理屏障的双重作用。从小肠近端到结肠远端，杯状细胞的数量不断增加。小肠和大肠上皮被主要由 MUC2 分泌的黏液覆盖，覆盖小肠的黏液为单层黏液，而结肠则为双层黏液，与上皮相邻的内层阻止细菌接触黏膜，而靠肠腔的外层则是肠道微生物群的栖息地 [35]。

多种因素，如免疫细胞、饮食和细菌，影响杯状细胞的分化。与常规饲养的小鼠相比，无菌小鼠肠道杯状细胞表达较少的 MUC2，仅有极薄的黏液层。将无菌小鼠暴露于常规环境中可增强 RELMβ 和 MUC2 的表达，黏液层显著增厚 [38]。微生物群通过与吲哚相关的分泌因子发挥作用，促进杯状细胞分化，并通过外源性芳香烃受体调节肠道内稳态，以增加 IL-10 的表达，逆转老年小鼠的衰老 [39]。

四、肠道巨噬细胞

肠道是体内最大的巨噬细胞池，哺乳动物胃肠道内有大量的组织巨噬细胞[7，40]。巨噬细胞是哺乳动物上皮下固有层最丰富的细胞成分。巨噬细胞在肠道不同部位的数量与微生物负荷密切相关，在大肠中最高，而在无菌小鼠的肠道中最低[6]。巨噬细胞与其他免疫细胞发挥协同作用，有助于在对共生微生物和食物抗原产生免疫耐受与对有害微生物或毒素产生免疫抵抗之间保持平衡。巨噬细胞通过捕获和清除穿过上皮屏障的细菌，同时发挥上皮更新所需的持续吞噬作用，可维持黏膜内环境的稳定。巨噬细胞是保护性免疫的重要组成部分，其功能缺陷与炎性肠病（inflammatory bowel diseases，IBD）的病理过程相关。巨噬细胞特异性缺乏 IL-10 受体，会导致小鼠发生严重的自发性结肠炎。

肠道常驻巨噬细胞起源于胚胎期的卵黄囊或肝脏，可长期存在直至成年。一方面，这些常驻巨噬细胞在出生后逐渐被骨髓来源的单核细胞衍生的细胞取代。另一方面，循环而来的单核细胞与肠道常驻巨噬细胞形成竞争，这些巨噬细胞能够根据所处位置的信号分子进行分化、增殖并占据常驻巨噬细胞的生态位[41]。成人肠道中，单核细胞衍生的细胞不断补充消耗的巨噬细胞。在分化过程中，单核细胞失去 Ly6C 表达，而其他巨噬细胞表面标志物如 MHC Ⅱ、F4/80、CD11c 和 CX3CR1 表达上调。巨噬细胞也存在于黏膜下层，在维持黏膜下层血管的完整性方面发挥重要作用[7，40]。

众所周知，巨噬细胞能迅速适应组织环境。最初的研究中，肠道巨噬细胞是从整个组织所提取，对肠壁不同层的巨噬细胞种群并未进行常规的组织或表型分离。最近的研究表明，在人和小鼠的肠道内，巨噬细胞可以根据其解剖位置来定义，大致分为固有层巨噬细胞和肌层巨噬细胞。与小鼠其他组织内的巨噬细胞相比，肠道巨噬细胞被认为是通过一个被称为“单核细胞瀑布”的过程不断被循环的单核细胞所取代，这种逐步分化也存在于人类。Ly6C^{hi} CD64^{-} CX3CR1^{int} MHC Ⅱ$^{-}$ 单核细胞（P1）进入肠道后通过 3 个阶段（P2 ～ P4）分化，转变为成熟的 CD64^{+}C^{+}3CR1^{hi} MHC Ⅱ^{hi} 巨噬细胞[40]。循环来源巨噬细胞的分化动力学在小肠和大肠之间存在差异，这表明环境因素，如微生物组，对巨噬细胞可能具有区域特异性[41]。

肠道巨噬细胞表达高水平的集落刺激因子 1 受体（colony stimulating factor 1 receptor，CSF1R），遗传或药物途径可阻断 CSF1 或 CSF1R，导致肠道中巨噬细胞的缺乏。除局部生长因子和募集细胞因子的作用外，微生物群是肠道巨噬细胞募集和分化的关键因素[40]。此外，常驻巨噬细胞在组织内显示出有规律的分布模式，这表明它们可以感知邻近的细胞，并可能被周围细胞排斥。这与细菌的群体感应机制类似。细菌群体感应系统主要用来调节种群密度，而巨噬细胞群体感应的机制尚不清楚，可能涉及营养因子的浓度差异。低水平的 CSF1 对巨噬细胞的存活是必要的，而高水平的 CSF1 能够诱导巨噬细胞增殖[41]。

1959 年，Palay 和 Karlin 利用电子显微镜首先发现了固有层巨噬细胞的存在。这些巨噬细胞位于肠上皮，捕获肠腔抗原，吞噬死亡细胞，并清除穿过上皮层的病原体。固有层巨噬细胞主要由来源于单核细胞瀑布的巨噬细胞组成，与肠道深层组织相比，常驻巨噬细胞是次要的来源。人肠道固有层巨噬细胞表达 CD64^{+} CD11c^{+} MHC Ⅱ^{hi}，在小鼠中则表达 CX3CR1^{hi}。活体成像显示，这些巨噬细胞有快速而动态的膜皱褶和伪足延伸[40]。

固有层巨噬细胞具有高度吞噬能力。在摄入食物抗原或无害的共生细菌后，它们不会释放炎性介质或一氧化氮，从而防止其他免疫细胞诱导免疫反应（这可能会导致附带损伤）并维持组织稳态[41]。固有层巨噬细胞将捕获的抗原呈递给 $CD103^+$ 树突状细胞，后者迁移到肠系膜淋巴结以诱导抗原特异性调节性 T 细胞[42]。$FoxP3^+$ 调节性 T 细胞参与对食物抗原的免疫耐受，其在肠道内的扩增依赖于 IL-10。固有层巨噬细胞产生 IL-10 依赖于肠道微生物群的刺激[41]。此外，肠道常驻巨噬细胞表达一种促进凋亡细胞吞噬的受体 CD36，具有高度吞噬功能，表现出较强的杀菌活性，而且不会引发明显的炎症。因此，局部巨噬细胞是对抗破坏上皮屏障的共生细菌的防火墙。它们不会表达高水平的共刺激分子，如 CD80、CD86 或 CD40，但表达胞质模式识别受体，具有较强的抗菌活性。

此外，常驻肠道巨噬细胞不仅可吞噬局部细菌和死亡细胞，维持肠道内环境的稳定，而且还调节上皮细胞的完整性。固有层巨噬细胞表达转录因子过氧化物酶体增殖物激活受体 -γ（peroxisome proliferators-activated receptors γ，PPAR-γ），并通过此途径抑制促炎基因表达以预防局部炎症反应。因此，巨噬细胞是参与正常生理过程中组织重塑和维持共生微生物免疫耐受的细胞亚群[6]。派尔集合淋巴结中存在 $Tim4^+$ 和 $Tim4^-CD4^+$ 巨噬细胞亚群，已证明该细胞能有效吞噬病毒颗粒和凋亡细胞，并呈现出抗菌和抗病毒的特征[43]。最近的研究表明，CSF1R 依赖性巨噬细胞亚群与隐窝上皮密切相关。这些巨噬细胞的耗竭导致 Lgr5 肠干细胞减少，潘氏细胞分化异常和杯状细胞密度异常[44]。

五、肠道 T 淋巴细胞

T 细胞是肠黏膜含量最丰富的白细胞之一，在哺乳动物的肠道免疫中发挥着重要作用，它们的缺失会导致严重细胞免疫功能缺陷，如人类免疫缺陷病毒感染[6]。健康动物的 T 细胞群主要由 1 型 T 辅助细胞（type 1 T helper，Th1）和 2 型 T 辅助细胞（type 2 T helper，Th2）组成。克罗恩病与 Th1 细胞因子相关，而溃疡性结肠炎则主要与 Th2 相关[6]。调节性 T 细胞（regulatory T-cells，Treg）和 17 型辅助性 T 细胞（type 17 T helper，Th17）是后续发现的 T 细胞亚群。

Th17 细胞的特征是表达转录因子维 A 酸相关孤核受体 γt（retinoic acid-related orphan receptor gamma t，RORγt）、IL23R 和 C-C 趋化因子受体 6。IL-1β、IL-6、IL-21、IL-23 和 TGF-β 可影响 Th17 细胞的分化[6]。Th17 细胞可分泌促炎细胞因子 IL-17A、IL-17F、IL-21、IL-22、IL-26 以及趋化因子 CCL20，Th17 细胞在肠道炎症，尤其是克罗恩病中起重要作用[6]。

Th17 细胞通过协调中性粒细胞内流和合成 IL-17、IL-22，维持上皮屏障的完整性或修复上皮屏障，中和病原体和共生微生物[6]。肠道黏膜上皮细胞表达 Th17 细胞因子的受体，促进紧密连接的形成、抗菌肽的产生，以及黏液的产生[6]。Th17 细胞既参与炎症的发生发展，也参与维持肠道屏障的完整性。小肠固有层 Th17 细胞的分化需要特定细菌参与。鼠柠檬酸杆菌诱导 Th17 细胞表现促炎特性，该 Th17 细胞主要参与有氧糖酵解，与炎症效应细胞有关。分节丝状菌诱导的 Th17 细胞没有促炎特性，它在维持肠道内环境稳定、黏膜结构发育、IgA 诱导和出生后肠道免疫功能成熟中发挥重要作用[10]。

肠道淋巴细胞持续暴露于食物和微生物抗原，进化出独特的机制以维持肠道屏障的完整性和免疫稳态。肠上皮作为不可渗透的屏障将机体与外界环境隔开。肠上皮细胞之间的淋巴细胞称为上皮内淋巴细胞。由于这一特定的位置，上皮内淋巴细胞可直接接触肠上皮细胞，并与肠腔内的抗原直接接触。因此，上皮内淋巴细胞具有广泛的调节和效应能力，包括预防病原体侵袭和维持耐受性以防止广泛的组织损伤。上皮内淋巴细胞几乎完全是 T 细胞，其数量甚至超过了脾脏中的数量。根据其发育和成熟的不同路径，上皮内淋巴细胞可以分为两个亚群：“传统型”（或“a 型”）肠道 T 细胞，表达 TCRαβ 以及 CD4 或 CD8αβ；“非传统型”（或“b 型”）肠道 T 细胞，表达 TCRαβ 或 TCRγδ，通常也表达 CD8αα 二聚体（表 3-1）。除了上皮内淋巴细胞，肠道 T 细胞也可以存在于固有层中。与上皮内淋巴细胞不同的是，固有层淋巴细胞来自于传统的 T 细胞，这些 T 细胞经历了常规的胸腺发育，在次级淋巴器官中被启动，并迁移到具有效应记忆表型的固有层[45]。

表 3-1　两种不同的肠道 T 细胞

	传统型（a 型）	非传统型（b 型）
特征性表达标记	$CD4^+$ $TCR\alpha\beta^+$ 或 $CD8\alpha\beta^+$ $TCR\alpha\beta^+$	$CD8\alpha\alpha^+$ 或 $CD8\alpha\alpha^-$ 和 $CD4^-$ $CD8\alpha\beta^-$ $TCR\alpha\beta^+$ 或 $TCR\gamma\delta^+$
位置	上皮内淋巴细胞和固有层淋巴细胞	主要在上皮内淋巴细胞
功能	防御感染（细胞毒性 T 细胞、Th1、Th2 和 Th17）、促炎（Th17）；对共生菌和食物抗原的耐受性（PTregs）	免疫调节、抗感染、维持肠道内稳态、对肠道抗原的耐受性、增强自身抗原特异性
发育及成熟	在胸腺中发育，循环并迁移到肠相关淋巴组织中，被抗原激活成为效应记忆细胞，驻留肠道	在胸腺中被选择，直接引导到肠道

由于来源于胸腺和肠道环境的影响，这些肠 T 细胞具有不同的表型和功能。大多数肠道 T 细胞在外周淋巴器官成熟。这些细胞获得肠道归巢受体后迁移到肠道。离开胸腺后，幼稚 T 细胞通过循环迁移到肠道相关淋巴组织。在肠道相关淋巴组织中，如派尔集合淋巴结和肠系膜淋巴结，幼稚的 $CD4^+T$ 和 $CD8\alpha\beta^+T$ 细胞由抗原呈递细胞启动，并通过上调肠道归巢分子，如整合素 α4β7、趋化因子受体 CCR9、活化标记 CD44、黏附分子 LFA-1 和极晚期抗原 -4（VLA-4）而获得迁移到肠道组织的能力（图 3-3）。然后，这些 T 细胞被趋化因子吸引，通过与肠细胞分泌的平行配体的相互作用进入肠道。

T 细胞上的趋化因子受体决定了它们在肠道中的不同位置。例如，CCL25 和 CCL28 分别由小肠上皮细胞和结肠细胞分泌。表达相应趋化因子受体的 T 细胞，即 CCR9（CCL25 的受体）或 CCR10（CCL28 的受体），被血管内皮细胞吸引并迁移进入肠道。在人类和小鼠，小肠中几乎所有的 T 细胞都表达 CCR9。在趋化因子引导 T 细胞迁移的同时，表达在 T 细胞表面的整合素，如 α4β7，与表达在内皮细胞和上皮细胞上的黏附分子，如黏膜寻址素细胞黏附分子 1 相互作用，产生对组织的黏附和扩散效应。有趣的是，小肠中超过 90% 的淋巴细胞 α4β7 整合素呈阳性，而 α4β7 的缺陷会导致肠道相关淋巴组织的形成中断[45]。

环境因素可能通过抗原呈递诱导 T 细胞适应肠道微环境。传统的 $CD4^+T$ 细胞通过感知肠道中的特定环境，分化为不同的辅助性 T 细胞亚群（Th1、Th2、Th17 和诱导型 Treg 细胞）。肠道屏障中的树突状细胞和肠上皮细胞等抗原呈递细胞可以调节 T 细胞的分化，以应对来自肠腔的各种刺激。例如，短链脂肪酸刺激上皮内 $CD103^+$ 树突状细胞分泌 TGF-β、维 A 酸和吲哚胺 2，3- 双加氧酶，从而诱导外周 Treg 细胞分化（图 3-3）。当 $CX3CR1^+CD103^+$ 树突状细胞与细菌黏附在一起时，微生物抗原被呈递给 T 细胞以启动 Th17 的分化，从而诱导免疫应答以保护宿主免受侵犯。

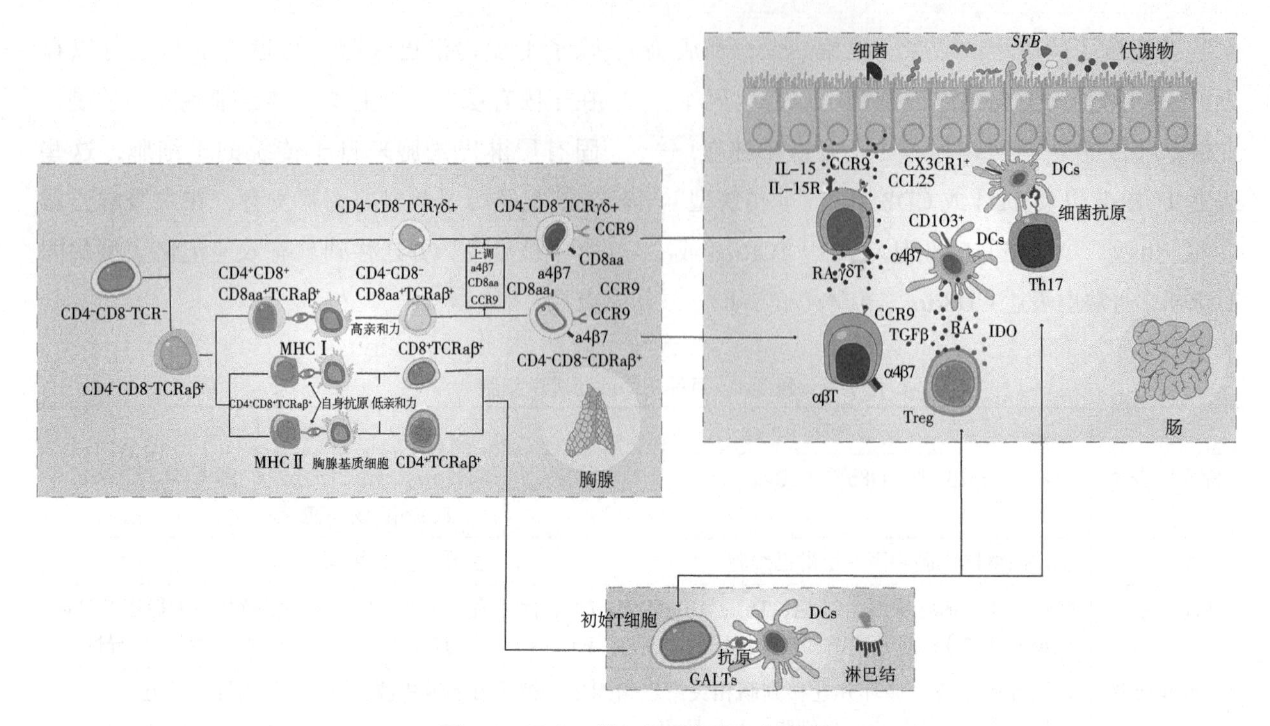

图 3-3　肠道 T 淋巴细胞的发育与成熟

肠道 T 细胞可分为诱导的“传统”（或“a 型”）肠道 T 细胞或“非传统”（或“b 型”）肠道 T 细胞。传统的肠道 T 细胞表达 TCRαβ 和 CD4 或 CD8αβ，并作为 TCR2 的辅助受体。非传统肠道 T 细胞表达 TcRαβ 或 TcRγδ，通常也表达 CD8αα 同源二聚体。传统的 T 细胞来源于胸腺中的 $CD4^-CD8^-$（DN）祖细胞，发展为 SP $CD4^+T$（MHC Ⅰ）细胞或 $CD8^+T$ 细胞（MHC Ⅱ），随后迁移到外周淋巴器官，如淋巴结，在那里它们遇到抗原并获得激活的效应器表型，从而驱动它们迁移到肠道。或者，胸腺中未成熟的三阴性胸腺细胞（$CD4^-CD8^-TCR^-$）分化为双阴性（$CD4^-CD8^-$）、TCRγδ 阳性或 TCRαβ 阳性的肠道 T 细胞前体。CCR9. 趋化因子受体 9；MHC Ⅰ. 主要组织相容性复合体 - Ⅰ；MHC Ⅱ. 主要组织相容性复合体 - Ⅱ；*SFB*. 肠道分节丝状菌；RA. 维甲酸；CCL25. 趋化因子 25；IDO. 双加氧酶；DCS. 树突状细胞；Treg. 调节性 T 细胞；TCR.T 细胞受体；TGF. 转化生长因子；GALT. 肠道相关淋巴组织

维 A 酸是从饮食中提取的一种维生素 A 代谢物，它是肠道归巢受体的重要诱导物。维 A 酸促进 T 细胞表面 α4β7 整合素和 CCR9 的表达，从而促进这些细胞在小肠的驻留。此外，树突状细胞呈递的肠道微生物和食物抗原可形成功能多样的特异性 T 细胞群，具有显著的可塑性，可转分化为具有其他特征甚至功能相反的 T 细胞，如炎症性 Th17 细胞可以转化为调节性 Tr1 细胞。非传统肠道 T 细胞和传统肠道 T 细胞都具有对病原体的免疫防御功能，同时保持对食物和共生体的免疫耐受性 [45]。无菌小鼠的固有层中几乎没有 Th17 细胞，但接种分节丝状菌的小鼠表现出 Th17 细胞的分化和数量增加，进而诱导 Th1 细胞数量增加 [10]。

六、肠道树突状细胞

树突状细胞是塑造免疫系统的关键调节器细胞。在黏膜组织中，树突状细胞起着感知感染的监视作用，同时也是刺激初始 T 细胞分化的主要抗原呈递细胞。它们在天然免疫和适应性免疫系统之间架起桥梁，以维持肠道免疫环境的动态平衡 [46]。树突状细胞是功能最强的专业抗原呈递细胞。与巨噬细胞不同，树突状细胞可以通过激活幼稚 T 细胞来启动初级免疫反应，并调节促炎或耐受性免疫反应 [6]。固有层和肠道相关淋巴组织驻留的树突状细胞在维持对共生微生物和食物的耐受性及产生针对病原体的保护性免疫反应方面都有特别的作用 [28]。此外，树突状细胞在体内和体外都能打

开上皮间紧密连接，并能直接穿过上皮摄取抗原。根据整合素 CD11c 和 CD103 的表面表达情况，可将肠道树突状细胞分为几个亚群[7]，分别是 $CD103^{+}CD11b^{-}$，$CD103^{+}CD11b^{+}$、$CD103^{-}CD11b^{+}$ 和 $CD103^{-}CD11b^{-}$ 树突状细胞[47]。

最近，为了更好地区分树突状细胞和巨噬细胞，引入了 CD24 和 Sirp α。树突状细胞主要位于固有层和黏膜相关淋巴组织，与上皮层有密切的相互作用。杯状细胞可以将较小的可溶性抗原从肠腔转移至 $CD103^{+}$ 树突状细胞。肠细胞分泌的趋化因子 TLR 配体，可引起固有层树突状细胞向上皮细胞迁移。上皮细胞和基质细胞可分泌多种诱导树突状细胞耐受的因子，如 RA、TGF-β、PGE-2 和 TSLP。建立肠道耐受对于预防肠道疾病（如 IBD）至关重要，而调节黏膜树突状细胞为预防传染病提供了潜在的治疗策略[7]。

树突状细胞可根据环境变化来表达 PRR，以此感应微生物的产物。树突状细胞能够诱导 Th1、Th2 或 Treg 的分化，产生相应免疫反应，这主要取决于它们先前暴露于何种细胞因子和（或）微生物配体。一旦树突状细胞迁移到黏膜下层，它们就可通过树突高效地采集肠道内容物，然后通过上皮层或 M 细胞捕获和处理抗原。在生理条件下，树突状细胞具有调节作用，可以阻止针对食物抗原和肠道微生物群的免疫反应。它可通过各种信号分子（胸腺基质淋巴细胞生成素、IL-10 及 TGF-β）和维 A 酸来调节免疫反应。在维 A 酸存在的情况下，肠道树突状细胞能够利用将维生素 A 转化为维 A 酸的酶，促进 Treg 细胞和分泌 IgA 的 B 细胞分化。

肠道树突状细胞与其他组织树突状细胞的区别在于：①减少 PRR 的表达；②减少共刺激分子的表达，从而减少抗原呈递；③增加抗炎细胞因子 IL-10 的产生；④有利于抗原特异性 Treg 细胞和分泌 IgA 的 B 细胞分化；⑤通过在 Treg 细胞和分泌 IgA 的 B 细胞中表达肠道归巢标记来诱导免疫耐受。肠道树突状细胞上述特征的任何改变都会导致对微生物群的异常免疫反应，可能导致 IBD 的发生[6]。

树突状细胞是一种骨髓来源的抗原呈递细胞，由两个主要亚群组成：经典树突状细胞（conventional dendritic cell，cDC）和浆细胞样树突状细胞（plasmacytoid dendritic cell，pDC）。它们在发育上既不同于组织驻留的巨噬细胞，也不同于单核细胞来源的巨噬细胞，但与这些细胞有许多共同的表型标记。肠道经典树突状细胞存在于有组织结构的淋巴组织中，包括派尔集合淋巴结、引流淋巴结，以及小肠和结肠的固有层。固有层经典树突状细胞包括 cDC1 和 cDC2，它们可以在小鼠和人类中分别通过 X-C 基序趋化因子受体 1 或信号调节蛋白 α（SIRPα/CD172a）的表达来确定。在其他物种中，则使用其他的细胞表面标记进一步确定肠道经典树突状细胞亚群。在小鼠中，cDC1 是 $CD103^{+}CD11b^{-}$，而 cDC2 包括 $CD103^{-}CD11b^{+}$ 细胞和肠道特异性 $CD103^{+}CD11b^{+}$ 群体。cDC2 中两个 $CD11b^{+}$ 群体密切相关，$CD103^{-}CD11b^{+}$cDC 在 TGFβ 的作用下产生 $CD103^{+}CD11b^{+}$ 细胞。小肠中 $CD103^{+}CD11b^{+}$ 的含量高于结肠中 $CD103^{+}CD11b^{+}$ 的含量。人类肠道中存在相当数量的经典树突状细胞，这些细胞通常是基于 CD103 与 SIRPα 相结合确定的，而不是 CD11b 的表达来确定的[48]。

为防御肠道病原体及建立对无害抗原的调节反应，树突状细胞可连续采集肠道内容物。在派尔集合淋巴结中，滤泡相关上皮中的 M 细胞将细菌和其他微粒内化，并将它们传递给下层的树突状细胞。派尔集合淋巴结中的经典树突状细胞还可以捕获易位 IgA 免疫复合物，并通过 M 细胞特异性的跨细胞孔延伸树突。经典树突状细胞交叉呈递从被感染的上皮细胞中捕获的病毒抗原。回肠 $CD103^{+}$ 经典树突状细胞从上皮样本中呈递可溶性和颗粒性抗原。IgG 免疫复合物可通过小肠杯状细胞转运的低分子可溶性物质和上皮逆向转运，将抗原传递给经典树突状细胞。非迁移性 CX3CR1^{+} 巨噬细胞通过跨上皮过程将小肠肠腔内的抗原传递给迁移性 CD103^{+} 经典树突状细胞 [48]。

肠道淋巴组织中活化的淋巴细胞可通过表达特异性黏附分子和趋化因子受体，如整合素 α4β7 与肠血管内皮细胞表达的黏膜地址素细胞黏附分子 -1 结合，促进淋巴细胞进入结肠和小肠黏膜。趋化因子受体 CCR9 是进入小肠所必需的物质，其配体 CCL25 可在小肠中表达 [48]。

感染可改变树突状细胞诱导的反应平衡，有利于其效应机制。CD103^{+} 小鼠树突状细胞通过 p38MAPK 信号调节初始 T 细胞向 Treg 和 Th1 分化的平衡，在缺乏 p38MAPK 的情况下，视黄醛脱氢酶 2 的表达和 Treg 的产生会减少，Th1 应答增强。肠道树突状细胞诱导的 T 细胞失衡可能是由于致病因素的直接刺激、常驻树突状细胞群体环境的改变或招募不同的促炎树突状细胞所致。肠经典树突状细胞可表达模式识别受体，并对微生物产物产生应答。小鼠和人树突状细胞亚群差异表达 Toll 样受体，提示树突状细胞亚群具有直接识别特定微生物的特异性。小鼠 CD103^{+}CD11b^{+} 树突状细胞表达 TLR5，被鞭毛蛋白激活后，诱导 Th17 应答的能力增强。CD103^{+}CD11b^{+} 树突状细胞的激活可促进 IL-6 和 IL-23 的产生，从而促进 Th17 的发育和抗菌肽 Reg Ⅲ γ 的产生。在体内使用 TLR7 激动剂可激活 CD103^{+}CD11b^{-} 树突状细胞亚群迁移，并可引起效应性 CD8^{+} T 细胞对交叉呈递抗原产生应答 [48]。

在大多数组织中，接触微生物产物就足以将未成熟的经典树突状细胞转化为成熟的细胞，从而产生强有力的效应反应。然而，接触来自共生微生物群的 PAMPs 在健康的肠道中也很常见，因此也可能需要第二种信号。事实上，即使在成熟经典树突状细胞的高水平共刺激分子表达的情况下，小鼠 CD103^{+} 小肠经典树突状细胞也能诱导 Treg。第二种信号的性质尚不清楚，但含有 IgA 的免疫复合物（通常局限于管腔，在受损组织中大量存在）可以增强树突状细胞的促炎活性 [48]。

肠道经典树突状细胞在黏膜中停留数天，它们被调教后获得调节特性。缺乏稳态调节作用的经典树突状细胞可能促进产生过强的效应 T 细胞免疫反应。暴露于 TLR 配体或炎症细胞因子后的经典树突状细胞周转加快，停留时间缩短，并增强经典树突状细胞产生的免疫反应。在骨髓发育过程中，肠道经典树突状细胞的功能也可能受到远程影响。肠道炎症会改变造血功能，影响单核细胞的发育，进而影响由单核细胞衍生的肠道细胞群体 [48]。

七、固有淋巴细胞

固有淋巴细胞（innate lymphoid cell，ILC）是近年来在黏膜免疫系统中发现的先天性免疫细胞，其特征是缺乏特异性抗原受体，在调节肠上皮细胞屏障完整性、免疫、炎症和组织修复中起着重要作用 [6]。ILC 不表达 T 细胞和 B 细胞上表达的各种类型的抗原受体，它们主要是组织驻留细胞，并深入整合到宿主组织中 [49]。适应性淋巴细胞在淋巴器官中数量最多，在初级和次级淋巴组织中相对较少。因此，它们的存在一直被忽视。然而，ILC 在外周组织的位置，特别是在肺、皮肤和肠道的屏障表面，为组织受侵的早期防御反应提供了优势。ILC 是快速反应的细胞，它们在激活后的数小时内产生细胞因子，这与原始的适应性淋巴细胞启动、扩增、分化和进入组织所需的时间形成了鲜明的对比 [50]。肠道中的 ILC 可以抵御许多类型的病原体，如病毒、细胞内和细胞外细菌、真菌和寄生虫。ILC 作用迅速，可释放细胞因子激活先天免疫细胞、获得性免疫细胞和肠道上皮细胞 [51]。

ILC富集于肠道等黏膜组织中，可对来自肠道微环境的信号（如细胞因子、警报素和其他炎性和非炎性刺激）做出反应，以启动适当的免疫反应并维持组织的动态平衡[52-54]。此外，ILC表面表达特定的受体和配体，进一步调节其功能[55, 56]。ILC已被鉴定出多种亚型，主要有3类。第1组ILC（ILC1）包括非细胞毒性ILC1细胞和细胞毒性NK细胞。传统的NK细胞最早于1975年被发现，与其他类型的ILC相比，其研究历史较长[57, 58]。ILC1受T-bet调节，可产生IFN-γ、GM-CSF、颗粒酶和穿孔素，对IL-12、IL-18或其他刺激因素（病原体或肿瘤）产生免疫反应。它们可与Th1细胞协同对抗细胞内微生物，如病毒、细菌或寄生虫的感染[59-62]。

第2组ILC（ILC2）与Th2类似，表达GATA3，能在IL-25、IL-33和胸腺基质淋巴细胞生成素的刺激下产生IL-4、IL-5、IL-13、IL-9和双调节素。ILC2在对抗大型细胞外寄生虫和过敏原的免疫反应中发挥重要作用。它们产生的抗菌肽可促进组织损伤修复[63, 64]。最近*Science*期刊上发表的一项研究中，利用小鼠模型和先进的成像技术监测ILC的激活和运动，发现ILC2来源于肠道，可通过淋巴管和血液循环迁移到其他器官抵抗蠕虫感染。ILC2的转运部分依赖于1-磷酸鞘氨醇[65]。

第3组ILC（ILC3）对应Th17细胞，可表达RORγt、淋巴毒素α和β、IL-17、IL-22、GM-CSF和TNF-α。它们可以被IL-23、IL-1β或NCR配体激活，并能对抗细菌和真菌等细胞外微生物感染[62]。表3-2将ILC分类为ILC1、ILC2和ILC3，有助于理解ILC分类和功能。在免疫应答过程中，ILC的功能和分化较为复杂。ILC的异质性和可塑性在人和小鼠的研究中都已被证实。组织驻留的T-bet^{+}ILC1可能有4个来源：ILC祖细胞、暴露于IL-12和IL-1β的转化的ILC2、暴露于IL-2、IL-15和IL-23转化的ILC3及暴露于TGF-β的NK细胞[66, 67]。此外，还有淋巴组织诱导样细胞（lymphoid tissue inducer-like cell，LTi）也属于ILC的范畴[49]，LTi对淋巴器官形成至关重要。ILC亚群之间有显著的可塑性，这取决于它们接受的环境刺激，环境因素可直接影响它们的效应器功能和免疫反应。虽然ILC对淋巴器官生成和组织内稳态至关重要，但它们也参与自身免疫性疾病或炎症驱动的癌变发展[68]。

表3-2　肠道内固有淋巴细胞

类型	表面标记物		刺激因子	调节转化因子	特异释放因子	在肠道屏障中的作用
	小鼠	人类				
ILC1	CD160 NKp46 NK1.1	CD103 CD160 CD56 NKp46 NKp44	IL-12, IL-15	T-bet	IFN-γ, TNF-α, GM-CSF, 颗粒酶, 穿孔素	防御病毒、病原体
ILC2	IL17RB IL-33R CD25 CD127	IL17RB IL-33R CD25 CD127 CRTh2	IL-25, IL-33, TSLP	GATA3	IL-4, IL-5, IL-13, IL-9, 双调蛋白	驱除蠕虫
ILC3	NKp46	NKp44	IL-1β, IL-23, NCR ligand	RORγt	IL-17, IL-22, GM-CSF, TNF-α	防御细菌、真菌

高维细胞仪分析显示，人类结直肠肿瘤内NK细胞和上皮内ILC1高度浸润[68]。它们在受刺激后分泌细胞毒性分子，如颗粒酶、穿孔素[69, 70]及细胞因子，如IFN-γ，并参与细胞毒性抗肿瘤反应[68]（图3-4）。NK细胞具有细胞毒性功能，类似于细胞毒性CD8^{+}T细胞。ILC1具有有限的细胞毒性，类似于Th1细胞[51]。上皮内ILC1的研究表明，它们在克罗恩病和大肠癌患者肠道中积聚。这些上皮内ILC1可被趋化因子和细胞因子激活，如：肠上皮细胞产生的IL-18或树突状细胞和单核细胞产生的IL-12[68]。此外，NK细胞还可以通过分泌细胞因子（如IFN-γ和TNF-α）和其他生长因子来调节包括树突状细胞在内的多种细胞类型。NK细胞能迅速杀伤细胞，这种杀伤能力受激活剂（如NKG2D、NKp30、NKp44、2B4、DNAM-1和CD16）[71-75]和MHC-Ⅰ类分子介导的抑制性受体（如各种Ly49受体、白细胞免疫球蛋白样受体、KLRG1）的调节[76]。在IL-12、IL-15或IL-18的刺激下，NK细胞和ILC1分泌TNF和IFN-γ，并依赖于T-box（T-bet）等关键转录因子的调节[59, 77-79]。

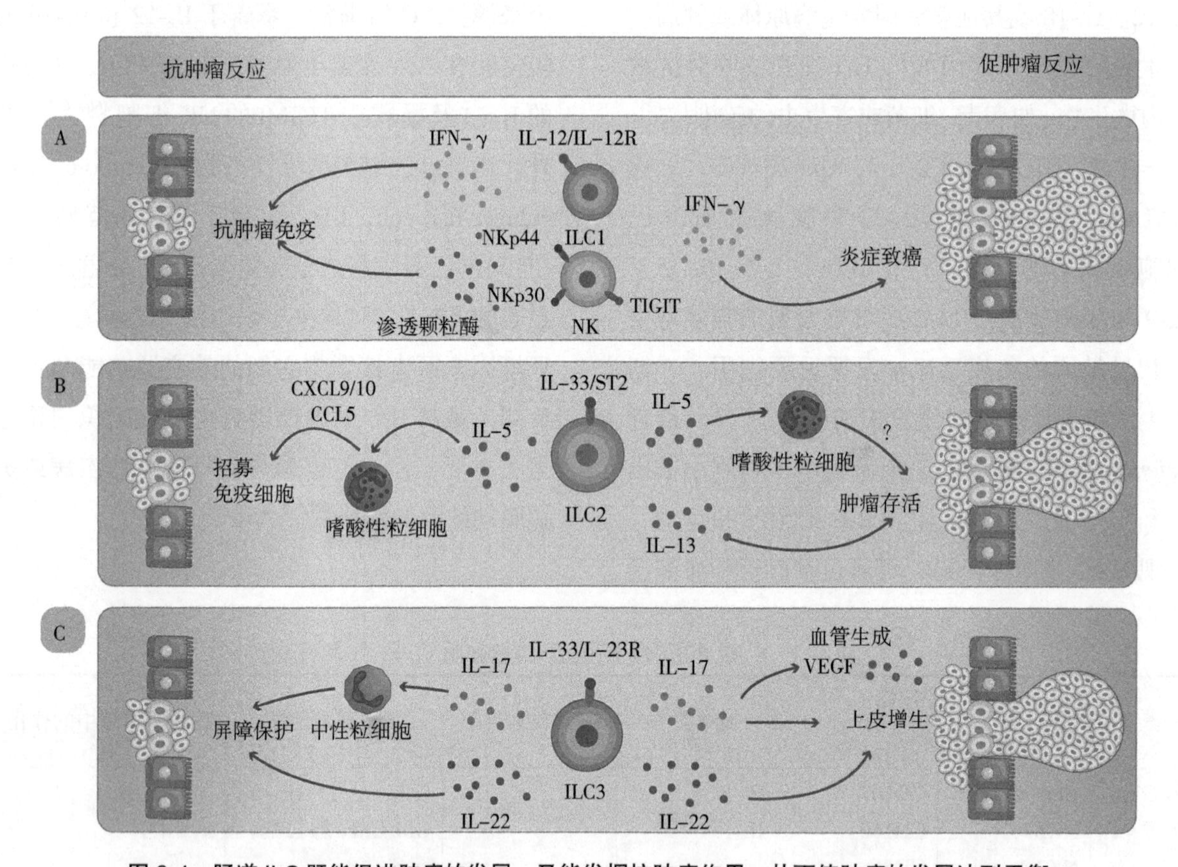

图3-4　肠道ILC既能促进肿瘤的发展，又能发挥抗肿瘤作用，从而使肿瘤的发展达到平衡

NK细胞和上皮内ILC1（A）、ILC2（B）和ILC3（C）通过其激活或抑制受体信号调节这些免疫细胞的功能。ILC通过分泌细胞因子和细胞毒分子，调节肿瘤微环境，抑制或促进大肠癌的发生和发展。ILC. 固有淋巴细胞；CXCL. 趋化因子配体；TIGIT.T细胞免疫球蛋白和ITIM结构域蛋白；VEGF. 血管内皮生长因子；CCL5. 趋化因子5

ILC2保护宿主免受蠕虫入侵，参与伤口愈合和过敏反应。ILC2s在一定条件下还能产生IL-4、IL-6、IL-9和双调蛋白[51]。与Th2相似，转录因子GATA-3也作为ILC2s的关键转录因子[80, 81]。ILC2存在于不同的解剖部位，包括脂肪组织、脾脏、鼻组织、肺、肠道和皮肤。ILC2的发育需要维A酸相关的孤核受体α、T细胞因子1和Notch信号[51]。ILC2在包括肺癌、胃癌等肿瘤中的作用已经被证实[82, 83]。然而，它们在结直肠癌中的作用还知之甚少。

事实上，与肠道中的其他ILC亚群相比，它们出现的频率相对较低。与健康人相比，IBD患者黏膜样本中ILC2的出现频率较高[84]。结直肠癌中的ILC2表达更高水平的ICOS和CD69[69, 70]。这些表型不同的ILC2可能是IL-4、IL-5和IL-13的潜在来源，因为结直肠癌小鼠血清中IL-4和IL-13水平升高[85]。IL-13可通过自分泌或旁分泌方式直接作用于肿瘤上皮细胞，促进其存活和进展[68]。在结直肠癌中，IL-13Rα2受体的高表达与更晚的疾病分期、淋巴结转移和生存率降低有关[86]。相反，有研究发现术前血清IL-13降低是结直肠癌患者预后较差的预测指标[87]。这些不一致的研究结果表明，IL-13的表达及其预测值可能与环境因素相关。IL-5在结直肠癌中的作用尚未明确。尽管如此，它对嗜酸性粒细胞的招募、扩增和激活发挥重要作用[68]。反之，一些细胞分泌化学诱导剂，如CXCL9、CXCL10和CCL5，可促进$CD8^+$T细胞向肿瘤部位募集[88]（表3-3）。

表3-3　固有免疫细胞在促进或抑制结直肠肿瘤发生中的作用

免疫细胞类型	预测	
	抑制肿瘤	促进肿瘤
NK细胞	高细胞毒性 肿瘤始动细胞的识别	浸润性NK细胞减少 激活受体水平降低(如NKp44、NKp30) 抑制性受体表达增加（如TIGIT） 低细胞毒性
1型固有淋巴细胞（ILC1）	产生与抗肿瘤免疫相关的干扰素-γ和细胞毒分子	积聚在发炎的组织中 IFN-γ促进炎症
2型固有淋巴细胞（ILC2）	术前高血清IL-13 IL-5募集嗜酸性粒细胞	局部IL-13促进肿瘤上皮细胞存活和增殖 溃疡性结肠炎患者IL-5升高
3型固有淋巴细胞（ILC3）	IL-17募集中性粒细胞保护组织屏障 IL-22促进伤口愈合并保护肠上皮细胞免受遗传毒性应激诱导的DNA损伤	IL-17的过度产生促进炎症和血管生成，并破坏肠上皮屏障 IL-22过度分泌对上皮细胞增殖的调节作用

ILC3是一种异质细胞群，参与抵抗细菌和真菌感染、调控共生菌以及淋巴组织的发育和修复[51]。ILC3通过分泌IL-22促进肠道内环境稳定，诱导T细胞耐受，保护肠道免受感染[4]。作为肠道黏膜免疫的重要一员，在生理稳态下，ILC3可维护肠上皮及黏液完整性，促进肠免疫系统的发育和完善，诱导适应性免疫耐受，维持肠道共生菌稳态，抵御病原体入侵。若ILC3s调控异常，则可促进适应性免疫应答，诱导菌群失调，导致肠道病理损伤[9]。激活后的ILC3可分泌IL-17A、IL-22、TNF和GM-CSF[51]。维A酸相关孤核受体γt和芳香烃受体是促进ILC3发育、维持功能的关键转录因子[89]。

炎症是肿瘤进展的关键驱动力，炎性肠病与结直肠癌之间的关系已被证实。在结直肠癌发展早期，肠道屏障被破坏导致炎症加剧，通透性增加，从而允许微生物及其产物渗透到肿瘤微环境中。研究表明，具核梭杆菌可以塑造肿瘤微环境促进结直肠癌发生发展[90-92]。活化的树突状细胞和巨噬细胞可产生具有调节ILC3功能和可塑性的关键细胞因子IL-23[68]。此外，IL-12、IL-18、TGF-β或Notch配体也可影响ILC3s的可塑性[79, 93, 94]。表达IL-22的ILC3在IL-12或TGF-β存在的情况下，获得ILC1的特征，并转分化为产生IFN-γ的细胞[94]。已发现IL-23在小鼠和人类结直肠癌中过表达，在小鼠癌症模型中，IL-23缺乏可防止肿瘤的发生[68]。相反，长期注射IL-23的小鼠可诱导产生ILC3依赖性十二指肠腺瘤[95]。此外，局部IL-23表达可促进结肠固有层ILC3的积聚和激活[96]。

肠道ILC3表达较高水平的活化受体ICOS和CD69，可促进下游信号细胞因子IL-17和IL-22的产生，这些细胞因子与结直肠癌的发病机制有关[97]。研究表明，小鼠和人结直肠癌组织中IL-17的表达均较高[97-100]。ILC3产生高水平的IL-17，其他免疫性和非免疫性肠道细胞，如Th17细胞、γδT细胞和潘氏细胞，也表达这种细胞因子[101]。在肠腺瘤发展到癌的过程中，IL-17转录增加[102]。此外，IL-17还可通过诱导血管内皮生长因子（vascular endothelial growth factor，VEGF）的表达，减轻细胞周期抑制，从而促进上皮细胞增殖和肠道炎症，增加血管生成[68]。IL-17是一种参与大肠癌发病的促炎细胞因子，同时也可能起到保护作用[68]。高水平的IL-17可以改善患者的总体生存率[99]，而中和IL-17可增加肠道通透性，加重克罗恩病和葡聚糖硫酸钠介导的结肠炎[103，104]。分节丝状菌可诱导ILC3产生IL-22，并促进Th17细胞分化和IL-17的表达[10]。然而，胃肠道ILC3也有一定的负面作用，它可导致IL-17和IFN-γ水平升高，与肝螺杆菌（*helicobacter hepaticus*）诱导的结肠炎有关。此外，ILC3产生大量的IL-22，可造成上皮细胞破坏，导致肠道损伤[4]。

第三节　肠道免疫功能与机体健康

肠道相关淋巴组织是人类最重要的淋巴网络[105]。位于小肠黏膜和黏膜下层的派尔集合淋巴结被含有M细胞的“滤泡相关上皮”所覆盖。这些细胞吞噬抗原，将抗原从肠腔运送到肠道树突状细胞，进一步激活T细胞[105]。随后，T细胞返回肠道淋巴组织，以发挥其效应功能。同时，M细胞启动黏膜sIgA的产生和体液免疫反应。故派尔集合淋巴结直接介导肠道菌群与体液免疫及细胞免疫反应之间的相互作用。另一方面，肠道致病菌能够侵入和破坏滤泡相关上皮，特别是M细胞，干扰T细胞的分化和免疫反应，细菌通过不同机制黏附于肠上皮，侵袭M细胞以及其他肠上皮细胞[105]。

肠道相关淋巴组织也含有一个免疫耐受性树突状细胞亚群，即$CD103^+$树突状细胞，参与Foxp3的表达和初始$CD4^+$T细胞向Treg细胞转化。共生菌群通过激活特定受体或自身代谢产物（如短链脂肪酸）刺激$CD103^+$树突状细胞。丁酸可刺激肠道树突状细胞，促进IL-10释放和Treg细胞极化来维护肠道免疫耐受，同时维持分泌IL-10的$CD4^+$T细胞、分泌IL-17的Th17细胞和$CD4^+$Th1效应细胞之间的平衡。此外，短链脂肪酸可减轻产生脂多糖的革兰阴性菌引起的过度炎症反应[105]。因此肠道免疫功能与机体健康密切相关[106，107]。

一、肠道免疫与胃肠道稳态

肠道稳态是宿主肠道黏膜和免疫屏障、肠道微生物、营养物质和代谢产物等相互作用而形成的动态平衡状态。肠道稳态由宿主免疫系统、肠腔抗原负荷和上皮屏障之间的复杂相互作用所制约[108]。肠道稳态对肠道的正常生理功能，如营养吸收、能量代谢及抵御肠道感染等具有重要意义，而肠道稳态失衡与很多严重疾病密切相关，如肠道感染、肥胖、2型糖尿病、炎性肠病、自闭综合征等。

肠道T细胞与肠腔内环境的相互作用是维持肠道免疫稳态的关键。通过分泌可溶性因子或直接与肠道内其他细胞的相互作用，T细胞可维持上皮层的完整性，清除已感染的细胞，促进B细胞产生IgA，甚至产生各种细胞因子。

因此，T细胞能高效有序地应对病原体[45]，对维持局部组织的内环境稳定起着至关重要的作用。在原发感染期间，无论是病毒、细菌还是寄生虫，一些记忆T细胞可长期保留在组织中，并获得特殊的表型[109, 110]，成为组织驻留记忆性T细胞（tissue- resident memory T cell，TRM）。

最近两项基于小鼠单细胞转录组学的研究表明，肠道TRM群体似乎存在相当大的异质性[111, 112]。当再次感染与上次相同的病原体后，TRM细胞能提供捷径，明显缩短机体适应性免疫应答的耗时，即由抗原呈递细胞（antigen presenting cell，APC）处理抗原，APC向次级淋巴组织迁移，T细胞识别，协同刺激与随后的激活、增殖以及效应T细胞的再循环和迁移到受感染组织[113]。与抗原结合后，TRM细胞可以直接增殖，分泌促炎细胞因子，如IFN-γ或TNF-α和趋化因子，并通过分泌颗粒酶B和穿孔素介导细胞毒性，直接清除感染细胞[114]。

潘氏细胞是高度特化的分泌上皮细胞。潘氏细胞产生的致密颗粒含有丰富的抗菌肽和免疫调节蛋白，其功能可调节肠道菌群的组成[115, 116]。多项研究表明，双硫腙诱导的潘氏细胞减少后，小肠在72小时内便会及时补充潘氏细胞群[117]。潘氏细胞来源的α-防御素在建立及维持肠道微生态的平衡方面起着重要作用。

小鼠α-防御素是一种非活化的前体肽，经过金属蛋白酶7（matrix metalloproteinase 7，MMP7）的切割后被活化。DEFA5转基因小鼠和MMP7缺失小鼠两种动物模型验证了α-防御素在肠道免疫中的相关作用。DEFA5转基因小鼠过表达人α-防御素5，而MMP7缺失小鼠不能产生有活性的α-防御素。在这两个小鼠模型中，潘氏细胞的效应因子，如溶菌酶、编码α-防御素的Defa1和Defcr4等在mRNA的表达水平以及肠菌的总数都没有明显的变化，而两组肠道菌群的组成却发生了改变。其中DEFA5转基因小鼠中厚壁菌的比例显著降低，而拟杆菌的比例显著升高，但在MMP7缺失小鼠中却得到了相反的结果。这说明α-防御素5调节肠道菌群不同菌属的组成比例，但不会影响肠道菌的总数。同时在DEFA5转基因小鼠的肠道中，几乎检测不到分节丝状菌的存在，它是厚壁菌的重要成员。此外，在DEFA5转基因小鼠中，固有层中Th17细胞的比例和数目也受到了相应的影响。这些都说明潘氏细胞来源的α-防御素能影响肠道共生菌的组成与肠道稳态。

为了使肠道达到稳态，免疫系统必须对来自食物和共生微生物的抗原保持耐受性。而肠道巨噬细胞在这一方面发挥着重要作用。它可以表现出耐受表型，通常表现为对TLR配体低反应。它还可控制效应Th细胞和Treg细胞之间的平衡，维持稳态环境。因此，巨噬细胞是缓解致病性免疫反应和维持肠道内稳态的关键细胞[118]。肠道巨噬细胞能产生多种细胞因子和其他可溶性因子以维持肠道稳态，其中包括前列腺素E2（PGE2），它可使局部的巨噬细胞刺激肠道隐窝中的上皮干细胞增殖，从而维持上皮屏障的完整性[119]。位于肌层和浆膜中的巨噬细胞可与神经系统相互作用，从而调节肠蠕动，确保食糜的运动[119]。在葡聚糖硫酸钠诱导的结肠炎模型中，上皮细胞的修复过程需要MyD88信号介导，也证明巨噬细胞对维持肠道稳态的重要性[120]。

肠黏膜还可通过黏膜免疫系统维持肠道内稳态。一方面，肠道黏膜免疫系统可保护宿主免受病原体的侵袭，并能清除有害的非自身抗原。另一方面，它又允许共生细菌在肠道内共生，且饮食中的非自身抗原可通过肠道黏膜表面进入人体。因此，维持肠道稳态需要肠道黏膜免疫网络在主动免疫（针对病原体和有害的非自身抗原）和免疫耐受（针对共生微生物和饮食抗原）之间保持动态平衡[121, 122]。肠黏膜免疫网络在复杂的外界环境中能巧妙地维持肠道内环境的稳定，其主要依赖于肠道相关淋巴组织、分泌型 IgA、抗菌肽（如防御素）、黏膜免疫细胞（如 Th1、Th2、Th17、Tfh 和 Treg 细胞）、细胞因子（如 IL-10）、趋化因子（如 CCR9）和共生细菌。

肠道内稳态主要是通过协调对病原体（如细菌、病毒和真菌）的物理、化学和免疫防御系统，以及对饮食抗原和共生细菌的免疫耐受来维持。在防御机制方面，小肠黏膜表面通过解剖学和免疫学防御机制，对病原体入侵和有害抗原进入机体具有特异性和非特异性保护作用。作为非特异性防御机制，肠上皮细胞被黏液层和抗菌肽（如防御素和溶菌酶）覆盖，由此组成的物理和化学屏障是抵御病原体入侵的第一道防线[121]。

sIgA是肠道黏膜免疫系统特有的主要抗体，在肠道黏膜组织的免疫防御中发挥着巨大作用。它的主要功能是对病原体和毒素进行特异性防御。此外，已被证实 sIgA 可建立并维持健康的肠道菌群[123]。肠道产生特异性 sIgA 的首要条件是病原体、共生菌和抗原穿过上皮黏膜屏障进入肠道相关淋巴组织，包括启动肠道抗原特异性免疫反应的中心场所——派尔集合淋巴结[121]。

IL-10 是一种重要的抗炎细胞因子，对微生物抗原的免疫应答起着负调节作用，它在维持肠道微生物免疫动态平衡方面尤为重要[124]。天然和获得性免疫系统的多数细胞均可产生 IL-10，这些细胞包括单核细胞、巨噬细胞、树突状细胞、肥大细胞、嗜酸性粒细胞、中性粒细胞、自然杀伤细胞、B 细胞和 T 细胞。胃肠道是免疫系统和外界环境之间最大的界面。由于每天暴露于食物抗原和微生物，胃肠道需要特定和高度调节的免疫反应来诱导和维持肠道内环境的动态平衡状态[125]。

IL-10 在维持肠道内稳态中起关键作用。通过使用遗传性缺乏 IL-10（$IL10^{-/-}$）或 IL-10 受体通路（$IL10rb^{-/-}$）的小鼠进行研究发现，IL-10 信号的缺失都会导致严重的自发性肠炎，尤其是结肠炎[126]。当饲养在无菌环境时，这些小鼠不会患结肠炎。抗生素治疗可减轻 $IL10^{-/-}$ 小鼠的疾病严重程度，表明肠道炎症源于对肠道微生物群的过度免疫反应[126]。自从开展对 IL-10 缺乏的小鼠的初步研究后，人们开始了解到 IL-10 促进肠道内稳态的多种机制[126]。肠道微生物群的过度免疫反应导致 IL-10 下调，从而促进 IBD 的发展[127]。IL-10 可预防无害的肠道抗原（特别是微生物群）所产生的超敏反应，从而避免肠道炎症的发生[128]。

二、肠道免疫与血糖稳态

1 型糖尿病是一种自身免疫性疾病，其发生与胰岛 B 细胞自身抗体的产生有关。胰岛自身免疫通常发生在生命早期。虽然约 60% 的 1 型糖尿病风险归因于遗传因素，但环境风险因素，如饮食、早期感染或分娩方式被认为是发病的重要因素[129]。人体胃肠道是一个富集免疫细胞的场所[8]。1 型糖尿病患者 T 细胞介导的自身免疫反应可导致胰岛 B 细胞破坏。胃肠道和胰腺之间有密切的免疫学联系，食物或肠道微生物群抗原可能会引发胰腺相关的免疫反应[130, 131]。

1 型糖尿病是一种由炎症介导的慢性代谢性疾病。肠道微生物群失调可通过对致病性 T 淋巴细胞和调节性 T 淋巴细胞的多种作用促进自身免疫性疾病的发生[45, 132, 133]。此外，肠道 T 细胞可向胰岛异常迁移，致使胰岛 B 细胞受损，从而诱发 1 型糖尿病[45, 134]。在 1 型糖尿病的动物模型和人类患者中，都发现 α4β7 整合素表达的 T 细胞浸润胰岛。此外，使用针对 α4β7 整合素或内皮标志物黏膜地址素细胞黏附分子 -1 的抗体，可显著减轻非肥胖糖尿病小鼠 1 型糖尿病的发生。这一发现表明，胰岛中的自身反应性 T 细胞可能是从肠道被诱导并迁移而来。

然而，另一项研究发现，肠道中诱导的 T 细胞也可预防 1 型糖尿病的发展。肠道诱导的 1 型调节性 T 细胞通过分泌 IL-10 迁移到胰岛，并控制疾病进展 [45]。肠道微生物群控制免疫系统的一个重要机制是通过形成短链脂肪酸（short-chain fatty acid，SCFA），如丁酸、乙酸和丙酸 [134]。大多数乙酸和丙酸产生菌属于拟杆菌门，而丁酸产生菌属于厚壁菌门，尤其是梭菌簇Ⅳ和XⅣ a。在怀孕的母鼠中，来自微生物的 SCFAs 会影响其后代的免疫系统。此外，SCFAs 可被胚胎的 G 蛋白偶联受体 41（GPR41）和 GPR43 感知，影响代谢和神经系统的产前发育，这对预防成年代谢综合征至关重要 [135]。

在 T 淋巴细胞中，SCFA 激活 G 蛋白偶联受体（GPR41/GPR43）信号通路，抑制组蛋白去乙酰化酶，并通过增强哺乳动物雷帕霉素靶蛋白（mammalian target of rapamycin，mTOR）复合物的活性诱导代谢改变，从而抑制炎症的级联反应，减少炎症细胞因子如 IL-10 和 IFN-γ 的产生 [136, 137]。近期有研究发现，新诊断的 1 型糖尿病患者不仅拥有与短链脂肪酸产生相关的独特肠道微生物群，还表现出 IgA 介导的相关免疫改变，此研究为探讨肠道菌群诱导 1 型糖尿病 IgA 免疫应答提供了新思路 [138]。

三、肠道免疫与骨代谢

骨代谢包括骨的吸收和形成，存在于人类生长发育的全过程 [139]。骨量从出生就开始增加，成年后则达到顶峰。成年人中后期骨量的维持与年龄、性别和种族等因素相关，而这些因素很大程度上又受到生活方式的影响 [140]。钙和维生素 D 的吸收对保持骨骼健康尤为重要，目前发现其他营养素也参与维持骨骼健康。例如，益生菌可降低肠道内的 pH 而促进肠道对钙的吸收 [141]。

T 淋巴细胞是一组异质性细胞，拥有丰富的细胞亚群，不同亚群可以分泌不同细胞因子。Th17 细胞是 $CD4^{+}$T 细胞的一个亚群，它在激活天然免疫机制中起重要作用，包括诱导上皮细胞产生抗菌肽、募集中性粒细胞等功能。肠道炎症激活的 Th17 细胞可向骨基质迁移，产生 IL-17，增强局部炎症，导致 TNF-α、IL-1 等炎性细胞因子升高，从而增强 NF-κB 受体活化因子配体的表达，激活破骨细胞前体细胞，促进破骨细胞分化 [142]。Th17 细胞对雌激素缺乏性骨丢失至关重要，它具有促进破骨细胞生成的作用。女性血清 IL-17 的升高与骨质疏松症密切相关 [143, 144]。因此去除 IL17 或使用抗 IL17 抗体可防止骨丢失。

另一个与骨代谢相关的 T 淋巴细胞系是调节性 T 细胞。Treg 细胞稳定存在于肠道黏膜中，对肠道和全身免疫系统均有显著影响。Treg 细胞可分泌 IL-4、IL-10 和 TGF-β，通过细胞毒性 T 淋巴细胞相关抗原 4 介导的信号通路来抑制 T 细胞激活，这样可以减少 RANKL 和其他细胞因子的表达，抑制破骨细胞的分化，减少骨吸收，促进骨形成，最终达到缓解骨质疏松的目标 [145]。此外，Treg 细胞的生成受雌激素的影响较大，雌激素能直接增加 Treg 细胞的相对数量，因此，补充雌激素能防止卵巢切除术后引起的骨丢失 [139]。

在细胞表面，肠道固有免疫系统通过 PRR 识别细菌，如 TLR 家族等信号通路。而在细胞质中，由 NLRs、NOD1 和 NOD2 负责对细菌的进一步识别。通过与细菌多肽结合后，它们能吸引一种很常见的蛋白激酶，即受体相互作用蛋白 2，该蛋白能激活 NF-κB 信号通路，从而引起趋化因子和细胞因子基因表达 [139]。NOD1 存在于多种细胞中。它主要是通过识别存在于革兰阴性菌中的肽聚糖来产生促炎信号 [146]。NOD2 广泛表达于非造血细胞、骨髓源性细胞和淋巴细胞。NOD2 可与革兰阳性和阴性细菌的所有类型的肽聚糖结合，随后激活 NLR 来诱导炎症反应。敲除 NOD1 和 NOD2 后，小鼠肠道菌群的变化对 TNF-α 和 RANKL 的表达及骨密度的变化均无影响，无菌小鼠骨量的变化取决于 NOD1 和 NOD2 这两种蛋白的作用，在 NOD2 基因缺陷的小鼠中，骨吸收和破骨细胞数量显著减少[139]。

四、肠道免疫与感染

肠黏膜是人体中表面积最大的器官，它与含有共生菌和致病微生物的肠道生态系统相互作用。由于肠道免疫系统的协调性和高效性，肠道能区别对待不同的微生物，如对共生菌产生免疫耐受，而致病微生物则被清除。肠道共生菌从出生时开始定植于人体，微生物群与肠黏膜形成互利关系，它与肠道上皮和肠道相关淋巴组织形成复杂的肠道屏障，这个庞大而多样的微生物群是肠道抵御病原细菌入侵的防御机制之一。这些共生关系与模式识别受体的存在，与肠道菌群对暴露在其中的抗原结构的免疫调节能力紧密相关[105]。

T 淋巴细胞可以通过产生增强屏障功能的细胞因子，或招募其他免疫保护和免疫调节细胞来清除被感染的细胞，从而保护宿主。Th1/Th17 所介导的免疫失调对共生菌的失调与肠道炎症性疾病有关。此外，对 Treg 反应的过度极化会减弱肠道对感染的免疫反应。鼠柠檬酸杆菌感染小鼠，是研究肠道病原体感染的主要啮齿动物模型。研究表明，在细胞外鼠柠檬酸杆菌感染的情况下，Th17 分泌的细胞因子，如 IL17A、IL17F 和 IL22 对肠道起到了保护作用[45]。此外，还有多种 T 细胞参与了肠道免疫。有研究报道，经口感染细胞内致病菌李斯特菌可诱导肠道 $CD8^{+}$T 细胞反应。同时小鼠 $CD4^{+}$T 细胞的耗竭能持续增加攻击性感染后的细菌负荷，这表明记忆性 $CD4^{+}$T 细胞在控制肠道细胞内病原体感染中起着关键作用[45]。

上皮内淋巴细胞能降解失调的肠道上皮细胞，也通过细胞因子影响健康的肠上皮细胞再生。上皮内淋巴细胞也是抵御病毒感染的前沿防线。既往研究表明，上皮内淋巴细胞在呼吸道及肠道病毒的感染过程中会升高。新生儿发生肠道病毒（柯萨奇病毒、轮状病毒和诺如病毒）感染时症状更严重，而成人肠道病毒感染中炎症反应相对较轻，这可能是由于新生儿出生时和断奶期间，上皮内淋巴细胞的数量较低，肠道的免疫屏障尚不成熟所致[147]。上皮内淋巴细胞的激活能促进肠上皮中Ⅰ/Ⅲ型干扰素受体依赖的 IFN 应答基因上调。另外，激活的上皮内淋巴细胞介质在体外能保护细胞免受病毒感染。

上皮内淋巴细胞的活化可有效抑制诺如病毒感染[27]。诺如病毒（norovirus，NoV）感染是引起散发性和流行性胃肠炎的主要病因。完整的免疫系统是有效控制 NoVs 感染的关键[148]。既往研究提示，促炎和抗炎细胞因子水平在 NoVs 感染的症状期达到顶峰，表明 NOVs 感染能迅速激活宿主免疫系统[131]。病毒进入细胞后，通常被多种宿主细胞因子感知。Toll 样受体是细胞传感机制的重要组成部分。RIG-Ⅰ和 MDA-5 等 RIG 样解旋酶位于细胞质中，它们能识别外源 RNA。已经证明，解旋酶 MDA-5 可通过诱导 IFN 产生来控制小鼠诺如病毒（murine norovirus，MNoV）感染[149]。MDA-5 的激活需要血红素氧化 IRP2 泛素连接酶 1 或 HOIL1，它是线性泛素链组装复合物的一部分，是多种免疫信号通路的重要调节因子[148]。MNoV 感染能诱导转录因子 IFN 调节因子 3（IRF-3）和 IRF-7 活化，最终诱导 I 型 IFN（IFN-α 和 IFN-β）的产生[148]。

小鼠的 IFN 反应能抑制 MNoV 的复制。虽然两种类型的 IFN 都能抑制病毒蛋白的翻译，但 IFN-γ 以双链 RNA 活化蛋白激酶依赖的方式发挥作用，而 IFN-α 则独立发挥作用。IFN-γ 能通过调节自噬蛋白和诱导干扰素诱导的 GTP 酶来破坏细胞质中的 MNoV 复合物。Ⅲ型干扰素（IFN-λ）在体内也可参与 MNoV 感染的抑制过程。即使在没有获得性免疫的情况下，腹腔注射 IFN-λ 也能清除持续的 MNoV 感染。IFN-λ 受体在肠上皮细胞上的表达是抗 NOVs 活性所必需的 [148]。

人类的获得性免疫反应对免疫记忆非常重要，有助于防止人诺如病毒再感染。人诺如病毒的保护性自身免疫的持续时间可能比最初认为的更长，可持续 1 ～ 4 年。$CD4^+$ 和 $CD8^+$ T 细胞是有效清除肠道和肠道淋巴结原发性急性 MNoV 感染所必需的细胞 [148]。

第四节　小结与展望

肠道屏障系统既要吸收营养物质，同时要识别有益或有害的信号、阻止病原体和毒素的侵袭以维持肠道内环境的稳定。肠道屏障包含黏膜机械屏障、微生物屏障、化学屏障及免疫屏障等多层结构，各屏障之间相互协作，共同维护肠道内环境稳态。肠黏膜的免疫屏障包括多种免疫细胞类型，是人体最大的免疫器官，这些免疫细胞在不同体内外信号刺激下，既能激活对病原微生物的免疫反应，同时保持对食物抗原和共生细菌的免疫耐受性。在多种因素调节下，免疫反应和免疫耐受之间保持精细的平衡状态，以维持肠道内环境及全身的免疫稳态。

针对改善肠道屏障功能的治疗能否改善胃肠道或系统性疾病的临床表现，尚未得到证实。现有的研究提示，胃肠道疾病中肠道屏障有望作为未来治疗的靶点。针对肠漏和肠道功能紊乱，补充益生菌或肠道菌群移植可能是很有潜力的治疗手段。此外，肠道免疫功能的检测仍然缺乏特异性和敏感性，更缺乏有效的评估标准。目前可用的多种测试方法有各自不同的测量终点，临床意义和相关性也尚不明确。肠道菌群和黏膜免疫系统都是肠道屏障的重要组成部分，对肠道稳态的维持有不可取代的作用。以肠道菌群为切入点的免疫调节治疗有一定的临床价值，值得进一步深入研究。

参考文献

[1] Russler-Germain EV, Rengarajan S, Hsieh CS. Antigen-specific regulatory t-cell responses to intestinal microbiota. Mucosal Immunology, 2017, 10(6): 1375-1386.

[2] Lopetuso LR, Scaldaferri F, Bruno G, et al. The therapeutic management of gut barrier leaking: The emerging role for mucosal barrier protectors. European Review for Medical and Pharmacological Sciences, 2015, 19(6): 1068-1076.

[3] Broom LJ, Kogut MH. Gut immunity: Its development and reasons and opportunities for modulation in monogastric production animals. Animal Health Research Reviews, 2018, 19(1): 46-52.

[4] Cardoso-Silva D, Delbue D, Itzlinger A, et al. Intestinal barrier function in gluten-related disorders. Nutrients, 2019, 11(10):14.

[5] Camilleri M, Lyle BJ, Madsen KL, et al. Role for diet in normal gut barrier function: Developing guidance within the framework of food-labeling regulations. American Journal of Physiology Gastrointestinal and Liver Physiology, 2019, 317(1): G17-G39.

[6] Chassaing B, Kumar M, Baker MT, et al. Mammalian gut immunity. Biomedical Journal, 2014, 37(5): 246-258.

[7] Fan H, Wang A, Wang Y, et al. Innate lymphoid cells: Regulators of gut barrier function and immune homeostasis. Journal of Immunology Research, 2019, 2019: 2525984.

[8] Endesfelder D, Engel M, Zu Castell W. Gut immunity and type 1 diabetes: A melange of microbes, diet, and host interactions? Current Diabetes Reports, 2016, 16(7): 60.

[9] Wen Z, Chen D, Bian J. Role of group 3 innate lymphoid cells in intestinal barrier. Zhonghua Wei Zhong Bing Ji Jiu Yi Xue, 2019, 31(2): 252-256.

[10] Peng J, Tang Y, Huang Y. Gut health: The results of microbial and mucosal immune interactions in pigs. Animal Nutrition, 2021, 7(2): 282-294.

[11] Chandwe K, Kelly P. Colostrum therapy for human gastrointestinal health and disease. Nutrients, 2021,

13(6):34.

[12] Suarez LJ, Arboleda S, Angelov N, et al. Oral versus gastrointestinal mucosal immune niches in homeostasis and allostasis. Frontiers in Immunology, 2021, 12: 705206.

[13] Suzuki T. Regulation of the intestinal barrier by nutrients: The role of tight junctions. Animal Science Journal, 2020, 91(1): e13357.

[14] Cui Y, Wang Q, Chang R, et al. Intestinal barrier function-non-alcoholic fatty liver disease interactions and possible role of gut microbiota. Journal of Agricultural and Food Chemistry, 2019, 67(10): 2754-2762.

[15] Garcia-Hernandez V, Quiros M, Nusrat A. Intestinal epithelial claudins: Expression and regulation in homeostasis and inflammation. Annals of the New York Academy of Sciences, 2017, 1397(1): 66-79.

[16] Grosheva I, Zheng D, Levy M, et al. High-throughput screen identifies host and microbiota regulators of intestinal barrier function. Gastroenterology, 2020, 159(5): 1807-1823.

[17] Buckley A, Turner JR. Cell biology of tight junction barrier regulation and mucosal disease. Cold Spring Harbor Perspectives in Biology, 2018, 10(1):123.

[18] Allert S, Forster TM, Svensson CM, et al. Candida albicans-induced epithelial damage mediates translocation through intestinal barriers. mBio, 2018, 9(3):44.

[19] Hensley-Mcbain T, Berard AR, Manuzak JA, et al. Intestinal damage precedes mucosal immune dysfunction in siv infection. Mucosal Immunology, 2018, 11(5): 1429-1440.

[20] Poeta M, Pierri L, Vajro P. Gut-liver axis derangement in non-alcoholic fatty liver disease. Children (Basel, Switzerland), 2017, 4(8):23.

[21] Dai X, Wang B. Role of gut barrier function in the pathogenesis of nonalcoholic fatty liver disease. Gastroenterology Research and Practice, 2015, 2015: 287348.

[22] Lopez-Santiago R, Sanchez-Argaez AB, De Alba-Nunez LG, et al. Immune response to mucosal brucella infection. Frontiers in Immunology, 2019, 10: 1759.

[23] Doulberis M, Kotronis G, Gialamprinou D, et al. Non-alcoholic fatty liver disease: An update with special focus on the role of gut microbiota. Metabolism: Clinical and Experimental, 2017, 71: 182-197.

[24] Pral LP, Fachi JL, Correa RO, et al. Hypoxia and hif-1 as key regulators of gut microbiota and host interactions. Trends in Immunology, 2021, 42(7): 604-621.

[25] Calleja-Conde J, Echeverry-Alzate V, Buhler KM, et al. The immune system through the lens of alcohol intake and gut microbiota. International Journal of Molecular Sciences, 2021, 22(14):16.

[26] Wang RX, Lee JS, Campbell EL, et al. Microbiota-derived butyrate dynamically regulates intestinal homeostasis through regulation of actin-associated protein synaptopodin. Proceedings of the National Academy of Sciences of the United States of America, 2020, 117(21): 11648-11657.

[27] Viggiano D, Ianiro G, Vanella G, et al. Gut barrier in health and disease: Focus on childhood. European Review for Medical and Pharmacological Sciences, 2015, 19(6): 1077-1085.

[28] Sun T, Nguyen A, Gommerman JL. Dendritic cell subsets in intestinal immunity and inflammation. Journal of Immunology, 2020, 204(5): 1075-1083.

[29] Hooper LV. Epithelial cell contributions to intestinal immunity. Advances in Immunology, 2015, 126: 129-172.

[30] Allaire JM, Crowley SM, Law HT, et al. The intestinal epithelium: Central coordinator of mucosal immunity. Trends in Immunology, 2018, 39(9): 677-696.

[31] Adolph TE, Mayr L, Grabherr F, et al. Paneth cells and their antimicrobials in intestinal immunity. Current pharmaceutical design, 2018, 24(10): 1121-1129.

[32] Zhou CB, Fang JY. The role of pyroptosis in gastrointestinal cancer and immune responses to intestinal microbial infection. Biochimica et Biophysica Acta Reviews on Cancer, 2019, 1872(1): 1-10.

[33] 张馨文，王海方，刘志华．潘氏细胞在肠道稳态中的作用．生命科学，2016, 28(2): 275-282.

[34] Brown EM, Sadarangani M, Finlay BB. The role of the immune system in governing host-microbe interactions in the intestine. Nature Immunology, 2013, 14(7): 660-667.

[35] Mccauley HA, Guasch G. Three cheers for the goblet cell: Maintaining homeostasis in mucosal epithelia. Trends in Molecular Medicine, 2015, 21(8): 492-503.

[36] Birchenough GM, Johansson ME, Gustafsson JK, et al. New developments in goblet cell mucus secretion and function. Mucosal Immunology, 2015, 8(4): 712-719.

[37] Corfield AP. The interaction of the gut microbiota with the mucus barrier in health and disease in human. Microorganisms, 2018, 6(3):171.

[38] Yang S, Yu M. Role of goblet cells in intestinal barrier and mucosal immunity. Journal of Inflammation Research, 2021, 14: 3171-3183.

[39] Powell DN, Swimm A, Sonowal R, et al. Indoles from the commensal microbiota act via the ahr and il-10 to tune the cellular composition of the colonic epithelium during aging. Proceedings of the National Academy of Sciences of the United States of America, 2020, 117(35): 21519-21526.

[40] Muller PA, Matheis F, Mucida D. Gut macrophages: Key players in intestinal immunity and tissue physiology. Current Opinion in Immunology, 2020, 62: 54-61.

[41] Viola MF, Boeckxstaens G. Niche-specific functional heterogeneity of intestinal resident macrophages. Gut, 2021, 70(7): 1383-1395.

[42] Mazzini E, Massimiliano L, Penna G, et al. Oral tolerance can be established via gap junction transfer of fed antigens from cx3cr1(+) macrophages to cd103(+) dendritic cells. Immunity, 2014, 40(2): 248-261.

[43] Bonnardel J, Da Silva C, Henri S, et al. Innate and adaptive immune functions of peyer's patch monocyte-derived cells. Cell Reports, 2015, 11(5): 770-784.

[44] Sehgal A, Donaldson DS, Pridans C, et al. The role of csf1r-dependent macrophages in control of the intestinal stem-cell niche. Nature Communications, 2018, 9(1): 1272.

[45] Ma H, Tao W, Zhu S. T lymphocytes in the intestinal mucosa: Defense and tolerance. Cellular & molecular Immunology, 2019, 16(3): 216-224.

[46] Chang SY, Ko HJ, Kweon MN. Mucosal dendritic cells shape mucosal immunity. Experimental & Molecular Medicine, 2014, 46: e84.

[47] Becker M, Guttler S, Bachem A, et al. Ontogenic, phenotypic, and functional characterization of xcr1(+) dendritic cells leads to a consistent classification of intestinal dendritic cells based on the expression of xcr1 and sirpalpha. Frontiers in Immunology, 2014, 5: 326.

[48] Stagg AJ. Intestinal dendritic cells in health and gut inflammation. Frontiers in Immunology, 2018, 9: 2883.

[49] Vivier E, Artis D, Colonna M, et al. Innate lymphoid cells: 10 years on. Cell, 2018, 174(5): 1054-1066.

[50] Kotas ME, Locksley RM. Why innate lymphoid cells? Immunity, 2018, 48(6): 1081-1090.

[51] Bostick JW, Zhou L. Innate lymphoid cells in intestinal immunity and inflammation. Cellular and Molecular Life Sciences, 2016, 73(2): 237-252.

[52] Almeida FF, Jacquelot N, Belz GT. Deconstructing deployment of the innate immune lymphocyte army for barrier homeostasis and protection. Immunological Reviews, 2018, 286(1): 6-22.

[53] Seillet C, Jacquelot N. Sensing of physiological regulators by innate lymphoid cells. Cellular & molecular Immunology, 2019, 16(5): 442-451.

[54] Jacquelot N, Luong K, Seillet C. Physiological regulation of innate lymphoid cells. Frontiers in Immunology, 2019, 10: 405.

[55] Taylor S, Huang Y, Mallett G, et al. Pd-1 regulates klrg1(+) group 2 innate lymphoid cells. The Journal of Experimental Medicine, 2017, 214(6): 1663-1678.

[56] Bi J, Tian Z. Nk cell dysfunction and checkpoint immunotherapy. Frontiers in Immunology, 2019, 10: 1999.

[57] Middendorp S, Nieuwenhuis EE. Nkt cells in mucosal immunity. Mucosal Immunology, 2009, 2(5): 393-402.

[58] Ma C, Han M, Heinrich B, et al. Gut microbiome-mediated bile acid metabolism regulates liver cancer via nkt cells. Science, 2018, 360(6391):eaan5931.

[59] Fuchs A, Vermi W, Lee JS, et al. Intraepithelial type 1 innate lymphoid cells are a unique subset of il-12- and il-15-responsive ifn-gamma-producing cells. Immunity, 2013, 38(4): 769-781.

[60] Artis D, Spits H. The biology of innate lymphoid cells. Nature, 2015, 517(7534): 293-301.

[61] Han L, Wang XM, Di S, et al. Innate lymphoid cells: A link between the nervous system and microbiota in intestinal networks. Mediators of Inflammation, 2019, 2019: 1978094.

[62] Bruchard M, Ghiringhelli F. Deciphering the roles of innate lymphoid cells in cancer. Frontiers in Immunology, 2019, 10: 656.

[63] Oliphant CJ, Hwang YY, Walker JA, et al. Mhcii-mediated dialog between group 2 innate lymphoid cells and cd4(+) t cells potentiates type 2 immunity and promotes parasitic helminth expulsion. Immunity, 2014, 41(2): 283-295.

[64] Wagner M, Moro K, Koyasu S. Plastic heterogeneity of innate lymphoid cells in cancer. Trends in Cancer, 2017, 3(5): 326-335.

[65] Huang Y, Mao K, Chen X, et al. S1p-dependent interorgan trafficking of group 2 innate lymphoid cells supports host defense. Science, 2018, 359(6371): 114-119.

[66] Lim AI, Li Y, Lopez-Lastra S, et al. Systemic human ilc precursors provide a substrate for tissue ilc differentiation. Cell, 2017, 168(6): 1086-1100.

[67] Colonna M. Innate lymphoid cells: Diversity, plasticity, and unique functions in immunity. Immunity, 2018, 48(6): 1104-1117.

[68] Huang Q, Cao W, Mielke LA, et al. Innate lymphoid cells in colorectal cancers: A double-edged sword.

Frontiers in Immunology, 2019, 10: 3080.

[69] Simoni Y, Fehlings M, Kloverpris HN, et al. Human innate lymphoid cell subsets possess tissue-type based heterogeneity in phenotype and frequency. Immunity, 2018, 48(5): 1060.

[70] De Vries NL, Van Unen V, Ijsselsteijn ME, et al. High-dimensional cytometric analysis of colorectal cancer reveals novel mediators of antitumour immunity. Gut, 2020, 69(4): 691-703.

[71] Vivier E, Tomasello E, Baratin M, et al. Functions of natural killer cells. Nature Immunology, 2008, 9(5): 503-510.

[72] Martinet L, Smyth MJ. Balancing natural killer cell activation through paired receptors. Nature Reviews Immunology, 2015, 15(4): 243-254.

[73] Martinet L, Ferrari De Andrade L, Guillerey C, et al. Dnam-1 expression marks an alternative program of nk cell maturation. Cell Reports, 2015, 11(1): 85-97.

[74] Yokoyama WM, Plougastel BF. Immune functions encoded by the natural killer gene complex. Nature Reviews Immunology, 2003, 3(4): 304-316.

[75] Huntington ND, Nutt SL, Carotta S. Regulation of murine natural killer cell commitment. Frontiers in Immunology, 2013, 4: 14.

[76] Nicholson SE, Keating N, Belz GT. Natural killer cells and anti-tumor immunity. Molecular Immunology, 2019, 110: 40-47.

[77] Klose CSN, Flach M, Mohle L, et al. Differentiation of type 1 ilcs from a common progenitor to all helper-like innate lymphoid cell lineages. Cell, 2014, 157(2): 340-356.

[78] Robinette ML, Fuchs A, Cortez VS, et al. Transcriptional programs define molecular characteristics of innate lymphoid cell classes and subsets. Nature Immunology, 2015, 16(3): 306-317.

[79] Bernink JH, Peters CP, Munneke M, et al. Human type 1 innate lymphoid cells accumulate in inflamed mucosal tissues. Nature Immunology, 2013, 14(3): 221-229.

[80] Mjosberg J, Bernink J, Golebski K, et al. The transcription factor gata3 is essential for the function of human type 2 innate lymphoid cells. Immunity, 2012, 37(4): 649-659.

[81] Hoyler T, Klose CS, Souabni A, et al. The transcription factor gata-3 controls cell fate and maintenance of type 2 innate lymphoid cells. Immunity, 2012, 37(4): 634-648.

[82] Ikutani M, Yanagibashi T, Ogasawara M, et al. Identification of innate il-5-producing cells and their role in lung eosinophil regulation and antitumor immunity. Journal of Immunology, 2012, 188(2): 703-713.

[83] Bie Q, Zhang P, Su Z, et al. Polarization of ilc2s in peripheral blood might contribute to immunosuppressive microenvironment in patients with gastric cancer. Journal of Immunology Research, 2014, 2014: 923135.

[84] Forkel M, Van Tol S, Hoog C, et al. Distinct alterations in the composition of mucosal innate lymphoid cells in newly diagnosed and established crohn's disease and ulcerative colitis. Journal of Crohn's & colitis, 2019, 13(1): 67-78.

[85] Zhou Y, Ji Y, Wang H, et al. Il-33 promotes the development of colorectal cancer through inducing tumor-infiltrating st2l(+) regulatory t cells in mice. Technology in Cancer Research & Treatment, 2018, 17: 1533033818780091.

[86] Barderas R, Bartolome RA, Fernandez-Acenero MJ, et al. High expression of il-13 receptor alpha2 in colorectal cancer is associated with invasion, liver metastasis, and poor prognosis. Cancer Research, 2012, 72(11): 2780-2790.

[87] Saigusa S, Tanaka K, Inoue Y, et al. Low serum interleukin-13 levels correlate with poorer prognoses for colorectal cancer patients. International Surgery, 2014, 99(3): 223-229.

[88] Carretero R, Sektioglu IM, Garbi N, et al. Eosinophils orchestrate cancer rejection by normalizing tumor vessels and enhancing infiltration of cd8(+) t cells. Nature Immunology, 2015, 16(6): 609-617.

[89] Qiu J, Zhou L. Aryl hydrocarbon receptor promotes rorgammat(+) group 3 ilcs and controls intestinal immunity and inflammation. Seminars in Immunopathology, 2013, 35(6): 657-670.

[90] Brennan CA, Garrett WS. Fusobacterium nucleatum-symbiont, opportunist and oncobacterium. Nature Reviews Microbiology, 2019, 17(3): 156-166.

[91] Hu L, Liu Y, Kong X, et al. Fusobacterium nucleatum facilitates m2 macrophage polarization and colorectal carcinoma progression by activating tlr4/nf-kappab/s100a9 cascade. Frontiers in Immunology, 2021, 12: 658681.

[92] Chen T, Li Q, Wu J, et al. Fusobacterium nucleatum promotes m2 polarization of macrophages in the microenvironment of colorectal tumours via a tlr4-dependent mechanism. Cancer Immunology Immunotherapy, 2018, 67(10): 1635-1646.

[93] Viant C, Rankin LC, Girard-Madoux MJ, et al. Transforming growth factor-beta and notch ligands

act as opposing environmental cues in regulating the plasticity of type 3 innate lymphoid cells. Science Signaling, 2016, 9(426): ra46.

[94] Cella M, Gamini R, Secca C, et al. Subsets of ilc3-ilc1-like cells generate a diversity spectrum of innate lymphoid cells in human mucosal tissues. Nature Immunology, 2019, 20(8): 980-991.

[95] Chan IH, Jain R, Tessmer MS, et al. Interleukin-23 is sufficient to induce rapid de novo gut tumorigenesis, independent of carcinogens, through activation of innate lymphoid cells. Mucosal Immunology, 2014, 7(4): 842-856.

[96] Chen L, He Z, Slinger E, et al. Il-23 activates innate lymphoid cells to promote neonatal intestinal pathology. Mucosal Immunology, 2015, 8(2): 390-402.

[97] Kirchberger S, Royston DJ, Boulard O, et al. Innate lymphoid cells sustain colon cancer through production of interleukin-22 in a mouse model. The Journal of Experimental Medicine, 2013, 210(5): 917-931.

[98] Cui G, Yang H, Zhao J, et al. Elevated proinflammatory cytokine il-17a in the adjacent tissues along the adenoma-carcinoma sequence. Pathology Oncology Research, 2015, 21(1): 139-146.

[99] Lin Y, Xu J, Su H, et al. Interleukin-17 is a favorable prognostic marker for colorectal cancer. Clinical & Translational Oncology, 2015, 17(1): 50-56.

[100] Liu J, Duan Y, Cheng X, et al. Il-17 is associated with poor prognosis and promotes angiogenesis via stimulating vegf production of cancer cells in colorectal carcinoma. Biochemical and Biophysical Research Communications, 2011, 407(2): 348-354.

[101] Cua DJ, Tato CM. Innate il-17-producing cells: The sentinels of the immune system. Nature Reviews Immunology, 2010, 10(7): 479-489.

[102] Cui G, Yuan A, Goll R, et al. Il-17a in the tumor microenvironment of the human colorectal adenoma-carcinoma sequence. Scandinavian Journal of Gastroenterology, 2012, 47(11): 1304-1312.

[103] Maxwell JR, Zhang Y, Brown WA, et al. Differential roles for interleukin-23 and interleukin-17 in intestinal immunoregulation. Immunity, 2015, 43(4): 739-750.

[104] Lee JS, Tato CM, Joyce-Shaikh B, et al. Interleukin-23-independent il-17 production regulates intestinal epithelial permeability. Immunity, 2015, 43(4): 727-738.

[105] Iacob S, Iacob DG. Infectious threats, the intestinal barrier, and its trojan horse: Dysbiosis. Frontiers in Microbiology, 2019, 10: 1676.

[106] Fekete E, Allain T, Siddiq A, et al. Giardia spp. And the gut microbiota: Dangerous liaisons. Frontiers in Microbiology, 2020, 11: 618106.

[107] Kinashi Y, Hase K. Partners in leaky gut syndrome: Intestinal dysbiosis and autoimmunity. Frontiers in Immunology, 2021, 12: 673708.

[108] Lei-Leston AC, Murphy AG, Maloy KJ. Epithelial cell inflammasomes in intestinal immunity and inflammation. Frontiers in Immunology, 2017, 8: 1168.

[109] Romagnoli PA, Sheridan BS, Pham QM, et al. Il-17a-producing resident memory gammadelta t cells orchestrate the innate immune response to secondary oral listeria monocytogenes infection. Proceedings of the National Academy of Sciences of the United States of America, 2016, 113(30): 8502-8507.

[110] Steinfelder S, Rausch S, Michael D, et al. Intestinal helminth infection induces highly functional resident memory cd4(+) t cells in mice. European Journal of Immunology, 2017, 47(2): 353-363.

[111] Milner JJ, Toma C, He Z, et al. Heterogenous populations of tissue-resident cd8(+) t cells are generated in response to infection and malignancy. Immunity, 2020, 52(5): 808-824.

[112] Kurd NS, He Z, Louis TL, et al. Early precursors and molecular determinants of tissue-resident memory cd8(+) t lymphocytes revealed by single-cell rna sequencing. Sci Immunology, 2020, 5(47):231.

[113] Paap EM, Muller TM, Sommer K, et al. Total recall: Intestinal trm cells in health and disease. Frontiers in Immunology, 2020, 11: 623072.

[114] Bartolome-Casado R, Landsverk OJB, Chauhan SK, et al. Cd4(+) t cells persist for years in the human small intestine and display a th1 cytokine profile. Mucosal Immunology, 2021, 14(2): 402-410.

[115] Lueschow SR, Mcelroy SJ. The paneth cell: The curator and defender of the immature small intestine. Frontiers in Immunology, 2020, 11: 587.

[116] Nikolenko VN, Oganesyan MV, Sankova MV, et al. Paneth cells: Maintaining dynamic microbiome-host homeostasis, protecting against inflammation and cancer. Bioessays, 2021, 43(3): e2000180.

[117] Lueschow SR, Stumphy J, Gong H, et al. Loss of murine paneth cell function alters the immature intestinal microbiome and mimics changes seen in neonatal necrotizing enterocolitis. PloS One, 2018, 13(10): e0204967.

[118] Moreira Lopes TC, Mosser DM, Goncalves R. Macrophage polarization in intestinal inflammation

and gut homeostasis. Inflammation Research, 2020, 69(12): 1163-1172.

[119] 范慧宁，张靖，朱金水．巨噬细胞在肠道稳态及炎性反应中的作用．国际消化病杂志．2019, 39(01): 21-25.

[120] Malvin NP, Seno H, Stappenbeck TS. Colonic epithelial response to injury requires myd88 signaling in myeloid cells. Mucosal Immunology, 2012, 5(2): 194-206.

[121] Tokuhara D, Kurashima Y, Kamioka M, et al. A comprehensive understanding of the gut mucosal immune system in allergic inflammation. Allergol International, 2019, 68(1): 17-25.

[122] Takahashi D, Kimura S, Hase K. Intestinal immunity: To be, or not to be, induced? That is the question. International Immunology, 2021,12:111-113.

[123] Mathias A, Pais B, Favre L, et al. Role of secretory iga in the mucosal sensing of commensal bacteria. Gut microbes, 2014, 5(6): 688-695.

[124] Nguyen HD, Aljamaei HM, Stadnyk AW. The production and function of endogenous interleukin-10 in intestinal epithelial cells and gut homeostasis. Cellular and Molecular Gastroenterology and Hepatology, 2021, 12(4):1343-1352.

[125] Rutz S, Ouyang W. Regulation of interleukin-10 expression. Advances in Experimental Medicine and Biology, 2016, 941: 89-116.

[126] Neumann C, Scheffold A, Rutz S. Functions and regulation of t cell-derived interleukin-10. Seminars in Immunology, 2019, 44: 101344.

[127] Fay NC, Muthusamy BP, Nyugen LP, et al. A novel fusion of il-10 engineered to traffic across intestinal epithelium to treat colitis. Journal of Immunology, 2020, 205(11): 3191-3204.

[128] Kole A, Maloy KJ. Control of intestinal inflammation by interleukin-10. Current Topics in Microbiology and Immunology, 2014, 380: 19-38.

[129] Knip M, Siljander H. The role of the intestinal microbiota in type 1 diabetes mellitus. Nature Reviews Endocrinology, 2016, 12(3): 154-167.

[130] Paun A, Yau C, Danska JS. Immune recognition and response to the intestinal microbiome in type 1 diabetes. Journal of Autoimmunity, 2016, 71: 10-18.

[131] Cutler AJ, Oliveira J, Ferreira RC, et al. Capturing the systemic immune signature of a norovirus infection: An n-of-1 case study within a clinical trial. Wellcome Open Research, 2017, 2: 28.

[132] Davis-Richardson AG, Triplett EW. A model for the role of gut bacteria in the development of autoimmunity for type 1 diabetes. Diabetologia, 2015, 58(7): 1386-1393.

[133] Durazzo M, Ferro A, Gruden G. Gastrointestinal microbiota and type 1 diabetes mellitus: The state of art. Journal of Clinical Medicine, 2019, 8(11):12-15.

[134] Luu M, Visekruna A. Short-chain fatty acids: Bacterial messengers modulating the immunometabolism of t cells. European Journal of Immunology, 2019, 49(6): 842-848.

[135] Kimura I, Miyamoto J, Ohue-Kitano R, et al. Maternal gut microbiota in pregnancy influences offspring metabolic phenotype in mice. Science, 2020, 367(6481):44.

[136] Gurung M, Li Z, You H, et al. Role of gut microbiota in type 2 diabetes pathophysiology. eBio Medicine, 2020, 51: 102590.

[137] Sroka-Oleksiak A, Mlodzinska A, Bulanda M, et al. Metagenomic analysis of duodenal microbiota reveals a potential biomarker of dysbiosis in the course of obesity and type 2 diabetes: A pilot study. Journal of Clinical Medicine, 2020, 9(2):127.

[138] Huang J, Pearson JA, Peng J, et al. Gut microbial metabolites alter iga immunity in type 1 diabetes. The Journal of Clinical Investigation, 2020, 5(10):12-14.

[139] Li L, Rao S, Cheng Y, et al. Microbial osteoporosis: The interplay between the gut microbiota and bones via host metabolism and immunity. Microbiology Open, 2019, 8(8): e00810.

[140] Weaver CM, Gordon CM, Janz KF, et al. The national osteoporosis foundation's position statement on peak bone mass development and lifestyle factors: A systematic review and implementation recommendations. Osteoporosis International, 2016, 27(4): 1281-1386.

[141] Wallace TC, Marzorati M, Spence L, et al. New frontiers in fibers: Innovative and emerging research on the gut microbiome and bone health. Journal of the American College of Nutrition, 2017, 36(3): 218-222.

[142] Li C, Pi G, Li F. The role of intestinal flora in the regulation of bone homeostasis. Frontiers in Cellular and Infection Microbiology, 2021, 11: 579323.

[143] Molnar I, Bohaty I, Somogyine-Vari E. Il-17a-mediated srank ligand elevation involved in postmenopausal osteoporosis. Osteoporosis International, 2014, 25(2): 783-786.

[144] Zhang J, Fu Q, Ren Z, et al. Changes of serum cytokines-related th1/th2/th17 concentration in patients with postmenopausal osteoporosis.

Gynecological Endocrinology, 2015, 31(3): 183-190.

[145] Hsu E, Pacifici R. From osteoimmunology to osteomicrobiology: How the microbiota and the immune system regulate bone. Calcified Tissue International, 2018, 102(5): 512-521.

[146] Clarke TB, Davis KM, Lysenko ES, et al. Recognition of peptidoglycan from the microbiota by nod1 enhances systemic innate immunity. Nature Medicine, 2010, 16(2): 228-231.

[147] Williams AM, Probert CS, Stepankova R, et al. Effects of microflora on the neonatal development of gut mucosal t cells and myeloid cells in the mouse. Immunology, 2006, 119(4): 470-478.

[148] Hassan E, Baldridge MT. Norovirus encounters in the gut: Multifaceted interactions and disease outcomes. Mucosal Immunology, 2019, 12(6): 1259-1267.

[149] Mccartney SA, Thackray LB, Gitlin L, et al. Mda-5 recognition of a murine norovirus. PLoS Pathogens, 2008, 4(7): e1000108.

第 4 章
肠道菌群与机体的跨界信号交流

第一节　跨界信号交流概述

细胞间的化学通信可协调它们的行为。在原核生物中，这种化学通信被称为群体感应；而真核细胞通过激素进行信号交流。研究发现，细菌可以产生包括群体感应信号在内的化学分子，向真核细胞传递信息。而宿主也可以产生激素等物质向细菌细胞发送信号。这种跨物种的信号交流可介导细菌、哺乳动物和植物宿主之间的共生和致病关系，称为跨界信号交流。

细菌与其宿主共同进化了几十亿年，通过这种进化，它们相互暴露于彼此产生和释放的信号分子。因此，不难理解微生物与宿主之间存在跨界信号交流过程[1]。本章节中，我们把微生物与真核生物（尤其是哺乳动物）间的跨界信号交流进行归纳总结，大致可分为以下几种方式：利用群体感应系统进行跨界信号交流，利用激素进行跨界信号交流，通过细胞胞外囊泡进行跨界信号交流，以及其他信号分子如吲哚等进行跨界信号交流。

第二节　细菌与真核生物通过群体感应系统对话

一、肠道菌群的群体感应系统

（一）群体感应系统概述

细菌生活在多种微生物群落中，由于单个细菌细胞的作用是微弱的，因而细菌进化出可以协同作用的机制，使其作为一个整体发挥远超单个个体的作用。就像我们人类使用语言进行交流一样，群居的细菌也可以通过分泌和感知特定的物质和信号分子，与生物和非生物因子相互作用[2]。Greenberg 和他的同事们称这种细菌间的通信过程为群体感应（quorum sensing，QS），又称密度感应，是一种细胞与细胞间的通信系统，既可调节相同种群内细菌间的交流，也可调节不同种群细菌间的交流[3]。群体感应系统通过合成称为自体诱导物（autoinducers，AIs）的小分子而起作用，这些分子在细胞内合成（贯穿细菌的整个生长周期）并释放到周围环境中，通过一系列信号转导机制来调控细菌种群的行为。AIs 在群体感应中起着至关重要的作用，它是细菌在群体范围内交流和同步特定行为的信号，从而使细菌作为一个整体而不是单一个体发挥作用[4]。一般来说，群体感应调控细菌致病性和毒力大致包括以下步骤：①群体感应信号分子的合成；②信号分子释放到环境中；③高细胞密度下感知信号分子并与膜受体结合；④受体激活后形成复合物与致病性基因启动子区结合；⑤调控致病性基因的转录。细菌产生并释放 AIs，其浓度随着特定部位细胞种群密度的增加而升高，当信号分子浓度随着细菌种群密度达到一定阈值时，细菌细胞内的受体与 AIs 结合，触发信号转导级联，导致群体范围内基因表达的变化[5]，进而调节细菌生物学行为，如细菌发光、毒力因子表达、消毒剂耐受性、孢子形成、毒素产生、运动、生物膜形成和药物抵抗等[6]。无论是革兰阴性菌还是革兰阳性菌，都存在群体感应现象，然而其传递信息的信号分子是不同的。一般来说，各种 AIs 大致可分为以下几种类型：①革兰阴性菌普遍利用的酰化高丝氨酸内酯（acylated homoserine lactones，AHLs），也称 AI-1；②革兰阳性菌利用自诱导肽（auto-inducing peptides，AIPs）；③用于微生物群落种间通信的 AI-2；④出血性大肠埃希菌中发现的 AI-3[3, 7]。

（二）革兰阴性细菌 AHLs 群体感应系统

革兰阴性菌普遍使用 AHLs 作为信号分子。AHLs 由一个含有 4～18 个碳的酰基侧链和一个高丝氨酸内酯环构成[8]。在革兰阴性细菌中，不同细菌甚至是相同细菌由于不同的合成系统，合成的 AHLs 结构也有所不同。分子结构的差异取决于酰基侧链上碳（4～18 个）的数量，第三碳上取代氢原子的取代基的差异（-H，-OH 或 -oxo），以及酰基侧链中是否存在不饱和双键[9]。这些差异导致 AHLs 分子结构和分泌途径的多样性。短侧链 AHL（酰基侧链上碳原子＜8 个，C4-8-HSL）可直接穿透细胞膜并在合成后释放到周围环境中，而长侧链 AHL（酰基侧链上碳原子＞8 个，C10-18-HSL）只能通过活性外排途径释放，例如 3-oxo-C12-HSL 通过活性 mexAB-oprM 编码的 mexAB-oprM 泵从膜中导出[10]。

AHL 信号分子由 LuxI 基因及其同源基因编码的产物催化合成，它以 S- 腺苷蛋氨酸（S-adenosyl methionine，SAM）和酰基 - 酰基载体蛋白（acyl-ACP）作为底物合成。LuxR 基因编码 LuxR 结合蛋白，是一种诱导应答的转录激活因子。AHL 与 LuxR 结合后，发生二聚或多聚，并且多聚产物与靶基因上游调控区结合，激活或抑制靶基因的表达[2, 11]。LuxI/LuxR 系统是革兰阴性菌中常见的群体感应调控系统，最早是在一种名为哈维氏弧菌（*Vibrio fischeri*）的海洋菌的生物发光现象中发现的，并且只在高密度菌群时发光而在低密度时不发光，该系统是目前研究最为深入和广泛的一类群体感应系统。该系统具有物种特异性，介导相同细菌种群内的交流，并且能调节多种细菌的各种表型特征，如生物发光、运动、生物膜形成、毒力等。这些机制在铜绿假单胞菌（*Pseudomonas aeruginosa*）中研究得较为深入，AHLs 参与铜绿假单胞菌感染过程及调节定植黏附能力等。

（三）革兰阳性细菌 AIPs 群体感应系统

以小分子寡肽为代表的信号分子，即 AIPs，是革兰阳性菌的主要群体感应信号分子。AIP 为 5～87 个氨基酸组成的线性或环状肽[12]。AIP 前体分子通常由先导肽修饰而成，它可以通过 ATP 结合盒转运蛋白在体内进行运输，并促进其分泌到细胞外[13]。当外界分泌的信号分子 AIPs 达到一定浓度时，便与细胞膜上的跨膜双组分信号转导系统（two component signal transduction system，TXSTS）组氨酸传感器激酶伴侣受体蛋白结合[14]，通过磷酸化 / 去磷酸化级联反应，将 AIPs 传递到细胞内并结合启动子，启动转录和翻译后修饰，调节基因表达[15]。

该群体感应系统中部分细菌应用另一种信号传导方式，即 RRNPP 系统，与双组分系统一样，细菌合成前肽，加工成熟的 AIP 通过该系统从细胞中输出。然而，在该系统中，AIP 不与膜结合受体结合，而是被信号接收细胞导入胞内，与细胞内受体 / 反应调节器形成复合物，从而调节相关基因[16]。

（四）AI-2 群体感应系统

以 AI-2 为代表的这种信号分子，是一种呋喃酰硼酸二酯化合物，为非特异性自体诱导物，是介导革兰阴性菌和革兰阳性菌之间种内和种间通信的一种“通用”信号[17-19]。细菌不仅可以接收同种细菌释放的 AI-2 信号，还可以接收不同种群细菌释放的 AI-2 信号，并调节相应基因表达。目前已知超过 80 余种革兰阳性及革兰阴性细菌可以产生 AI-2 信号分子[20, 21]。AI-2 为具有不同构象的类似小分子的混合物，迄今为止，已发现两种 AI-2 信号分子的结构：（2S，4S）-2- 甲基 -2，3，3，4- 四氢呋喃硼酸酯和（2R，4S）-2- 甲基 -2，3，3，4- 四氢呋喃[18]。基于 AI-2 的群体感应系统最早是 20 世纪 90 年代在革兰阴性杆菌 *V. harveyi* 中发现的[22]。

迄今为止，发现了 AI-2 合成的两种途径。第一，经典 AI-2 合成路径。依赖于细菌 LuxS 基因编码合成的 LuxS 蛋白酶，具体过程如下：主要由 S- 腺苷蛋氨酸通过三个酶促步骤产生：①以 SAM 作为甲基供体产生中间产物 S-腺苷同型半胱氨酸（S-adenosylhomocysteine，SAH）；② SAH 被核苷酶 Pfs 水解为 S- 核糖同型半胱氨酸（S-ribosylhomocysteine，SRH）和腺嘌呤；③由 LuxS 基因编码合成的 LuxS 蛋白酶催化 SRH 裂解为 4，5- 二羟基 2，3- 戊二酮（4，5-dihydroxy-2，3-pentanedione，DPD）和同型半胱氨酸，其中 DPD 不稳定，为 AI-2 的前体物质，可发生自身环化现象，经过化学键重排作用，形成有活性的 AI-2 信号分子[23]。第二，除了公认的合成途径外，在嗜热菌中还发现了另一种生物合成途径：在没有 luxs 而只有 Pfs 的情况下，嗜热菌利用 SAH 水解酶将 SAH 裂解为腺苷和同型半胱氨酸。腺苷被核苷磷酸化酶转化为核糖 -1- 磷酸（核糖 -1-P），然后被磷酸糖突变酶异构化为核糖 -5- 磷酸（核糖 -5-P）。最后，核糖 -5-P 通过热诱导转化为 DPD 和 AI-2[24]。

AI-2 信号分子的浓度同样随细菌密度增加而升高，当体外浓度达到阈值时，则发生内化，进而启动一系列信号转导机制。目前发现了 3 种 AI-2 受体，包括弧菌 LuxP 蛋白、大肠埃希菌和鼠伤寒沙门菌 LsrB 蛋白，以及放线菌群 RbsB 蛋白，表明细菌具有识别 AI-2 分子的不同受体。

AI-2 与 LuxP 的结合包含由两种蛋白质形成的复合物，即：与 AI-2 结合的周质蛋白 LuxP 和传感器激酶 / 磷酸酶 LuxQ[25]。AI-2 与 LuxP 和 LuxQ 蛋白复合物结合后，引发了与 LuxQ 蛋白和磷酸化 LuxU 蛋白相关的去磷酸化级联反应，导致发光调节蛋白 LuxO 的去磷酸化。去磷酸化的 LuxO 进而激活转录激活因子 LuxR，而激活的 LuxR 启动操纵子 luxCDABEGH 的转录，从而诱导生物发光[26-28]。

在弧菌属、大肠埃希菌和沙门菌等细菌中还发现了 LuxP 的同系物，包括 LuxS 调节蛋白 B（LsrB）[20, 29]。在这个系统中，LsrB 主要与 AI-2 内化有关。AI-2 与受体蛋白 LsrB 结合后内化，进而被 AI-2 激酶（LsrK）磷酸化，磷酸化的 AI-2 与转录调节蛋白 LsrR 结合，启动 LsrACDBFGE 操纵子的转录，进而加速 AI-2 的摄取[30, 31]。尽管 AI-2 在群体感应系统中的作用有待进一步研究，但目前研究表明，AI-2 介导的群体感应途径参与了多种生理功能，包括细胞信号转导、代谢、毒性因子释放、应激反应等[32]。另一重要作用是，AI-2 能影响肠道主要菌群的平衡，促进肠道内稳态的恢复[33]。

由于 AI-2 具有同时介导革兰阳性及革兰阴性菌群体感应的这种通用性，针对 AI-2 的研究也越来越多。目前 AI-2 的检测有以下几种方法。①生物工程菌株：常作为敏感、快速检测自体诱导物的一种方法。迄今为止，LuxN（AHL 受体编码基因）突变的 *V. harveyi* BB170 和 LuxN+LuxS 突变的 *V. harveyi* MM32 已被用作 AI-2 的报告菌株，其中 *V. harveyi* BB170

使用更为广泛[34]。②基于荧光共振能量转移（fluorescence resonance energy transfer，FRET）的生物感受系统：基于 FRET 的生物传感系统由融合蛋白质 LuxP-EGFP（增强绿色荧光蛋白）和荧光团 7- 二乙氨基 -3-[N-（2- 马来酰亚胺乙基）氨基甲酰] 香豆素 {7-diethylamino-3-[N-（2-maleimidoethyl）carbamoyl]coumarin，MDCC} 组成。MDCC 的激发触发 FRET 和 EGFP 荧光。荧光发射强度随 AI-2 浓度的增加而降低。在所有检测方法中，该系统对 AI-2 的检测灵敏度最高，可用于临床[35]。③荧光生物传感器：荧光染料修饰的 AI-2 受体蛋白 LuxP 和 LsrB 对 BAI-2 和 AI-2 的结合分别表现出明显的荧光变化。与工程菌 *V. harveyi* 相比，该方法具有检测速度快、检测范围宽、不受糖和 DPD 类似物干扰等优点。此外，还可用于实时监测 AI-2 的产生，这有助于揭示不同物种 AI-2 产生的精确时间变化和差异[36]。④液相色谱法和气相色谱法：采用气相色谱 - 质谱联用法对生物样品中的 DPD 或 AI-2 进行分析[37]。

（五）AI-3 群体感应系统

在研究肠出血性大肠埃希菌（*enterohaemorrhagic E.coli*，EHEC）毒力基因表达的过程中，细胞间通信涉及一种新的自体诱导物，称为 AI-3。AI-3 系统最初是在被革兰阴性菌感染的动物肠组织中发现的。在肠出血性大肠埃希菌感染发病机制研究中，发现了一种芳香化合物，其结构未知，为 EHEC 的代谢产物。并证明 AI-3 及其类似物来源于苏氨酸脱氢酶产物和部分 tRNA 合成酶反应，它们分布在多种革兰阴性和革兰阳性细菌中[38]。该系统的诸多功能仍在探索中。

AI-3 信号分子与肾上腺素 / 去甲肾上腺素具有交叉作用，肾上腺素 / 去甲肾上腺素 /AI-3 信号常作为微生物与其宿主之间的跨界化学信号交流系统。这些信号由双组分系统 QseBC 感知，而细菌膜受体 QseC 则通过自身磷酸化和磷酸化 QseB 反应调节器启动复杂的磷酸化信号级联，激活第二个双组分系统 QseE/QseF 的表达。QseE/QseF 双组分系统也参与毒力基因的表达，它能感应肾上腺素、磷酸盐和硫酸盐[39, 40]。QseC 通过直接与这些信号结合，特异性地感知 AI-3、肾上腺素和去甲肾上腺素。

Sperandio 等的研究表明，肠出血性大肠埃希菌能感知胃肠道微生物产生的 AI-3，从而激活由肠细胞清除位点（locus of enterocyte effacement，LEE）编码的毒力基因表达。致病岛 LEE 编码一个Ⅲ型分泌系统，该系统向真核宿主细胞内注射细菌的效应器。LEE 基因的表达导致“黏附和擦拭性损伤”（attaching and effacing，AE）病变的形成。AE 病变的特征是微绒毛的破坏和肌动蛋白细胞骨架的重新排列，从而形成吸附细菌的基座[41]。

二、群体感应系统对菌群的调控作用

（一）调节毒力

细菌毒力的关键调控中心便是群体感应，这在抗细菌毒力治疗中研究较为深入[42]。过去几十年的研究表明，在植物、动物和人类的细菌病原体中，部分毒力因子的产生是由群体感应控制的。

铜绿假单胞菌是一种人类机会性病原体，能够引起严重的多重耐药感染[43]。群体感应在该菌的毒力中起着关键作用，鉴于该细菌作为人类病原体的重要性，铜绿假单胞菌已成为群体感应研究的模式生物之一[44]。转录组研究表明，LasI/LasR 和 RhlI/RhlR 系统控制了铜绿假单胞菌近 10% 的基因组的表达，其中 254 个基因（包括几个毒力基因）是由 AHLs 诱导的[45]。铜绿假单胞菌群体感应信号影响的毒力因子包括 LasA 蛋白酶（破坏上皮屏障；由 3-oxo-C12-HSL 控制）、LasB 弹性蛋白酶（降解基质蛋白，如胶原；由 3-oxo-C12-HSL- 和 BHL 控制）、碱性蛋白酶（降解宿主防御蛋白；由 3-oxo-C12-HSL 控制）、鼠李糖脂（引起免疫细胞坏死；由 BHL 控制）、绿脓杆菌素（参与免疫逃避；由 3-oxo-C12-HSL-、BHL- 和 PQS 控制）和 LecA 凝集素（促进定植；由 PQS 控制）[44]。

哈维氏弧菌是水生动物的主要病原体，对全球水产养殖业造成重大损失[46]。哈维氏弧菌包含 3 个群体感应系统，有 3 种不同类型的信号分子：HAI-1、AI-2 和 CAI-1；能够控制主要调节因子 LuxR 和 AphA 产生的信号转导级联反应。哈维氏弧菌的群体感应系统可控制不同毒力相关表型的表达，包括Ⅲ型分泌系统[47]、铁载体[48]、几丁质酶[49]、磷脂酶[50]、vhp 金属蛋白酶[51]和鞭毛运动[52]。

单核细胞增生李斯特菌，简称单增李斯特菌，是一种普遍存在的食源性病原体，可引起动物和人类李斯特菌病。单增李斯特菌分泌的 AIs 分子可被群体中的相邻细胞感应到。研究证实，依赖 AgrBDCA 操纵子调控的群体感应机制与单增李斯特菌的生物膜形成、侵袭、毒力及多种基因表达有关。ΔAgrD 突变细菌的生物膜形成显著减少，侵入结肠上皮细胞 Caco-2 减少，细胞壁内质素 A 水平降低，毒力基因（hlyA、actA、plcA、prfA 和 inlA）的表达发生改变导致 ΔAgrD 突变体的毒力受损[53]。

（二）影响生物膜形成

最近研究表明，人类细菌感染大多与生物膜有关，生物膜的形成是临床细菌感染难以治愈和复发的重要原因之一[54]。因此了解细菌生物膜的形成过程有助于我们对临床细菌感染采取新的治疗措施。细菌附着在惰性或有生命的物体表面，一起形成生物膜[55]。据目前的认识，生物膜是地球上细菌生命的主要形式[56]。生物膜由细菌及其包裹细菌多糖、蛋白质和细胞外 DNA 组成的细胞外基质构成[57, 58]。细菌生物膜的形成一般分为 3 个阶段[59]。①黏附：生物膜形成的初始步骤便是细菌附着在物体表面[60]。此时，细菌可利用其附属物（如鞭毛）附着到物体表面，同时也可利用物理作用，如范德华力 / 静电相互作用等附着到物体表面[61]。②成熟：细菌黏附于物体表面后，接下来就是生物膜的成熟过程，开始分泌细胞外多糖物质（extracellular polysaccharide substance，EPS），EPS 是生物膜中的主要组分[62]。成熟生物膜的结构可以是扁平的或均匀的，也可以是高度结构化的，其特征是在细胞外基质中有空隙和细胞塔[63]。③聚集、溶解或分散：各种细菌的单个细胞能够主动离开生物膜。据推测，这种扩散过程可以使细菌在新的表面定植并重新启动生物膜的发育过程[64, 65]。

生物膜中细菌群落释放多种细胞间和细胞内信号分子，直接影响生物膜的种群和动态结构[66]，这些生物活性物质包括自体诱导物、D-氨基酸和代谢物的小信号分子等[67]。生物膜的形成涉及数百个生物膜相关基因的调控，包括与应激反应、群体感应、运动、细胞表面附属物、代谢和运输有关的基因[68]。在生物膜形成过程中，群体感应系统起着重要的调控作用，常通过调控细菌鞭毛、胞外多糖、黏附素等物质的合成基因而影响生物膜的形成。

群体感应系统可调节革兰阴性菌和革兰阳性菌生物膜的形成。以 AHL 为信号分子的革兰阴性菌，由信号分子和相应的信号分子受体组成的群体感应系统调控其生物膜的形成。费氏弧菌的 LuxI/LuxR 系统为经典的 AHL-QS 系统[69] LuxI 催化合成自体诱导剂 N-（3- 氧代己基）-L-高丝氨酸内酯（N-（3-oxo-hexanoyl）-L-homoserine lactone，3-oxo-C6-HSL），LuxR 与 3-oxo-C6-HSL 形成复合物后导致广泛的基因表达激活。基于 AHL 的群体感应系统已被证明能影响革兰阴性菌液化沙雷菌（*Serratia liquefaciens*）的生物膜成熟[70]。铜绿假单胞菌群体感应系统有两个信号系统：Lasl/LasR 和 rhlI/rhlI。Lasl/rhlI、LasR 和 rhlR 基因分别编码不同的信号分子合成酶和信号分子受体[71]。信号分子浓度随着细菌密度的增加而升高，当信号分子浓度达到一定阈值时，便与相应的信号分子受体结合并激活受体。激活的受体随后激活相关的转录调节因子，合成胞外多糖、毒力因子等，从而形成细菌生物膜[72]。

在革兰阳性菌中，群体感应系统以寡肽为信号分子调控生物膜的形成。不同类型的细菌具有不同的寡肽信号分子和群体感应调控途径。链球菌的双组分系统是组氨酸蛋白激酶和应答调节蛋白，而金黄色葡萄球菌的群体感应系统是高度保守的[73]。WalK/WalR 也称为 YycG/YycF 双组分系统，它直接调节生物膜的形成[74]。在表皮葡萄球菌中，群体感应系统相关的综合调节因子 sarA 与生物膜的形成密切相关，是表皮葡萄球菌生物膜形成的阳性调节因子[75]。此外，金黄色葡萄球菌 Agr（辅助基因调节）系统中，AgrD 编码 AIP 信号序列的前肽[76]。金黄色葡萄球菌信号肽酶 B 对信号肽进行切割，AgrB 对尾部进行切割，然后将成熟的 AIP 分泌到细胞外环境中。AIP 与传感器跨膜组氨酸激酶 AgrC 结合，其同源反应调节因子 AgrA 被磷酸化，随后与 P2 和 P3 启动子区结合。这使得 RNA 核酸聚合酶Ⅲ（RNA Ⅲ）的产生能够进一步触发 AIP 的合成，同时诱导 RNA Ⅲ调节与毒力、生物膜形成和其他过程相关的基因[77]。这种群体感应信号在生物膜形成中起着重要作用。在霍乱弧菌和金黄色葡萄球菌中，细胞密度的增加抑制了生物膜的形成[78, 79]，而铜绿假单胞菌中群体感应回路（LasI/R 和 RhlI/R）的激活则促进了生物膜的形成[80]。

AI-2 作为细菌间的“通用语言”，在细菌种内和种间对生物膜的调节作用也有较多研究。在牙菌斑中，数百种细菌构成了混合型生物膜，其中格氏链球菌（*Streptococcus gordonii*）是主要的定植菌，该菌在牙齿上产生的 AI-2 可诱导其他细菌如牙龈卟啉单胞菌连续定植[81]。口腔中含有 300 多种不同的细菌，它们在口咽的各部位都有定植，牙齿和黏膜上皮细胞都有由微生物形成的牙菌斑，为更多的细菌定植提供了有利条件，形成小的微生物群。几种口腔细菌都能产生 AI-2，这些微生物产生的 AI-2 信号分子参与其他微生物群落的生物膜形成和基因调控，可导致口腔环境中微生物种群的变化[82]。另外，AI-2 增多会导致大肠埃希菌生物膜形成增加[30]。此外，肺炎克雷伯菌的 AI-2 也可以促进早期生物膜的形成[83]。

三、肠道菌群群体感应分子诱导宿主应答

原核生物和真核生物共存了数百万年。据估计，人体有 10^{13} 个细胞和 10^{14} 个细菌。真核生物与原核生物之间的关系是动态变化的，这些相互作用既可能有益，也可能有害。人类与肠道微生物群保持着共生关系，这对营养吸收和先天性免疫系统的发育至关重要[84]。肠道微生物群在人类健康中扮演着多种重要角色，从调节新陈代谢到影响免疫系统，甚至促进肠道神经系统的成熟[85]。

原核生物和真核生物期间，互相暴露在对方产生和释放的信号中。微生物细胞和它们的宿主通过一系列信号分子相互交流，这种细胞间跨界信号传递涉及一些小分子物质，如真核生物产生的激素和细菌产生的激素样化学物质等。最近的证据表明，群体感应不仅限于细菌间的通信，还介导微生物与其宿主之间的通信。基于脂质或脂肪酸的信号机制在真核生物中也有广泛的特征。数十种脂质激素已经被证明具有数百种生物学功能。这些脂质分子包括二十碳烯酸家族成员及脂质类激素和类固醇激素。脂质类激素在细胞膜上自由扩散，并与核激素受体家族成员相互作用，改变转录调控。细菌的 AHLs 和真核生物的脂质信号分子在化学上是相似的，并且有相似的作用方式。因而有学

者猜想，细菌群体感应信号亦可使细菌信号传递到真核细胞并介导转录调节[86]；相反，宿主激素可以向细菌细胞发出信号，从而调节细菌的生物学活动[87]。这种界间信号传递称为跨界信号交流。最近发现的细菌信号分子和宿主激素之间的互相作用关系开辟了一个新的研究领域，有助于更好地理解病原体与宿主之间的关系（图 4-1）。

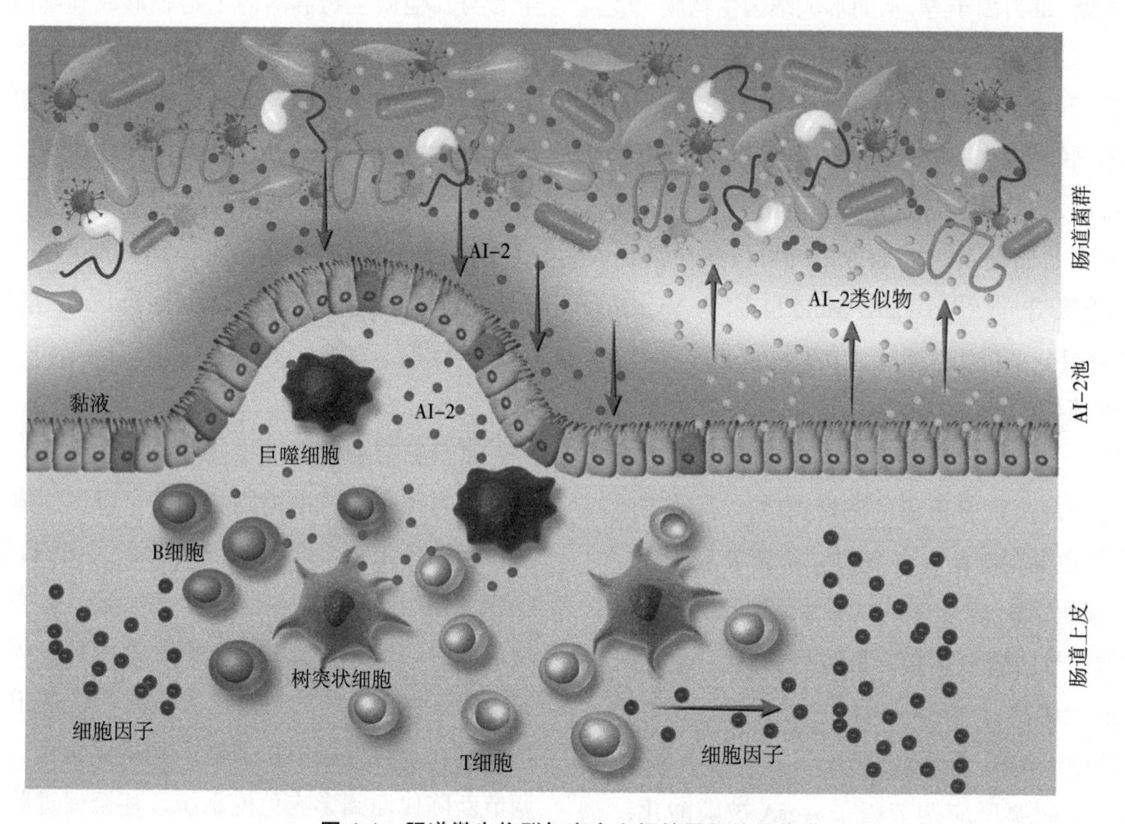

图 4-1　肠道微生物群与宿主之间的界间信号传递

肠道微生物群的群体感应分子（AI-2）通过与 T 淋巴细胞、巨噬细胞、树突状细胞和中性粒细胞等免疫细胞作用，从而与宿主肠道对话。宿主肠道分泌的激素可以作用于肠道细菌，并通过 AI-2 类似物干扰和淬灭细菌的群体感应，从而影响肠道细菌的生物学行为。引自：Li Q，Y Ren，X Fu，Inter-kingdom signaling between gut microbiota and their host. Cell Mol Life Sci, 2019, 76（12）: 2383-2389.

某些植物中的变形菌通过 LuxR 转录因子亚家族成员来识别植物信号，以此来感知宿主环境，这种跨界交流对于其互利共生和致病性调节都很重要。在豆科植物和固氮根瘤菌及根瘤菌属农杆菌及其宿主中，涉及植物源分子和细菌蛋白受体的跨界信号系统已经得到了很好的研究和利用[88]。AHLs 是革兰阴性菌群体中协调群体行为的信号分子。最近的证据表明，它们在植物生长和防御反应中发挥作用。3-oxo-C6-HSL 增强了植物的耐盐性，依赖 ABA 和不依赖 ABA 的信号途径以及 SOS 信号可能参与了 3-oxo-C6-HSL 诱导的植物抗盐性过程。这些研究成果为研究植物与微生物之间的相互交流提供了新的视角[89]。

然而细菌与哺乳动物间的跨界信号交流仍在初期探索阶段。真核生物拥有多种微生物生态系统，构成了微生物群落[90]，物种间和跨物种界间的化学交流影响着肠道微生物群的物种组成[91]。群体感应在病原菌和共生菌与宿主的相互作用中起着关键作用[92]。最近的研究表明，N-（3- 氧代十二烷酰）-L- 高丝氨酸内酯［N-（3-oxododecanoyl）-L-homoserine lactone，3-oxo-C12-HSL］可以进入哺乳动物细胞并发挥作用[93]。早在 10 多年前，人们就知道 AHLs

可以在哺乳动物细胞中引起反应。这一现象最初是通过呼吸道上皮细胞观察到的，在纯化的AHL 刺激下，白细胞介素 -8（IL-8）的表达呈剂量依赖性增加[94]。

（一）AHL 群体感应分子

AHL 作为革兰阴性菌普遍使用的信号分子，其在跨界信号交流中的作用已被广泛研究。越来越多的证据表明，3-oxo-C12-HSL 在体内外都能调节真核细胞的信号转导和免疫应答[95]。3-oxo-C12-HSL 可自由进入哺乳动物细胞，并且在细胞内具有活性[93, 96]。根据目前的研究成果，可将 AHLs 对真核哺乳动物的作用分为抑制宿主免疫反应、诱导细胞凋亡、促进炎症反应、促进抗肿瘤活性等方面（表 4-1）。

表 4-1　原核生物 AHLs 对真核生物的作用

信号分子	真核组织或细胞	功能	参考文献
3-oxo-C12-HSL	人成纤维细胞（L828）和支气管上皮细胞(16HBE)	诱导核因子 NF-κB 和激活蛋白 AP-2 的表达；在 mRNA 水平和蛋白水平都产生大量 IL-8	[100, 101]
3-oxo-C12-HSL	人多形核中性粒细胞	诱导多形核中性粒细胞在体外定向迁移	[102, 103]
3-oxo-C12-HSL	小鼠皮肤	细胞因子 IL-1α、IL-6、MIP2、MCP1、MIP1β、诱导蛋白 10 和 T 细胞活化基因 3 的 mRNAs 水平增加。并激活 T 细胞产生炎性细胞因子γ- 干扰素，促进 Th1 型微环境形成。通过 NF-κB 途径促进环氧合酶 2 的表达	[105]
3-oxo-C12-HSL	肾组织	诱导中性粒细胞迁移，引起肾组织炎症和细胞因子反应	[106]
3-oxo-C12-HSL	巨噬细胞	3-oxo-C12-HSL 可抑制脂多糖刺激的巨噬细胞增殖，并下调 IL-122 的产生	[108]
3-oxo-C12-HSL	巨噬细胞	诱导巨噬细胞 P388D1 凋亡	[109]
3-oxo-C12-HSL	淋巴细胞	抑制淋巴细胞增殖和肿瘤坏死因子 α 的产生	[108]
3-oxo-C12-HSL	淋巴细胞	抑制 Th1 和 Th2 细胞分化和增殖	[111]
3-oxo-C12-HSL	淋巴细胞	影响 T 细胞和 T 细胞介导的免疫应答，并抑制脾脏抗原呈递细胞激活 T 细胞的能力；3-oxo-C12-HSL 可抑制对钙离子载体及佛波酯对 T 细胞的激活作用	[111]
3-oxo-C12-HSL	淋巴细胞	3-oxo-C12-HSL 抑制 T 细胞增殖和 IL-2 释放	[112]
3-oxo-C12-HSL	淋巴细胞	抑制有丝分裂原和抗原刺激的 T 淋巴细胞的增殖和功能（细胞因子的产生）	[113]
3-oxo-C12-HSL	淋巴细胞，树突状细胞	抑制正常树突状细胞和 T 细胞的活化和增殖，下调树突状细胞共刺激分子的表达；在混合淋巴细胞树突状细胞反应中，IL-4、IL-10、TNF-α 的分泌增加，IFN-γ 和 IL-6 降低。群体感应分子诱导树突状细胞和 $CD4^+$ T 细胞凋亡	[114]
3-oxo-C12-HSL	巨噬细胞，中性粒细胞和单核细胞株	诱导巨噬细胞，中性粒细胞和单核细胞株 U-937 和 P388D1 发生凋亡，并呈剂量依赖性和孵育时间依赖性增加	[115]
3-oxo-C12-HSL	小鼠成纤维细胞和人血管内皮细胞	诱导凋亡并调节免疫介质，通过内质网钙动员诱导细胞凋亡	[93, 116]
3-oxo-C12-HSL	乳腺癌细胞系	抑制 Akt/PKB 通路，抑制信号转导和转录激活因子 3 的活性	[117]
3-oxo-C12-HSL	结直肠癌和前列腺癌细胞	下调胸苷酸合成酶，增强化疗药物 5- 氟尿嘧啶和紫杉醇的疗效	[118]
3-oxo-C12-HSL	人鼻窦上皮细胞	AHL 与人类苦味受体 T2R38 结合，激活钙信号和一氧化氮信号转导，增加黏液纤毛清除率，进而对抗吸入病原体，加强呼吸道物理防御能力	[119]

1. 促进炎症反应　炎症是先天免疫系统对有害刺激（如病原体、受损细胞或刺激物）的复杂生物反应。通常表现为对入侵微生物病原体产生固有免疫反应，通过一组复杂的信号转导启动炎症反应过程，包括诱导转录因子 NF-κB，NF-κB 通过调节细胞生长、分化、细胞存活 / 凋亡、炎症和免疫等多种生物学过程，在宿主应激反应中发挥着关键作用[97]。先天免疫系统在病原微生物入侵时被激活，转录因子 NF-κB 的激活对于调节先天性免疫应答、消除入侵的病原微生物至关重要。例如，真核细胞膜上 Toll 样受体 4（Toll-like receptor 4，TLR4）识别革兰阴性菌的脂多糖（lipopolysaccharide，LPS），导致核转录因子 NF-κB 的快速激活，促进编码促炎性细胞因子的基因表达。研究发现，在铜绿假单胞菌中，信号分子 3-oxo-C12-HSL 通过选择性地抑制哺乳动物细胞 NF-κB 的活化，削弱 TLR4 依赖的先天免疫反应，从而促进持续感染[95]。早期体外实验研究表明，3-oxo-C12-HSL 在低浓度（＜ 10μmol/L）时发挥免疫抑制 / 抗炎作用，在高浓度时（20μmol/L 及以上）表现为促炎作用[98]。

3-oxo-C12-HSL 可影响宿主细胞的反应，包括炎症因子的上调。大部分囊性纤维化患者是由肺气道铜绿假单胞菌感染引起的，铜绿假单胞菌可在囊性纤维化患者的肺内形成生物膜，痰中可检测到 3-oxo-C12-HSL 分子[99]。囊性纤维化患者的肺组织表现为强烈的炎症反应，体外研究表明：3-oxo-C12-HSL 可通过诱导核因子 NF-κB 和激活蛋白 AP-2 的表达，诱导人成纤维细胞（L828）和支气管上皮细胞（16HBE）在 mRNA 水平和蛋白水平都产生大量 IL-8，而 IL-8 可诱导中性粒细胞的趋化[100, 101]。另外，3-oxo-C12-HSL 在 10 ～ 100μmol/L 剂量下还可直接诱导人多形核中性粒细胞在体外定向迁移（趋化）[102, 103]。多形核中性粒细胞是抵抗细菌感染的"一线防御"细胞，其趋化性是宿主防御的关键。因此，3-oxo-C12-HSL 通过对中性粒细胞的趋化，促进机体炎症反应。适量的 3-oxo-C12-HSL 刺激，宿主可通过识别细菌的 3-oxo-C12-HSL 分子，诱发炎症反应清除病原微生物；当过量的 3-oxo-C12-HSL 作用于机体免疫细胞时，可能诱发强烈的免疫反应，同时对机体组织细胞产生损伤。同时有研究表明，肺暴露于中性粒细胞产生的炎症介质如弹性蛋白酶和活性氧，会导致组织破坏、阻塞性肺疾病，最终导致肺衰竭[104]。

将 3-oxo-C12-HSL 注射到小鼠皮肤中，导致细胞因子 IL-1α、IL-6、巨噬细胞炎症蛋白 2（macrophage inflammatory protein 2，MIP2）、单核细胞趋化蛋白 1（monocyte chemoattractant protein 1，MCP1）、MIP1β、诱导蛋白 10 和 T 细胞活化基因 3 的 mRNAs 水平增加，并激活 T 细胞产生 γ- 干扰素（interferon-gamma，IFN-γ），促进以 IFN-γ 为代表的 Th1 型微环境形成[105]。3-oxo-C12-HSL 还可以通过 NF-κB 途径促进环氧合酶 2（cyclooxygenase 2，COX-2）的表达。COX-2 酶对花生四烯酸转化为前列腺素很重要，与水肿、炎症浸润、发热和疼痛有关。膜相关前列腺素 E（prostaglandin E，PGE）合成酶及其产物 PGE2 水平的升高表明，3-oxo-C12-HSLs 可作为脂质介导免疫反应的有效激活剂。前列腺素代谢物在结构和化学特性上与 3-oxo-C12-HSL 相似，具有与 3-oxo-C12-HSL 相同的 COX2 激活表型；这表明它们可能通过共同受体发挥作用[105]。这些炎症介质的产生能诱导大量白细胞聚集，造成组织破坏。

为研究群体感应信号分子在尿路感染中诱导炎症的作用，研究人员采用能产生群体感应信号分子的铜绿假单胞菌标准亲本株（PAO1）和缺乏群体感应信号分子的等基因单、双突变株感染小鼠。结果发现，与缺乏群体感应信号分子的菌株比较，小鼠感染 PAO1 后，体内 AHLs 增多，并且中性粒细胞迁移更明显，并引起肾组织炎症和细胞因子反应[106]。

研究人员试图寻找真核细胞上 AHLs 的受体，他们发现，3-oxo-C12-HSL 作用于真核细胞时，过氧化物酶体增殖物激活受体（peroxisome proliferator-activated receptor，PPAR）的两个成员 PPARβδ 和 PPARγ 以剂量依赖的方式表达增强。PPAR 是抑制促炎症基因表达的转录因子。即使在 PPARγ 激动剂罗格列酮存在的情况下，3-oxo-C12-HSL 也能抑制 PPARγ 的转录活性，而 PPARβδ 的活性增加。这些实验表明 3-oxo-C12-HSL 是 PPARβδ 的配体[107]，也就是说，PPARβδ 可能是真核生物细胞上针对 AHLs 的受体。

2. *抑制宿主免疫反应*　研究表明，3-oxo-C12-HSL 也可以直接调节宿主的免疫反应。例如铜绿假单胞菌的群体感应信号分子 3-oxo-C12-HSL，能影响包括免疫细胞在内的多种哺乳动物细胞的功能。

（1）巨噬细胞：巨噬细胞是先天性免疫系统的重要成分，它们是从组织中的单核细胞分化出来的，在中性粒细胞的刺激下到达感染部位。巨噬细胞具有先天免疫细胞的特性和启动适应性免疫机制的特性。针对微生物抗原，它们表现出较强的吞噬活性，产生抗菌活性氧、一氧化氮、与组织修复有关的鸟氨酸、促炎细胞因子，刺激其他免疫细胞对病原体产生反应。1998 年，科学家首次证明 AHLs 可抑制宿主免疫反应，3-oxo-C12-HSL 可抑制脂多糖刺激的巨噬细胞增殖，并下调 IL-12 的产生[108]。另外一项研究发现，3-oxo-C12-HSL 及其类似物可诱导巨噬细胞 P388D1 凋亡，并发现除了 L- 高丝氨酸内酯结构，oxo 基团在酰基侧链中的位置对细胞凋亡的诱导活性也至关重要[109]。其中含有 11 ～ 13 个碳的侧链并含有 3- 氧基或 3- 羟基的结构的 AHL，具有最佳的免疫抑制活性[110]。随后的研究表明，3-oxo-C12-HSL 在体外可抑制细胞因子的产生，且对 INF-γ 的抑制作用大于 IL-4[111]。

（2）T 细胞：研究发现，3-oxo-C12-HSL 能抑制淋巴细胞增殖和肿瘤坏死因子 α（tumor necrosis factor alpha，TNF-α）的产生[108]。另一项研究发现，3-oxo-C12-HSL 可抑制 TH1 和 TH2 细胞的分化和增殖[111]。3-oxo-C12-HSL 抑制了 T 细胞大多数细胞因子的产生，并且可抑制抗原呈递细胞的功能。在体内外试验中，3-oxo-C12-HSL 均能影响 T 细胞和 T 细胞介导的免疫应答，并抑制脾脏抗原呈递细胞（APCs）激活 T 细胞的能力。另外，3-oxo-C12-HSL 可抑制钙离子载体和佛波酯对 T 细胞的激活作用[111]。采用 T 细胞受体和 CD28（共刺激分子）激活人 T 细胞，结果发现 3-oxo-C12-HSL 能抑制 T 细胞增殖和 IL-2 释放[112]。因而有学者推断，铜绿假单胞菌可能通过这种抑制作用，导致宿主的持续感染状态。

另外，3-oxo-C12-HSL 还可抑制有丝分裂原和抗原刺激的 T 淋巴细胞的增殖和功能（细胞因子的产生）[113]。除此之外，群体感应信号分子还可抑制正常树突状细胞和 T 细胞的活化和增殖，下调树突状细胞共刺激分子的表达；在混合淋巴细胞树突状细胞反应（mixed lymphocyte dendritic cell reaction，MLDCR）中，导致 IL-4、IL-10 和 TNF-α 的分泌增加，而 IFN-γ 和 IL-6 分泌降低。群体感应信号分子可诱导树突状细胞和 $CD4^+$ T 细胞凋亡，但不诱导 $CD8^+$ T 细胞凋亡[114]。

3. *促进凋亡*　最近研究表明，3-oxo-C12-HSL 能诱导多种哺乳动物细胞凋亡，包括中性粒细胞、巨噬细胞等多种细胞，这提示了细菌介导的细胞凋亡的另一个机制。

首次观察到这种凋亡现象的研究是在加入 3-oxo-C12-HSL 处理后，诱导了巨噬细胞、中性粒细胞和单核细胞株 U-937 和 P388D1 发生凋亡，并呈剂量依赖性和孵育时间依赖性增加[115]。

除免疫细胞外，3-oxo-C12-HSL 还能诱导小鼠成纤维细胞和人血管内皮细胞凋亡并调节免疫介质的表达。这种作用与内质网中动员的胞质钙水平的增加有关[93]。此外有研究表明，磷脂酶C抑制剂能阻断钙的释放，从而阻断3-oxo-C12-HSL的促凋亡作用，但不影响3-oxo-C12-HSL的免疫调节作用[116]。

4. 抗肿瘤活性　细胞生存信号可通过磷脂酰肌苷3激酶和蛋白激酶B等脂质依赖性激酶传递。近年来，信号转导和转录激活蛋白（signal transducer and activator of transcription，STAT）被认为是多种细胞增殖的正向调节因子。这些途径的持续激活与肿瘤细胞的生长有关，而抑制它们可以阻止细胞增殖、促进凋亡细胞死亡。

在乳腺癌细胞系中，3-oxo-C12-HSL能部分抑制Akt/PKB通路，并抑制STAT3的活性，而利用药物抑制STAT3途径导致了与加入3-oxo-C12-HSL相同的表型。此外，加入活性STAT3蛋白可抑制3-oxo-C12-HSL的促凋亡活性[117]。在结直肠癌和前列腺癌细胞中，3-oxo-C12-HSL能下调胸苷酸合成酶，并增强常用肿瘤化疗药物5-氟尿嘧啶和紫杉醇的疗效[118]。

5. 清除病原体　人类苦味受体T2R38是上呼吸道固有防御的重要组成部分，它可以与革兰阴性菌分泌的酰基高丝氨酸内酯群体感应分子相互作用。此外，人鼻窦上皮细胞中的T2R38激活能刺激钙和NO信号，增加黏液纤毛清除率，进而对抗吸入病原体，加强呼吸道物理防御功能[119]。

（二）AIP群体感应分子

细菌群体感应分子作为调控细菌生物学行为的信号分子，可能在细菌的黏膜定植中发挥作用[120]。AIP为部分革兰阳性菌用于群体感应交流的信号分子，现有研究表明它可介导细菌与宿主间的跨界信号交流。有关AIP与宿主间的跨界信号交流研究目前较少。研究发现以下AIP：群体感应肽Phr0662由芽孢杆菌类合成[121]，EntF代谢物来自于屎肠球菌，EDF衍生肽由大肠埃希菌合成[122]，均能促进结肠癌细胞株HCT-8/E11的侵袭作用，此外Phr0662还能促进血管生成[123]（表4-2）。

表4-2　原核生物AIP、AI-2及AI-3对真核生物的作用

信号分子	真核组织或细胞	功能	参考文献
群体感应肽Phr0662，EntF,EDF衍生肽	结肠癌细胞株HCT-8/E11	促进结肠癌细胞株HCT-8/E11侵袭作用，Phr0662可促进血管生成	[123]
DPD	牙龈上皮细胞	促进细胞增殖	[124-126]
AI-2	HCT-8结肠癌细胞	促进炎性细胞因子IL-8的表达	[128]
AI-2	肺泡壁细胞	细胞损伤	[129]
AI-2	U937来源的巨噬细胞	通过TNFSF9/TRAF1/p-AKT/IL-1β信号途径增强巨噬细胞的迁移率和促进M1极化	[131]
AI-2	结肠癌	与肿瘤免疫相关	[137]
AI-2	鸡Ⅱ型肺细胞	诱导鸡Ⅱ型肺细胞的损伤过程	[130]
AI-3	宫颈癌HeLa细胞	促进黏附和擦拭性损伤	[138]

（三）AI-2群体感应分子

许多革兰阴性菌及革兰阳性菌都可产生信号分子AI-2，作为种间通信的语言，AI-2也能介导原核生物与真菌间，以及原核生物与哺乳动物间的跨界信号交流（表4-2）。

在口腔中，抵抗细菌的首道防线是口腔上皮。上皮细胞通过分泌抗菌肽和酶形成一道屏障，阻止细菌持续作用于口腔上皮，此外还产生趋化因子与免疫细胞交流[124]。口腔细菌还可通过LuxS/AI-2信号与宿主细胞进行跨界交流。牙龈假单胞菌LuxS突变菌株诱导牙周膜成纤维细胞炎症反应的能力下降，表明牙龈假单胞菌

LuxS 信号在与牙周膜成纤维细胞的直接相互作用中起重要作用[125]。另外，使用 AI-2 的前体物质 DPD 及其类似物作用于牙龈上皮细胞，结果发现，浓度为 100nmol/L 的丁基 -DPD 能促进牙龈角质形成细胞增殖；10μmol/L 的乙基 -DPD 作用于牙龈角质形成细胞，可使 MMP-9 活性升高[126]。已知 MMP-9 具有改变细胞紧密连接功能的能力[127]，因此 DPD 可能通过改变细胞紧密连接功能而与真核生物上皮细胞进行跨界信号交流。

为研究 AI-2 对真核生物的直接作用，将 AI-2 直接作用于 HCT-8 结肠癌细胞后，炎性细胞因子 IL-8 的表达在 6 小时和 12 小时都有显著上调，随后在 24 小时显著下调[128]。已知 IL-8 在招募中性粒细胞中起重要作用，因此 AI-2 可能通过促进促炎因子表达而招募免疫细胞，促进炎症反应。Bryan 等将肺泡细胞暴露于浓度为 50μmol/L 的 AI-2 后，进行微阵列研究，发现有 5 个基因的表达变化超过 2 倍（DPYD，TAGLN，COL11A1，ANK2，AK094929）[129]。另外，将鸡Ⅱ型肺细胞与禽致病性大肠埃希菌 APEC-O78 共培养时，发现 AI-2 还参与禽致病性大肠埃希菌诱导鸡Ⅱ型肺细胞损伤的过程[130]。研究发现，具核梭杆菌产生的 AI-2 能促进 U937 来源巨噬细胞的迁移，并促进其表型发生 M1 极化，其机制可能是 AI-2 激活 TNFSF9/TRAF1/p-AKT/IL-1β 信号通路，而 TNFSF9/IL-1β 在结直肠癌表达增高与 AI-2 浓度及患者生存率相关[131]。

细菌生物膜广泛存在于人类牙根表面、呼吸道上皮细胞、炎症性肠病肠上皮表面和大肠肿瘤表面[125, 132-134]，而结肠生物膜的存在可能与散发性结直肠癌的风险增加有关[135, 136]。我们探索了 AI-2 浓度与人类结直肠癌（colorectal cancer，CRC）的相关性，通过检测 CRC 患者粪便、癌组织和唾液标本中 AI-2 的浓度，发现 CRC 患者结肠癌组织和粪便中 AI-2 浓度均显著高于正常人，且浓度随 CRC 进展显著升高。此外，在 CRC 组织中，AI-2 浓度与 $CD3^+T$ 细胞数量呈正相关，与 CD4/CD8 比值呈负相关，表明 AI-2 可能参与了 CRC 的肿瘤免疫过程[137]。

（四）AI-3 群体感应分子

AI-3 同样参与细菌的种间通信，然而近年研究发现它对真核细胞也具有潜在作用。由于编码肠出血性大肠埃希菌Ⅲ型分泌系统基因的转录是通过 LuxS/AI-2 群体感应系统激活的，学者检测了 LuxS 突变对Ⅲ型分泌的影响，发现 LuxS 突变株在抑制Ⅲ型分泌后，仍然能够在宫颈癌 HeLa 细胞上产生黏附和擦拭性损伤。结果表明上皮细胞和细菌细胞之间存在某种信号传导，从而激活了Ⅲ型分泌，最终发现对宫颈癌 HeLa 细胞产生此作用的是肠出血性大肠埃希菌产生的 AI-3 信号分子[138]（表 4-2）。

综上表明，细菌群体感应信号分子能影响肠上皮细胞和肠道免疫功能。肠道上皮细胞和免疫细胞可以“监听”居住在消化道管腔中细菌群落之间的通信，并根据这些信号改变细菌的生物学行为[139]。进一步研究细菌群体感应分子对肠上皮细胞的影响将有助于理解微生物相关性疾病，调控跨界信号交流网络可能为临床提供新的抗感染方法。

四、宿主通过调控群体感应系统与微生物对话

真核生物拥有多种微生物生态系统，构成了微生物群落[90]。例如哺乳动物皮肤、口腔和肠道中的细菌群落[140]。越来越多的证据表明，细菌物种间和跨物种间的交流决定了肠道微生物群的物种组成[91]。哺乳动物的组织和细胞已经进化出多种防御机制，以防止细菌感染。这些物质包括蛋白质和肽类，如溶菌酶和防御素，宿主的免疫系统通过分泌 α- 防御素等抗菌物质来控制微生物群的定位和组成，或者杀死细菌[141-143]。此外，宿主还分泌特定的抗生物膜防

御分子，包括乳铁蛋白，它能螯合铁并阻止铜绿假单胞菌生物膜的形成[144]。

近来研究发现，宿主也可以通过多种方式影响微生物的群体感应系统，进而调控微生物群的组成和功能。

（一）影响群体感应分子活性的理化因素

居住在肠黏膜表面的微生物群落受宿主多种因素的影响，包括先天免疫成分、黏液成分和饮食[145, 146]。宿主的肠道微环境会影响AI-2的活性，包括pH、胆汁酸、温度、渗透压。①pH对AI-2信号具有较大影响：在两个鼠李糖乳杆菌菌株中（LGG，BFE5264），两个菌株在pH为5.0时对AI-2信号有较强的上调作用，但在pH为8.0时仅对LGG产生的AI-2信号有显著的下调作用。植物乳杆菌菌株产AI-2能力在低pH（pH 5.0）和高pH（pH 8.0）时均下调。②胆汁酸亦会影响AI-2的活性。鼠李糖乳杆菌菌株对胆汁酸表现出浓度依赖性反应，随着胆汁酸浓度的增加，菌株的AI-2活性增加。而植物乳杆菌的AI-2活性由于胆汁酸的存在被抑制。③温度对鼠李糖乳杆菌的AI-2活性也有显著影响。当温度从37℃降到25℃时，AI-2的活性可降低到50%左右。④渗透压的改变也影响着AI-2的活性。随着NaCl浓度的增加，即渗透压增加，鼠李糖乳杆菌和植物乳杆菌菌株的AI-2活性都增加[147]。这意味着宿主肠道微环境的改变会影响菌群AIs的活性，从而影响肠道微生物群的密度和结构，进而反过来影响宿主健康。

（二）真核生物产生物质干扰或淬灭群体感应信号分子

哺乳动物细胞可能进化出干扰细菌群体感应信号的机制，以此防御细菌感染的。许多有机体，包括细菌、真菌、植物和哺乳动物，都可以产生淬灭酶来干扰和灭活AHL等信号分子，以此来削弱细菌病原体的毒性。这些真核生物产物在结构上似乎与AHL无关，但就其功能状态而言，它们通常被称为AHL干扰物[148]。

许多哺乳动物细胞系能使3-oxo-C12-HSL失活，为降低细菌毒性提供了可能的机制。真核生物可以产生抑制细菌群体感应的酶，例如，淡水水螅产生氧化还原酶[149]，将水螅的主要定植细菌曲线杆菌产生的自体诱导分子3-oxo-C12-HSL还原为3-OH-C12-HSL[150]。3-OH-C12-HSL分子促进曲线杆菌定植于宿主，然而，只有3-oxo-C12-HSL自体诱导剂可以激活曲线杆菌的一个关键表型，这个表型与鞭毛基因、运动性及不同宿主有关。因此，水螅通过产生氧化还原酶影响细菌信号分子，从而捕获曲线杆菌。红藻*Delisea pulchra*可产生卤代呋喃酮，作为群体感应受体拮抗剂，通过与LuxR型AHL受体结合，卤代呋喃酮破坏受体的稳定性，导致加速受体降解，从而干扰重要的细菌通信调节过程，如生物膜的形成[151]。

铜绿假单胞菌是一种环境中普遍存在的机会致病菌，可引起多种感染，包括烧伤感染和慢性耳道感染、膀胱感染等[152]，它可利用两种群体感应分子3-oxo-C12-HSL和C4-HSL来控制细胞外毒力因子的产生和生物膜的形成。哺乳动物的呼吸道可通过多种防御机制保护自己免受细菌感染，包括抗菌肽、黏液纤毛清除和吞噬细胞。近年研究发现，哺乳动物细胞还可以调节细菌群体感应系统，从而保护自已免受细菌侵袭。人类气道上皮可以能选择性水解并灭活铜绿假单胞菌产生的高丝氨酸内酯自体诱导物3-oxo-C12-HSL，其降解活性由热敏感因子介导，该因子为钙依赖性酯酶的对氧磷酶（paraoxonase，PON）[153]。PONs是一个独特的、高度保守的钙依赖性酯酶家族，由PON1、PON2和PON3组成[154]。血清PON1水解3-oxo-C12-HSL的内酯环，PON1基因敲除小鼠血清水解3-oxo-C12-HSL活性减弱[155]。敲除小鼠的气管上皮细胞PON2基因，导致对3-oxo-C12-HSL的降解能力下降，而PON2的过度表达则促进了3-oxo-C12-HSL的降解[156, 157]。宿主免疫细胞也能表达 PON2，水解 3-oxo-C12-HSL。宿主还能产生其他物质影响群体感应信号，从而抵御病原体入侵。慢性感染性伤口通常有多种微生物，金黄色葡萄球菌和铜绿假单胞菌是两种最常见的物种。虽然在体外铜绿假单胞菌很容易杀死金黄色葡萄球菌，但在慢性感染性伤口中，这两个物种可以长期共存。在慢性感染性伤口中，宿主因子如血清白蛋白，可隔离 3-oxo-C12-HSL 自体诱导因子，从而抑制铜绿假单胞菌 LasR 依赖的群体感应行为，导致铜绿假单胞菌无法产生杀死金黄色葡萄球菌的毒力因子[158]。

宿主细胞还能通过 Nox2 和 ApoB 干扰细菌的群体感应进行防御行为。Nox2 是一种 NADPH 氧化酶，主要在吞噬细胞中表达，可产生活性氧。载脂蛋白 B（apolipoprotein B，ApoB）是低密度脂蛋白（low-density lipoprotein，LDL）的一个组成部分，血清中 LDL 浓度较高。Nox2 和 ApoB 都能拮抗群体感应信号中的自身诱导肽，激活金黄色葡萄球菌（*Staphylococcus aureus*）的毒力基因表达[159, 160]。ApoB 直接与 AIP 结合后，阻止其与受体的相互作用，从而抑制金黄色葡萄球菌群体感应介导的行为[160]。Nox2 能氧化 LDL，促进 ApoB 的构象变化，并调节其与 AIPs 的结合[161]。

此外，植物能产生黄酮类柚皮苷，它能抑制群体感应调节的铜绿假单胞菌毒力因子的产生。研究表明，柚皮苷通过与其生理激活剂 3-oxo-C12-HSL 竞争，能直接与群体感应调节因子 LasR 结合，从而产生抑制效应[162]。

（三）真核生物产生物质激活细菌群体感应系统

原核细胞通过自体诱导物发出信号，AIs 是一种激素样化合物，使微生物在群体水平上协调行动。近年来大量研究表明，一些病原菌能够识别宿主信号分子，从而激活致病因子的产生；另一方面，真核生物产生的某些物质也能激活细菌群体感应系统，导致细菌毒力增强（表 4-3）。

表 4-3　真核生物信号分子对于原核生物的影响

真核生物信号分子	功能	参考文献
氧化还原酶	将曲线杆菌产生的自身诱导分子 3-oxo-C12-HSL 还原为 3-OH-C12-HSL	[150]
卤代呋喃酮	群体感应受体拮抗剂，通过与 LuxR 型 AHL 受体结合，破坏受体的稳定性，加速降解受体	[151]
对氧磷酶	水解并灭活铜绿假单胞菌产生的高丝氨酸内酯自体诱导物 3-oxo-C12-HSL	[155-157]
血清白蛋白	隔离 3-oxo-C12-HSL 自体诱导因子，从而抑制铜绿假单胞菌 LasR 依赖的群体感应行为	[158]
载脂蛋白 B	与金黄色葡萄球菌寡肽自体诱导剂结合，阻止其与受体的相互作用，从而抑制金黄色葡萄球菌的群体感应	[159, 160]
黄酮类柚皮苷	通过与其生理激活剂 3-oxo-C12-HSL 竞争，直接与群体感应调节因子 LasR 结合而产生抑制作用	[162]
阿片类物质	铜绿假单胞菌毒力增强	[166]
膜菱形丝氨酸蛋白酶（Rho）	在真核生物中 Rho 与 Providencia stuartii 菌株的 Aar 蛋白具有相似性，可扰乱其群体感应信号	[167, 168]
肾上腺素和去甲肾上腺素	激活肠出血性大肠埃希菌毒力基因的表达，促进其在人类结肠黏膜的定植，并能调节细菌的新陈代谢等	[169]
去甲肾上腺素	诱导铜绿假单胞菌毒力基因的表达	[170]
AI-2 模拟物	影响细菌基因表达	[171]

暴露于生理或创伤应激的动物，随后的细菌感染已被证明会导致死亡率增加[163]，这与免疫功能受损和细菌清除能力减弱有关[164]。大量证据表明，在宿主应激期间释放的化合物直接激活了某些机会性致病菌。在宿主应激期间，多个组织部位可释放内源性阿片类物质，内源性阿片类物质广泛分布于受神经支配的肠黏膜内，并在神经、免疫和肠上皮细胞应激期间发挥多重作用[165]。因此，阿片类物质可能作为宿主源性细菌信号分子，发挥跨界信号交流激素的作用。研究发现，铜绿假单胞菌能感知宿主应激时释放的阿片类化合物，并将该类化合物与群体感应系统进行整合，导致铜绿假单胞菌毒力增强[166]。

细菌与宿主之间的跨界通信也会影响细菌定植。最近有研究表明，人类肠道上皮细胞能模仿细菌的群体感应与微生物群进行通信。上皮细胞产生一种模拟 AI-2 的分子称为 AI-2 模拟物，以应对细菌的相互应答或介导紧密连接的破坏。这种 AI-2 模拟物由细菌 AI-2 受体（LuxP/LsrB）检测并激活细菌的群体感应，例如鼠伤寒沙门菌等肠道病原体。如果上皮细胞直接或间接暴露于这些细菌，这种模拟 AI-2 的活性就会被诱导，这表明细菌分泌的某种成分刺激了 AI-2 模拟物的产生。这些发现揭示了哺乳动物 AI-2 在界间信号传递中的潜在作用。据推测，利用这种通用的种间 AI-2 自体诱导物作为模拟物，而不是种特异性的自体诱导物，可以使宿主与肠道中的大量细菌发生相互作用[171]。

细胞因子是免疫系统的主要信号分子，是病原体与宿主作用的主要靶点。几种细胞因子［如白细胞介素 -1（IL-1）、白细胞介素 -2（IL-2）和粒 - 巨噬细胞集落刺激因子］能促进致病性大肠埃希菌的生长，表明这些细胞因子可调控细菌的行为[172]。IL-1 可与致病性大肠埃希菌和鼠疫耶尔森菌（鼠疫的病原体）结合[173]。IL-1 与鼠疫耶尔森菌荚膜上的 F1 抗原结合，这种结合通过鼠疫耶尔森菌荚膜抗原 F1 组装蛋白（Caf1A）实现[173]。Caf1A 与人 IL-1 受体具有显著的同源性。此外，一些革兰阴性细菌，包括大肠埃希菌、志贺菌和肠沙门菌与促炎细胞因子 TNF-α 结合后，自身毒力会发生改变，但这些细菌识别 TNF-α 的确切机制尚不清楚。与 TNF-α 结合后的志贺菌，对上皮细胞的侵袭性明显增加[174]。铜绿假单胞菌外膜蛋白 OprF 可识别γ- 干扰素，导致群体感应基因 rhlI 的激活，从而合成信号分子 C4-HSL，促进凝集素 PA-1 的表达及铜绿假单胞菌素产生，进而导致肠黏膜屏障的破坏[175]。

真核生物通过产生激素或细胞因子等物质干扰细菌群体感应系统（模拟或淬灭），调节细菌细胞间通信及其生物学行为，影响细菌毒力、黏附能力、生物膜形成等。这些研究结果表明，宿主通过多种机制干预细菌的群体感应系统，具有调控细菌生物学行为的能力，这种跨界交流有助于二者之间保持平衡。目前该领域的研究尚缺乏深度，未来仍有广阔的研究前景。

第三节　宿主与微生物通过激素进行跨界信号交流

真核生物利用激素调节多种生理功能并维持内环境平衡。哺乳动物激素有三大类：蛋白质（或肽）、类固醇（脂激素的一个亚类）和氨基酸衍生物。激素的结构决定其受体的位置，胺和肽类激素不能穿过细胞膜与细胞表面受体（如受体激酶和 G- 蛋白偶联受体）结合，而类固醇激素可以穿过脂质膜，并与细胞内受体结合。所有这些激素都参与和微生物之间的跨界信号传递。受体激酶是具有内在酪氨酸或苏氨酸激酶活性的细胞表面受体，当与激素细胞外氨基末端结合时被激活。这种激活导致细胞内下游蛋白的募集和磷酸化，从而启动信号级联。

一、肾上腺素和去甲肾上腺素

肾上腺素和去甲肾上腺素是人体内含量最丰富的儿茶酚胺，对肠道生理有重要作用，包括调节肠道平滑肌收缩、黏膜下血流量及氯化物和钾的分泌[176]。肠道内常驻微生物群及病原体可影响结肠中肾上腺素 / 去甲肾上腺素的浓度[177]。此外，T 细胞、巨噬细胞和中性粒细胞也可产生并分泌肾上腺素和去甲肾上腺[178]。

肾上腺素和去甲肾上腺素被认为可调节细菌生长和毒力基因的表达。细菌细胞不表达肾上腺素能受体，但许多研究表明它们对肾上腺素和（或）去甲肾上腺素能做出反应。这些研究对象大多为居住在人体胃肠道的细菌，而胃肠道存在较多肾上腺素和（或）去甲肾上腺素[179]。细菌能感应肾上腺素和去甲肾上腺素并做出反应，如人类结肠内的肠出血性大肠埃希菌主要定植在近端结肠和横结肠，而这些部位肠细胞高表达 α_2 肾上腺素能受体[180]。研究发现，肠出血性大肠埃希菌能够识别肾上腺素和去甲肾上腺素，并激活 LEE 基因和鞭毛及细菌毒素基因的转录，导致毒力基因的激活和表达，促进其在人类结肠黏膜的定植，并能调节细菌的新陈代谢等多种生物学活动。同时去甲肾上腺素参与了肠杆菌素的诱导表达和通过铁载体帮助铁的吸收，这可能这是去甲肾上腺素诱导大肠埃希菌生长的机制[169, 181, 182]。

肾上腺素和去甲肾上腺素能与群体感应分子 AI-3 产生相互串扰，这是由一种叫作 QseC 的传感器组氨酸激酶介导的。这种膜修饰的传感器激酶能特异性地感知细菌激素样化合物 AI-3 及宿主激素（肾上腺素和去甲肾上腺素），并直接与这些信号分子结合[183]。这可能是肾上腺素和去甲肾上腺素调节肠出血性大肠埃希菌毒力的一种机制。

在外科手术创伤期间，去甲肾上腺素释放到肠内会诱导铜绿假单胞菌毒力基因的表达，从而导致肠源性败血症[170]。肠出血性大肠埃希菌中的 AI-3 信号可能被肾上腺素和去甲肾上腺素所取代，这表明 AI-3 可能也参与了跨界信号转导[138]。在其他情况下，群体感应系统是通过宿主产生的信号分子与细菌上的特定膜受体结合而激活的，例如宿主产生的 IFNγ 与铜绿假单胞菌的 OprF 外膜蛋白结合，调节铜绿假单胞菌毒力[175]。

二、肽类激素

肽类激素由多种激素构成，细菌能感知其中的几种激素。全世界范围内，有超过 50% 的人群胃内有幽门螺杆菌定植。这种细菌可以引起十二指肠溃疡和胃癌[184]。胃泌素是一种由胃细胞分泌的肽类激素，能刺激胃酸的释放[185]，并促进幽门螺杆菌生长。胃泌素类似物五肽胃泌素和另一种肽类激素胆囊收缩素也可产生类似效应。尽管幽门螺杆菌中胃泌素的感受器尚未被确认，然而胃泌素类似物可以阻断这种反应，这种现象可能提示细菌上存在这种激素的受体。值得注意的是，这种反应仅限于人胃泌素，这可能解释了为什么幽门螺杆菌特异性地感染人类[186]。相反，另一种胃肽类激素，生长抑素，可抑制胃酸分泌，抑制幽门螺杆菌生长。生长抑素已被证明能直接与细菌结合，而这种结合能被抗生长抑素抗体阻断，这再次表明幽门螺杆菌存在尚未被发现的肽类激素受体[187]。此外，幽门螺杆菌感染也会改变这两种激素的水平，分别升高胃泌素和降低生长抑素的水平[188, 189]。

类鼻疽伯克霍尔德菌（*Burkholderia pseu-*

domallei）是类鼻疽的病因，而类鼻疽最常见的危险因素是糖尿病。类鼻疽伯克霍尔德菌及其相关种属洋葱伯克霍尔德氏菌（*B. cepacia*）能结合胰岛素（胰腺分泌的肽激素，负责调节血糖水平）并对其做出反应，通过降低磷脂酶C和酸性磷酸酶的活性来增加自身生长和毒力[190]。然而，目前尚未发现该细菌上的胰岛素受体。

细菌感受到的另一类肽类激素是利钠肽，它参与血液的渗透调节。这个家族包含3种结构相关的肽：心钠素、脑钠肽和C型钠尿肽，它们在哺乳动物细胞中通过膜结合的鸟苷酸环化酶受体发出信号。脑钠肽和C型钠尿肽可增强铜绿假单胞菌和荧光性假单胞菌（*P. fluorescen*）的细胞毒性和修饰脂多糖。这种LPS修饰使其比未修饰的LPS对真核细胞更加具有致死性。这些激素通过两种胞质腺苷酸环化酶增加假单胞菌的cGMP和cAMP水平。在铜绿假单胞菌中，环核苷酸是一类重要的细胞内信号分子，能促进细菌生物膜的形成和毒力性状的表达。cAMP与毒力因子调节因子Vfr结合后，细菌毒力基因被激活，促进急性铜绿假单胞菌感染，而cGMP信号抑制急性感染相关基因，促进生物膜形成相关基因的表达，从而导致慢性感染[191，192]。

表皮生长因子（epidermal growth factor，EGF）是一种小分子蛋白质，常被动物细胞用作信号传递物质。表皮生长因子受体（epidermal growth factor receptor，EGFR），这种受体酪氨酸激酶对宿主-微生物的交流很重要，EGFR存在于细胞表面，与表皮生长因子结合后激活[193]。EGF的分泌需要经过真核细胞膜菱形体（rhomboid，RHO）蛋白水解激活。原核生物中存在RHO同系物，如由斯氏普罗威登斯菌（*Providencia stuartii*）编码的Aar蛋白，由此微生物和宿主间通信便建立了联系。在黑腹果蝇中，Aar能在功能上替代RHO；此外，真核生物RHO可以逆转斯氏普罗威登斯菌Aar突变菌株的功能[167，168]。以上研究表明，真核RHO和原核Aar可以互换，这提示EGF可能参与了真核生物与原核生物间的跨界信号交流。然而，目前还不清楚斯氏普罗威登斯菌是否能对真核细胞的EGF信号做出反应。

三、类固醇激素

类固醇激素控制哺乳动物的新陈代谢、免疫、性发育和许多其他生理功能。它们通常由胆固醇合成，是一类可以通过细胞膜的脂质。它们的受体通常位于细胞核内或细胞质内。在筛选根癌农杆菌（*Agrobacterium tumefaciens*）和铜绿假单胞菌基于AHL的群体感应抑制剂的研究中，Beury-Cirou等发现，三种哺乳动物类固醇激素（雌酮、雌三醇和雌二醇）可作为群体感应抑制剂，这些激素能减少AHL的积累，并降低依赖于群体感应系统的基因表达[194]。

四、抗菌肽

天然免疫系统可通过产生阳离子抗菌肽（cationic antimicrobial peptide，CAMP）发挥防御功能，由CAMPs形成的黏膜屏障被认为是保护宿主免受致病性肠道感染的关键。CAMPs与细菌膜相互作用并破坏细菌膜。一些细菌病原体通过改变脂多糖的脂质A，以防止CAMP与其外膜结合。其中最典型的一个例子是细胞内病原体肠道沙门菌，它可以通过脂质A修饰而产生CAMP抗性[195]。

第四节　宿主与微生物通过细胞外囊泡进行跨界信号交流

细胞外囊泡是由细胞产生的球形碎片，直径 10～1000nm，由脂质双层分隔，包含蛋白质、脂质、核酸和许多其他细胞成分。大量研究表明，细胞外囊泡在细胞 - 细胞信号传导中发挥重要作用，并且是跨界信号交流的重要机制，包括宿主和病原体之间的通信。由于许多细菌存在于肠道上皮细胞的黏液中，不需要直接的细胞 - 细胞接触，因此细胞外囊泡介导的细胞间通信是宿主 - 病原体相互作用的一个重要方面 [196]。细胞外囊泡通过各种机制进入细胞，一旦进入靶细胞，就会释放出其中的蛋白质、小 RNA（sRNA），转运 RNA（tRNA）和微小 RNA（miRNA）等，这些 RNA 通过调控靶基因的表达，调节受体细胞基因的表达和功能 [197]。

一、细菌分泌的胞外囊泡将 RNA 片段传递给哺乳动物宿主

外膜囊泡（outer membrane vesicle，OMV）是革兰阴性细菌分泌的细胞外囊泡，为 10～300nm，其内容物包括细胞周质和细胞中心质成分 [198]。革兰阴性细菌也分泌由细菌外膜和内膜独立分隔的外膜 - 内膜囊泡（O-IMV）[199]。革兰阳性细菌也分泌细胞外囊泡，但由于革兰阳性细菌不含外膜，这些细胞外囊泡通常被描述为细菌细胞外小泡（bacterial extracellular vesicle，BEV）[200]。OMVs 和 BEVs 介导细菌和哺乳动物细胞之间 sRNA 和 tRNA 片段的跨界转移，而不需要细胞直接接触 [201]。铜绿假单胞菌位于肺上皮细胞上的黏液层，分泌的 OMVs 通过黏液扩散，与气道上皮细胞顶膜的脂筏融合，并将 23 nt tRNA 片段（sRNA-52320）传递到宿主细胞。sRNA-52320 通过 LPS 靶向激活的 MAPK 信号通路中的多个基因，抑制对细菌感染的免疫应答，从而减少 IL-8 的分泌和中性粒细胞向小鼠肺部的迁移 [202]。因此，sRNA-52320 可促进铜绿假单胞菌在免疫功能低下的个体中建立慢性肺部感染。牙周病病原体放线共生放线埃希菌（*Aggregatibacter actinomycetemcomitans*）产生的 OMV 中含有 sRNAs，它们可以穿过血脑屏障，通过激活巨噬细胞的 NF-κB 信号通路，刺激 TNF-α 的产生，促进脑内促炎细胞因子的分泌 [203]。大肠埃希菌分泌的 OMVs 中的 sRNAs 被转移到膀胱上皮细胞中，能抑制 LPS 诱导的 IL-1a 分泌 [204]。幽门螺杆菌分泌的 OMVs 中含两种 sRNAs（sR-2509025 和 sR-989262），可减少人胃腺癌细胞由 LPS 刺激产生的 IL-8[205]。

二、哺乳动物细胞分泌的胞外囊泡向细菌传递 miRNA

研究表明，真核生物的 miRNA（长约 22nt）能影响细菌的基因表达和表型 [206, 207]。粪便 miRNAs 主要由肠道上皮细胞（IEC）和 Hopx+ 细胞产生。肠道微生物群可以利用粪便中的 miRNAs 来改变自身基因的表达 [208]。粪便 miRNA 对肠道微生物群的调控不同于传统的 miRNA 调控，后者主要包括 mRNA 转录后抑制，导致 mRNA 翻译效率降低 [209]。并且，大多数可检测的粪便 miRNAs 以细胞外囊泡的形式存在，这种形式使它们更稳定。

在肠道疾病中，具核梭杆菌与大肠埃希菌是与 CRC 密切相关的微生物。Liu 等首次证明，宿主粪便中的 miRNAs 可以通过改变细菌的相对丰度来调节肠道微生物群的组成。事实上，与野生型小鼠相比，肠上皮细胞特异性 miRNA 缺陷小鼠的细菌在属水平的多样性增加。他们证实，miR-515-5p 提高了具核梭杆菌 16S rRNA/23S rRNA 转录物的比例，miR-1226-5p

和 miR-4747-3p 分别上调了大肠埃希菌中 YegH 和 RNaseP 的 mRNA 水平；miR-1224-5p 和 miR-663 分别下调了大肠埃希菌中 rutA 和 FucO 的 mRNA 水平。YegH 属于与细胞膜相关的黄素腺嘌呤二核苷酸结合蛋白，是一种黄素蛋白氧化还原酶；随着 RNaseP 表达的增强，YegH 促进细菌的代谢和生长。RutA 是一种参与嘧啶代谢（降解和生物合成）的嘧啶单加氧酶，FucO 是一种参与碳水化合物代谢的乳醛还原酶。这些 miRNAs 转染到具核梭杆菌和大肠埃希菌内，可以影响细菌体内 miRNA 的丰度及细菌的生长和代谢[208]。miR-21 可通过影响肠道微生物群，加重葡聚糖硫酸钠诱导的结肠炎[210]。在蚊虫体内，miR-275 的缺失可导致肠道菌群减少；维持 miR-275 水平对多种肠道功能至关重要[211]。Ji 等发现，炎性肠病（inflammatory bowel disease，IBD）患者粪便中含有不同表达水平的 miRNA，它们通过调节某些细菌的生长影响 IBD 的进展。IBD 中差异表达的 miRNA 包括 miR-199a、miR-1226、miR-548a 和 miR-515-5p。它们影响病原菌具核梭杆菌、大肠埃希菌，以及益生菌片段丝状菌（SFB）的生长，最终可导致 IBD 的发生[212]。其中，miR-199a-5p 抑制 SFB 的增殖，但不影响具核梭杆菌和大肠埃希菌。相反，miR-1226 和 miR-515-5p 抑制具核梭杆菌和大肠埃希菌的生长，而促进 SFB 的增殖。此外，miR-548ab 抑制具核梭杆菌和大肠埃希菌的生长，但对 SFB 的生长没有影响。这些数据表明，粪便中 miRNAs 的差异表达能影响肠道细菌的生长。此外，miRNAs 可以调控不同种类细菌的相对丰度。因此，miRNAs 是肠道微生物群落组成和动态变化的重要“影响者”。

科学家还发现，口服 miR-30d 可改善实验性自身免疫性脑脊髓炎，这种作用是通过增加肠道中嗜黏蛋白阿克曼菌的丰度，从而增加调节性 T 细胞实现的[213]。在最近的一项研究中，原代培养的人气道上皮细胞分泌的细胞外囊泡可以将 miRNA let-7b-5p 传递给铜绿假单胞菌，从而减少生物膜形成所必需的几种蛋白质的丰度，最终减少具有抗生素抗性的生物膜形成。同时，miRNA let-7b-5p 也可靶向调控另一种机会性肺病原菌洋葱伯克霍尔德菌（*Burkholderia cenocepacia*）中相应的生物膜基因同源序列，表明 miRNA let-7b-5p 减少生物膜形成和提高抗生素敏感性的能力并不局限于铜绿假单胞菌[214]。

综上所述，上述研究表明，由 sRNA、tRNA 片段和含有 miRNA 的细胞外囊泡介导的界间通信是双向的：含有 miRNA 的真核细胞外囊泡能调节细菌的基因表达，反之，细菌 OMVs 和 BEVs 分泌的 sRNA 和 tRNA 片段能影响真核细胞的基因表达和功能。考虑到微生物群组成的变化与多种人类疾病有关，以 miRNAs 为靶标可能有助于调控微生物群平衡。此外，微生物可能主动或选择性地吸收不同的 miRNAs，这些 miRNAs 反过来又促进微生物的生长。以上这些研究均证实了原核微生物与真核生物间通过 miRNAs 进行跨物种交流的科学假设。

第五节　宿主与微生物跨界信号交流的其他分子

细胞间的化学信号传递是协调种群行为的有效途径。经过多年的研究，人们逐渐认识到，许多物质可用作跨物种间的化学信号交流，并且与物种营养获取、毒力及宿主防御等方面有重要的调节作用。哺乳动物和细菌细胞可利用这些信号来快速适应环境变化。

一、吲哚

吲哚是一种典型的氮 - 杂环芳香族化合物，

广泛存在于自然环境中。许多细菌及植物可产生吲哚，因此，吲哚及其衍生物广泛存在于原核和真核生物群落中。近年来的研究发现，吲哚可介导细胞间、种群间和不同物种界间的信号交流。吲哚及其衍生物通过抑制群体感应和毒力因子的产生，从而抑制多种耐药病原体的致病能力。昆虫的行为也可受吲哚调控。此外，科学家们发现，吲哚可调节动物体内氧化应激、肠道炎症和激素分泌，并能影响人类疾病，如炎症、神经系统和代谢性疾病[215]。

（一）吲哚在细菌中的产生及作用

许多革兰阳性和革兰阴性细菌可利用色氨酸酶（TnaA）将色氨酸代谢合成吲哚。大肠埃希菌和霍乱弧菌液体培养上清中吲哚浓度可达 0.5mmol/L[216, 217]。在小鼠、大鼠和人肠道中，可检测到的吲哚浓度高达 1.1mmol/L[218]。

吲哚可调节革兰阳性菌株的孢子形成[219]；调节大肠埃希菌的质粒稳定性[220]、细胞分裂[221]、抗生素耐受性[222]和毒力[223]，以及大肠埃希菌和霍乱弧菌的生物膜形成[216, 217]。

具体来说，吲哚可以调节大肠埃希菌的耐药性形成，这主要是通过药物外排泵和氧化应激保护机制发挥作用[222]。另一方面，大肠埃希菌在抗生素作用下能增加吲哚的生物合成[224]。吲哚还增加了细菌对抗生素的耐药性，这可能与氧化应激和噬菌体休克途径有关[225]。然而，也有研究认为吲哚可降低细菌耐药性：即吲哚可通过磷酸二酯酶 DosP（降低激活 TnaA 所需的环磷酸腺苷浓度）和毒素蛋白 YafQ（切割 TnaA mRNA）来降低耐药性的形成[226]。得出这种不同结论可能是由于研究中使用的检测方法不同所致。高浓度的吲哚（＞ 1mmol/L）可通过调节膜电位抑制大肠埃希菌细胞分裂[221]。最初认为，吲哚的摄入与排出仅依赖于 AcrEF-TolC 和 Mtr 转运蛋白。然而，最近发现，吲哚在大肠埃希菌中可不依赖于 AcrEF-TolC 和 Mtr 而通过细胞膜转运[227]。

大肠埃希菌和霍乱菌株是目前吲哚研究的主要对象。近年来，吲哚在其他吲哚产生菌中的作用也逐渐被科学家们所认识。据报道，海洋病原体迟钝爱德华菌（*Edwardsiella tarda*）中色氨酸酶的表达增加可促进生物膜和抗生素抗性形成[228]；而在鳗弧菌（*Vibrio anguillarum*）中，吲哚可抑制生物膜的形成和毒力[229]。

（二）吲哚对人类疾病的影响

吲哚不仅可调节细菌的毒力、生物膜形成、抗生素耐受性和群体感应，还能影响真核生物的生命活动。动物不能合成吲哚，但能用嗅觉系统感知吲哚，如苍蝇、蝴蝶等能感受到吲哚，并调节它们的行为[230]。最近研究表明，吲哚可能通过影响氧化应激[231]、肠道炎症[232]和激素分泌[233]对人体健康产生有益的影响。

吲哚在产吲哚细菌中是稳定存在的，但许多非产吲哚细菌和真核生物可以利用多种加氧酶（如单加氧酶、双加氧酶和 P450 家族成员）修饰或降解吲哚[234]。因此，吲哚衍生物广泛存在于原核和真核生物群落中。

在小鼠和人体肠道中能检测到的吲哚浓度为 250 ～ 1100μmol/L，然而动物细胞不能合成吲哚，因而这些吲哚主要是由肠道细菌产生的。研究人员在血液、外周组织、尿液甚至脑组织中检测到浓度高达 10 ～ 200μmol/L 的吲哚及其衍生物，如吲哚硫酸酯、吲哚 -3- 丙酸、靛红和 5- 羟基吲哚[231, 235]。然而目前它们在人体中的作用研究较少。现有研究表明：吲哚及其衍生物可能影响人类疾病，如细菌感染、肠道炎症、神经系统疾病、糖尿病、癌症等。

在人类肠上皮细胞中，吲哚和 7- 羟基吲哚能增加上皮细胞紧密连接，降低细胞炎症因子水平[218]。在无菌小鼠模型中，口服吲哚可增加结肠上皮细胞间紧密连接和黏附性[232]。这些结果表明，吲哚和吲哚产生菌在建立上皮屏障和预防肠道炎症方面发挥了有益作用。最近的研究表明，肠道微生物组在 IBD 中有不可或缺的作用，或许，吲哚及其衍生物在IBD这一类肠道微生物相关疾病中也具有潜在作用。

吲哚能调节肠内分泌细胞分泌肠促胰岛素肽GLP-1[233]，因此，吲哚或产生吲哚的微生物群可能影响代谢性疾病，如2型糖尿病。葡萄糖通过抑制色氨酸酶的分解代谢抑制了吲哚的生物合成[236]，表明高糖饮食能抑制吲哚的产生，而高色氨酸蛋白质饮食能促进人体肠道中吲哚的产生[233]。

十字花科蔬菜中的吲哚-3-甲醇和3，3′-二吲哚甲烷已被证明对多种癌症具有抗癌活性[237]。美国国家癌症研究所已开始在癌症患者中对这两种药物进行临床试验。吲哚-3-甲醇已被证明可调节多种癌细胞系中的细胞信号，包括Akt、NF-κB、JNK、MEK和p53[238]；3，3′-二吲哚甲烷是一种有效的辐射保护剂和缓解剂，通过刺激DNA修复和NF-κB生存信号发挥作用[239]。因此，吲哚及其衍生物被认为对多种癌症具有治疗潜力，值得进一步研究。

二、腺嘌呤核苷三磷酸（ATP）

（一）真核细胞与原核细胞中的胞外ATP（eATP）

ATP不仅是细胞内普遍存在的能量载体，而且能分泌到细胞外（eATP）作为细胞外信号分子发挥重要作用[240]。作为细胞外的信号分子，ATP与血小板活化和凝血[241，242]、炎症[243]，以及多种神经系统疾病有关[244]。

众所周知，ATP是多细胞生物细胞间通信的信使，其作用在过去几十年里已经被广泛研究。然而，不仅是多细胞生物，研究表明：单细胞生物如酵母或细菌也能分泌或使用ATP作为细胞外信号分子[245]。Ivanova等于2005年首次在细菌中发现了ATP，他们在86株海洋细菌的培养上清中，发现了革兰阴性α-变形杆菌和γ-变形杆菌能分泌eATP，测得的浓度为0.2～7mmol/L。而革兰阳性菌产生的ATP要比α-变形杆菌和γ-变形杆菌少得多[246]。

（二）真核细胞与原核细胞通过eATP进行跨界信号交流

细菌ATP已被证明在不同器官系统中可作为一种跨界信号分子发挥作用。哺乳动物肠道相关淋巴组织派尔斑中的滤泡辅助性T细胞（T follicular helper，Tfh）可受细菌ATP的调控。Tfh是B细胞发育及产生IgA的关键细胞，这个过程部分是通过P2X7介导的凋亡发挥作用[247]。Proietti等在动物实验中，用含或不含水解ATP的三磷酸腺苷双磷酸酶质粒的大肠埃希菌及肠道沙门菌灌胃，发现细菌ATP可诱导P2X7介导的Tfh凋亡，随后限制B细胞分泌IgA抗体，从而降低口服疫苗的效率[248]。Atarashi等发现，在腹腔和直肠内予以1.25 mg（2.3μmol/L）ATPγS后，$CD70^{high}CD11c^{Low}$肠固有层细胞产生细胞因子（IL-6，IL-23，TGF-β）增加；另外在无菌小鼠中，予以ATP刺激，促使原始$CD4^{+}$T细胞向辅助性T细胞（TH17）的分化增多[249]。Abbasian等认为，尿路致病菌分泌的ATP作为一种毒力因子，能促进Ca^{2+}内流，影响使膀胱收缩的尿上皮细胞，从而参与急性尿失禁和过度活动性膀胱疾病[250]。

细菌产生的eATP不仅可以直接作用于真核生物，还可以刺激真核细胞产生ATP增加。Alvarez等发现，Caco-2细胞在大肠埃希菌刺激后，培养上清中ATP浓度上升了6倍；大肠埃希菌刺激后的大鼠空肠段管腔分泌ATP增加了2倍[251]。研究还表明，大肠埃希菌的成孔毒素α-溶血素和聚集放线菌的白细胞毒素A可导致红细胞的ATP分泌增加[252]。

此外，细菌也可以利用宿主来源的eATP。Mempin等发现，向在平台期生长的大肠埃希菌K12菌株中添加10或100μmol/L ATP，其存活时间可提高7天[253]。

这种作用表明，ATP对于细菌生长是有

利的，细菌入侵宿主细胞，可能利用细胞内的ATP作为营养物质或信号分子。也有研究表明，eATP能促进牙周病中生物膜的形成，并促进细菌从具核梭杆菌生物膜中分离出来，形成新的克隆菌落[254]。

ATP是细胞间嘌呤信号系统的主要参与者。越来越多的证据表明，细菌等单核生物也能利用这种通信系统。细菌以依赖于葡萄糖、呼吸链等方式分泌ATP。此外，细菌也可以利用宿主细胞产生的eATP。eATP已被证明是一种跨界交流的信号分子，特别是在肠道及口腔黏膜部位。然而，这方面还需深入研究，尤其是细菌如何利用宿主细胞ATP产生有利于自身生存的环境，从而增强致病性。阐明这些问题，将为细菌感染的防治策略提供新的解决方法。

第六节　跨界信号交流的研究前景

一、改善细菌耐药

细菌耐药性的产生代表着一场巨大的全球健康危机，也是当今人类面临的最严重威胁之一。一些细菌对几乎所有抗生素都产生了耐药性。因此，迫切需要新的抗菌剂来克服耐药菌。为了满足新治疗方法的迫切需要，研究人员进行了大量研究工作，其中一些通过抑制耐药机制（如β-内酰胺酶抑制剂抗生素佐剂的作用）成功地产生了针对耐药革兰阴性菌的抗菌活性。另一个较有研究前景的方向是一些具有抗菌活性的天然衍生药物，如噬菌体、DCAP{2-[(3-(3，6-dichloro-9H-carbazol-9-yl)-2-hydroxypropyl)amino]-2-(hydroxymethyl)propane-1，3-diol}、ODLs（odilorhabdins）、肽基苯并咪唑、群体感应抑制剂，以及金属抗菌剂[255]。细菌耐药性是植物、动物和人类中正在出现的阻碍抗菌药物治疗的障碍之一。细菌可以通过产生生物膜这种方式有利于其自身生存，而生物膜导致常规抗生素无法穿透生物膜发挥杀菌抑菌效应。群体感应系统作为调控细菌生物膜形成的信号系统，因而针对群体感应系统的药物可能是解决这一问题的策略之一[256]。

工程菌通过群体感应调节来感应和杀死病原体。Hwang等通过使用具有益生菌作用的大肠埃希菌菌株Nissle 1917，使其通过AI（3-oxo-C12-HSL）调节的运动寻找病原体，并诱导病原体自溶（由赖氨酸E7驱动）和释放抗铜绿假单胞菌毒素pyocin c5，由此杀死病原体而发挥有益作用[257，258]。该工程菌对铜绿假单胞菌肠道感染具有体内预防和治疗作用。

群体感应在条件致病菌铜绿假单胞菌感染过程中起着重要作用。研究发现，群体感应抑制剂可以在不杀死细菌的情况下干扰感染发生的初级阶段，因此群体感应抑制剂被认为是抗感染治疗的新途径。肉桂醛是一种铜绿假单胞菌生物膜抑制剂和生物膜分散剂。学者将肉桂醛和黏菌素联合应用，结果发现这两者具有协同作用，而肉桂醛与羧苄青霉素、妥布霉素或红霉素联合应用则无此效应。进一步研究发现，肉桂醛通过抑制群体感应信号分子及其受体（lasB、rhlA和pqsA）的表达而表现出群体感应抑制剂的活性。因此，可将肉桂醛与黏菌素联合运用，来治疗铜绿假单胞菌感染[259]。

二、微生物可能通过群体感应促进疾病发生

细菌群体感应分子，无论是革兰阴性菌的AHLs分子，还是革兰阳性菌的AIPs分子，或

者通用分子 AI-2，均可与真核细胞生物间进行跨界信号交流，它们具有诱导宿主细胞产生炎症反应，抑制宿主免疫等功能。以往的观点认为，真核生物与原核生物各自使用自己的语言，将二者作为独立的个体；现在的研究表明，真核生物来源的信号分子与原核生物来源的信号分子可能具有相似性，能被彼此所监听和感知，并对其产生反应。目前，AHLs 对真核生物作用的研究已经较多，然而 AIPs 和 AI-2 的作用仍需大量研究来证明。这种跨界信号相互作用的研究可能为疾病发病机制的研究提供新的视角和观念。

三、利用群体感应调节肠道菌群以治疗疾病

研究发现，抗生素处理引起的微生物群失衡小鼠，予以大肠埃希菌后增加了肠道 AI-2 水平，影响了微生物群的整体结构，特别有利于厚壁菌门的生长。这表明 AI-2 可以影响肠道内主要菌群的平衡，并促进肠道内稳态的恢复[33]。然而，哺乳动物的肠道微生物之间相互作用的网络十分复杂，群体感应分子可以对这些菌群进行调控，改善肠道菌群失衡状态。因此，目前学者们正在尝试使用各种益生菌或化学试剂来调节肠道菌群，改变菌群结构，使其尽量恢复健康状态，从而治疗各种菌群失衡导致的疾病。这种新的治疗途径具有广阔的发展前景，仍需要大量研究进行深入探索。

参考文献

[1] Pacheco AR, Sperandio V. Inter-kingdom signaling: chemical language between bacteria and host. Current Opinion in Microbiology, 2009, 12(2): 192-198.

[2] Papenfort K, Bassler BL. Quorum sensing signal-response systems in Gram-negative bacteria. Nature Reviews: Microbiology, 2016, 14(9): 576-588.

[3] Whiteley M, Diggle SP, Greenberg EP. Progress in and promise of bacterial quorum sensing research. Nature, 2017, 551(7680): 313-320.

[4] Kaper JB, Sperandio V. Bacterial cell-to-cell signaling in the gastrointestinal tract. Infection And Immunity, 2005, 73(6): 3197-3209.

[5] Ng WL, Bassler BL. Bacterial quorum-sensing network architectures. Annual Review of Genetics, 2009, 43: 197-222.

[6] Moreno-Gamez S, Sorg RA, Domenech A, et al. Quorum sensing integrates environmental cues, cell density and cell history to control bacterial competence. Nature Communications, 2017, 8(1): 854.

[7] Sintim HO, Smith JA, Wang J, et al. Paradigm shift in discovering next-generation anti-infective agents: targeting quorum sensing, c-di-GMP signaling and biofilm formation in bacteria with small molecules. Future Medicinal Chemistry, 2010, 2(6): 1005-1035.

[8] Galloway WR, Hodgkinson JT, Bowden SD, et al. Quorum sensing in Gram-negative bacteria: small-molecule modulation of AHL and AI-2 quorum sensing pathways. Chemical Reviews, 2011, 111(1): 28-67.

[9] Kumari A, Pasini P, Deo SK, et al. Biosensing systems for the detection of bacterial quorum signaling molecules. Analytical Chemistry, 2006, 78(22): 7603-7609.

[10] Pearson JP, Van Delden C, Iglewski BH. Active efflux and diffusion are involved in transport of Pseudomonas aeruginosa cell-to-cell signals. Journal of Bacteriology, 1999, 181(4): 1203-1210.

[11] Chbib C. Impact of the structure-activity relationship of AHL analogues on quorum sensing in Gram-negative bacteria. Bloorganic & Medicinal Chemistry, 2020, 28(3): 115282.

[12] Irie Y, Parsek MR. Quorum sensing and microbial biofilms. Current Topics in Microbiology and Immunology, 2008, 322: 67-84.

[13] Syvitski RT, Tian XL, Sampara K, et al. Structure-activity analysis of quorum-sensing signaling peptides from Streptococcus mutans. Journal of Bacteriology, 2007, 189(4): 1441-1450.

[14] Novick RP, Geisinger E. Quorum sensing in staphylococci. Annual Review of Genetics, 2008, 42: 541-564.

[15] Monnet V, Gardan R. Quorum-sensing regulators in Gram-positive bacteria: ‘cherchez le peptide’. Molecular Microbiology, 2015, 97(2): 181-184.

[16] McBrayer DN, Cameron CD, Tal-Gan Y. Development and utilization of peptide-based quorum sensing

modulators in Gram-positive bacteria. Organic & Biomolecular Chemistry, 2020, 18(37): 7273-7290.

[17] Chen X, Schauder S, Potier N, et al. Structural identification of a bacterial quorum-sensing signal containing boron. Nature, 2002, 415(6871): 545-549.

[18] Zhao J, Quan C, Jin L, et al. Production, detection and application perspectives of quorum sensing autoinducer-2 in bacteria. Journal of Biotechnology, 2018, 268: 53-60.

[19] Pereira CS, Thompson JA, Xavier KB. AI-2-mediated signalling in bacteria. FEMS Microbiology Reviews, 2013, 37(2): 156-181.

[20] Miller ST, Xavier KB, Campagna SR, et al. Salmonella typhimurium recognizes a chemically distinct form of the bacterial quorum-sensing signal AI-2. Molecular Cell, 2004, 15(5): 677-687.

[21] Wang S, Payne GF, Bentley WE. Quorum sensing communication: molecularly connecting cells, Their neighbors, and even devices. Annual Review of Chemical and Biomolecular Engineering, 2020, 11: 447-468.

[22] Bassler BL, Wright M, Silverman MR. Multiple signalling systems controlling expression of luminescence in Vibrio harveyi: sequence and function of genes encoding a second sensory pathway. Molecular Microbiology, 1994, 13(2): 273-286.

[23] Schauder S, Shokat K, Surette MG, et al. The LuxS family of bacterial autoinducers: biosynthesis of a novel quorum-sensing signal molecule. Molecular Microbiology, 2001, 41(2): 463-476.

[24] Nichols JD, Johnson MR, Chou CJ, et al. Temperature, not LuxS, mediates AI-2 formation in hydrothermal habitats. FEMS Microbiology Ecology, 2009, 68(2): 173-181.

[25] Bhattacharyya M, Vishveshwara S. Functional correlation of bacterial LuxS with their quaternary associations: interface analysis of the structure networks. BMC Structural Biology, 2009, 9: 8.

[26] Vendeville A, Winzer K, Heurlier K, et al. Making 'sense' of metabolism: autoinducer-2, LuxS and pathogenic bacteria. Nature Reviews: Microbiology, 2005, 3(5): 383-396.

[27] Freeman JA, Lilley BN, Bassler BL. A genetic analysis of the functions of LuxN: a two-component hybrid sensor kinase that regulates quorum sensing in Vibrio harveyi. Molecular Microbiology, 2000, 35(1): 139-149.

[28] Freeman JA, Bassler BL. Sequence and function of LuxU: a two-component phosphorelay protein that regulates quorum sensing in Vibrio harveyi. Journal of Bacteriology, 1999, 181(3): 899-906.

[29] Kaur A, Capalash N, Sharma P. Communication mechanisms in extremophiles: Exploring their existence and industrial applications. Microbiological Research, 2019, 221: 15-27.

[30] Herzberg M, Kaye IK, Peti W, et al. YdgG (TqsA) controls biofilm formation in Escherichia coli K-12 through autoinducer 2 transport. Journal of Bacteriology, 2006, 188(2): 587-598.

[31] Servinsky MD, Terrell JL, Tsao CY, et al. Directed assembly of a bacterial quorum. ISME Journal, 2016, 10(1): 158-169.

[32] Di Cagno R, De Angelis M, Calasso M, et al. Proteomics of the bacterial cross-talk by quorum sensing. Journal of Proteomics, 2011, 74(1): 19-34.

[33] Bivar Xavier K. Bacterial interspecies quorum sensing in the mammalian gut microbiota. Comptes Rendus Biologies, 2018, 341(5): 297-299.

[34] Blair WM, Doucette GJ. The Vibrio harveyi bioassay used routinely to detect AI-2 quorum sensing inhibition is confounded by inconsistent normalization across marine matrices. Journal of Microbiological Methods, 2013, 92(3): 250-252.

[35] Raut N, Joel S, Pasini P, et al. Bacterial autoinducer-2 detection via an engineered quorum sensing protein. Analytical Chemistry, 2015, 87(5): 2608-2614.

[36] Zhu J, Pei D. A LuxP-based fluorescent sensor for bacterial autoinducer II. ACS Chemical Biology, 2008, 3(2): 110-119.

[37] Thiel V, Vilchez R, Sztajer H, et al. Identification, quantification, and determination of the absolute configuration of the bacterial quorum-sensing signal autoinducer-2 by gas chromatography-mass spectrometry. ChemBioChem, 2009, 10(3): 479-485.

[38] Kim CS, Gatsios A, Cuesta S, et al. Characterization of autoinducer-3 structure and biosynthesis in E. coli. ACS Central Science, 2020, 6(2): 197-206.

[39] Moreira CG, Sperandio V. The epinephrine/norepinephrine/autoinducer-3 interkingdom signaling system in escherichia coli O157:H7. Advances in Experimental Medicine and Biology, 2016, 874: 247-261.

[40] Tobias NJ, Brehm J, Kresovic D, et al. New vocabulary for bacterial communication. ChemBioChem, 2020, 21(6): 759-768.

[41] Kaper JB, Nataro JP, Mobley HL. Pathogenic Escherichia coli. Nature Reviews: Microbiology, 2004, 2(2): 123-140.

[42] LaSarre B, Federle MJ. Exploiting quorum sensing to confuse bacterial pathogens. Microbiology and Molecular Biology Reviews, 2013, 77(1): 73-111.

[43] Lister PD, Wolter DJ, Hanson ND. Antibacterial-resistant Pseudomonas aeruginosa: clinical impact and complex regulation of chromosomally encoded resistance mechanisms. Clinical Microbiology Reviews, 2009, 22(4): 582-610.

[44] Lee J, Zhang L. The hierarchy quorum sensing network in Pseudomonas aeruginosa. Protein & Cell, 2015, 6(1): 26-41.

[45] Schuster M, Greenberg EP. A network of networks: quorum-sensing gene regulation in Pseudomonas aeruginosa. International Journal of Medical Microbiology, 2006, 296(2-3): 73-81.

[46] Defoirdt T, Boon N, Sorgeloos P, et al. Alternatives to antibiotics to control bacterial infections: luminescent vibriosis in aquaculture as an example. Trends in Biotechnology, 2007, 25(10): 472-479.

[47] Ruwandeepika HA, Karunasagar I, Bossier P, et al. Expression and quorum sensing regulation of type III secretion system genes of vibrio harveyi during infection of gnotobiotic brine shrimp. PLoS One, 2015, 10(12): e0143935.

[48] Lilley BN, Bassler BL. Regulation of quorum sensing in Vibrio harveyi by LuxO and sigma-54. Molecular Microbiology, 2000, 36(4): 940-954.

[49] Defoirdt T, Darshanee Ruwandeepika HA, Karunasagar I, et al. Quorum sensing negatively regulates chitinase in Vibrio harveyi. Environmental Microbiology Reports, 2010, 2(1): 44-49.

[50] Natrah FM, Ruwandeepika HA, Pawar S, et al. Regulation of virulence factors by quorum sensing in Vibrio harveyi. Veterinary Microbiology, 2011, 154(1-2): 124-129.

[51] Ruwandeepika HA, Bhowmick PP, Karunasagar I, et al. Quorum sensing regulation of virulence gene expression in Vibrio harveyi in vitro and in vivo during infection of gnotobiotic brine shrimp larvae. Environmental Microbiology Reports, 2011, 3(5): 597-602.

[52] Yang Q, Defoirdt T. Quorum sensing positively regulates flagellar motility in pathogenic Vibrio harveyi. Environmental Microbiology, 2015, 17(4): 960-968.

[53] Riedel CU, Monk IR, Casey PG, et al. AgrD-dependent quorum sensing affects biofilm formation, invasion, virulence and global gene expression profiles in Listeria monocytogenes. Molecular Microbiology, 2009, 71(5): 1177-1189.

[54] France MT, Cornea A, Kehlet-Delgado H, et al. Spatial structure facilitates the accumulation and persistence of antibiotic-resistant mutants in biofilms. Evolutionary Applications, 2019, 12(3): 498-507.

[55] Kolter R, Greenberg EP. Microbial sciences: the superficial life of microbes. Nature, 2006, 441(7091): 300-302.

[56] Tolker-Nielsen T. Biofilm development. Microbiology Spectrum, 2015, 3(2): MB-0001-2014.

[57] Flemming HC, Wingender J. The biofilm matrix. Nature reviews: microbiology, 2010, 8(9): 623-633.

[58] Hoiby N, Bjarnsholt T, Moser C, et al. ESCMID guideline for the diagnosis and treatment of biofilm infections 2014. Clinical Microbiology And Infection, 2015, 21 Suppl 1: S1-25.

[59] Parsek MR, Greenberg EP. Sociomicrobiology: the connections between quorum sensing and biofilms. Trends in Microbiology, 2005, 13(1): 27-33.

[60] O'Toole G, Kaplan HB, Kolter R. Biofilm formation as microbial development. Annual Review of Microbiology, 2000, 54: 49-79.

[61] Jamal M, Ahmad W, Andleeb S, et al. Bacterial biofilm and associated infections. Journal of the Chinese Medical Association, 2018, 81(1): 7-11.

[62] Roy R, Tiwari M, Donelli G, et al. Strategies for combating bacterial biofilms: A focus on anti-biofilm agents and their mechanisms of action. Virulence, 2018, 9(1): 522-554.

[63] Costerton JW, Lewandowski Z, Caldwell DE, et al. Microbial biofilms. Annual Review of Microbiology, 1995, 49: 711-745.

[64] Hall-Stoodley L, Costerton JW, Stoodley P. Bacterial biofilms: from the natural environment to infectious diseases. Nature Reviews: Microbiology, 2004, 2(2): 95-108.

[65] Parsek MR, Fuqua C. Biofilms 2003: emerging themes and challenges in studies of surface-associated microbial life. Journal of Bacteriology, 2004, 186(14): 4427-4440.

[66] Giaouris E, Heir E, Desvaux M, et al. Intra- and inter-species interactions within biofilms of important foodborne bacterial pathogens. Frontiers in Microbiology, 2015, 6: 841.

[67] Kostakioti M, Hadjifrangiskou M, Hultgren SJ. Bacterial biofilms: development, dispersal, and therapeutic strategies in the dawn of the postantibiotic era. Cold Spring Harbor Perspectives in Medicine, 2013, 3(4): a010306.

[68] Domka J, Lee J, Bansal T, et al. Temporal gene-

expression in Escherichia coli K-12 biofilms. Environmental Microbiology, 2007, 9(2): 332-346.

[69] Fuqua WC, Winans SC, Greenberg EP. Quorum sensing in bacteria: the LuxR-LuxI family of cell density-responsive transcriptional regulators. Journal of Bacteriology, 1994, 176(2): 269-275.

[70] Labbate M, Queck SY, Koh KS, et al. Quorum sensing-controlled biofilm development in Serratia liquefaciens MG1. Journal of Bacteriology, 2004, 186(3): 692-698.

[71] An SQ, Murtagh J, Twomey KB, et al. Modulation of antibiotic sensitivity and biofilm formation in Pseudomonas aeruginosa by interspecies signal analogues. Nature Communications, 2019, 10(1): 2334.

[72] Zhao X, Yu Z, Ding T. Quorum-sensing regulation of antimicrobial resistance in bacteria. Microorganisms, 2020, 8(3): 425.

[73] McCluskey J, Hinds J, Husain S, et al. A two-component system that controls the expression of pneumococcal surface antigen A (PsaA) and regulates virulence and resistance to oxidative stress in Streptococcus pneumoniae. Molecular Microbiology, 2004, 51(6): 1661-1675.

[74] Dubrac S, Boneca IG, Poupel O, et al. New insights into the WalK/WalR (YycG/YycF) essential signal transduction pathway reveal a major role in controlling cell wall metabolism and biofilm formation in Staphylococcus aureus. Journal of Bacteriology, 2007, 189(22): 8257-8269.

[75] Gimza BD, Larias MI, Budny BG, et al. Mapping the global network of extracellular protease regulation in staphylococcus aureus. mSphere, 2019, 4(5): e00676-00619.

[76] Zhang L, Gray L, Novick RP, et al. Transmembrane topology of AgrB, the protein involved in the post-translational modification of AgrD in Staphylococcus aureus. Journal of Biological Chemistry, 2002, 277(38): 34736-34742.

[77] Queck SY, Jameson-Lee M, Villaruz AE, et al. RNAIII-independent target gene control by the agr quorum-sensing system: insight into the evolution of virulence regulation in Staphylococcus aureus. Molecular Cell, 2008, 32(1): 150-158.

[78] Boles BR, Horswill AR. Agr-mediated dispersal of Staphylococcus aureus biofilms. PLoS Pathogens, 2008, 4(4): e1000052.

[79] Hammer BK, Bassler BL. Quorum sensing controls biofilm formation in Vibrio cholerae. Molecular Microbiology, 2003, 50(1): 101-104.

[80] Duan K, Surette MG. Environmental regulation of Pseudomonas aeruginosa PAO1 Las and Rhl quorum-sensing systems. Journal of Bacteriology, 2007, 189(13): 4827-4836.

[81] McNab R, Ford SK, El-Sabaeny A, et al. LuxS-based signaling in Streptococcus gordonii: autoinducer 2 controls carbohydrate metabolism and biofilm formation with Porphyromonas gingivalis. Journal of Bacteriology, 2003, 185(1): 274-284.

[82] Guo L, He X, Shi W. Intercellular communications in multispecies oral microbial communities. Frontiers in Microbiology, 2014, 5: 328.

[83] Balestrino D, Haagensen JA, Rich C, et al. Characterization of type 2 quorum sensing in Klebsiella pneumoniae and relationship with biofilm formation. Journal of Bacteriology, 2005, 187(8): 2870-2880.

[84] Belkaid Y, Hand TW. Role of the microbiota in immunity and inflammation. Cell, 2014, 157(1): 121-141.

[85] De Vadder F, Grasset E, Manneras Holm L, et al. Gut microbiota regulates maturation of the adult enteric nervous system via enteric serotonin networks. Proceedings of the National Academy of Sciences of the United States of America, 2018, 115(25): 6458-6463.

[86] Shiner EK, Rumbaugh KP, Williams SC. Inter-kingdom signaling: deciphering the language of acyl homoserine lactones. FEMS Microbiology Reviews, 2005, 29(5): 935-947.

[87] Hughes DT, Sperandio V. Inter-kingdom signalling: communication between bacteria and their hosts. Nature Reviews: Microbiology, 2008, 6(2): 111-120.

[88] Coutinho BG, Mevers E, Schaefer AL, et al. A plant-responsive bacterial-signaling system senses an ethanolamine derivative. Proceedings of the National Academy of Sciences of the United States of America, 2018, 115(39): 9785-9790.

[89] Zhao Q, Yang XY, Li Y, et al. N-3-oxo-hexanoyl-homoserine lactone, a bacterial quorum sensing signal, enhances salt tolerance in Arabidopsis and wheat. Botanical Studies, 2020, 61(1): 8.

[90] McFall-Ngai M, Hadfield MG, Bosch TC, et al. Animals in a bacterial world, a new imperative for the life sciences. Proceedings of the National Academy of Sciences of the United States of America, 2013, 110(9): 3229-3236.

[91] Hsiao A, Ahmed AM, Subramanian S, et al. Members of the human gut microbiota involved in recovery from Vibrio cholerae infection. Nature, 2014, 515(7527): 423-426.

[92] Boyer M, Wisniewski-Dye F. Cell-cell signalling in bacteria: not simply a matter of quorum. FEMS Microbiology Ecology, 2009, 70(1): 1-19.

[93] Williams SC, Patterson EK, Carty NL, et al. Pseudomonas aeruginosa autoinducer enters and functions in mammalian cells. Journal of Bacteriology, 2004, 186(8): 2281-2287.

[94] DiMango E, Zar HJ, Bryan R, et al. Diverse Pseudomonas aeruginosa gene products stimulate respiratory epithelial cells to produce interleukin-8. Journal of Clinical Investigation, 1995, 96(5): 2204-2210.

[95] Kravchenko VV, Kaufmann GF, Mathison JC, et al. Modulation of gene expression via disruption of NF-kappaB signaling by a bacterial small molecule. Science, 2008, 321(5886): 259-263.

[96] Ritchie AJ, Whittall C, Lazenby JJ, et al. The immunomodulatory Pseudomonas aeruginosa signalling molecule N-(3-oxododecanoyl)-L-homoserine lactone enters mammalian cells in an unregulated fashion. Immunology and Cell Biology, 2007, 85(8): 596-602.

[97] Kravchenko VV, Kaufmann GF. Bacterial inhibition of inflammatory responses via TLR-independent mechanisms. Cellular Microbiology, 2013, 15(4): 527-536.

[98] Cooley M, Chhabra SR, Williams P. N-Acylhomoserine lactone-mediated quorum sensing: a twist in the tail and a blow for host immunity. chemistry & biology, 2008, 15(11): 1141-1147.

[99] Singh PK, Schaefer AL, Parsek MR, et al. Quorum-sensing signals indicate that cystic fibrosis lungs are infected with bacterial biofilms. Nature, 2000, 407(6805): 762-764.

[100] Smith RS, Fedyk ER, Springer TA, et al. IL-8 production in human lung fibroblasts and epithelial cells activated by the Pseudomonas autoinducer N-3-oxododecanoyl homoserine lactone is transcriptionally regulated by NF-kappa B and activator protein-2. Journal of Immunology, 2001, 167(1): 366-374.

[101] Palfreyman RW, Watson ML, Eden C, et al. Induction of biologically active interleukin-8 from lung epithelial cells by Burkholderia (Pseudomonas) cepacia products. Infection And Immunity, 1997, 65(2): 617-622.

[102] Zimmermann S, Wagner C, Muller W, et al. Induction of neutrophil chemotaxis by the quorum-sensing molecule N-(3-oxododecanoyl)-L-homoserine lactone. Infection And Immunity, 2006, 74(10): 5687-5692.

[103] Hansch GM, Prior B, Brenner-Weiss G, et al. The Pseudomonas quinolone signal (PQS) stimulates chemotaxis of polymorphonuclear neutrophils. Journal of Applied Biomaterials & Functional Materials, 2014, 12(1): 21-26.

[104] Polverino E, Rosales-Mayor E, Dale GE, et al. The Role of neutrophil elastase inhibitors in lung diseases. Chest, 2017, 152(2): 249-262.

[105] Smith RS, Harris SG, Phipps R, et al. The Pseudomonas aeruginosa quorum-sensing molecule N-(3-oxododecanoyl)homoserine lactone contributes to virulence and induces inflammation in vivo. Journal of Bacteriology, 2002, 184(4): 1132-1139.

[106] Gupta RK, Chhibber S, Harjai K. Quorum sensing signal molecules cause renal tissue inflammation through local cytokine responses in experimental UTI caused by Pseudomonas aeruginosa. Immunobiology, 2013, 218(2): 181-185.

[107] Jahoor A, Patel R, Bryan A, et al. Peroxisome proliferator-activated receptors mediate host cell proinflammatory responses to Pseudomonas aeruginosa autoinducer. Journal of Bacteriology, 2008, 190(13): 4408-4415.

[108] Telford G, Wheeler D, Williams P, et al. The Pseudomonas aeruginosa quorum-sensing signal molecule N-(3-oxododecanoyl)-L-homoserine lactone has immunomodulatory activity. Infection And Immunity, 1998, 66(1): 36-42.

[109] Horikawa M, Tateda K, Tuzuki E, et al. Synthesis of Pseudomonas quorum-sensing autoinducer analogs and structural entities required for induction of apoptosis in macrophages. Bioorganic & Medicinal Chemistry Letters, 2006, 16(8): 2130-2133.

[110] Chhabra SR, Harty C, Hooi DS, et al. Synthetic analogues of the bacterial signal (quorum sensing) molecule N-(3-oxododecanoyl)-L-homoserine lactone as immune modulators. Journal of Medicinal Chemistry, 2003, 46(1): 97-104.

[111] Ritchie AJ, Jansson A, Stallberg J, et al. The Pseudomonas aeruginosa quorum-sensing molecule N-3-(oxododecanoyl)-L-homoserine lactone inhibits T-cell differentiation and cytokine production by a mechanism involving an early step in T-cell activation. Infection And Immunity, 2005, 73(3): 1648-1655.

[112] Hooi DS, Bycroft BW, Chhabra SR, et al. Differential immune modulatory activity of Pseudomonas aeruginosa quorum-sensing signal molecules.

Infection And Immunity, 2004, 72(11): 6463-6470.

[113] Ritchie AJ, Yam AO, Tanabe KM, et al. Modification of in vivo and in vitro T- and B-cell-mediated immune responses by the Pseudomonas aeruginosa quorum-sensing molecule N-(3-oxododecanoyl)-L-homoserine lactone. Infection And Immunity, 2003, 71(8): 4421-4431.

[114] Boontham P, Robins A, Chandran P, et al. Significant immunomodulatory effects of Pseudomonas aeruginosa quorum-sensing signal molecules: possible link in human sepsis. Clinical Science (London, England: 1979), 2008, 115(11): 343-351.

[115] Tateda K, Ishii Y, Horikawa M, et al. The Pseudomonas aeruginosa autoinducer N-3-oxododecanoyl homoserine lactone accelerates apoptosis in macrophages and neutrophils. Infection And Immunity, 2003, 71(10): 5785-5793.

[116] Shiner EK, Terentyev D, Bryan A, et al. Pseudomonas aeruginosa autoinducer modulates host cell responses through calcium signalling. Cellular Microbiology, 2006, 8(10): 1601-1610.

[117] Li L, Hooi D, Chhabra SR, et al. Bacterial N-acylhomoserine lactone-induced apoptosis in breast carcinoma cells correlated with down-modulation of STAT3. Oncogene, 2004, 23(28): 4894-4902.

[118] Oliver CM, Schaefer AL, Greenberg EP, et al. Microwave synthesis and evaluation of phenacylhomoserine lactones as anticancer compounds that minimally activate quorum sensing pathways in Pseudomonas aeruginosa. Journal of Medicinal Chemistry, 2009, 52(6): 1569-1575.

[119] Lee RJ, Chen B, Redding KM, et al. Mouse nasal epithelial innate immune responses to Pseudomonas aeruginosa quorum-sensing molecules require taste signaling components. Innate Immunity, 2014, 20(6): 606-617.

[120] Macfarlane S, Woodmansey EJ, Macfarlane GT. Colonization of mucin by human intestinal bacteria and establishment of biofilm communities in a two-stage continuous culture system. Applied and Environmental Microbiology, 2005, 71(11): 7483-7492.

[121] Perego M, Brannigan JA. Pentapeptide regulation of aspartyl-phosphate phosphatases. Peptides, 2001, 22(10): 1541-1547.

[122] Kolodkin-Gal I, Hazan R, Gaathon A, et al. A linear pentapeptide is a quorum-sensing factor required for mazEF-mediated cell death in Escherichia coli. Science, 2007, 318(5850): 652-655.

[123] Wynendaele E, Verbeke F, D'Hondt M, et al. Crosstalk between the microbiome and cancer cells by quorum sensing peptides. Peptides, 2015, 64: 40-48.

[124] Gursoy UK, Kononen E. Understanding the roles of gingival beta-defensins. Journal of Oral Microbiology, 2012, 4.

[125] Scheres N, Lamont RJ, Crielaard W, et al. LuxS signaling in Porphyromonas gingivalis-host interactions. Anaerobe, 2015, 35(Pt A): 3-9.

[126] Elmanfi S, Ma X, Sintim HO, et al. Quorum-sensing molecule dihydroxy-2,3-pentanedione and its analogs as regulators of epithelial integrity. Journal of Periodontal Research, 2018, 53(3): 414-421.

[127] Giebel SJ, Menicucci G, McGuire PG, et al. Matrix metalloproteinases in early diabetic retinopathy and their role in alteration of the blood-retinal barrier. Laboratory Investigation, 2005, 85(5): 597-607.

[128] Zargar A, Quan DN, Carter KK, et al. Bacterial secretions of nonpathogenic Escherichia coli elicit inflammatory pathways: a closer investigation of interkingdom signaling. MBio, 2015, 6(2): e00025.

[129] Bryan A, Watters C, Koenig L, et al. Human transcriptome analysis reveals a potential role for active transport in the metabolism of Pseudomonas aeruginosa autoinducers. Microbes and Infection, 2010, 12(12-13): 1042-1050.

[130] Cui ZQ, Wu ZM, Fu YX, et al. Autoinducer-2 of quorum sensing is involved in cell damage caused by avian pathogenic Escherichia coli. Microbial Pathogenesis, 2016, 99: 247-252.

[131] Wu J, Li K, Peng W, et al. Autoinducer-2 of Fusobacterium nucleatum promotes macrophage M1 polarization via TNFSF9/IL-1beta signaling. International Immunopharmacology, 2019, 74: 105724.

[132] Vidal JE, Howery KE, Ludewick HP, et al. Quorum-sensing systems LuxS/autoinducer 2 and Com regulate Streptococcus pneumoniae biofilms in a bioreactor with living cultures of human respiratory cells. Infection And Immunity, 2013, 81(4): 1341-1353.

[133] Trier JS. Mucosal flora in inflammatory bowel disease: Intraepithelial bacteria or endocrine epithelial cell secretory granules? Gastroenterology, 2002, 123(3): 955; author reply 956.

[134] Dejea CM, Wick EC, Hechenbleikner EM, et al. Microbiota organization is a distinct feature of proximal colorectal cancers. Proceedings of the

National Academy of Sciences of the United States of America, 2014, 111(51): 18321-18326.

[135] Chen T, Li Q, Zhang X, et al. TOX expression decreases with progression of colorectal cancers and is associated with CD4 T-cell density and Fusobacterium nucleatum infection. Human Pathology, 2018, 79: 93-101.

[136] Yu J, Chen Y, Fu X, et al. Invasive Fusobacterium nucleatum may play a role in the carcinogenesis of proximal colon cancer through the serrated neoplasia pathway. International Journal of Cancer, 2016, 139(6): 1318-1326.

[137] Li Q, Peng W, Wu J, et al. Autoinducer-2 of gut microbiota, a potential novel marker for human colorectal cancer, is associated with the activation of TNFSF9 signaling in macrophages. Oncoimmunology, 2019, 8(10): e1626192.

[138] Sperandio V, Torres AG, Jarvis B, et al. Bacteria-host communication: the language of hormones. Proceedings of the National Academy of Sciences of the United States of America, 2003, 100(15): 8951-8956.

[139] Li Q, Ren Y, Fu X. Inter-kingdom signaling between gut microbiota and their host. Cellular And Molecular Life Sciences, 2019, 76(12): 2383-2389.

[140] Sender R, Fuchs S, Milo R. Revised estimates for the number of human and bacteria cells in the body. PLoS Biology, 2016, 14(8): e1002533.

[141] Hooper LV, Littman DR, Macpherson AJ. Interactions between the microbiota and the immune system. Science, 2012, 336(6086): 1268-1273.

[142] Boman HG. Antibacterial peptides: basic facts and emerging concepts. Journal of Internal Medicine, 2003, 254(3): 197-215.

[143] Ganz T. Defensins: antimicrobial peptides of innate immunity. Nature Reviews: Immunology, 2003, 3(9): 710-720.

[144] Farnaud S, Evans RW. Lactoferrin--a multifunctional protein with antimicrobial properties. Molecular Immunology, 2003, 40(7): 395-405.

[145] Ley RE, Hamady M, Lozupone C, et al. Evolution of mammals and their gut microbes. Science, 2008, 320(5883): 1647-1651.

[146] Sommer F, Backhed F. The gut microbiota--masters of host development and physiology. Nature Reviews: Microbiology, 2013, 11(4): 227-238.

[147] Yeo S, Park H, Ji Y, et al. Influence of gastrointestinal stress on autoinducer-2 activity of two Lactobacillus species. FEMS Microbiology Ecology, 2015, 91(7): fiv065.

[148] Zhang LH, Dong YH. Quorum sensing and signal interference: diverse implications. Molecular Microbiology, 2004, 53(6): 1563-1571.

[149] Bosch TC. Cnidarian-microbe interactions and the origin of innate immunity in metazoans. Annual Review of Microbiology, 2013, 67: 499-518.

[150] Pietschke C, Treitz C, Foret S, et al. Host modification of a bacterial quorum-sensing signal induces a phenotypic switch in bacterial symbionts. Proceedings of the National Academy of Sciences of the United States of America, 2017, 114(40): E8488-E8497.

[151] Harder T, Campbell AH, Egan S, et al. Chemical mediation of ternary interactions between marine holobionts and their environment as exemplified by the red alga Delisea pulchra. Journal of Chemical Ecology, 2012, 38(5): 442-450.

[152] Smith RS, Iglewski BH. P. aeruginosa quorum-sensing systems and virulence. Current Opinion in Microbiology, 2003, 6(1): 56-60.

[153] Chun CK, Ozer EA, Welsh MJ, et al. Inactivation of a Pseudomonas aeruginosa quorum-sensing signal by human airway epithelia. Proceedings of the National Academy of Sciences of the United States of America, 2004, 101(10): 3587-3590.

[154] Draganov DI, La Du BN. Pharmacogenetics of paraoxonases: a brief review. Naunyn-Schmiedebergs Archives of Pharmacology, 2004, 369(1): 78-88.

[155] Ozer EA, Pezzulo A, Shih DM, et al. Human and murine paraoxonase 1 are host modulators of Pseudomonas aeruginosa quorum-sensing. FEMS Microbiology Letters, 2005, 253(1): 29-37.

[156] Stoltz DA, Ozer EA, Ng CJ, et al. Paraoxonase-2 deficiency enhances Pseudomonas aeruginosa quorum sensing in murine tracheal epithelia. American Journal of Physiology Lung Cellular and Molecular Physiology, 2007, 292(4): L852-860.

[157] Teiber JF, Horke S, Haines DC, et al. Dominant role of paraoxonases in inactivation of the Pseudomonas aeruginosa quorum-sensing signal N-(3-oxododecanoyl)-L-homoserine lactone. Infection And Immunity, 2008, 76(6): 2512-2519.

[158] Smith AC, Rice A, Sutton B, et al. Albumin inhibits pseudomonas aeruginosa quorum sensing and alters polymicrobial interactions. Infection And Immunity, 2017, 85(9): e00116-00117.

[159] Rothfork JM, Timmins GS, Harris MN, et al. Inactivation of a bacterial virulence pheromone

by phagocyte-derived oxidants: new role for the NADPH oxidase in host defense. Proceedings of the National Academy of Sciences of the United States of America, 2004, 101(38): 13867-13872.

[160] Peterson MM, Mack JL, Hall PR, et al. Apolipoprotein B is an innate barrier against invasive staphylococcus aureus infection. Cell Host & Microbe, 2008, 4(6): 555-566.

[161] Hall PR, Elmore BO, Spang CH, et al. Nox2 modification of LDL is essential for optimal apolipoprotein B-mediated control of agr type III Staphylococcus aureus quorum-sensing. PLoS Pathogens, 2013, 9(2): e1003166.

[162] Hernando-Amado S, Alcalde-Rico M, Gil-Gil T, et al. Naringenin inhibition of the pseudomonas aeruginosa quorum sensing response is based on its time-dependent competition with N-(3-Oxo-dodecanoyl)-L-homoserine lactone for LasR binding. Frontiers in Molecular Biosciences, 2020, 7: 25.

[163] Wu L, Zaborina O, Zaborin A, et al. High-molecular-weight polyethylene glycol prevents lethal sepsis due to intestinal Pseudomonas aeruginosa. Gastroenterology, 2004, 126(2): 488-498.

[164] Vallejo R, de Leon-Casasola O, Benyamin R. Opioid therapy and immunosuppression: a review. American Journal of Therapeutics, 2004, 11(5): 354-365.

[165] Sternini C, Patierno S, Selmer IS, et al. The opioid system in the gastrointestinal tract. Neurogastroenterology and Motility, 2004, 16 Suppl 2: 3-16.

[166] Zaborina O, Lepine F, Xiao G, et al. Dynorphin activates quorum sensing quinolone signaling in Pseudomonas aeruginosa. PLoS Pathogens, 2007, 3(3): e35.

[167] Gallio M, Sturgill G, Rather P, et al. A conserved mechanism for extracellular signaling in eukaryotes and prokaryotes. Proceedings of the National Academy of Sciences of the United States of America, 2002, 99(19): 12208-12213.

[168] Stevenson LG, Strisovsky K, Clemmer KM, et al. Rhomboid protease AarA mediates quorum-sensing in Providencia stuartii by activating TatA of the twin-arginine translocase. Proceedings of the National Academy of Sciences of the United States of America, 2007, 104(3): 1003-1008.

[169] Burton CL, Chhabra SR, Swift S, et al. The growth response of Escherichia coli to neurotransmitters and related catecholamine drugs requires a functional enterobactin biosynthesis and uptake system. Infection And Immunity, 2002, 70(11): 5913-5923.

[170] Alverdy J, Holbrook C, Rocha F, et al. Gut-derived sepsis occurs when the right pathogen with the right virulence genes meets the right host: evidence for in vivo virulence expression in Pseudomonas aeruginosa. Annals of Surgery, 2000, 232(4): 480-489.

[171] Ismail AS, Valastyan JS, Bassler BL. A host-produced autoinducer-2 mimic activates bacterial quorum sensing. Cell Host & Microbe, 2016, 19(4): 470-480.

[172] Porat R, Clark BD, Wolff SM, et al. Enhancement of growth of virulent strains of Escherichia coli by interleukin-1. Science, 1991, 254(5030): 430-432.

[173] Zav'yalov VP, Chernovskaya TV, Navolotskaya EV, et al. Specific high affinity binding of human interleukin 1 beta by Caf1A usher protein of Yersinia pestis. FEBS Letters, 1995, 371(1): 65-68.

[174] Luo G, Niesel DW, Shaban RA, et al. Tumor necrosis factor alpha binding to bacteria: evidence for a high-affinity receptor and alteration of bacterial virulence properties. Infection And Immunity, 1993, 61(3): 830-835.

[175] Wu L, Estrada O, Zaborina O, et al. Recognition of host immune activation by Pseudomonas aeruginosa. Science, 2005, 309(5735): 774-777.

[176] Horger S, Schultheiss G, Diener M. Segment-specific effects of epinephrine on ion transport in the colon of the rat. American Journal of Physiology, 1998, 275(6): G1367-1376.

[177] Asano Y, Hiramoto T, Nishino R, et al. Critical role of gut microbiota in the production of biologically active, free catecholamines in the gut lumen of mice. American Journal of Physiology Gastrointestinal and Liver Physiology, 2012, 303(11): G1288-1295.

[178] Flierl MA, Rittirsch D, Huber-Lang M, et al. Catecholamines-crafty weapons in the inflammatory arsenal of immune/inflammatory cells or opening pandora's box? Molecular Medicine, 2008, 14(3-4): 195-204.

[179] Lyte M, Ernst S. Catecholamine induced growth of gram negative bacteria. Life Sciences, 1992, 50(3): 203-212.

[180] Valet P, Senard JM, Devedjian JC, et al. Characterization and distribution of alpha 2-adrenergic receptors in the human intestinal mucosa. Journal of Clinical Investigation, 1993, 91(5): 2049-2057.

[181] Lyte M, Erickson AK, Arulanandam BP, et al. Norepinephrine-induced expression of the K99 pilus adhesin of enterotoxigenic Escherichia

coli. Biochemical and Biophysical Research Communications, 1997, 232(3): 682-686.

[182] Freestone PP, Lyte M, Neal CP, et al. The mammalian neuroendocrine hormone norepinephrine supplies iron for bacterial growth in the presence of transferrin or lactoferrin. Journal of Bacteriology, 2000, 182(21): 6091-6098.

[183] Clarke MB, Hughes DT, Zhu C, et al. The QseC sensor kinase: a bacterial adrenergic receptor. Proceedings of the National Academy of Sciences of the United States of America, 2006, 103(27): 10420-10425.

[184] Noto JM, Zackular JP, Varga MG, et al. Modification of the gastric mucosal microbiota by a strain-specific helicobacter pylori oncoprotein and carcinogenic histologic phenotype. mBio, 2019, 10(3): e00955-00919.

[185] Schubert ML. Physiologic, pathophysiologic, and pharmacologic regulation of gastric acid secretion. Current Opinion in Gastroenterology, 2017, 33(6): 430-438.

[186] Chowers MY, Keller N, Tal R, et al. Human gastrin: a Helicobacter pylori--specific growth factor. Gastroenterology, 1999, 117(5): 1113-1118.

[187] Yamashita K, Kaneko H, Yamamoto S, et al. Inhibitory effect of somatostatin on Helicobacter pylori proliferation in vitro. Gastroenterology, 1998, 115(5): 1123-1130.

[188] Joseph IM, Kirschner D. A model for the study of Helicobacter pylori interaction with human gastric acid secretion. Journal of Theoretical Biology, 2004, 228(1): 55-80.

[189] Maciorkowska E, Panasiuk A, Kondej-Muszynska K, et al. Mucosal gastrin cells and serum gastrin levels in children with Helicobacter pylori infection. Advances in Medical Sciences, 2006, 51: 137-141.

[190] Kanai K, Kondo E, Kurata T. Affinity and response of Burkholderia pseudomallei and Burkholderia cepacia to insulin. Southeast Asian Journal of Tropical Medicine and Public Health, 1996, 27(3): 584-591.

[191] Veron W, Lesouhaitier O, Pennanec X, et al. Natriuretic peptides affect Pseudomonas aeruginosa and specifically modify lipopolysaccharide biosynthesis. FEBS Journal, 2007, 274(22): 5852-5864.

[192] Blier AS, Veron W, Bazire A, et al. C-type natriuretic peptide modulates quorum sensing molecule and toxin production in Pseudomonas aeruginosa. Microbiology (Reading), 2011, 157(Pt 7): 1929-1944.

[193] Moghal N, Sternberg PW. Multiple positive and negative regulators of signaling by the EGF-receptor. Current Opinion in Cell Biology, 1999, 11(2): 190-196.

[194] Beury-Cirou A, Tannieres M, Minard C, et al. At a supra-physiological concentration, human sexual hormones act as quorum-sensing inhibitors. PLoS One, 2013, 8(12): e83564.

[195] Goto R, Miki T, Nakamura N, et al. Salmonella Typhimurium PagP- and UgtL-dependent resistance to antimicrobial peptides contributes to the gut colonization. PLoS One, 2017, 12(12): e0190095.

[196] Koeppen K, Hampton TH, Jarek M, et al. A novel mechanism of host-pathogen interaction through sRNA in bacterial outer membrane vesicles. PLoS Pathogens, 2016, 12(6): e1005672.

[197] Nahui Palomino RA, Vanpouille C, Costantini PE, et al. Microbiota-host communications: bacterial extracellular vesicles as a common language. PLoS Pathogens, 2021, 17(5): e1009508.

[198] Elmi A, Watson E, Sandu P, et al. Campylobacter jejuni outer membrane vesicles play an important role in bacterial interactions with human intestinal epithelial cells. Infection And Immunity, 2012, 80(12): 4089-4098.

[199] Delgado L, Baeza N, Perez-Cruz C, et al. Cryo-transmission electron microscopy of outer-inner membrane vesicles naturally secreted by Gram-negative pathogenic bacteria. Bio-protocol, 2019, 9(18): e3367.

[200] Resch U, Tsatsaronis JA, Le Rhun A, et al. A two-component regulatory system impacts extracellular membrane-derived vesicle production in group a streptococcus. mBio, 2016, 7(6): e00207-00216.

[201] Ahmadi Badi S, Bruno SP, Moshiri A, et al. Small RNAs in outer membrane vesicles and their function in host-microbe interactions. Frontiers in Microbiology, 2020, 11: 1209.

[202] Bomberger JM, Ye S, Maceachran DP, et al. A Pseudomonas aeruginosa toxin that hijacks the host ubiquitin proteolytic system. PLoS Pathogens, 2011, 7(3): e1001325.

[203] Han EC, Choi SY, Lee Y, et al. Extracellular RNAs in periodontopathogenic outer membrane vesicles promote TNF-alpha production in human macrophages and cross the blood-brain barrier in mice. FASEB Journal, 2019, 33(12): 13412-13422.

[204] Dauros-Singorenko P, Hong J, Swift S, et al. Effect of the extracellular vesicle RNA cargo from

uropathogenic Escherichia coli on bladder cells. Frontiers in Molecular Biosciences, 2020, 7: 580913.

[205] Zhang H, Zhang Y, Song Z, et al. sncRNAs packaged by Helicobacter pylori outer membrane vesicles attenuate IL-8 secretion in human cells. International Journal of Medical Microbiology, 2020, 310(1): 151356.

[206] Munhoz da Rocha IF, Amatuzzi RF, Lucena ACR, et al. Cross-kingdom extracellular vesicles EV-RNA communication as a mechanism for host-pathogen interaction. Frontiers in Cellular and Infection Microbiology, 2020, 10: 593160.

[207] Mathieu M, Martin-Jaular L, Lavieu G, et al. Specificities of secretion and uptake of exosomes and other extracellular vesicles for cell-to-cell communication. Nature Cell Biology, 2019, 21(1): 9-17.

[208] Liu S, da Cunha AP, Rezende RM, et al. The host shapes the gut microbiota via fecal microRNA. Cell Host & Microbe, 2016, 19(1): 32-43.

[209] Fabian MR, Sonenberg N, Filipowicz W. Regulation of mRNA translation and stability by microRNAs. Annual Review of Biochemistry, 2010, 79: 351-379.

[210] Johnston DGW, Williams MA, Thaiss CA, et al. Loss of microRNA-21 influences the gut microbiota, causing reduced susceptibility in a murine model of colitis. Journal of Crohn's & Colitis, 2018, 12(7): 835-848.

[211] Zhao B, Lucas KJ, Saha TT, et al. MicroRNA-275 targets sarco/endoplasmic reticulum Ca2+ adenosine triphosphatase (SERCA) to control key functions in the mosquito gut. PLoS Genetics, 2017, 13(8): e1006943.

[212] Ji Y, Li X, Zhu Y, et al. Faecal microRNA as a biomarker of the activity and prognosis of inflammatory bowel diseases. Biochemical and Biophysical Research Communications, 2018, 503(4): 2443-2450.

[213] Liu S, Rezende RM, Moreira TG, et al. Oral administration of miR-30d from feces of MS patients suppresses MS-like symptoms in mice by expanding Akkermansia muciniphila. Cell Host & Microbe, 2019, 26(6): 779-794. e778.

[214] Koeppen K, Nymon A, Barnaby R, et al. Let-7b-5p in vesicles secreted by human airway cells reduces biofilm formation and increases antibiotic sensitivity of P. aeruginosa. Proceedings of the National Academy of Sciences of the United States of America, 2021, 118(28): e2105370118.

[215] Lee JH, Wood TK, Lee J. Roles of indole as an interspecies and interkingdom signaling molecule. Trends in Microbiology, 2015, 23(11): 707-718.

[216] Mueller RS, Beyhan S, Saini SG, et al. Indole acts as an extracellular cue regulating gene expression in Vibrio cholerae. Journal of Bacteriology, 2009, 191(11): 3504-3516.

[217] Lee J, Jayaraman A, Wood TK. Indole is an inter-species biofilm signal mediated by SdiA. BMC Microbiology, 2007, 7: 42.

[218] Bansal T, Alaniz RC, Wood TK, et al. The bacterial signal indole increases epithelial-cell tight-junction resistance and attenuates indicators of inflammation. Proceedings of the National Academy of Sciences of the United States of America, 2010, 107(1): 228-233.

[219] Kim YG, Lee JH, Cho MH, et al. Indole and 3-indolylacetonitrile inhibit spore maturation in Paenibacillus alvei. BMC Microbiology, 2011, 11: 119.

[220] Chant EL, Summers DK. Indole signalling contributes to the stable maintenance of Escherichia coli multicopy plasmids. Molecular Microbiology, 2007, 63(1): 35-43.

[221] Chimerel C, Field CM, Pinero-Fernandez S, et al. Indole prevents Escherichia coli cell division by modulating membrane potential. Biochimica et Biophysica Acta, 2012, 1818(7): 1590-1594.

[222] Lee HH, Molla MN, Cantor CR, et al. Bacterial charity work leads to population-wide resistance. Nature, 2010, 467(7311): 82-85.

[223] Hirakawa H, Kodama T, Takumi-Kobayashi A, et al. Secreted indole serves as a signal for expression of type III secretion system translocators in enterohaemorrhagic Escherichia coli O157:H7. Microbiology (Reading), 2009, 155(Pt 2): 541-550.

[224] Han TH, Lee JH, Cho MH, et al. Environmental factors affecting indole production in Escherichia coli. Research in Microbiology, 2011, 162(2): 108-116.

[225] Vega NM, Allison KR, Khalil AS, et al. Signaling-mediated bacterial persister formation. Nature Chemical Biology, 2012, 8(5): 431-433.

[226] Hu Y, Kwan BW, Osbourne DO, et al. Toxin YafQ increases persister cell formation by reducing indole signalling. Environmental Microbiology, 2015, 17(4): 1275-1285.

[227] Pinero-Fernandez S, Chimerel C, Keyser UF, et al. Indole transport across Escherichia coli membranes. Journal of Bacteriology, 2011, 193(8): 1793-1798.

[228] Han Y, Yang CL, Yang Q, et al. Mutation of

tryptophanase gene tnaA in Edwardsiella tarda reduces lipopolysaccharide production, antibiotic resistance and virulence. Environmental Microbiology Reports, 2011, 3(5): 603-612.

[229] Li X, Yang Q, Dierckens K, et al. RpoS and indole signaling control the virulence of Vibrio anguillarum towards gnotobiotic sea bass (Dicentrarchus labrax) larvae. PLoS One, 2014, 9(10): e111801.

[230] Lindh JM, Kannaste A, Knols BG, et al. Oviposition responses of Anopheles gambiae s.s. (Diptera: Culicidae) and identification of volatiles from bacteria-containing solutions. Journal of Medical Entomology, 2008, 45(6): 1039-1049.

[231] Wikoff WR, Anfora AT, Liu J, et al. Metabolomics analysis reveals large effects of gut microflora on mammalian blood metabolites. Proceedings of the National Academy of Sciences of the United States of America, 2009, 106(10): 3698-3703.

[232] Shimada Y, Kinoshita M, Harada K, et al. Commensal bacteria-dependent indole production enhances epithelial barrier function in the colon. PLoS One, 2013, 8(11): e80604.

[233] Chimerel C, Emery E, Summers DK, et al. Bacterial metabolite indole modulates incretin secretion from intestinal enteroendocrine L cells. Cell Reports, 2014, 9(4): 1202-1208.

[234] Ensley BD, Ratzkin BJ, Osslund TD, et al. Expression of naphthalene oxidation genes in Escherichia coli results in the biosynthesis of indigo. Science, 1983, 222(4620): 167-169.

[235] Crumeyrolle-Arias M, Tournaire MC, Rabot S, et al. 5-hydroxyoxindole, an indole metabolite, is present at high concentrations in brain. Journal of Neuroscience Research, 2008, 86(1): 202-207.

[236] Wyeth FJ. The Effects of acids, alkalies, and sugars on the growth and indole formation of bacillus coli: a report to the medical research committee. Biochemical Journal, 1919, 13(1): 10-24.

[237] Biersack B, Schobert R. Indole compounds against breast cancer: recent developments. Current Drug Targets, 2012, 13(14): 1705-1719.

[238] Choi HS, Cho MC, Lee HG, et al. Indole-3-carbinol induces apoptosis through p53 and activation of caspase-8 pathway in lung cancer A549 cells. Food And Chemical Toxicology, 2010, 48(3): 883-890.

[239] Fan S, Meng Q, Xu J, et al. DIM (3,3’-diindolylmethane) confers protection against ionizing radiation by a unique mechanism. Proceedings of the National Academy of Sciences of the United States of America, 2013, 110(46): 18650-18655.

[240] Burnstock G. Purinergic signalling. British Journal of Pharmacology, 2006, 147 Suppl 1: S172-181.

[241] Fagerberg SK, Patel P, Andersen LW, et al. Erythrocyte P2X1 receptor expression is correlated with change in haematocrit in patients admitted to the ICU with blood pathogen-positive sepsis. Critical Care (London, England), 2018, 22(1): 181.

[242] Wang Y, Ouyang Y, Liu B, et al. Platelet activation and antiplatelet therapy in sepsis: A narrative review. Thrombosis Research, 2018, 166: 28-36.

[243] Idzko M, Ferrari D, Eltzschig HK. Nucleotide signalling during inflammation. Nature, 2014, 509(7500): 310-317.

[244] Burnstock G. Historical review: ATP as a neurotransmitter. Trends in Pharmacological Sciences, 2006, 27(3): 166-176.

[245] Fountain SJ. Primitive ATP-activated P2X receptors: discovery, function and pharmacology. Frontiers in Cellular Neuroscience, 2013, 7: 247.

[246] Ivanova EP, Alexeeva YV, Pham DK, et al. ATP level variations in heterotrophic bacteria during attachment on hydrophilic and hydrophobic surfaces. International Microbiology, 2006, 9(1): 37-46.

[247] Proietti M, Cornacchione V, Rezzonico Jost T, et al. ATP-gated ionotropic P2X7 receptor controls follicular T helper cell numbers in Peyer’s patches to promote host-microbiota mutualism. Immunity, 2014, 41(5): 789-801.

[248] Proietti M, Perruzza L, Scribano D, et al. ATP released by intestinal bacteria limits the generation of protective IgA against enteropathogens. Nature Communications, 2019, 10(1): 250.

[249] Atarashi K, Nishimura J, Shima T, et al. ATP drives lamina propria T(H)17 cell differentiation. Nature, 2008, 455(7214): 808-812.

[250] Abbasian B, Shair A, O’Gorman DB, et al. Potential role of extracellular ATP released by bacteria in bladder infection and contractility. mSphere, 2019, 4(5): e00439-00419.

[251] Alvarez CL, Corradi G, Lauri N, et al. Dynamic regulation of extracellular ATP in Escherichia coli. Biochemical Journal, 2017, 474(8): 1395-1416.

[252] Skals M, Bjaelde RG, Reinholdt J, et al. Bacterial RTX toxins allow acute ATP release from human erythrocytes directly through the toxin pore. Journal of Biological Chemistry, 2014, 289(27): 19098-19109.

[253] Mempin R, Tran H, Chen C, et al. Release of

extracellular ATP by bacteria during growth. BMC Microbiology, 2013, 13: 301.

[254] Ding Q, Tan KS. The danger signal extracellular ATP is an inducer of Fusobacterium nucleatum biofilm dispersal. Frontiers in Cellular and Infection Microbiology, 2016, 6: 155.

[255] Breijyeh Z, Jubeh B, Karaman R. Resistance of Gram-negative bacteria to current antibacterial agents and approaches to resolve it. Molecules, 2020, 25(6): 1340.

[256] Mostafa I, Abbas HA, Ashour ML, et al. Polyphenols from salix tetrasperma impair virulence and inhibit quorum sensing of pseudomonas aeruginosa. Molecules, 2020, 25(6): 1341.

[257] Hwang IY, Tan MH, Koh E, et al. Reprogramming microbes to be pathogen-seeking killers. ACS Synthetic Biology, 2014, 3(4): 228-237.

[258] Hwang IY, Koh E, Wong A, et al. Engineered probiotic Escherichia coli can eliminate and prevent Pseudomonas aeruginosa gut infection in animal models. Nature Communications, 2017, 8: 15028.

[259] Topa SH, Palombo EA, Kingshott P, et al. Activity of cinnamaldehyde on quorum sensing and biofilm susceptibility to antibiotics in pseudomonas aeruginosa. Microorganisms, 2020, 8(3): 455.

第5章
肠道微生物群与机体免疫

第一节　肠道微生物群对机体免疫功能的影响

肠道菌群在宿主免疫调节、控制肠黏膜炎症中发挥着至关重要的作用。肠道黏膜免疫系统是维持肠道黏膜屏障完整性和调节肠道炎症的基础[1]，涉及肠上皮细胞、固有层淋巴细胞和黏膜下集合淋巴结等[2]。肠道黏膜在耐受大量微生物和膳食抗原的同时，也能产生快速有效的抗感染免疫反应[1]，限制组织损伤，并通过形成特异的长期适应性免疫反应来预防或降低继发性感染的严重程度[3]。肠道微生物组在功能上并不独立于宿主黏膜，它可以通过微生物相关分子模式（microbe-associated molecular pattern，MAMP）和微生物相关代谢物来影响黏膜屏障的完整性并调节宿主免疫[1]，通过改变宿主免疫系统的结构和功能，重塑免疫微环境，促进或干扰特定疾病的发展[4]。

一、微生物群决定宿主免疫系统的发育和免疫应答

人类的胃肠道包含约100万亿个细菌，因此成为微生物与宿主免疫系统之间相互作用的主要场所[5, 6]。微生物群与人类共同进化了数百万年，两者之间具有确切的互惠关系。在这种关系中，微生物群有助于宿主的许多生理功能，同时，宿主也为微生物提供营养和栖息地。除了有助于食物的消化和发酵，微生物群对于防御病原体也至关重要。病原体能争夺养分和黏附位点，有些微生物甚至通过分泌抗菌肽来积极对抗病原体。稳定的微生物群和黏液层对于防止病原菌感染宿主至关重要[7]。最近，无菌动物实验表明，生命早期的微生物群定植对于免疫系统的最佳发育非常重要。在没有微生物群的情况下，肠道黏膜免疫发育不良[8]。

微生物群对宿主进行生理调节的途径之一，是消化道中食物残渣经微生物厌氧发酵产生的多种代谢产物。结肠中细菌发酵纤维的主要产物是短链脂肪酸（short-chain fatty acid，SCFA），包括乙酸盐、丙酸盐和丁酸盐，它们不仅是肠道菌群和肠上皮细胞的重要能源，并且在宿主生理和免疫方面具有多种调节功能，通常和抗菌肽一起被认为是有益代谢产物[9]。它们可以穿过肠上皮细胞，并能够与宿主细胞相互作用，从而影响宿主的免疫应答和疾病风险[10]。

肠道微生物定植和膳食纤维含量会影响结肠中SCFA的浓度。富含纤维的饮食可以促进水解纤维素和木聚糖的细菌生长，包括普雷沃菌属、木聚糖杆属和产丁酸的普拉梭菌[9, 11]。普拉梭菌和其他产生SCFA的细菌可以保护宿主免受炎症和非感染性结肠疾病的侵害。有报道

发现，低浓度的普拉梭菌与克罗恩疾病（Crohn's disease，CD）和其他炎性肠病（inflammatory bowel disease，IBD）相关[12, 13]。SCFAs 是组蛋白去乙酰化酶（histone deacetylase，HDAC）的有效抑制剂，它们可能会促进具有免疫耐受、抗炎特性的细胞表型的表达，这对于维持免疫稳态至关重要。微生物通过表观遗传控制免疫系统的一个有力的证据是丁酸盐对 Treg 细胞分化的调节。幼稚 $CD4^{+}$ T 细胞在 Treg 诱导分化条件下与丁酸盐共培养后，从培养物纯化的 $Foxp3^{-}$ 细胞中观察到了 Foxp3 启动子和 CNS1 增强子处 H3K27-Ac 表达显著增加，这可能会增强 Foxp3 的诱导，从而促进胸腺外 Treg 细胞的生成[14]。

此外，SCFA 还通过加强肠上皮细胞的屏障功能来增强防御机制。体内研究结果表明，定植有产 SCFA 的多形拟杆菌或普拉梭菌的无菌小鼠会诱导杯状细胞分化和黏液产生[15]。体外实验中，SCFA 能促进肠上皮杯状细胞中黏蛋白基因的转录[16]。同样，SCFA 有助于肠上皮细胞中紧密连接蛋白的合成，并促进长双歧杆菌的定植，该菌株能产生大量乙酸盐，从而可以防止致命的致病性大肠埃希菌 O157：H7 感染。这表明 SCFA 可以增强肠上皮细胞的完整性，并抑制致命毒素从肠腔进入全身循环[17, 18]。

微生物群组成的变化会影响许多生理过程（包括发育、代谢和免疫细胞功能），并与多种疾病的易感性有关，但其如何调节免疫稳态的机制尚不清楚。有研究发现，特定细菌物种的存在可以通过促进某些亚型淋巴细胞的发育来改变免疫反应。例如，分节丝状菌诱导 IL-17 和 IL-22 的产生，有利于小鼠 Th17 细胞的生成[19]，而无菌小鼠的梭状芽孢杆菌、ASF 菌（altered Schaedler flora，一种特定共生菌组合，包括乳酸杆菌、拟杆菌、弯枝菌属和梭菌属）和脆弱拟杆菌的重建，促进了结肠中 $IL\text{-}10^{+}$Treg 细胞的募集[20, 21]。此外，识别微生物刺激对于免疫调节机制很重要。有证据表明，在非肥胖糖尿病小鼠中，一种识别微生物刺激的多种固有免疫受体的衔接子——髓样分化初级应答基因 88（myeloid differentiation factor 88，MyD88）的缺失，可以改变微生物的组成并降低糖尿病的严重程度，而微生物群的消耗导致糖尿病的快速发展[22]。Yoshiyuki Mishima 等也发现，肠道常驻细菌以 MyD88/TLR2/PI3K 依赖的方式，激活产生 IL-10 的肠道 B 细胞，抑制攻击性免疫反应并维持黏膜稳态[23]。

目前，有一些学派提出：微生物组和肠道免疫系统不仅维持局部免疫，也是维持全身免疫调节的关键，而且微生态失调会促进触发自身免疫性疾病的免疫应答效应[24]。稳定的微生物群有利于维持良好的免疫系统平衡，令人担忧的是，抗生素作为儿童最常用的处方药之一，破坏了婴儿肠道中脆弱的生态系统，并可能在以后的生活中增加自身炎症性疾病的风险[25, 26]。摄入抗生素会破坏肠道微生物组，增加宿主对过度增殖的机会致病菌的敏感性。支持这一假说的研究表明，在幼年小鼠中使用抗生素会导致微生物群改变，诱发促炎性免疫反应，增加了炎症性疾病的风险[27]。在新生小鼠中，抗生素治疗可使梭菌属细菌耗竭，减少了肠固有层 $ROR\gamma t^{+}$ 固有淋巴细胞（innate lymphoid cell，ILC）和 T 细胞诱导 IL-22 的产生，导致对食物过敏原的敏感性增强[28]。生命早期使用低剂量青霉素会引起微生物群的短暂变动和持续的代谢改变，并影响免疫相关基因在回肠的表达[29]。

此外，分节丝状菌与某些自身炎症性疾病有关。动物模型显示，分节丝状菌可加剧自身免疫性脑炎和鼠关节炎[30, 31]。溃疡性结肠炎和克罗恩病患者的某些炎症部位也可检测到分节丝状菌，在给予青霉素预处理根除分节丝状菌后，可降低 Th17 的表达以及小鼠对葡聚糖硫酸钠诱导的结肠炎的敏感性[32]。另一方面，分节丝状菌的定植对于防御细菌病原体有重要作用。抗生素可减少分节丝状菌和Th17细胞的数量，降低了机体对肠道病原体柠檬酸杆菌感染的抵抗力。而分节丝状菌在小鼠的定植与炎症和抗菌防御相关基因的表达增加相关，并导致对鼠柠檬酸杆菌的抵抗力更强[19]。这些研究表明，通过微调肠道菌群，某些IBD或其他自身免疫性疾病患者可能获得更良好的免疫状态。

微生物群影响全身各个系统，如何进行协作以确保人类健康是目前面临的科学难题。科学家们齐心协力并开展了一系列研究计划，例如2007年启动的人类微生物组计划，目的是描述和鉴定人类微生物群的组成，以及他们在健康和疾病中的作用[33]。尽管做出了这些努力，但是仍然很难定义是什么构成了健康的微生物群，以及微生物与宿主免疫系统是如何共同进化的[34]。由于多种因素，包括个体的表观遗传变异、饮食、地理位置、生活方式，均能够影响微生物的组成[35-37]，因此特定细菌是如何影响个体的研究变得非常复杂。

此外，肠道微生物群的组成在整个生命进程中都会发生变化[38]，为了设计有效的疾病预防疗法，确定干预肠道微生物组成的特定时间范围也很重要。有研究表明，从受孕到2岁的这段时间，通过饮食、抗生素、益生菌、益生元和粪便微生物群移植等干预措施调节微生物群，是预防或治疗营养不良的时间窗[39]。因此，进一步明确特定的微生物如何调节免疫反应，以及有效的干预时期，将有助于我们通过靶向调节微生物群的方法，达到预防和治疗某些疾病的目的。

二、生命早期的微生物群会影响生命后期的免疫力

在婴儿时期，肠道菌群有较高的多样性和变异性；随着年龄增长，多样性降低，并趋于稳定[40]（图5-1）。生命早期的微生物群在形成耐受性免疫功能、防止针对自身抗原和非自身抗原（例如过敏原）的不良炎症反应方面具有重要作用。多种因素参与调节了生命早期微生物群的发育，包括饮食、抗生素、分娩方式、母乳喂养和感染[41]。

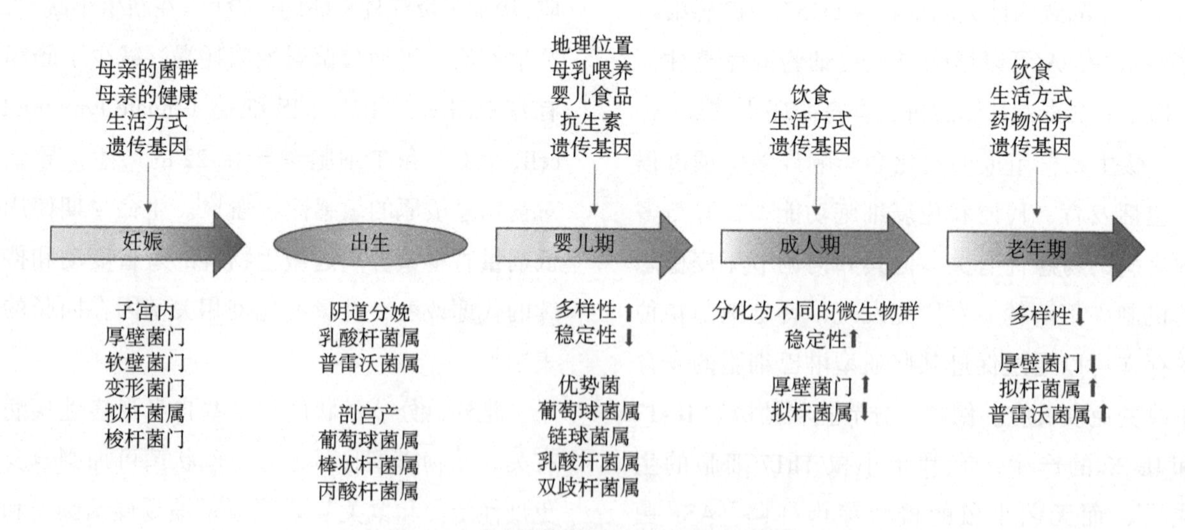

图5-1　人类生命周期中微生物群的变化

子宫不是无菌环境，在胎盘、胎膜、脐带血和胎粪中都存在细菌。婴儿肠道细菌的定植取决于分娩方式，并从中获得独特的细菌群落。幼儿肠道菌群的组成经常变化，非常多样化且不稳定，随着年龄的增长，肠道菌群变得更稳定。与年轻人相比，老年人肠道菌群多样性降低，并且产生短链脂肪酸的细菌减少，分泌内毒素的革兰阴性细菌增加

近年来，基于培养的研究确定了在胎盘[42]、胎膜[43]、脐带血[44]和胎粪中存在细菌，无菌子宫的观念已经被颠覆[45]。人类胎盘中含有低丰度但代谢丰富的微生物群，主要包括厚壁菌门、软壁菌门、变形菌门、拟杆菌属和梭杆菌门，这与口腔微生物组相似[42]。有研究从一位健康妇女母乳中分离出来的屎肠球菌，采用基因标记后口服接种至怀孕小鼠，标记菌株可以从接种动物的羊水中分离并检测出来[44]。因此，有学者认为肠道微生物定植过程可能是在产前由胎盘和羊水中的微生物群启动的[46]。最新的研究发现，胎儿肠道含有调节性 T 细胞和 $PLZF^+CD161^+$ 记忆 T 细胞，它们在子宫内对细菌、自身和母体抗原有反应而被激活[47]。此外，实验鼠模型表明，孕期注射和（或）雾化脂多糖（lipopolysaccharides，LPS）、鼻内细菌暴露，或补充益生菌，可以减轻后代的过敏性气道炎症，诱导黏膜耐受性增加，并抑制变应原诱导 IgE 产生、嗜酸性气道炎症和气道高反应性[48-50]。

在哺乳动物中，婴儿根据其分娩方式的不同，接触不同的细菌，这决定了婴儿肠道微生物定植的差异化。阴道分娩的婴儿会在其母亲的阴道微生物区（乳酸杆菌、普雷沃菌或克雷伯菌属）中获得细菌群落。相比之下，剖宫产婴儿的肠道细菌群落与母亲皮肤表面的细菌类似（金黄色葡萄球菌、棒状杆菌、丙酸杆菌）[51, 52]。剖宫产与肠道微生物多样性降低、拟杆菌门定植延迟和出生后前 2 年 Th1 反应减少有关[53]。此外，伴随潜在致病性肠球菌属和克雷伯菌属的增多，剖宫产分娩的婴儿在出生后的第 1 年有更多的呼吸道感染事件和更高的抗生素需求[54]。除了细菌种类的不同，剖宫产分娩的婴儿肠道微生物数量也低于经阴道分娩出生的婴儿，并与免疫耐受、T 细胞应答和特定细胞因子的显著降低有关[4]。上皮耐受性仅在阴道分娩的小鼠中观察到，而在剖宫产中则没有。因此，新生儿肠道微生物定植似乎可以调节生命早期的免疫系统。

在婴儿成长过程中，饮食对微生物群的组成有很大影响。配方奶和母乳喂养的儿童将接受不同的营养、微生物和免疫成分。母乳因含有 IgA、乳铁蛋白和防御素，对婴儿的免疫防御很重要，可以防止感染，并在生命的最初几天为免疫系统的成熟做出贡献[55, 56]。此外，母乳可能影响后代的免疫耐受性和过敏反应，其方式主要分为两种：一种是以非抗原特异性方式促进免疫调节机制，另一种是诱导抗原特异性免疫耐受[57]。研究人员还发现，人母乳中含有影响肠道神经元存活和神经突触生长的神经营养因子和细胞因子[58]。重要的是，与普遍的看法相反，母乳并不是无菌的，其中含有共生菌和潜在的益生菌。母乳中发现的细菌成分会因地理位置和母亲饮食的不同而变化，研究表明，母乳中主要存在有葡萄球菌、链球菌、乳杆菌和双歧杆菌[59]。迄今为止收集的数据表明，缺乏母乳喂养可能会增加婴儿的健康风险。例如，未接受母乳的早产儿出现坏死性小肠结肠炎（necrotizing enterocolitis，NEC）的风险增加[60]。此外，非母乳喂养也会增加婴儿许多疾病的发生率，例如传染病、肠胃炎、肺炎、儿童肥胖、2 型糖尿病、白血病和婴儿猝死综合征[61]。

三、菌群失调对机体免疫的影响

肠道菌群失调是指肠道微生物群的组成和功能发生变化，直接影响肠道中免疫系统稳态，与黏膜屏障功能障碍和炎症反应有关[1, 4, 62]。虽然肠道菌群失调是某些炎症性疾病的标志，但菌群失调也会反过来触发肠道菌群免疫调节机制。最近的研究发现，溃疡性结肠炎患者肠道内具核梭杆菌丰度增加，并通过激活结肠中的 AKT2 信号通路诱导巨噬细胞 M1 极化，并分泌大量的炎症因子，如 TNF-α、IFN-γ 和 MCP-1，进而招募更多的巨噬细胞进入炎症部位，破坏黏膜屏障，并促进细菌移位[63]。在健康人的肠道中，少量易位的共生细菌可通过拟杆菌属多糖诱导的Th1和Th17细胞，以及黏膜上黏附的分节丝状菌，促进Th17细胞的分化作用来清除[64]。但是大量细菌入侵会持续激活Toll样受体（Toll-like Receptors，TLRs），并引起促炎细胞因子的过表达，从而破坏肠道上皮并导致慢性肠道炎症[65]。令人惊讶的是，在MyD88缺陷型小鼠中发现，高水平的分节丝状菌可以保护非肥胖糖尿病小鼠免于发展为糖尿病[66, 67]。

受干扰的微生物群也会影响先天免疫系统的成熟，因为肠道细菌本身就是该过程的驱动力。如果没有微生物群，中性粒细胞和树突状细胞的功能会受到损害，前者表现为对病原体杀伤力减弱，后者影响Ⅰ型干扰素的分泌[68, 69]。另外，骨髓中的骨髓细胞发育也将会延迟[70]，这种延迟会削弱全身性感染的清除率，并增加对过敏的敏感性[70, 71]。通过对微生物群进行干扰，可能具有类似的有害作用。例如，在早期发育过程中，使用抗生素治疗的小鼠IL-4分泌增加，Treg细胞数量减少，并且在生命后期更容易出现气道高反应性和结肠炎[72]。在婴幼儿时期，抗生素治疗引起的持续性改变与成年后的炎性肠病、哮喘和特应性皮炎相关[73, 74]。肠道的非炎性稳态可以被宿主免疫系统和肠道菌群所动摇，它们之间相互作用的失衡会增加免疫性疾病的风险。

总而言之，肠道免疫系统和共生微生物的失衡，导致了微生物群失调、肠道屏障完整性受损和对共生微生物的促炎免疫反应，这反过来又会加剧不平衡的程度[75]。

四、肠道菌群失调和疾病

肠道菌群失调可影响多种疾病的病程和严重程度。典型的例子包括炎性肠病[76]、1型糖尿病[77]、乳糜泻[78]和心血管疾病[79]。在此，我们将重点讨论影响人类不同生命阶段的3种疾病，即新生儿坏死性小肠结肠炎、成人的结直肠癌和老年人的艰难梭菌相关性腹泻。

（一）坏死性小肠结肠炎

坏死性小肠结肠炎是一种暴发性肠道炎症，在早产儿中最常见，死亡率可高达30%[80]。坏死性小肠结肠炎的特征表现为严重的炎症、肠积气和肠坏死[81]。尽管其发病机制尚不清楚，但早产、剖宫产、细菌感染、抗生素的长期使用、免疫力不成熟和微生物失调等多种危险因素会增加坏死性小肠结肠炎的发生率[81, 82]。其中，微生物失调和炎症被认为在坏死性小肠结肠炎的发病机制中起关键作用[82]。

在早产儿发展为坏死性小肠结肠炎之前，微生物失调的特点是变形杆菌的相对丰度增加，而厚壁菌门和拟杆菌门的相对丰度降低[83]。由于TLR4在早产儿肠道中过表达，增加的变形菌门通过LPS与这些特定受体相互作用，可能触发肠道中的过度免疫反应，导致肠道损伤和坏死。但变形杆菌相对丰度的增加是坏死性小肠结肠炎的原因还是结果，目前尚不明确[81]。那么，通过调节微生物组成的方式，是否能够预防坏死性小肠结肠炎并改善预后？

有证据表明，纯母乳喂养与仅用配方奶粉喂养相比，坏死性小肠结肠炎的发生率至少降低了6倍[84]。为此，美国儿科学会建议在出生后立即用母乳喂养早产婴儿[85]。母乳降低这种风险的潜在机制可能是多方面的，最近的许多研究主要集中在母乳中的免疫调节成分，以及它们是如何作用于肠道微生物群[86]。母乳中含量丰富的人乳寡糖，例如二唾液酸乳-N-四糖，与母乳喂养婴儿较低的坏死性小肠结肠炎风险相关[87]，人乳寡糖到达肠道后促进双歧杆菌的生长[81]。体外实验还发现，相较于葡萄糖和乳糖，在人乳寡糖上生长的婴儿双歧杆菌和短双歧杆菌更能增加细胞间的紧密连接以及抗炎细胞因子 IL-10 的分泌，并减少促炎细胞因子 TNF-α 的表达，下调趋化因子活性[81]。

此外，补充益生菌也可以降低早产儿坏死性小肠结肠炎的风险和死亡率[88, 89]。并且，多菌株益生菌似乎是预防极低出生体重儿坏死性小肠结肠炎、降低死亡率行而有效的办法[90]。虽然缺乏有效的证据支持其在预防坏死性小肠结肠炎中的临床疗效，但益生元可以抑制有害微生物的定植并促进双歧杆菌的生长，从而调节肠道微生物的组成[91]。值得注意的是，由于新生儿免疫系统较弱且不成熟，服用益生菌并非没有风险。未来我们还需要进一步的研究，以控制菌株和剂量，量化早产儿益生菌给药的风险和益处[82]。

（二）结直肠癌

越来越多的研究表明肠道菌群失调与结直肠癌之间的关系[92]。在所有癌症病例中，约有 15% 与病毒或细菌感染有关[93]。微生物病原体可以通过导致 DNA 损伤的物质，例如一氧化氮或活性氧，直接影响肿瘤发生，也可以通过产生促炎微环境间接影响肿瘤发生[94]。

肠道微生物群中的几种细菌，如幽门螺杆菌、大肠埃希菌、具核梭杆菌、粪肠球菌，被认为与结直肠癌的发生有关[95]。例如，致癌性幽门螺杆菌感染会导致慢性炎症，上皮细胞中的 β-catenin 信号传导失调，从而促进胃黏膜的恶性转化[96]。同样在结肠中，幽门螺杆菌感染会增加腺癌风险[97]，尤其是毒力因子细胞毒素相关基因 A（CagA）阳性的菌株与癌症的发展有关[97]。与健康人相比，大肠癌患者的微生物群组成显著不同，表现为厚壁菌门和梭杆菌门的丰度增加[98, 99]。在大肠癌患者中，梭杆菌占肠道细菌的 10% 左右，而在健康个体中不到 0.1%。微生物群中的这些变化可以形成具有更高遗传毒性和致癌潜力的肠道群落。具核梭杆菌是一种在肿瘤组织中高度丰富的物种[100]，细胞表面蛋白 FadA 是其关键毒力因子[101]。FadA 可以调节细菌的黏附和侵袭，还可以通过调节 E-cadherin/β-catenin 途径，诱导炎症并抑制宿主免疫，从而影响结直肠癌的发生发展[102]。

肠道菌群失调可能通过影响固有免疫反应，更确切地说是可通过 MyD88 的激活，触发肿瘤的发展。在腺瘤性大肠息肉病（APC）基因杂合突变小鼠中，接头蛋白 MyD88 的信号传导在自发肿瘤发展中发挥了关键作用。MyD88 依赖的信号通路控制肠道肿瘤发生的几个关键修饰基因的表达，并在自发和诱导的肿瘤发展中发挥关键作用[103]。相反，MyD88 信号在氧化偶氮甲烷 / 葡聚糖硫酸钠所致结肠炎相关癌症的发展中具有保护作用[104]。MyD88 信号通路的第二分支是炎性小体衍生的 IL-18，这可能解释了 MyD88 信号在不同癌症模型中的矛盾作用。缺乏保护性和促进组织修复的 IL-18，以及由此导致无法治愈的化学诱导的上皮损害，可能会增强上皮细胞的突变率和腺瘤形成，从而使 MyD88 缺乏症的保护作用失衡[105-107]。

与 MyD88 相比，TLR4 信号在几种癌症模型中显示出一致的促肿瘤作用[108]，大肠癌患者中 TLR4 信号转导增加[109]。在肠道损伤的急性反应期，LPS 作用于 TLR4 并激活下游的信号级联反应，导致 COX-2 转录增强及 PGE2 产量增加，有益于肠上皮细胞增殖并减少细胞凋亡[108, 110]。但是，当 LPS 刺激超过正常水平并引起慢性 TLR4 激活时，相同的机制也可能促进结直肠肿瘤的形成和生长[111, 112]。有趣的是，肥胖个体的炎症状态增加与大肠癌的高风险相关[113]。在 IBD 中也观察到 TLR4 激活升高[114]，炎性肠病的两种主要形式是溃疡性结肠炎和克罗恩病，是结肠炎相关大肠癌的危险因素[115]。TLR/MyD88 通路在微生物诱导的结肠炎相关癌症的发展中起着非常重要的作用。动物实验发现，在没有 MyD88 的情况下，IL-10（-/-）自发性结肠炎的小鼠无法发展为致癌物诱导的肿瘤[116]。

微生物群的特定变化促进了结直肠癌的形成。肉类是含硫氨基酸的主要来源，可以增加肠道中能利用硫酸盐的细菌的数量，例如脱硫弧菌、脱硫杆菌。这些细菌产生的硫化氢减少了黏液的形成，抑制了 DNA 的甲基化并增加了活性氧的产生 [117]。同样，单一细菌物种也可以促进肿瘤生长。脆弱拟杆菌通过增加 β-catenin 核信号传导来刺激细胞增殖 [118]，并通过活性氧来破坏 DNA[119]。研究发现，在炎性肠病和结直肠癌患者中，黏膜相关聚酮化合物合酶 pks^+ 大肠埃希菌的比例很高，其产生的聚酮化合物 - 肽基因毒素 Colibactin 在结肠炎易感的 IL-10 缺陷小鼠中促进了浸润性癌的发展 [120]。Colibactin 导致 DNA 双链断裂和 DNA 修复不完全，从而导致基因组不稳定 [121]。

微生物群也可以赋予抗结直肠癌的保护作用并防止癌变，改变微生物群或针对特定的免疫成分可能是开发结直肠癌新治疗方案的有效策略。微生物代谢产物丁酸盐激活受体 GPR109a，触发细胞保护性 IL-18 的产生，并通过 IL-10 诱导 Treg 细胞分化，同时抑制促炎性 Th17 细胞的形成 [122]。益生菌（如双歧杆菌和乳杆菌）创造了有利的微环境，不仅减少了炎症，而且减少了结直肠癌的发生。特别是与益生元合用时，双歧杆菌属和乳杆菌属可减少小鼠和大鼠结肠异常隐窝病灶的发生 [123]。

（三）艰难梭菌相关性腹泻

艰难梭菌是一种革兰阳性、产芽孢厌氧菌 [124]，可引起一系列疾病，从轻度腹泻到严重并发症 [125]。新生儿胃肠系统是其生长和繁殖的自然栖息地，且不会出现活动性感染。但随着新生儿的成长，其他微生物进入肠道成为优势物种，艰难梭菌的定植便受到了限制 [126]。对老年人而言，肠道中双歧杆菌和某些厚壁菌门、拟杆菌门减少，变形杆菌增加 [127]，以及频繁使用抗生素，更多的医源性暴露和共存疾病的发展，可能增加了对艰难梭菌的易感性 [128]。

正常肠道微生物群限制外来细菌整合到现有群体中的能力被称为定植抗性 [27]。研究发现，肠道微生物群可以通过竞争营养素、阻止进入肠黏膜的黏附位点、产生抑制物质和刺激宿主的免疫系统来提供定植抗性，防止病原体的感染 [129]。已知的风险因素，包括抗生素、高龄、炎性肠病和免疫抑制剂，其共同的特征是肠道微生物生态系统被破坏，表现为多样性减少、某些细菌丰度的改变及定植抗性的丧失 [125]。当肠道微生物失衡，内源或外源性病原体孢子出芽并生长，随后穿透黏液层并黏附在肠细胞上，定植于胃肠道 [130]。艰难梭菌释放的毒素 TcdA 和 TcdB 修饰上皮细胞骨架的肌动蛋白，使细胞骨架解体，抑制细胞分裂和膜运输，导致肠上皮细胞的破坏并诱导炎症反应 [124]。当 TcdA 和 TcdB 毒素在宿主肠道中释放时，其活性足以引发疾病 [131]。

体外研究表明，初级胆汁酸刺激艰难梭菌孢子的萌发；而次级胆汁酸能够抑制病原体的生长，并降低艰难梭菌在肠道的定植 [124]。镰孢梭菌（*Clostridium scindens*，C.scindens）是一种具有 7-α- 脱羟基酶活性的共生细菌，除了能够将初级胆汁酸转化为次级胆汁酸，并限制艰难梭菌的萌发和生长之外，可能通过产生色氨酸衍生的抗生素影响艰难梭菌的生存能力 [132]。其中，肠道微生物群 - 胆汁酸相互作用可能是通过影响 FXR-FGF 通路与艰难梭菌感染的发病机制相关 [133]。据报道，肠道微生物产生的一种次级胆汁酸，3β- 羟基脱氧胆酸，通过法尼醇 X 受体（farnesoid X receptor，FXR）作用促进外周调节性 T 细胞的产生 [134]，但其对艰难梭菌感染的潜在影响目前尚不清楚 [133]。此外，某些微生物可以通过消耗艰难梭菌生长所需的营养来阻止艰难梭菌的繁殖。碳水化合物，例如来源于肠道黏蛋白的 N- 乙酰氨基葡萄糖和 N- 乙酰神经氨酸（N-acetylneuraminic acid，Neu5Ac），是支持艰难梭菌生长的重要营养素 [135]。这些碳水化合物优先被其他肠道微生物代谢，从而限制了艰难梭菌的繁殖，进而限制了疾病的发展[136]。

艰难梭菌感染是抗生素相关性腹泻的主要原因，并且 22% ～ 32% 的患者出现对抗生素无效的长期或反复感染[137]。目前，粪便微生物移植已成为复发或难治性艰难梭菌感染的一种有效的治疗方法[137, 138]。粪便微生物移植通过输注来自健康供体的微生物群，治愈了 90% 以上的难治性艰难梭菌感染的患者，而传统的万古霉素治疗仅改善了 30% 的病例[139]。在动物模型中，艰难梭菌感染与抗生素治疗相结合的研究发现，艰难梭菌的增殖与毛螺菌科的减少和肠杆菌科的水平提高有关[140-142]。在患有艰难梭菌相关性腹泻的老年受试者中，肠杆菌科的水平也有所升高[143]。研究发现，粪便微生物移植恢复了复发性艰难梭菌感染者肠道微生物和胆汁酸的组成，且这个过程与回肠 FXR 信号的激活相关，表现为 FGF19 增加和 FGF21 表达减少[138]。

此外，关于恢复肠道的微生物平衡，益生菌制剂已经被标准化，并作为对抗复发性艰难梭菌感染的营养药物，进入了临床试验阶段，例如 RBX2660 和 SER-109[144]。Kevin Chen 等还通过基因修饰布拉酵母菌，使其分泌一种特异性抗体，有效而广泛地中和 TcdA 及 TcdB，在小鼠疾病模型中显示出对原发性和复发性艰难梭菌感染预防和治疗的保护作用[145]。

五、微生物群和免疫反应：孰先孰后

许多炎症性疾病，如炎性肠病，与微生物群的失衡有关，这表明微生物失调、炎症和病因之间存在联系[146]。人们试图阐明炎性肠病发展过程中微生物群与炎症的因果关系，但存在许多限制。例如，大多数已发表的研究是基于横断面，而非前瞻性纵向队列研究，不能提供微生物失调与发病相关的时间[147]。目前为止，仍不清楚微生物失调是导致疾病的原因，还是免疫功能紊乱的结果[146, 148, 149]。

Floris Imhann 等发现，对 IBD 有高遗传风险的健康个体，其肠道微生物群发生了显著变化，如产丁酸盐的罗氏菌属（*Roseburia spp*）减少[150]。全基因组关联研究还发现了与 IBD 易感性相关的基因，例如 NOD2 和 ATG16L1，其编码的重要成分用于感知和适应肠道微生物组的变化，并影响宿主防御和免疫机制[148]。此外，儿科风险分层研究在新诊断的、未接受过治疗的克罗恩病儿童粪便和黏膜活检中发现了微生态失调，这表明微生态失调可能先于临床疾病出现，且独立于长期炎症和药物治疗[147]。而在另一项克罗恩病患儿的前瞻性研究中发现，炎症、抗生素暴露和饮食作为环境压力源，能够独立地影响不同微生物分类群。并且，随着抗肿瘤坏死因子或全肠内营养治疗后肠道炎症的减少，微生物失调也减少[151]。

炎性肠病通过环境因素、微生物组、免疫和遗传因素之间的相互作用而发展[148]。因此，IBD 菌群失调和炎症之间的联系比因果关系更具动态性。有人推测：肠道免疫系统的表观遗传变化可能为促炎微生物创造环境，从而增加对 IBD 的易感性；随后，由环境因素导致的微生物变化可能进一步破坏微生物与肠道免疫的动态平衡，从而形成恶性循环，最终导致 IBD（图 5-2）[146]。

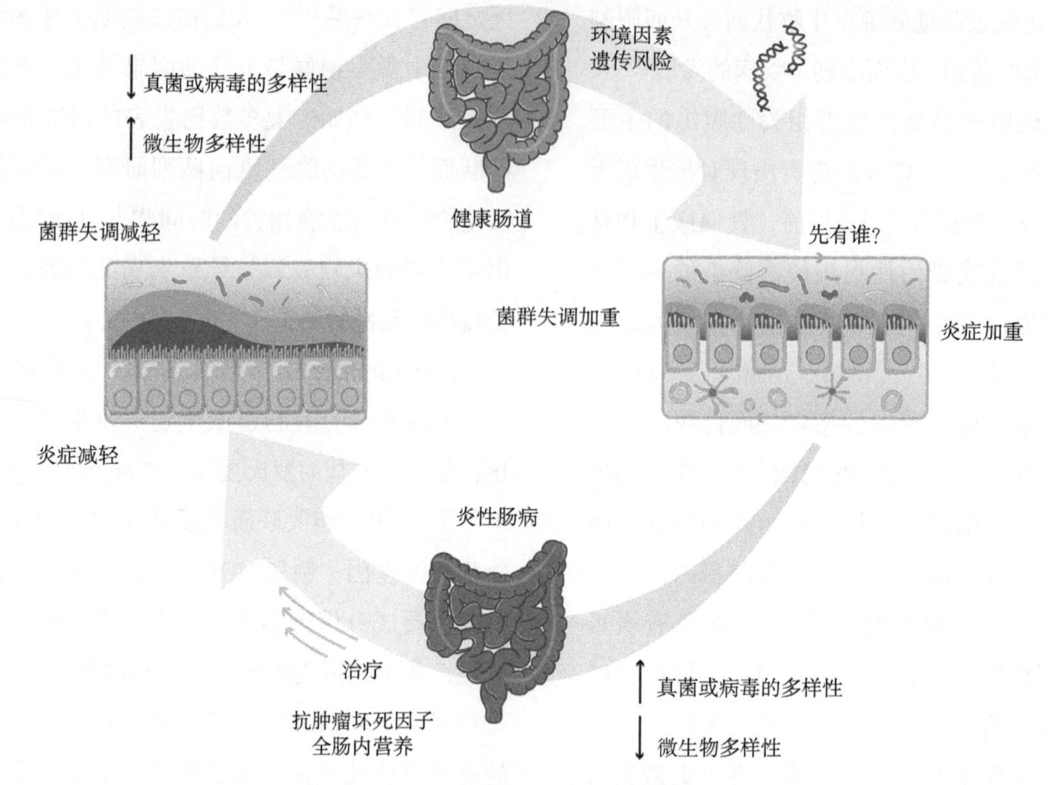

图 5-2　肠道菌群失调和免疫反应

在环境因素影响下，具有遗传倾向（例如 NOD2、IL-10R 或 ATG16L1 突变）的健康个体可能发生肠道菌群失调和炎症，但两者的先后顺序尚不清楚。一旦发生微生态失调和炎症反应的正反馈循环，就会导致炎性肠病。用抗肿瘤坏死因子或全肠内营养治疗后，炎症减轻，随后菌群失调好转，再次回到肠道稳态

第二节　肠道微生物与免疫的相互作用

在数百万年的进化中，微生物群和宿主免疫系统之间形成了相互依赖的关系：先天和适应性免疫的成熟和功能依赖于共生微生物，而免疫系统也能够影响微生物生态系统[152]。肠道菌群深刻地影响宿主免疫系统的表型和功能[153, 154]。例如，无菌小鼠存在许多免疫缺陷，包括肠道中 T 细胞分化和募集受损[20, 155]、IgA 的产生和分泌减少[156]、抗微生物肽的产生减少[157]、全身性 IgE 升高[158]。重要的是，可以通过应用来自鼠和人类供体的粪便微生物群，或其衍生的细菌菌株、真菌或病毒来逆转这些缺陷[20, 159]。了解这种微生物与宿主之间相互作用的分子机制是开发新型药物的基础，这些药物可以有针对性地塑造免疫功能，从而有益于宿主健康。

一、胃肠道菌群和免疫系统发育

黏膜免疫系统在其结构和功能上具有高度复杂性和异质性，在保护机体免受微生物病原体侵害方面起着关键作用[160]，并且在细菌菌落定居于肠道之后发生了重大变化[161]。其中，共生微生物是免疫系统成熟所必需的，免疫系统需要“学习”才能区分共生细菌和致病菌[162]。来自小肠上皮和淋巴样细胞的 Toll 样受体参与了这种差异识别，负责肠道黏膜免疫系统的正常发育。一方面，TLR 抑制炎症反应的发生，促进对共生微生物群的免疫耐受。另一方面，TLR 识别不同的微生物相关分子模式，包含各种细菌抗原，如荚膜多糖和脂多糖、鞭毛蛋白和未甲基化的细菌 DNA CpG 基序，并触发天然

肠道免疫[163]。TLR 识别微生物的这些特定分子成分后，将信号传递到细胞中激活 NF-κ B，从而诱导促炎细胞因子的过度释放，包括 IL-1、IL-6、TNF-α 和 IL-12[164]。在生命的最初几周内 TLR 活性降低，可能使稳定的肠道细菌得以发展。此外，共生微生物群激活 TLR 对于维持上皮稳态、保护肠道免受损伤至关重要[165]。另外，NOD 样受体（NOD-like receptors，NLR）识别各种微生物特定分子并触发炎症小体的装配，这些炎症小体可以作为损伤相关模式的传感器。NLRP6 炎症体缺乏与免疫反应改变（例如 IL-18 水平降低）、菌群失调和肠道增生（结肠隐窝增生，末端回肠隐窝 / 绒毛比值改变，派尔集合淋巴结生发中心增大）有关[166, 167]。

胃肠道菌群已显示出调节中性粒细胞功能和迁移的作用[21]，并影响 T 细胞群体分化为不同类型的辅助细胞（Th）：Th1，Th2 和 Th17 或调节性 T 细胞[163]。Th17 细胞是 $CD4^+$T 细胞的子集，它们分泌多种细胞因子（IL-22、IL-17A 和 IL-17F），对免疫稳态和炎症有显著影响[168]。Th1 和 Th2 细胞在分化后具有稳定的分泌特征，而 Th17 细胞则保留了不同特征和功能的细胞因子表达[169]。研究表明，从共生细菌脆弱拟杆菌中纯化出的荚膜多糖能抑制 IL-17 的产生，并保护结肠黏膜免受细菌抗原引发的炎症反应，刺激 $CD4^+$ T 淋巴细胞产生 IL-10 。另一方面，结肠环境也刺激了新生 $CD4^+$T 细胞向外周来源的调节性 T 细胞分化和增殖[170]。Treg 细胞是免疫耐受的关键介质，限制了不适当的高炎症反应[171]，其功能障碍可导致自身免疫疾病[172]。

分泌型免疫球蛋白 A（secretory immunoglobulin A，sIgA）在局部免疫反应中起着至关重要的作用，被认为是对抗病原体和毒素的第一道防线。sIgA 与肠道相关淋巴样组织（gut-associated lymphoid tissue，GALT）中的微折叠细胞（microfold cells，也称 M 细胞）相互作用，并将抗原传递给黏膜的树突状细胞、进而 T 细胞活化，最终在肠系膜淋巴结中发生 B 细胞类别转换重组，以诱导抗原特异性黏膜和系统免疫[173]。通过调节免疫显性决定簇诱导低水平的 IgA，被认为是共生细菌在肠道环境中获得优势的一种机制[167]。包括 TGF-β、IL-4、IL-10、IL-5 和 IL-6 在内的多种细胞因子能刺激 IgA 的产生，其中一些细胞因子，尤其是 IL-10 和 TGF-β 对于维持黏膜耐受性至关重要，因此证明了 sIgA 的产生、免疫和肠道稳态之间的联系[174]。

此外，宿主 - 共生菌群的相互作用触发了上皮的抗菌反应，释放多种抗菌肽，包括胰岛再生源蛋白Ⅲ γ（regenerating islet-derived protein Ⅲ γ，Reg Ⅲ γ）、α- 防御素和血管生成素[157, 175, 176]。这些抗菌肽减少了潜在病原微生物的数量，并对随后的异常免疫反应提供了保护作用。例如，多形拟杆菌触发了抗菌肽的产生，该抗菌肽可靶向作用于其他肠道微生物。表达人类肠道 α- 防御素基因 DEFA5 的小鼠，肠道分节丝状细菌明显减少，产生 IL-17 的固有层 T 细胞也减少[176]。研究发现，在克罗恩病患者中防御素的表达降低，可能是导致宿主微生物失调和炎性肠病炎症反应增强的原因[177]。

总之，固有免疫系统成熟期间微生物的异常发育导致免疫耐受性下降，继而导致自身免疫性疾病和炎症性疾病的发展[178]。同时，微生物产物也可能会诱导免疫应答，从而导致组织损伤和慢性炎症，特别是在黏膜损伤后。当肠道微生物群失调时，机体将调整免疫反应，通过特异性的 IgA 效应、固有免疫效应因子（如防御素），或非特异性效应（如腹泻），消除不需要的微生物群落，并为有益微生物的重新定植做准备，以促进微生物群重新配置到最佳状态[179]。

二、肠道菌群对局部免疫的影响

肠道菌群与免疫系统的发育和调节密切相

关。在局部，肠道微生物 - 宿主共生关系对于屏障强化、营养吸收、抵抗肠道病原体及黏膜免疫系统的发展和维持至关重要[152]。例如，肠黏液层是宿主与肠道内环境接触的第一道屏障，也是抵抗肠道病原体的第一道防线，主要由杯状细胞合成和分泌的黏液糖蛋白组成。杯状细胞能通过基础及调节分泌途径，生成黏蛋白来维持和更新肠黏液层，肠道菌群也是黏液的组成、厚度及黏液屏障通透性功能形成的关键因素。研究发现，无菌小鼠的杯状细胞数量和大小都明显少于常规饲养的小鼠，使得无菌小鼠的黏液层更薄[180, 181]。无菌小鼠中还存在其他胃肠道免疫缺陷的表现：①肠系膜淋巴结（mesenteric lymph nodes，MLN）变小和内皮微静脉形态异常，淋巴细胞结合不良[182]；②缺少生发中心的派尔集合淋巴结减少[183]；③肠固有层中缺乏淋巴滤泡，但是存在新生的隐窝斑，它们在微生物定植后会发展为功能性的独立淋巴滤泡[184, 185]。这些局部免疫缺陷伴随着肠固有层 $CD4^+$ T 细胞、浆细胞数量减少，以及 IgA 产生减少，从而导致肠屏障功能进一步受损[186, 187]。此外，Ivanov 等[188]的研究表明，肠道中 $IL\text{-}17^+$ Th17 细胞和 $Foxp3^+$ Tregs 之间的平衡，需要共生细菌的信号，并且依赖于肠道微生物群的组成。

三、肠道菌群对全身免疫的影响

全身固有免疫调节也受共生微生物的影响。肠道微生物在骨髓和外周粒细胞 - 巨噬细胞祖细胞水平上对骨髓生成有刺激作用，对树突状细胞、巨噬细胞和中性粒细胞的功能也有作用。在许多情况下，这些系统性作用归因于循环中细菌衍生分子，即微生物相关分子模式或病原体相关分子模式（pathogen-associated molecular pattern，PAMP），例如脂多糖、肽聚糖或鞭毛蛋白。当其被识别时，固有免疫细胞上的模式识别受体（pattern-recognition receptors，PRRs）可以通过 MyD88 依赖性途径发出信号，从而增强全身固有免疫细胞的反应能力[152]。细菌代谢物，例如短链脂肪酸，与刺激树突状细胞在骨髓中的产生及其吞噬能力有关[189]。共生细菌的存在也会刺激全身适应性免疫，特别是远处非黏膜淋巴样组织的正常发育，例如脾和外周淋巴结。在无菌小鼠的这些器官中，B 细胞滤泡和 T 细胞区域发育不良，血清中 IgG 水平降低[190, 191]。与无菌动物相比，脆弱拟杆菌定植期间通过荚膜多糖 A 的免疫调节活性，纠正全身性 T 细胞缺乏和 Th1/Th2 失衡，并指导淋巴器官发育[192]。

四、肠道菌群对固有免疫的作用

（一）上皮细胞

肠上皮细胞构成了机体和外部环境之间的边界。无菌小鼠的研究表明，肠腔内的微生物定植影响肠上皮细胞的代谢、增殖、存活、屏障功能和与免疫细胞的交流[193]。尽管被认为不是固有免疫系统的真正细胞，但肠上皮细胞同样表达先天免疫受体，以识别微生物相关分子模式，并启动下游的信号级联反应[194]。这些信号受体包括 Toll 样受体、RIG-I 样受体、NOD 样受体等[195, 196]。

肠上皮细胞先天免疫受体的表达和微生物识别后的主动信号转导对于肠道稳态至关重要。其上皮特异性缺失会导致上皮屏障的破坏，使组织易感自发性炎症。TLR 信号传导中涉及的成分证明了这一点，包括髓样分化初级应答蛋白 MyD88、TNF 受体相关因子 6 和 NF-κ B 必需调节剂（NEMO）[197-199]，以及细胞死亡的协调者，例如与受体相互作用的丝氨酸 / 苏氨酸蛋白激酶 1，与 FAS 相关的死亡域蛋白和 caspase-8[200-202]。最近的研究表明，结肠炎中 TLR4 信号通路的增加可驱动双氧化酶 2 的表

达和上皮细胞 H_2O_2 的产生，从而导致肿瘤的发生 [203]。小肠潘氏细胞中高度表达的核苷酸结合寡聚化结构域蛋白 2 由微生物肽聚糖激活，并诱导包括细胞因子分泌、自噬诱导、细胞内囊泡运输、上皮再生和抗菌肽产生在内的细胞反应，从而影响微生物群的组成 [204-206]。上皮细胞 NOD1 对趋化因子 CCL20 介导的肠内孤立淋巴滤泡的产生和细菌定植都很重要 [185]。

肠上皮细胞中的模式识别受体对防御病原体感染也很重要。NOD 样受体 NLRC4 在感知细胞内鞭毛蛋白或 3 型细菌分泌系统成分后形成炎症体，驱动被感染细胞排出病原体，以限制其在上皮内的增殖 [207]。NLRC4 炎症小体已被证明在防御多种革兰阴性细菌病原体，如假单胞菌属、沙门菌和志贺菌中起重要作用 [208]。微生物群相关代谢物牛磺酸、组胺和精胺通过调节 NLRP6 炎症小体的信号传导、上皮细胞 IL-18 的分泌和下游抗菌肽（antimicrobial peptide，AMP）来塑造宿主 - 微生物群界面 [209]。NLRP6 缺乏会导致杯状细胞的自噬缺陷、黏液分泌减少 [210]，微生物群组成和功能失衡、微生物分布改变，以及宿主对肠道感染的易感性增加 [166, 210]。此外，NLRP6 被描述为肠道抗病毒免疫的调节剂 [211]，这表明它可能在控制微生物组的病毒部分起作用。其他受体，包括代谢物感应受体——GPR43 和 GPR109A [212]、干扰素诱导基因 HIN-200 家族受体——AIM2[196]，通过整合微生物信号触发炎症体激活，以调节 IL-18 的水平。而这些受体的基因缺失与肠道炎症的易感性增加有关，这突出了上皮细胞 IL-18 在协调肠道宿主 - 微生物界面中的重要作用。

有趣的是，微生物对肠上皮细胞影响远远超出了这些细胞的经典免疫功能，共生微生物定植似乎在肠上皮细胞代谢中起重要作用。微生物群来源 SCFA 的上皮代谢可能是黏膜“生理性缺氧”的主要决定因素，它们促进肠上皮耗氧量达到缺氧诱导因子（hypoxia-inducible factor，HIF）稳定的程度，并增加屏障保护性 HIF 靶基因的表达，连接微生物、代谢和黏膜固有免疫 [213]。吲哚——微生物代谢产物之一，通过孕烷 X 受体促进屏障功能 [214]，并增加胰高血糖素样肽 1 的分泌。由于微生物群本身在组成和功能上会发生有规律的变化 [215]，这表明微生物对固有免疫系统的影响程度可能会在一天中产生明显的波动。微生物在肠上皮细胞中诱导的 TLR 信号转导通过昼夜节律的协调来驱动肠道激素的产生 [216]。

总之，肠上皮细胞将微生物信号整合到由黏液和抗菌肽组成的宿主 - 微生物界面的编排中，并动态调节细胞代谢。

（二）肠道微生物对骨髓细胞的调节

肠道菌群极大地影响了肠道免疫系统。髓样细胞，如中性粒细胞和巨噬细胞，通常是感染的第一个免疫应答者。尽管骨髓生成发生在骨髓中，但多项研究表明，肠道菌群对骨髓造血作用和肠道巨噬细胞功能具有广泛的影响（图 5-3）[217]。

1. 中性粒细胞　经广谱抗生素治疗的小鼠，骨髓干细胞和祖细胞数量减少 [218]。由于缺乏肠道菌群，无菌小鼠的固有免疫出现缺陷，对病原体的早期免疫反应受损。而无菌小鼠的复杂微生物群再定植可以恢复骨髓造血功能和对单核细胞增生性李斯特菌全身感染的抵抗力 [70]。从机制上讲，微生物相关分子模式被证明可以维持中性粒细胞的稳态，并通过 Toll 样受体信号转导调节中性粒细胞，对抗细菌感染 [219, 220]。此外，微生物群通过 TLR4 和 MyD88 依赖机制，诱导产 IL-17 的 ILC3 在肠道中的积累，促进骨髓中粒细胞集落刺激因子（granulocyte colony-stimulating factor，G-CSF）介导的粒细胞生成，这对抵抗新生小鼠的大肠埃希菌败血症至关重要 [221]。另一方面，微生物群介导的中性粒细胞活化会增加循环中活化的中性粒细胞数量，它们分泌促炎细胞因子和颗粒蛋白酶，破坏组织并加剧疾病。在镰状细胞疾病和内毒素诱导的脓毒血症小鼠模型中，肠道微生物群的耗竭显著减少了循环中性粒细胞数量，减轻了炎症相关的器官损伤[222]。在全身性大肠埃希菌感染期间，宿主模式识别受体识别细菌成分，通过TLR4和NOD1协同诱导G-CSF，动员造血干细胞进入脾脏，并成熟为中性粒细胞以抵抗感染[223]。肠道微生物群衍生的配体可能在骨髓祖细胞动员到各个器官中发挥相似的作用，在这些器官中它们产生成熟的骨髓细胞以维持组织稳态。总之，从骨髓中髓系祖细胞的粒细胞生成，到成熟中性粒细胞对感染的应答，以及中性粒细胞的衰老，肠道微生物的微生物分子似乎对中性粒细胞具有深远而持久的影响[217]。

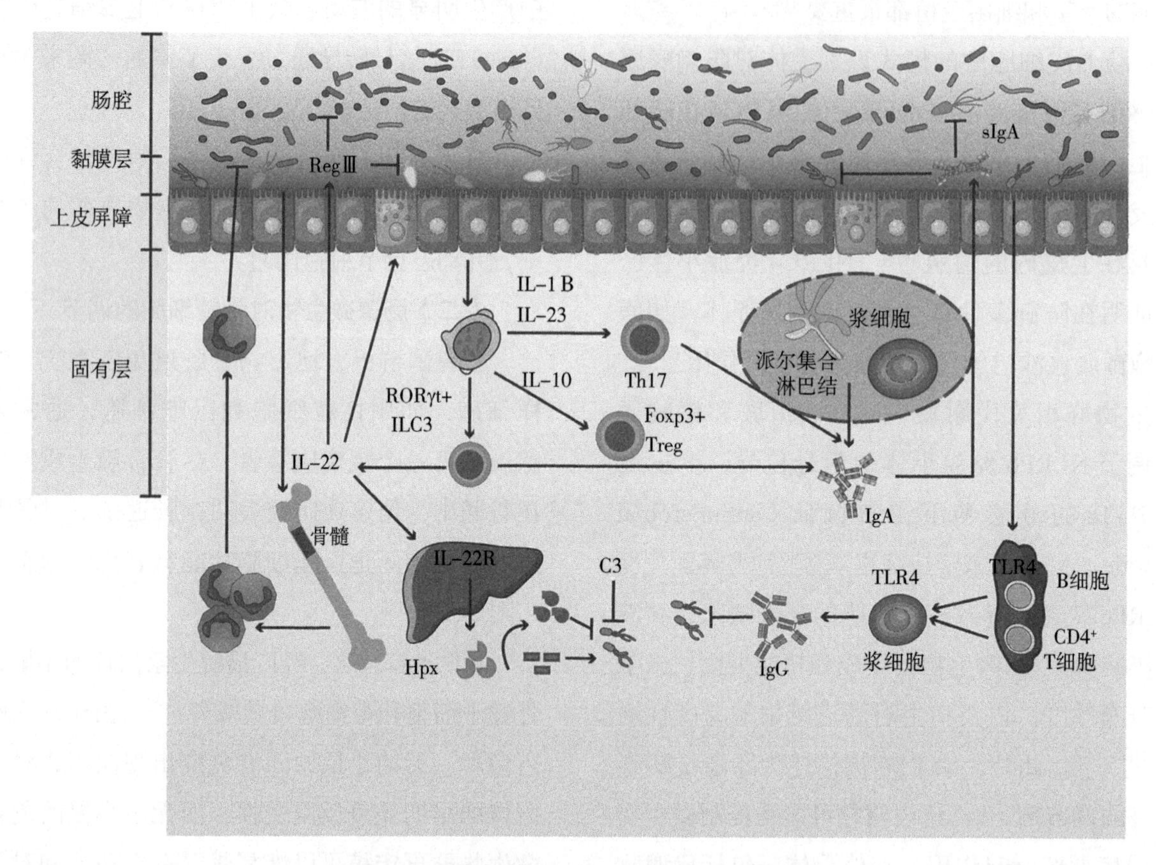

图 5-3　肠道微生物群对宿主免疫反应的影响

通过Toll样受体，来自肠道共生细菌的分子促进骨髓中嗜中性粒细胞生成和感染后中性粒细胞的动员。肠道共生细菌的存在对于肠道巨噬细胞中pro-IL-1的表达很重要。在肠道损伤过程中，肠道共生细菌通过NLRP3炎症体将其裂解为成熟的IL-1，从而促进适当的炎症反应。来自肠道巨噬细胞的IL-1、IL-23和IL-6对促进黏膜Th17细胞应答很重要，而肠道巨噬细胞的IL-10和微生物短链脂肪酸参与了稳态条件下的肠道Tregs的发育。肠道菌群的存在对于诱导T细胞依赖性和独立产生IgA抗体至关重要。其中大多数IgA抗体对肠道共生体具有特异性，并被分泌到肠腔中，以侵袭性细菌为目标，防止它们穿过上皮屏障。在肠道柠檬酸杆菌或艰难梭菌感染期间，由3型固有淋巴细胞产生的IL-22可以诱导肝细胞产生血红素结合蛋白和补体C3，分别抑制全身性移位细菌的生长和促进其清除。在稳态条件下，部分革兰阴性肠道共生菌诱导机体产生能够识别细菌表面抗原的IgG抗体，如某些革兰阴性病原体上表达的胞壁质脂蛋白，从而有助于宿主抵御肠道共生体或病原体的全身感染。

RegⅢ. 胰岛再生源蛋白Ⅲ；TLR4.toll样受体4；RORγt. 维甲酸相关孤儿核受体γt；Hpx. 血红素结合蛋白；Foxp3. 叉头状转录因子P3；ILC3.3型固有淋巴细胞；Treg. 调节性T细胞

2. 肠嗜酸性粒细胞　嗜酸性粒细胞是源自骨髓多能造血干细胞的重要白细胞亚型。在稳态条件下，嗜酸性粒细胞驻留在肠道中，通过与肠道微生物交流来维持免疫反应的平衡[224]。研究表明，不同组织中的嗜酸性粒细胞，其功能具有显著差异[217]。例如，与血液和肺中的嗜酸性粒细胞相比，肠道嗜酸性粒细胞很少脱颗粒，且寿命更长 [225, 226]。小肠嗜酸性粒细胞分泌高水平的 IL-1 受体拮抗剂（IL-1Ra）并发挥抗炎功能，而大肠嗜酸性粒细胞会产生 TNF-α 和 IL-13 并促进结肠炎 [227]。

肠道微生物可以调节嗜酸性粒细胞的数量和功能。与 SPF 小鼠相比，无菌小鼠的小肠嗜酸性粒细胞增多更明显，并表现出更多的细胞质突起和更少的颗粒含量。这种变化与调节嗜酸性粒细胞趋化和存活的肠道信号（包括 IL-3、VEGF、IL-11 和 CXCL9）有关，并且能被复杂的微生物定植完全逆转。而在嗜酸性粒细胞缺乏的小鼠中，黏膜下层胶原集聚，并且表现出对食物抗原的过敏反应性降低 [228]。据报道，细菌能够引发嗜酸性粒细胞的多种功能反应，包括吞噬作用、颗粒衍生蛋白的释放、活性氧的产生 [229]。来自革兰阴性菌的脂多糖激活 IL-5 或 IFN-γ，触发肠道嗜酸性粒细胞以依赖活性氧的方式释放线粒体 DNA，防止细菌入侵宿主 [230]。此外，肠道嗜酸性粒细胞及其上游细胞调节者（如 ILC2）表达游离脂肪酸受体，并对微生物代谢物产生反应 [231]。

阐明嗜酸性粒细胞 - 微生物组的相互作用还需要进一步的研究，可以考虑将微生物组改造作为塑造组织嗜酸性粒细胞表型的策略 [231]。

3. 肠道巨噬细胞　在人体的每种组织中都有常驻巨噬细胞，它们可以吞噬和杀死感染性病原体，从而执行免疫哨兵和稳态功能 [232]。巨噬细胞是稳态条件下肠道固有层中最丰富的单核吞噬细胞 [233]。研究发现，卵黄囊和胎肝衍生的巨噬细胞存在于新生儿肠道中，但在出生后不久被 Ly6C^{hi} 血液单核细胞所取代，并分化为成熟的抗炎巨噬细胞。这一过程需要 CCR2 依赖性经典单核细胞的募集和共生微生物群的参与 [234]。

稳态条件下，大多数驻留在小鼠结肠中的巨噬细胞表达高水平的 CX3CR1，具有强烈的吞噬作用和 MHCII^{hi}，但对 TLR 刺激（例如脂多糖）有抗性，并产生 IL-10[235]。共生微生物群触发肠道巨噬细胞产生细胞因子 IL - 1β，从而促进 ILC3 细胞释放粒细胞 - 巨噬细胞集落刺激因子（granulocyte-macrophage colony-stimulating factor，GM-CSF），而后者又反作用于巨噬细胞，影响其数量和功能。GM-CSF 的缺乏会导致结肠中巨噬细胞的数量减少，产 IL-10 和维 A 酸的能力降低 [236]。此外，微生物代谢物丁酸盐可以通过抑制组蛋白脱乙酰酶活性作用于肠道巨噬细胞，导致脂多糖诱导的促炎介质下调，包括 NO、IL-6 和 IL-12，维持对肠道微生物群的耐受性 [237]。

微生物群在调节肠道巨噬细胞介导的炎症中起着关键作用 [238]。它们可能是通过 TLR-MyD88 信号通路与肠道巨噬细胞中的 pro-IL-1β 诱导联系起来 [239]。肠道常驻巨噬细胞组成性地表达 NLRC4 和 pro-IL-1β，并在致病菌（如沙门菌）感染后分泌大量的 IL-1β，从而引发炎症反应 [240]。在肠道损伤时，肠杆菌科，特别是奇异变形杆菌可以诱导新招募的 Ly6C^{hi} 单核细胞释放 NLRP3 依赖性 IL-1β，从而促进肠道内炎症 [241]。有证据表明，LPS 通过 NF-κB 途径上调芳香烃受体（aryl hydrocarbon receptor，AHR）表达，经 Src-STAT3 信号通路促进小鼠炎性巨噬细胞中 IL-10 的产生，但目前尚不清楚该途径是否与体内肠道巨噬细胞相关 [242]。

综上所述，肠道菌群在控制循环单核细胞募集、补充肠道常驻巨噬细胞，以及响应损伤或感染而引起的肠道巨噬细胞功能变化方面起着复杂的作用 [217]。

4. 固有淋巴样细胞 3　固有淋巴细胞是一群先天免疫细胞，是固有免疫系统的重要组成部分，主要分为 3 个亚组：ILC1、ILC2 和 ILC3。其中，ILC3 在被细胞外细菌或真菌病原体感染期间产生 IL-17、IL-22 和 GM-CSF[243]。ILC3 是稳态下小肠固有层中的主要 ILC 亚群 [244]，越来越多的证据支持其在肠道免疫中的重要作用[245]。例如，对肠道中非T细胞依赖性IgA合成起关键作用的分离淋巴滤泡，其形成受RORγt$^+$淋巴组织诱导细胞和基质细胞之间相互作用的调节[246]。

关于ILC3的发育是否需要肠道共生细菌，部分研究表明，在无菌或Myd88$^{-/-}$小鼠中，产IL-22的ILC3减少；但也有学者发现，在无菌小鼠中ILC3的数量正常或增加[244]。肠道共生菌通过ILC3上的模式识别受体直接发出信号，或者通过调节肠髓样细胞和上皮细胞间接影响ILC3的功能[217]。人类CD127$^+$淋巴组织诱导细胞样ILC可以直接感知细菌成分，激活TLR2信号通路，诱导IL-2以自分泌方式增强IL-22的产生[247]。肠道共生菌在稳态条件下可以促进肠道巨噬细胞中IL-1β的表达，从而诱导RORγt$^+$ ILC产生IL-22[248]。树突状细胞通过识别MAMP在微生物群和ILC3之间建立联系。例如，鼠柠檬酸杆菌感染导致树突状细胞产生IL-23，进而诱导ILC3产生IL-22。由于ILC3表达LTα1β2，而树突状细胞表达LTβR，从而建立了一个正反馈回路，进一步促进IL-23的产生[249]。此外，色氨酸可以被肠道微生物代谢，其代谢产物可以作为AhR配体，特异性调节ILC3的功能[250]。由于Card9（-/-）小鼠的微生物群色氨酸代谢受损，不能产生作为AHR配体的代谢产物，导致ILC产生IL-22减少[251]。因此，我们认为固有淋巴细胞在没有微生物的情况下正常发育，但其正常的功能依赖于共生微生物的定植[162]。

ILCs在肠道感染中的作用已经得到证实：ILC3s参与对柠檬酸杆菌、艰难梭菌、肠道沙门菌、单核细胞增生李斯特菌的免疫应答[245]。它通过IL-22诱导肠上皮干细胞增殖并产生AMPs[252]、促进上皮蛋白的岩藻糖基化[253]，以此加强屏障功能来抑制微生物的入侵。表达CCR6的淋巴组织诱导细胞样ILC3通过表达MHC Ⅱ，下调共生细菌特异性CD4$^+$ T细胞反应，从而限制自发性炎症[254]。Castellanos等还展示了微生物诱导CX3CR1$^+$单核吞噬细胞释放TNF样配体1A，促进ILC3屏障免疫中的保护作用，并揭示了TL1A诱导ILC3上OX40L表达在驱动慢性T细胞结肠炎中的致病作用[255]。另外，研究还强调了肠道ILC3通过IL-22在介导宿主防御全身性感染中的重要性。在肠道艰难梭菌感染中，IL-22通过诱导表达IL-22R的肝细胞产生补体C3，促进移位肠道细菌的清除[256]。而在鼠柠檬酸杆菌感染过程中，ILC3产生的IL-22会系统性地诱导血红素结合蛋白的产生，从而限制血红素的可利用性，抑制因肠道屏障受损而全身性移位的细菌生长[257]。

五、肠道菌群对适应性免疫的影响

（一）肠道微生物对T细胞反应的调节

稳态条件下，小肠固有层包含有两个与稳态调节和发育相关的CD4$^+$ T细胞群，即Th17细胞和Foxp3$^+$调节性T细胞[258]。效应性T细胞和调节性T细胞之间的微妙平衡，介导对病原体的免疫反应并限制过度的免疫激活[75]。在无菌小鼠中，肠道相关Th17细胞和Treg细胞发育受损[217]，表明共生微生物在T细胞亚群的发育和外周形成过程中起着重要的作用[259]。此外，微生物相关代谢产物也参与调节适应性免疫[259]。

1. Th17细胞　Th17细胞是CD4$^+$效应T细胞的分支，可产生大量的IL-17，在宿主防御病原体及自身免疫疾病的发展中起着重要的作用[260]。微生物群，尤其是分节丝状菌，影响Th17细胞的分化[261]。肠道免疫系统通过识别与微生物的物理相互作用来启动Th17细胞反应[261]。分节丝状菌黏附至小肠上皮细胞，增强了血清淀粉样蛋白A的表达和释放，从而增强CX3CR1$^+$吞噬细胞中IL-1β和IL-23的产生，这两种细胞因子协同促进 ILC3 产生 IL-22。另一方面，IL-22 又反过来增强血清淀粉样蛋白 A 介导的 IL-1β 生成，最终上调 RORgt⁺CD4⁺T 细胞中 IL-17 的产生 [217，261，262]。最新研究表明，小鼠共生细菌在骨髓来源的树突状细胞中通过 TLR4 信号触发 Th17 诱导细胞因子的分泌，而 Th17 细胞分泌的细胞因子（如 IL-6、IL-1b、IL-23 和 IL-10）数量取决于共生细菌的免疫原性 [263]。此外，微生物相关代谢产物——戊二酸盐增强了 mTOR 活性，从而诱导 Th17 细胞表达 IL-10[264]。吲哚 -3- 乙酸也是一种由肠道微生物群产生的吲哚衍生物，可以作为 6- 甲酰基吲哚并 [3，2-b] 咔唑（6-Formylindolo[3，2-b] carbazole，FICZ）内源性合成的底物 [265]。FICZ 是一种内源性 AhR 配体，而 AhR 在产生全功能 Th17 细胞中起着关键作用 [259]。

研究表明，丙酸杆菌菌株 P. UF1 对无菌小鼠的定植与新生小鼠肠道 Th17 细胞的增加和坏死性小肠结肠炎的缓解相关 [266]。近年来，肠道微生物群在调节免疫检查点抑制剂治疗反应中的作用引起了广泛关注。通过接受抗 PD-1 免疫疗法无反应黑色素瘤患者的肠道微生物群移植，可以促进无菌小鼠脾脏和肿瘤中 RORγT⁺ Th17 细胞的增加 [267]。无法诱导有效的抗肿瘤免疫反应的原因，可能部分涉及微生物对肿瘤微环境中 Th17 细胞的激活 [268]。因此，可以利用共生细菌与 Th17 细胞之间的联系来解决涉及 Th17 异常反应的疾病 [217]。

2. Treg 细胞　在肠道，Foxp3⁺ Treg 细胞主要位于固有层中，并且能够以微生物群依赖的方式迁移到上皮细胞转化为 Foxp3⁻ CD8αα⁺CD4⁺ 上皮内淋巴细胞 [269]。Treg 细胞可以在胸腺中与常规的 CD4⁺ T 细胞一起分化；也可以在外周组织中通过诱导 Foxp3 在常规 CD4⁺ T 细胞中表达以响应抗原刺激而产生，被称为 pTreg。其中，来自共生微生物群的抗原对于 pTreg 诱导至关重要 [270]。共生细菌，例如梭杆菌属 [20] 和脆弱拟杆菌 [271]，能够促进肠道 Foxp3⁺ Treg 细胞的增殖和分化。另一方面，脆弱拟杆菌的荚膜多糖 A 激活 Foxp3⁺ Treg 细胞上 TLR2 信号通路，诱导黏膜免疫耐受，促进稳态条件下的微生物定植 [272]。

细菌代谢产物短链脂肪酸，已经被证明能够调节结肠 Treg 细胞的数量和功能 [273]。例如，丁酸盐通过表观遗传控制 Foxp3 基因的表达，诱导肠道中 Treg 细胞的产生 [274]。丙酸盐特异性诱导 Foxp3⁺ 产 IL-10 的 Treg 细胞，其作用依赖于编码 GPR43 的基因——Ffar2 的表达 [273]。此外，在维持 Foxp3+ Treg 细胞稳态中起关键作用的水溶性维生素 B9（叶酸），可以由几种细菌合成，包括双歧杆菌和乳酸杆菌 [250]。次级胆汁酸 3β- 羟基脱氧胆酸通过作用于树突状细胞来降低其免疫刺激特性，从而增加 Foxp3 的诱导。树突状细胞中 FXR 的缺失增强了 Treg 细胞的生成 [134]。

另外，RORγt⁺ Treg 细胞是结肠中一个独特的 Treg 群体，转录因子 RORγt 的表达由肠道微生物群诱导 [75]。但诱导 RORγt⁺ Treg 发育的分子介质目前仍不清楚，SCFA 可能是其中之一，且 RORγt 在 Th17 细胞和结肠 Treg 细胞的诱导可能遵循不同的途径 [275]。

（二）肠道菌群对 B 细胞免疫反应的调节

B 细胞在黏膜微生物群和宿主免疫系统的对话中起着关键作用，它们不仅能分化为抗体分泌细胞，还可以作为抗原呈递细胞并产生大量的细胞因子 [276]。据报道，肠道相关淋巴组织至少包含人体 80% 的浆细胞和 90% 的 sIgA[277]。除骨髓以外，小肠固有层也是 B 细胞抗原受体（B cell receptor，BCR）多样化和 B 细胞发育的场所。研究发现，肠道微生物组可以诱导全身性，包括肠道固有层的祖 B 细胞水平增加。并且，固有层和骨髓 Rag2⁺ B 细胞具有明显不同的 Vκ 组成，这可能是因为肠道和骨髓中 BCR 编辑的差异所致，并与共生微生物定植相关[278]。最近的研究表明，细菌可以将 L-氨基酸转化为 D-氨基酸，作为细胞壁的基本成分和生态系统中的信号分子。后者，比如 D-丙氨酸，通过刺激 M1 巨噬细胞，促进肠道幼稚 B 细胞存活[279]。

微生物群的定植会显著刺激 B 细胞并诱导免疫球蛋白产生。有证据表明，肠道黏膜和全身免疫系统可以通过微生物黏膜暴露或全身暴露的任一途径来启动，而暴露于微生物抗原的途径和顺序决定了 B 细胞免疫反应的免疫球蛋白谱和功能结果。其中，微生物黏膜暴露剂量增加后，主要针对细胞表面抗原的 IgA 谱没有扩大，且连续黏膜暴露产生了有限的重叠储备和 IgA 结合特异性的降低[280]。来自不同小鼠的派尔斑生发中心持续表达公共 BCR 克隆型，大部分具有典型的 CDR3 序列。它们依赖或独立于肠道微生物群产生，在派尔斑生发中心经历亲和力成熟后形成高度选择的抗体，以靶向特定的细菌群体来促进宿主 - 微生物的共生，并通过与微生物表面多糖的交叉反应保护机体免受病原体的侵害[281]。

产生 IL-10 的 B 细胞，也称为调节性 B 细胞（regulatory B cells，Bregs），具有抑制自身免疫性炎症的能力。Bregs 的诱导和分化依赖于先天免疫受体的信号传导[276]。研究表明，肠道常驻细菌通过 TLR2、MyD88 和 PI3K 途径激活肠道产 IL-10 的 B 细胞，从而减少结肠 T 细胞的活化，并维持黏膜稳态[23]。Elizabeth C Rosser 等在关节炎模型中证明了，肠道微生物群通过驱动 IL-1β 和 IL-6 的产生，促进了脾脏和肠系膜淋巴结中 Breg 细胞的分化，从而抑制过度炎症[282]。Qinghui Mu 等也发现，通过口服细菌 DNA 介导的 Breg 细胞扩增，能够抑制 MRL/lpr 小鼠模型中自身免疫性疾病的发作[283]。此外，细菌还可以通过产生乙酸盐来诱导产 IL-10 的 B 细胞的分化，其过程与乙酰辅酶 A 增加引起的代谢变化有关[284]。

1. IgA　IgA 是黏膜免疫系统中的主要抗体类型，其中，sIgA 是黏膜 IgA 的主要有效形式。据估计，成人每天约有 3g sIgA 分泌到肠腔中[285]。在共生细菌定居之前，无菌小鼠中黏膜分泌 IgA 的细胞明显减少，在新生儿肠道中则无法检测到，表明肠道共生细菌可能提供关键的刺激信号，以诱导黏膜 IgA 的产生[217]。GALT 的 IgA 反应包括依赖于 T 细胞和不依赖于 T 细胞的分化途径。T 细胞非依赖性应答似乎产生结合共生细菌的多反应性 IgA 抗体，而 T 细胞依赖性 IgA 应答由肠道病原体触发[276]。研究表明，T 细胞依赖性 IgA 的诱导需要 B 细胞表面的 CD40 与其衍生自 T 细胞的配体 CD40L 之间的相互作用，从而产生高亲和力的抗原特异性 IgA 来中和病原体[285]。另一项研究还证明了 Th17 细胞也是高亲和力、T 细胞依赖性 IgA 产生所需的关键亚群[286]。肠道 T 细胞非依赖性途径需要来自跨膜激活因子和 CAML 相互作用因子（transmembrane activator and CAML interactor，TACI），一种主要由 B 细胞表达的先天 CD40 配体样因子 BAFF/APRIL 受体的激活信号，且不依赖于 CD40。其诱导的 sIgA 反应只针对肠道微生物群的一部分，而不影响整体组成。值得注意的是，TACI 不参与肠相关淋巴组织中 T 细胞依赖性 IgA 的诱导[287]。

肠道微生物群影响肠道淋巴细胞组织中 IgA 的多样性和产 sIgA 浆细胞的积累。据报道，哺乳动物肠道表面 IgA 的多样性与肠道分类群的多样性是平行的[75]。共生细菌分节丝状菌刺激了孤立淋巴滤泡和第三级淋巴组织的出生后发育。并且，这些淋巴滤泡和第三级淋巴组织取代了派尔氏斑，成为肠道 IgA 和分节丝状菌特异性 Th17 反应的诱导位点[183]。由于滤泡辅助性 T 细胞（T follicular helper，Tfh）表达高水平的 ATP 门控 P2X7 受体，它限制了派尔氏斑中 Tfh 的扩增和生发中心反应。而肠道微生物群通过产生细胞外 ATP，限制了针对肠道病原体的保护性 IgA 的产生 [288]。此外，SCFAs 可以通过调节 B 细胞代谢和基因表达来促进 T 细胞非依赖性 IgA 的产生 [75]。然而，微生物对高亲和力和低亲和力 sIgA 分布的影响尚不确定 [285]。

2. IgE　IgE 是过敏反应的典型介质 [289]。IgE^+ 浆细胞主要存在于骨髓或淋巴组织中，并且在过敏性疾病儿童血液中表达 IgE 的记忆 B 细胞和浆细胞增多。而最近的研究表明，一些 IgE 可能起源于肠道 [290]。有证据表明，IgE 的产生受肠道微生物群的影响。例如，无菌小鼠或多样性较低微生物群定植的小鼠表现出血清高 IgE 水平。而在生命早期的关键发育窗口，微生物丰度的增加可以抑制这种现象 [291]。Cahenzli 等报道，新生无菌小鼠的 B 细胞以 CD4 T 细胞和 IL-4 依赖的方式在黏膜部位向 IgE 进行类别转换 [158]。当黏膜部位暴露于不同微生物时，可以在触发 IgA 类别转换的同时，下调 IgE 至基线水平，并且强调了婴儿期作为重要的机会窗口期 [217]。最新的研究表明，接受健康婴儿细菌定植的无菌小鼠免受牛奶过敏原 β- 乳球蛋白诱导的过敏反应，包括特异性 IgE 的产生，并确定了粪厌氧棒杆菌在其中的保护作用 [292]。Madeleine Wyss 等还发现，小鼠可以通过定植具有指定特征的细菌物种的组合来抑制 IgE 水平。这些特征包括存在于生命早期、产乙酸盐的细菌，以及具有 IgA 免疫原性 [291]。

目前，关于肠道微生物群对降低 IgE 反应的年龄特异性的潜在机制仍然难以捉摸 [217]。并且，抑制高水平的 IgE 是仅依赖于某些微生物种类，还是微生物丰度的累加效应，仍有待确定。因此，我们还需要进一步研究其他细菌特征和代谢物对 IgE 的调节作用，以开发基于微生物的治疗方法来预防人类过敏性疾病 [291]。

3. IgG　IgG 是人血清中最丰富的免疫球蛋白类型 [293]。与肠道共生细菌能够在 GALTs 中局部诱导 IgA 相反，人们认为高亲和力、抗原特异性 IgG 抗体的诱导发生在肠外器官，比如脾脏 [217]。另外，也有报道在人肠道中发现了 $IgG1^+$ 和 $IgG2^+$ 浆细胞，健康成年小鼠的派尔氏斑和肠系膜淋巴结中也存在产 IgG2b 和 IgG3 的 B 细胞群 [294]。

研究显示，在幼年小鼠中没有检测到抗阴沟肠杆菌，一种革兰阳性肠道共生细菌的血清 IgG [295]。而在克罗恩病或结肠炎患者中，经常有较高滴度的肠道共生细菌 IgG 抗体，这被认为是肠道细菌因肠道屏障破坏而导致全身移位所致 [296]。那么，在稳态条件下是否存在对共生细菌的全身性 IgG 反应？ Melody Y Zeng 等发现，稳态条件下的部分肠道共生革兰阴性菌能够全身传播以诱导 IgG 反应，并确定了胞壁质脂蛋白，一种在革兰阴性肠杆菌种上大量表达的分泌外膜蛋白，是稳态条件下诱导小鼠和健康人产生血清 IgG 的主要抗原。此外，研究还证明了 $CD4^+$ T 细胞和 B 细胞上 TLR4 在整个过程中的重要性 [297]。目前，导致胃肠道内局部产生 IgG 的信号尚未完全确定 [294]。大肠埃希菌 LPS 已被证明可以触发 BCR/TLR4 双重信号转导，通过 NF-κ B 通路，诱导胞苷脱氨酶和 T 细胞非依赖性类别转换 DNA 重组，从而产生 IgG2b 和 IgG3 [298]，这可能有助于 GALT 中 IgG 的生成。

肠道病原体可促进炎症反应，并诱导肠道相关的黏膜体液免疫。细菌抗原可通过淋巴系统到达肠系膜和远端淋巴器官，这导致记忆 B 细胞和浆细胞产生 IgG 抗体 [299]。成人和哺乳动物肠上皮细胞表达新生的 Fc 受体，一种介导 IgG 双向转运的载体。它在介导 IgG 分泌的同时，通过将 IgG/ 抗原复合物转运回肠黏膜，经由树突状细胞处理并呈递给区域淋巴组织，启动适应性免疫反应 [300]，从而在防御上皮相关病原体，例如柠檬酸杆菌中起重要作用 [301]。树突状细胞表达 SCFA 受体 FFA2，它的激活导致 B 细胞激活因子和视黄酸调节蛋白醛脱氢酶 1 家族成员 ALDH1A2 的释放，从而触发浆细胞分化基因的表达，进而产生 IgA 和 IgG。并且，SCFA-GPR43 相互作用促进了强大的抗体反应，增强了宿主对柠檬酸杆菌感染的抵抗力[302]。此外，已有证据表明，革兰阴性细菌抗原诱导的 IgG 反应，可以通过 Fc-γR 介导的免疫保护作用对抗全身性大肠埃希菌和沙门菌感染[294]。

六、免疫系统对肠道菌群的调节

微生物群对免疫系统具有显著的影响。同时，免疫系统在控制和塑造微生物的组成方面也起着重要的作用。两者之间的相互作用是复杂的、动态的[303]。在这里，我们主要介绍免疫系统如何调节肠道菌群。

（一）固有免疫对肠道菌群的调节

先天免疫系统通过感知微生物组变化，将信号传递给宿主，以适应组织生理水平，并调整微生物组成和功能。来自人类和小鼠的遗传证据表明，微生物组成随时间和个体之间的变化而不同，而先天免疫系统在其中起着重要作用[162]。

Toll 样受体是病原体和共生体固有免疫识别的关键介质[165]，参与宿主对病原体的防御，调节共生微生物的数量，并维持组织的完整性[303]。研究表明，TLR5 可以感知细菌鞭毛蛋白，在缺乏 TLR5 的小鼠中肠道菌群发生改变[304]。有趣的是，微生物群中的这些改变可能使人容易患肥胖症、代谢综合征和结肠炎。这表明 TLR5 可能是维持“健康”微生物群组成的必要条件[305]。George MH Birchenough 等发现，杯状细胞可以通过激活 TLR-MyD88 依赖的 Nox/DUOX ROS 合成下游的 NLRP6 炎性小体。这会触发杯状细胞分泌钙离子依赖的黏蛋白 MUC2，并产生细胞间隙连接信号来诱导隐窝中相邻的杯状细胞分泌 MUC2，从而驱逐细菌，通过防止细菌侵入内黏液层来保护结肠隐窝[306]。然而，关于 TLR 在控制微生物群组成中的作用仍存在一些争议。有研究表明，TLR 缺陷型小鼠中微生物群的改变很大程度上是由母体传递给后代而非先天免疫缺陷造成[307]。

Nod 样受体是感染和组织应激的胞质传感器，能够影响微生物群的组成。例如，缺乏 Nod1 或 Nod2 的小鼠都可以出现微生物群的改变[185, 204]。研究发现，Nod2$^{-/-}$ 小鼠的小肠上皮表现出异常，包括杯状细胞产生黏液减少。这与上皮内淋巴细胞产生过多的 γ- 干扰素有关，并且依赖于先天免疫信号分子 MyD88[206]。NLRP6 是形成炎性小体的 NLR 家族成员，可在肠细胞和杯状细胞中表达，在调节肠道稳态、防御感染等方面发挥关键作用[308]。先天性免疫系统通过涉及 NLRP6、ASC 和 caspase-1 的机制，导致 pro-IL18 的裂解来调节肠道微生物。并且，NLRP6 及炎性小体成分 ASC 和 caspase-1 的缺失形成了一种可传播的、更具致病性的微生物群[166]。有证据表明，NLRP6 缺乏可减少黏蛋白颗粒释放和由此产生的黏液层，导致杯状细胞自噬功能丧失。这使得 NLRP6 基因缺陷的小鼠更容易受到肠道病原体的感染，并且更不能维持微生物的稳态[210]。然而，在另一项研究中，Joana K. Volk 等认为，NLRP6 炎性小体在结肠内黏液层的形成或功能中并不是必需的[309]。

此外，MyD88 是识别微生物的多种先天免疫受体的信号适配器，MyD88 缺陷的小鼠同样表现出微生物组成的改变[22]。据报道，MyD88 控制着潘氏细胞中几种 AMP 的表达，包括主要针对革兰阳性菌的 Reg Ⅲ γ，它可以限制黏膜表面相关细菌的数量[198]。另外，潘氏细胞功能失调或者遗传诱导的 AMP 表达改变也会触发微生物群改变和菌群失调[176, 310]。

这些证据表明，固有免疫系统在调节肠道菌群组成和维持肠道菌群稳态方面起着重要作用。

（二）适应性免疫对肠道菌群的作用

1. T 细胞　有学者提出，适应性免疫系统的发展，部分是为了与复杂的微生物群共生[311]。

实际上，缺乏适应性免疫的小鼠，其微生物群会发生变化，这表明适应性免疫在调节微生物群的组成中起着重要的作用[312]。研究表明，T 细胞缺陷型小鼠的微生物群发生了变化[313]。T 细胞影响微生物群组成的主要机制是作为分泌 sIgA 的 B 细胞助手而发挥作用。另一方面，T 细胞似乎也可以直接影响微生物群的组成，例如通过诱导潘氏细胞中 AMP 表达，尽管目前缺乏充足的证据[313]。

研究发现，$Foxp3^{+}$ T 细胞通过生发中心依赖或非依赖的方式抑制炎症，并支持派尔氏斑生发中心的 IgA 选择，从而有助于肠道微生物群的多样化。$Foxp3^{+}$ T 细胞可以通过分化为 T 滤泡调节和滤泡辅助性 T 细胞，从而发挥对派尔氏斑生发中心和 IgA 合成的调节作用。它也可以通过阻止 $Foxp3^{-}$ T 细胞的扩增及其细胞因子的过度产生，从而以不依赖于生发中心的方式控制炎症，调节微生物的组成[314]。据报道，在具有针对鞭毛蛋白的 TCR 转基因小鼠中，IgA 的产生在很大程度上取决于 Treg 细胞[315]。Treg 细胞特异性 c-Maf 缺乏导致肠内产 IgA 的浆细胞和产 IL-17A-IL-22 的 Th17 细胞增多，以及明显的微生物失调[316]。

此外，Th17 细胞在派尔氏斑中获得 Tfh 表型，并诱导产生 IgA 的 B 细胞发育。缺乏 Th17 细胞的小鼠在霍乱毒素免疫后无法产生抗原特异性 IgA 反应，这证明了 Th17 细胞是高亲和力 T 细胞依赖性 IgA 产生所需的关键亚群[286]。最新的研究表明，对肠道细菌的反应性是人类 $CD4^{+}$ T 细胞库的正常特性。其对共生体的反应可能通过产生屏障保护性细胞因子和提供大量对病原体起反应的 T 细胞来维持肠道内稳态。循环和肠道常驻的肠道微生物反应性 T 细胞显示出 Th17 和 Th1 细胞特征，并且在某些情况下会产生 IL-10。与循环细胞群相比，肠道常驻细胞表现出明显的 Th17 倾向，这在 IBD 中更为突出[317]。

2. B 细胞和 IgA　B 细胞是肠道内稳态的关键介质，通过产生大量的分泌型 IgA 抗体来响应共生体。其中，以 T 细胞依赖的方式产生的 IgA 在肠道微生物群的形成中起着更重要的作用[303]。IgA 通过促进共生微生物的定植，以及中和侵袭性病原体来塑造肠道微生物组[316]。

在缺乏活化诱导的胞苷脱氨酶的小鼠中发现了 IgA 在塑造微生物群中的潜在作用，这些小鼠无法进行类别转换重组或体细胞超突变[318]。特别是，这些小鼠表现出分节丝状菌的增生，并可以通过重建肠道 IgA 产生来纠正这一缺陷[319]。因此，似乎 IgA 对分节丝状菌的反应限制了该物种的生长。研究发现，抑制性共受体程序性细胞死亡 -1（programmed cell death–1，PD-1）缺乏导致表型改变的 Tfh 细胞数量过多，从而使派尔氏斑生发中心 IgA 前体细胞选择失调。因此，降低了 PD-1 缺陷小鼠产生的 IgA 结合细菌的能力，可导致肠道微生物群落的改变[320]。此外，由于 T 细胞中 MyD88 信号的丢失会导致 Tfh 细胞和产生 IgA 的 B 细胞减少，T 细胞内缺乏 MyD88 的动物不能控制与黏膜相关的细菌群落，从而出现微生态失调[321]。不可忽视的是，IgA 对发育过程中微生物群的组成也很重要。γ- 变形菌是新生小鼠中的优势菌门，在正常成年小鼠的微生物群中受到抑制，但可在缺乏 IgA 的小鼠肠道中持续定植，从而导致持续的肠道炎症，并增加了对肠道损伤模型的易感性[322]。

微生物群特异性的 IgA 还可以通过减少对微生物群的炎症反应来促进宿主 - 微生物群的共生[323]。这种减少的炎症反应主要是通过“免疫排斥”来介导，其机制为 sIgA 通过识别病毒和细菌表面的多种抗原受体以及蛋白质，将肠腔中的各种抗原交叉连接，从而延迟或消除其黏附或穿透上皮的固有潜能[324]。IgA 减少对微生物群的炎症反应的另一种机制可能是通过 IgA 涂层对细菌基因表达的直接影响。Tyler

C.Cullender 等通过体外实验证明，鞭毛特异性 IgG 抑制了细菌的运动并下调鞭毛基因的表达，并且认为 IgA 可能具有同样的反应[325]。

七、微生物群和免疫系统：人类疾病动物模型的经验教训

动物模型已经被广泛用于证明肠道中宿主 - 微生物的相互作用，其中无菌动物和悉生动物是理解宿主与微生物相互作用的关键工具[180, 326]。由于具有高度的遗传相似性，并且在微生物分类水平上也与人类具有一定的相似性，在所有非人类模型中，小鼠是迄今为止研究肠道微生物群最常见且最有效的模型[327]。这些动物模型表明，肠道菌群广泛地影响宿主的生物过程，是驱动小鼠黏膜和系统免疫发育的关键因素[328]，并且能够影响某些疾病的发展[329]。例如，许多结肠炎的啮齿类动物模型，包括转基因人类 HLA B27 等位基因的大鼠和缺乏 TCRβ 的小鼠，都会出现肠道炎症[329, 330]。而在部分动物模型中，疾病的严重程度取决于肠道菌群的组成，且该表型可以通过菌群移植转移到非基因修饰的小鼠上[166, 331]。

特异性细菌通过其在宿主组织的定植或产生的代谢物，能够不同程度地影响宿主免疫，这在生物模型和常规小鼠模型中得到了证明[332-334]。例如，分节丝状细菌通过与回肠中的肠上皮细胞附着，导致强烈的 Th17 和 IgA 反应，从而限制其扩增[183]。梭菌通过产生大量的 SCFA 促进小鼠 $CD4^+$ Treg 的诱导[14, 274]。最近的研究还确定了驻留在小鼠结肠黏液中的细菌物种，它们能够促进 $CD8^+$ T 细胞反应，从而防止感染并增强抗肿瘤免疫力[155]。目前，共生特异性 TCR 转基因技术和 MHC Ⅱ 四聚体技术已被用于研究抗共生 T 细胞反应在调节宿主免疫 - 微生物联系方面的作用[335, 336]。通过这些方法，Mo Xu 等发现了病原体依赖型 IBD 是由微生物反应性 T 细胞驱动的[336]。

微生物代谢物是肠道菌群与肠道和全身免疫系统进行交流的关键物质[337]。其中，短链脂肪酸已经被证明是饮食 - 宿主 - 微生物相互作用的有效介体[338]。有证据表明，SCFAs 刺激 GPR43 对于某些炎症反应的正常消退是必要的。例如，GPR43（-/-）小鼠在结肠炎模型中表现出炎症加重或未消退[339]。此外，SCFA 能够影响远端部位，例如调节肺部炎症、自身免疫和造血功能[189, 340, 341]。然而，这些发现与人类的相关性有待进一步研究，但一些有希望的数据已经开始出现：口服丁酸盐后，人单核细胞中出现了某些可检测到的抗炎特征[342]。因此，科学家们提出了“饮食微生物群假说”，即工业化饮食通过改变肠道微生物群和代谢产物而导致炎症性疾病增加[343-345]。在工业化社会中，成人每日膳食纤维摄入量远低于推荐水平，而饮食中纤维含量的增加与人类纵向研究中较低的死亡风险相关[346, 347]。动物研究也证明了缺乏膳食纤维的肠道微生物群会降解结肠黏液屏障，导致宿主对病原体的易感性增高[348]。

了解微生物代谢物与宿主之间的关系，需要微生物高度定植在无菌小鼠模型的背景下，结合微生物和宿主的分子遗传学，来发现或验证其在宿主中的作用。研究表明，产芽孢梭菌通过产生的吲哚 -3- 丙酸激活上皮细胞孕烷 X 受体，增强肠道屏障功能[214]。而使用不产吲哚 -3- 丙酸的产芽孢梭菌 fldC 突变体接种的无菌小鼠，表现为肠道通透性增加[349]。目前，无菌小鼠已被广泛用作研究宿主 - 微生物相互作用的动物模型，以及具有特定疾病状态的人类肠道菌群或粪便移植的接受者。通过微生物群移植转移宿主表型或疾病状态，不仅可以提供微生物群因果关系的支持，还可以将小鼠用作临床前实验模型[350, 351]。微生物群已显示可转移肥胖、营养不良和结肠炎的疾病状态[352, 353]。尽管有营养干预，但这些方法在证明微生物群与宿主营

养不良中的因果关系方面尤其具有启发性，并提供了成功的治疗方法[352, 354, 355]。使用人类衍生的菌株和群落定植的小鼠，将是测试基于人类研究中确定的候选宿主 - 微生物途径是否可以使用动物模型进行验证和研究的重要手段。

动物模型为我们理解宿主免疫 - 肠道微生物之间的相互作用提供了重要的依据。然而，将这些原理扩展到人类的健康和疾病却是一个挑战[356]。有研究发现，小鼠经常以和人类明显不同的方式对实验干预做出反应，模仿人类遗传疾病而创建的小鼠品系通常具有与人类不同的表型[357]。另外，肠道微生物群和宿主之间的串扰是宿主特异性的，因此在小鼠模型中的观察并不完全适用于人类[356]。解释动物模型研究的关键将是基于人类的研究和介入试验。

八、优先考虑微生物与免疫的关系：从人类干预研究中吸取的教训

人类在遗传学、肠道微生物群的复杂性、个性化和动态性方面彼此不同，并且每个人的微生物群对干预的反应也是个体化的。因此，要确定与人类健康相关的微生物 - 免疫通路，那么基于人的研究必不可少。

迄今为止，大多数微生物组研究都是横断面的，在单个时间点进行采样，而具有明确干预措施的纵向研究受到限制。在包括人类微生物组计划和美国肠道计划在内的大型人群中，已经做出了巨大的努力来进行微生物群横截面分析[358, 359]。这些项目提供的数据和见解，可为微生物组研究提供巨大的资源，例如揭示个体间变异和多样性水平。由于要考虑的相关因素很多，因此将微生物群的各个方面与其他宿主和环境因素相关联一直是一个挑战。为了建立因果关系的最终目标，需要大量具有高质量元数据的个体来解开在此类观察研究中共同变化的因素。纵向干预研究提供了多个优势，包括建立因果关系，并且由于每个人都可以充当自己的控制对象（即基线状态），因此能够大幅度地减少阐明微生物群与宿主参数之间关系所需的受试者数量。因此，随着时间的推移，在受试者中测得的对干预的反应可以与微生物组或微生物代谢产物的改变相匹配，以增加识别因果相互作用的可能性。

多项纵向饮食干预研究指出，饮食能改变微生物群以及宿主健康[360-362]。高纤维饮食和益生元补充剂的摄入，导致肠道微生物群短链脂肪酸产生增加，并改善了 2 型糖尿病的指标[362]。相比之下，已确定基于微生物群特征的个性化饮食与一般饮食建议相比，可以更大程度地改善餐后血糖反应[361]。为了研究食物摄入量、肠道微生物群与代谢和炎症表型之间的时间关系，研究根据肥胖个体中微生物群的基因丰度确定了不同的亚组。其中，微生物基因丰度降低的个体表现出更明显的代谢障碍和低度炎症[363]。将人群范围内的饮食干预措施和个性化饮食干预措施结合起来，利用微生物群和免疫的相互作用，可能在人类疾病的预防和治疗中发挥更积极的作用。

为了在有限的研究中识别微生物群和免疫相互作用，应该使用高维测量来获取微生物群和免疫系统的深入资料[364, 365]。微生物群的高维分析工具，包括基于测序和培养的方法以及高通量代谢组学；免疫系统分析的辅助工具包括细胞计数（CyTOF）、RNAseq、抗体序列测序和高倍血清蛋白组学[366-369]。该领域面临的一个重大挑战是创建用户友好的、直观的方法来整合和分析多组数据，这可能推动该方法的应用和发展，类似于微生物生态学定量研究支持广泛的微生物群落分析和可视化[370]。

探测微生物群与宿主相互作用的其他方法包括检查微生物群的哪些成分被 IgA 包被（IgA-Seq），或检查对共生细菌的全身抗体反应[371, 372]。这两种基于抗体的方法都可以识别

与宿主免疫系统密切相互作用的细菌，从而导致适应性免疫反应。基于 IgA-Seq 的研究表明，IgA 对特定分类群的应答在很大程度上依赖于 T 细胞，且非 T 细胞依赖性应答也扮演着重要角色[373]。这些工具结合使用，可以生成大量的免疫和微生物群特征，然后将其输入机器学习模型，以识别微生物群和免疫系统中的其他模式，例如预测特定结果的元素或状态的变化[374]。

基于人的研究有助于确定改善微生物群和免疫状态的候选干预措施，以及在人类中发生的一系列可能的微生物 - 免疫相互作用。但在人类受试者中，进一步测试和分析受到限制，故而需要转向动物和体外模型，以测试由人类数据提供的假设。该方法可以包括从人类受试者到无菌小鼠的粪便微生物群转移实验或在候选相互作用中鉴定出的单个微生物定植。在定植的同时，可以调整小鼠的饮食或其他环境条件，以测试给定的假设或复制基于人体的试验（即高脂肪、低纤维、基于地理位置的试验）[352, 375]。此外，如果某些微生物代谢物被鉴定为免疫调节剂，则可以通过遗传或作为底物的形式引入肠道，以评估其对小鼠免疫系统的作用，或直接作用于体外的人免疫细胞[349, 376]。

此外，最近的研究已经确定了某些人体特有的生物过程，无法在其他动物中建模，例如新陈代谢和药物功效测试。因此，人类类器官，一种干细胞衍生的 3D 培养系统的出现受到了广泛关注，并为研究人类疾病和补充动物模型提供了独特的机会[377]。目前，类器官技术已经能够阐明胃肠上皮内模式识别受体的差异和区域特异性表达和功能，而整个胃肠道内复杂调节系统的整体功能后果仍不清楚，必须在未来解决[378]。

九、优先考虑微生物群落与免疫的关系：向祖先微生物群落学习的经验教训

近几十年来，与医学的现代化和进步相一致，包括抗生素的使用、卫生条件的改善、剖宫产和食品生产的工业化，微生物群经历了实质性的重塑，并提出了“微生物群不全综合征”一词来描述微生物类群和相关功能的丧失。在进化意义上，人类基因组“预期”这些微生物信号的丢失可能导致重要系统的失调，包括免疫功能、新陈代谢和中枢神经系统功能[379]。纳入进化论视角可能会推动阐明重要的人类微生物 - 免疫相互作用。大多数研究，例如人类微生物组计划和人体肠道基因组学，分别提供了 300 和 124 人的 16S rRNA 和宏基因组学数据，以及工业化人群肠道菌群的详细视图[380, 381]。2010 年，一项对居住在布基纳法索农村地区的 14 名儿童的研究表明，该微生物群的成分与西方微生物群不同，这提示绘制人类微生物群的图谱需要对生活在不同生活方式下的人群进行采样[11]。研究发现了许多易受影响或与人类工业化社会负相关的分类单元，又称为 VANISH，它揭示了由城市化决定的微生物群落变化的全球模式[382, 383]。生活方式的改变，可能使工业化微生物群落成为一种不同于塑造人类基因组的状态[344]。

随着时间的流逝，全球疾病负担已从传染性转向非传染性慢性病。其中，许多疾病的原因与免疫失调和慢性炎症相关[384]。这些疾病包括糖尿病、肥胖症、心血管疾病和癌症，工业化社会对其有着不同程度的影响。而在非工业化、猎人聚集的人群中，成人预期寿命约为 72 岁，肥胖和心血管疾病的负担估计低于 5%[385]。许多研究将具有传统生活方式的人群，如狩猎 - 采集者和农民，视为了解微生物群系状态的窗口[382, 386]，观察生活方式的梯度可以突出与生活方式相关的特定微生物和功能。一项对尼泊尔人口（其中一些人是狩猎采集者，另一些人处于向自给农业过渡的不同阶段）的研究确定了微生物群组成的梯度和特定类群，其流行率和相对丰度的下降反映了生活方式的改变[387]。

工业化社会的发展，选择了一种“工业化”

的微生物群，而这种微生物群并不十分适合促进人类健康。一个假设是，微生物类群和相关功能的缺失导致工业化人口的免疫系统失调和对炎症疾病的易感性增加。因此，为了在短期内改善或重建肠道微生物群与人类基因组之间的兼容性，可以重建工业化微生物群中缺乏或罕见的、已被证明对宿主有益的"祖先"物种，包括 VANISH 分类群[379]。VANISH 类群和未被充分代表的功能元件作为免疫健康或失调的候选驱动因素，将在小鼠、体外，并最终在人类中进行探索。

第三节　微生物群介导免疫调节的生物学机制

这些将肠道微生物组与免疫治疗功效联系起来的发现，只是划破了这种复杂相互作用的表面。确定生物学机制对于靶向微生物群的治疗以优化患者反应至关重要。在此，以微生物群介导的抗肿瘤免疫为例展开讨论。

在探讨微生物群介导的抗肿瘤免疫调节的可能机制时，出现了两个普遍的问题。第一个问题从胃肠道向肿瘤或引流淋巴结传递信号的信使的本质是什么？它能够进入循环系统以便进入远处的肿瘤部位，并且可以分类为微生物群或宿主衍生的细胞或分子。第二个问题是信使赋予肿瘤内免疫力的本质是什么？免疫抑制作用可通过增强调节功能或直接抑制抗肿瘤免疫来实现，而通过减轻调节功能或刺激抗肿瘤 T 细胞应答可以达到免疫刺激作用。微生物群介导的对肿瘤生长和免疫疗法功效的确切机制才刚刚开始被理解[388]。

一、以活细菌或 MAMP/PAMP 作为信使

已经在通常认为是无菌的胃肠外组织中鉴定出了共生细菌。Leore T Geller 等在人胰腺导管腺癌（pancreatic ductal adenocarci-noma，PDAC）的肿瘤微环境（tumor microenviron-ment，TME）中发现了细菌，并表明细菌从十二指肠逆行迁移到胰腺可能是 PDAC 相关细菌的来源。此外，这些细菌还被证明通过代谢药物的活性形式降低吉西他滨的化疗功效[389]。就对免疫功能的影响而言，细菌易位进入 MLN 和脾脏会产生特异于易位物种的 Th1 记忆反应[390]。在细菌易位的情况下，活细菌可以进入脾脏、淋巴结或肿瘤，同时提供外来抗原和佐剂（MAMPs/PAMPs），从而启动强大的免疫反应。因此，TME 中的 T 细胞交叉反应或旁观者激活可能导致杀伤肿瘤细胞的作用[388]。

研究表明，具核梭杆菌可能通过激活 β- 连环蛋白途径促进结直肠癌的发展和进展，并通过募集免疫抑制髓样细胞和抑制自然杀伤细胞等功能增强肿瘤免疫逃避。结直肠癌组织中的具核梭杆菌的丰度较高与肿瘤浸润性 CD3 T 细胞密度较低有关，这表明了具核梭杆菌在减弱抗肿瘤免疫中的作用[391]。另外，肠道共生菌可以通过 PRR 向 GALT 和 MLN 内的免疫细胞发送信号。PRR 可以识别细菌、真菌、病毒的 MAMPs 特征，并启动炎症细胞因子信号，以招募额外的免疫细胞亚群并消除感染。然而，长时间的 PRR 信号可导致慢性炎症和组织损伤从而促进肿瘤的发展[392]。

二、细菌和肿瘤抗原的交叉反应性，增强了抗原性

一些数据表明，细菌和肿瘤细胞可以共享 T 细胞表位的机械作用[393, 394]。针对细菌抗原的交叉反应性 T 细胞，可能通过 $CD4^+$ T 细胞辅助或 $CD8^+$ T 细胞直接杀伤而发挥抗肿瘤作用。在一项临床前研究中，脆弱拟杆菌反应性 $CD4^+$ T 细胞的过继转移可增强无菌小鼠的肿瘤控制能

力，并恢复其抗 CTLA-4 功效[395]。研究发现，从接受免疫检查点封锁治疗的患者中分离外周免疫细胞，通过体外嗜黏蛋白阿克曼菌刺激后测定 T 细胞 IFN-γ 的产生。结果发现，IFN-γ 的释放高于中位数，并且与延长的无进展生存期相关[393]。PDAC 长期生存者的肿瘤内和循环 T 细胞对新抗原和交叉反应的微生物表位都具有特异性，这表明在新抗原质量模型的背景下嵌入微生物同源性可以帮助创建一个免疫原性新抗原的有效替代品[394]。

三、MAMP / PAMP 的辅助性

微生物来源的 MAMP 或 PAMP 可以穿过黏膜屏障并进入循环系统。例如，健康人的血清含有能够激活一系列 TLR 和 NOD 受体的刺激物[396]。在癌症的背景下，细菌 LPS 在全身照射后异常进入循环系统，从而增强了小鼠模型中过继性 T 细胞疗法的活性[397]。另外，来自细菌的核酸也已被证明可以作为天然佐剂[398]。尤其是富含原核生物未甲基化的 CpG 二核苷酸通常是 TLR9 的有效激活剂。

这些促炎微生物产物可以触发固有免疫细胞，例如树突状细胞的部分活化。这种条件下的抗原呈递细胞可能具有增强启动抗肿瘤 T 细胞的能力。研究发现，双歧杆菌处理的小鼠显示出显著增强的肿瘤控制能力，伴随着外周肿瘤特异性 T 细胞的增加，以及肿瘤内抗原特异性 $CD8^+$ T 细胞的积累。而从双歧杆菌喂养的小鼠中分离的树突状细胞，表现为与抗肿瘤免疫相关的基因表达增加和活化 T 细胞的能力增强[399]。此外，与免疫检查点封锁治疗反应相关的转移性黑色素瘤患者粪杆菌属的富集，也与肿瘤中抗原加工和呈递的增加有关[267]。

四、微生物代谢物作为信使

肠道细菌会产生各种生物活性分子作为其代谢的副产物。这些代谢物可对宿主表现出多种作用，包括调节免疫系统[337]。其中，SCFA 是已知能影响宿主免疫力的微生物代谢物之一[400]，例如乙酸盐、丁酸盐和丙酸盐。这些代谢物除了是肠上皮细胞的主要能量来源[401]，也能影响细胞因子的产生[402]、影响巨噬细胞和树突状细胞功能[189]，以及 B 细胞的类别转换[403]。SCFA 还可以抑制组蛋白脱乙酰基酶，促进 Treg 细胞的分化[14]。另外，通过模仿人类信号分子，SCFA 还可以充当 G 蛋白偶联受体的配体[404]。SCFA 产生菌丰度较高的患者，如双歧杆菌和粪杆菌属，似乎对免疫治疗更为敏感，这进一步支持了 SCFA 在肿瘤免疫调节中的重要性[391]。丁酸盐还可通过下调硫氧还蛋白 -1 并升高活性氧水平来抑制结肠癌细胞系的上皮间充质转化、细胞增殖和迁移[405]。研究发现，代谢物肌苷在大肠癌小鼠模型中通过假长双歧杆菌改变免疫治疗的有效性。肌苷单独诱导 $CD4^+$ T 细胞中 TH1 调节基因的表达，而加入免疫检查点抑制剂可导致 IFN-γ 产生的同时增加。而检查点阻断疗法和肌苷的体内抗肿瘤活性需要通过 CpG 及树突状细胞的 IL-12 协同刺激[392]。与宿主免疫相关的其他细菌代谢物还包括视黄酸和共代谢物，例如多胺和芳香烃受体配体[406]。这些小分子可以通过充当信号分子、表观遗传调节剂和代谢转换来影响免疫力，并最终形成抗肿瘤免疫力。

五、宿主细胞因子作为信使

肠道细菌可能通过局部诱导可溶性免疫调

节因子并在全身进行扩散，例如增加 I 型干扰素、IL-12 和 TNF-α 的产生，或降低免疫抑制细胞因子，如 TGF-β 的产生，从而改变 TME 或引流淋巴结中关键免疫亚群的激活阈值，在免疫治疗的背景下导致增强的适应性免疫反应 [388]。在癌症模型中，小鼠口服嗜黏蛋白阿克曼菌以 IL-12 依赖的方式改善了 PD-1 阻断的功效 [393]。对预先暴露于抗生素的小鼠给予沙氏别样杆菌（*Alistipes shahii*，A. shahii）管饲，重建了肿瘤相关骨髓细胞产生 TNF 的能力，并影响肿瘤对 CpG- 寡聚脱氧核苷酸反应的能力 [407]。此外，IL-17 和 IL-10 被认为是各种致癌作用的主要免疫调节剂，而微生物群在调节这些细胞因子中的作用表明它们参与了这一过程 [408]。研究发现，一种名为 Prohep 的新型益生菌混合物通过减少从肠道募集 Th17 细胞，导致肿瘤微环境中 IL-17 的水平降低，从而削弱了肝肿瘤中的血管生成，抑制了肿瘤的生长 [409]。另外，有研究表明，TNF、IL-1、IL-6、IL-10、IL-17 和干扰素在大肠癌的发生和发展中起着重要作用 [410]。

六、免疫细胞作为信使

树突状细胞是关键的微生物传感器，可将先天性与适应性免疫联系起来，对于在 TME 中塑造 T 细胞反应也至关重要。微生物信号可能仅需要在固有层和 MLN 局部发挥作用以驱动树突状细胞功能，而随后传递给 TME 的免疫调节作用可能是由树突状细胞本身或下游的 T 细胞进行的。已经显示出各种固有免疫细胞能够在稳定状态下从肠道固有层转移至脾脏和外周淋巴结 [411]。在肠道屏障受损与完好无损的情况下，树突状细胞的微生物感测机制可能不同。活细菌或微生物相关分子模式可能通过受损的肠道屏障易位至循环系统，随后被树突状细胞上的 PRR 识别，并影响下游先天和适应性免疫。这种潜在的机制可能有助于肠道炎症情况下微生物群介导的抗肿瘤免疫调节。另一方面，在肠道屏障完整的情况下，黏膜树突状细胞通过各种机制不断采集抗原，包括在上皮细胞之间延伸树突 [412]，以及通过杯状细胞通道 [413] 或微折叠细胞获得蛋白质 [104]。载有细菌抗原的树突状细胞可以诱导对共生细菌的免疫耐受；可以引发细菌抗原反应性 T 细胞，并在某些情况下可能与肿瘤抗原发生交叉反应 [393, 394]；还可能在抗肿瘤反应中提供旁观者帮助。

有研究表明，环磷酰胺改变了小肠中微生物群的组成，并导致细菌易位到次级淋巴器官中。在那里，这些细菌诱导特定的“致病性”Th17（pathogenic Th17 cells，pTh17）细胞和记忆 Th1 免疫反应的产生。而 pTh17 细胞的过继转移部分恢复了环磷酰胺的抗肿瘤功效 [390]。此外，免疫抑制细胞，如调节性 T 细胞和浆细胞样树突状细胞（plasmacytoid dendritic cells，pDCs），在肿瘤免疫逃逸中起着至关重要的作用。研究发现，增加的 BDCA2$^+$ pDCs 和 Foxp3$^+$ Tregs 可能受胃黏膜微生物群的调节，并且两者都参与了胃癌的免疫抑制微环境 [414]。Tregs 对肠道细菌感染的反应限制了局部微环境中 IL-2 的可用性，允许 Th17 发育以促进产肠毒素脆弱拟杆菌触发的肿瘤形成 [415]。

考虑到共生菌 - 宿主之间相互作用的复杂性，微生物组的多样性及个体间的差异性，很可能多种模式会共同影响微生物群对免疫治疗的功效。此外，微生物组的相对贡献还需要与影响免疫疗法疗效的其他因素结合在一起，包括遗传因素和肿瘤细胞内在的致癌性改变 [416, 417]。要确定所有这些因素的相对贡献及对人类健康的影响最大的因素，还需要在癌症患者中进行更严谨的实验设计，以证实源自动物模型的假设。

第四节 肠道微生物与肠外器官的联系

一、共同黏膜免疫系统

黏膜免疫系统包含了广泛分布的胃肠道、呼吸道、泌尿生殖道黏膜及一些外分泌腺体处的淋巴组织[418]。1975 年，有研究通过证明含有 IgA 的支气管相关淋巴组织（bronchus-associated lymphoid tissue，BALT）淋巴细胞可以重新填充支气管和肠黏膜，提出了基于黏膜部位的共同黏膜免疫系统（common mucosal immune system，CMIS）的概念[419]。基于在动物模型和人类中的广泛研究，有可靠的证据表明，来自肠相关淋巴组织分泌 IgA 的致敏浆细胞前体能够扩散到肠道、其他黏膜相关组织和外分泌腺[420]。黏膜致敏淋巴细胞产生局部黏膜反应的同时，部分通过淋巴细胞归巢至其他黏膜组织参与反应，形成一个功能上相关联的系统，称之为共同黏膜免疫系统[421]。

随后，越来越多的研究结果也支持了共同黏膜免疫系统的存在，并对其功能和机制做了进一步的探索。微生物和环境抗原通过具有胞饮和吞噬作用的微折叠细胞进入诱导位点，与抗原呈递细胞相互作用。骨髓中的 B 细胞进入诱导部位，在 T 细胞、上皮细胞和树突细胞的影响下，表达 sIgA 产生局部免疫反应。同时，致敏后的 T 和 B 细胞通过胸导管进入血流，基于归巢受体与效应部位相应配体的相互作用，被分布到其他黏膜免疫组织。最后产生特定的 IgA，以防止病原体附着在黏膜上[422, 423]。

（一）肺 - 肠轴

消化道和呼吸道虽然在解剖结构和功能上不同，但具有相同的胚胎起源和黏膜免疫系统的共同成分，如上皮屏障、黏膜下淋巴组织、IgA 和防御素的产生，以及天然淋巴细胞和树突状细胞的存在[424]。肠道微生物群，尤其是微生态失调，是影响肺部疾病（如肺炎、哮喘和肺癌等）发展的重要因素，尽管其潜在的机制仍不明确，但越来越多的流行病学和实验研究强调了肠道微生物和肺部之间的相互作用，被称为“肠 - 肺轴”[425, 426]。这个轴允许内毒素、微生物代谢物、细胞因子和激素进入血液，连接肠和肺[425]。不仅肠道疾病会出现肺部症状，肺部疾病也会伴随胃肠道症状[427, 428]。

肠道微生物群影响免疫反应和肺部炎症的机制正在被广泛研究（图 5-4）。其相互关系主要有两种：循环系统中的细菌产物和肠、肺淋巴结中的免疫细胞[429]。已有研究表明，上皮细胞和免疫细胞吸收内皮细胞的信号，形成局部细胞因子微环境，从而导致远端部位免疫反应的改变[430]。越来越多的证据表明，肠道微生物相关代谢物影响骨髓造血前体细胞的迁移和肺部炎症的消退[431]。

1. 哮喘　研究已经发现，肠道微生物与过敏性疾病关系密切，生命早期肠道微生物的异常可能是引起过敏和哮喘的关键因素。而免疫在哮喘发病机制中起着至关重要的作用，如调节性 T 细胞亚群和 TLR 的参与，因此推测肠道微生物与变态反应之间存在联系[428]。在生命早期，肠道菌群的多样性可能通过调节 Th1/Th2 平衡来预防哮喘的气道炎症[432]。研究发现，哮喘患儿微生物组中观察到较低数量的嗜黏蛋白阿克曼菌和普拉梭菌。这两种细菌可能通过其分泌的代谢物诱导了抗炎细胞因子 IL-10，并阻止了像 IL-12 这样的促炎细胞因子的分泌来抑制炎症[433]。随着外周血中炎症因子水平的升高，哮喘儿童出现肠道微生物失调和胃肠不适症状的可能性将会增加[428]。

已发表的研究还讨论了短链脂肪酸在免疫系统调节和过敏性疾病发展中的潜在作用。哮喘患者粪便中脂肪酸总量、乙酸、丁酸、丙酸等特异酸的绝对浓度及异构酸含量均显著降低。肠道生态失调所致 SCFAs 改变使免疫过程向影响肺支气管系统状态的 Th2 免疫反应转移[434]。

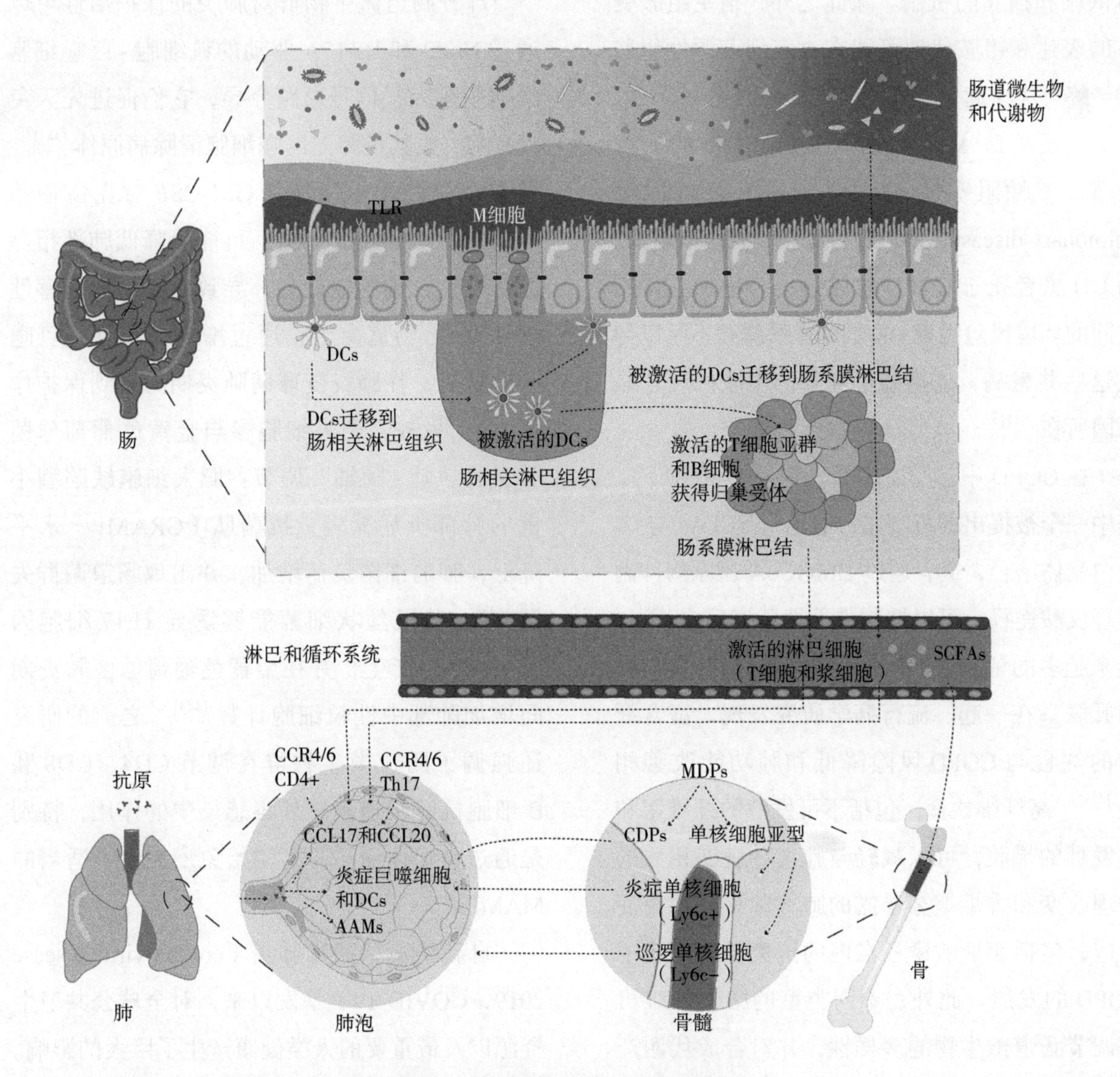

图 5-4 肠道微生物群与肺免疫的关系

肠道细菌与肺免疫之间的相互作用由多种因素介导，包括病原体相关分子模式、模式识别受体、短链脂肪酸、肠道完整性和固有层免疫细胞。在正常状态下，树突状细胞通过微折叠细胞活性、树突延伸和肠道屏障功能从管腔中连续取样，这决定了细菌 /PAMP 易位。在树突状细胞取样后，这些细胞迁移到肠道相关淋巴样组织，然后迁移到肠系膜淋巴结，根据肠道微生物群释放的某些细胞因子来调节淋巴细胞（T 和 B 细胞）的分化和归巢。活化的 T 细胞和 B 细胞通过循环分布在肺部。此外，微生物暴露后肺部产生的 CCL20 和 CCL17 的水平有助于 T 细胞亚群的印记。短链脂肪酸可以渗入骨髓，通过影响树突状细胞祖细胞向炎症或抗炎免疫细胞的分化来影响肺免疫。肺部的炎性巨噬细胞和树突状细胞来源于树突状细胞前体和 Ly6c+ 炎性单核细胞；活化巨噬细胞是肺部的抗炎免疫细胞，来源于 Ly6c⁻ 单核细胞。DCs. 树突状细胞；TLR.toll 样受体；CCR. 趋化因子受体；CCL. 趋化因子配体；CDPs. 普通树突状细胞前体细胞；MDPs. 巨噬树突细胞祖细胞；AAMs. 选择性激活的巨噬细胞；SCFAs. 短链脂肪酸

组胺作为影响机体免疫的重要免疫调节剂，尤其是在与过敏反应相关的许多急性症状表现突出。有证据表明，哮喘患者肠道微生物群中组胺产生菌的水平增加，并且分泌大量组胺的摩氏摩根菌升高水平与哮喘的严重程度呈正相关。推测可能是由于成人哮喘患者体内细菌产生组胺水平升高，从而降低了宿主自体产生组胺驱动过敏反应所需的水平[435]。肠道内细菌分

泌的组胺对肺部炎症反应有显著影响，其对远处器官（如肺）的免疫调节作用可能涉及 H_2 受体依赖和独立的机制。除此之外，宿主组胺受体的表达和组胺代谢可能在决定细菌源性组胺的最终反应中起关键作用 [436]。

2. 慢性阻塞性肺疾病　基于人群的研究发现，慢性阻塞性肺疾病（chronic obstructive pulmonary disease，COPD）患者 IBD 患病率增加，而 CD 患者死于 COPD 的风险也会增加。仅用共同的环境风险因素 - 香烟烟雾暴露并不足以导致这些共患病，连接肠道和肺的串联机制才是原因所在 [437]。

在 COPD 中已经描述了“肠 - 肺 - 肝轴”。其中一个被提出的机制是肝脏通过 SCFAs 与 G-蛋白受体结合，或者通过 HMGCoA 还原酶抑制甲羟戊酸途径，可以抑制先天性免疫反应 [438]。越来越多的证据表明，肠道通过饮食因素与肺和肝联系在一起。流行病学研究发现，富含纤维的饮食与 COPD 风险降低和肺功能改善相关 [428]。高纤维饮食，包括不可发酵的纤维素和可发酵的果胶，可以减轻与肺气肿进展相关的病理改变和香烟烟雾暴露的肺气肿小鼠的炎症反应，包括亚油酸途径在内的抗炎机制阻止了 COPD 的发展。此外，不同类型的膳食纤维可以调节肠道微生物的多样性，并对合成代谢产生不同的影响，包括 SCFAs、胆汁酸和鞘磷脂的生成。这些结果表明，高纤维饮食对肺气肿的保护作用可能取决于微生物群 - 代谢 - 免疫相互作用 [426]。

3. 呼吸道感染　近年来，人们对肠道微生物群与呼吸道感染之间的联系在小鼠和人类中进行了探索。研究发现，活动性结核病患者的肠道微生物群主要以 SCFAs 产生菌及相关代谢途径的显著减少为特征 [428]，从而影响免疫细胞并增加结核分枝杆菌增殖 [439]。抗生素诱导的肠道微生态失调小鼠，其肺树突状细胞表现出巨噬细胞诱导的 C 型凝集素低表达，使得激活幼稚 CD4 T 细胞的能力降低，结核分枝杆菌得以存活 [440]。

部分肠道微生物群对肺炎的保护活性可以通过 Nod2 和 IL-17A 驱动的粒细胞 - 巨噬细胞集落刺激因子信号通路介导，它们促进先天免疫反应，尤其是肺泡巨噬细胞清除病原体 [441]。肠道感染的小鼠，肺部 GM-CSF 对克雷伯菌属感染的反应显著降低，可能会降低肺部招募循环中性粒细胞的能力，导致对感染的易感性增加 [424]。肠道微生物通过增强初级肺泡巨噬细胞功能，在肺炎链球菌肺炎期间起到保护作用 [442]。另外，肥大细胞参与金黄色葡萄球菌肺炎的“肺 - 肠轴”调节：肥大细胞缺陷型小鼠的肺部炎症减轻，抗菌肽（CRAMP）水平降低，肺的细菌负荷增加，并出现肠道菌群失调 [443]。分段丝状细菌能够诱导 Th17 细胞因子 IL-22 的产生，并在金黄色葡萄球菌肺炎期间增加肺部中性粒细胞计数 [444]。之前的研究还强调了肠道微生物群在调节 CD4、CD8 和 B 细胞抗呼吸道流感病毒感染中的作用，特别是通过炎症体激活，为初始免疫提供了适当的 MAMPs[445]。

自新型冠状病毒肺炎（corona virus disease 2019，COVID-19）暴发以来，对全球公共卫生资源以及最重要的人类健康产生了巨大的影响。调查分析发现，新型冠状病毒感染除了典型的呼吸道症状，许多患者还有胃肠道表现 [446]。其原因可能是由于病毒直接感染胃肠道细胞 [447]，也有可能是因为病毒从肺部进入血液，然后从血液传播到肠道 [448]。新型冠状病毒肺炎胃肠道表现延长，与肠道微生物多样性和丰度降低、免疫异常和 SARSCoV-2 清除延迟相关。呼吸道黏膜和肠道微生物群之间的双向相互作用，即“肠 - 肺轴”，被认为参与了新型冠状病毒的病理免疫反应 [431]。

新型冠状病毒通过 S 蛋白与细胞表面血管紧张素转换酶 2（angiotensin-converting enzyme

2，ACE2）受体结合进入宿主。这种受体在呼吸道和胃肠道中高度表达，并且可以调节肠道中的氨基酸的转运、微生态和肠道炎症 [447]。另外，新冠病毒 E 蛋白导致 PALS1 功能受损，干扰紧密连接的形成，使肠屏障完整性遭到破坏，可能导致漏肠综合征、细菌移位及免疫激活 [449]。多项研究表明，新型冠状病毒肺炎病患者的肠道微生物组成发生了改变 [450]，例如乳酸杆菌属、双歧杆菌属 [451]，以及产丁酸盐的细菌减少 [452]。中重度新型冠状病毒肺炎住院患者粪便微生物组有显著的变化，其特征是机会性病原体的富集和有益共生菌的减少，并且新型冠状病毒肺炎严重程度与某些细菌基线丰度相关。这些肠道微生物失调甚至在病毒清除后仍然存在。此外，已被证明可以下调小鼠结肠中 ACE2 表达的部分拟杆菌属与粪便新型冠状病毒负荷呈显著负相关 [453]。事实上，由于新型冠状病毒的进入与 ACE2 的表达水平有关，下调 ACE2 的表达水平，可以减少 ACE2-Ang1-7-Mas 途径，并进一步减少新型冠状病毒感染期间的肺损伤 [445]。

另一方面，ACE2 可以通过调节氨基酸转运蛋白 B0AT1 的表达控制色氨酸的转运，进而影响抗菌肽的产生。由于 ACE2 下调，肠道对色氨酸的吸收减少，并减少了抗菌肽的分泌，导致病原体存活增加和肠道微生态失调 [454]。研究发现，色氨酸缺乏是新型冠状病毒肺炎的一个突出的特征和重要驱动因素，犬尿氨酸途径的改变损害了免疫系统的负向调节能力，从而促进了过度炎症和细胞因子风暴综合征的发生 [455]。

（二）肠 - 阴道轴

“肠 - 阴道轴”是指肠道微生物群可能通过调节免疫信号的流动对生殖器疾病的病程产生深远影响 [456]。肠道微生物群、免疫、阴道微生物群和激素之间的平衡对于女性生殖道的生理状态至关重要 [457]。而肠道微生物失调会导致健康微生物群产生的抗肿瘤物质（如短链脂肪酸）减少，肠道屏障的改变激活 toll 样受体而促进慢性炎症状态，并导致代谢和激素失调。这些失衡所带来的影响不仅限于胃肠道，还会影响妇科肿瘤的发生 [458]。

早在 2001 年，Kutteh 等通过研究发现，口服和直肠注射相同剂量的疫苗后，阴道和宫颈液中的 IgA、IgG 显著增加，并且初次接种诱导的特异性抗体可通过随后的直肠给药得到增强 [459]。从人类粪便中分离出来的长双歧杆菌 NK49 通过调节肠道微生物群，抑制 NF-κB 相关 TNF-α 的表达，可显著减轻阴道加德纳菌诱发的阴道病 [460]。目前为止，我们对“肠 - 阴道轴”的研究还较少 [456]。毫无疑问，随着对“肠 - 阴道轴”的深入探索，为治疗女性生殖系统疾病提供了新思路：益生菌的应用 [461]、粪便微生物群移植 [457]、新一代抗人乳头瘤病毒口服疫苗 [456]，可能在未来发挥显著的作用。

（三）肠 - 乳腺轴

通过口服或吸入抗原后，特异的 sIgA 在远离诱导位点的腺体分泌物中同时出现 [420]。乳腺是普通黏膜免疫系统的一个组成部分，表现出其他黏膜部位所没有的特征：没有大量且丰富多样的细菌定植，也没有类似于肠道派尔斑的黏膜诱导位点 [423]。女性在妊娠和哺乳期间，机体会产生一系列的变化，例如妊娠晚期特有的低度炎症和肠道通透性增加 [462]，这影响了肠道吸收代谢、微生物群和免疫系统 [463]，并在妊娠晚期利于细菌和免疫细胞的生理易位 [464]。此外，细菌代谢物也通过血流从肠道大量转运到乳房 [57]。

来自胃肠道的细菌可以通过涉及单核 - 免疫细胞的机制易位，通过“肠 - 乳腺”途径迁移到乳腺，并且随后在母乳喂养的新生儿胃肠道中定植 [464]。肠道 B 细胞迁移到乳腺的过程中，乳腺的归巢作用尚未完全明确，可能与黏膜地址素细胞黏附分子 -1、肠道归巢受体 α4β7 整合素和黏膜相关 CCL28/CCR10 有关。此外，乳腺和乳汁中免疫细胞种类和数量不同，这表明可能存在免疫细胞向乳汁迁移的选择性机制 [57]。

母乳在母亲和婴儿之间提供了复杂的营养、免疫、神经内分泌和微生物整合，并在生命早期抗感染、诱导免疫耐受和预防过敏方面起作用 [57]。乳源性免疫调节剂，例如转化生长因子 β、维生素 A、IGF-1 和 IgG，与维持肠道屏障、抗原跨肠道转移、树突状细胞抗原呈递及诱导调节性 T 细胞水平相关 [57]。由于早产儿肠道的易感性和高反应性，使婴儿肠道容易发生免疫反应改变、血管损伤和微生物群失调，是导致坏死性小肠结肠炎的根本原因 [465]。最近研究表明，人乳中所含的外泌体通过减少炎症和肠上皮损伤，以及恢复肠道紧密连接蛋白，在预防坏死性小肠结肠炎中发挥有益的作用 [466]。

（四）肠 - 眼 - 泪腺轴

干燥综合征（Sjögren syndrome，SS）是一种影响外分泌腺（如唾液腺和泪腺）的自身免疫性疾病，在泪腺和唾液腺中表现出 $CD4^+$ T 细胞和 B 细胞的浸润，并导致腺泡细胞功能丧失 [467]。有临床研究表明干燥综合征表现出显著的肠道菌群改变，并且肠道生态失调与眼干燥症的严重程度部分相关 [468]。无菌小鼠的产酸拟杆菌肠道定植会增加泪腺中的 IgA 转录物水平，其中 IL-1β 对 sIgA 在眼相关淋巴组织和眼表的表达至关重要，并推测在“眼 - 肠轴”中存在一个肠道来源的 B 细胞循环 [469]。有证据表明，无菌环境能够加速泪腺淋巴细胞浸润、腺体破坏和产 IFN-γ 细胞的积累，而通过肠道微生物组重建可以减少致病 Th1 细胞的产生，这表明肠道菌群移植有可能成为干燥综合征的新疗法 [470]。这些发现强调了“肠 - 眼 - 泪腺轴”的存在 [467, 471]。此外，尚需进一步研究对眼表有影响的肠道微生物群，以及肠道生态失调影响泪腺、唾液腺淋巴浸润的机制，这或许将为干燥综合征的发病机制提供更多的信息，以及为使用益生菌治疗开辟新的可能 [467, 471]。

二、脑 - 肠轴

微生物群、肠道和大脑形成一种联系，即“脑 - 肠轴”，使肠道与中枢神经系统同步，并改变行为和大脑免疫稳态 [472]。研究发现，许多大脑或精神疾病都可能与肠道息息相关，比如脑梗死、精神分裂症、阿尔茨海默病、自闭症、重度抑郁障碍、焦虑、双相型障碍、神经性厌食症、成瘾、多动症、创伤后应激障碍、强迫症、帕金森病、多发性硬化、癫痫、颅内感染等 [473]。大脑和肠道可以通过多种途径进行沟通，包括肠道神经系统（ENS）、自主神经系统、免疫系统、肠道微生物代谢物、神经递质以及下丘脑 - 垂体 - 肾上腺轴等途径 [472, 474]。脑肠轴和微生物群之间相互作用的机制主要包括调节肠屏障和连接完整性，调节传入神经、神经递质的产生、表达及周转，调节细菌代谢产物与黏膜免疫等 [475]。而其中免疫系统与机体稳态息息相关，并参与多种疾病的发生与发展。

（一）脑卒中

脑卒中是全球成年人死亡的第三大原因，也是造成成年人永久残疾的最常见原因。其中缺血性脑卒中——脑梗死占比达 70%。而 T 细胞向大脑的浸润，以及细胞因子 IL-23 和 IL-17，在脑梗死及其伴随的神经功能缺损的演变中起着关键作用 [476]。研究表明，小鼠效应 T 细胞从肠道运输到大脑，定位于软脑膜中，并通过分泌白介素 -17 增强缺血性神经炎症，导致脑实质中趋化因子产生增加，以及中性粒细胞等炎症细胞的细胞毒性浸润 [477]。恢复肠道微生物群的健康和平衡可以增强对卒中患者的治疗效果 [478]。

（二）精神分裂症

精神分裂症是一种以精神错乱、冷漠、社交退缩和认知障碍为特征的精神疾病 [474]。研究表明，各种炎症因子是精神分裂症和其他精神病的危险因素[479]。而肠道微生物可以调节各种炎症因子的变化，如促炎因子IL-8、IL-1或抗炎因子IL-10和TGF-β。精神分裂症患者血清促炎因子水平高于正常对照组，并且血清炎症标志物水平与精神分裂症患者的临床症状群呈明显正相关[480]。同时肠道菌群也可产生有害物质，破坏肠道上皮屏障，导致有神经毒性的细菌产物和蛋白质进入循环系统[481]。研究发现，精神分裂症患者体内肠道炎症标志物抗酿酒酵母的抗体浓度明显升高[482]。这些都表明肠道微生物群可能对精神分裂症的发生与发展起着关键作用。

（三）阿尔茨海默病

阿尔茨海默病（Alzheimer disease，AD）是一种进行性神经退行性疾病，其特征为存在细胞外的淀粉样β肽聚集体以及神经内神经原纤维的缠结[483]。所以调节β-淀粉样蛋白的体内平衡失调是治疗阿尔茨海默病的一个有效且有说服力的方法。有证据表明，大脑和肠道之间双向交流的障碍可能导致神经炎症。例如，肠道微生物群的细菌菌落自然释放大量的脂多糖、淀粉样蛋白和其他促炎分子，这些分子从肠道渗漏并逐渐在全身和大脑积累，因而可能导致与年龄相关的炎症变化和神经退行性变[484]。而另一方面，据推测，从肠道微生物释放的功能性淀粉样蛋白通过交叉接种、引发先天免疫和激活神经炎症来触发神经元蛋白质的错误折叠[485]。

肠道微生物群可以通过多种机制改变阿尔茨海默病中β-淀粉样蛋白体内平衡，而这些途径可以为进一步探索肠道微生物群用于β-淀粉样蛋白靶向治疗提供可能性。肠道屏障受损或肠道渗漏时，肠道菌群分泌的脂多糖和淀粉样蛋白可进一步加剧肠道的渗漏，并可增加细胞因子和其他促炎症分子的水平，如与阿尔茨海默病直接相关的IL-17A和IL-22。这些细胞因子可通过胃肠道和血脑屏障进入大脑并进一步触发免疫原性反应、活性氧释放和toll样受体2/1、CD14和NF-κB的信号传导。而这些因素在神经退行性变中起关键作用。同时大脑中淀粉样蛋白的存在通过激活小胶质细胞和星形胶质细胞触发炎症[486]。另外，小胶质细胞的形态、成熟和功能受到来自肠道微生物群的SCFAs、细菌产物或其他代谢物的调节[487]。因此，肠道微生物代谢产物可能通过影响小胶质细胞功能和增强神经炎症来影响阿尔茨海默病的病理过程[472]。

大脑与肠道的相互沟通包括神经、代谢、激素、免疫等多个方面，任何一个组成部分的改变都可能导致两个系统的功能改变。肠道微生物的正常生态平衡在维持这种关系中起着重要作用。目前许多神经系统疾病的发病机制仍不清楚，而脑-肠轴微生物群的发现，为尚未明确致病机制的神经系统疾病的研究提供了新的研究方向[474]。

三、肝-肠轴

肝脏血液供应的近75%来自门静脉，这确保了其与肠道微生物及其代谢物的相互作用[488]。在过去，关于肠道微生物群稳态与肝脏疾病之间的密切关系已有描述：肠道生态失调与一系列肝脏疾病有关，同时肝脏疾病也可以改变肠道微生物群[489]。1998年，Marshall提出了“肠-肝轴”的概念[490]，它是胃肠道和肝脏通过门静脉循环形成的一个密切的解剖和功能双向作用的结果，涉及代谢、免疫、神经内分泌信号等多种成分[491]。目前，大多数研究主要集中在慢性肝病，如非酒精性脂肪性肝病（NAFLD）、自身免疫性肝病、肝硬化和肝癌等[492]。

（一）急性肝损伤

急性肝损伤（acute liver injury，ALI）是

一种严重威胁患者生命和健康的常见疾病，药物过量或化学品暴露是引起急性肝损伤的主要原因[492]。在美国和欧洲，尤以对乙酰氨基酚（APAP）所致急性肝损伤最常见[493]。

最近的证据表明，肠道屏障功能障碍，以及肠道微生物群的变化，包括它们的丰度、多样性和代谢物，在APAP诱导的肝损害中发挥着重要作用[494]。在化学诱导的急性肝损伤模型中，肠道微生物群调节急性肝损伤的机制主要概括为以下三点：首先，肠道微生物通过细菌代谢产物和化学毒素的转化直接影响肝细胞；其次，微生物和MAMPs通过受损的肠黏膜屏障激活免疫细胞，向肝脏释放大量趋化因子和促炎细胞因子，诱发肝脏炎症；最后，肠道菌群失调或化学毒物直接引起肠黏膜屏障损伤，增加细菌和MAMPs向肝脏易位，介导肝损伤[492]。

肠道微生物组可以通过TLR信号传导调节MYC依赖的转录程序，协调星状细胞、内皮细胞和Kupffer细胞的激活[495]。临床队列研究及动物实验表明，肠道生态失调的个体对APAP引起的肝损伤可能会更敏感。从生态失调小鼠到野生型小鼠的肠道菌群移植增加了野生型小鼠APAP引起的肝损伤，并使单核细胞极化偏向Ly6C^{hi}炎症表型[496]。而含有芽孢杆菌孢子的益生菌补充剂（MSB）对APAP过量引起的大鼠急性肝损伤具有类似于肝保护剂的保护作用。表现为MSB预处理的个体血清AST、ALT、促炎细胞因子（TNF-α、IL-1β）、ZO-1和总抗氧化能力及肝细胞坏死显著降低[493]。

（二）慢性肝病

随着对“肝-肠轴”的认识，肠道生态失调已被确定为先天免疫反应的重要驱动因素，加速了慢性肝病的进展[496]。非酒精性脂肪性肝病（non-alcoholic fatty liver disease，NAFLD）是世界范围内肝脏疾病的主要原因，并在不久的将来成为终末期肝病的主要原因[489]。NAFLD包括多种肝病，从脂肪变性到非酒精性脂肪性肝炎、肝纤维化和肝硬化[488]，并且增加患肝细胞癌的风险[497]。

肠道微生物群参与了NAFLD中肝脏脂肪变性、炎症和纤维化，以及肝癌的发生过程[488, 489]。NAFLD肠道菌群失调表现为α多样性较低，β多样性的改变。此外，NAFLD及其严重程度与细菌编码炎性产物的基因丰度也有关[498]。肠道菌群失调及SCFAs的显著减少引起肠道黏膜屏障的损伤，导致病原菌及其代谢产物经“肠-肝轴”移位，引发肝脏慢性炎症[499]。LPS-LBP复合物与CD14结合，与TLR协同刺激丝裂原活化蛋白激酶（MAPK）、JNK、p38和NF-κ B通路，导致促炎细胞因子的转录，包括TNF、IL-1和IL-6[488]。PAMPs使炎症体激活，导致促炎细胞因子（如IL-1β和IL-18）的表达，并通过caspase-1的激活促进细胞凋亡和纤维化[500]。另外，肠道微生物相关代谢物在肝脏疾病中的作用也是不容忽视的。肠道微生物群依赖的代谢物色胺和吲哚-3-乙酸减少了巨噬细胞中脂肪酸和脂多糖刺激的促炎细胞因子产生，并抑制了细胞向趋化因子的迁移。同时I3A还减弱了肝细胞中脂质负荷下的炎症反应[501]。

在代偿性肝硬化中，屏障功能障碍的特征与其病因类似，而失代偿期肝硬化的肠道屏障破坏则与病因无关[502]。随着肝硬化的进展，肠道微生态失调导致肝硬化大鼠肠黏膜的上皮内淋巴细胞和固有层淋巴细胞呈现出免疫失调的促炎模式，其特征为活化淋巴细胞的增加、功能向Th1型转换，伴随Th17的减少，并与肝硬化严重程度相关[503]。肝硬化患者分泌到肠道的胆汁酸减少会导致产生严重的生态失调和大量致病菌。而另一方面，肠道生态失调又会促进肠道炎症、肠道屏障损伤和肝脏炎症，进一步抑制肝脏分泌胆汁酸[502]。

此外，肠道微生物群通过调节不同因素影响肝癌，包括胆汁酸、免疫检查点抑制剂和Toll

样受体等[504]。肠道微生物利用胆汁酸作为信使，通过胆汁酸 /CXCL16/CXCR6 途径控制趋化因子依赖的肝脏 NKT 细胞的聚集，介导肿瘤抑制[505]。脂磷壁酸与脱氧胆酸协同增强肝星状细胞的衰老相关分泌表型（senescence-associated secretory phenotype，SASP），通过 TLR2 上调 SASP 和 COX2 的表达并介导 PGE2 的产生，进而通过 PTGER4 受体抑制抗肿瘤免疫，导致肝癌进展[506]。

越来越多的证据表明，“肝 - 肠轴”能够影响肝脏疾病的进展[504]。那么，我们是否能够通过调节“肝 - 肠轴”，进而影响疾病的转归？研究表明，通过应用抗生素、益生菌和益生元调节肠道微生物群，可以改善肠道屏障，并减缓肝脏炎症 - 纤维化 - 肝硬化 - 癌症的进展[504]。已经报道了利福昔明可以通过减少革兰阴性菌，导致 LPS-TLR4 通路下调，缓解肝脏炎症状态和门静脉高压[507]；万古霉素通过去除介导初级胆汁酸向次级胆汁酸转化的细菌，可以诱导肝 NKT 细胞积累并抑制肝肿瘤生长[505]。

新型益生菌混合物 Prohep 可以增加肠道中的抗炎细菌和代谢物，限制肠道中过量 Th17 细胞的产生，并减少 Th17 从肠道向肿瘤部位的募集，下调肝脏肿瘤中 IL-17 和其他促血管生成基因，以减少血管生成来限制肿瘤生长[409]。戊糖片球菌 CGMCC 7049 能够逆转肠道生态失调、增加 SCFA 水平、降低促炎细胞因子和趋化因子（包括 IL-5、TNF-α、G-CSF、MIP-1α、KC、MCP-1）的循环水平来缓解乙醇诱导的肝损伤[508]。益生元牛奶低聚糖可以增加产丁酸盐的细菌丰度，而合生元（婴儿双歧杆菌加牛奶低聚糖）在西方饮食喂养的胆汁酸受体 FXR 敲除小鼠中抑制了肝脏炎症，减少肝脏脂肪沉积和损伤，起到改善非酒精性脂肪性肝炎的作用[509]。最近的研究还发现，菊粉可以调节肠道微生物群，并通过“肠 - 肝轴”抑制 LPS-TLR4-Mψ-NF-κB-NLRP3 炎症途径，调节炎性 Mψs 的激活及其极化，降低了促炎细胞因子水平，对非酒精性脂肪性肝病发挥有益作用[499]。此外，还包括粪菌移植、噬菌体、FXR 激动剂、工程益生菌、肠道微生物对靶向治疗的调节等，尚需我们进一步的研究[502]。

四、肠 - 肾轴

肠道微生物群不断与宿主的重要器官系统交流，如大脑、骨髓、血管、肾脏、免疫系统和自主神经系统等，这种交流有助于宿主的体内平衡和健康[510]。肠道微生物群与肾脏进行双向交流的过程，即“肠 - 肾轴”。

肠道微生物群的主要作用有：①通过恢复蛋白质紧密连接结构、上调黏蛋白基因以及与病原菌竞争结合上皮细胞等机制维持肠上皮屏障[511, 512]。②抑制涉及 TLR 信号传导的肠道炎症[513]。③通过分子信号维持正常 T 细胞功能[192]。同时肠道微生物代谢物如短链脂肪酸在维持肠道稳态方面也有多种作用。短链脂肪酸的主要作用包括调控肠上皮细胞增殖和凋亡之间的平衡、诱导肠上皮细胞分泌内源性抗菌肽及 Treg 细胞的分化、调节细胞因子的产生和维持肠道屏障功能[514]。而这些作用与肾脏疾病的发生与发展息息相关。

肠 - 肾轴可分为代谢依赖型和免疫型。代谢依赖途径主要由肠道微生物群产生的代谢物介导。而在免疫途径中，免疫系统的成分（例如淋巴细胞、细胞因子）在肠道和肾脏之间的沟通中起着关键作用[515]。肠道微生物群不仅可以调节肠道免疫细胞的激活，还调节骨髓中免疫祖细胞的分布[218, 516]。骨髓排出的成熟免疫细胞在肠道中由外周淋巴器官中（如肠道相关淋巴组织）的肠道微生物群激活。肠道黏膜表面的微生物与肠道上皮细胞接触，通过其代谢产物改变肠道通透性，并通过与黏膜表面细胞上表达的 TLR 相互作用调节宿主免疫系统，从而在调节肠道相关淋巴组织中发挥作用[517]。另外，微生物代谢物短链脂肪酸可通过降低MCP-1和IL-1β来减轻肾脏炎症[518]。短链脂肪酸还抑制缺氧后肾小球系膜细胞和肾近端小管上皮细胞中由高葡萄糖和脂多糖诱导的活性氧生成[519]。

肠道微生物群与慢性肾病之间存在许多致病联系，如肠道通透性增加、微生物群产生的潜在肾病毒素超载，以及某些可能具有肾脏保护特性的微生物群衍生分子（如丁酸盐和维生素K）水平降低[520]。在代谢方面，尿毒症影响微生物群的组成和代谢，同时重要的尿毒症毒素，如氧化三甲胺（Trimethylamine-N-oxide，TMAO）、对甲酚硫酸盐、吲哚氧基硫酸盐和吲哚-3乙酸等也来源于微生物代谢。此外，肠道生态失调可能会破坏上皮屏障，最终使宿主暴露于内毒素（如脂多糖）的机会增加[515]。这为通过改变饮食、改变微生物群、减少微生物群产生的尿毒症毒素、增加毒素排泄或靶向特定的尿毒症毒素来进行治疗干预提供了可能性。在免疫方面，研究表明，慢性肾病与局部和全身炎症相关，而全身炎症与生存率和肾脏预后呈明显相关[521, 522]。炎症的驱动因素包括细胞因子和炎症介导的肾清除率降低及炎症介质的产生增加[523]。同时肠道微生物群释放的尿毒症毒素也是慢性肾脏病慢性炎症的主要来源[524]。进一步了解肠道微生物群改变对慢性肾脏病进展的影响，无论是通过促进与肠道通透性增加相关的炎症，还是通过产生饮食来源的肾毒性尿毒症毒素或其前体，或者通过降低肾保护分子的可用性，都有可能极大地影响慢性肾脏病的管理，并为减缓慢性肾脏病进展的新治疗方法铺平道路[510]。

五、肠－心轴

大多数心血管疾病的危险因素，包括衰老、肥胖、某些饮食模式和久坐不动的生活方式，都可能导致肠道微生态失调。相反，肠道微生态失调也可能影响心血管疾病的发展，包括冠状动脉疾病、高血压和心力衰竭[525-527]。尽管其潜在的机制我们尚不明确，肠道微生物组通过直接和间接的方式影响动脉粥样硬化的所有危险因素，因此在心血管疾病中发挥着重要的作用[527]。既往的研究发现，细菌DNA存在于动脉粥样硬化斑块中，而肠道可能是其潜在的来源[528]。此外，心血管疾病可能还与肠道微生物组成、肠道屏障功能障碍、代谢性内毒素血症及肠道微生物相关代谢物（如TMAO、SCFAs、胆汁酸）相关[529, 530]。

随着对肠道微生物群和疾病之间关系的理解加深，有学者提出了心力衰竭的“心-肠轴”假说：心力衰竭心排血量减少，引起肠道缺血/淤血，使得肠道通透性增加，致病菌得到生长，产生TMAO增加；同时，细菌移位使得循环内毒素增加，单核-巨噬细胞活化，潜在的炎症细胞因子激活，导致内皮功能障碍，从而加重心力衰竭[531, 532]。TMAO对动脉粥样硬化的影响，可能是通过诱导细胞表达两种清道夫受体（CD36和清道夫受体A），起到抑制胆固醇反向转运和巨噬细胞清除胆固醇的作用，导致泡沫细胞形成的增加[526]。TMAO还可以通过NF-κB信号通路促进细胞因子和黏附分子的激活，如IL-6、COX-2、ICAM1和E-选择素，导致血管内皮细胞白细胞募集和动脉粥样硬化[533]。动物研究发现，嗜黏蛋白阿克曼菌可以增加紧密连接蛋白的表达水平，降低肠道通透性，从而减少血液中LPS水平，显著降低了MCP-1、TNF-α和ICAM-1的表达和巨噬细胞的主动脉浸润，从而减少动脉粥样硬化损伤[534]。有证据表明，血压与肠道微生物多样性、丰度密切相关，并且受厚壁菌门和拟杆菌门比值（F/B）的影响；SCFAs的升高血压和降低血压作用分别通过与Olfr78和GPR41结合来介导。高盐饮食减少了肠道乳酸杆菌，并伴随着Th17细胞和血压的升高[535]。

心房颤动是最常见的心律失常，目前与肠道生态失调两者之间关系的研究还较少。研究发现，TMAO 可以通过 p65 NF-κB 信号的激活增加右前神经节丛中炎症细胞因子（IL-1β、IL-6 和 TNF-α）的表达，进一步激活心脏自主神经系统并促进心房颤动的诱发[536]。在心房颤动患者中，与 IFN-γ、IL-17 和 IL-22 相关、具有促炎特性的瘤胃球菌属过表达；与心血管保护代谢物呈负相关的一种革兰阴性厌氧菌——韦荣球菌富集；而产丁酸盐的普拉梭菌减少[537]。果糖摄入过量会导致肠道微生态失调，进而增加肠道通透性并激活宿主炎症，诱导 NF-κ B/NLRP3 信号的激活，细胞因子 TNF-α、TGF-β、NF-κB 和 IL-6 的分泌促进心脏炎症和心律失常的发生[538]。

因此，肠道微生物群是治疗心血管疾病的一个新的潜在治疗靶点，目前正在探索对人体进行干预研究，包括通过饮食影响肠道菌群、益生菌、益生元、针对肠道微生物相关代谢物（例如 TMAO 抑制剂）、肠道菌群移植等微生物靶向疗法[526, 527, 530]。

六、肠 - 皮肤轴

研究发现，一些常见的炎症性皮肤病伴有肠道菌群失调，例如特应性皮炎、银屑病、酒渣鼻和寻常痤疮。这些证据表明，皮肤与肠道之间可能存在联系，即“肠 - 皮肤轴”[539]。两者之间的串联涉及代谢系统、免疫系统、神经系统和内分泌系统，但其完整的机制尚需进一步研究[540, 541]。肠道微生物可以通过影响 T 细胞对各种免疫刺激的分化来促进皮肤的稳态应变；肠道微生物相关代谢产物 SCFAs 也被认为在确定某些皮肤微生物分布中起重要作用，而这种分布将影响皮肤免疫防御机制[541, 542]。

IBD 患者中 7% ～ 11% 患有银屑病，而在银屑病患者血浆中含有肠道细菌的 DNA[540]；并且两者表现出一些共同特征，包括免疫反应（Th17、Th1 和 Treg）和炎性细胞因子（TNF-α、IFN-γ、IL-12、IL-17 和 IL-23）以及嗜黏蛋白阿克曼菌丰度的降低[543]。肠道微生物群可能是通过调节 Th17 细胞和促炎反应，从而影响银屑病发展[544]。最近，有研究总结了“肠 - 脑 - 皮肤轴”中肠道微生物分泌的几种神经递质（多巴胺、5- 羟色胺、γ- 氨基丁酸）作为免疫调节剂在银屑病中的作用[545]。多巴胺可以抑制活化的 T 细胞，并有可能通过调节这种通信，靶向抑制银屑病中 T 细胞介导的炎症。此外，高选择性 D1 样受体激动剂非诺多泮，也能够抑制银屑病患者活化的 T 细胞，并减少趋化因子 SDF-1/CXCL12 和炎性细胞因子（TNF-α、IL-1β、IL-6）的分泌，以及 T 细胞活化蛋白 / 标记物：CD69、CD28 和 IL-2 的表达[546]。γ- 氨基丁酸可以抑制 IFN-γ 和 IL-17 的产生以及抗原呈递细胞上 MHC Ⅱ和 CD80 的表达，并对免疫调节分子如 Foxp3+、IL-10、TGF-β、CTLA4 和 SIRP-α 的表达表现出很强的促进作用[547]。

特应性皮炎，也称为湿疹，是一种慢性过敏性皮肤病[544]。研究发现，出生时对微生物刺激反应的 IL-10 产生受损可能与特应性皮炎发生风险增加有关[548]。在特应性皮炎患者中，普拉梭菌亚种水平上的生物失调和肠上皮炎症，导致肠上皮屏障受损，最终导致皮肤中对过敏原的异常 Th2 型免疫反应[549]。当出现“肠漏”时，免疫原性分子（包括饮食抗原、细菌毒素和病原体）的渗透增加，可能在皮肤中累积，从而扰乱表皮屏障，导致慢性皮肤炎症和持续的免疫反应[539, 541]。而口服清酒乳杆菌 WIKIM30 可以显著降低特应性皮炎样皮损和血清 IgE 和 IL-4 水平，同时降低外周淋巴结 $CD4^+$ T 细胞和 B 细胞的数量以及 Th2 细胞因子（IL-4、IL-5 和 IL-13）水平，增加肠系膜淋巴结 Treg 分化和 IL-10 分泌[550]。

总之，越来越多的证据支持“肠-皮肤轴”的存在，调节肠道微生物群以改善皮肤状况似乎是一种可行的方法[539]。

七、肠-骨轴

骨骼是一个动态的器官，在生命的各个阶段都会经历不断的重塑。这一重塑过程需要破骨细胞和成骨细胞两种细胞的协同作用。破骨细胞进行骨吸收的作用，而成骨细胞参与新骨的形成[551]。破骨细胞分化受前破骨细胞生成细胞因子调节，主要包括RANKL、肿瘤坏死因子、IL-6、IL-1和IL-17[552]。肠道微生物群可以影响宿主免疫系统的发育和功能，其释放的代谢物以及微生物和免疫细胞之间的直接接触刺激了肠道内皮屏障的免疫系统。免疫细胞（包括T细胞和树突状细胞）与肠道内壁的微生物菌群相互作用，并迁移到淋巴结，激活促炎或抗炎免疫反应。这些细胞也可以释放可溶性促炎或抗炎介质或细胞因子到循环系统中，然后调节全身骨重塑[553]。与常规饲养的动物相比，无菌雌性C57BL/6小鼠的骨密度增加，股骨干骺端小梁骨体积分数增加39%，破骨细胞表面减少，矿化表面增加[554]。

骨质疏松症最常见的原因是雌激素缺乏，其机制主要是增加破骨细胞的形成、活性和寿命，其驱动因素是骨髓中炎症细胞因子IL-1、IL-17、TNF-α和破骨细胞生成因子RANKL的产生。其中TNF-α刺激骨吸收的关键机制包括增强RANKL活性和诱导Th17细胞[555]。在体内，Th17细胞在炎症状态下可增强人类破骨细胞分化。肠道炎症部位激活的Th17细胞产生高水平的破骨细胞生成因子RANKL、IL-17和TNF-α。它们迁移到骨髓后可通过产生细胞因子和上调间充质细胞的RANKL表达来增强破骨细胞分化[556]。而这些过程需要肠道微生物的参与[557]。

肠道微生物群还可以通过将微生物相关分子模式（如脂多糖、肽聚糖、鞭毛蛋白）引入体循环来影响远处的器官[558]。在骨骼中，MAMPs通过刺激骨、软骨和滑膜的常驻免疫细胞上的先天免疫受体对骨重塑造成直接影响，包括TLR2（对肽聚糖有反应）、TLR4（对脂多糖有反应）和TLR5（对鞭毛蛋白有反应）[553，559]，从而导致骨关节炎的发生。一般认为全身低度慢性炎症是骨关节炎发展的主要诱因之一，通常与肥胖有关[560]。紧密连接蛋白和肠屏障完整性的变化可能使革兰阴性菌成分脂多糖进入体循环[561]。而血清脂多糖和脂多糖结合蛋白水平与影像学骨关节炎和关节症状的严重程度之间存在关联[562]。这表明肠道微生物群可通过增加脂多糖的产生促进损伤诱导的骨关节炎[563]。同时微生物群还可能通过诱导IGF-1促进骨形成[516]，并且还可能影响类固醇激素、甲状旁腺激素或维生素D代谢物的水平[564]。

研究表明，益生菌可以减少几种促炎和溶骨性细胞因子的表达，如TNF-α和IL-1β[565]。罗伊乳杆菌、鼠李糖乳杆菌和副干酪乳杆菌可显著降低破骨细胞生成和骨吸收[557]。罗伊乳杆菌抑制了小鼠促炎细胞因子TNF介导的骨吸收，并降低了骨髓中$CD4^+$ T细胞百分比[566，567]。

尽管目前肠道微生物群与骨骼或关节的交流机制并不完全明确，但动物模型研究报告的结果强烈提示肠道微生物的失衡正在驱动炎性关节病，这表明靶向改变微生物群可以帮助我们找到预防或治疗骨骼相关疾病的有效策略[563]。

第五节　小结与展望

在过去的 10 年里，科学家们对微生物和免疫系统相互作用的研究，使人们更好地理解了它们的分子基础及其重要性，但在解开人类健康和疾病中微生物组 - 免疫系统相互作用方面，仍存在许多未知和挑战：①发病前或在疾病早期阶段，微生物群和免疫力之间的直接因果关系尚未确定；②病毒、真菌等以前被忽视的微生物及其对宿主免疫的影响，成为未来研究中一个重要且具有挑战性的课题；③还需进一步阐明微生物群和免疫系统是如何在环境触发因素和宿主遗传学的背景下相互作用的；④研究所使用的小鼠，具有与人类不同的微生物群，因此其转化潜力和可重复性有限；⑤人类的个体间变异性和相关的复杂性加大了研究的难度 [303]。

健康的人体肠道微生物生态系统是维持全身有序功能的必要条件，而肠道生态失调已被证明与多种病理状况有关 [568]，恢复失调的肠道微生物组已逐渐成为一种有效的辅助手段和治疗方法 [569]。目前，使用益生菌、益生元、饮食改变、抗生素和肠道菌群移植是调节肠道微生物群有效且公认的方法 [568]。然而，想要开发这些基于微生物群的疗法并将其转化为临床实践，还需要加强对微生物组 - 宿主免疫之间复杂相互作用的理解，以及标准化、严格和公正的临床前和临床干预研究 [303]。此外，人体肠道微生物生态系统的复杂性需要寻求新的策略以更合适的方式来调控。一些新兴的有潜力的手段，例如工程纳米材料、噬菌体，以及抗菌肽、非抗生素药物、疫苗和免疫球蛋白的使用还需要进一步的探索 [568]。

共生微生物群对宿主免疫系统有着深远的影响，而深入了解微生物介导的免疫调节机制，识别精确的免疫刺激和免疫抑制细菌菌株或途径，有助于提高治疗方法的精确度，从而避免不良后果 [392]。药物微生物组学领域为减轻药物副作用和提高治疗效益开辟了一条新途径。全面了解基于微生物组的药动学和药效学或可将微生物组治疗转化为临床。为了实现这一点，可采用简化方法和系统调查相结合的方法，将代谢组学和转录组学的数据整合后绘制药物代谢相关的代谢途径，然后进行系统级建模。在相关模型系统中进行实践，并进行根本原因和稳健性分析，可能是提高我们了解药物生物学效率的一种实用方法，进而开发微生物组疗法，这或将改变我们医疗保健行业的未来 [570]。因此，免疫疗法的未来可能会将直接的、基于药物的免疫调节与微生物组和代谢组修饰相结合，以共同靶向疾病分子病因的微生物和宿主成分 [162]。

参考文献

[1] Sanders DJ, Inniss S, Sebepos-Rogers G, et al. The role of the microbiome in gastrointestinal inflammation. Bioscience Reports, 2021, 41(6): BSR20203850.

[2] Shi N, Li N, Duan X, et al. Interaction between the gut microbiome and mucosal immune system. Military Medical Research, 2017, 4: 14.

[3] Scharen OP, Hapfelmeier S. Robust microbe immune recognition in the intestinal mucosa. Genes And Immunity, 2021, 22(5-6): 268-275.

[4] Zhou B, Yuan Y, Zhang S, et al. Intestinal flora and disease mutually shape the regional immune system in the intestinal tract. Frontiers in Immunology, 2020, 11: 575.

[5] Caricilli AM, Castoldi A, Câmara NO. Intestinal barrier: A gentlemen's agreement between microbiota and immunity. World Journal of Gastrointestinal Pathophysiology, 2014, 5(1): 18-32.

[6] Hara N, Alkanani AK, Ir D, et al. The role of the intestinal microbiota in type 1 diabetes. Clinical Immunology, 2013, 146(2): 112-119.

[7] Takiishi T, Fenero CIM, Camara NOS. Intestinal barrier and gut microbiota: Shaping our immune responses throughout life. Tissue Barriers, 2017, 5(4): e1373208.

[8] Sommer F, Bäckhed F. The gut microbiota--masters of host development and physiology. Nature Reviews: Microbiology, 2013, 11(4): 227-238.

[9] Kumar M, Babaei P, Ji B, et al. Human gut microbiota and healthy aging: Recent developments and future prospective. Nutrition and Healthy Aging, 2016, 4(1): 3-16.

[10] Cummings JH, Pomare EW, Branch WJ, et al. Short chain fatty acids in human large intestine, portal, hepatic and venous blood. Gut, 1987, 28(10): 1221-1227.

[11] De Filippo C, Cavalieri D, Di Paola M, et al. Impact of diet in shaping gut microbiota revealed by a comparative study in children from Europe and rural Africa. Proceedings of the National Academy of Sciences of the United States of America, 2010, 107(33): 14691-14696.

[12] Martinez-Medina M, Aldeguer X, Gonzalez-Huix F, et al. Abnormal microbiota composition in the ileocolonic mucosa of Crohn's disease patients as revealed by polymerase chain reaction-denaturing gradient gel electrophoresis. Inflammatory Bowel Diseases, 2006, 12(12): 1136-1145.

[13] Frank DN, St Amand AL, Feldman RA, et al. Molecular-phylogenetic characterization of microbial community imbalances in human inflammatory bowel diseases. Proceedings of the National Academy of Sciences of the United States of America, 2007, 104(34): 13780-13785.

[14] Arpaia N, Campbell C, Fan X, et al. Metabolites produced by commensal bacteria promote peripheral regulatory T-cell generation. Nature, 2013, 504(7480): 451-455.

[15] Wrzosek L, Miquel S, Noordine ML, et al. Bacteroides thetaiotaomicron and Faecalibacterium prausnitzii influence the production of mucus glycans and the development of goblet cells in the colonic epithelium of a gnotobiotic model rodent. BMC Biology, 2013, 11: 61.

[16] Gaudier E, Jarry A, Blottière HM, et al. Butyrate specifically modulates MUC gene expression in intestinal epithelial goblet cells deprived of glucose. American Journal of Physiology Gastrointestinal and Liver Physiology, 2004, 287(6): G1168-G1174.

[17] Fukuda S, Toh H, Hase K, et al. Bifidobacteria can protect from enteropathogenic infection through production of acetate. Nature, 2011, 469(7331): 543-547.

[18] Valenzano MC, DiGuilio K, Mercado J, et al. Remodeling of tight junctions and enhancement of barrier integrity of the CACO-2 intestinal epithelial cell layer by micronutrients. PloS One, 2015, 10(7): e0133926.

[19] Ivanov, II, Atarashi K, Manel N, et al. Induction of intestinal Th17 cells by segmented filamentous bacteria. Cell, 2009, 139(3): 485-498.

[20] Atarashi K, Tanoue T, Shima T, et al. Induction of colonic regulatory T cells by indigenous Clostridium species. Science, 2011, 331(6015): 337-341.

[21] Geuking MB, Cahenzli J, Lawson MA, et al. Intestinal bacterial colonization induces mutualistic regulatory T cell responses. Immunity, 2011, 34(5): 794-806.

[22] Wen L, Ley RE, Volchkov PY, et al. Innate immunity and intestinal microbiota in the development of Type 1 diabetes. Nature, 2008, 455(7216): 1109-1113.

[23] Mishima Y, Oka A, Liu B, et al. Microbiota maintain colonic homeostasis by activating TLR2/MyD88/PI3K signaling in IL-10-producing regulatory B cells. Journal of Clinical Investigation, 2019, 129(9): 3702-3716.

[24] Vaarala O. Is the origin of type 1 diabetes in the gut? Immunology and Cell Biology, 2012, 90(3): 271-276.

[25] Carstens LE, Westerbeek EA, van Zwol A, et al. Neonatal antibiotics in preterm infants and allergic disorders later in life. Pediatric Allergy and Immunology, 2016, 27(7): 759-764.

[26] Mulder B, Pouwels KB, Schuiling-Veninga CC, et al. Antibiotic use during pregnancy and asthma in preschool children: the influence of confounding. Clinical And Experimental Allergy, 2016, 46(9): 1214-1226.

[27] Littman DR, Pamer EG. Role of the commensal microbiota in normal and pathogenic host immune responses. Cell Host & Microbe, 2011, 10(4): 311-323.

[28] Stefka AT, Feehley T, Tripathi P, et al. Commensal bacteria protect against food allergen sensitization. Proceedings of the National Academy of Sciences of the United States of America, 2014, 111(36): 13145-13150.

[29] Cox LM, Yamanishi S, Sohn J, et al. Altering the intestinal microbiota during a critical developmental window has lasting metabolic consequences. Cell, 2014, 158(4): 705-721.

[30] Lee YK, Menezes JS, Umesaki Y, et al. Proinflammatory T-cell responses to gut microbiota promote experimental autoimmune encephalomyelitis. Proceedings of the National Academy of Sciences of the United States of America, 2011, 108 Suppl 1(Suppl 1): 4615-4622.

[31] Lee H, Jin BE, Jang E, et al. Gut-residing microbes alter the host susceptibility to autoantibody-mediated arthritis. Immune Network, 2014, 14(1): 38-44.

[32] Jin S, Zhao D, Cai C, et al. Low-dose penicillin exposure in early life decreases Th17 and the susceptibility to DSS colitis in mice through gut microbiota modification. Scientific Reports, 2017, 7: 43662.

[33] Turnbaugh PJ, Ley RE, Hamady M, et al. The human microbiome project. Nature, 2007, 449(7164): 804-810.

[34] Proctor L. Priorities for the next 10 years of human microbiome research. Nature, 2019, 569(7758): 623-625.

[35] Shock T, Badang L, Ferguson B, et al. The interplay between diet, gut microbes, and host epigenetics in health and disease. Journal of Nutritional Biochemistry, 2021, 95: 108631.

[36] Deschasaux M, Bouter KE, Prodan A, et al. Depicting the composition of gut microbiota in a population with varied ethnic origins but shared geography. Nature Medicine, 2018, 24(10): 1526-1531.

[37] McBurney MI, Davis C, Fraser CM, et al. Establishing what constitutes a healthy human gut microbiome: State of the science, regulatory considerations, and future directions. Journal of Nutrition, 2019, 149(11): 1882-1895.

[38] Abdelsalam NA, Ramadan AT, ElRakaiby MT, et al. Toxicomicrobiomics: the human microbiome vs. pharmaceutical, dietary, and environmental xenobiotics. Frontiers in Pharmacology, 2020, 11: 390.

[39] Robertson RC, Manges AR, Finlay BB, et al. The human microbiome and child growth - first 1000 days and beyond. Trends in Microbiology, 2019, 27(2): 131-147.

[40] Qin J, Li R, Raes J, et al. A human gut microbial gene catalogue established by metagenomic sequencing. Nature, 2010, 464(7285): 59-65.

[41] Rodríguez JM, Murphy K, Stanton C, et al. The composition of the gut microbiota throughout life, with an emphasis on early life. Microbial Ecology in Health and Disease, 2015, 26: 26050.

[42] Aagaard K, Ma J, Antony KM, et al. The placenta harbors a unique microbiome. Science Translational Medicine, 2014, 6(237): 237ra265.

[43] Steel JH, O'Donoghue K, Kennea NL, et al. Maternal origin of inflammatory leukocytes in preterm fetal membranes, shown by fluorescence in situ hybridisation. Placenta, 2005, 26(8-9): 672-677.

[44] Jimenez E, Fernandez L, Marin ML, et al. Isolation of commensal bacteria from umbilical cord blood of healthy neonates born by cesarean section. Current Microbiology, 2005, 51(4): 270-274.

[45] Ardissone AN, de la Cruz DM, Davis-Richardson AG, et al. Meconium microbiome analysis identifies bacteria correlated with premature birth. PloS One, 2014, 9(3): e90784.

[46] Collado MC, Rautava S, Aakko J, et al. Human gut colonisation may be initiated in utero by distinct microbial communities in the placenta and amniotic fluid. Scientific Reports, 2016, 6: 23129.

[47] Silverstein RB, Mysorekar IU. Group therapy on in utero colonization: seeking common truths and a way forward. Microbiome, 2021, 9(1): 7.

[48] Gerhold K, Avagyan A, Seib C, et al. Prenatal initiation of endotoxin airway exposure prevents subsequent allergen-induced sensitization and airway inflammation in mice. Journal Of Allergy And Clinical Immunology, 2006, 118(3): 666-673.

[49] Blümer N, Herz U, Wegmann M, et al. Prenatal lipopolysaccharide-exposure prevents allergic sensitization and airway inflammation, but not airway responsiveness in a murine model of experimental asthma. Clinical And Experimental Allergy, 2005, 35(3): 397-402.

[50] Conrad ML, Ferstl R, Teich R, et al. Maternal TLR signaling is required for prenatal asthma protection by the nonpathogenic microbe Acinetobacter lwoffii F78. Journal of Experimental Medicine, 2009, 206(13): 2869-2877.

[51] Dominguez-Bello MG, Costello EK, Contreras M, et al. Delivery mode shapes the acquisition and structure of the initial microbiota across multiple body habitats in newborns. Proceedings of the National Academy of Sciences of the United States of America, 2010, 107(26): 11971-11975.

[52] Kristensen K, Henriksen L. Cesarean section and disease associated with immune function. Journal Of Allergy And Clinical Immunology, 2016, 137(2): 587-590.

[53] Jakobsson HE, Abrahamsson TR, Jenmalm MC, et al. Decreased gut microbiota diversity, delayed Bacteroidetes colonisation and reduced Th1 responses in infants delivered by caesarean section. Gut, 2014, 63(4): 559-566.

[54] Reyman M, van Houten MA, van Baarle D, et al. Impact of delivery mode-associated gut microbiota dynamics on health in the first year of life. Nature Communications, 2019, 10(1): 4997.

[55] Brandtzaeg P. The mucosal immune system and its integration with the mammary glands. Journal of Pediatrics, 2010, 156(2 Suppl): S8-S15.

[56] Lönnerdal B. Infant formula and infant nutrition: bioactive proteins of human milk and implications for composition of infant formulas. American Journal of Clinical Nutrition, 2014, 99(3): 712s-717s.

[57] Rodriguez JM, Fernandez L, Verhasselt V. The gutbreast axis:programming health for life. Nutrients, 2021, 13(2): 606.

[58] Fichter M, Klotz M, Hirschberg DL, et al. Breast milk contains relevant neurotrophic factors and cytokines for enteric nervous system development. Molecular Nutrition & Food Research, 2011, 55(10): 1592-1596.

[59] Fernández L, Langa S, Martín V, et al. The microbiota of human milk in healthy women. Cellular and Molecular Biology (Noisy-Le-Grand), 2013, 59(1): 31-42.

[60] McGuire W, Anthony MY. Donor human milk versus formula for preventing necrotising enterocolitis in preterm infants: systematic review. Archives of Disease in Childhood: Fetal and Neonatal Edition, 2003, 88(1): F11-F14.

[61] Stuebe A. The risks of not breastfeeding for mothers and infants. Reviews in Obstetrics and Gynecology, 2009, 2(4): 222-231.

[62] Levy M, Kolodziejczyk AA, Thaiss CA, et al. Dysbiosis and the immune system. Nature Reviews: Immunology, 2017, 17(4): 219-232.

[63] Liu L, Liang L, Liang H, et al. Fusobacterium nucleatum aggravates the progression of colitis by regulating M1 macrophage polarization via AKT2 pathway. Frontiers in Immunology, 2019, 10: 1324.

[64] Wang Y, Yin Y, Chen X, et al. Induction of intestinal th17 cells by flagellins from segmented filamentous bacteria. Frontiers in Immunology, 2019, 10: 2750.

[65] Karczewski J, Poniedziałek B, Adamski Z, et al. The effects of the microbiota on the host immune system. Autoimmunity, 2014, 47(8): 494-504.

[66] Burrows MP, Volchkov P, Kobayashi KS, et al. Microbiota regulates type 1 diabetes through Toll-like receptors. Proceedings of the National Academy of Sciences of the United States of America, 2015, 112(32): 9973-9977.

[67] Larsson E, Tremaroli V, Lee YS, et al. Analysis of gut microbial regulation of host gene expression along the length of the gut and regulation of gut microbial ecology through MyD88. Gut, 2012, 61(8): 1124-1131.

[68] Clarke TB, Davis KM, Lysenko ES, et al. Recognition of peptidoglycan from the microbiota by Nod1 enhances systemic innate immunity. Nature Medicine, 2010, 16(2): 228-231.

[69] Ganal SC, Sanos SL, Kallfass C, et al. Priming of natural killer cells by nonmucosal mononuclear phagocytes requires instructive signals from commensal microbiota. Immunity, 2012, 37(1): 171-186.

[70] Khosravi A, Yanez A, Price JG, et al. Gut microbiota promote hematopoiesis to control bacterial infection. Cell Host & Microbe, 2014, 15(3): 374-381.

[71] Hill DA, Siracusa MC, Abt MC, et al. Commensal bacteria-derived signals regulate basophil hematopoiesis and allergic inflammation. Nature Medicine, 2012, 18(4): 538-546.

[72] Zeissig S, Blumberg RS. Life at the beginning: perturbation of the microbiota by antibiotics in early life and its role in health and disease. Nature Immunology, 2014, 15(4): 307-310.

[73] Ng SC, Bernstein CN, Vatn MH, et al. Geographical variability and environmental risk factors in inflammatory bowel disease. Gut, 2013, 62(4): 630-649.

[74] Murk W, Risnes KR, Bracken MB. Prenatal or early-life exposure to antibiotics and risk of childhood asthma: a systematic review. Pediatrics, 2011, 127(6): 1125-1138.

[75] Wang L, Zhu L, Qin S. Gut microbiota modulation on intestinal mucosal adaptive immunity. Journal of Immunology Research, 2019, 2019: 4735040.

[76] Cammarota G, Ianiro G, Cianci R, et al. The involvement of gut microbiota in inflammatory bowel disease pathogenesis: potential for therapy. Pharmacol Therapeut, 2015, 149: 191-212.

[77] Scott FW, Pound LD, Patrick C, et al. Where genes meet environment-integrating the role of gut luminal contents, immunity and pancreas in type 1 diabetes. Translational Research, 2017, 179: 183-198.

[78] Losurdo G, Principi M, Iannone A, et al. The interaction between celiac disease and intestinal microbiota. Journal of Clinical Gastroenterology, 2016, 50 Suppl 2, Proceedings from the 8th Probiotics, Prebiotics & New Foods for Microbiota and Human Health meeting held in Rome, Italy on September 13-15, 2015: S145-S147.

[79] Tang WH, Hazen SL. The contributory role of gut microbiota in cardiovascular disease. Journal of Clinical Investigation, 2014, 124(10): 4204-4211.

[80] Rich BS, Dolgin SE. Necrotizing enterocolitis. Pediatrics in Review, 2017, 38(12): 552-559.

[81] Masi AC, Stewart CJ. The role of the preterm intestinal microbiome in sepsis and necrotising enterocolitis.

Early Human Development, 2019, 138: 104854.
[82] Baranowski JR, Claud EC. Necrotizing enterocolitis and the preterm Infant microbiome. Advances in Experimental Medicine and Biology, 2019, 1125: 25-36.
[83] Pammi M, Cope J, Tarr PI, et al. Intestinal dysbiosis in preterm infants preceding necrotizing enterocolitis: a systematic review and meta-analysis. Microbiome, 2017, 5(1): 31.
[84] Lucas A, Cole TJ. Breast milk and neonatal necrotising enterocolitis. Lancet, 1990, 336(8730): 1519-1523.
[85] Mosca F, Gianni ML. Human milk: composition and health benefits. La Pediatria Medica e Chirurgica, 2017, 39(2): 155.
[86] Granger CL, Embleton ND, Palmer JM, et al. Maternal breastmilk, infant gut microbiome and the impact on preterm infant health. Acta Paediatrica, 2021, 110(2): 450-457.
[87] Autran CA, Kellman BP, Kim JH, et al. Human milk oligosaccharide composition predicts risk of necrotising enterocolitis in preterm infants. Gut, 2018, 67(6): 1064-1070.
[88] Underwood MA. Arguments for routine administration of probiotics for NEC prevention. Current Opinion in Pediatrics, 2019, 31(2): 188-194.
[89] Olsen R, Greisen G, Schroder M, et al. Prophylactic probiotics for preterm infants: a systematic review and meta-analysis of observational studies. Neonatology, 2016, 109(2): 105-112.
[90] Chang HY, Chen JH, Chang JH, et al. Multiple strains probiotics appear to be the most effective probiotics in the prevention of necrotizing enterocolitis and mortality: An updated meta-analysis. PloS One, 2017, 12(2): e0171579.
[91] Garg BD, Balasubramanian H, Kabra NS. Physiological effects of prebiotics and its role in prevention of necrotizing enterocolitis in preterm neonates. Journal of Maternal-Fetal & Neonatal Medicine, 2018, 31(15): 2071-2078.
[92] Schwabe RF, Jobin C. The microbiome and cancer. Nature Reviews: Cancer, 2013, 13(11): 800-812.
[93] Kuper H, Adami HO, Trichopoulos D. Infections as a major preventable cause of human cancer. Journal of Internal Medicine, 2000, 248(3): 171-183.
[94] Irrazábal T, Belcheva A, Girardin SE, et al. The multifaceted role of the intestinal microbiota in colon cancer. Molecular Cell, 2014, 54(2): 309-320.
[95] Jaye K, Li CG, Bhuyan DJ. The complex interplay of gut microbiota with the five most common cancer types: From carcinogenesis to therapeutics to prognoses. Critical Reviews in Oncology/Hematology, 2021, 165: 103429.
[96] Franco AT, Israel DA, Washington MK, et al. Activation of beta-catenin by carcinogenic Helicobacter pylori. Proceedings of the National Academy of Sciences of the United States of America, 2005, 102(30): 10646-10651.
[97] Wang F, Sun MY, Shi SL, et al. Helicobacter pylori infection and normal colorectal mucosa-adenomatous polyp-adenocarcinoma sequence: a meta-analysis of 27 case-control studies. Colorectal Disease, 2014, 16(4): 246-252.
[98] Gao Z, Guo B, Gao R, et al. Microbiota disbiosis is associated with colorectal cancer. Frontiers in Microbiology, 2015, 6: 20.
[99] Ahn J, Sinha R, Pei Z, et al. Human gut microbiome and risk for colorectal cancer. Journal of the National Cancer Institute, 2013, 105(24): 1907-1911.
[100] Castellarin M, Warren RL, Freeman JD, et al. Fusobacterium nucleatum infection is prevalent in human colorectal carcinoma. Genome Research, 2012, 22(2): 299-306.
[101] Wu J, Li Q, Fu X. Fusobacterium nucleatum contributes to the carcinogenesis of colorectal cancer by inducing inflammation and suppressing host immunity. Translational Oncology, 2019, 12(6): 846-851.
[102] Guo P, Tian Z, Kong X, et al. FadA promotes DNA damage and progression of Fusobacterium nucleatum-induced colorectal cancer through up-regulation of chk2. Journal of Experimental and Clinical Cancer Research, 2020, 39(1): 202.
[103] Rakoff-Nahoum S, Medzhitov R. Regulation of spontaneous intestinal tumorigenesis through the adaptor protein MyD88. Science, 2007, 317(5834): 124-127.
[104] Salcedo R, Worschech A, Cardone M, et al. MyD88-mediated signaling prevents development of adenocarcinomas of the colon: role of interleukin 18. Journal of Experimental Medicine, 2010, 207(8): 1625-1636.
[105] Zaki MH, Vogel P, Body-Malapel M, et al. IL-18 production downstream of the Nlrp3 inflammasome confers protection against colorectal tumor formation. Journal of Immunology, 2010, 185(8): 4912-4920.
[106] Chen GY, Liu M, Wang F, et al. A functional role for Nlrp6 in intestinal inflammation and tumorigenesis. Journal of Immunology, 2011, 186(12): 7187-7194.

[107] Saleh M, Trinchieri G. Innate immune mechanisms of colitis and colitis-associated colorectal cancer. Nature Reviews: Immunology, 2011, 11(1): 9-20.

[108] Fukata M, Chen A, Vamadevan AS, et al. Toll-like receptor-4 promotes the development of colitis-associated colorectal tumors. Gastroenterology, 2007, 133(6): 1869-1881.

[109] Wang EL, Qian ZR, Nakasono M, et al. High expression of Toll-like receptor 4/myeloid differentiation factor 88 signals correlates with poor prognosis in colorectal cancer. British Journal of Cancer, 2010, 102(5): 908-915.

[110] Fukata M, Chen A, Klepper A, et al. Cox-2 is regulated by Toll-like receptor-4 (TLR4) signaling: Role in proliferation and apoptosis in the intestine. Gastroenterology, 2006, 131(3): 862-877.

[111] Fukata M, Abreu MT. TLR4 signalling in the intestine in health and disease. Biochemical Society Transactions, 2007, 35(Pt 6): 1473-1478.

[112] Yu LC, Wang JT, Wei SC, et al. Host-microbial interactions and regulation of intestinal epithelial barrier function: From physiology to pathology. World Journal of Gastrointestinal Pathophysiology, 2012, 3(1): 27-43.

[113] Calle EE, Kaaks R. Overweight, obesity and cancer: epidemiological evidence and proposed mechanisms. Nature Reviews: Cancer, 2004, 4(8): 579-591.

[114] Cario E, Podolsky DK. Differential alteration in intestinal epithelial cell expression of toll-like receptor 3 (TLR3) and TLR4 in inflammatory bowel disease. Infection And Immunity, 2000, 68(12): 7010-7017.

[115] So J, Tang W, Leung WK, et al. Cancer risk in 2621 chinese patients with inflammatory bowel disease: a population-based cohort study. Inflammatory Bowel Diseases, 2017, 23(11): 2061-2068.

[116] Uronis JM, Mühlbauer M, Herfarth HH, et al. Modulation of the intestinal microbiota alters colitis-associated colorectal cancer susceptibility. PloS One, 2009, 4(6): e6026.

[117] Azcárate-Peril MA, Sikes M, Bruno-Bárcena JM. The intestinal microbiota, gastrointestinal environment and colorectal cancer: a putative role for probiotics in prevention of colorectal cancer? American Journal of Physiology Gastrointestinal and Liver Physiology, 2011, 301(3): G401-G424.

[118] Wu S, Morin PJ, Maouyo D, et al. Bacteroides fragilis enterotoxin induces c-Myc expression and cellular proliferation. Gastroenterology, 2003, 124(2): 392-400.

[119] Goodwin AC, Destefano Shields CE, Wu S, et al. Polyamine catabolism contributes to enterotoxigenic Bacteroides fragilis-induced colon tumorigenesis. Proceedings of the National Academy of Sciences of the United States of America, 2011, 108(37): 15354-15359.

[120] Arthur JC, Perez-Chanona E, Mühlbauer M, et al. Intestinal inflammation targets cancer-inducing activity of the microbiota. Science, 2012, 338(6103): 120-123.

[121] Cuevas-Ramos G, Petit CR, Marcq I, et al. Escherichia coli induces DNA damage in vivo and triggers genomic instability in mammalian cells. Proceedings of the National Academy of Sciences of the United States of America, 2010, 107(25): 11537-11542.

[122] Singh N, Gurav A, Sivaprakasam S, et al. Activation of Gpr109a, receptor for niacin and the commensal metabolite butyrate, suppresses colonic inflammation and carcinogenesis. Immunity, 2014, 40(1): 128-139.

[123] Kriegel MA, Sefik E, Hill JA, et al. Naturally transmitted segmented filamentous bacteria segregate with diabetes protection in nonobese diabetic mice. Proceedings of the National Academy of Sciences of the United States of America, 2011, 108(28): 11548-11553.

[124] Perez-Cobas AE, Moya A, Gosalbes MJ, et al. Colonization resistance of the gut microbiota against clostridium difficile. Antibiotics-Basel, 2015, 4(3): 337-357.

[125] Abbas A, Zackular JP. Microbe-microbe interactions during Clostridioides difficile infection. Current Opinion in Microbiology, 2020, 53: 19-25.

[126] Sehgal K, Khanna S. Gut microbiome and Clostridioides difficile infection: a closer look at the microscopic interface. Therapeutic Advances in Gastroenterology, 2021, 14: 1756284821994736.

[127] Claesson MJ, Cusack S, O'Sullivan O, et al. Composition, variability, and temporal stability of the intestinal microbiota of the elderly. Proceedings of the National Academy of Sciences of the United States of America, 2011, 108 Suppl 1: 4586-4591.

[128] Seekatz AM, Young VB. Clostridium difficile and the microbiota. Journal of Clinical Investigation, 2014, 124(10): 4182-4189.

[129] Rosa R, Donskey CJ, Munoz-Price LS. The intersection between colonization resistance, antimicrobial stewardship, and clostridium difficile.

Current Infectious Disease Reports, 2018, 20(8): 27.

[130] Deneve C, Janoir C, Poilane I, et al. New trends in Clostridium difficile virulence and pathogenesis. International Journal of Antimicrobial Agents, 2009, 33 Suppl 1: S24-S28.

[131] Kelly CP, Becker S, Linevsky JK, et al. Neutrophil recruitment in Clostridium difficile toxin A enteritis in the rabbit. Journal of Clinical Investigation, 1994, 93(3): 1257-1265.

[132] Kang JD, Myers CJ, Harris SC, et al. Bile acid 7alpha-dehydroxylating gut bacteria secrete antibiotics that inhibit clostridium difficile: role of secondary bile acids. Cell Chemical Biology, 2019, 26(1): 27-34, e24.

[133] Mullish BH, Allegretti JR. The contribution of bile acid metabolism to the pathogenesis of clostridioides difficile infection. Therapeutic Advances in Gastroenterology, 2021, 14: 17562848211017725.

[134] Campbell C, McKenney PT, Konstantinovsky D, et al. Bacterial metabolism of bile acids promotes generation of peripheral regulatory T cells. Nature, 2020, 581(7809): 475-479.

[135] Ng KM, Ferreyra JA, Higginbottom SK, et al. Microbiota-liberated host sugars facilitate post-antibiotic expansion of enteric pathogens. Nature, 2013, 502(7469): 96-99.

[136] Wilson KH, Perini F. Role of competition for nutrients in suppression of clostridium difficile by the colonic microflora. Infection And Immunity, 1988, 56(10): 2610-2614.

[137] Baunwall SMD, Lee MM, Eriksen MK, et al. Faecal microbiota transplantation for recurrent clostridioides difficile infection: An updated systematic review and meta-analysis. EClinical Medicine, 2020, 29-30: 100642.

[138] Monaghan T, Mullish BH, Patterson J, et al. Effective fecal microbiota transplantation for recurrent clostridioides difficile infection in humans is associated with increased signalling in the bile acid-farnesoid X receptor-fibroblast growth factor pathway. Gut Microbes, 2019, 10(2): 142-148.

[139] van Nood E, Vrieze A, Nieuwdorp M, et al. Duodenal infusion of donor feces for recurrent clostridium difficile. New England Journal of Medicine, 2013, 368(5): 407-415.

[140] Buffie CG, Jarchum I, Equinda M, et al. Profound alterations of intestinal microbiota following a single dose of clindamycin results in sustained susceptibility to Clostridium difficile-induced colitis. Infection And Immunity, 2012, 80(1): 62-73.

[141] Lawley TD, Clare S, Walker AW, et al. Targeted restoration of the intestinal microbiota with a simple, defined bacteriotherapy resolves relapsing Clostridium difficile disease in mice. PLoS Pathogens, 2012, 8(10): e1002995.

[142] Reeves AE, Theriot CM, Bergin IL, et al. The interplay between microbiome dynamics and pathogen dynamics in a murine model of Clostridium difficile Infection. Gut Microbes, 2011, 2(3): 145-158.

[143] Rea MC, O'Sullivan O, Shanahan F, et al. Clostridium difficile carriage in elderly subjects and associated changes in the intestinal microbiota. Journal of Clinical Microbiology, 2012, 50(3): 867-875.

[144] Tortajada-Girbes M, Rivas A, Hernandez M, et al. Alimentary and pharmaceutical approach to natural antimicrobials against clostridioides difficile gastrointestinal infection. Foods, 2021, 10(5): 1124.

[145] Chen K, Zhu Y, Zhang Y, et al. A probiotic yeast-based immunotherapy against clostridioides difficile infection. Science Translational Medicine, 2020, 12(567): eaax4905.

[146] Fritsch J, Abreu MT. The microbiota and the immune response: what is the chicken and what is the egg? Gastrointestinal Endoscopy Clinics of North America, 2019, 29(3): 381-393.

[147] Ni J, Wu GD, Albenberg L, et al. Gut microbiota and IBD: causation or correlation? Nature Reviews: Gastroenterology & Hepatology, 2017, 14(10): 573-584.

[148] Lee M, Chang EB. Inflammatory bowel diseases (IBD) and the microbiome-searching the crime scene for clues. Gastroenterology, 2021, 160(2): 524-537.

[149] Alshehri D, Saadah O, Mosli M, et al. Dysbiosis of gut microbiota in inflammatory bowel disease: Current therapies and potential for microbiota-modulating therapeutic approaches. Bosnian Journal of Basic Medical Sciences, 2021, 21(3): 270-283.

[150] Imhann F, Vich Vila A, Bonder MJ, et al. Interplay of host genetics and gut microbiota underlying the onset and clinical presentation of inflammatory bowel disease. Gut, 2018, 67(1): 108-119.

[151] Lewis JD, Chen EZ, Baldassano RN, et al. Inflammation, Antibiotics, and Diet as Environmental Stressors of the Gut Microbiome in Pediatric Crohn's Disease. Cell Host & Microbe, 2015, 18(4): 489-500.

[152] Gorjifard S, Goldszmid RS. Microbiota-myeloid

cell crosstalk beyond the gut. Journal of Leukocyte Biology, 2016, 100(5): 865-879.

[153] Honda K, Littman DR. The microbiota in adaptive immune homeostasis and disease. Nature, 2016, 535(7610): 75-84.

[154] Ost KS, Round JL. Communication between the microbiota and mammalian immunity. Annual Review of Microbiology, 2018, 72: 399-422.

[155] Tanoue T, Morita S, Plichta DR, et al. A defined commensal consortium elicits CD8 T cells and anti-cancer immunity. Nature, 2019, 565(7741): 600-605.

[156] Umesaki Y, Setoyama H, Matsumoto S, et al. Differential roles of segmented filamentous bacteria and clostridia in development of the intestinal immune system. Infection And Immunity, 1999, 67(7): 3504-3511.

[157] Cash HL, Whitham CV, Behrendt CL, et al. Symbiotic bacteria direct expression of an intestinal bactericidal lectin. Science, 2006, 313(5790): 1126-1130.

[158] Cahenzli J, Koller Y, Wyss M, et al. Intestinal microbial diversity during early-life colonization shapes long-term IgE levels. Cell Host & Microbe, 2013, 14(5): 559-570.

[159] Atarashi K, Tanoue T, Oshima K, et al. Treg induction by a rationally selected mixture of Clostridia strains from the human microbiota. Nature, 2013, 500(7461): 232-236.

[160] Wright PF, Ackerman ME, Brickley EB. Mucosal Immunity: The Forgotten Arm of the Immune System. Journal of the Pediatric Infectious Diseases Society, 2019, 8(1): 53-54.

[161] Kabat AM, Srinivasan N, Maloy KJ. Modulation of immune development and function by intestinal microbiota. Trends in Immunology, 2014, 35(11): 507-517.

[162] Thaiss CA, Zmora N, Levy M, et al. The microbiome and innate immunity. Nature, 2016, 535(7610): 65-74.

[163] Francino MP. Early development of the gut microbiota and immune health. Pathogens, 2014, 3(3): 769-790.

[164] Fu W, Zhao J, Liu X, et al. The roles of the TLR/NFkappaB signaling pathway in the mutual interactions between the lung and the large intestine. Molecular Medicine Reports, 2018, 18(2): 1387-1394.

[165] Rakoff-Nahoum S, Paglino J, Eslami-Varzaneh F, et al. Recognition of commensal microflora by toll-like receptors is required for intestinal homeostasis. Cell, 2004, 118(2): 229-241.

[166] Elinav E, Strowig T, Kau AL, et al. NLRP6 inflammasome regulates colonic microbial ecology and risk for colitis. Cell, 2011, 145(5): 745-757.

[167] Clemente JC, Ursell LK, Parfrey LW, et al. The impact of the gut microbiota on human health: an integrative view. Cell, 2012, 148(6): 1258-1270.

[168] Rossi M, Bot A. The Th17 cell population and the immune homeostasis of the gastrointestinal tract. International Reviews of Immunology, 2013, 32(5-6): 471-474.

[169] Weaver CT, Elson CO, Fouser LA, et al. The Th17 pathway and inflammatory diseases of the intestines, lungs, and skin. Annual Review of Pathology, 2013, 8: 477-512.

[170] Tanoue T, Atarashi K, Honda K. Development and maintenance of intestinal regulatory T cells. Nature Reviews: Immunology, 2016, 16(5): 295-309.

[171] Collins CB, Aherne CM, Kominsky D, et al. Retinoic acid attenuates ileitis by restoring the balance between T-helper 17 and T regulatory cells. Gastroenterology, 2011, 141(5): 1821-1831.

[172] Wu HJ, Wu E. The role of gut microbiota in immune homeostasis and autoimmunity. Gut Microbes, 2012, 3(1): 4-14.

[173] Boyaka PN. Inducing Mucosal IgA: A Challenge for Vaccine Adjuvants and Delivery Systems. Journal of Immunology, 2017, 199(1): 9-16.

[174] Mantis NJ, Rol N, Corthesy B. Secretory IgA's complex roles in immunity and mucosal homeostasis in the gut. Mucosal Immunology, 2011, 4(6): 603-611.

[175] Hooper LV, Stappenbeck TS, Hong CV, et al. Angiogenins: a new class of microbicidal proteins involved in innate immunity. Nature Immunology, 2003, 4(3): 269-273.

[176] Salzman NH, Hung K, Haribhai D, et al. Enteric defensins are essential regulators of intestinal microbial ecology. Nature Immunology, 2010, 11(1): 76-83.

[177] Chamaillard M, Dessein R. Defensins couple dysbiosis to primary immunodeficiency in Crohn's disease. World Journal of Gastroenterology, 2011, 17(5): 567-571.

[178] Shreiner AB, Kao JY, Young VB. The gut microbiome in health and in disease. Current Opinion in Gastroenterology, 2015, 31(1): 69-75.

[179] Gordon JI, Dewey KG, Mills DA, et al. The

human gut microbiota and undernutrition. Science Translational Medicine, 2012, 4(137): 137ps112.

[180] Smith K, McCoy KD, Macpherson AJ. Use of axenic animals in studying the adaptation of mammals to their commensal intestinal microbiota. Seminars in Immunology, 2007, 19(2): 59-69.

[181] Deplancke B, Gaskins HR. Microbial modulation of innate defense: goblet cells and the intestinal mucus layer. American Journal of Clinical Nutrition, 2001, 73(6): 1131S-1141S.

[182] Manolios N, Geczy CL, Schrieber L. High endothelial venule morphology and function are inducible in germ-free mice: a possible role for interferon-gamma. Cellular Immunology, 1988, 117(1): 136-151.

[183] Lecuyer E, Rakotobe S, Lengline-Garnier H, et al. Segmented filamentous bacterium uses secondary and tertiary lymphoid tissues to induce gut IgA and specific T helper 17 cell responses. Immunity, 2014, 40(4): 608-620.

[184] Pabst O, Herbrand H, Friedrichsen M, et al. Adaptation of solitary intestinal lymphoid tissue in response to microbiota and chemokine receptor CCR7 signaling. Journal of Immunology, 2006, 177(10): 6824-6832.

[185] Bouskra D, Brezillon C, Berard M, et al. Lymphoid tissue genesis induced by commensals through NOD1 regulates intestinal homeostasis. Nature, 2008, 456(7221): 507-510.

[186] Macpherson AJ, Martinic MM, Harris N. The functions of mucosal T cells in containing the indigenous commensal flora of the intestine. Cellular And Molecular Life Sciences, 2002, 59(12): 2088-2096.

[187] Macpherson AJ, Hunziker L, McCoy K, et al. IgA responses in the intestinal mucosa against pathogenic and non-pathogenic microorganisms. Microbes And Infection, 2001, 3(12): 1021-1035.

[188] Ivanov, II, Littman DR. Segmented filamentous bacteria take the stage. Mucosal Immunology, 2010, 3(3): 209-212.

[189] Trompette A, Gollwitzer ES, Yadava K, et al. Gut microbiota metabolism of dietary fiber influences allergic airway disease and hematopoiesis. Nature Medicine, 2014, 20(2): 159-166.

[190] Bauer H, Horowitz RE, Levenson SM, et al. The response of the lymphatic tissue to the microbial flora. Studies on germfree mice. American Journal of Pathology, 1963, 42: 471-483.

[191] Benveniste J, Lespinats G, Adam C, et al. Immunoglobulins in intact, immunized, and contaminated axenic mice: study of serum IgA. Journal of Immunology, 1971, 107(6): 1647-1655.

[192] Mazmanian SK, Liu CH, Tzianabos AO, et al. An immunomodulatory molecule of symbiotic bacteria directs maturation of the host immune system. Cell, 2005, 122(1): 107-118.

[193] Soderholm AT, Pedicord VA. Intestinal epithelial cells: at the interface of the microbiota and mucosal immunity. Immunology, 2019, 158(4): 267-280.

[194] Pott J, Hornef M. Innate immune signalling at the intestinal epithelium in homeostasis and disease. EMBO Reports, 2012, 13(8): 684-698.

[195] Wells JM, Rossi O, Meijerink M, et al. Epithelial crosstalk at the microbiota-mucosal interface. Proceedings of the National Academy of Sciences of the United States of America, 2011, 108 Suppl 1: 4607-4614.

[196] Hu S, Peng L, Kwak YT, et al. The DNA Sensor AIM2 Maintains Intestinal Homeostasis via Regulation of Epithelial Antimicrobial Host Defense. Cell Reports, 2015, 13(9): 1922-1936.

[197] Nenci A, Becker C, Wullaert A, et al. Epithelial NEMO links innate immunity to chronic intestinal inflammation. Nature, 2007, 446(7135): 557-561.

[198] Vaishnava S, Yamamoto M, Severson KM, et al. The antibacterial lectin RegIIIgamma promotes the spatial segregation of microbiota and host in the intestine. Science, 2011, 334(6053): 255-258.

[199] Vlantis K, Polykratis A, Welz PS, et al. TLR-independent anti-inflammatory function of intestinal epithelial TRAF6 signalling prevents DSS-induced colitis in mice. Gut, 2016, 65(6): 935-943.

[200] Dannappel M, Vlantis K, Kumari S, et al. RIPK1 maintains epithelial homeostasis by inhibiting apoptosis and necroptosis. Nature, 2014, 513(7516): 90-94.

[201] Gunther C, Martini E, Wittkopf N, et al. Caspase-8 regulates TNF-alpha-induced epithelial necroptosis and terminal ileitis. Nature, 2011, 477(7364): 335-339.

[202] Welz PS, Wullaert A, Vlantis K, et al. FADD prevents RIP3-mediated epithelial cell necrosis and chronic intestinal inflammation. Nature, 2011, 477(7364): 330-334.

[203] Burgueno JF, Fritsch J, Gonzalez EE, et al. Epithelial TLR4 signaling activates duox2 to induce microbiota-driven tumorigenesis. Gastroenterology, 2021, 160(3): 797-808. e6.

[204] Couturier-Maillard A, Secher T, Rehman A, et al. NOD2-mediated dysbiosis predisposes mice to transmissible colitis and colorectal cancer. Journal of Clinical Investigation, 2013, 123(2): 700-711.

[205] Nigro G, Rossi R, Commere PH, et al. The cytosolic bacterial peptidoglycan sensor Nod2 affords stem cell protection and links microbes to gut epithelial regeneration. Cell Host & Microbe, 2014, 15(6): 792-798.

[206] Ramanan D, Tang MS, Bowcutt R, et al. Bacterial sensor Nod2 prevents inflammation of the small intestine by restricting the expansion of the commensal Bacteroides vulgatus. Immunity, 2014, 41(2): 311-324.

[207] Sellin ME, Muller AA, Felmy B, et al. Epithelium-intrinsic NAIP/NLRC4 inflammasome drives infected enterocyte expulsion to restrict Salmonella replication in the intestinal mucosa. Cell Host & Microbe, 2014, 16(2): 237-248.

[208] Hu B, Elinav E, Huber S, et al. Inflammation-induced tumorigenesis in the colon is regulated by caspase-1 and NLRC4. Proceedings of the National Academy of Sciences of the United States of America, 2010, 107(50): 21635-21640.

[209] Levy M, Thaiss CA, Zeevi D, et al. Microbiota-Modulated Metabolites Shape the Intestinal Microenvironment by Regulating NLRP6 Inflammasome Signaling. Cell, 2015, 163(6): 1428-1443.

[210] Wlodarska M, Thaiss CA, Nowarski R, et al. NLRP6 inflammasome orchestrates the colonic host-microbial interface by regulating goblet cell mucus secretion. Cell, 2014, 156(5): 1045-1059.

[211] Wang P, Zhu S, Yang L, et al. Nlrp6 regulates intestinal antiviral innate immunity. Science, 2015, 350(6262): 826-830.

[212] Macia L, Tan J, Vieira AT, et al. Metabolite-sensing receptors GPR43 and GPR109A facilitate dietary fibre-induced gut homeostasis through regulation of the inflammasome. Nature Communications, 2015, 6: 6734.

[213] Kelly CJ, Zheng L, Campbell EL, et al. Crosstalk between Microbiota-Derived Short-Chain Fatty Acids and Intestinal Epithelial HIF Augments Tissue Barrier Function. Cell Host & Microbe, 2015, 17(5): 662-671.

[214] Venkatesh M, Mukherjee S, Wang H, et al. Symbiotic bacterial metabolites regulate gastrointestinal barrier function via the xenobiotic sensor PXR and Toll-like receptor 4. Immunity, 2014, 41(2): 296-310.

[215] Zarrinpar A, Chaix A, Yooseph S, et al. Diet and feeding pattern affect the diurnal dynamics of the gut microbiome. Cell Metabolism, 2014, 20(6): 1006-1017.

[216] Mukherji A, Kobiita A, Ye T, et al. Homeostasis in intestinal epithelium is orchestrated by the circadian clock and microbiota cues transduced by TLRs. Cell, 2013, 153(4): 812-827.

[217] Pickard JM, Zeng MY, Caruso R, et al. Gut microbiota: Role in pathogen colonization, immune responses, and inflammatory disease. Immunological Reviews, 2017, 279(1): 70-89.

[218] Josefsdottir KS, Baldridge MT, Kadmon CS, et al. Antibiotics impair murine hematopoiesis by depleting the intestinal microbiota. Blood, 2017, 129(6): 729-739.

[219] Fiedler K, Kokai E, Bresch S, et al. MyD88 is involved in myeloid as well as lymphoid hematopoiesis independent of the presence of a pathogen. American Journal of Blood Research, 2013, 3(2): 124-140.

[220] Balmer ML, Schurch CM, Saito Y, et al. Microbiota-derived compounds drive steady-state granulopoiesis via MyD88/TICAM signaling. Journal of Immunology, 2014, 193(10): 5273-5283.

[221] Deshmukh HS, Liu Y, Menkiti OR, et al. The microbiota regulates neutrophil homeostasis and host resistance to Escherichia coli K1 sepsis in neonatal mice. Nature Medicine, 2014, 20(5): 524-530.

[222] Zhang D, Chen G, Manwani D, et al. Neutrophil ageing is regulated by the microbiome. Nature, 2015, 525(7570): 528-532.

[223] Burberry A, Zeng MY, Ding L, et al. Infection mobilizes hematopoietic stem cells through cooperative NOD-like receptor and Toll-like receptor signaling. Cell Host & Microbe, 2014, 15(6): 779-791.

[224] Upparahalli Venkateshaiah S, Manohar M, Kandikattu HK, et al. Experimental Modeling of Eosinophil-Associated Diseases. Methods in Molecular Biology, 2021, 2241: 275-291.

[225] Carlens J, Wahl B, Ballmaier M, et al. Common gamma-chain-dependent signals confer selective survival of eosinophils in the murine small intestine. Journal of Immunology, 2009, 183(9): 5600-5607.

[226] Verjan Garcia N, Umemoto E, Saito Y, et al. SIRPalpha/CD172a regulates eosinophil homeostasis. Journal of Immunology, 2011, 187(5): 2268-2277.

[227] Yang BG, Seoh JY, Jang MH. Regulatory eosinophils

in inflammation and metabolic disorders. Immune Network, 2017, 17(1): 41-47.

[228] Jimenez-Saiz R, Anipindi VC, Galipeau H, et al. Microbial regulation of enteric eosinophils and its impact on tissue remodeling and Th2 immunity. Frontiers in Immunology, 2020, 11: 155.

[229] Ravin KA, Loy M. The Eosinophil in Infection. Clinical Reviews in Allergy & Immunology, 2016, 50(2): 214-227.

[230] Yousefi S, Gold JA, Andina N, et al. Catapult-like release of mitochondrial DNA by eosinophils contributes to antibacterial defense. Nature Medicine, 2008, 14(9): 949-953.

[231] Masterson JC, Menard-Katcher C, Larsen LD, et al. Heterogeneity of intestinal tissue eosinophils: potential considerations for next-generation eosinophil-targeting strategies. Cells, 2021, 10(2): 426.

[232] Davies LC, Taylor PR. Tissue-resident macrophages: then and now. Immunology, 2015, 144(4): 541-548.

[233] Gross M, Salame TM, Jung S. Guardians of the gut-murine intestinal macrophages and dendritic cells. Frontiers in Immunology, 2015, 6: 254.

[234] Bain CC, Bravo-Blas A, Scott CL, et al. Constant replenishment from circulating monocytes maintains the macrophage pool in the intestine of adult mice. Nature Immunology, 2014, 15(10): 929-937.

[235] Bain CC, Scott CL, Uronen-Hansson H, et al. Resident and pro-inflammatory macrophages in the colon represent alternative context-dependent fates of the same Ly6Chi monocyte precursors. Mucosal Immunology, 2013, 6(3): 498-510.

[236] Mortha A, Chudnovskiy A, Hashimoto D, et al. Microbiota-dependent crosstalk between macrophages and ILC3 promotes intestinal homeostasis. Science, 2014, 343(6178): 1249288.

[237] Chang PV, Hao L, Offermanns S, et al. The microbial metabolite butyrate regulates intestinal macrophage function via histone deacetylase inhibition. Proceedings of the National Academy of Sciences of the United States of America, 2014, 111(6): 2247-2252.

[238] Scott NA, Mann ER. Regulation of mononuclear phagocyte function by the microbiota at mucosal sites. Immunology, 2020, 159(1): 26-38.

[239] Shaw MH, Kamada N, Kim YG, et al. Microbiota-induced IL-1beta, but not IL-6, is critical for the development of steady-state TH17 cells in the intestine. Journal of Experimental Medicine, 2012, 209(2): 251-258.

[240] Franchi L, Kamada N, Nakamura Y, et al. NLRC4-driven production of IL-1beta discriminates between pathogenic and commensal bacteria and promotes host intestinal defense. Nature Immunology, 2012, 13(5): 449-456.

[241] Seo SU, Kamada N, Munoz-Planillo R, et al. Distinct commensals induce interleukin-1beta via NLRP3 Inflammasome in Inflammatory monocytes to promote intestinal inflammation in response to injury. Immunity, 2015, 42(4): 744-755.

[242] Zhu J, Luo L, Tian L, et al. Aryl hydrocarbon receptor promotes IL-10 expression in inflammatory macrophages through src-STAT3 signaling pathway. Frontiers in Immunology, 2018, 9: 2033.

[243] Kim M, Kim CH. Colonization and effector functions of innate lymphoid cells in mucosal tissues. Microbes And Infection, 2016, 18(10): 604-614.

[244] Kim CH, Hashimoto-Hill S, Kim M. Migration and tissue tropism of innate lymphoid cells. Trends in Immunology, 2016, 37(1): 68-79.

[245] Panda SK, Colonna M. Innate lymphoid cells in mucosal immunity. Frontiers in Immunology, 2019, 10: 861.

[246] Tsuji M, Suzuki K, Kitamura H, et al. Requirement for lymphoid tissue-inducer cells in isolated follicle formation and T cell-independent immunoglobulin A generation in the gut. Immunity, 2008, 29(2): 261-271.

[247] Crellin NK, Trifari S, Kaplan CD, et al. Regulation of cytokine secretion in human CD127(+) LTi-like innate lymphoid cells by Toll-like receptor 2. Immunity, 2010, 33(5): 752-764.

[248] Sonnenberg GF, Artis D. Innate lymphoid cell interactions with microbiota: implications for intestinal health and disease. Immunity, 2012, 37(4): 601-610.

[249] Cording S, Medvedovic J, Cherrier M, et al. Development and regulation of RORgammat(+) innate lymphoid cells. FEBS Letters, 2014, 588(22): 4176-4181.

[250] Wang G, Huang S, Wang Y, et al. Bridging intestinal immunity and gut microbiota by metabolites. Cellular And Molecular Life Sciences, 2019, 76(20): 3917-3937.

[251] Lamas B, Richard ML, Leducq V, et al. CARD9 impacts colitis by altering gut microbiota metabolism of tryptophan into aryl hydrocarbon receptor ligands. Nature Medicine, 2016, 22(6): 598-605.

[252] Lindemans CA, Calafiore M, Mertelsmann AM, et al. Interleukin-22 promotes intestinal-stem-cell-mediated epithelial regeneration. Nature, 2015, 528(7583): 560-564.

[253] Goto Y, Obata T, Kunisawa J, et al. Innate lymphoid cells regulate intestinal epithelial cell glycosylation. Science, 2014, 345(6202): 1254009.

[254] Beck K, Ohno H, Satoh-Takayama N. Innate Lymphoid Cells: Important Regulators of Host-Bacteria Interaction for Border Defense. Microorganisms, 2020, 8(9): 1342.

[255] Castellanos JG, Woo V, Viladomiu M, et al. Microbiota-Induced TNF-like Ligand 1A Drives Group 3 Innate Lymphoid Cell-Mediated Barrier Protection and Intestinal T Cell Activation during Colitis. Immunity, 2018, 49(6): 1077-1089. e1075.

[256] Hasegawa M, Yada S, Liu MZ, et al. Interleukin-22 regulates the complement system to promote resistance against pathobionts after pathogen-induced intestinal damage. Immunity, 2014, 41(4): 620-632.

[257] Sakamoto K, Kim YG, Hara H, et al. IL-22 Controls Iron-Dependent Nutritional Immunity Against Systemic Bacterial Infections. Science Immunology, 2017, 2(8): eaai8371.

[258] Ivanov, II, Frutos Rde L, Manel N, et al. Specific microbiota direct the differentiation of IL-17-producing T-helper cells in the mucosa of the small intestine. Cell Host & Microbe, 2008, 4(4): 337-349.

[259] Wojciech L, Tan KSW, Gascoigne NRJ. Taming the Sentinels: Microbiome-Derived Metabolites and Polarization of T Cells. International Journal of Molecular Sciences, 2020, 21(20): 7740.

[260] Miossec P, Kolls JK. Targeting IL-17 and TH17 cells in chronic inflammation. Nature Reviews: Drug Discovery, 2012, 11(10): 763-776.

[261] Atarashi K, Tanoue T, Ando M, et al. Th17 Cell Induction by Adhesion of Microbes to Intestinal Epithelial Cells. Cell, 2015, 163(2): 367-380.

[262] Sano T, Huang W, Hall JA, et al. An IL-23R/IL-22 Circuit Regulates Epithelial Serum Amyloid A to Promote Local Effector Th17 Responses. Cell, 2016, 164(1-2): 324.

[263] Michaelis L, Treß M, Löw H-C, et al. Gut commensal-induced IκBζ expression in dendritic cells Influences the Th17 Response. Frontiers in Immunology, 2021, 11: 612336.

[264] Luu M, Pautz S, Kohl V, et al. The short-chain fatty acid pentanoate suppresses autoimmunity by modulating the metabolic-epigenetic crosstalk in lymphocytes. Nature Communications, 2019, 10(1): 760.

[265] Smirnova A, Wincent E, Vikstrom Bergander L, et al. Evidence for new light-independent pathways for generation of the endogenous aryl hydrocarbon receptor agonist FICZ. Chemical Research in Toxicology, 2016, 29(1): 75-86.

[266] Colliou N, Ge Y, Sahay B, et al. Commensal propionibacterium strain UF1 mitigates intestinal inflammation via Th17 cell regulation. Journal of Clinical Investigation, 2017, 127(11): 3970-3986.

[267] Gopalakrishnan V, Spencer CN, Nezi L, et al. Gut microbiome modulates response to anti-PD-1 immunotherapy in melanoma patients. Science, 2018, 359(6371): 97-103.

[268] Hurtado CG, Wan F, Housseau F, et al. Roles for Interleukin 17 and Adaptive Immunity in Pathogenesis of Colorectal Cancer. Gastroenterology, 2018, 155(6): 1706-1715.

[269] Sujino T, London M, Hoytema van Konijnenburg DP, et al. Tissue adaptation of regulatory and intraepithelial CD4(+) T cells controls gut inflammation. Science, 2016, 352(6293): 1581-1586.

[270] Kraj P, Ignatowicz L. The mechanisms shaping the repertoire of CD4(+) Foxp3(+) regulatory T cells. Immunology, 2018, 153(3): 290-296.

[271] Round JL, Mazmanian SK. Inducible Foxp3+ regulatory T-cell development by a commensal bacterium of the intestinal microbiota. Proceedings of the National Academy of Sciences of the United States of America, 2010, 107(27): 12204-12209.

[272] Round JL, Lee SM, Li J, et al. The Toll-like receptor 2 pathway establishes colonization by a commensal of the human microbiota. Science, 2011, 332(6032): 974-977.

[273] Smith PM, Howitt MR, Panikov N, et al. The microbial metabolites, short-chain fatty acids, regulate colonic Treg cell homeostasis. Science, 2013, 341(6145): 569-573.

[274] Furusawa Y, Obata Y, Fukuda S, et al. Commensal microbe-derived butyrate induces the differentiation of colonic regulatory T cells. Nature, 2013, 504(7480): 446-450.

[275] Sefik E, Geva-Zatorsky N, Oh S, et al. MUCOSAL IMMUNOLOGY. Individual intestinal symbionts induce a distinct population of RORgamma(+) regulatory T cells. Science, 2015, 349(6251): 993-997.

[276] Botia-Sanchez M, Alarcon-Riquelme ME, Galicia

G.B cells and microbiota in autoimmunity. International Journal of Molecular Sciences, 2021, 22(9): 4846.

[277] Brandtzaeg P, Farstad IN, Johansen FE, et al. The B-cell system of human mucosae and exocrine glands. Immunological Reviews, 1999, 171: 45-87.

[278] Wesemann DR, Portuguese AJ, Meyers RM, et al. Microbial colonization influences early B-lineage development in the gut lamina propria. Nature, 2013, 501(7465): 112-115.

[279] Suzuki M, Sujino T, Chiba S, et al. Host-microbe cross-talk governs amino acid chirality to regulate survival and differentiation of B cells. Science Advances, 2021, 7(10): eabd6480.

[280] Li H, Limenitakis JP, Greiff V, et al. Mucosal or systemic microbiota exposures shape the B cell repertoire. Nature, 2020, 584(7820): 274-278.

[281] Chen H, Zhang Y, Ye AY, et al. BCR selection and affinity maturation in Peyer's patch germinal centres. Nature, 2020, 582(7812): 421-425.

[282] Rosser EC, Oleinika K, Tonon S, et al. Regulatory B cells are induced by gut microbiota-driven interleukin-1beta and interleukin-6 production. Nature Medicine, 2014, 20(11): 1334-1339.

[283] Mu Q, Edwards MR, Swartwout BK, et al. Gut Microbiota and Bacterial DNA Suppress Autoimmunity by Stimulating Regulatory B Cells in a Murine Model of Lupus. Frontiers in Immunology, 2020, 11: 593353.

[284] Daien CI, Tan J, Audo R, et al. Gut-derived acetate promotes B10 cells with antiinflammatory effects. JCI Insight, 2021, 6(7): e144156.

[285] Li Y, Jin L, Chen T. The Effects of Secretory IgA in the Mucosal Immune System. Biomed Research International, 2020, 2020: 2032057.

[286] Hirota K, Turner JE, Villa M, et al. Plasticity of Th17 cells in Peyer's patches is responsible for the induction of T cell-dependent IgA responses. Nature Immunology, 2013, 14(4): 372-379.

[287] Grasset EK, Chorny A, Casas-Recasens S, et al. Gut T cell-independent IgA responses to commensal bacteria require engagement of the TACI receptor on B cells. Science Immunology, 2020, 5(49): eaat7117.

[288] Proietti M, Perruzza L, Scribano D, et al. ATP released by intestinal bacteria limits the generation of protective IgA against enteropathogens. Nature Communications, 2019, 10(1): 250.

[289] Gould HJ, Sutton BJ. IgE in allergy and asthma today. Nature Reviews: Immunology, 2008, 8(3): 205-217.

[290] Campbell E, Hesser LA, Nagler CR. B cells and the microbiota: a missing connection in food allergy. Mucosal Immunology, 2021, 14(1): 4-13.

[291] Wyss M, Brown K, Thomson CA, et al. Using Precisely Defined in vivo Microbiotas to Understand Microbial Regulation of IgE. Frontiers in Immunology, 2019, 10: 3107.

[292] Feehley T, Plunkett CH, Bao R, et al. Healthy infants harbor intestinal bacteria that protect against food allergy. Nature Medicine, 2019, 25(3): 448-453.

[293] Vidarsson G, Dekkers G, Rispens T. IgG subclasses and allotypes: from structure to effector functions. Frontiers in Immunology, 2014, 5: 520.

[294] Castro-Dopico T, Clatworthy MR. IgG and Fcgamma Receptors in Intestinal Immunity and Inflammation. Frontiers in Immunology, 2019, 10: 805.

[295] Macpherson AJ, Gatto D, Sainsbury E, et al. A primitive T cell-independent mechanism of intestinal mucosal IgA responses to commensal bacteria. Science, 2000, 288(5474): 2222-2226.

[296] Macpherson A, Khoo UY, Forgacs I, et al. Mucosal antibodies in inflammatory bowel disease are directed against intestinal bacteria. Gut, 1996, 38(3): 365-375.

[297] Zeng MY, Cisalpino D, Varadarajan S, et al. Gut Microbiota-Induced Immunoglobulin G Controls Systemic Infection by Symbiotic Bacteria and Pathogens. Immunity, 2016, 44(3): 647-658.

[298] Pone EJ, Zhang J, Mai T, et al. BCR-signalling synergizes with TLR-signalling for induction of AID and immunoglobulin class-switching through the non-canonical NF-kappaB pathway. Nature Communications, 2012, 3: 767.

[299] Amadou Amani S, Lang ML. Bacteria That Cause Enteric Diseases Stimulate Distinct Humoral Immune Responses. Frontiers in Immunology, 2020, 11: 565648.

[300] Yoshida M, Claypool SM, Wagner JS, et al. Human neonatal Fc receptor mediates transport of IgG into luminal secretions for delivery of antigens to mucosal dendritic cells. Immunity, 2004, 20(6): 769-783.

[301] Yoshida M, Kobayashi K, Kuo TT, et al. Neonatal Fc receptor for IgG regulates mucosal immune responses to luminal bacteria. Journal of Clinical Investigation, 2006, 116(8): 2142-2151.

[302] Bolognini D, Dedeo D, Milligan G. Metabolic and inflammatory functions of short-chain fatty acid receptors. Current Opinion in Endocrine and Metabolic Research, 2021, 16: 1-9.

[303] Zheng D, Liwinski T, Elinav E. Interaction between

microbiota and immunity in health and disease. Cell Research, 2020, 30(6): 492-506.

[304] Vijay-Kumar M, Aitken JD, Carvalho FA, et al. Metabolic syndrome and altered gut microbiota in mice lacking Toll-like receptor 5. Science, 2010, 328(5975): 228-231.

[305] Chassaing B, Ley RE, Gewirtz AT. Intestinal epithelial cell toll-like receptor 5 regulates the intestinal microbiota to prevent low-grade inflammation and metabolic syndrome in mice. Gastroenterology, 2014, 147(6): 1363-1377. e1317.

[306] Birchenough GM, Nystrom E , Johansson ME, et al. A sentinel goblet cell guards the colonic crypt by triggering Nlrp6-dependent Muc2 secretion. Science, 2016, 352(6293): 1535-1542.

[307] Ubeda C, Lipuma L, Gobourne A, et al. Familial transmission rather than defective innate immunity shapes the distinct intestinal microbiota of TLR-deficient mice. Journal of Experimental Medicine, 2012, 209(8): 1445-1456.

[308] Guo H, Gibson SA, Ting JPY. Gut microbiota, NLR proteins, and intestinal homeostasis. Journal of Experimental Medicine, 2020, 217(10): e20181832.

[309] Volk JK, Nystrom EEL, van der Post S, et al. The Nlrp6 inflammasome is not required for baseline colonic inner mucus layer formation or function. Journal of Experimental Medicine, 2019, 216(11): 2602-2618.

[310] Salzman NH, Bevins CL. Dysbiosis--a consequence of Paneth cell dysfunction. Seminars in Immunology, 2013, 25(5): 334-341.

[311] Lee YK, Mazmanian SK. Has the microbiota played a critical role in the evolution of the adaptive immune system? Science, 2010, 330(6012): 1768-1773.

[312] Zhang H, Sparks JB, Karyala SV, et al. Host adaptive immunity alters gut microbiota. ISME Journal, 2015, 9(3): 770-781.

[313] Kato LM, Kawamoto S, Maruya M, et al. The role of the adaptive immune system in regulation of gut microbiota. Immunological Reviews, 2014, 260(1): 67-75.

[314] Kawamoto S, Maruya M, Kato LM, et al. Foxp3(+) T cells regulate immunoglobulin a selection and facilitate diversification of bacterial species responsible for immune homeostasis. Immunity, 2014, 41(1): 152-165.

[315] Cong Y, Feng T, Fujihashi K, et al. A dominant, coordinated T regulatory cell-IgA response to the intestinal microbiota. Proceedings of the National Academy of Sciences of the United States of America, 2009, 106(46): 19256-19261.

[316] Neumann C, Blume J, Roy U, et al. c-Maf-dependent Treg cell control of intestinal TH17 cells and IgA establishes host-microbiota homeostasis. Nature Immunology, 2019, 20(4): 471-481.

[317] Hegazy AN, West NR, Stubbington MJT, et al. Circulating and Tissue-Resident CD4(+) T Cells With Reactivity to Intestinal Microbiota Are Abundant in Healthy Individuals and Function Is Altered During Inflammation. Gastroenterology, 2017, 153(5): 1320-1337. e1316.

[318] Fagarasan S, Muramatsu M, Suzuki K, et al. Critical roles of activation-induced cytidine deaminase in the homeostasis of gut flora. Science, 2002, 298(5597): 1424-1427.

[319] Suzuki K, Meek B, Doi Y, et al. Aberrant expansion of segmented filamentous bacteria in IgA-deficient gut. Proceedings of the National Academy of Sciences of the United States of America, 2004, 101(7): 1981-1986.

[320] Kawamoto S, Tran TH, Maruya M, et al. The inhibitory receptor PD-1 regulates IgA selection and bacterial composition in the gut. Science, 2012, 336(6080): 485-489.

[321] Kubinak JL, Petersen C, Stephens WZ, et al. MyD88 signaling in T cells directs IgA-mediated control of the microbiota to promote health. Cell Host & Microbe, 2015, 17(2): 153-163.

[322] Mirpuri J, Raetz M, Sturge CR, et al. Proteobacteria-specific IgA regulates maturation of the intestinal microbiota. Gut Microbes, 2014, 5(1): 28-39.

[323] Peterson DA, McNulty NP, Guruge JL, et al. IgA response to symbiotic bacteria as a mediator of gut homeostasis. Cell Host & Microbe, 2007, 2(5): 328-339.

[324] Corthesy B. Multi-faceted functions of secretory IgA at mucosal surfaces. Frontiers in Immunology, 2013, 4: 185.

[325] Cullender TC, Chassaing B, Janzon A, et al. Innate and adaptive immunity interact to quench microbiome flagellar motility in the gut. Cell Host & Microbe, 2013, 14(5): 571-581.

[326] Gordon HA, Pesti L. The gnotobiotic animal as a tool in the study of host microbial relationships. Bacteriological Reviews, 1971, 35(4): 390-429.

[327] Cahana I, Iraqi FA. Impact of host genetics on gut microbiome: Take-home lessons from human and mouse studies. Animal Models And Experimental

Medicine, 2020, 3(3): 229-236.

[328] Cebra JJ. Influences of microbiota on intestinal immune system development. American Journal of Clinical Nutrition, 1999, 69(5): 1046S-1051S.

[329] Taurog JD, Richardson JA, Croft JT, et al. The germfree state prevents development of gut and joint inflammatory disease in HLA-B27 transgenic rats. Journal of Experimental Medicine, 1994, 180(6): 2359-2364.

[330] Dianda L, Hanby AM, Wright NA, et al. T cell receptor-alpha beta-deficient mice fail to develop colitis in the absence of a microbial environment. American Journal of Pathology, 1997, 150(1): 91-97.

[331] Garrett WS, Lord GM, Punit S, et al. Communicable ulcerative colitis induced by T-bet deficiency in the innate immune system. Cell, 2007, 131(1): 33-45.

[332] Ahern PP, Faith JJ, Gordon JI. Mining the human gut microbiota for effector strains that shape the immune system. Immunity, 2014, 40(6): 815-823.

[333] Faith JJ, Ahern PP, Ridaura VK, et al. Identifying gut microbe-host phenotype relationships using combinatorial communities in gnotobiotic mice. Science Translational Medicine, 2014, 6(220): 220ra211.

[334] Geva-Zatorsky N, Sefik E, Kua L, et al. Mining the Human Gut Microbiota for Immunomodulatory Organisms. Cell, 2017, 168(5): 928-943. e11.

[335] Chai JN, Peng Y, Rengarajan S, et al. Helicobacter species are potent drivers of colonic T cell responses in homeostasis and inflammation. Science Immunology, 2017, 2(13): eaal5068.

[336] Xu M, Pokrovskii M, Ding Y, et al. c-MAF-dependent regulatory T cells mediate immunological tolerance to a gut pathobiont. Nature, 2018, 554(7692): 373-377.

[337] Rooks MG, Garrett WS. Gut microbiota, metabolites and host immunity. Nature Reviews: Immunology, 2016, 16(6): 341-352.

[338] Koh A, De Vadder F, Kovatcheva-Datchary P, et al. From Dietary Fiber to Host Physiology: Short-Chain Fatty Acids as Key Bacterial Metabolites. Cell, 2016, 165(6): 1332-1345.

[339] Maslowski KM, Vieira AT, Ng A, et al. Regulation of inflammatory responses by gut microbiota and chemoattractant receptor GPR43. Nature, 2009, 461(7268): 1282-1286.

[340] Marino E, Richards JL, McLeod KH, et al. Gut microbial metabolites limit the frequency of autoimmune T cells and protect against type 1 diabetes. Nature Immunology, 2017, 18(5): 552-562.

[341] Trompette A, Gollwitzer ES, Pattaroni C, et al. Dietary Fiber Confers Protection against Flu by Shaping Ly6c(-) Patrolling Monocyte Hematopoiesis and CD8(+) T Cell Metabolism. Immunity, 2018, 48(5): 992-1005. e1008.

[342] Cleophas MCP, Ratter JM, Bekkering S, et al. Effects of oral butyrate supplementation on inflammatory potential of circulating peripheral blood mononuclear cells in healthy and obese males. Scientific Reports, 2019, 9(1): 775.

[343] Maslowski KM, Mackay CR. Diet, gut microbiota and immune responses. Nature Immunology, 2011, 12(1): 5-9.

[344] Sonnenburg ED, Sonnenburg JL. Starving our microbial self: the deleterious consequences of a diet deficient in microbiota-accessible carbohydrates. Cell Metabolism, 2014, 20(5): 779-786.

[345] Thorburn AN, Macia L, Mackay CR. Diet, metabolites, and "western-lifestyle" inflammatory diseases. Immunity, 2014, 40(6): 833-842.

[346] King DE, Mainous AG, 3rd, Lambourne CA. Trends in dietary fiber intake in the United States, 1999-2008. Journal of the Academy of Nutrition and Dietetics, 2012, 112(5): 642-648.

[347] Park Y, Subar AF, Hollenbeck A, et al. Dietary fiber intake and mortality in the NIH-AARP diet and health study. Archives of Internal Medicine, 2011, 171(12): 1061-1068.

[348] Desai MS, Seekatz AM, Koropatkin NM, et al. A Dietary Fiber-Deprived Gut Microbiota Degrades the Colonic Mucus Barrier and Enhances Pathogen Susceptibility. Cell, 2016, 167(5): 1339-1353. e1321.

[349] Dodd D, Spitzer MH, Van Treuren W, et al. A gut bacterial pathway metabolizes aromatic amino acids into nine circulating metabolites. Nature, 2017, 551(7682): 648-652.

[350] Faith JJ, Rey FE, O'Donnell D, et al. Creating and characterizing communities of human gut microbes in gnotobiotic mice. ISME Journal, 2010, 4(9): 1094-1098.

[351] Turnbaugh PJ, Hamady M, Yatsunenko T, et al. A core gut microbiome in obese and lean twins. Nature, 2009, 457(7228): 480-484.

[352] Blanton LV, Charbonneau MR, Salih T, et al. Gut bacteria that prevent growth impairments transmitted by microbiota from malnourished children. Science, 2016, 351(6275): 10.1126/science.aad3311 aad3311.

[353] Britton GJ, Contijoch EJ, Mogno I, et al. Microbiotas

from Humans with Inflammatory Bowel Disease Alter the Balance of Gut Th17 and RORgammat(+) Regulatory T Cells and Exacerbate Colitis in Mice. Immunity, 2019, 50(1): 212-224. e214.

[354] Blanton LV, Barratt MJ, Charbonneau MR, et al. Childhood undernutrition, the gut microbiota, and microbiota-directed therapeutics. Science, 2016, 352(6293): 1533.

[355] Trehan I, Goldbach HS, LaGrone LN, et al. Antibiotics as part of the management of severe acute malnutrition. Malawi Medical Journal, 2016, 28(3): 123-130.

[356] Nguyen TL, Vieira-Silva S, Liston A, et al. How informative is the mouse for human gut microbiota research? Disease Models & Mechanisms, 2015, 8(1): 1-16.

[357] Perlman RL. Mouse models of human disease: An evolutionary perspective. Evolution Medicine and Public Health, 2016, 2016(1): 170-176.

[358] Lloyd-Price J, Mahurkar A, Rahnavard G, et al. Strains, functions and dynamics in the expanded Human Microbiome Project. Nature, 2017, 550(7674): 61-66.

[359] McDonald D, Hyde E, Debelius JW, et al. American Gut: an Open Platform for Citizen Science Microbiome Research. mSystems, 2018, 3(3): e00031-00018.

[360] Lewis JD, Albenberg L, Lee D, et al. The Importance and Challenges of Dietary Intervention Trials for Inflammatory Bowel Disease. Inflammatory Bowel Diseases, 2017, 23(2): 181-191.

[361] Zeevi D, Korem T, Zmora N, et al. Personalized Nutrition by Prediction of Glycemic Responses. Cell, 2015, 163(5): 1079-1094.

[362] Zhao L, Zhang F, Ding X, et al. Gut bacteria selectively promoted by dietary fibers alleviate type 2 diabetes. Science, 2018, 359(6380): 1151-1156.

[363] Cotillard A, Kennedy SP, Kong LC, et al. Dietary intervention impact on gut microbial gene richness. Nature, 2013, 500(7464): 585-588.

[364] Schirmer M, Smeekens SP, Vlamakis H, et al. Linking the Human Gut Microbiome to Inflammatory Cytokine Production Capacity. Cell, 2016, 167(4): 1125-1136. e1128.

[365] Thomas S, Rouilly V, Patin E, et al. The Milieu Interieur study - an integrative approach for study of human immunological variance. Clinical Immunology, 2015, 157(2): 277-293.

[366] Hasin Y, Seldin M, Lusis A. Multi-omics approaches to disease. Genome Biology, 2017, 18(1): 83.

[367] Papalexi E, Satija R. Single-cell RNA sequencing to explore immune cell heterogeneity. Nature Reviews: Immunology, 2018, 18(1): 35-45.

[368] Olsen LR, Pedersen CB, Leipold MD, et al. Getting the Most from Your High-Dimensional Cytometry Data. Immunity, 2019, 50(3): 535-536.

[369] Spitzer MH, Nolan GP. Mass Cytometry: Single Cells, Many Features. Cell, 2016, 165(4): 780-791.

[370] Caporaso JG, Kuczynski J, Stombaugh J, et al. QIIME allows analysis of high-throughput community sequencing data. Nature Methods, 2010, 7(5): 335-336.

[371] Paun A, Yau C, Meshkibaf S, et al. Association of HLA-dependent islet autoimmunity with systemic antibody responses to intestinal commensal bacteria in children. Science Immunology, 2019, 4(32): eaau8125.

[372] Planer JD, Peng Y, Kau AL, et al. Development of the gut microbiota and mucosal IgA responses in twins and gnotobiotic mice. Nature, 2016, 534(7606): 263-266.

[373] Palm NW, de Zoete MR, Cullen TW, et al. Immunoglobulin A coating identifies colitogenic bacteria in inflammatory bowel disease. Cell, 2014, 158(5): 1000-1010.

[374] Zmora N, Zilberman-Schapira G, Suez J, et al. Personalized Gut Mucosal Colonization Resistance to Empiric Probiotics Is Associated with Unique Host and Microbiome Features. Cell, 2018, 174(6): 1388-1405. e1321.

[375] Sonnenburg ED, Smits SA, Tikhonov M, et al. Diet-induced extinctions in the gut microbiota compound over generations. Nature, 2016, 529(7585): 212-215.

[376] Skelly AN, Sato Y, Kearney S, et al. Mining the microbiota for microbial and metabolite-based immunotherapies. Nature Reviews: Immunology, 2019, 19(5): 305-323.

[377] Kim J, Koo BK, Knoblich JA. Human organoids: model systems for human biology and medicine. Nature Reviews: Molecular Cell Biology, 2020, 21(10): 571-584.

[378] Kayisoglu O, Schlegel N, Bartfeld S. Gastrointestinal epithelial innate immunity-regionalization and organoids as new model. Journal of Molecular Medicine (Berlin, Germany), 2021, 99(4): 517-530.

[379] Sonnenburg ED, Sonnenburg JL. The ancestral and industrialized gut microbiota and implications for human health. Nature Reviews Microbiology, 2019,

17(6): 383-390.

[380] Costello EK, Stagaman K, Dethlefsen L, et al. The application of ecological theory toward an understanding of the human microbiome. Science, 2012, 336(6086): 1255-1262.

[381] Qin J, Li R, Raes J, et al. A human gut microbial gene catalogue established by metagenomic sequencing. Nature, 2010, 464(7285): 59-65.

[382] Smits SA, Leach J, Sonnenburg ED, et al. Seasonal cycling in the gut microbiome of the Hadza hunter-gatherers of Tanzania. Science, 2017, 357(6353): 802-806.

[383] Vangay P, Johnson AJ, Ward TL, et al. US Immigration Westernizes the Human Gut Microbiome. Cell, 2018, 175(4): 962-972. e910.

[384] Lozano R, Naghavi M, Foreman K, et al. Global and regional mortality from 235 causes of death for 20 age groups in 1990 and 2010: a systematic analysis for the Global Burden of Disease Study 2010. Lancet, 2012, 380(9859): 2095-2128.

[385] Pontzer H, Wood BM, Raichlen DA. Hunter-gatherers as models in public health. Obesity Reviews, 2018, 19 Suppl 1: 24-35.

[386] Fragiadakis GK, Smits SA, Sonnenburg ED, et al. Links between environment, diet, and the hunter-gatherer microbiome. Gut Microbes, 2019, 10(2): 216-227.

[387] Jha AR, Davenport ER, Gautam Y, et al. Gut microbiome transition across a lifestyle gradient in Himalaya. PLoS Biology, 2018, 16(11): e2005396.

[388] Fessler J, Matson V, Gajewski TF. Exploring the emerging role of the microbiome in cancer immunotherapy. Journal for ImmunoTherapy of Cancer, 2019, 7(1): 108.

[389] Geller LT, Barzily-Rokni M, Danino T, et al. Potential role of intratumor bacteria in mediating tumor resistance to the chemotherapeutic drug gemcitabine. Science, 2017, 357(6356): 1156-1160.

[390] Viaud S, Saccheri F, Mignot G, et al. The intestinal microbiota modulates the anticancer immune effects of cyclophosphamide. Science, 2013, 342(6161): 971-976.

[391] Song M, Chan AT. Environmental Factors, Gut Microbiota, and Colorectal Cancer Prevention. Clinical Gastroenterology and Hepatology, 2019, 17(2): 275-289.

[392] Matson V, Chervin CS, Gajewski TF. Cancer and the Microbiome-Influence of the Commensal Microbiota on Cancer, Immune Responses, and Immunotherapy. Gastroenterology, 2021, 160(2): 600-613.

[393] Routy B, Le Chatelier E, Derosa L, et al. Gut microbiome influences efficacy of PD-1-based immunotherapy against epithelial tumors. Science, 2018, 359(6371): 91-97.

[394] Balachandran VP, Łuksza M, Zhao JN, et al. Identification of unique neoantigen qualities in long-term survivors of pancreatic cancer. Nature, 2017, 551(7681): 512-516.

[395] Vétizou M, Pitt JM, Daillère R, et al. Anticancer immunotherapy by CTLA-4 blockade relies on the gut microbiota. Science, 2015, 350(6264): 1079-1084.

[396] Thaiss CA, Levy M, Grosheva I, et al. Hyperglycemia drives intestinal barrier dysfunction and risk for enteric infection. Science, 2018, 359(6382): 1376-1383.

[397] Paulos CM, Wrzesinski C, Kaiser A, et al. Microbial translocation augments the function of adoptively transferred self/tumor-specific CD8+ T cells via TLR4 signaling. Journal of Clinical Investigation, 2007, 117(8): 2197-2204.

[398] Hall JA, Bouladoux N, Sun CM, et al. Commensal DNA limits regulatory T cell conversion and is a natural adjuvant of intestinal immune responses. Immunity, 2008, 29(4): 637-649.

[399] Sivan A, Corrales L, Hubert N, et al. Commensal Bifidobacterium promotes antitumor immunity and facilitates anti-PD-L1 efficacy. Science, 2015, 350(6264): 1084-1089.

[400] Morrison DJ, Preston T. Formation of short chain fatty acids by the gut microbiota and their impact on human metabolism. Gut Microbes, 2016, 7(3): 189-200.

[401] den Besten G, van Eunen K, Groen AK, et al. The role of short-chain fatty acids in the interplay between diet, gut microbiota, and host energy metabolism. Journal of Lipid Research, 2013, 54(9): 2325-2340.

[402] Iraporda C, Errea A, Romanin DE, et al. Lactate and short chain fatty acids produced by microbial fermentation downregulate proinflammatory responses in intestinal epithelial cells and myeloid cells. Immunobiology, 2015, 220(10): 1161-1169.

[403] White CA, Pone EJ, Lam T, et al. Histone deacetylase inhibitors upregulate B cell microRNAs that silence AID and Blimp-1 expression for epigenetic modulation of antibody and autoantibody responses. Journal of Immunology, 2014, 193(12): 5933-5950.

[404] Cohen LJ, Esterhazy D, Kim SH, et al. Commensal

bacteria make GPCR ligands that mimic human signalling molecules. Nature, 2017, 549(7670): 48-53.

[405] Hanus M, Parada-Venegas D, Landskron G, et al. Immune System, Microbiota, and Microbial Metabolites: The Unresolved Triad in Colorectal Cancer Microenvironment. Frontiers in Immunology, 2021, 12: 612826.

[406] Levy M, Thaiss CA, Elinav E. Metabolites: messengers between the microbiota and the immune system. Genes & Development, 2016, 30(14): 1589-1597.

[407] Iida N, Dzutsev A, Stewart CA, et al. Commensal bacteria control cancer response to therapy by modulating the tumor microenvironment. Science, 2013, 342(6161): 967-970.

[408] Khan AA, Sirsat AT, Singh H, et al. Microbiota and cancer: current understanding and mechanistic implications. Clinical & Translational Oncology, 2021: 1-10.

[409] Li J, Sung CY, Lee N, et al. Probiotics modulated gut microbiota suppresses hepatocellular carcinoma growth in mice. Proceedings of the National Academy of Sciences of the United States of America, 2016, 113(9): E1306-1315.

[410] Heo G, Lee Y, Im E. Interplay between the Gut Microbiota and Inflammatory Mediators in the Development of Colorectal Cancer. Cancers, 2021, 13(4): 734.

[411] Morton AM, Sefik E, Upadhyay R, et al. Endoscopic photoconversion reveals unexpectedly broad leukocyte trafficking to and from the gut. Proceedings of the National Academy of Sciences of the United States of America, 2014, 111(18): 6696-6701.

[412] Arques JL, Hautefort I, Ivory K, et al. Salmonella induces flagellin- and MyD88-dependent migration of bacteria-capturing dendritic cells into the gut lumen. Gastroenterology, 2009, 137(2): 579-587, 587.e571-572.

[413] McDole JR, Wheeler LW, McDonald KG, et al. Goblet cells deliver luminal antigen to CD103+ dendritic cells in the small intestine. Nature, 2012, 483(7389): 345-349.

[414] Ling Z, Shao L, Liu X, et al. Regulatory T Cells and Plasmacytoid Dendritic Cells Within the Tumor Microenvironment in Gastric Cancer Are Correlated With Gastric Microbiota Dysbiosis: A Preliminary Study. Frontiers in Immunology, 2019, 10: 533.

[415] Geis AL, Fan H, Wu X, et al. Regulatory T-cell Response to Enterotoxigenic Bacteroides fragilis Colonization Triggers IL17-Dependent Colon Carcinogenesis. Cancer Discovery, 2015, 5(10): 1098-1109.

[416] Queirolo P, Morabito A, Laurent S, et al. Association of CTLA-4 polymorphisms with improved overall survival in melanoma patients treated with CTLA-4 blockade: a pilot study. Cancer Investigation, 2013, 31(5): 336-345.

[417] Peng W, Chen JQ, Liu C, et al. Loss of PTEN Promotes Resistance to T Cell-Mediated Immunotherapy. Cancer Discovery, 2016, 6(2): 202-216.

[418] 张欣悦，高永翔，谢怡敏. 黏膜免疫系统研究进展. 中药与临床，2015, 6(5): 64-68.

[419] Clancy RL, Cripps AW. An Oral Whole-Cell Killed Nontypeable Haemophilus influenzae Immunotherapeutic For The Prevention Of Acute Exacerbations Of Chronic Airway Disease. International Journal of Chronic Obstructive Pulmonary Disease, 2019, 14: 2423-2431.

[420] Mestecky J. The common mucosal immune system and current strategies for induction of immune responses in external secretions. Journal of Clinical Immunology, 1987, 7(4): 265-276.

[421] Date Y, Ebisawa M, Fukuda S, et al. NALT M cells are important for immune induction for the common mucosal immune system. International Immunology, 2017, 29(10): 471-478.

[422] He Y, Wen Q, Yao F, et al. Gut-lung axis: The microbial contributions and clinical implications. Critical Reviews in Microbiology, 2017, 43(1): 81-95.

[423] Mestecky J. The Mammary Gland as an Integral Component of the Common Mucosal Immune System. Nestle Nutrition Institute Workshop Series, 2020, 94: 27-37.

[424] Trivedi S, Grossmann AH, Jensen O, et al. Intestinal Infection Is Associated With Impaired Lung Innate Immunity to Secondary Respiratory Infection. Open Forum Infectious Diseases, 2021, 8(6): ofab237.

[425] Zhou A, Lei Y, Tang L, et al. Gut Microbiota: the Emerging Link to Lung Homeostasis and Disease. Journal of Bacteriology, 2021, 203(4): e00454-00420.

[426] Jang YO, Kim OH, Kim SJ, et al. High-fiber diets attenuate emphysema development via modulation of gut microbiota and metabolism. Scientific Reports, 2021, 11(1): 7008.

[427] Ubags NDJ, Marsland BJ. Mechanistic insight into

the function of the microbiome in lung diseases. European Respiratory Journal, 2017, 50(3): 1602467.

[428] Zhang D, Li S, Wang N, et al. The Cross-Talk Between Gut Microbiota and Lungs in Common Lung Diseases. Frontiers in Microbiology, 2020, 11: 301.

[429] Tan JY, Tang YC, Huang J. Gut Microbiota and Lung Injury. Advances in Experimental Medicine and Biology, 2020, 1238: 55-72.

[430] Barcik W, Boutin RCT, Sokolowska M, et al. The Role of Lung and Gut Microbiota in the Pathology of Asthma. Immunity, 2020, 52(2): 241-255.

[431] de Oliveira GLV, Oliveira CNS, Pinzan CF, et al. Microbiota Modulation of the Gut-Lung Axis in COVID-19. Frontiers in Immunology, 2021, 12: 635471.

[432] Qian LJ, Kang SM, Xie JL, et al. Early-life gut microbial colonization shapes Th1/Th2 balance in asthma model in BALB/c mice. BMC Microbiology, 2017, 17(1): 135.

[433] Demirci M, Tokman HB, Uysal HK, et al. Reduced Akkermansia muciniphila and Faecalibacterium prausnitzii levels in the gut microbiota of children with allergic asthma. Allergologia et Immunopathologia, 2019, 47(4): 365-371.

[434] Ivashkin V, Zolnikova O, Potskherashvili N, et al. Metabolic activity of intestinal microflora in patients with bronchial asthma. Clinics and Practice, 2019, 9(1): 1126.

[435] Barcik W, Pugin B, Westermann P, et al. Histamine-secreting microbes are increased in the gut of adult asthma patients. Journal Of Allergy And Clinical Immunology, 2016, 138(5): 1491-1494. e1497.

[436] Barcik W, Pugin B, Bresco MS, et al. Bacterial secretion of histamine within the gut influences immune responses within the lung. Allergy, 2019, 74(5): 899-909.

[437] Raftery AL, Tsantikos E, Harris NL, et al. Links Between Inflammatory Bowel Disease and Chronic Obstructive Pulmonary Disease. Frontiers in Immunology, 2020, 11: 2144.

[438] Young RP, Hopkins RJ, Marsland B. The Gut-Liver-Lung Axis. Modulation of the Innate Immune Response and Its Possible Role in Chronic Obstructive Pulmonary Disease. American Journal of Respiratory Cell and Molecular Biology, 2016, 54(2): 161-169.

[439] Osei Sekyere J, Maningi NE, Fourie PB. Mycobacterium tuberculosis, antimicrobials, immunity, and lung-gut microbiota crosstalk: current updates and emerging advances. Annals of the New York Academy of Sciences, 2020, 1467(1): 21-47.

[440] Negi S, Pahari S, Bashir H, et al. Gut Microbiota Regulates Mincle Mediated Activation of Lung Dendritic Cells to Protect Against Mycobacterium tuberculosis. Frontiers in Immunology, 2019, 10: 1142.

[441] Brown RL, Sequeira RP, Clarke TB. The microbiota protects against respiratory infection via GM-CSF signaling. Nature Communications, 2017, 8(1): 1512.

[442] Schuijt TJ, Lankelma JM, Scicluna BP, et al. The gut microbiota plays a protective role in the host defence against pneumococcal pneumonia. Gut, 2016, 65(4): 575-583.

[443] Liu C, Yang L, Han Y, et al. Mast cells participate in regulation of lung-gut axis during Staphylococcus aureus pneumonia. Cell Proliferation, 2019, 52(2): e12565.

[444] Gauguet S, D'Ortona S, Ahnger-Pier K, et al. Intestinal Microbiota of Mice Influences Resistance to Staphylococcus aureus Pneumonia. Infection And Immunity, 2015, 83(10): 4003-4014.

[445] Ahmadi Badi S, Tarashi S, Fateh A, et al. From the Role of Microbiota in Gut-Lung Axis to SARS-CoV-2 Pathogenesis. Mediators of Inflammation, 2021, 2021: 6611222.

[446] Cheung KS, Hung IFN, Chan PPY, et al. Gastrointestinal Manifestations of SARS-CoV-2 Infection and Virus Load in Fecal Samples From a Hong Kong Cohort: Systematic Review and Meta-analysis. Gastroenterology, 2020, 159(1): 81-95.

[447] Neurath MF. COVID-19 and immunomodulation in IBD. Gut, 2020, 69(7): 1335-1342.

[448] Lin L, Lu L, Cao W, et al. Hypothesis for potential pathogenesis of SARS-CoV-2 infection-a review of immune changes in patients with viral pneumonia. Emerging Microbes & Infections, 2020, 9(1): 727-732.

[449] Megyeri K, Dernovics A, Al-Luhaibi ZII, et al. COVID-19-associated diarrhea. World Journal of Gastroenterology, 2021, 27(23): 3208-3222.

[450] Gautier T, David-Le Gall S, Sweidan A, et al. Next-Generation Probiotics and Their Metabolites in COVID-19. Microorganisms, 2021, 9(5): 941.

[451] 徐凯进，蔡洪流，沈毅弘，等 . 2019 冠状病毒病 (COVID-19) 诊疗浙江经验，浙江大学学报（医学版），2020, 49(2): 147-157.

[452] Howell MC, Green R, McGill AR, et al. SARS-CoV-

2-Induced Gut Microbiome Dysbiosis: Implications for Colorectal Cancer. Cancers, 2021, 13(11): 2676.

[453] Zuo T, Zhang F, Lui GCY, et al. Alterations in Gut Microbiota of Patients With COVID-19 During Time of Hospitalization. Gastroenterology, 2020, 159(3): 944-955. e948.

[454] Rajput S, Paliwal D, Naithani M, et al. COVID-19 and Gut Microbiota: A Potential Connection. Indian Journal of Clinical Biochemistry, 2021, 36(3): 1-12.

[455] Qin WH, Liu CL, Jiang YH, et al. Gut ACE2 expression, tryptophan deficiency and inflammatory responses: the potential connection that should not be ignored during SARS-CoV-2 infection. Cellular and Molecular Gastroenterology and Hepatology, 2021, 12(4): 1514-1516. e1514.

[456] Taghinezhad SS, Keyvani H, Bermudez-Humaran LG, et al. Twenty years of research on HPV vaccines based on genetically modified lactic acid bacteria: an overview on the gut-vagina axis. Cellular And Molecular Life Sciences, 2021, 78(4): 1191-1206.

[457] Quaranta G, Sanguinetti M, Masucci L. Fecal Microbiota Transplantation: A Potential Tool for Treatment of Human Female Reproductive Tract Diseases. Frontiers in Immunology, 2019, 10: 2653.

[458] Borella F, Carosso AR, Cosma S, et al. Gut Microbiota and Gynecological Cancers: A Summary of Pathogenetic Mechanisms and Future Directions. ACS Infectious Diseases, 2021, 7(5): 987-1009.

[459] Kutteh WH, Kantele A, Moldoveanu Z, et al. Induction of specific immune responses in the genital tract of women after oral or rectal immunization and rectal boosting with Salmonella typhi Ty 21a vaccine. Journal of Reproductive Immunology, 2001, 52(1-2): 61-75.

[460] Kim DE, Kim JK, Han SK, et al. Lactobacillus plantarum NK3 and Bifidobacterium longum NK49 Alleviate Bacterial Vaginosis and Osteoporosis in Mice by Suppressing NF-kappaB-Linked TNF-alpha Expression. Journal of Medicinal Food, 2019, 22(10): 1022-1031.

[461] Jahanshahi M, Maleki Dana P, Badehnoosh B, et al. Anti-tumor activities of probiotics in cervical cancer. Journal of Ovarian Research, 2020, 13(1): 68.

[462] Gosalbes MJ, Compte J, Moriano-Gutierrez S, et al. Metabolic adaptation in the human gut microbiota during pregnancy and the first year of life. EBioMedicine, 2019, 39: 497-509.

[463] Koren O, Goodrich JK, Cullender TC, et al. Host remodeling of the gut microbiome and metabolic changes during pregnancy. Cell, 2012, 150(3): 470-480.

[464] Rodriguez JM. The origin of human milk bacteria: is there a bacterial entero-mammary pathway during late pregnancy and lactation? Advances in Nutrition, 2014, 5(6): 779-784.

[465] Zeng R, Wang J, Zhuo Z, et al. Stem cells and exosomes: promising candidates for necrotizing enterocolitis therapy. Stem Cell Research & Therapy, 2021, 12(1): 323.

[466] He S, Liu G, Zhu X. Human breast milk-derived exosomes may help maintain intestinal epithelial barrier integrity. Pediatric Research, 2021, 90(2): 366-372.

[467] Trujillo-Vargas CM, Schaefer L, Alam J, et al. The gut-eye-lacrimal gland-microbiome axis in Sjogren Syndrome. Ocular Surface, 2020, 18(2): 335-344.

[468] Moon J, Choi SH, Yoon CH, et al. Gut dysbiosis is prevailing in Sjogren's syndrome and is related to dry eye severity. PloS One, 2020, 15(2): e0229029.

[469] Kugadas A, Wright Q, Geddes-McAlister J, et al. Role of Microbiota in Strengthening Ocular Mucosal Barrier Function Through Secretory IgA. Investigative Ophthalmology and Visual Science, 2017, 58(11): 4593-4600.

[470] Zaheer M, Wang C, Bian F, et al. Protective role of commensal bacteria in Sjogren Syndrome. Journal of Autoimmunity, 2018, 93: 45-56.

[471] Moon J, Yoon CH, Choi SH, et al. Can Gut Microbiota Affect Dry Eye Syndrome? International Journal of Molecular Sciences, 2020, 21(22): 8443.

[472] Doifode T, Giridharan VV, Generoso JS, et al. The impact of the microbiota-gut-brain axis on Alzheimer's disease pathophysiology. Pharmacological Research, 2021, 164: 105314.

[473] Cryan JF, O'Riordan KJ, Cowan CSM, et al. The Microbiota-Gut-Brain Axis. Physiological Reviews, 2019, 99(4): 1877-2013.

[474] Zhu X, Han Y, Du J, et al. Microbiota-gut-brain axis and the central nervous system. Oncotarget, 2017, 8(32): 53829-53838.

[475] Arneth BM. Gut-brain axis biochemical signalling from the gastrointestinal tract to the central nervous system: gut dysbiosis and altered brain function. Postgraduate Medical Journal, 2018, 94(1114): 446-452.

[476] Shichita T, Sugiyama Y, Ooboshi H, et al. Pivotal role of cerebral interleukin-17-producing gammadelta T cells in the delayed phase of ischemic brain injury.

Nature Medicine, 2009, 15(8): 946-950.

[477] Benakis C, Brea D, Caballero S, et al. Commensal microbiota affects ischemic stroke outcome by regulating intestinal γδ T cells. Nature Medicine, 2016, 22(5): 516-523.

[478] Singh V, Roth S, Llovera G, et al. Microbiota Dysbiosis Controls the Neuroinflammatory Response after Stroke. Journal of Neuroscience, 2016, 36(28): 7428-7440.

[479] Muller N. Inflammation in Schizophrenia: Pathogenetic Aspects and Therapeutic Considerations. Schizophrenia Bulletin, 2018, 44(5): 973-982.

[480] Hope S, Ueland T, Steen NE, et al. Interleukin 1 receptor antagonist and soluble tumor necrosis factor receptor 1 are associated with general severity and psychotic symptoms in schizophrenia and bipolar disorder. Schizophrenia Research, 2013, 145(1-3): 36-42.

[481] Hornig M. The role of microbes and autoimmunity in the pathogenesis of neuropsychiatric illness. Current Opinion in Rheumatology, 2013, 25(4): 488-795.

[482] Severance EG, Alaedini A, Yang S, et al. Gastrointestinal inflammation and associated immune activation in schizophrenia. Schizophrenia Research, 2012, 138(1): 48-53.

[483] Minter MR, Zhang C, Leone V, et al. Antibiotic-induced perturbations in gut microbial diversity influences neuro-inflammation and amyloidosis in a murine model of Alzheimer's disease. Scientific Reports, 2016, 6: 30028.

[484] Zhao Y, Lukiw WJ. Microbiome-generated amyloid and potential impact on amyloidogenesis in Alzheimer's disease (AD). Journal of Natural Sciences, 2015, 1(7): e138.

[485] Friedland RP. Mechanisms of molecular mimicry involving the microbiota in neurodegeneration. Journal of Alzheimer's Disease, 2015, 45(2): 349-362.

[486] Pistollato F, Sumalla Cano S, Elio I, et al. Role of gut microbiota and nutrients in amyloid formation and pathogenesis of Alzheimer disease. Nutrition Reviews, 2016, 74(10): 624-634.

[487] Erny D, Hrab ě de Angelis AL, Jaitin D, et al. Host microbiota constantly control maturation and function of microglia in the CNS. Nature Neuroscience, 2015, 18(7): 965-977.

[488] Li R, Mao Z, Ye X, et al. Human Gut Microbiome and Liver Diseases: From Correlation to Causation. Microorganisms, 2021, 9(5): 1017.

[489] Fianchi F, Liguori A, Gasbarrini A, et al. Nonalcoholic Fatty Liver Disease (NAFLD) as Model of Gut-Liver Axis Interaction: From Pathophysiology to Potential Target of Treatment for Personalized Therapy. International Journal of Molecular Sciences, 2021, 22(12): 6485.

[490] Marshall JC. The gut as a potential trigger of exercise-induced inflammatory responses. Canadian Journal of Physiology and Pharmacology, 1998, 76(5): 479-484.

[491] Milosevic I, Vujovic A, Barac A, et al. Gut-Liver Axis, Gut Microbiota, and Its Modulation in the Management of Liver Diseases: A Review of the Literature. International Journal of Molecular Sciences, 2019, 20(2): 395.

[492] Chen T, Li R, Chen P. Gut Microbiota and Chemical-Induced Acute Liver Injury. Frontiers in Physiology, 2021, 12: 688780.

[493] Neag MA, Catinean A, Muntean DM, et al. Probiotic Bacillus Spores Protect Against Acetaminophen Induced Acute Liver Injury in Rats. Nutrients, 2020, 12(3): 632.

[494] Dey P. The role of gut microbiome in chemical-induced metabolic and toxicological murine disease models. Life Sciences, 2020, 258: 118172.

[495] Kolodziejczyk AA, Federici S, Zmora N, et al. Acute liver failure is regulated by MYC- and microbiome-dependent programs. Nature Medicine, 2020, 26(12): 1899-1911.

[496] Schneider KM, Elfers C, Ghallab A, et al. Intestinal Dysbiosis Amplifies Acetaminophen-Induced Acute Liver Injury. Cellular and Molecular Gastroenterology and Hepatology, 2021, 11(4): 909-933.

[497] Kanwal F, Kramer JR, Mapakshi S, et al. Risk of Hepatocellular Cancer in Patients With Non-Alcoholic Fatty Liver Disease. Gastroenterology, 2018, 155(6): 1828-1837. e1822.

[498] Schwimmer JB, Johnson JS, Angeles JE, et al. Microbiome Signatures Associated With Steatohepatitis and Moderate to Severe Fibrosis in Children With Nonalcoholic Fatty Liver Disease. Gastroenterology, 2019, 157(4): 1109-1122.

[499] Bao T, He F, Zhang X, et al. Inulin Exerts Beneficial Effects on Non-Alcoholic Fatty Liver Disease via Modulating gut Microbiome and Suppressing the Lipopolysaccharide-Toll-Like Receptor 4-Mpsi-Nuclear Factor-kappaB-Nod-Like Receptor Protein 3 Pathway via gut-Liver Axis in Mice. Frontiers in Pharmacology, 2020, 11: 558525.

[500] Friedman SL, Neuschwander-Tetri BA, Rinella M, et al. Mechanisms of NAFLD development and therapeutic strategies. Nature Medicine, 2018, 24(7): 908-922.

[501] Krishnan S, Ding Y, Saedi N, et al. Gut Microbiota-Derived Tryptophan Metabolites Modulate Inflammatory Response in Hepatocytes and Macrophages. Cell Reports, 2018, 23(4): 1099-1111.

[502] Albillos A, de Gottardi A, Rescigno M. The gut-liver axis in liver disease: Pathophysiological basis for therapy. Journal of Hepatology, 2020, 72(3): 558-577.

[503] Munoz L, Borrero MJ, Ubeda M, et al. Intestinal Immune Dysregulation Driven by Dysbiosis Promotes Barrier Disruption and Bacterial Translocation in Rats With Cirrhosis. Hepatology, 2019, 70(3): 925-938.

[504] Zhang C, Yang M, Ericsson AC. The Potential Gut Microbiota-Mediated Treatment Options for Liver Cancer. Frontiers in Oncology, 2020, 10: 524205.

[505] Ma C, Han M, Heinrich B, et al. Gut microbiome-mediated bile acid metabolism regulates liver cancer via NKT cells. Science, 2018, 360(6391): eaan5931.

[506] Loo TM, Kamachi F, Watanabe Y, et al. Gut Microbiota Promotes Obesity-Associated Liver Cancer through PGE2-Mediated Suppression of Antitumor Immunity. Cancer Discovery, 2017, 7(5): 522-538.

[507] Giannelli V, Di Gregorio V, Iebba V, et al. Microbiota and the gut-liver axis: bacterial translocation, inflammation and infection in cirrhosis. World Journal of Gastroenterology, 2014, 20(45): 16795-16810.

[508] Jiang XW, Li YT, Ye JZ, et al. New strain of Pediococcus pentosaceus alleviates ethanol-induced liver injury by modulating the gut microbiota and short-chain fatty acid metabolism. World Journal of Gastroenterology, 2020, 26(40): 6224-6240.

[509] Jena PK, Sheng L, Nagar N, et al. Synbiotics Bifidobacterium infantis and milk oligosaccharides are effective in reversing cancer-prone nonalcoholic steatohepatitis using western diet-fed FXR knockout mouse models. Journal of Nutritional Biochemistry, 2018, 57: 246-254.

[510] Castillo-Rodriguez E, Fernandez-Prado R, Esteras R, et al. Impact of Altered Intestinal Microbiota on Chronic Kidney Disease Progression. Toxins, 2018, 10(7): 300.

[511] Vaziri ND, Goshtasbi N, Yuan J, et al. Uremic plasma impairs barrier function and depletes the tight junction protein constituents of intestinal epithelium. American Journal of Nephrology, 2012, 36(5): 438-443.

[512] Sabatino A, Regolisti G, Brusasco I, et al. Alterations of intestinal barrier and microbiota in chronic kidney disease. Nephrology Dialysis Transplantation, 2015, 30(6): 924-933.

[513] Cario E, Gerken G, Podolsky DK. Toll-like receptor 2 controls mucosal inflammation by regulating epithelial barrier function. Gastroenterology, 2007, 132(4): 1359-1374.

[514] Yang T, Richards EM, Pepine CJ, et al. The gut microbiota and the brain-gut-kidney axis in hypertension and chronic kidney disease. Nature Reviews Nephrology, 2018, 14(7): 442-456.

[515] Evenepoel P, Poesen R, Meijers B. The gut-kidney axis. Pediatric Nephrology, 2017, 32(11): 2005-2014.

[516] Yan J, Herzog JW, Tsang K, et al. Gut microbiota induce IGF-1 and promote bone formation and growth. Proceedings of the National Academy of Sciences of the United States of America, 2016, 113(47): E7554-E7563.

[517] Coppo R. The gut-kidney axis in IgA nephropathy: role of microbiota and diet on genetic predisposition. Pediatric Nephrology, 2018, 33(1): 53-61.

[518] Shakeel M. Recent advances in understanding the role of oxidative stress in diabetic neuropathy. Diabetes & Metabolic Syndrome, 2015, 9(4): 373-378.

[519] Huang W, Guo HL, Deng X, et al. Short-Chain Fatty Acids Inhibit Oxidative Stress and Inflammation in Mesangial Cells Induced by High Glucose and Lipopolysaccharide. Experimental and Clinical Endocrinology and Diabetes, 2017, 125(2): 98-105.

[520] Fernandez-Prado R, Esteras R, Perez-Gomez MV, et al. Nutrients Turned into Toxins: Microbiota Modulation of Nutrient Properties in Chronic Kidney Disease. Nutrients, 2017, 9(5): 489.

[521] Kooman JP, Dekker MJ, Usvyat LA, et al. Inflammation and premature aging in advanced chronic kidney disease. American Journal of Physiology: Renal Physiology, 2017, 313(4): F938-F950.

[522] Perez-Gomez MV, Sanchez-Niño MD, Sanz AB, et al. Targeting inflammation in diabetic kidney disease: early clinical trials. Expert Opinion On Investigational Drugs, 2016, 25(9): 1045-1058.

[523] Castillo-Rodríguez E, Pizarro-Sánchez S, Sanz AB, et al. Inflammatory Cytokines as Uremic Toxins: "Ni Son Todos Los Que Estan, Ni Estan Todos Los Que Son". Toxins, 2017, 9(4): 114.

[524] Lau WL, Kalantar-Zadeh K, Vaziri ND. The Gut as a

Source of Inflammation in Chronic Kidney Disease. Nephron, 2015, 130(2): 92-98.

[525] Battson ML, Lee DM, Weir TL, et al. The gut microbiota as a novel regulator of cardiovascular function and disease. Journal of Nutritional Biochemistry, 2018, 56: 1-15.

[526] Jin M, Qian Z, Yin J, et al. The role of intestinal microbiota in cardiovascular disease. Journal of Cellular and Molecular Medicine, 2019, 23(4): 2343-2350.

[527] Novakovic M, Rout A, Kingsley T, et al. Role of gut microbiota in cardiovascular diseases. World Journal of Cardiology, 2020, 12(4): 110-122.

[528] Koren O, Spor A, Felin J, et al. Human oral, gut, and plaque microbiota in patients with atherosclerosis. Proceedings of the National Academy of Sciences of the United States of America, 2011, 108 Suppl 1: 4592-4598.

[529] Rajendiran E, Ramadass B, Ramprasath V. Understanding connections and roles of gut microbiome in cardiovascular diseases. Canadian Journal of Microbiology, 2021, 67(2): 101-111.

[530] Forkosh E, Ilan Y. The heart-gut axis: new target for atherosclerosis and congestive heart failure therapy. Open Heart, 2019, 6(1): e000993.

[531] Harikrishnan S. Diet, the Gut Microbiome and Heart Failure. Card Fail Rev, 2019, 5(2): 119-122.

[532] Kamo T, Akazawa H, Suzuki JI, et al. Novel Concept of a Heart-Gut Axis in the Pathophysiology of Heart Failure. Korean Circulation Journal, 2017, 47(5): 663-669.

[533] Seldin MM, Meng Y, Qi H, et al. Trimethylamine N-Oxide Promotes Vascular Inflammation Through Signaling of Mitogen-Activated Protein Kinase and Nuclear Factor-kappaB. Journal of the American Heart Association, 2016, 5(2): e002767.

[534] Li J, Lin S, Vanhoutte PM, et al. Akkermansia Muciniphila Protects Against Atherosclerosis by Preventing Metabolic Endotoxemia-Induced Inflammation in Apoe-/- Mice. Circulation, 2016, 133(24): 2434-2446.

[535] Wilck N, Matus MG, Kearney SM, et al. Salt-responsive gut commensal modulates TH17 axis and disease. Nature, 2017, 551(7682): 585-589.

[536] Yu L, Meng G, Huang B, et al. A potential relationship between gut microbes and atrial fibrillation: Trimethylamine N-oxide, a gut microbe-derived metabolite, facilitates the progression of atrial fibrillation. International Journal of Cardiology, 2018, 255: 92-98.

[537] Zuo K, Li J, Li K, et al. Disordered gut microbiota and alterations in metabolic patterns are associated with atrial fibrillation. Gigascience, 2019, 8(6): giz058.

[538] Cheng WL, Li SJ, Lee TI, et al. Sugar Fructose Triggers Gut Dysbiosis and Metabolic Inflammation with Cardiac Arrhythmogenesis. Biomedicines, 2021, 9(7): 728.

[539] Szanto M, Dozsa A, Antal D, et al. Targeting the gut-skin axis-Probiotics as new tools for skin disorder management? Experimental Dermatology, 2019, 28(11): 1210-1218.

[540] Ahlawat S, Asha, Sharma KK. Gut-organ axis: a microbial outreach and networking. Letters in Applied Microbiology, 2021, 72(6): 636-668.

[541] Park DH, Kim JW, Park HJ, et al. Comparative Analysis of the Microbiome across the Gut-Skin Axis in Atopic Dermatitis. International Journal of Molecular Sciences, 2021, 22(8): 4228.

[542] Salem I, Ramser A, Isham N, et al. The Gut Microbiome as a Major Regulator of the Gut-Skin Axis. Frontiers in Microbiology, 2018, 9: 1459.

[543] Tan L, Zhao S, Zhu W, et al. The Akkermansia muciniphila is a gut microbiota signature in psoriasis. Experimental Dermatology, 2018, 27(2): 144-149.

[544] Ellis SR, Nguyen M, Vaughn AR, et al. The Skin and Gut Microbiome and Its Role in Common Dermatologic Conditions. Microorganisms, 2019, 7(11): 550.

[545] Chen G, Chen ZM, Fan XY, et al. Gut-Brain-Skin Axis in Psoriasis: A Review. Dermatology and Therapy, 2021, 11(1): 25-38.

[546] Keren A, Gilhar A, Ullmann Y, et al. Instantaneous depolarization of T cells via dopamine receptors, and inhibition of activated T cells of Psoriasis patients and inflamed human skin, by D1-like receptor agonist: Fenoldopam. Immunology, 2019, 158(3): 171-193.

[547] Bajic SS, Dokic J, Dinic M, et al. GABA potentiate the immunoregulatory effects of Lactobacillus brevis BGZLS10-17 via ATG5-dependent autophagy in vitro. Scientific Reports, 2020, 10(1): 1347.

[548] Suzuki S, Campos-Alberto E, Morita Y, et al. Low Interleukin 10 Production at Birth Is a Risk Factor for Atopic Dermatitis in Neonates with Bifidobacterium Colonization. International Archives of Allergy and Immunology, 2018, 177(4): 342-349.

[549] Song H, Yoo Y, Hwang J, et al. Faecalibacterium prausnitzii subspecies-level dysbiosis in the human

gut microbiome underlying atopic dermatitis. Journal of Allergy And Clinical Immunology, 2016, 137(3): 852-860.

[550] Kwon MS, Lim SK, Jang JY, et al. Lactobacillus sakei WIKIM30 Ameliorates Atopic Dermatitis-Like Skin Lesions by Inducing Regulatory T Cells and Altering Gut Microbiota Structure in Mice. Frontiers in Immunology, 2018, 9: 1905.

[551] Quach D, Britton RA. Gut Microbiota and Bone Health. Advances in Experimental Medicine and Biology, 2017, 1033: 47-58.

[552] Jones D, Glimcher LH, Aliprantis AO. Osteoimmunology at the nexus of arthritis, osteoporosis, cancer, and infection. Journal of Clinical Investigation, 2011, 121(7): 2534-2542.

[553] Hernandez CJ, Guss JD, Luna M, et al. Links Between the Microbiome and Bone. Journal of Bone And Mineral Research, 2016, 31(9): 1638-1646.

[554] Sjögren K, Engdahl C, Henning P, et al. The gut microbiota regulates bone mass in mice. Journal of Bone And Mineral Research, 2012, 27(6): 1357-1367.

[555] Hsu E, Pacifici R. From Osteoimmunology to Osteomicrobiology: How the Microbiota and the Immune System Regulate Bone. Calcified Tissue International, 2018, 102(5): 512-521.

[556] Ciucci T, Ibáñez L, Boucoiran A, et al. Bone marrow Th17 TNFα cells induce osteoclast differentiation, and link bone destruction to IBD. Gut, 2015, 64(7): 1072-1081.

[557] Li JY, Chassaing B, Tyagi AM, et al. Sex steroid deficiency-associated bone loss is microbiota dependent and prevented by probiotics. Journal of Clinical Investigation, 2016, 126(6): 2049-2063.

[558] Potgieter M, Bester J, Kell DB, et al. The dormant blood microbiome in chronic, inflammatory diseases. FEMS Microbiology Reviews, 2015, 39(4): 567-591.

[559] Lorenzo D, GianVincenzo Z, Carlo Luca R, et al. Oral-Gut Microbiota and Arthritis: Is There an Evidence-Based Axis? Journal of Clinical Medicine, 2019, 8(10): 1753.

[560] Livshits G, Kalinkovich A. Hierarchical, imbalanced pro-inflammatory cytokine networks govern the pathogenesis of chronic arthropathies. Osteoarthritis and Cartilage, 2018, 26(1): 7-17.

[561] Cani PD, Bibiloni R, Knauf C, et al. Changes in gut microbiota control metabolic endotoxemia-induced inflammation in high-fat diet-induced obesity and diabetes in mice. Diabetes, 2008, 57(6): 1470-1481.

[562] Huang ZY, Stabler T, Pei FX, et al. Both systemic and local lipopolysaccharide (LPS) burden are associated with knee OA severity and inflammation. Osteoarthritis and Cartilage, 2016, 24(10): 1769-1775.

[563] Kalinkovich A, Livshits G. A cross talk between dysbiosis and gut-associated immune system governs the development of inflammatory arthropathies. Seminars in Arthritis and Rheumatism, 2019, 49(3): 474-484.

[564] Charles JF, Ermann J, Aliprantis AO. The intestinal microbiome and skeletal fitness: Connecting bugs and bones. Clinical Immunology, 2015, 159(2): 163-169.

[565] Mbalaviele G, Novack DV, Schett G, et al. Inflammatory osteolysis: a conspiracy against bone. Journal of Clinical Investigation, 2017, 127(6): 2030-2039.

[566] Britton RA, Irwin R, Quach D, et al. Probiotic L. reuteri treatment prevents bone loss in a menopausal ovariectomized mouse model. Journal of Cellular Physiology, 2014, 229(11): 1822-1830.

[567] McCabe LR, Irwin R, Schaefer L, et al. Probiotic use decreases intestinal inflammation and increases bone density in healthy male but not female mice. Journal of Cellular Physiology, 2013, 228(8): 1793-1798.

[568] Jimenez-Avalos JA, Arrevillaga-Boni G, Gonzalez-Lopez L, et al. Classical methods and perspectives for manipulating the human gut microbial ecosystem. Critical Reviews In Food Science And Nutrition, 2021, 61(2): 234-258.

[569] Dixit K, Chaudhari D, Dhotre D, et al. Restoration of dysbiotic human gut microbiome for homeostasis. Life Sciences, 2021, 278: 119622.

[570] Sharma A, Das P, Buschmann M, et al. The Future of Microbiome-Based Therapeutics in Clinical Applications. Clinical Pharmacology and Therapeutics, 2020, 107(1): 123-128.

第 6 章 肠道菌群代谢物对机体免疫功能的影响

第一节 概 述

哺乳动物胃肠道及相关黏膜免疫系统含有大量原核和真核生物来源的代谢产物，这些代谢产物在真核生物的发育和生理中起着重要的作用[1]。结肠含有未消化的外源性饮食成分，以及由微生物和宿主相互作用产生的内源性化合物，肠道菌群对这些成分进行厌氧发酵，产生大量多样化的代谢产物。肠道单层上皮细胞构成宿主和微生物之间的黏膜界面，使肠道菌群代谢产物能够接触宿主细胞并与之相互作用，从而影响宿主免疫反应和疾病的发生[2]。越来越多的证据表明，饮食和肠道菌群代谢产物通过调节代谢、免疫和炎症，从而调控宿主的病理生理过程[3]。胃肠道的不同部分寄生着不同比例的肠道菌群，因而在胃肠道的不同位置存在独特的细菌代谢产物。肠道菌群将摄入的营养物质转化为代谢产物，这些代谢产物能作用于肠道菌群和宿主细胞，并在其间发挥信使作用[4]。

哺乳动物胃肠道中有 10^{13} 个以上的共生细菌，它们是维持肠道生理稳定和机体动态平衡的关键因素。肠道菌群不仅是消化、吸收和储存食物所必需的，对宿主免疫系统的发育也是必需的，特别是在调节肠道免疫系统的稳态方面必不可少[5]。肠道菌群对宿主代谢和免疫的调节主要通过肠腔与上皮表面之间小分子（肠道菌群代谢产物）的交换。肠道菌群合成、调节和降解多种代谢产物，作为宿主新陈代谢的功能补充，特别是宿主无法代谢的膳食成分[6]。大多数代谢产物以两种方式产生：①与饮食有关的微生物产物；②与饮食无关的微生物产物。粪便和尿液中，超过 50% 的代谢产物来自肠道菌群或由肠道菌群修饰[5]。

最新研究表明，共生菌的代谢产物通过直接和间接机制，在调节宿主先天性和适应性免疫细胞的稳态和功能方面发挥了关键作用[7]。小分子代谢物可以直接通过黏膜层扩散，通过代谢产物特异性受体触发上皮信号，从而激活信号转导途径和转录程序，控制细胞的分化、增殖、成熟和效应功能[5]。尽管已知肠道菌群以多种方式调节宿主生理功能，但许多已知的代谢产物尚未进行功能表征，而且，目前对参与这些相互作用的分子机制仍知之甚少。在本章中，我们总结了多种肠道菌群代谢产物的产生和作用，并着重介绍了这些代谢产物如何影响和调节宿主的免疫功能，从而维持宿主健康。

第二节　肠菌代谢物介导宿主 - 细菌相互作用

一、肠菌代谢物作为细菌和宿主之间通信的信使

肠道菌群是宿主内的微生物器官，由不同的细菌种群组成，具有相互交流及与宿主细胞之间进行交流的能力[8]。肠道菌群与宿主共同进化，具有特定的遗传和代谢特性。例如，肠道菌群包含厌氧菌，它们分解宿主摄入的多糖，为自身生长和宿主提供碳和能量来源[4]。肠道菌群对宿主生理功能，尤其是免疫系统具有很大影响。最近的研究揭示了宿主 - 微生物群相互作用的分子基础，它是由细胞间相互作用以及多种生物活性小分子（代谢产物）的产生、修饰和传感介导的[1]。

结肠细菌通过多种代谢途径向结肠腔内释放特异性代谢产物。细菌的正常周转也会从死细胞中释放中间代谢产物进入结肠腔。代谢产物的产生和再释放取决于各种因素，包括宿主饮食成分和在结肠中定植的细菌菌株。这些代谢物大多数只对细菌附近的宿主细胞（例如结肠上皮细胞、黏膜免疫细胞、肠道神经元）产生局部影响，但有一些代谢物可能进入系统循环，对远处的细胞产生影响[9]。此外，结肠细菌及其代谢产物可能通过影响肠黏膜固有层中的免疫细胞，导致这些免疫细胞在生物学和免疫学上发生改变，然后传播到远处的器官产生影响[10, 11]。新的证据表明，宿主免疫系统可以识别肠道菌群代谢产物，对这些小分子代谢物的识别会影响宿主免疫反应以及肠道疾病的发生[12]。

肠道菌群可以产生无数的代谢产物，一般来说，这些代谢产物可大致分为 3 类：①由微生物降解食物产生的代谢物；②由宿主产生的代谢物前体，被微生物修饰后产生的代谢物；③肠道微生物从头合成的代谢物[13]。这些代谢产物通过向细菌和宿主细胞发出信号来促进肠道内稳态，特别是调节某些免疫细胞亚群，从而产生有利于细菌定植的环境[14]。微生物代谢物因为在体积上比细菌小，可以进入全身循环，再加上微生物群的可塑性，为利用微生物及其产生的代谢物来治疗局部和全身性疾病提供了可能[15]。此外，肠道菌群代谢物不局限于胃肠道，还能到达宿主其他组织，如肺、脑部等[16]。代谢物对宿主细胞的作用，如促进肠上皮再生和屏障完整性或调节黏膜免疫稳态，取决于靶细胞类型[17]。因此，肠道微生物群的组成影响宿主的发育、健康和疾病。

最近，许多研究集中于调节肠道菌群及其代谢物以改善宿主健康和预防或治疗疾病。因此，我们聚焦于肠道微生物群产生的不同种类的代谢物及其在调节宿主健康和疾病中的作用。这些代谢产物作为肠道菌群和宿主之间通信的信使，深入了解这种化学语言可以为治疗人类各种疾病提供新策略。

二、肠菌代谢物介导宿主 – 肠菌相互作用的分子靶点

表 6-1 列出了结肠腔内的细菌代谢物，这些代谢物已被证明在结肠上皮、黏膜免疫细胞以及远处器官产生显著的生物学效应。表 6-1 还确定了这些代谢物各自的宿主分子靶点。基于他们的结构和来源不同，我们将这些代谢物分为以下 4 组。

表 6-1　细菌代谢物及其在宿主中的分子靶点

代谢产物	分子靶点
短链羧酸盐	
乙酸盐、丙酸盐、丁酸盐	GPR41、GPR43、HDACs
丁酸盐	GPR109A、HDACs
乳酸	GPR81，NDRG3
琥珀酸盐	GPR91、TETs、PHD2
色氨酸代谢产物	
吲哚，吲哚 -3- 醛	AhR
吲哚 -3- 乙酸	AhR
吲哚 -3- 丙酸	PXR
脂质和脂质代谢物	
三甲胺	TAAR5
脱氧胆酸、石胆酸	FXR、PXR
共轭亚油酸和亚麻酸	PPARα、PPARγ
10- 羟基 - 顺式 -12- 十八烯酸甲酯	GPR40
细菌细胞壁成分	
脂多糖	TLR4
肽聚糖	NOD1、NOD2
多糖 A	TLR2

GPR. G 蛋白偶联受体；HDACs. 组蛋白去乙酰化酶；NDRG. N-myc 下游调节基因；PHD2. 脯氨酰羟化酶 2；AhR：芳香烃受体；PXR. 孕烷 X 受体；FXR. 法尼醇 X 受体；TAAR5. 微量胺相关受体 5；PPAR. 过氧化物酶体增殖物激活受体；TLR. Toll 样受体；NOD. 核苷酸结合寡聚化结构域蛋白

第一组（短链羧酸盐）由乙酸盐、丙酸盐、丁酸盐和乳酸单羧酸盐及琥珀酸二羧酸盐组成。这些代谢物主要是由细菌对膳食纤维发酵而产生，但也有一些是在正常饮食中发现的（例如酸奶中的乳酸）。这些代谢产物的分子靶点包括细胞表面 G 蛋白偶联受体（GPR41、GPR43、GPR109A、GPR81 和 GPR91）、细胞内酶（组蛋白去乙酰化酶、TETs、脯氨酰羟化酶 2）和细胞内信号成分（N-myc 下游调节基因）[18]。

第二组由色氨酸代谢物（吲哚、吲哚 -3- 醛、吲哚 -3- 乙酸和吲哚 -3- 丙酸）组成。这些代谢物是由结肠细菌通过代谢细菌合成的色氨酸或从饮食中产生的。它们的分子靶点是选择性核受体（芳香烃受体和孕烷 X 受体）[18]。

第三组由脂质代谢产物组成，为细菌中的内源性脂质和（或）膳食脂质代谢而来。这些代谢物包括有机阳离子三甲胺、次级胆汁酸、脱氧胆酸和石胆酸，以及多不饱和脂肪酸。在宿主细胞中，这些脂质代谢产物通过核受体（法尼醇 X 受体，孕烷 X 受体，过氧化物酶体增殖物激活受体 α 和过氧化物酶体增殖物激活受体 γ）以及细胞表面受体（GPR40 和微量胺相关受体 5）发出信号 [18]。

第四组主要由细菌细胞壁成分（脂多糖、肽聚糖和多糖 A）组成。这些代谢物通过细胞表面受体（TLR4）和细胞内受体（TLR2，NOD1 和 NOD2）发挥作用。

第三节　各种肠菌代谢物对宿主的影响

一、短链脂肪酸

短链脂肪酸（short-chain fatty acids，SCFAs）是一组烷基链短于六个碳原子的脂肪酸化合物 [19]。SCFAs（包括乙酸盐，丙酸盐和丁酸盐）源自膳食纤维（如菊粉）的细菌发酵，这些膳食纤维在小肠中无法吸收 [20]。结肠中 SCFAs 的比例受多种因素的影响，包括肠道菌群组成、代谢物底物和宿主状态 [4]。乙酸盐和丙酸盐主要由拟杆菌属产生，而丁酸盐主要由厚壁菌门产生 [21]。通常 SCFAs 对肠道有益，例如，增强肠屏障、为肠道上皮细胞提供丰富的能量以及抑制炎症 [19]。除此之外，SCFAs 还具有多种调节功能，它们对宿主生理和免疫的影响不断被发现 [2]（图 6-1）。

SCFAs 调节免疫反应的途径主要分为细胞内和细胞外途径：细胞内和细胞外信号通路，包括 G 蛋白偶联受体（G Protein-Coupled

Receptors，GPCRs）和组蛋白去乙酰化酶（histone deacetylase，HDACs）[22]。细胞外途径被证明涉及上皮细胞表面配体，包括GPR41（也称为FFAR3）、GPR43（也称为FFAR2）和GPR109A（也称为HCAR2）[13]。GPR41识别丙酸盐和丁酸盐，GPR43识别所有SCFAs，GPR109A只识别丁酸盐[23]。GPR41在外周血单个核细胞、树突状细胞和多形核中性粒细胞以及脾脏、淋巴结、骨髓、肺、小肠中均有表达。相反，GPR43的表达受到限制，因为它主要位于肠道和特定的免疫群体，如外周血单个核细胞、多形核中性粒细胞、单核细胞和淋巴细胞[24]。GPR109A在肠道细胞、巨噬细胞、单核细胞、中性粒细胞、树突状细胞、脂肪细胞和郎格罕细胞中表达[25]。

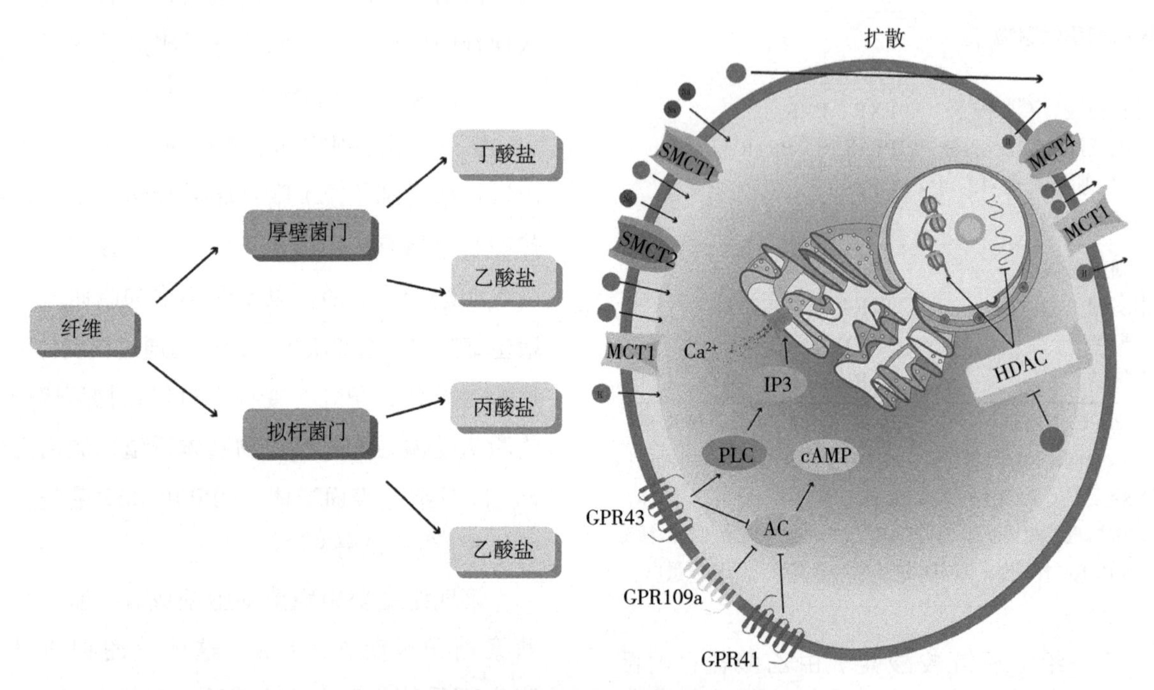

图6-1　短链脂肪酸的形成、转运和作用机制

SCFAs（包括乙酸盐、丙酸盐和丁酸盐）由细菌对膳食纤维发酵而产生。SCFAs的吸收通过3种机制发生：被动扩散、电中性和生电性摄取。SCFAs的电荷决定其吸收是通过被动扩散还是通过载体机制进行。例如，当SCFAs以质子化形式存在时，SCFAs的吸收主要为被动扩散；在生理条件下，这是SCFAs转运的主要机制。相反，阴离子形式的SCFAs的吸收依赖于载体介导，这种吸收通过4种转运蛋白进行。MCT1和MCT4是电中性转运蛋白，它们依赖氢，而SMCT1和SMCT2则依赖钠，分别是生电性转运蛋白和电中性转运蛋白。SCFAs通过刺激3种G蛋白偶联受体（GPR41、GPR43和GPR109a），以及作为组蛋白去乙酰化酶抑制剂而发挥作用。HDAC.组蛋白去乙酰化酶；AC.腺苷酸环化酶；cAMP.环磷酸腺苷；PLC.磷脂酶C；IP3.三磷酸肌醇；MCT.单羧酸转运体；SMCT.Na^+偶联单羧酸转运蛋白；GPR.G蛋白偶联受体

对于SCFAs诱导中性粒细胞趋化以及调节性T细胞（regulatory T-cells，Treg）的扩增，GPR43表达是必需的[2]。研究发现，$GPR43^{-/-}$小鼠缺乏GPR43的表达会严重改变SCFAs介导的结肠Treg细胞的扩增[26]。在肠外，SCFA-GPR43的相互作用可以通过减少中性粒细胞的趋化性以及炎症基因的表达，降低细菌引起的早产风险[27]，并通过调节尿酸盐晶体的炎性小体的组装和免疫细胞的清除，从而下调痛风相关的炎症[2]。SCFAs对宿主生理性的GPR43依赖性作用也延伸到中枢神经系统。小胶质细胞是中枢神经系统的常驻巨噬细胞，其成熟和功能依赖于肠道菌群，而维持小胶质细胞的稳态需要SCFAs和GPR43[28]。在GPR43缺陷型（$GPR43^{-/-}$）和无菌小鼠中观察到炎症调节失调，这种状态表明GPR43与SCFAs的结合是免疫调

节中至关重要的一个环节[13]。此外，一些研究发现，在 GPR43$^{-/-}$ 和 GPR41$^{-/-}$ 小鼠中表现出较低的炎症反应和较慢的细菌清除，表明 GPR43 和 GPR41 可能具有抗炎及清除致病菌的作用[13]。

GPR41 对丙酸盐的亲和力最高，其次为丁酸盐，乙酸盐较弱[24]。GPR41 和 GPR43 通过与 SCFAs 结合参与能量稳态的调节。缺乏 GPR41 会增加小鼠的体脂质量，这表明 SCFAs 通过激活 GPR41 促进能量消耗和减少肥胖[29]。

与 GPR41 或 GPR43 不同，GPR109A 只与丁酸盐结合。GPR109A 是丁酸盐的选择性受体，在结肠上皮细胞面向管腔的顶膜上表达，其表达与丁酸盐一起沿肠道从空肠到结肠逐渐增加[18]。GPR109A 缺失小鼠更容易出现实验性结肠炎和结肠癌。研究表明，丁酸盐受体在黏膜免疫细胞上表达，这些免疫细胞上的受体以及结肠上皮细胞上的受体有助于防止结肠炎症和癌变[23]。缺乏 GPR109A 的小鼠的结肠巨噬细胞和树突状细胞诱导幼稚 T 细胞分化为分泌 IL-10 的 Treg 细胞的能力存在缺陷，导致结肠 Treg 细胞普遍减少，其机制仍不明确[23]。

与它们在肠内的局部作用类似，循环中的 SCFAs 通过降低肺内 IL-4、IL-5、IL-13 和 IL-17a 的水平来减轻过敏性气道炎症的严重性，这可能是通过增加骨髓中巨噬细胞和树突状细胞的生成来实现的[30]。因此，SCFAs 及其 GPCRs 是共生微生物群调节炎症反应的一种途径，但这些代谢物的敏感细胞和免疫介导机制值得进一步研究。

SCFAs 驱动的 HDACs 抑制作用促进产生耐受性、抗炎性的细胞表型，这对于维持免疫稳态是至关重要的，这种现象支持了微生物群可以作为宿主生理的表观遗传调节剂的概念。外周血单个核细胞和中性粒细胞暴露于 SCFAs 中，与暴露于 HDACs 抑制剂、失活 NF-κB 和下调促炎症细胞因子（如肿瘤坏死因子）的生成相似[31]。此外，研究发现 SCFAs 通过抑制 HDACs，导致巨噬细胞促炎因子减少[32, 33]。这些结果表明，SCFAs 通过抑制 HDACs，从而抑制 NF-κB，使促炎症因子基因表达下调，进而抑制肠道炎症反应。SCFAs 还通过抑制 HDACs 影响外周 T 细胞功能，特别是 Treg 细胞。HDACs 抑制剂可以在体内改变 Treg 细胞的数量和功能。这表明 SCFAs 在免疫系统发育和疾病预防中的表观遗传潜力。

在几种炎性疾病的动物模型中，已经证明了 HDACs 抑制剂的免疫调节作用和治疗益处，这提示我们，SCFAs 可能是具有治疗潜力的药物。SCFAs 是一种半衰期短、代谢迅速的挥发性化合物，其对 HDACs 的抑制作用具有浓度依赖性。只有高浓度的 SCFAs 才足以干扰 HDACs 功能，并且影响可能需要特定的转运体[34]。SCFAs 也可以通过 GPCRs 依赖机制间接抑制 HDACs，GPCRs 的 SCFAs 特异性不同，SCFAs 的效力也不同[34]。因此，SCFAs 是否直接或间接阻断 HDACs 活性受到多种因素的影响，包括浓度、转运体、受体以及所涉及的细胞和（或）组织类型。因此，需要更多的研究来探讨 SCFAs 在健康和疾病中的免疫调节功能和治疗潜力。

然而，SCFAs 也会加剧疾病。一项测量囊性纤维化患者痰液中 SCFAs 浓度的研究发现，SCFAs 介导的中性粒细胞募集和持续存在加重了炎症反应，促进了铜绿假单胞菌的生长。因此，SCFAs 的免疫调节作用取决于内环境和细胞类型。细胞特异性和组织特异性 GPCRs 的存在及其不同的代谢物感应能力使宿主能够调节炎症从而控制炎症。

二、色氨酸代谢物

色氨酸是合成蛋白质的原料之一，但是肠菌可以直接利用该氨基酸产生许多具有免疫学意义的代谢产物。目前研究最多的 3 种与宿主微生物群相互作用有关的色氨酸代谢途径如下（图 6-2）：①肠道微生物群直接将色氨酸转化为多种分子，包括芳香烃受体（aryl hydrocarbon receptor，AhR）的配体；②免疫细胞和上皮细胞通过吲哚胺 2，3- 双加氧酶途径的犬尿氨酸代谢途径；③肠嗜铬细胞通过色氨酸（tryptophan，Trp）羟化酶 1 产生 5- 羟色胺的途径 [35]。吲哚、吲哚衍生物和色胺是肠道中主要的微生物色氨酸代谢产物 [36]。吲哚作为色氨酸代谢物的主要物质之一，被认为是重要的中间信号分子，可用于生物膜的形成以及调节细菌的运动性和抵抗非吲哚产生物种的入侵，例如肠沙门菌和铜绿假单胞菌 [37]。大多数吲哚衍生物被认为是 AhR 的配体，如吲哚 -3- 乙醛和吲哚 -3- 醛。

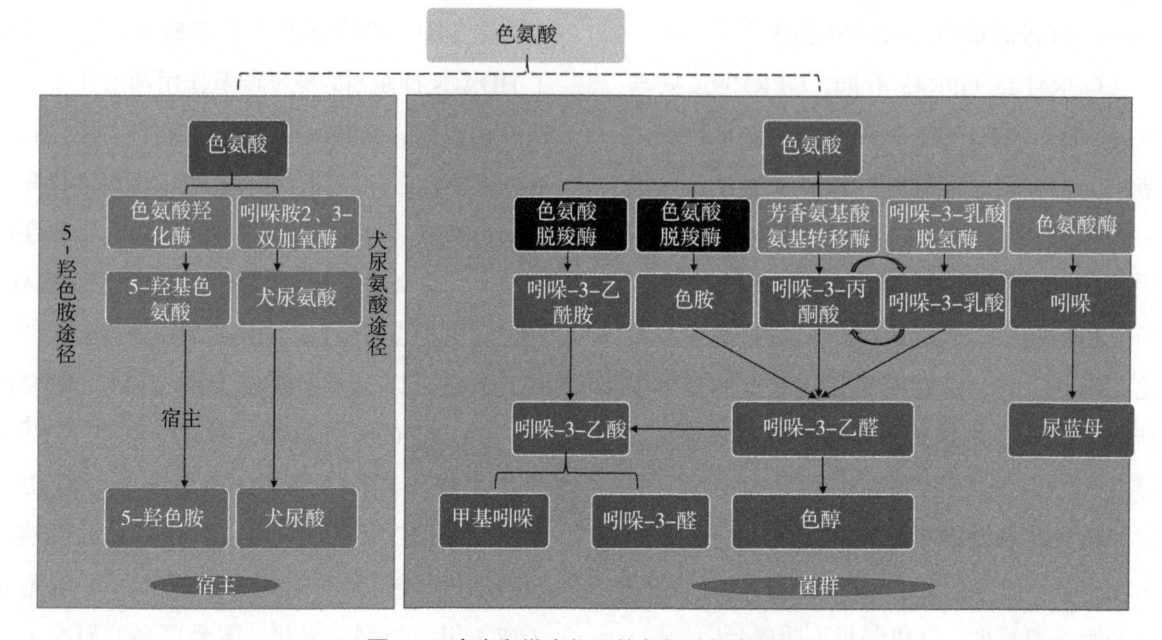

图 6-2　宿主和微生物群的色氨酸代谢途径

在微生物代谢产物中，吲哚和吲哚酸衍生物是肠道内主要的色氨酸微生物代谢产物，肠道微生物产生不同的代谢产物，这些代谢产物是细菌产生催化酶的基础。犬尿氨酸途径和 5- 羟色胺途径是宿主色氨酸代谢的主要途径

在结肠炎小鼠模型中，膳食补充色氨酸和色氨酸代谢产物可减少结肠炎的严重程度 [38]。相反，饲喂缺乏色氨酸饮食的小鼠更容易患上结肠炎 [38]。核受体 AhR 和孕烷 X 受体（pregnane X receptor，PXR）是细菌衍生的色氨酸代谢产物的分子靶标 [39]。细菌衍生的色氨酸代谢产物通过对 AhR 和 PXR 的激活产生抗炎、抑制肿瘤作用 [40]，以及维持上皮屏障和上皮内淋巴细胞正常功能 [41]。此外，PXR 还调节细胞增殖、肿瘤转移和肠道炎症 [42]；因此，激活该受体的色氨酸代谢物可能会影响结肠炎症和结肠癌的发展和严重程度。

此外，胃肠道中 AhR 的配体通过诱导罗伊乳酸杆菌的扩张和抑制致病菌的生长而影响肠道微生物组成 [43, 44]。3 型固有淋巴细胞（group 3 innate lymphoid cell，ILC3）中 AhR 信号的激活诱导 IL-22 的产生，从而促进抗菌肽的分泌。通过这种方式，AhR 的配体可以保护宿主免受白念珠菌的致病性感染 [21]。小鼠体内缺乏 AhR 或缺乏 AhR 的配体会导致肠道微生物紊乱，从而使动物更容易受到啮齿柠檬酸杆菌和单核细胞增生性李斯特菌的感染 [44, 45]。

最近研究发现，吲哚能减轻脂多糖诱导的小鼠肝脏炎症和胆固醇的代谢改变 [46]。这项研究表明，吲哚可以改善肝脏的炎症性疾病。同样，其他吲哚衍生物也得到了广泛的关注，包括吲哚 -3- 醛、吲哚 -3- 乙酸和吲哚 -3- 丙酸。这些分子通过作用于宿主细胞中的核受体而发挥重要的生物学功能。吲哚醛主要由吲哚丙酮酸在芳香族氨基酸转氨酶催化下合成。作为 AhR 的配体，吲哚醛诱导 IL-22 的产生，从而预防念珠菌病和对黏膜损伤提供保护。此外，最近研究表明，吲哚醛诱导 IL-10 受体的表达，这表明吲哚醛具有 IL-10 信号依赖性的抗炎机制 [47]。

研究表明，代谢综合征患者的吲哚 -3- 乙酸水平降低。例如，肥胖者的粪便样本比非肥胖者显示出更低的吲哚 -3- 乙酸水平 [48]。此外，在高脂肪饮食喂养的小鼠中，观察到结肠内吲哚 -3- 乙酸浓度的降低，表现出葡萄糖不耐受和肝脏脂肪变性。此外，吲哚 -3- 乙酸剂量依赖性地降低脂多糖诱导的 TNF-α、MCP-1 和 IL-1β 等促炎症细胞因子，从而产生抗炎作用。吲哚 -3- 丙酸作为 PXR 的配体，也被证明可以抑制肠道促炎症细胞因子并增强屏障功能 [49]。作为一种自由基清除剂，吲哚 -3- 丙酸可以保护大脑神经元和减少肝微粒体膜的氧化应激损伤 [50, 51]。有趣的是，最近的研究发现吲哚 -3- 丙酸通过增加血浆胰岛素水平降低血糖，从而改善葡萄糖代谢 [52]。因此，吲哚 -3- 丙酸可能具有治疗胰岛素抵抗相关代谢功能障碍性疾病的潜力。

一般来说，色氨酸衍生物对人体几乎无害，并可能与宿主发生器官特异性或物种特异性的相互作用。与宿主内源性色氨酸代谢相对简单的背景不同，肠道环境中细菌色氨酸代谢极其复杂，许多具有能代谢色氨酸的菌株仍然未知。因此，需要对色氨酸代谢细菌进行更多的研究。

三、苯丙氨酸代谢物

大多数情况下，饮食中的苯丙氨酸在苯丙氨酸羟化酶（PAH）的帮助下被消化成酪氨酸，然后参与黑色素代谢。苯丙氨酸在苯丙氨酸解氨酶的帮助下转化为苯丙酮酸和苯乙酸（phenylacetic acid，PAA）。苯丙酮尿症患者由于 PAH 活性降低，导致苯丙氨酸毒性代谢物蓄积。一些研究人员描述了肠道共生菌生孢梭菌产生苯丙氨酸代谢产物的途径。该物种可以通过 fldH、fldBC 和 acdA 编码的酶将苯丙氨酸代谢为相应的丙酸衍生物苯丙酸，也可以通过 porA 编码的酶将苯丙氨酸代谢为 PAA[53]。

最近的一项研究表明，PAA 作为肠道微生物群产生的代谢物苯乙酰谷氨酰胺（phenylacetyl glutamine，PAGln）和苯乙酰甘氨酸（phenylacetylglycine，PAGly）的前体。此外，PAGln 与心血管疾病（cardiovascular disease，CVD）和重大心血管不良事件（心肌梗死、脑卒中或死亡）相关 [54]。值得一提的是，PAGln 和 PAGly 的产生具有种属特异性，这意味着在不同的菌群和不同生物肠道中，PAA 可能是不同苯丙氨酸衍生物的前体。所有这些结果表明，重要的是微生物拥有的酶，而不是微生物本身。因此，创造一种能将苯丙氨酸的有毒代谢物转化为有益代谢物的工程菌，以其特殊的酶比例，使人们可以预防某些酶丢失引起的疾病，如苯丙酮尿症，同时在没有“后方”威胁的情况下享受食物，这将是一个有前景的课题。

四、支链氨基酸

最丰富的支链氨基酸（branched-chain amino acid，BCAA）是缬氨酸、异亮氨酸和亮氨酸，是植物、真菌和细菌合成蛋白质的必需氨基酸，尤其是肠道微生物群的成员。它们通过调节蛋白质合成、葡萄糖和脂质代谢、胰岛素抵抗、肝细胞增殖和免疫，在维持哺乳动物体内平衡方面发挥关键作用 [55]。在棕色脂肪组织中 BCAA 分解代谢对于控制产热至关重要。它们通过 SLC25A44 转运蛋白出现在线粒体中，有助于改善代谢状态 [56]。此外，向小鼠补充 BCAA 混合物可促进健康的微生物群增加，肠杆菌科减少 [57]。

然而，BCAA 潜在的作用是有争议的。全身BCAA水平升高与肥胖和糖尿病有关[58]。在遗传肥胖小鼠中，BCAA的积累会诱导胰岛素抵抗[59]。肠道微生物群是BCAA水平的调节剂，因为它可以产生和使用BCAA。普雷沃菌和普通拟杆菌是BCAA的有效生产者，它们的数量与BCAA水平和胰岛素抵抗呈正相关。同时，在胰岛素抵抗患者中，能够吸收BCAA的细菌丰度减少，例如穗状丁酸弧菌和惰性真杆菌[60]。因此，需要进一步研究以更准确地阐明BCAA在代谢紊乱发病机制中的作用。

五、胆汁酸

细菌代谢物不仅来源于膳食成分，还来源于通过宿主代谢分泌到肠腔的底物[61]。初级胆汁酸，如胆酸和鹅去氧胆酸，来源于肝脏中的胆固醇分解作用，在与甘氨酸（人体）或牛磺酸（小鼠）结合后，由胆囊进一步分泌到肠道中[62]。在回肠末端，近95%的胆汁酸被主动重吸收，剩下的则成为结肠细菌进行生物转化反应的底物[63]。这种转化是由于结肠中的特定细菌能够去除胆酸和鹅去氧胆酸中的羟基；这些细菌修饰的胆汁酸被称为次级胆汁酸（如脱氧胆酸和石胆酸）[18]。这些胆汁酸被吸收进入门静脉循环，被肝细胞吸收，并在胆汁中重新分泌。因此，正常胆汁中不仅含有初级胆汁酸，还含有细菌衍生的次级胆汁酸[18]。

微生物群在胆汁酸肠肝循环中起着两个关键作用：将初级胆汁酸转化为次级胆汁酸，以及将结合型胆汁酸去结合生成游离型胆汁酸以供排泄[64]。总的来说，微生物在调节宿主胆汁酸循环中起着至关重要的作用[5]。值得注意的是，细菌可以在功能上改变胆汁酸种类，同时胆汁酸也可以调节细菌的生长。最近的一项研究表明，饮食中饱和脂肪酸的摄入与牛磺酸结合胆汁酸的生成之间存在联系，这种胆汁酸可以促进有害细菌的生长，从而增加对肠道炎症的易感性，表明饮食、共生肠道细菌、胆汁酸代谢和炎症之间存在复杂的相互作用[65]。

初级和次级胆汁酸是消化和吸收膳食脂肪和脂溶性维生素所必需的[3]。此外，胆汁酸在各种疾病中有重要的治疗价值，包括神经系统疾病、脂肪肝、肝纤维化、胆结石、原发性胆汁性肝硬化、胰腺炎和炎症性肠病等[66-68]。胆汁酸通过核受体PXR和法尼醇X受体（farnesoid X receptor，FXR）发挥作用，并通过PXR/FXR介导信号传导、影响靶细胞的基因表达[69]。FXR是核激素受体家族的一员，广泛存在于回肠和肝脏，调节多种代谢途径中的基因表达。调节FXR的配体已成为预防和治疗脂肪肝和相关代谢紊乱疾病的潜在策略[70]。缺乏FXR的小鼠表现出对葡萄糖的耐受性降低及对胰岛素的敏感性降低[71]。相比之下，胆酸激活FXR后，通过抑制与肝脏糖异生相关的多个基因的表达来降低血糖水平[71]。

除了FXR外，膜受体胆汁酸G蛋白偶联受体5（Takeda G protein-coupled receptor 5，TGR5）是另一种典型的胆汁酸受体[29]。该受体是在2002年发现的[72]，它在免疫细胞（单核细胞和巨噬细胞）、肌肉、脊髓、脂肪细胞和肠神经系统中高度表达[73]。这种受体与肠道巨噬细胞和单核细胞显著相关，因为它可以识别初级和次级胆汁酸，从而引发大量的代谢和免疫效应，所有这些都受到饮食和肠道微生物群相互作用的高度调节。石胆酸和脱氧胆酸作为次级胆汁酸，是TGR5最有效的配体[74]。在小鼠中，TGR5缺乏导致葡聚糖硫酸钠和2，4，6-三硝基苯磺酸引起的结肠炎恶化[75]。在小鼠巨噬细胞中，TGR5的特异性缺失导致胰岛素抵抗[76]。

此外，TGR5调节肠黏膜L细胞的胰高血糖素样肽-1（glucagon-like peptide-1，GLP-1）释放，在维持体内葡萄糖的平衡方面具有关键作用[77]。胆汁酸与TGR5结合并抑制NF-κB的激活，导致炎症介质如TNF-α、IL-1β、IL-6和NOS2

的表达降低[64]。最近的一项研究揭示了 TGR5 抗炎症作用的机制，胆汁酸通过 TGR5-cAMP-PKA 轴抑制 NLRP3 炎性小体激活。TGR5 诱导的 PKA 激酶激活导致 NLRP3 泛素化，这与 NLRP3 在单个残基 Ser291 上的磷酸化有关。这种磷酸化作用阻止了 NLRP3 炎性小体的激活，并随后将 IL-1β 前体裂解为 IL-1β。这项研究表明，胆汁酸和 TGR5 结合可阻断 NLRP3 依赖性脂多糖诱导的系统性炎症、明矾诱导的腹膜炎症和 2 型糖尿病相关的炎症[78]。有趣的是，给小鼠补充胆汁酸以 TGR5 依赖的方式促进棕色脂肪组织的能量消耗[79]。

迄今为止，胆汁酸代谢物对肠道微生物群的影响尚无深入的研究。一些研究表明，胆汁酸可以诱导细菌 DNA 和膜损伤，但细菌对胆汁酸诱导的损伤具有保护和修复机制[80, 81]。此外，肠道胆汁酸的缺乏会促进细菌从肠道向内脏组织转移。

六、三甲胺 -N- 氧化物

一般认为，正常的结肠菌群有助于结肠健康，并且由于宿主和细菌的共生关系已经进化了数百万年，这种共生关系不会损害其中的任何一方。但令人惊讶的是，最近的研究却有相反的发现[18]。胆碱通常以磷脂酰胆碱（卵磷脂）的形式存在于饮食中，肉碱是肉类的重要组成部分。肠道微生物将胆碱、磷脂酰胆碱和左旋肉碱代谢产生三甲胺，被宿主肝脏氧化成三甲胺 -N- 氧化物（Trimethylamine-N-oxide，TMAO）[82]。TMAO 由肝脏中黄素单加氧酶（flavin-containing monooxygenase，FMO）催化形成，被认为是早期代谢综合征新的生物标志物[83]。TMAO 与动脉粥样硬化和心肌梗死等心脏事件有关，这为富含脂肪和红肉的饮食与动脉粥样硬化风险增加之间提供了分子联系[84]。

最近的小鼠研究发现了这种心血管毒素的细菌来源，该研究表明动脉粥样硬化的风险可以通过肠道微生物移植来传播[85]。研究发现，TMAO 可在体外增加血小板聚集和与胶原的黏附，并在体内增强血栓形成[86]。这些影响可以通过增加饮食中的胆碱来复制，并被肠道菌群组成显著影响[86]。TMAO 的血浆水平受不同因素的影响，例如饮食、肠道微生物群、药物等。TMAO 的循环水平与动脉粥样硬化斑块大小和心血管事件密切相关[87]。几项荟萃分析发现，血浆 TMAO 水平与 CVD 和死亡风险之间存在关联[88]。循环中 TMAO 水平最高四分位数的患者比最低四分位数的患者出现主要不良心血管事件的风险更高[84]。TMAO 的水平还与斑块形成、患者心血管事件的长期风险和不良预后相关[89]。然而，这种心血管毒素的分子靶点仍然未知。L- 肉碱代谢不仅限于心血管健康，还与胰岛素抵抗相关[90]。研究发现，细菌胆碱分解代谢能产生 TMAO 和诱导胆碱缺乏状态，进而通过调节 DNA 甲基化加重代谢紊乱[91]。

七、多胺

多胺，如腐胺、亚精胺和精胺，存在于几乎所有的活细胞中，是广泛的生物功能的组成部分，包括基因转录和翻译以及细胞生长和死亡[2]。多胺是一种小的聚阳离子分子，来自食物或由肠道微生物群合成[92]。几乎所有的食品都含有多胺，包括大豆、绿茶叶、小麦胚芽、蘑菇、橙子和肉等。大多数食物来源的多胺在小肠中被吸收[93]。而结肠中大量的多胺，特别是腐胺和亚精胺，主要由肠道微生物群产生，包括拟杆菌属和梭杆菌属[94]。

哺乳动物的多胺合成包括精氨酸酶 1（将精氨酸转化为鸟氨酸）、限速酶鸟氨酸脱羧酶（从鸟氨酸合成腐胺）以及相互转化腐胺、亚精胺和精胺的顺序酶。由于其反应性，宿主通过生物合成、分解代谢、吸收和流出机制以及生物合成酶的转录、翻译和降解，从而使得细胞内多胺浓度受到严格调控。与宿主多胺代谢不同，细菌利用组成型或诱导型氨基酸脱羧酶产生多胺[95]。两种主要的肠道微生物：大肠埃希菌和粪肠球菌，可以通过混合途径从精氨酸生物合成腐胺[96, 97]。精氨酸被大肠埃希菌转化为胍丁胺[96]，然后被粪肠球菌代谢成腐胺[96]。

此外，产生酸性化合物的双歧杆菌能加速腐胺的产生[98]。亚精胺的产生由至少两种肠道细菌介导，大肠埃希菌和拟杆菌属[99, 100]。多胺对于肠道微生物群（主要是变形菌门和厚壁菌门）的多种功能至关重要，包括细胞信号、抗应激性、RNA 和蛋白质合成[101]。此外，多胺通过促进自噬作用提高细菌寿命[102]。几种细菌病原体有赖于多胺，从而在宿主体内产生毒性及提高存活，包括幽门螺杆菌、山夫登堡沙门菌等[95]。此外，霍乱弧菌和鼠疫耶尔森菌形成生物膜也需要多胺[84, 85]。

多胺对宿主细胞的功能也很重要，但由于多胺在低浓度下导致细胞生长缺陷以及在高浓度下的毒性和致癌作用，使得多胺和鸟氨酸脱羧酶水平失调。肠道含有高浓度的多胺，肠上皮细胞（IEC）依赖于这个储存库来维持它们的快速周转和高增殖率。多胺也有助于增强肠上皮细胞黏膜屏障的完整性。体外研究表明，多胺可以刺激细胞间连接蛋白的产生，包括闭合蛋白、带状闭合蛋白 1 和 E- 钙黏蛋白（又称钙黏蛋白 1），这些蛋白对调节细胞旁通透性和增强上皮屏障功能至关重要。此外，给幼鼠多胺可诱导小肠黏液和分泌型 IgA 的产生，而喂食缺乏多胺的饲料会导致肠黏膜发育不良[2]。这些观察结果表明，宿主 - 微生物多胺合成是肠道微生物组的一项重要功能，它在生命早期就已获得，是出生后胃肠道发育所必需的。

多胺代谢在调节免疫中起着重要作用，必须受到宿主的严格控制。精氨酸酶 1 和一氧化氮合酶分别作用于精氨酸产生多胺或一氧化氮，是平衡免疫反应效应的重要酶。巨噬细胞向典型（M1）促炎表型的极化会导致一氧化氮合酶诱导亚型的激活、促炎细胞因子的产生和细胞毒性的增强。精胺可通过抑制鸟氨酸脱羧酶的表达和促炎细胞因子的合成来抑制 M1 巨噬细胞的激活，而不会改变抗炎 TGFβ 和 IL-10（均为巨噬细胞抑制剂）的合成，从而保护宿主免受巨噬细胞过度激活引起的严重炎症损伤[103]。益生菌双歧杆菌 LK512 导致肠蠕动和结肠多胺水平升高，这与结肠促炎因子 TNF-α 和 IL-6 水平降低相关[104]。由此可以假设，控制饮食和提供益生菌可能有助于改变结肠多胺代谢而有益于宿主健康。值得注意的是，在健康小鼠中，双歧杆菌 LK512 诱导了对氧化应激的抵抗力和延长寿命，这取决于微生物多胺合成的增强[105]。

此外，研究发现，随着年龄的增长，多胺水平会降低，并且在许多神经退行性疾病中其水平发生改变[106]。多胺还调节全身和黏膜适应性免疫。接受富含多胺母乳的幼犬显示上皮内 $CD8^+$T 细胞和固有层 $CD4^+$T 细胞加速成熟，以及脾脏 B 细胞的早熟增加[107]。

越来越多的证据支持多胺异常生物合成在致癌和肿瘤免疫中的作用。与所有高度增殖的细胞一样，癌细胞需要多胺来满足持续快速生长的需求。与健康人相比，许多癌症患者的尿液和血液中的多胺水平增加，宿主和肠道微生物群对多胺代谢的失调可能导致结直肠癌（colorectal cancer，CRC）[108]。一项比较结肠癌和正常组织样本的代谢组学筛查显示，癌细胞中产生的多胺可以促进细菌生物膜的生长，反过来，生物膜中细菌产生的多胺可以促进和加强癌症的发展[109]。经抗生素治疗后，切除的 CRC 组织没有生物膜，与生物膜阳性组织相比，特异性多胺代谢物 N1，N12- 二乙酰精胺的水平降低[109]。因此，宿主源性多胺和细菌源性多胺可能协同作用，促进 CRC 的发生。N1，N12- 二乙酰精胺可能是生物膜相关肿瘤的潜在生物标志物。

多胺也与皮肤癌和激素相关的癌症有关，包括乳腺癌和前列腺癌[108]。通过抑制鸟氨酸脱羧酶活性减少多胺后，可以以 T 细胞依赖的方式阻止肿瘤生长，这表明减少肿瘤内的多胺可以逆转肿瘤微环境中的免疫抑制[110]。因此，多胺代谢在肿瘤发生中具有重要作用，使得多胺途径成为抗癌治疗和化学预防的一个很有前景的靶点。多胺的复杂调节对宿主和微生物细胞的功能至关重要，需要进一步研究，以了解宿主和微生物的多胺代谢是如何影响宿主健康和疾病的。

八、丙酮酸和乳酸

丙酮酸是由细菌发酵膳食纤维产生的，并进一步还原生成乳酸。细菌代谢产物丙酮酸盐和乳酸诱导小肠 CX3CR1$^+$ 单核细胞将树突伸入肠腔，以捕获管腔抗原并促进抗原特异性免疫反应，从而形成对沙门菌感染的抵抗力[111]。

乳酸在结肠腔中的浓度很高。它可能是由不同细菌菌株（尤其是乳酸杆菌）的新陈代谢引起的，也可能来自饮食（例如酸奶）。乳酸以往被认为是与能量产生有关的代谢终产物和糖异生的碳源，但最近被认为是重要的信号分子[112]。最近研究发现，乳酸是 GPR81 的激动剂[113]，虽然它对免疫系统的影响尚不清楚，但乳酸通过抑制脂肪分解对新陈代谢有着实质性的影响[113]。GPR81 可在乳酸菌的益生菌效应中发挥作用，乳酸菌可诱导分泌型 IgA 产生、抗原摄取和细胞因子产生[114]，从而对机体起保护作用。微生物产生的乳酸也有利于肠屏障的维持，因为它通过 GPR81 依赖的机制促进肠干细胞的分化[115]。GPR81 不是乳酸的唯一作用方式。最近发现 N-myc 下游调节基因家族成员 3（N-Myc Downstream-Regulated Gene 3，NDRG3），一种乳酸靶向的细胞内信号分子，能激活涉及 RAF 和 ERK 的下游信号通路[116]。细菌来源的乳酸和膳食乳酸通过结肠上皮细胞顶膜上的 H^+ 偶联和 Na^+ 偶联单羧酸转运蛋白，进入结肠上皮细胞，但目前尚不清楚新发现的细胞内靶点 NDRG3 是否在结肠上皮细胞中表达，是否与乳酸的生物学效应有关。

九、琥珀酸盐

琥珀酸盐是结肠腔中存在的一种细菌代谢物[117]。这种代谢物是 DNA 去甲基化酶和缺氧诱导因子 1α（hypoxia-inducible factor 1α，HIF1α）脯氨酰羟化酶（prolyl hydroxylase domain，PHD）的 TET 家族的抑制剂[118]；因此，琥珀酸盐可能通过增强 DNA 甲基化来影响结肠上皮细胞的表观遗传特征，并通过增加转录因子 HIF1α 的水平来影响缺氧信号。琥珀酸还可以通过 GPR91 影响细胞功能[119]。此外，琥珀酸盐能促进艰难梭菌的增殖。抗生素治疗引起的菌群失调，会增加局部肠腔琥珀酸水平，艰难梭菌将其作为代谢源转化为丁酸盐，导致细菌生长[120]。多形拟杆菌引起的琥珀酸水平增加也通过上调转录因子 Cra，促进肠出血性大肠埃希菌 O157：H7 的致病性，Cra 可以积极调节与毒力相关的基因[121]。然而，目前关于琥珀酸盐与结肠健康的相关性的研究很少。

十、雌马酚

雌马酚可由人和动物胃肠道中的产雌马酚细菌通过代谢大豆异黄酮类物质（主要是大豆苷元）转化而成。无菌动物和缺乏丰富肠道菌群的新生婴儿无法形成雌马酚，这表明肠道微生物群参与其形成[4]。在人类中，许多肠道微生物群负责生产雌马酚，如双歧杆菌、屎肠球菌、乳杆菌、瘤胃球菌和链球菌[122]。肠道菌群可以产生 β- 葡萄糖苷酶，使大豆异黄酮去糖基化以释放可利用的大豆苷元，大豆苷元可以被整个肠细胞吸收或进一步转化为雌马酚[123]。雌马酚能抑制艰难梭菌的生长和孢子形成，说明它具有抗菌活性[124]。此外，雌马酚还调节某些肠道微生物群的生长速度，如脆弱拟杆菌和普拉梭菌[125]。

十一、化合物 K

化合物 K 是肠道微生物群转化膳食成分后形成的代谢产物[126，127]。它是人参皂苷 Rb1 的微生物代谢产物，人参皂苷 Rb1 是一种 20（S）- 原人参二醇型人参皂苷，也是人参的一种成分[128]。人参皂苷 Rb1 膜渗透性低，易降解[129]。因此，它的生物效应通过肠道微生物代谢而增强[128]。人参皂苷 Rb1 被肠道微生物（如乳酸杆菌、双歧杆菌、拟杆菌和嗜热链球菌）通过脱糖基化和水解进行代谢[128]。不易消化的食品成分和益生元能刺激化合物 K 的形成和吸收，主要是通过增加糖苷酶活性和促进产生化合物 K 的肠道微生物（如乳酸杆菌和拟杆菌）的生长[130]。此外，化合物 K 可通过促进产生 SCFAs 的细菌（乳球菌属和梭状芽孢杆菌属）生长，从而改变肠道微生物群的组成[131]。由于肠道微生物群参与多种有效代谢物的产生，调节肠道菌群组成可能是调节微生物代谢物水平的另一种方法。

十二、类胡萝卜素和酚类化合物

类胡萝卜素和多酚类物质都是植物次级代谢产物，具有潜在的抗氧化和抗炎作用。类胡萝卜素是脂溶性四萜化合物，在蔬菜和水果中以色素的形式存在[132]，多酚是含有一个以上与苯环相连的羟基的分子[133]。虽然类胡萝卜素和多酚的来源主要是植物，但肠道微生物群的贡献是非常值得关注的。膳食类胡萝卜素和多酚的生物可利用性和生物利用度可以被肠道微生物群所改善，从而增加其对人类健康的有益程度[134]。肠道微生物群通过微生物酶将肠道内容物降解为简单分子，从而促进类胡萝卜素和多酚的生成[135]。除了作为自由基清除剂和抗流感分子的直接作用外，类胡萝卜素和多酚已被证明是 AhR 和 PXR 的激动剂，从而维持肠道内环境的稳定、糖代谢正常以及提高对炎症性疾病的抵抗力[136]。

十三、烟酸

烟酸是一种细菌产物，可以通过结合 GPR109A 来抑制炎症和预防肿瘤的发生。在一项体内研究中，通过 GPR109A 信号传导显示，烟酸通过诱导巨噬细胞和树突状细胞的抗炎活性，以及促进 Treg 细胞和产生 IL-10 的 T 细胞分化，从而抑制炎症和肿瘤进展[23]。在溃疡性结肠炎的小鼠模型中证明了这一研究结果，烟酸可以通过介导 GPR109A 来降低结肠炎的严重程度[137]。然而，遗憾的是，我们在文献中没有找到任何相关的有力证据或例子来支持烟酸对结直肠癌的影响。

十四、咪唑丙酸

咪唑丙酸是肠道微生物群利用组氨酸产生的代谢物。最近的研究发现，咪唑丙酸在 2 型糖尿病中增加，并与胰岛素抵抗有关[138]。在肝脏，咪唑丙酸似乎通过哺乳动物的雷帕霉素复合物 1（Mammalian target of rapamycin complex 1，mTORC1）靶点影响胰岛素信号通路。在人群中对咪唑丙酸的检测发现，咪唑丙酸水平与代谢和生活方式有关[60]。糖尿病前期和糖尿病的受试者，以及在该队列中有低细菌基因丰度和拟杆菌属 2 肠型的受试者中，咪唑丙酸水平升高。咪唑丙酸水平与低度炎症标志物之间存在关联。此外，血清咪唑丙酸水平和不健康饮食有关。因此，在 2 型糖尿病患者中，肠道微生物群可能转向产生咪唑丙酸，从而影响宿主炎症和代谢[60]。

十五、精胺和组胺

精胺和组胺由宿主和肠道微生物产生，并通过抑制 NLRP6 炎症体，降低 IL-18 水平从而降低肠道上皮屏障完整性[139]。

第四节　肠菌代谢物和宿主免疫系统

一、肠菌代谢物与宿主免疫系统

肠道微生物群有助于调节宿主许多生理功能，如营养稳态、能量消耗和免疫功能等[140]。肠道微生物群与宿主免疫系统的复杂相互作用始于出生时，微生物启动免疫发育，免疫系统随后协调肠道微生物群的组成[141]。

新近研究表明，微生物群在宿主的免疫发育和免疫调节中发挥了关键作用。这可能是由小分子代谢物通过信号传导驱动的，并对宿主免疫和生理产生深远的影响[142]。如上所述，许多微生物群相关代谢物具有生物活性，如 SCFAs、吲哚和次级胆汁酸，这些代谢物可与宿主中相应的传感系统产生反应[143]。生物活性代谢物影响先天和适应性免疫细胞的成熟、激活、极化和分泌细胞因子等功能，从而调节抗炎或促炎反应（图 6-3）。在接下来的章节中，我们将概述微生物群相关代谢物对肠道免疫细胞亚群及其功能的影响。

二、肠菌代谢物与宿主先天性免疫的相互作用

1. *肠上皮细胞*　肠道在保护宿主免受环境影响和调节营养吸收方面起着至关重要的作用。肠黏膜表面主要由肠上皮细胞构成，肠上皮细胞形成了重要的物理屏障，将共生菌和肠道固有层分隔开来[144]。尽管肠上皮细胞不是典型的先天免疫细胞，但它们是黏膜免疫系统的关键部分。事实上，肠上皮细胞有大量先天免疫相关受体，它们通过识别配体促进肠道内环境稳定[145]。

SCFAs 是肠上皮细胞的能量底物，特别是丁酸盐，它可以调节肠上皮细胞的能量代谢[146]。SCFAs 通过激活 GPCRs 参与调节肠上皮细胞的增殖功能[147-149]。例如，当给无菌小鼠或抗生素治疗的小鼠施用 SCFAs 的混合物时，变薄的肠黏膜厚度增加，减少的肠上皮细胞增殖和周转能力得以恢复，抵抗病毒及细菌入侵的能力加强；这种效应可能是 SCFAs 通过 GPR41 或 GPR43 激活 MEK-ERK 信号来介导的[148]。此外，SCFAs 诱导 GPR43 激活后，通过激活 mTOR 和 STAT3 信号，促进肠上皮细胞抗微生物肽（如 Reg Ⅲ和 β- 防御素）的产生，进而增强抵抗病毒及细菌入侵的能力[150]。

SCFAs 还能促进上皮杯状细胞中黏蛋白基因的转录。研究发现，在无菌小鼠中接种几种产生 SCFAs 的细菌，可以诱导杯状细胞的分化和黏液的分泌[151, 152]。丁酸盐还诱导肠上皮细胞分化并增强细胞凋亡，但涉及的详细机制尚未阐明[4]。此外，丁酸盐还抑制肠干细胞，可能是通过抑制 HDACs 和增强细胞周期负调控因子活性来发挥作用[153]。虽然丁酸盐能抑制肠干细胞，但肠细胞通过对丁酸盐的利用而减少了丁酸盐到达肠干细胞，从而保护肠干细胞免受丁酸盐抑制作用的影响。炎症体是肠上皮细胞的一种重要的先天免疫感应复合物，可产生细胞因子 IL-18 和下游抗菌肽并分泌黏液，从而调节宿主和微生物群之间的相互作用[154]。丁酸盐通过激活肠上皮细胞中的 GPR43 或 GPCR109A 来影响 NLRP3 炎症体，促进下游炎症细胞因

子 IL-18 的表达，从而促进上皮修复和维持屏障功能[139，140]。

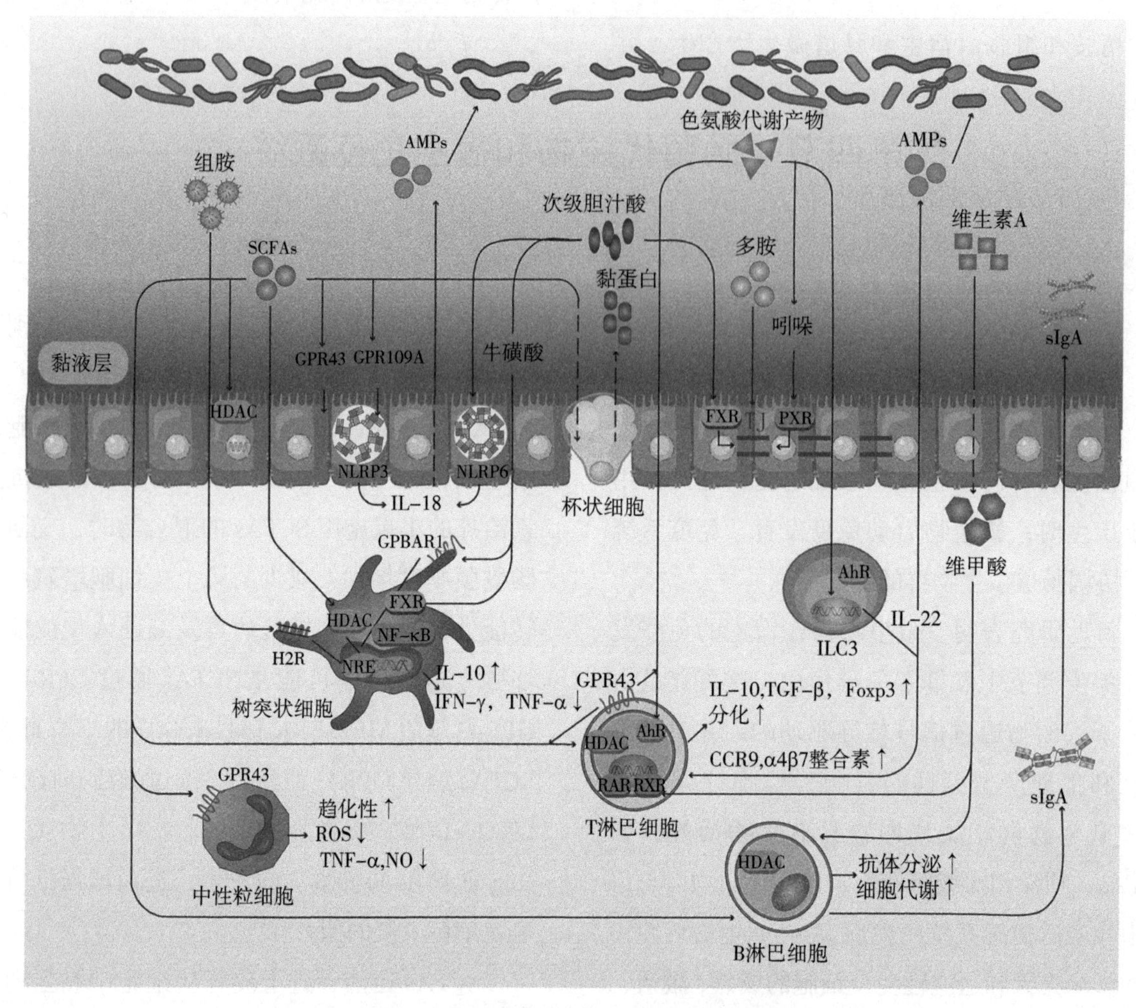

图 6-3 肠道菌群代谢物对免疫细胞的影响

来自微生物或宿主的代谢物参与复杂的宿主 - 微生物相互作用。许多微生物群相关代谢物具有生物活性，如短链脂肪酸、吲哚和次级胆汁酸等，这些代谢物可以通过信号通路与宿主中相应的传感平台反应。生物活性代谢物影响先天性和适应性免疫细胞的成熟、激活、极化和效应功能，从而调节抗炎或促炎反应。AMPs. 抗菌肽；HDAC. 组蛋白去乙酰化酶；NLRP3. 核苷酸结合寡聚化结构域样受体蛋白 3；NLRP6. 核苷酸结合寡聚化结构域样受体蛋白 6；FXR. 法尼醇 X 受体；PXR. 孕烷 X 受体；sIgA. 分泌型免疫球蛋白 A；H2R. 组胺受体 2；NRE. 负调控元件；NF-κB. 核转录因子 -κ B；GPBAR1.G 蛋白偶联胆汁酸受体 1；AhR. 芳香烃受体；RAR. 维甲酸受体；RXR. 维甲酸 X 受体；Foxp3. 叉头转录因子 3；CCR9. 趋化因子受体 9；ILC3.3 型固有淋巴细胞；ROS. 活性氧；SCFAs. 短链脂肪酸；GPR.G 蛋白偶联受体；IL. 白细胞介素；TNF-α. 肿瘤坏死因子 α；TGF. 转化生长因子

IL-10RA 表达与肠黏膜屏障形成直接相关，丁酸盐通过 IL-10RA 依赖性机制抑制紧密连接蛋白 -2 的表达，从而增强肠上皮屏障功能[155]。丁酸盐还能加速肠上皮细胞的线粒体耗氧率，稳定缺氧诱导因子及其靶基因，增强屏障功能[156]。此外，丁酸盐通过激活 AMP 活化蛋白激酶加速带状闭合蛋白 1 的组装，以降低肠黏膜的通透性[157]。此外，研究表明，丁酸盐通过转录因子 SP1 上调肠上皮细胞中 TGFβ 的产生，从而维持肠上皮内稳态[158]。SCFAs 作为 HDACs 抑制剂，抑制脂多糖诱导的 NLRP3 炎症体的激活和自噬，从而增强肠屏障功能，并保护肠屏障免受破坏[159]。根据最近的一项研究，SCFAs 诱导的 HDACs 抑制还通过激活 p21 活化激酶 1 和乳脂肪球表皮生长因子 8 蛋白，从而诱导肠上皮细胞迁移来促进伤口愈合[4]。

吲哚是一种微生物介导的氨基酸代谢物，通过 PXR 促进上皮屏障功能，有助于加强紧密连接。吲哚增加了肠黏膜紧密连接形成、黏蛋白产生和抑制活化转录因子有关的基因表达，减少了促炎症细胞因子（如 IL-8）的产生，同时增加了抗炎症细胞因子（如 IL-10）的产生[160]，从而减轻炎症反应、保护肠道黏膜屏障。

肠道微生物群产生的胆汁酸被认为是消化和吸收膳食脂肪和胆固醇的必要物质，尤其是在小肠中。在肠道中，胆汁酸激活核受体 FXR 或 TGR5，进而发挥其调节作用[161, 162]；而次级胆汁酸与 FXR 结合后，还参与了调节上皮细胞完整性和微生物组成[5]。FXR 缺乏小鼠表现出肠上皮屏障破坏和肠道稳态平衡失调[163]。例如，微生物胆汁酸衍生物脱氧胆酸和石胆酸与 FXR 结合并调节肠上皮细胞完整性[163]。此外，脱氧胆酸（生理浓度）和 FXR 之间的相互作用可以抑制肠上皮细胞中的 EGFR/Src/ERK 途径，导致肠上皮细胞增殖减少[164]。

然而，胆汁酸也可以与激活蛋白 1 和 c-Myc 信号通路相互作用，增加结肠癌细胞的增殖及转移等生物学行为[165]。此外，闪烁梭菌可以介导原发性胆汁酸向继发性胆汁酸转化，它定植后会通过抑制肝窦内皮细胞产生趋化因子 CXCL16 和减少自然杀伤 T 细胞在肝脏的蓄积而诱导肝肿瘤生长[166]。肠内分泌 L 细胞是肠上皮细胞的一个分支，它也通过 FXR 和 TGR5 感知胆汁酸，后者调节 GLP-1 的分泌。胆汁酸激活 TGR5 后，细胞内环磷酸腺苷（cyclic adenosine monophosphate，cAMP）和钙水平增加，从而促进 GLP-1 的分泌[167]。此外，去结合形式的胆汁酸可能会增加结肠细胞的周转[168]。

多胺，主要是腐胺，增加肠上皮细胞的 DNA 合成。这种效应可能是通过调节酶活性来实现的，而酶活性是启动 DNA 合成或刺激起始因子所必需的[4]。多胺诱导 Rac1 活化，并依次激活 RhoA 和 Cdc42，这对肌动蛋白聚合非常重要，导致肠上皮细胞迁移[4]。已有研究表明，多胺活化的 Rac1 与 PLCγ1 形成复合物，增加钙离子的流入，促进肠上皮细胞迁移[169]。肠上皮细胞从隐窝向绒毛的迁移对肠黏膜的组织完整性以及受损肠黏膜的修复或愈合至关重要。此外，多胺通过增加闭合蛋白、带状闭合蛋白 1 和 E- 钙黏蛋白的产生，也有助于增强肠上皮细胞屏障的完整性[170, 171]。多胺也可能参与肠道成熟；例如，亚精胺和精胺诱导半乳糖基转移酶活性和糖蛋白半乳糖基化，这对肠上皮细胞刷状缘膜中黏蛋白和大多数酶的合成非常重要[4]。此外，精胺可能通过减少 NLRP6 炎性体组装，从而抑制肠上皮细胞产生 IL-18 和抗菌肽。

雌马酚通过促进抗氧化基因的表达，增强抗氧化酶活性，可以保护肠上皮细胞免受氧化损伤[172]。此外，雌马酚还可以保持紧密连接的完整性，并抑制肠上皮细胞产生 IL-8[173]。肠内分泌 L 细胞表达特定的雌马酚受体、GPR30 和雌激素受体。雌马酚与受体之间的相互作用导致细胞内 Ca^{2+} 和肌动蛋白重组水平升高，从而抑制 GLP-1 分泌[174]。

通过对 Caco-2 细胞（一种人克隆结肠腺癌细胞）体外研究表明，人参皂苷代谢产物 K 通过 Na^+ / 葡萄糖共转运蛋白 1（Na^+/glucose cotransporter 1，SGLT1）介导葡萄糖摄取，这对改善肠道炎症有重要意义。化合物 K 诱导表皮生长因子受体（epidermal growth factor receptor，EGFR）磷酸化，这在功能上是环磷酸腺苷反应元件结合蛋白（cAMP response element binding protein，CREB）和 CREB 结合蛋白结合 SGLT1 启动子所必需的，将 SGLT1 染色质转变为激活状态，从而导致肠上皮细胞摄取葡萄糖[175]。化合物 K 还抑制脂多糖激活的人结直肠癌细胞系分泌 IL-8，提示具有抗炎作用[176]。另外，化合物 K 还可以将细胞周期阻滞在 G1 期，从而抑制细胞生长，诱导活性氧生成，通过调节线粒体依赖的凋亡和丝裂原活化蛋白激酶（mitogen-activated protein kinase，MAPK）途径，导致细胞凋亡[176，177]。

2. *固有淋巴细胞* 与适应性淋巴细胞相比，固有淋巴细胞是相对罕见的先天性淋巴细胞，但是它们在黏膜相关组织的屏障表面上大量存在[178，179]。最近，一些研究报道了特定的微生物代谢物可以调节固有淋巴细胞，并且大多数研究集中在ILC3[180]，它产生细胞因子IL-22，是IL-10细胞因子家族的一员。IL-22的缺乏与宿主感染和代谢紊乱有关[181]。此外，IL-22促进抗菌肽的表达，以限制共生细菌（如肠道分节丝状菌）的定植，诱导表面蛋白的岩藻糖基化，以促进有益细菌的定植，并增强杯状细胞的增殖以分泌黏蛋白[182，183]。

AhR在黏膜免疫中起着关键作用。核受体RORγt对固有淋巴细胞的发育至关重要，并且RORγt$^+$固有淋巴细胞大量存在于肠黏膜固有层中[184，185]。事实上，在没有AhR的情况下，RORγt$^+$固有淋巴细胞经历更高的凋亡率，产生更少的IL-22。RORγt与AhR相互作用，刺激后者在IL-22位点结合，促进转录。如上所述，色氨酸分解代谢产物可以特异性调节固有淋巴细胞。色氨酸可以被微生物和宿主转化为活性物质，并且大多数这些代谢物是AhR激动剂。乳酸杆菌代谢色氨酸产生的代谢物是AhR的配体，这些配体有助于抵抗白念珠菌和肠道分节丝状菌的定植。例如，在Card9$^{-/-}$小鼠肠道中的乳酸杆菌数量较低，这导致色氨酸代谢受损，从而降低了AhR配体的产生。这种缺陷进一步损害了IL-22的产生，增加了小鼠对葡聚糖硫酸钠诱导的结肠炎的易感性。此外，在炎性肠病患者的粪便中产生AhR激动剂的能力，尤其是那些来自微生物群的激动剂的能力受到损害。一般来说，3型固有淋巴细胞群体和IL-22产生的平衡，在很大程度上依赖于产生AhR激动剂的能力；因此，这些微生物来源的色氨酸代谢物对于抵抗致病菌定植和维持肠道免疫稳态至关重要。

3. *树突状细胞* 树突状细胞是抗原呈递细胞，在连接先天免疫系统和获得性免疫系统中发挥关键作用。树突状细胞的免疫调节作用依赖于产生细胞因子和诱导幼稚CD4$^+$T淋巴细胞的分化[186]。树突状细胞的激活、存活和成熟主要受其微环境内的局部因素影响，如微生物成分、细胞因子和代谢物[187]。

SCFAs，主要是丁酸盐，通过控制趋化因子或趋化剂受体的表达，调节免疫细胞向肠道的募集。例如，丙酸盐和丁酸盐通过减少人单核细胞来源的树突状细胞产生的几种趋化因子，包括CCL3、CCL4、CCL5、CXCL9、CXCL10和CXCL11，来抑制白细胞迁移[188]。SCFAs作为HDACs抑制剂，当暴露于树突状细胞时，SCFAs驱动的HDACs抑制对于抑制NF-κB和TNF-α至关重要[189]。丙酸盐和丁酸盐通过其转运蛋白Slc5a8抑制骨髓前体分化为树突状细胞[190，191]，树突状细胞暴露于丁酸盐时，炎症细胞因子IL-12和IFN-γ的产生减少，IL-10和IL-23的产生增加[192]。一些证据也表明丁酸盐具有调节树突状细胞表达抗原和诱导T细胞分化的能力。例如，用丁酸盐处理树突状细胞降低了共刺激蛋白的表达，包括CD40、CD80、CD86和主要组织相容性复合体Ⅱ类分子[193]。

组胺是由组氨酸通过脱羧酶转化而来的生物胺[194]。组胺通过组胺受体2（histamine receptor 2，H2R）调节树突状细胞对微生物配体的反应，然后增强树突状细胞抗原呈递能力[5]。给H2R敲除小鼠服用产生组胺的鼠李糖乳杆菌菌株，降低了微生物的免疫调节活性[5]。另外，罗伊氏乳杆菌来源的组胺抑制TNF-α的产生，并抑制MEK/细胞外信号调节激酶MAPK信号。

次级胆汁酸作为G蛋白偶联胆汁酸受体1（G protein-coupled bile acid receptor 1，GPBAR1，又称为TGR5）的激动剂，能调节树突状细胞的功能[195，196]。这些受体与抗炎症反应有关，包括抑制NF-κB活性和NF-κB依赖性转录[197]。GPBAR1 信号激活导致 cAMP-PKA 调节的 STAT1 和 NF-κB 的抑制，GPBAR1 的激活导致细胞内 cAMP 的积累，进而促进 CD14$^+$ 单核细胞分化为 CD209$^+$ 树突状细胞[5]。

4. 中性粒细胞　中性粒细胞是聚集在侵入部位的第一个效应细胞，在那里它们杀死并消化细菌[5]。在黏膜感染或炎症的情况下，中性粒细胞将通过黏膜固有层进入感染或炎症部位。中性粒细胞的迁移取决于多种因素，如趋化因子和整合素[5]。迁移到指定位置后，活化的中性粒细胞产生一氧化氮、趋化因子（如 IL-8）和促炎症因子（如 TNF-α 和 IL-1β），导致炎症反应和黏膜损伤[198]。一些报道表明，SCFAs 是中性粒细胞的有效激活剂[199]，体外实验已经证明 SCFAs 通过结合 GPR43 诱导中性粒细胞趋化性。反过来，GPR43 依赖性信号导致 p38MAPK 的磷酸化和活化[198]，p38MAPK 磷酸化被认为是趋化性的主要决定因素[200]。

与正常骨髓来源的中性粒细胞（bone marrow-derived neutrophils，BMDNs）相比，Gpr43$^{-/-}$ BMDNs 在用乙酸盐处理时不显示趋化性[200]。此外，与野生型结肠炎小鼠相比，用 SCFAs（乙酸盐或丁酸盐）处理的 GPR43$^{-/-}$ 结肠炎小鼠显示结肠中性粒细胞含量降低[200]。SCFAs 对 HDACs 活性的抑制也抑制了中性粒细胞 NF-κ B 的活化和一氧化氮的产生。暴露于 HDACs 抑制剂后观察到类似的结果。尽管如此，SCFAs 对于活性氧产生有一定作用，乙酸促进活性氧产生，丁酸抑制活性氧产生，丙酸则没有影响[201]。中性粒细胞产生的活性氧有助于抵御微生物群，调节细胞内信号活性，并调节炎症过程[202]。

三、肠菌代谢物与宿主适应性免疫的相互作用

共生微生物对免疫系统的影响不仅包括先天性免疫反应，还包括适应性免疫反应。例如，无菌小鼠有一个不成熟的适应性免疫系统[203]。尽管肠道菌群代谢物对免疫系统发育影响的分子机制在很大程度上仍是未知的，但目前已有研究证明几种代谢物能调节适应性免疫细胞的功能，特别是 CD4$^+$T 和 B 淋巴细胞的功能。

很多观察结果揭示了特定代谢物的功能及其机制。例如，SCFAs 促进结肠中 Treg 细胞的 GPR15 依赖性归巢[204]。丁酸盐抑制 HDACs 和过氧化物酶体增殖物激活的受体 γ，导致促炎症细胞因子表达减少[205]。此外，丁酸盐以浓度依赖的方式影响 T 细胞增殖。低水平的丁酸盐刺激 T 细胞增殖，高水平的丁酸盐抑制 T 细胞增殖，这可能与细胞周期进程的改变有关。精胺和亚精胺除了促进 B 细胞早期成熟（诱导 IgM$^+$ 细胞百分比增加）外，还可加速 CD8$^+$T 细胞的成熟，增加上皮内自然杀伤细胞的存在，并提高固有层中成熟 CD4$^+$T 淋巴细胞的百分比[4]。

此外，化合物 K 可抑制 T 细胞激活信号，包括主要组织相容性复合体Ⅱ类分子和共刺激分子，如 CD80 和 CD86，从而抑制 T 细胞启动和减少活化 T 细胞的数量[206]。最后，化合物 K 可能具有细胞毒性，因为它在体外降低 T 细胞的存活[207]。总之，大多数代谢物发挥了维持组织内稳态所必需的抗炎症作用。我们将在下面具体讨论肠道菌群代谢物与宿主适应性免疫功能的相互作用。

1. CD4$^+$ T 淋巴细胞　如前所述，共生物种有助于调节 CD4$^+$T 细胞极化，使之成为特定的 CD4$^+$T 细胞亚群，包括辅助性 T 细胞（Th1、Th2、Th17 细胞）和 Treg 细胞。一般来说，肠道的免疫稳态和耐受性构成了促炎的 Th 细胞和抗炎的 Treg 细胞之间的平衡。

（1）辅助性 T 细胞：抗原呈递细胞分泌的 IFN-γ 作用于原始 Th 细胞的信号转导过程，并作用于转录激活因子 -1（signal transducers and activators of transcription1，STAT-1）。STAT-1 激活特异性转录因子 T-bet，促使原始 Th 细胞分化

为 Th 1 细胞[5]。人和小鼠 T 细胞上的石胆酸和维生素 D 受体之间的相互作用降低了 ERK1/2 磷酸化，抑制了 Th1 细胞激活，这是由 Th1 细胞相关细胞因子的生成减少和 STAT-1 磷酸化的降低所决定的[4, 208]。Th1 细胞应答可由微生物代谢产物调节。与食用不饱和脂肪（来自红花油）的小鼠相比，那些食用饱和脂肪（来自牛奶）的小鼠表现出沃氏嗜胆菌显著富集。沃氏嗜胆菌的扩增诱导了促炎性 Th1 细胞应答，并增加 $IL-10^{-/-}$ 小鼠中的结肠炎发生率[15]。因此，细菌的致病能力可以通过饮食代谢物的产生来调节[15]。

Th1 和 Th2 细胞在生理情况下是动态平衡的，并被相互分泌的细胞因子调节。Th1 细胞分泌的 IFN-γ 抑制 Th2 细胞的极化，而 Th2 细胞产生的 IL-4 抑制 Th1 细胞的增殖。Th1/Th2 细胞的失衡与多种自身免疫性疾病和炎症性疾病有关，如炎性肠病[209]。细胞膜或细胞内不同受体的激活可能导致不同的活性。例如，淋巴细胞 H1R 的激活促进 Th1 细胞极化，而 H2R 的激活抑制 Th1 和 Th2 细胞极化[194]。

组胺对依赖受体的 T 淋巴细胞有重要的调节作用。Th17 细胞是除 Th1 和 Th2 细胞之外的 $CD4^{+}$T 细胞极化，Th1 和 Th2 细胞也是促炎症效应的 Th 细胞。Th17 细胞主要通过分泌 IL-17、IL-23 和 IL-22 等细胞因子发挥作用，促炎症介质 IL-17 和 IL-23 可激活 NF-κ B 和相关的炎症信号通路，并阻断抗炎细胞因子 IL-10 的表达[210]。

AhR 的配体是 T 细胞分化的关键调节剂。例如，激活的 AhR 结合 STAT-1 和 STAT-5，抑制 Th17 细胞的发育[211]。此外，吲哚通过诱导 Treg 细胞的扩增、功能和稳定性，来调控 Treg 细胞 /Th17 细胞系的比例，同时抑制 STAT-3 和 RORγt- 介导的 Th17 细胞发育[4]。致病性 Th17 细胞中的 AhR 表达降低，导致组织损伤。体外研究表明，AhR 的配体吲哚 -3- 乳酸抑制小鼠 Th17 细胞的极化[5]。除了微生物代谢产物对 $CD4^{+}$T 细胞功能的直接调节之外，微生物群信号诱导的肠上皮细胞、树突状细胞或巨噬细胞产生的细胞因子也在 $CD4^{+}$T 细胞功能中发挥重要作用[212, 213]。

（2）调节性 T 细胞：在肠道淋巴细胞中，许多淋巴细胞协同工作，抵御传染性病原体，并维持肠上皮黏膜屏障完整性。然而，效应 T 细胞或髓系细胞对食物和常驻共生微生物群产生的无限制的免疫应答，对肠道健康和免疫稳态是有害的[214]。为了维持免疫稳态，免疫系统必须耐受来自食物和肠道微生物群的抗原，这些抗原部分依赖于诱导型 Treg 细胞。Treg 细胞的特征是 CD4、CD25 和 FOXP3 的表达以及抗炎症细胞因子 TGF-β 和 IL-10 的产生。肠道 Treg 细胞的发育受到来自微生物和宿主的几种代谢物的影响。

如上所述，SCFAs 对先天免疫系统有许多影响，并对 Treg 细胞生物学产生深远影响。研究表明，SCFAs 可影响 $FOXP3^{+}$/$IL-10^{+}$ 结肠 Treg 细胞的数量，并提高对结肠 Treg 细胞的调节[215]。丁酸盐是一种 HDACs 抑制剂，通过促进启动子和增强子保守的非编码序列 CNS1 的组蛋白乙酰化作用，对 FOXP3 位点进行表观遗传控制，从而增强基因表达以及效应子功能[216, 217]。进一步的研究表明，SCFAs 通过激活 GPR43 或 GPR41 刺激 Treg 细胞的增殖[218]，并通过抑制 HDACs（如 HDAC6 和 HDAC9）激活将原始 $CD4^{+}$ T 细胞分化为 Treg 细胞[215]。

在无菌和无特定病原体小鼠中，丙酸盐通过 GPR43 发出信号，特异性诱导产生 IL-10 的 $FOXP3^{+}$Treg 细胞增殖。T 细胞中的 HDACs 抑制增强了 p70 S6 激酶的乙酰化和 rS6 的磷酸化，并进一步调节与产生 $IL-10^{+}$T 细胞相关的 mTOR 途径[219]。有趣的是，不同的 SCFAs 通过不同的机制调控 Treg 细胞。乙酸盐和丙酸盐刺激结肠 Treg 细胞的扩增，而丁酸盐增加原始 T 细胞向 Treg 细胞的从头分化。除了直接调节 T 细胞反应外，SCFAs 还调节树突状细胞与 T 细胞相互作用，例如，通过 HDACs 的抑制作用抑制 NF-κB 中 RelB 的表达，并通过树突状细胞中 GPR109A 的激活诱导抗炎症基因的表达，从而导致 Treg 细胞分化。总之，这些结果表明，包括梭状芽孢杆菌在内的微生物群产生的 SCFAs，通过多种不同的机制来调控抗炎的 Treg 细胞 [15]。

维 A 酸（retinoic acids，RA），是一种通过醛脱氢酶催化膳食维生素 A 的代谢产物，在介导 Treg 细胞扩增中起关键作用 [220, 221]。与 TGF-β 结合，RA 诱导外周来源的 Treg 细胞（peripheral regulatory T cells，pTregs）分化 [222, 223]。此外，通过与 RA 受体和视黄醇 X 受体异源二聚体结合，RA 激活 TGFβ–SMAD 信号以促进 FOXP3 转录 [224, 225]。RA 在体外也能促进 RORγt^{+} FOXP3^{+}pTregs 的生成，抑制 RA 信号可阻止这些细胞在体内的发育 [226, 227]。RA 增加 pTregs 上肠道归巢标记物（如 CCR9 和 α4β7 整合素）的表达，这有助于原始 T 细胞迁移到不同的组织部位 [222, 228, 229]。其他调节肠道微生物群与宿主之间相互作用的维生素大多属于 B 族和 K 族，宿主无法合成，必须依赖共生微生物进行合成。例如，水溶性维生素 B_9 可由几种细菌（如双歧杆菌和乳酸杆菌）合成，在维持 Foxp3^{+}Treg 细胞内稳态方面起着关键作用 [230, 231]。

大多数色氨酸代谢物是 AhR 的配体。一些宿主来源的色氨酸代谢产物也可以作为 GPR35 和 GPR109A 激动剂，诱导结肠 Treg 细胞分化 [232]。罗伊乳杆菌衍生的 AhR 激动剂吲哚 -3-乳酸，可将上皮内 CD4^{+}T 细胞重编程为 CD4^{+} CD8α α^{+} T 细胞 [233]。据报道，多胺能加速黏膜固有层 CD4^{+}T 细胞成熟，以调节大鼠的适应性免疫 [234]。组胺除了影响树突状细胞分泌细胞因子外，还影响 Treg 细胞的功能 [235]；例如，激活 Treg 细胞的 H1R，导致细胞的抑制功能受到抑制，这与 CD25 和 FOXP3 的表达降低有关 [236]。

此外，叶酸通过上调抗凋亡因子 BCL-2 的表达，促进 FOXP3^{+}Treg 细胞的存活。喂食缺乏叶酸食物的小鼠在肠黏膜固有层中表现出 FOXP3^{+}Treg 细胞的数量减少，并且对结肠炎的易感性更高 [237, 238]。

2. B 细胞　在免疫调节和保护肠道免受微生物侵害方面，免疫球蛋白的分泌发挥着重要作用。分泌型免疫球蛋白 A（secretory immunoglobulin A，sIgA）是肠黏膜中最大的一类免疫球蛋白 [5]。sIgA 由黏膜固有层中的浆细胞（分化的 B 细胞）分泌，然后通过肠上皮进入管腔，在管腔中靶向微生物抗原，防止细菌移位和感染 [239]。膳食纤维摄入量与肠道 IgA 水平呈正相关 [5]。

胃肠道中抗体的产生和分泌依赖于微生物代谢物的感应。通过将 ^{13}C 标记的大肠埃希菌导入无菌小鼠体内，发现许多微生物代谢物，特别是氨基酸，能迅速渗透宿主几乎所有组织。研究发现，小肠是微生物代谢物渗透到宿主组织的主要部位，并且免疫球蛋白能加速微生物从小肠清除。

此外，SCFAs 能够通过抑制 HDACs 来调节 B 细胞基因表达，从而促进抗体分泌 [240]。进一步的研究表明，SCFAs 调节代谢传感器，以增强 B 细胞的氧化磷酸化、糖酵解和脂肪酸合成 [241]。这些功能增加了线粒体能量的产生，以促进 B 细胞的活化、分化和抗体的产生 [242]。这些观察结果表明，微生物代谢物，特别是 SCFAs，有助于抗体的产生，抗体可以同时加速病原体的清除和增强共生体的定植。与 SCFAs 在增强抗体生成中的作用相反，支链短链脂肪酸可以抑制 IgA 反应 [53]。

此外，维 A 酸还调节 B 细胞的活性，直接影响 B 细胞产生 IgA 的能力 [243]。研究发现，给予雌马酚的小鼠表现出更多的 B 细胞抗原特异性 IgE 产生 [4]。尽管如此，由于微生物代谢物在调节 B 细胞抗体产生中的作用仍不清楚，因此需要更多研究探索微生物代谢物对宿主抗体反应的调节作用[5]。

总之，大多数代谢物发挥了维持组织内稳态所必需的抗炎症作用。然而，高水平的代谢物可能会在宿主中产生负反馈；例如，长期的抗炎作用可能会增加感染疾病的风险。因此，保持肠道微生物群和微生物衍生代谢物的平衡对宿主健康至关重要。重要的是，微生物群与宿主的关系是相互作用的，微生物的代谢物会影响宿主细胞，宿主也可能影响微生物代谢物的存在。具体而言，宿主免疫系统影响微生物代谢物的存在；例如，分泌性抗体可以缩短微生物在小肠中的停留时间，并限制微生物代谢物对其他组织的渗透能力。因此，宿主免疫系统的改变可能会导致微生物群的失衡，从而改变微生物代谢产物的释放。肠道微生物群失衡或缺乏代表性代谢物可能导致宿主疾病，如下一节所述。

第五节　肠菌代谢物与人类疾病

基于前述，微生物代谢物水平和成分的改变可能会影响宿主免疫功能和新陈代谢，以及其他许多看似不相关的生理功能，因而与人类某些疾病的发生、发展密切相关。这些疾病包括炎性肠病、结直肠癌、糖尿病和非酒精性脂肪肝，甚至心血管和神经系统疾病等（图 6-4）。

一、肠菌代谢物与炎性肠病

炎性肠病（inflammatory bowel diseases，IBDs）是一组病因未明且与免疫相关的肠道炎症性疾病。克罗恩病和溃疡性结肠炎是 IBD 最常见的形式。许多研究已经证明 IBD 患者肠道菌群组成的变化，表明肠道菌群的变化以及相关代谢产物可能与这些疾病有关[244, 245]。此外，微生物代谢物的变化可能会导致代谢物与宿主细胞之间的相互作用受损，特别是肠上皮细胞和免疫细胞。

IBD 与色氨酸代谢产物的改变有关。IBD 患者血浆犬尿氨酸和犬尿烯酸水平升高[246]，血浆色氨酸浓度降低，外周血和结肠细胞中吲哚胺 2, 3- 双加氧酶（indoleamine 2, 3-dioxygenase，IDO 活性升高[247]。作为色氨酸降解途径的第一步，IDO1 可以将色氨酸和其他吲哚衍生物转化为犬尿氨酸[19]。细胞中 IDO1 的表达会根据疾病的活动而变化[248]，IDO1 的局部高表达可以产生抗炎作用，以抵消激活的 T 细胞对 IBD 结肠黏膜的破坏作用[249, 250]。有趣的是，植物来源的吲哚化合物已在传统草药中用于治疗 IBD，从而支持犬尿氨酸与 AhR 相互作用的重要性及其对免疫系统的作用[251, 252]。

在 IBD 中，增加的促炎细胞因子，包括干扰素、IL-1 和 IL-6，被证明可以诱导色氨酸分解代谢途径，从而降低血浆色氨酸水平，增加色氨酸分解代谢物水平[247]。在克罗恩病患者和炎性肠病动物模型中，炎症黏膜中 5- 羟色胺水平升高，这表明 5- 羟色胺在驱动 IBD 肠道炎症中发挥了重要作用[253]。几种特异性肠道细菌色氨酸代谢产物也参与了 IBD 的病理生理学[254]。在患有 IBD 的犬中，细菌色氨酸代谢物（吲哚乙酸酯和吲哚丙酸酯）明显减少，这些代谢物被认为在肠道中具有抗炎症功能[255]。在 IBD 患者中，粪便中的吲哚 -3- 乙酸（肠内抗炎功能）水平降低，这表明细菌色氨酸代谢降低可能有助于炎性肠病的发生。因此我们可以假设，通过恢复肠道内细菌色氨酸代谢来增加吲哚丙烯酸生成，以促进 IBD 患者的抗炎反应，这可能具有有益的治疗效果。

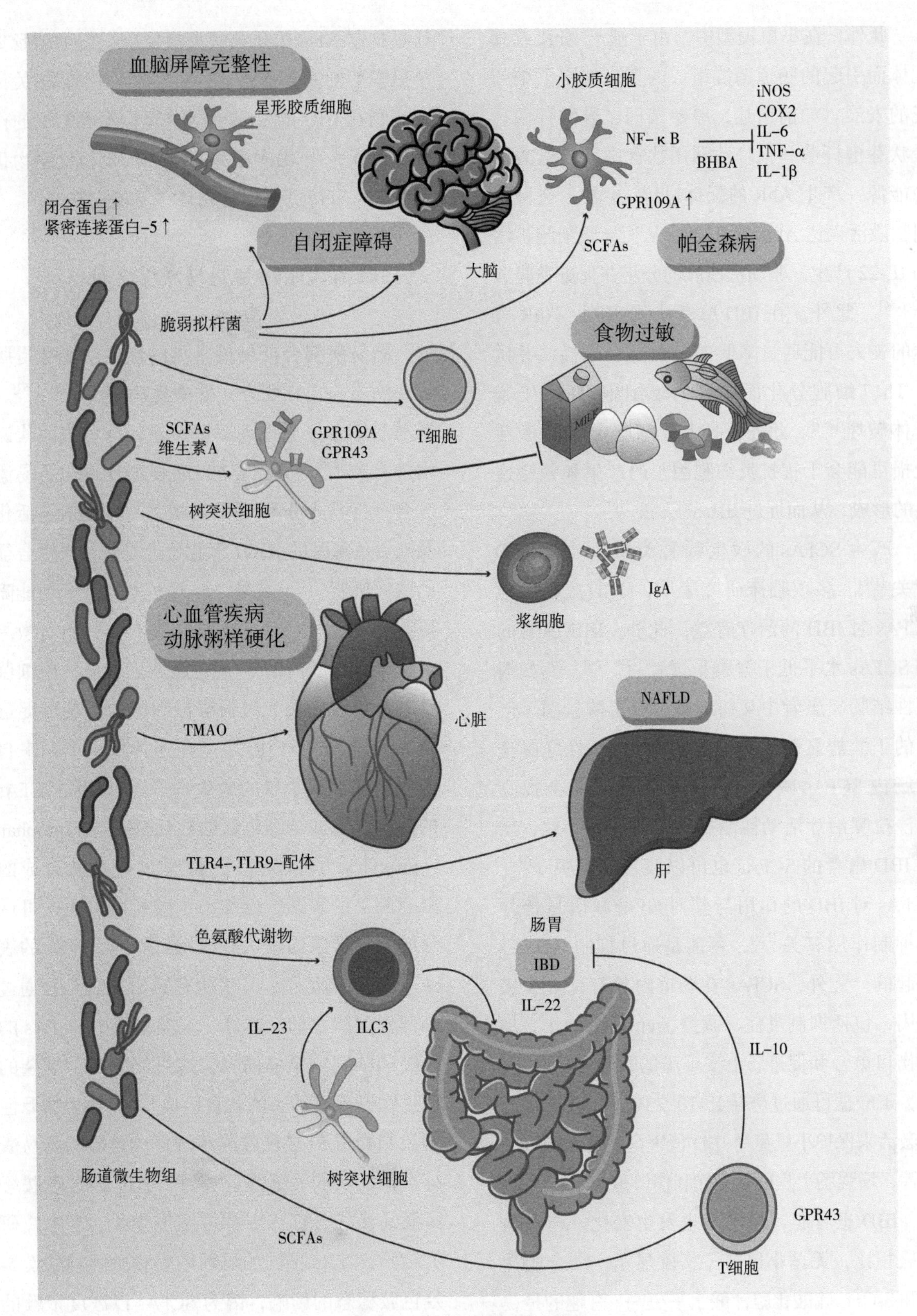

图 6-4　微生物代谢物与疾病

微生物代谢物水平改变与免疫介导和免疫相关性疾病有关。代谢物介导的疾病可能局限于胃肠道，如 IBD 或食物过敏，或影响全身远处器官，如大脑、心脏和肝脏。IBD. 炎性肠病；BHBA. β- 羟丁酸；iNOS. 诱导型一氧化氮合成酶；COX2. 环氧化酶 2；IgA. 免疫球蛋白 A；ILC3. 3 型固有淋巴细胞；GPR109A. G 蛋白偶联受体 109A；GPR43. G 蛋白偶联受体 43；SCFAs. 短链脂肪酸；TMAO. 三甲胺 -N- 氧化物；NAFLD. 非酒精性脂肪性肝病；NF-κB. 核因子 κB；IL. 白细胞介素；TLR.Toll 受体；TNF-α：肿瘤坏死因子 α

此外，在小鼠模型中，由于缺乏 AhR 或其配体而引起的肠道菌群组成的变化加剧了结肠炎的发展。如前所述，厚壁菌门、乳酸杆菌、梭状芽孢杆菌和拟杆菌利用饮食中的色氨酸作为能源，产生 AhR 的配体吲哚 -3- 醛。这种代谢物激活表达 AhR 的 ILC3，从而导致细胞扩增和 IL-22 产生，驱动抗菌肽的分泌并促进黏膜愈合 [256]。此外，在 IBD 患者中观察到，AhR 表达的丧失可能通过降低 Treg 细胞稳定性，并诱导 Th17 细胞分化而导致 Treg 细胞 /Th17 细胞群体破坏 [257]。此外，肠上皮细胞中 AhR 激活受损可能会干扰抗炎细胞因子的产生和紧密连接的形成，从而促进 IBD 的发展 [160]。

产生 SCFAs 的微生物种类减少也与 IBD 有关 [258]。多项临床研究表明，丁酸盐或其他 SCFAs 对 IBD 的治疗有益。此外，IBD 患者的肠 SCFAs 水平低于健康受试者 [259，260]。例如溃疡性结肠炎患者中某些细菌（产丁酸盐菌）产生的丁酸盐显著减少，从而导致溃疡性结肠炎结肠内 SCFAs 减少。口服丁酸盐可能会提高美沙拉嗪治疗活动性溃疡性结肠炎的疗效，增加 IBD 患者的 SCFAs 也可以改善结肠炎 [261]。SCFAs 对 IBD 的作用与其对 NF-κ B 信号传导的抑制作用有关 [262]，可能是通过抑制 HDACs 引起的。此外，SCFAs 在肠道内具有许多积极作用，包括抑制炎症、调控 Treg 细胞分化，加速伤口愈合和促进肠上皮屏障的完整性 [263]。另外，丁酸盐可通过诱导 IL-10 受体 α 亚基的表达和激活来保护小鼠免受 IBD 侵害，从而增加抗炎活性，增强肠上皮屏障功能和伤口愈合 [155，264]。

IBD 患者胆汁酸水平也发生变化。与普通小鼠相比，无菌小鼠模型实验显示，结合胆汁酸和 3-OH- 硫酸化胆汁酸水平更高。有趣的是，与健康人相比，IBD 患者的结合胆汁酸也增加了 [265]。胆汁酸具有抗菌作用。肠道微生物的稳态改变后，胆汁酸的合成和代谢受到影响，进而影响肠黏膜的稳态 [266，267]。此外，化合物 K 通过抑制 NF-κB 介导的炎症反应，促进结肠炎小鼠模型的恢复 [268]。总之，微生物及其相关代谢产物在 IBD 的发病机制中起着不可忽视的作用。然而，与 IBD 相关的许多微生物代谢物仍不明确，有待进一步研究。

二、肠菌代谢物与肠易激综合征

肠易激综合征是最常见的慢性功能性胃肠道疾病之一，在西方人群中患病率为 15%[269]。肠易激综合征通过犬尿氨酸途径与色氨酸代谢增加有关 [270]。犬尿氨酸 / 色氨酸比率与肠易激综合征症状严重程度呈正相关 [271]，IFN-γ 活化及随后色氨酸的 IDO1 氧化可能是肠易激综合征的致病机制 [271]。此外，5- 羟色胺能系统功能障碍与肠易激综合征的病理生理有关。肠易激综合征患者小肠内的 5- 羟色胺浓度较低，但细菌产物（如 SCFAs）对肠嗜铬细胞的影响可促进结 5- 羟色胺的产生 [272]。在遗传水平上，来自人源化和常规小鼠的微生物群通过刺激 SCFAs 的活性，增加结肠色氨酸羟化酶 1（Tryptophan hydroxylase 1，Tph1）（黏膜 5- 羟色胺合成的限速酶）的表达。例如，丁酸盐可以通过可诱导的锌指转录因子 ZBP-89 激活小鼠 Tph1 的表达 [273]。在肠道中，几乎所有的 SCFAs 都是通过细菌发酵产生的。此外，人类粪便中 SCFAs 的种类和浓度因个体而异，这可能归因于复杂的微生物群落或动态的饮食组成 [255]。微生物源性刺激和肠道 5- 羟色胺能系统可能是影响肠易激综合征症状的关键因素。为了通过 5- 羟色胺缓解肠易激综合征患者的胃肠道症状，首先应研究结肠 SCFAs 不同的局部浓度和比例对黏膜 5- 羟色胺稳态的影响，因为 SCFAs 以浓度依赖的方式影响 Tph1 的表达；如何通过操纵复杂的微生物群落来平衡黏膜 5- 羟色胺的稳态，是这一策略所面临的挑战。此外，SCFAs 与肠道 5- 羟色胺能系统之间的协同调节还有待进一步研究。

三、肠菌代谢物与结直肠癌

结直肠癌是导致死亡的第三大常见癌症。肠道微生物群已被证明通过微生物代谢物与CRC发生相关。通过分解肠道中的各种膳食残留物，肠道微生物群获得能量并产生多种代谢物来影响宿主生理。然而，其中一些代谢物是CRC的致癌因素[274]。大量研究表明，肠道菌群代谢物，如丁酸盐、丙酸盐、乙酸盐和烟酸等对CRC有保护作用，而次级胆汁酸、乳酸、TMAO、N-亚硝基化合物、乙醛、4-羟基苯乙酸（4-hydroxyphenylacetic acid，HO-PAA）、PAA、苯酚和细菌毒素等代谢物与CRC的发生有关。由此可见，宿主、肠道微生物群及其代谢物之间的相互作用，可以为预防和控制癌症提供一种全新的策略[275]。

如上所述，SCFAs在肠道具有抗炎作用，从而降低肿瘤形成的可能性[82]。SCFAs在结肠癌细胞的分化、生长和凋亡中起着重要作用[276]。这在流行病学上进一步被证实，因为高纤维含量的饮食与结肠癌的发病率呈负相关[276]。SCFAs对结肠癌细胞的抗肿瘤作用主要是诱导细胞凋亡[276]。研究表明，GPR43作为SCFAs的受体，其缺失可改变肠道微生物群的组成，降低肠道屏障完整性、过度激活树突状细胞和促进$CD8^{+}$T细胞失活，从而导致结肠肿瘤的发生[277]。另外，SCFAs还可以促进宿主产生抗体，SCFAs介导血浆B细胞分化，增加B细胞中乙酰辅酶A的水平，还可以控制代谢传感器，这可以导致糖酵解、脂肪酸合成、氧化磷酸化和抗体的产生，这些过程最终可以促进抗体产生[240]。同样，通过上调紧密连接蛋白的表达，SCFAs可以维持肠道屏障的完整性[278]。

关于SCFAs在CRC中作用，目前大多数研究主要集中在丁酸盐上。各种研究表明，丁酸盐能够抑制或增强CRC发生，因此关于丁酸盐在CRC中的作用尚未达成共识[279]。丁酸盐可调节肝脂肪生成和糖生成[280]，从而减少CRC、非酒精性脂肪肝等疾病[281]。丁酸盐诱导CRC细胞凋亡和自噬，从而抑制CRC细胞的增殖。丁酸盐激活GPR109A并抑制蛋白激酶B或Akt（PKB/Akt）和NF-κ B信号通路，以改善肠上皮屏障功能障碍和炎症反应[282]。研究发现，丁酸盐或烟酸以GPR109A依赖性方式诱导结肠上皮中的IL-18表达，从而防止炎症和癌变。此外，丁酸盐可通过激活GPR109A，促使$APC^{Min/+}$小鼠（$APC^{Min/+}$小鼠缺乏肿瘤抑制因子APC并自发发展为肠腺瘤）产生IL-18，从而预防肿瘤形成。在$APC^{Min/+}$小鼠中，先用抗生素耗竭肠道微生物，然后用GPR109A激动剂治疗，结果显示小鼠息肉减少，但在$GPR109A^{-/-}APC^{Min/+}$小鼠中没有发现这种情况。

相比之下，在缺乏DNA错配修复蛋白MSH2且易于形成肿瘤的$APC^{Min/+}MSH2^{-/-}$小鼠模型中，丁酸盐在较低浓度下驱动息肉形成和肠细胞转化，但在高浓度时与HDAC抑制相关的作用不明显，这表明较低浓度的丁酸盐可以驱动$APC^{Min/+}MSH2^{-/-}$肠细胞的增殖[279]。此外，研究发现，不同浓度的丁酸盐可增加不同类型CRC细胞株TLR4（一种与炎症信号相关的多功能受体）的表达。有趣的是，丁酸盐通过上调CRC细胞TLR4的表达，以及磷酸化NF-κ B和MAPKs信号，启动抗肿瘤免疫从而维持肠道稳态[283]。丁酸盐对细胞凋亡和DNA错配修复基因的激活与其能增加异柠檬酸脱氢酶1的表达、组蛋白乙酰化、DNA去甲基化以及伴随的α-酮戊二酸水平的提高有关，从而在CRC细胞株中产生抗肿瘤作用[284]。

丁酸盐可通过促进丙酮酸激酶异构体的去磷酸化和四聚化，重新编程CRC细胞代谢程序，抑制沃伯格效应，改变结肠肿瘤的能量代谢进而抑制其生长[285]。也有一些观点认为，可以利用丁酸盐重新平衡肠道生物失调来治疗CRC。高水平的丁酸盐可以抑制HDACs，并上调氧化偶氮甲烷/葡聚糖硫酸钠诱导的CRC小鼠肠道抗炎反应[278]。由于丁酸盐在维持肠道稳态、全身免疫系统和预防CRC中的有益功能，因此识别潜在的丁酸盐产生菌作为下一代益生菌（如产丁酸菌和普氏栖粪杆菌），并用来治疗CRC将是一个有趣的研究方向[286]。

丙酸盐在肠道和结肠癌中的抗炎作用与丁酸盐相当[287]。体内外研究均表明，丙酸盐和丁酸盐可以直接增加Tc17细胞和$CD8^+$CTLs相关基因的表达，促进抗肿瘤作用[288]。大量证据表明丙酸盐可诱导人结肠癌细胞凋亡。因此，有研究指出，丙酸盐通过促进活性氧的产生和减少葡萄糖氧化，从而引起Caco-2细胞凋亡[278]。丙酸盐还可以通过抑制NF-κ B信号通路和结肠细胞中IL-6的释放，从而发挥抗炎作用[278]。丙酸盐的另一个关键作用是通过诱导组蛋白乙酰化来调节结肠癌细胞的恶性增殖[278]。基于以上研究，我们可以推测，丙酸盐是防治CRC的合适候选药物。因此，产生丙酸盐的益生菌（如费氏丙酸杆菌）被认为是未来抑制CRC的潜在下一代益生菌[289]。

乙酸盐是体内产量最大的SCFAs之一，其次是丙酸盐和丁酸盐。与丁酸盐和丙酸盐相比，很少研究乙酸盐在CRC中的作用。因此，乙酸盐对CRC抗炎作用的确切机制尚不清楚。有研究表明，乙酸盐通过增加CRC细胞中抗炎细胞因子的表达，降低促炎症因子的水平，抑制NF-κ B信号通路的活化，从而抑制CRC的进展[278]。除此之外，乙酸盐可通过诱导DNA断裂和caspase-3活化，进而促进结肠癌细胞凋亡，减低其增殖反应，从而延缓结肠肿瘤的进展。然而，乙酸盐的凋亡作用可能受到一些抗凋亡剂如组织蛋白酶D在CRC细胞中过度表达的影响[290]。此外，较高水平的乙酸盐可以调节细胞代谢和刺激$CD8^+$T淋巴细胞，导致IFN-γ表达增加。这些结果为CRC的治疗开辟一个新的窗口[288]。

AhR的配体在诱导Treg细胞和肠道炎症中发挥着重要的调节作用，对胃肠道健康有重要影响[291]。SCFAs可以显著提高人CRC细胞系和小鼠结肠细胞中AhR相关基因的表达，证明SCFAs和AhR的配体在预防CRC方面具有协同作用[292]。此外，经犬尿氨酸途径代谢的色氨酸已被证明是一个潜在的免疫治疗靶点。犬尿氨酸途径在多种肿瘤类型中被激活；IDO1启动犬尿氨酸途径，并在原发肿瘤和CRC浸润的髓系来源细胞中表达[293]。IDO1似乎在癌症患者中被慢性激活[294]。IDO1在CRC中通常过表达，并常伴有色氨酸水平降低和犬尿氨酸途径代谢物水平升高[295]。值得注意的是，肿瘤微环境中的肿瘤细胞、基质细胞和（或）单核吞噬细胞上调IDO1或色氨酸2，3-双加氧酶（tryptophan 2，3-dioxygenase 2，TDO2），导致Trp分解代谢的激活，剥夺T细胞的必需氨基酸Trp，同时，产生对T细胞反应有毒的色氨酸代谢物[296]，或诱导Treg细胞分化或未成熟髓细胞的免疫抑制功能[296]。CMG017和CB548是IDO1和TDO2的两种双重抑制剂，已被证明能有效抑制犬尿氨酸途径，显示出良好的抗肿瘤疗效，具有良好的药理学特征，克服了对免疫检查点抑制剂的耐药性[297]。

研究发现，FXR对CRC发生发展具有抑制作用，表明FXR可以通过维持肠道胆汁酸稳态来减少炎症和控制CRC进展。但是较高浓度的胆汁酸与FXR的丧失相关，导致促肿瘤表型发生[298]。IL-8是肿瘤转移和生长的调节因子，它的水平在CRC患者中增加，从而增加了原发性CRC转移风险[299]。研究表明，石胆酸作为肿瘤启动子，通过刺激ERK1/2上调IL-8的表达，下调STAT3的磷酸化，从而促进CRC的侵袭转移[300]。进一步药理研究表明，石胆酸通过促进活性氧的产生和NF-κ B信号传导，从而上调IL-8的表达。然而，二甲双胍可以通过阻断活性氧的产生和NF-κ B信号传导来抑制

CRC 细胞的这种效应[301]。研究证明了石胆酸在尿激酶型纤溶酶原激活剂受体（urokinase-type plasminogen activator receptor，uPAR）表达中的作用，uPAR 可以促进人类 CRC 细胞的 MAPK 信号通路激活，从而导致 CRC 的进展[302]。

此外，体外实验结果证明脱氧胆酸可以激活 MAPK 信号通路，从而诱导 CRC 的发生[303]。脱氧胆酸通过激活 ERK1/2、聚 ADP 核糖聚合酶和 caspase-3 信号通路，增加 DNA 损伤和活性氧水平，降低视网膜母细胞瘤蛋白水平，导致细胞周期中凋亡途径下调和抑制肿瘤抑制因子[304]。脱氧胆酸可通过上调促炎细胞因子的表达，促进 $Apc^{min/+}$ 小鼠肠腺瘤和 CRC 的发生[305]。相比之下，二果糖酐Ⅲ作为益生元，可通过调节胆汁酸代谢减少大鼠体内脱氧胆酸的产生[306]。熊去氧胆酸是另一种胆汁酸，其化学结构与脱氧胆酸部分相似，通过阻止 COX-2 的产生和抑制脱氧胆酸效应发挥预防 CRC 的作用[278]。

乳酸是肠道微生物群产生的一种中间和（或）终末代谢产物，基于乳酸的血管生成作用，向肿瘤微环境输送营养物质和生长因子可能与肿瘤细胞糖酵解的功能有关[307]。因此，乳酸可以通过酸化 TME 来促进肿瘤侵袭[308]。此外，它可以刺激血管生成反应，将氧气、葡萄糖和其他营养物质转移到 CRC 细胞，从而诱导 CRC 细胞的增殖、侵袭和迁移[278]。高乳酸水平可通过诱导酸中毒来改变 TME，从而作为癌症代谢燃料并抑制宿主免疫功能[278]。

体外和离体研究结果表明，多酚类物质可导致盲肠中大量琥珀酸盐的产生，并通过抑制 CRC 细胞的生长和增殖而影响 CRC[309]。然而，最近的研究表明，琥珀酸可以通过介导琥珀酸受体（SUCNR1）信号传导促进体内肿瘤转移[310]。考虑到这一理论，了解微生物产生的琥珀酸如何调节肠道炎症将是一个有趣的问题。然而，关于琥珀酸对 CRC 作用仍然缺乏确切证据。

肠道微生物群发酵蛋白质产生大量氨基酸代谢物。硫化氢（hydrogen sulfide，H_2S）是一种氨基酸代谢产物，通常被用作结肠上皮细胞代谢的能量来源。一些研究总结了 H_2S 对 CRC 影响的证据。有趣的是，高蛋白饮食引起的高浓度 H_2S 可通过刺激硫化物 - 醌还原酶基因的表达，在结肠中形成适应性反应。因此，上皮层免受有害浓度的 H_2S 的影响而受到保护[311]。相反，有学者认为与硫化物氧化途径相关的酶在上皮 CRC 中丢失[312]。关于 H_2S 对 CRC 细胞炎症和增殖的刺激或抑制作用，一直存在相互矛盾的数据。有趣的是，细胞培养研究中发现了硫酸盐还原菌产生的 H_2S 的细胞毒性和遗传毒性作用。

有研究认为 Th1/ Th2 细胞失衡是由于 H_2S 暴露导致的，H_2S 暴露可抑制抗炎作用并激活参与 NF-κ B 信号通路的基因[313]。此外，低浓度 H_2S 可增加炎症相关基因如 iNos 和 IL-6 的表达，并抑制结肠细胞线粒体的耗氧[311]。然而，最近研究表明，低浓度 H_2S 对 CRC 细胞增殖的抑制作用是通过调控细胞凋亡和细胞周期来实现的。这些研究还证明，观察到的抗增殖作用是由于呼吸抑制和 H_2S 氧化导致的电子受体不足[314]。体外评估由氨基酸产生的肠道微生物群代谢物，这些代谢产物由氨基酸产生，例如 HO-PAA（由 L- 酪氨酸产生）[278]、PAA（由苯丙氨酸或酪氨酸产生）[278]、苯酚（由酪氨酸产生）[278]，乙醛（由乙醇或丙酮酸盐产生）[278]，证明 PAA 和苯酚的细胞毒性作用或 HO-PAA 和乙醛对结肠细胞的遗传毒性作用[315]。结果表明，HO-PAA 可降低线粒体复合物 I 活性和耗氧量。事实上，研究人体粪便样本中与氨基酸相关的肠道微生物群代谢物发现，HO-PAA 具有细胞毒性作用[316]。

微生物可以通过产生有毒的代谢物或致癌物来促进癌症的发生，从而增加疾病的发病率和易感性。细菌毒素是另一类 CRC 相关的代谢物，可由肠道微生物群如脆弱拟杆菌和大肠埃希菌产生。脆弱拟杆菌毒素通过激活 β- 连环蛋白信号和 E- 钙黏蛋白裂解、诱导 NF-κ B 通路、刺激 Th17 细胞免疫应答等多种机制，在 CRC 的发生发展中发挥重要作用 [278]。研究表明，产肠毒素型脆弱拟杆菌（*enterotoxigenic bacteroides fragilis*，ETBF）可通过刺激上皮屏障损伤和抑制黏膜免疫，在 CRC 的发生发展中发挥重要作用，因此 ETBF 可作为 CRC 检测的潜在标志物 [317]。越来越多的体内外证据表明，脆弱拟杆菌毒素（bacteroid fragilis toxin，BFT）可通过抑制 T 细胞增殖和刺激 Nos2 和 Arg1 基因的表达而引起 CRC[318]。基于这些证据，研究发现，BFT 可通过激活 IL-17 依赖性 NF-κB 和 STAT3 信号传导途径，触发结肠上皮细胞的致癌作用，这两种信号共同导致小鼠 CRC 发生 [319]。

最近关于 BFT 作用的流行病学观察表明，BFT 通过诱导基因突变、破坏 DNA 并最终破坏上皮细胞基因组，从而导致 CRC 的进展 [320]。基因毒素 colibactin 是具有遗传毒性的次级代谢产物，由大肠埃希菌等携带 pks 基因的微生物产生的。研究表明，它可以通过破坏 DNA 和阻滞细胞周期而导致肿瘤进展和癌变 [321]。大肠埃希菌的另一种毒素是细胞毒性坏死因子 1（cytotoxic necrotizing factor 1，CNF1），已证实其对 CRC 有影响。众所周知，这种毒素可以通过上调 ras 同源蛋白家族（Rho 家族）的鸟苷三磷酸酶，扰乱真核细胞周期并抑制细胞凋亡 [322]。此外，有研究评估了 CNF1 对人 CRC 细胞系的负面影响，结果发现这种毒素可以通过引发多倍体化和内复制，以及下调胞质分裂和有丝分裂来逆转细胞衰老 [323]。

有研究人员使用磁共振代谢组学对 CRC 患者粪便样本中的代谢物进行定量，发现戊酸盐、异戊酸盐、支链脂肪酸、苯乙酸和异丁酸盐在 CRC 患者中浓度较高，而糖、甲醇、氨基酸、胆酸、脱氧胆酸的含量在同一患者中较低 [324]。研究表明，酒石酸是 CRC 所涉及的代谢产物中浓度最高的 [325]。关于人类粪便代谢组学，最近的一项数据表明，黏膜内癌具有明显较高的脱氧胆酸、苯丙氨酸和支链氨基酸负荷。他们的结论是，CRC 患者肠道微生物群衍生代谢产物的动态变化是从早期阶段到Ⅲ / Ⅳ阶段，表明粪便样本代谢组学的评估可以揭示 CRC 患者肠道微生物群不同阶段的特异性表型 [326]。总的来说，越来越多的代谢组数据集为 CRC 的诊断提供了丰富的来源，但到目前为止，由于个体之间微生物群及其代谢物之间的高变异性，尚未确定用于检测 CRC 的通用生物标志物。

四、肠菌代谢物与肝癌

原发性肝癌和转移性肝癌目前是癌症相关死亡的主要原因。肝脏与肠道密切相关，并执行许多与消化、营养代谢和清除细菌代谢物有关的基本功能。肝脏通过门静脉暴露于肠道细菌成分及其代谢产物 [327]。肝脏是免疫细胞大量聚集的免疫器官，肠道微生物组的变化可能会影响肝脏免疫细胞的功能。

在原发性肝癌模型和转移性肝癌模型中，研究人员发现，改变肠道微生物组会导致肝脏 $CXCR6^+$ 自然杀伤 T 细胞（natural killer T cell，NKT）和效应记忆 $CD4^+$ 或 $CD8^+$ T 细胞的积累，与小鼠品系、性别或肝脏肿瘤的存在无关 [328]。NKT 细胞的激活状态在遇到载有抗原的肿瘤细胞时，会导致更高水平的 IFN-γ，从而促进抗肿瘤反应。NKT 缺陷小鼠的体内研究证实，NKT 细胞介导抑制肝脏肿瘤的生长 [328]。进一步的研究表明，NKT 细胞的积累受肝窦内皮细胞上 CXCL16（CXCR6 的唯一配体）的表达调节。初级胆汁酸增加 CXCL16 表达，而次级胆汁酸显示相反的效果。通过万古霉素治疗去除革兰阳性菌（其中含有介导初级胆汁酸转化为次级胆汁酸的细菌）后，足以诱导肝脏 NKT 细胞积聚并减少肝肿瘤生长。在肠道共生菌改变的小鼠中，给予喂养次级胆汁酸或胆汁酸代谢细菌的定植，可以逆转 NKT 细胞积累和肝脏肿瘤生长的抑制 [328]。

在原发性肝癌患者的非肿瘤肝组织中，初级胆汁酸鹅去氧胆酸水平与CXCL16表达相关，而与次级胆汁酸乙醇酸表达呈负相关，表明该发现可能适用于人类 [328]。这些研究结果表明，肠道微生物组利用胆汁酸作为信使来控制肝脏 NKT 细胞的积累，从而控制小鼠肝脏的抗肿瘤免疫。这些发现不仅在未来对人类癌症的治疗研究具有重要意义，而且还提供了肠道微生物组、代谢产物和肝脏免疫反应之间的联系。

五、肠菌代谢物与非酒精性脂肪性肝病

非酒精性脂肪性肝病（non-alcoholic fatty liver disease，NAFLD）包括单纯性脂肪变性、非酒精性脂肪性肝炎、纤维化和肝硬化。NAFLD 已成为主要的肝病，在全世界成人中患病率为 22%～29%[329]。NAFLD 与以向心性肥胖、胰岛素抵抗、高血压、高脂血症和血脂异常为特征的代谢综合征相关 [330]。时至今日，NAFLD 复杂的发病机制仍不清楚。随着研究的深入，人类肠道微生物群被认为是 NAFLD 的关键决定因素。除了肠道菌群组成的变化外，肠道菌群的组成及其代谢产物已成为调节 NAFLD 病理过程的关键因素。研究表明，肠道微生物群产生多种代谢物，通过门静脉与宿主肝细胞相互作用。这些物质包括 SCFAs、吲哚及其衍生物、三甲胺、次级胆汁酸等代谢物。肝脏对肠道细菌生物活性物质的反应机制与糖脂代谢调节、免疫信号转导和氧化还原稳态有关。阐明肠道微生物群产生的代谢物与肝脏之间相互作用的机制，将为 NAFLD 提供一个新的治疗靶点。

研究发现，SCFAs（乙酸盐和丙酸盐）有抗炎特性和抑制肝脏脂肪生成和脂质积聚的作用 [331]。来自动物模型的数据表明，丁酸盐通过调节肠道微生物群、肠道屏障功能、下调炎症信号和肝脏氧化损伤来减轻脂肪性肝炎 [332-335]。丁酸盐通过提高呼吸能力和脂肪酸氧化来增强线粒体功能 [336]。脂多糖在 NAFLD 的进展中也起重要作用 [337]，而丁酸盐通过增强紧密连接的表达，在维持肠道完整性方面起主要作用 [155]，因此，丁酸盐可能阻止脂多糖向肝脏的转运，从而间接抑制 NAFLD 的进展。乙酸盐和丁酸盐均可诱导 GLP-1 的分泌，GLP-1 可通过增加脂肪酸氧化，减少脂肪生成，改善肝脏糖代谢来预防 NAFLD[338，339]。与此不一致的是，最近的一项研究发现，NAFLD 患者肠道细菌产生的 SCFAs 水平较高，乙酸盐和丙酸盐的增加通过对循环免疫细胞的影响维持轻度炎症 [340]。

吲哚衍生物，如吲哚 -3- 乙酸通过直接作用于肝细胞，以 AhR 依赖的方式减轻细胞因子和游离脂肪酸介导的脂肪生成，可以改善肝脏的炎症性疾病。胆汁酸是通过核受体 FXR 起作用的信号分子。据报道，FXR 通过下调 LXR 和 SREBP-1C 的表达，减少肝脏中脂肪酸和三酰甘油的合成，从而减轻脂肪性肝炎 [341]。除了 FXR 外，TGR5 是另一种典型的胆汁酸受体 [29]。在肝组织中，TGR5 在库普弗细胞和内皮细胞中表达，具有调节肝脏炎症、糖代谢和提高胰岛素敏感性的功能。TGR5 通过抑制巨噬细胞的 NF-κB 信号传导和细胞因子的产生来减轻炎症反应 [342]。此外，TGR5 控制肠 L 细胞释放 GLP-1，从而在维持葡萄糖稳态方面具有关键作用。这些证据提示 TGR5 在调节炎症、糖代谢和能量平衡中具有重要作用。

TMAO 是肝脏中 FMO 催化的三甲胺（trimethylamine，TMA）氧化产物，被认为是早期代谢综合征的新生物标志物。TMAO 通过多种途径影响 NAFLD 的发生。首先，血 TMAO 的升高预示着 TMA 产生的增加，间接反映了胆碱和磷脂酰胆碱代谢的变化。尤其是胆碱缺乏会阻碍极低密度脂蛋白的合成和分泌，

从而导致肝内三酰甘油的积聚和脂肪变性[343]。其次，肝脏细胞色素P450家族7A1（CYP7A1）基因的表达与NAFLD患者血清TMAO水平呈正相关，该基因编码内质网膜蛋白，催化胆固醇转化为胆汁酸。然而，在正常生理条件下，补充TMAO的小鼠表现出胆汁酸的减少。这归因于肝脏中胆汁酸合成酶和转运蛋白的表达减少。因此，TMAO可能通过调节胆汁酸代谢和转运来调节NAFLD。最后，肠道微生物群介导的TMA/FMO3/TMAO通路调节胰岛素抵抗、糖脂代谢、胆固醇稳态和肝脏炎症[344, 345]，从而影响肝脏三酰甘油的积累和肝脏脂肪变性，增加NAFLD的发病风险。

肠道微生物群产生的关键代谢物，包括SCFAs、次级胆汁酸、吲哚及其衍生物、三甲胺等代谢物，在宿主代谢、免疫系统和氧化还原稳态中起调节作用，从而从根本上改变NAFLD的进展。尽管有关肠道微生物与肝脏疾病相互作用的研究取得了长足进展，但其潜在机制尚未完全阐明。微生物代谢物的靶向应用需要对其细胞受体和宿主的潜在信号通路进行全面的描述，这需要更广泛的临床和实验研究。探索肠道微生物群代谢物调节宿主细胞的机制，有利于预防肝脏疾病。靶向肠道微生物代谢物和相关的宿主细胞信号，将为NAFLD的干预提供新的策略。

六、肠菌代谢物与心血管疾病

心血管疾病（cardio vascular disease，CVD）是全球第一大死亡原因[346]。CVD包括多种疾病，如动脉粥样硬化、高血压、血小板功能亢进、脑卒中、高脂血症和心力衰竭。研究发现，微生物组分和功能的变化与CVD的发病机制有关。肠道微生物代谢物，如TMAO、SCFAs和胆汁酸代谢物与CVD之间存在关联。从临床角度来看，肠道微生物群及其代谢物可能是CVD治疗的目标[346]。

动脉壁粥样硬化斑块是由于脂质稳态失调导致含胆固醇低密度脂蛋白颗粒积聚，并伴有慢性炎症[347]。TMAO是一种代谢产物，在人体中由微生物代谢食物中的磷脂酰胆碱或左旋肉碱而产生，这些物质在肉类和高脂肪饮食中含量丰富。喂食左旋肉碱的小鼠表现为TMAO水平增加，导致动脉粥样硬化加剧。无菌小鼠或抗生素治疗的小鼠免受疾病的侵袭，从而将肠道菌群、营养和CVD的风险联系起来。在人类身上也观察到类似的发现，在摄入核素标记的食物后，血浆TMAO（和其他磷脂酰胆碱代谢物）随时间依赖性增加。这些增加的代谢物水平被抗生素治疗抑制，并在抗生素停药后再次出现。空腹血浆TMAO水平升高与主要心血管不良事件（如总死亡率、心肌梗死或脑卒中）显著相关。这些结果证明了这种微生物源性代谢物在CVD危险因素中的潜在中心作用[348]。

然而，TMAO及其前体在动脉粥样硬化发生中的作用并不是完全直接的。许多鱼类都含有大量TMAO，但长期以来，以鱼类为基础的饮食与降低CVD的风险有关。食用鱼类的心脏保护作用通常被认为是由ω-3脂肪酸介导的，这可能超过同时摄入TMAO的任何不利影响[349, 350]。胆碱通常被认为是TMAO的饮食来源，然而没有实质性证据表明胆碱摄入量与CVD风险之间存在显著关联[351]。喂食左旋肉碱导致ApoE$^{-/-}$小鼠循环中TMAO水平显著增加，但其与主动脉病变大小呈负相关[346]。膳食补充左旋肉碱还与CVD风险显著降低有关，这显然和血浆左旋肉碱升高与CVD风险增加相关的报道相矛盾[350, 352]。不同国家的人群研究表明，饮食中胆碱和甜菜碱的摄入与CVD的发病机制无关[346]。为了更好地了解膳食TMAO和左旋肉碱对CVD的直接影响，针对性的大规模研究将是必要的。

研究发现，TMAO浓度过高不会影响体外巨噬细胞对胆固醇的摄取或排出，虽然此前已经发现 TMAO 喂养小鼠的巨噬细胞清道夫受体 CD36 和 SR-A1 的上调 [353, 354]。已有报道称 TMAO 可抑制胆固醇逆向转运。研究发现，TMAO 通过抑制胆固醇逆向转运和改变巨噬细胞中胆固醇转运蛋白的活性来促进动脉粥样硬化 [355]。此外，TMAO 抑制肝脏胆汁酸合成酶 Cyp7a1 和 Cyp27a1 以及胆汁酸转运蛋白 Oatp1、Oatp4、Mrp2 和 Ntcp 的表达，导致胆汁酸相关通路紊乱并促进动脉粥样硬化 [356]。此外，有研究表明，FXR 可以通过调节肝脏 FMO3 的表达，控制胆汁酸代谢和 TMAO 的产生 [346]。FXR 通过抑制 $ApoE^{-/-}$ 小鼠中 CYP7A1 和 CYP8B1 的表达，保护小鼠免受动脉粥样硬化 [357]。TMAO 通过多种机制加速动脉粥样硬化，例如促进胆固醇流入、抑制胆固醇流出、阻断胆汁酸途径和引起血小板过度活跃。

最近的一项研究发现，TMAO 通过一种未知的机制，在内皮细胞和平滑肌细胞中诱导编码促炎、促动脉粥样硬化蛋白的几个基因的转录 [358]。TMAO 上调血管细胞黏附分子 -1 和活化蛋白激酶 C 和 NF-κ B 的表达。因此，TMAO 通过调节内皮细胞功能导致动脉粥样硬化，并增加单核细胞的黏附 [359]。这些发现证实了 TMAO 在 CVD 中的作用。除了对动脉粥样硬化的直接影响外，TMAO 还通过增强细胞内 Ca^{2+} 释放来增加血小板反应性，从而增加血栓事件和 CVD 的可能性。对于 TMAO 在动脉粥样硬化中的确切作用，以及通过靶向产生 TMAO 的细菌或酶来验证其治疗潜力，还需要进行更深入的研究。

研究发现，摄入含多胺的饮食与 CVD 风险呈负相关。多胺可能通过抑制促炎症细胞因子的产生和改善内皮细胞功能来改善 CVD。其他微生物代谢物，包括胆汁酸和化合物 K，也可能通过调节心血管功能来降低 CVD 的风险 [360]。胆汁酸可通过调节胆盐水解酶活性和激活胆汁酸受体，加速动脉粥样硬化的发展 [361]。细菌介导的胆盐水解酶活性可通过刺激胆固醇积聚、泡沫细胞形成和增加动脉粥样硬化斑块大小，促进动脉粥样硬化进展 [346]。此外，肠道微生物群通过其代谢物刺激肠道传入感觉纤维，或影响负责血压调节的靶器官如肾脏，从而调节血压 [362]。

补充乙酸盐可以改善心血管功能并防止 CVD 的发展 [363]。肠道微生物代谢物 SCFAs 调节 GPCRs 的活性，包括 GPR41、GPR43 和 GPR109A[364]。SCFAs 通过 GPCRs 调节血管紧张素 - 肾素系统合成肾素来调节血压 [346]。与野生型小鼠相比，GPR41 基因敲除小鼠的收缩压较高，SCFAs 通过激活内皮 GPR41 降低血压 [365]。SCFAs 调节嗅觉受体 78（Olfr78）活性，导致血管舒张从而在小鼠体内产生降压作用 [364]。这些研究表明，肠道微生物群可能通过 SCFAs 介导的机制在宿主血压中发挥重要作用。然而，SCFAs 作为高血压治疗靶点的潜力还需要进一步深入研究。

大量流行病学证据表明，内毒素血症提示血液中脂多糖水平高，是动脉粥样硬化的重要危险因素，并且是脂多糖与动脉粥样硬化疾病之间关联的纽带 [366]。据报道，慢性心力衰竭患者在急性水肿加重期间，血浆脂多糖和细胞因子浓度升高，表明内毒素可触发慢性心力衰竭患者免疫系统的激活 [366]。此外，血浆脂多糖水平可预测心房颤动患者的主要心血管不良事件 [367]。内毒素血症也参与肥胖和胰岛素抵抗机制 [368]，这与 CVD 密切相关。这些临床和基础研究清楚地表明，内毒素血症与 CVD 有关，可作为 CVD 治疗的有力靶点。此外，研究发现，使用拟杆菌可以降低脂多糖的活性和抑制动脉粥样硬化 [366]。脂多糖被认为是炎症性疾病中包括 CVD 的一种毒素，参与 CVD 发病和进展的病理生理。到目前为止，直接降低血液或粪便脂多糖水平的疗法还不存在。因此，开发管理内毒素血症或调节肠道微生物组成的临床应用措施，可能成为治疗 CVD 的新选择。

七、肠菌代谢物与慢性肾脏病

慢性肾脏病（chronic kidney disease，CKD）是一个全球性的公共卫生问题。CVD 是 CKD 患者早期死亡的主要原因[369]，此外，血管钙化是主要心血管不良事件的危险因素，尤其是 CKD 患者[370]。20 世纪 70 年代通过内镜检查首次证明，CKD 和非 CKD 患者的肠道微生物群都发生了显著改变。在 CKD 患者中，需氧菌和厌氧菌在十二指肠和空肠中大量定植[369]。此外，还观察到乳酸杆菌减少，肠杆菌科细菌增多[371]。此外，透析患者的粪便分析显示 SCFAs 丁酸水平下降。CKD 中肠道微生物群及其代谢物的改变被认为与 CKD 中血管钙化过程有关[372]。脂多糖和细菌 DNA 可直接诱导炎症和免疫反应，导致终末器官损伤[373]。肠道微生物群衍生的代谢物，如尿毒症毒素、TMAO、胆汁酸均通过调节血管表型、氧化应激和表观遗传学与 CKD 中的血管钙化相关[374]。因此，干预肠道微生物群衍生代谢物可能是抑制 CKD 中血管钙化的重要策略之一。

八、肠菌代谢物与神经系统疾病

尽管越来越多的研究发现，神经系统疾病以重要的免疫因素为特征，但对神经免疫调节的研究仍处于起步阶段。中枢神经系统中的常驻免疫细胞亚群介导中枢神经系统的免疫调节，这些细胞亚群影响疾病的进展，特别是小胶质细胞，即中枢神经系统的常驻髓样细胞。以前中枢神经系统被认为不受外周免疫系统和血液传播介质的影响，现在很明显地发现，在健康和疾病状态下，微生物代谢物可以进入大脑，影响大脑免疫[375]。

小胶质细胞是脑内驻留的巨噬细胞，对维持组织稳态和清除碎片、死亡细胞以及病原体起着至关重要的作用[376]，在脑发育中起着核心作用[377]。在小鼠模型中发现，小胶质细胞受到 SCFAs 的影响。在小鼠稳定状态下，肠道微生物群分泌的 SCFAs 调节小胶质细胞激活和成熟基因表达谱。无菌小鼠显示出小胶质细胞的整体缺陷、细胞数量改变和先天免疫反应受损。利用复杂的微生物群进行重新定植，恢复了小胶质细胞的部分特性，并且表明 SCFAs 调节了小胶质细胞的稳态。缺乏 GPR43 的小鼠出现小胶质细胞缺陷，这表明维持小胶质细胞内稳态需要 SCFAs 和 GPR43。丁酸盐和其他 SCFAs 从胃肠道通过门静脉循环流入，或通过远端结肠血供直接吸收到全身循环，并可通过血脑屏障进入中枢神经系统。

与喂食不溶性纤维的小鼠相比，喂食可溶性纤维饲料的小鼠表现出内毒素相关疾病的减少[378]。此外，无菌小鼠大脑中紧密连接蛋白（闭合蛋白和紧密连接蛋白 -5）减少，导致血脑屏障的通透性增加[379]。用产丁酸盐的酪丁酸梭状芽孢杆菌治疗，可以提高无菌小鼠大脑中的闭合蛋白和紧密连接蛋白 -5 水平，并将其血脑屏障通透性恢复到无特定病原体小鼠水平[379]。SCFAs 也可能刺激上皮细胞产生维 A 酸[380]，一种维生素 A 衍生代谢物，与 TGF-β 协同作用促进 Treg 细胞分化[381]。维甲酸除了增强 Treg 细胞分化外，还可以阻止 Th17 细胞分化，因此可能有改善神经炎症的作用[382]。

阿尔茨海默病是一种进行性神经退行性疾病，其特征是 β- 淀粉样蛋白（β-amyloid protein，Aβ）和 tau 的聚集和沉积。最近的研究表明，肠道微生物组在阿尔茨海默病的 Aβ 病理改变中起着关键作用[383]。一项临床研究发现，Aβ 负荷和血液 SCFAs 浓度之间存在相关性[384]。微生物来源的 SCFAs 有助于大脑中 Aβ 斑块沉积。无菌阿尔茨海默病小鼠模型的 SCFAs 浓度显著降低，Aβ 斑块负荷减少，补充 SCFAs 后足以模拟微生物组的作用并增加 Aβ 斑块负荷。小胶质细胞是对 SCFAs 有反应的关键细胞群，并且小胶质细胞载脂蛋白 e 的增加可能在淀粉样变的早期阶段加速 Aβ 的沉积 [383]。综上所述，微生物来源的 SCFAs 是沿着肠 - 脑轴的关键介质，可能通过调节小胶质细胞的表型促进 Aβ 沉积。这个发现将为微生物组 -SCFAs- 小胶质细胞轴的靶向治疗开辟新途径，以减少炎症对阿尔茨海默病发展的影响 [383]。然而，关于 SCFAs 浓度与阿尔茨海默病进展之间潜在联系的临床证据很少，需要进一步研究 [384]。

最近的一项研究表明，与健康对照组相比，帕金森病患者粪便 SCFAs 水平升高，这可能与 SCFAs 在帕金森病中的作用有关。帕金森病小鼠模型在缺乏肠道微生物组的情况下，补充 SCFAs 足以诱发神经炎症和疾病进展 [385]。研究发现，给予帕金森病大鼠模型 β- 羟基丁酸盐治疗改善了运动技能，并通过减少神经炎性反应保护了黑质纹状体神经元 [386]。最近发现 GPR109A 由小胶质细胞表达，并且在帕金森病患者的黑质中与小胶质细胞标记物共定位表达增加，因此可能参与了疾病发生 [386]。向中脑神经胶质细胞培养物中添加 β- 羟基丁酸盐，通过 GPR109A 抑制 NF-κB 途径，降低由脂多糖诱导的小胶质细胞激活，从而减少促炎症反应酶（COX-2，iNOS）和细胞因子（IL-6、TNF-α、IL-1b）[387]。

其他神经精神疾病，如自闭症也被认为涉及微生物群代谢物依赖性调节。易患自闭症的小鼠表现为生物失调、肠道通透性增加和血清代谢物改变 [388]。脆弱拟杆菌治疗可恢复自闭症的肠道通透性和神经症状，可能是通过调节 4- 乙基苯基硫酸酯，一种导致某些自闭症症状的代谢物。这些结果表明，肠道微生物群可能通过全身代谢物影响中枢神经系统的某些功能，包括神经炎症反应 [388]。这些影响中是否可能涉及中枢神经系统免疫反应的改变，以及是否在人类身上观察到类似的影响仍然未知。

在实验性自身免疫性脑脊髓炎（experimental autoimmune encephalomyelitis，EAE）小鼠模型中，肠道微生物来源的吲哚 -3- 甲醛、吲哚 -3- 丙酸和吲哚 -3- 硫酸酯通过 AhR 调节 T 细胞和树突状细胞，并且通过抑制星形胶质细胞中的 NF-κB 来抑制中枢神经系统炎症 [389]。在 EAE 小鼠中，色氨酸代谢物 6- 甲酰基吲哚并 [3，2-b] 咔唑（6-Formylindolo[3，2-b]carbazole，FICZ）增加小鼠产生 IL-17 和 IL-22 的 $CD4^+$T 细胞，从而加速 EAE 的发病和中枢神经系统的病理改变 [390]。此外，FICZ 可诱导脾脏中的 Th1 和 Th17 细胞，并伴有严重的 EAE 症状 [390]。相反，色氨酸衍生的 2-（1′ H- 吲哚 -3- 基羰基）-4- 噻唑羧酸甲酯 [2-（1' H-indole-3' -carbonyl）-thiazole-4-carboxylic acid methyl ester，ITE] 通过诱导 Treg 细胞来改善 EAE 症状 [391]。

所有这些研究都是注射外源性 AhR 的配体，而不是研究饮食或微生物源性内源性代谢物。从饮食色氨酸中产生各种吲哚代谢物，可刺激 AhR 活性。吲哚可通过激活星形胶质细胞中的 AhR，跨越血脑屏障，抑制促炎症反应。膳食中色氨酸缺乏或星形胶质细胞缺乏 AhR 将导致严重的 EAE 症状。在对照组小鼠中补充色氨酸可逆转这种效应，但在星形胶质细胞特异性 AhR 敲除小鼠中不能逆转。有趣的是，多发性硬化患者显示外周血中 AhR 的配体浓度降低，表明多发性硬化患者似乎在这些抗炎代谢物的产生、摄取或稳定性方面存在缺陷，从而导致 AhR 依赖性免疫调节水平的降低 [382]。

九、肠菌代谢物与脊柱关节炎

脊柱关节炎（spondyloarthropathies，SpA）在美国约 1% 的人口中发生 [392]。50% 的 SpA 患者可能有肠道感染，但没有明显的胃肠道症状（也称为亚临床感染）[393]。研究发现，SpA 患者体内色氨酸代谢相对减少，色氨酸代谢物水平较低。SpA 的色氨酸代谢物减少，在类风湿关节炎患者滑膜液中有类似的发现 [394]。在色氨酸代谢途径中，发现 SpA 患者粪便中有 21 种独特的色氨酸代谢物被还原。这些色氨酸代谢产物中有相当一部分与肠道免疫和肠道炎症有关，但哪些代谢产物在 SpA 亚临床炎症中起主导作用尚待研究。

十、肠菌代谢物与类风湿关节炎

类风湿关节炎（rheumatoid arthritis，RA）是一种自身免疫性疾病，其特征为滑膜关节的慢性炎症。炎症、新血管形成（血管生成）和骨吸收（破骨细胞生成）是参与关节炎关节损伤和畸形的三个关键过程。各种肠道微生物群衍生的代谢物与 RA 发病机制有关 [395]。在最近的一项研究中，发现与骨关节炎患者相比，RA 滑液中两种吲哚衍生物，即吲哚 -3- 乙酸和吲哚 -3- 乙酰醛减少。此外，恢复产生吲哚衍生物的细菌菌株，可改善佐剂诱发的关节炎，即 RA 大鼠模型 [396]。此外，吲哚衍生物的潜在益处是在结肠炎和 EAE 动物模型中改善疾病症状 [47]。吲哚 -3- 醛和吲哚 -3- 乙酸尽管结构相似，但这两种代谢物的生物活性却截然不同。吲哚 -3- 醛表现出促血管生成活性和促破骨细胞生成活性。相比之下，吲哚 -3- 乙酸对内皮细胞管形成表现出抗血管生成活性，但对破骨细胞生成没有影响 [395]。吲哚 -3- 醛和吲哚 -3- 乙酸都被鉴定作为 AhR 激动剂。使用 CH-223191（AhR 抑制剂）可抑制吲哚 -3- 乙酸的抗血管生成活性，但未能减轻吲哚 -3- 醛的影响。这些吲哚衍生物的相对生物利用度可能对 RA 进展以及可能具有相似细胞过程的其他疾病产生不同影响。然而，目前尚不清楚吲哚衍生物是否可以调节参与关节炎发病机制的细胞过程。

十一、肠菌代谢物与 1 型糖尿病

1 型糖尿病是一种以 T 淋巴细胞介导的胰岛 B 细胞进行性损伤为特征的全身性自身免疫性疾病。1 型糖尿病的发病机制涉及遗传因素、环境因素和自身免疫因素 [397]。

SCFAs 通过增加 $Foxp3^+CD4^+$Treg 细胞分化，从而促进抗炎反应并改善 1 型糖尿病症状。非肥胖糖尿病小鼠，其模型为 1 型糖尿病，喂养发酵为乙酸盐和丁酸盐的酰化淀粉，通过乙酸盐激活 GPR43 和丁酸盐诱导 $Foxp3^+CD4^+$Treg，导致肠道中的拟杆菌增多，降低疾病严重程度 [398]。丁酸盐通过诱导胰腺固有淋巴细胞分泌 IL-22，改善非肥胖糖尿病小鼠的 1 型糖尿病，并增加 β- 防御素 14 的表达。β- 防御素 14 是一种抗菌肽，通过刺激分泌 IL-4 的 B 细胞上的 TLR-2，从而诱导调节性巨噬细胞及保护性 $Foxp3^+CD4^+$Treg 细胞，进而改善 1 型糖尿病 [399]。

此外，SCFAs 通过促进胰腺 A 和 B 细胞中 Cathelicidin 相关抗菌肽（CRAMP）的表达，进一步阻止非肥胖糖尿病小鼠的 1 型糖尿病发展 [400]。并且，乙酸盐减少了促进胰岛自身抗原反应性 T 细胞增殖的边缘区 B 细胞数量，而丁酸盐增强了 Treg 细胞的功能，进一步减轻了非肥胖糖尿病小鼠的胰岛炎症和糖尿病症状 [401]。而且，SCFAs 可以增强肠道的完整性，降低糖尿病危险因子如 IL-21 的血清浓度。SCFAs 对 1 型糖尿病有潜在的预防作用 [402]。

十二、肠菌代谢物与肥胖

当能量摄入超过能量消耗时可产生肥胖，导致脂肪在脂肪细胞中过度堆积，脂肪组织炎症的发展伴随着促炎性脂肪因子的产生和分泌增加 [403]。大量研究表明，SCFAs 通过调节食欲、能量摄入、能量消耗和能量收获而与肥胖相关[404]。在一项啮齿动物研究中发现，SCFAs通过刺激肠内分泌细胞分泌 GLP-1 和多肽 YY（peptide YY，PYY）的分泌，从而降低体重[338]。一项针对人类肥胖患者的研究报道，口服丙酸盐增加了 GLP-1 和 PYY 的血浆浓度，这与食物摄入量减少相关[405]。与此类似，在肥胖小鼠中，丁酸盐通过能量消耗和脂质氧化而导致体重下降[406]。此外，另一种 AhR 的配体，吲哚，可以调节肠 L- 细胞释放 GLP-1，这可能有助于预防肥胖[407]。每日给予多胺，无论是亚精胺还是精胺，在饮食诱导的肥胖小鼠模型中都能有效诱导体重减轻并提高胰岛素敏感性[408]。肥胖与非酒精性脂肪肝和代谢性疾病等其他疾病密切相关，因此了解微生物代谢物如何影响肥胖至关重要。

十三、肠菌代谢物与 2 型糖尿病

2 型糖尿病的特点是由胰岛 B 细胞的胰岛素分泌受损和胰岛素介导的肝脏葡萄糖生成抑制引起的高血糖。许多证据表明，SCFAs 产生菌的定植和 SCFAs 的存在可能改善糖尿病，尤其是 2 型糖尿病[409, 410]。毫无疑问，SCFAs 对葡萄糖和能量稳态具有有益影响。SCFAs 可以通过激活结肠细胞中的 GPR43/41，分泌 PYY 或 GLP-1 到血浆中，进而改善 2 型糖尿病。GLP-1 能促进胰岛素的分泌，抑制胰高血糖素的分泌。PYY 可以改善外周组织的葡萄糖摄取和利用[411]。如前所述，SCFAs 通过促进胰岛素生成和依次调节葡萄糖代谢来改善 2 型糖尿病[338, 409, 410]。

除 SCFAs 外，胆汁酸还能诱导机体产生 GLP-1，GLP-1 最近被认为是 2 型糖尿病治疗的新靶点[167, 412]。研究发现，对胆汁酸有反应的 TGR5 可以微调能量平衡，作为 BA-TGR5-cAMP-D2 信号通路的一部分，可以靶向改善代谢[411]。此外，微生物色氨酸代谢物吲哚丙酸的水平与 2 型糖尿病的发生率呈负相关，与胰岛素分泌呈正相关，表明这种代谢物可能通过促进胰腺 B 细胞分泌胰岛素，而对 2 型糖尿病产生积极作用[413]。咪唑丙酸是一种微生物产生的组氨酸衍生代谢产物，在 2 型糖尿病患者中浓度较高[411]。进一步研究发现，咪唑丙酸通过 mTORC1 损害胰岛素信号，从而导致糖尿病[411]。目前，大众的饮食趋势已转向非纤维食品的消费增加，导致体重增加过多，产生微生物代谢物的底物供应不足。此外，超重或肥胖的人患 2 型糖尿病的风险也会增加。

第六节　小结与展望

随着研究的不断深入，肠道微生物代谢物被认为是人体生理学的重要组成部分，对免疫功能有着深远的影响。微生物代谢物是通过微生物 - 微生物、宿主 - 微生物相互作用而产生的，这种共同代谢在人类健康和疾病中发挥了重要作用。由于各种微生物代谢物在增强肠道屏障功能、抑制炎症反应、促进抗炎反应等方面发挥着相似的作用，因此对肠道微生物群及其相关代谢产物的调控对维持宿主健康具有重要意义。肠道微生物群及其分泌的代谢物参与调节宿主的许多代谢和免疫功能。微生物群分泌的代谢物如 SCFAs 能调节宿主的许多功能，主要是自噬、脂质代谢、免疫功能和抗氧化活性。

微生物群代谢物可能影响宿主免疫反应的许多方面，但共生菌调节宿主免疫的机制仍不清楚。因此，确定肠道微生物群如何调节宿主生理的分子机制，对于理解这些微生物在我们体内的作用至关重要，值得进一步研究。同样，需要进一步的研究来确定胃肠道代谢物的种类、分布和生物活性，以及它们如何影响局部和全

身炎症过程。将不同疾病情况下微生物群落结构与代谢组及其生物活性的相应变化相结合，可以明确微生物组影响宿主生理、病理生理甚至其自身群落功能的分子机制。人源化无菌动物可以更好地阐明人类特定的共生细菌种类及其相关代谢物在疾病发生和进展中的作用。

肠道微生物群产生的微生物源性代谢物作为化学信使，介导微生物与宿主之间的相互作用，对人类健康既有利又有害。这些代谢物已经被证明影响许多疾病的发生发展，包括炎性肠病、非酒精性脂肪肝、CVD、癌症、自身免疫性疾病和神经退行性疾病等。肠道微生物代谢组也被证明可以调节肠道病原体的定植抗性。

全面了解所有肠道微生物产生的小分子代谢物、它们的分子靶点及其生物学意义仍然是该领域的一个重要目标。因此在系统水平上，破译宿主 - 肠道微生物群相互作用的新方法处于该领域的前沿。结合宏基因组分析和血清代谢组学分析，能够发现代谢芳香族氨基酸（如组氨酸、缬氨酸、亮氨酸和异亮氨酸）的细菌与胰岛素敏感性和肥胖的改善相关 [414]。循环代谢物对宿主的影响也延伸到肠 - 脑轴，因为 4- 乙基苯酚硫酸盐可诱发自闭症障碍行为。此外，比较健康和患者粪便代谢组学，证明了哪些代谢物对于预防 IBD 很重要，以及在患有这种疾病的个体中，这些代谢物是如何改变的 [82]。例如，抗坏血酸最近被鉴定为与克罗恩病相关的生物活性微生物代谢物，并被证实通过靶向 T 细胞代谢对活化的效应 $CD4^+T$ 细胞表现出抑制作用 [82]。

另外，胸腺外 $Foxp3^+CD4^+Treg$ 细胞缺乏导致微生物暴露引发的 2 型免疫反应增强，并破坏了在定植期间边界栖息细菌的生态位建立，这与粪便和血清代谢组中脂质和氨基酸代谢的全身性变化相关 [415]。最近，研究人员用核素标记代谢前体来标记肠道中细菌产生的代谢物。这些方法利用高分辨率质谱法，对非重复 ^{13}C 标记的大肠埃希菌 HA107 在悉生鼠体内进行稳定核素示踪，可以区分 ^{12}C 宿主代谢组，以确定微生物产生的代谢产物的分布。与此类似，使用 ^{13}C 标记的葡萄糖确定母体抗体促进某些肠道微生物衍生的代谢物转移给后代 [82]。另一方面，目前已开发出新的方法以了解微生物代谢物对宿主的影响。代替传统的代谢组学方法，宏基因组数据可以通过生物信息学挖掘能够生物合成某些代谢物的操纵子 [82]。此外，分析现有宏基因组数据的新的计算方法已经能够识别代谢物水平和疾病结果之间的关联 [416]。最终，这些新的方法将提供更深入的理解，即宿主 - 肠道微生物相互作用是由微生物代谢物介导的。未来研究将包括发现新的微生物代谢物，鉴定产生代谢物的细菌种类，了解这些代谢物对宿主健康和疾病的影响。

参考文献

[1] Blacher E, Levy M, Tatirovsky E, et al. Microbiome-Modulated Metabolites at the Interface of Host Immunity. The Journal of Immunology, 2017, 198(2): 572-580.

[2] Rooks MG, Garrett WS. Gut microbiota, metabolites and host immunity. Nature Reviews Immunology, 2016, 16(6): 341-352.

[3] Melhem H, Kaya B, Ayata CK, et al. Metabolite-Sensing G Protein-Coupled Receptors Connect the Diet-Microbiota-Metabolites Axis to Inflammatory Bowel Disease. Cells, 2019, 8(5): 450.

[4] Sittipo P, Shim J, Lee Y. Microbial Metabolites Determine Host Health and the Status of Some Diseases. International Journal of Molecular Sciences, 2019, 20(21): 5296.

[5] Wang G, Huang S, Wang Y, et al. Bridging intestinal immunity and gut microbiota by metabolites. Cellular and Molecular Life Sciences, 2019, 76(20): 3917-3937.

[6] Nicholson JK, Holmes E, Kinross J, et al. Host-Gut Microbiota Metabolic Interactions. Science, 2012, 336(6086): 1262-1267.

[7] Hooper LV, Dan RL, JMA. Interactions between the microbiota and the immune system. Science, 2012,

336(6083): 1268-1273.

[8] Sittipo P, Lobionda S, Lee YK, et al. Intestinal microbiota and the immune system in metabolic diseases. Journal of Microbiology, 2018, 56(3): 154-162.

[9] Schroeder BO, Backhed F. Signals from the gut microbiota to distant organs in physiology and disease. Nature Medicine, 2016, 22(10): 1079-1089.

[10] Bhutia YD, Ganapathy V. Short, but Smart: SCFAs Train T Cells in the Gut to Fight Autoimmunity in the Brain. Immunity, 2015, 43(4): 629-631.

[11] Haghikia A, Jörg S, Duscha A, et al. Dietary Fatty Acids Directly Impact Central Nervous System Autoimmunity via the Small Intestine. Immunity, 2015, 43(4): 817-829.

[12] Zhou M, He J, Shen Y, et al. New Frontiers in Genetics, Gut Microbiota, and Immunity: A Rosetta Stone for the Pathogenesis of Inflammatory Bowel Disease. Biomed Research International, 2017, 2017: 1-17.

[13] Meng X, Zhou H, Shen H, et al. Microbe-metabolite-host axis, two-way action in the pathogenesis and treatment of human autoimmunity. Autoimmunity Reviews, 2019, 18(5): 455-475.

[14] Postler TS, Ghosh S. Understanding the Holobiont: How Microbial Metabolites Affect Human Health and Shape the Immune System. Cell Metabolism, 2017, 26(1): 110-130.

[15] Skelly AN, Sato Y, Kearney S, et al. Mining the microbiota for microbial and metabolite-based immunotherapies. Nature reviews. Immunology, 2019, 19(5): 305-323.

[16] Uchimura Y, Fuhrer T, Li H, et al. Antibodies Set Boundaries Limiting Microbial Metabolite Penetration and the Resultant Mammalian Host Response. Immunity, 2018, 49(3): 545-559.

[17] Shibata N, Kunisawa J, Kiyono H. Dietary and Microbial Metabolites in the Regulation of Host Immunity. Frontiers in Microbiology, 2017, 8: 2717.

[18] Bhutia YD, Ogura J, Sivaprakasam S, et al. Gut Microbiome and Colon Cancer: Role of Bacterial Metabolites and Their Molecular Targets in the Host. Current colorectal cancer reports, 2017, 13(2): 111-118.

[19] Dong L, Wang M, Guo J, et al. Role of intestinal microbiota and metabolites in inflammatory bowel disease. Chinese Medical Journal, 2019, 132(13): 1610-1614.

[20] Bilotta AJ, Cong Y. Gut microbiota metabolite regulation of host defenses at mucosal surfaces: implication in precision medicine. Precision Clinical Medicine, 2019, 2(2): 110-119.

[21] Levy M, Thaiss C, Elinav E. Metabolites: messengers between the microbiota and the immune system. Genes & Development, 2016, 30(14): 1589-1597.

[22] Chang PV, Hao L, Offermanns S, et al. The microbial metabolite butyrate regulates intestinal macrophage function via histone deacetylase inhibition. Proceedings of the National Academy of Sciences, 2014, 111(6): 2247-2252.

[23] Singh N, Gurav A, Sivaprakasam S, et al. Activation of Gpr109a, Receptor for Niacin and the Commensal Metabolite Butyrate, Suppresses Colonic Inflammation and Carcinogenesis. Immunity, 2014, 40(1): 128-139.

[24] Brown AJ, Goldsworthy SM, Barnes AA, et al. The orphan G proteincoupled receptors GPR41 and GPR43 are activated by propionate and other short chain carboxylic acids. Journal of Biological Chemistry, 2003, 278(13): 11312-11319.

[25] Koh A, De Vadder F, Kovatcheva-Datchary P, et al. From Dietary Fiber to Host Physiology: Short-Chain Fatty Acids as Key Bacterial Metabolites. Cell, 2016, 165(6): 1332-1345.

[26] Smith PM, Howitt MR, Panikov N, et al. The microbial metabolites, short-chain fatty acids, regulate colonic treg cell homeostasis. Science, 2013, 341(6145): 569-573.

[27] Voltolini C, Battersby S, Etherington SL, et al. A novel antiinflammatory role for the short-chain fatty acids in human labor. Endocrinology, 2012, 153(1): 395-403.

[28] Erny D, Hrabě De Angelis AL, Jaitin D, et al. Host microbiota constantly control maturation and function of microglia in the CNS. Nature Neuroscience, 2015, 18(7): 965-977.

[29] Ji Y, Yin Y, Li Z, et al. Gut Microbiota-Derived Components and Metabolites in the Progression of Non-Alcoholic Fatty Liver Disease (NAFLD). Nutrients, 2019, 11(8): 1712.

[30] Trompette A, Gollwitzer ES, Yadava K, et al. Gut microbiota metabolism of dietary fiber influences allergic airway disease and hematopoiesis. Nature Medicine, 2014, 20(2): 159-166.

[31] Vinolo MA, Rodrigues HG, Hatanaka E, et al. Suppressive effect of short-chain fatty acids on production of proinflammatory mediators by neutrophils. Journal of Nutritional Biochemistry, 2011, 22(9): 849-855.

[32] Chang PV, Hao L, Offermanns S, et al. The microbial metabolite butyrate regulates intestinal macrophage function via histone deacetylase inhibition. Proceedings

of the National Academy of Sciences, 2014, 111(6): 2247-2252.

[33] Trompette A, Gollwitzer ES, Yadava K, et al. Gut microbiota metabolism of dietary fiber influences allergic airway disease and hematopoiesis. Nature Medicine, 2014, 20(2): 159-166.

[34] Schilderink R, Verseijden C, de Jonge WJ. Dietary Inhibitors of Histone Deacetylases in Intestinal Immunity and Homeostasis. Frontiers in Immunology, 2013, 4: 226.

[35] Agus A, Planchais J, Sokol H. Gut Microbiota Regulation of Tryptophan Metabolism in Health and Disease. Cell Host & Microbe, 2018, 23(6): 716.

[36] Lamas B, Richard ML, Leducq V, et al. CARD9 impacts colitis by altering gut microbiota metabolism of tryptophan into aryl hydrocarbon receptor ligands. Nature Medicine, 2016, 22(6): 598-605.

[37] Li G, Young KD. Indole production by the tryptophanase TnaA in Escherichia coli is determined by the amount of exogenous tryptophan. Microbiology, 2013, 159(Pt_2): 402-410.

[38] Hashimoto T, Perlot T, Rehman A, et al. ACE2 links amino acid malnutrition to microbial ecology and intestinal inflammation. Nature, 2012, 487(7408): 477-481.

[39] Jin U, Lee S, Sridharan G, et al. Microbiome-Derived Tryptophan Metabolites and Their Aryl Hydrocarbon Receptor-Dependent Agonist and Antagonist Activities. Molecular Pharmacology, 2014, 85(5): 777-788.

[40] Xie G, Raufman J. Role of the Aryl Hydrocarbon Receptor in Colon Neoplasia. Cancers, 2015, 7(3): 1436-1446.

[41] Li Y, Innocentin S, Withers DR, et al. Exogenous stimuli maintain intraepithelial lymphocytes via aryl hydrocarbon receptor activation. Cell, 2011, 147(3): 629-640.

[42] Pondugula SR, Pavek P, SM. Pregnane X Receptor and Cancer: Context-Specificity is Key. Nuclear Receptor Research, 2016, 3: 101198.

[43] Zelante T, Iannitti RG, Cunha C, et al. Tryptophan Catabolites from Microbiota Engage Aryl Hydrocarbon Receptor and Balance Mucosal Reactivity via Interleukin-22. Immunity (Cambridge, Mass.), 2013, 39(2): 372-385.

[44] Qiu J, Heller JJ, Guo X, et al. The aryl hydrocarbon receptor regulates gut immunity through modulation of innate lymphoid cells. Immunity, 2012, 36(1): 92-104.

[45] Shi LZ, Faith NG, Nakayama Y, et al. The aryl hydrocarbon receptor is required for optimal resistance to Listeria monocytogenes infection in mice. Journal of immunology, 2007, 179(10): 6952-6962.

[46] Beaumont M, Neyrinck AM, Olivares M, et al. The gut microbiota metabolite indole alleviates liver inflammation in mice. FASEB journal : official publication of the Federation of American Societies for Experimental Biology, 2018, 32(12): 6681-6693.

[47] Alexeev EE, Lanis JM, Kao DJ, et al. Microbiota-Derived Indole Metabolites Promote Human and Murine Intestinal Homeostasis through Regulation of Interleukin-10 Receptor. American Journal of Pathology, 2018, 188(5): 1183-1194.

[48] Natividad JM, Agus A, Planchais J, et al. Impaired Aryl Hydrocarbon Receptor Ligand Production by the Gut Microbiota Is a Key Factor in Metabolic Syndrome. Cell Metabolism, 2018, 28(5): 737-749.

[49] Venkatesh M, Mukherjee S, Wang H, et al. Symbiotic bacterial metabolites regulate gastrointestinal barrier function via the xenobiotic sensor PXR and Toll-like receptor 4. Immunity, 2014, 41(2): 296-310.

[50] Hwang IK, Yoo K, Li H, et al. Indole-3-propionic acid attenuates neuronal damage and oxidative stress in the ischemic hippocampus. Journal of Neuroscience Research, 2009, 87(9): 2126-2137.

[51] Karbownik M, Reiter RJ, Garcia JJ, et al. Indole - 3 - propionic acid, a melatonin - related molecule, protects hepatic microsomal membranes from iron - induced oxidative damage: Relevance to cancer reduction. Journal of Cellular Biochemistry, 2001, 81(3): 507-513.

[52] Abildgaard A, Elfving B, Hokland M, et al. The microbial metabolite indole-3-propionic acid improves glucose metabolism in rats, but does not affect behaviour. Archives of Physiology and Biochemistry, 2018, 124(4): 306-312.

[53] Dodd D, Spitzer MH, Treuren WV, et al. A gut bacterial pathway metabolizes aromatic amino acids into nine circulating metabolites. Nature, 2017, 3(7682): e438.

[54] Nemet I, Saha PP, Gupta N, et al. A Cardiovascular Disease-Linked Gut Microbial Metabolite Acts via Adrenergic Receptors. Cell, 2020, 180(5): 862-877.

[55] Tajiri K, Shimizu Y. Branched-chain amino acids in liver diseases. Translational Gastroenterology & Hepatology, 2018, 3: 47.

[56] Yoneshiro T, Wang Q, Tajima K, et al. BCAA catabolism in brown fat controls energy homeostasis through SLC25A44. Nature, 2019, 572(7771):614-619.

[57] Yang Z, Huang S, Zou D, et al. Metabolic shifts and structural changes in the gut microbiota upon branched-chain amino acid supplementation in middle-aged mice.

Amino Acids, 2016, 48(12): 1-15.
[58] Zoltan A, Michael N. Branched Chain Amino Acids in Metabolic Disease. Current Diabetes Reports, 2018, 18(10): 76.
[59] Zhou M, Shao J, Wu CY, et al. Targeting BCAA Catabolism to Treat Obesity-Associated Insulin Resistance. Diabetes, 2019, 68(9): 1730-1746.
[60] Agus A, Clément K, Sokol H. Gut microbiota-derived metabolites as central regulators in metabolic disorders. Gut, 2021, 70(6): 1174-1182.
[61] Backhed F, Manchester JK, Semenkovich CF, et al. Mechanisms underlying the resistance to diet-induced obesity in germ-free mice. Proceedings of the National Academy of Sciences of the United States of America, 2007, 104(3): 979-984.
[62] Vamanu E, Pelinescu D, Sarbu I. Comparative Fingerprinting of the Human Microbiota in Diabetes and Cardiovascular Disease. Journal of Medicinal Food, 2016, 19(12): 1188-1195.
[63] Wahlström A, Sayin SI, Marschall H, et al. Intestinal Crosstalk between Bile Acids and Microbiota and Its Impact on Host Metabolism. Cell Metabolism, 2016, 24(1): 41-50.
[64] Brestoff JR, Artis D. Commensal bacteria at the interface of host metabolism and the immune system. Nature Immunology, 2013, 14(7): 676-684.
[65] Devkota S, Wang Y, Musch MW, et al. Dietary-fat-induced taurocholic acid promotes pathobiont expansion and colitis in Il10−/− mice. Nature, 2012, 487(7405): 104-108.
[66] Chow MD, Lee Y, Guo GL. The role of bile acids in nonalcoholic fatty liver disease and nonalcoholic steatohepatitis. Molecular Aspects of Medicine, 2017, 56: 34-44.
[67] McMillin M, DeMorrow S. Effects of bile acids on neurological function and disease. FASEB journal: official publication of the Federation of American Societies for Experimental Biology, 2016, 30(11): 3658-3668.
[68] Tiratterra E, Franco P, Porru E, et al. Role of bile acids in inflammatory bowel disease. Annals of Gastroenterology, 2018, 31(3): 266-272.
[69] Jonker JW, Liddle C, Downes M. FXR and PXR: Potential therapeutic targets in cholestasis. The Journal of Steroid Biochemistry and Molecular Biology, 2012, 130(3-5): 147-158.
[70] Zhang L, Xie C, Nichols RG, et al. Farnesoid X Receptor Signaling Shapes the Gut Microbiota and Controls Hepatic Lipid Metabolism. Msystems, 2016, 1(5): e16-e70.
[71] Ma K, Saha PK, Chan L, et al. Farnesoid X receptor is essential for normal glucose homeostasis. Journal of Clinical Investigation, 2006, 116(4): 1102-1109.
[72] Maruyama T, Miyamoto Y, Nakamura T, et al. Identification of membrane-type receptor for bile acids (M-BAR). Biochemical and Biophysical Research Communications, 2002, 298(5): 714-719.
[73] Duboc H, Taché Y, Hofmann AF. The bile acid TGR5 membrane receptor: From basic research to clinical application. Digestive and Liver Disease, 2014, 46(4): 302-312.
[74] Ding L, Yang L, Wang Z, et al. Bile acid nuclear receptor FXR and digestive system diseases. Acta Pharmaceutica Sinica B, 2015, 5(2): 135-144.
[75] Cipriani S, Mencarelli A, Chini MG, et al. The bile acid receptor GPBAR-1 (TGR5) modulates integrity of intestinal barrier and immune response to experimental colitis. Plos One, 2011, 6(10): e25637.
[76] Perino A, Pols TWH, Nomura M, et al. TGR5 reduces macrophage migration through mTOR-induced C/EBPβ differential translation. Journal of Clinical Investigation, 2014, 124(12): 5424-5436.
[77] Thomas C, Gioiello A, Noriega L, et al. TGR5-mediated bile acid sensing controls glucose homeostasis. Cell Metabolism, 2009, 10(3): 167-177.
[78] Guo C, Xie S, Chi Z, et al. Bile Acids Control Inflammation and Metabolic Disorder through Inhibition of NLRP3 Inflammasome. Immunity, 2016, 45(4): 802-816.
[79] Watanabe M, Houten SM, Mataki C, et al. Bile acids induce energy expenditure by promoting intracellular thyroid hormone activation. Nature (London), 2006, 439(7075): 484-489.
[80] Merritt ME, Donaldson JR. Effect of bile salts on the DNA and membrane integrity of enteric bacteria. Journal of Medical Microbiology, 2009, 58(12): 1533-1541.
[81] Prieto AI, Ramos-Morales F, Casadesús J. Repair of DNA Damage Induced by Bile Salts inSalmonella enterica. Genetics, 2006, 174(2): 575-584.
[82] Nicolas GR, Chang PV. Deciphering the Chemical Lexicon of Host–Gut Microbiota Interactions. Trends in Pharmacological Sciences, 2019, 40(6): 430-445.
[83] Barrea L, Annunziata G, Muscogiuri G, et al. Trimethylamine-N-oxide (TMAO) as Novel Potential Biomarker of Early Predictors of Metabolic Syndrome. Nutrients, 2018, 10(12): 1971.
[84] Tang WH, Wang Z, Levison BS, et al. Intestinal

microbial metabolism of phosphatidylcholine and cardiovascular risk. The New England journal of medicine, 2013, 368(17): 1575-1584.

[85] Gregory JC, Buffa JA, Org E, et al. Transmission of Atherosclerosis Susceptibility with Gut Microbial Transplantation. The Journal of biological chemistry, 2015, 290(9): 5647-5660.

[86] Zhu W, Gregory JC, Org E, et al. Gut Microbial Metabolite TMAO Enhances Platelet Hyperreactivity and Thrombosis Risk. Cell, 2016, 165(1): 111-124.

[87] Qi J, You T, Li J, et al. Circulating trimethylamine N - oxide and the risk of cardiovascular diseases: a systematic review and meta - analysis of 11 prospective cohort studies. Journal of Cellular and Molecular Medicine, 2017, 22(1): 185-194.

[88] Heianza Y, Ma W, Manson J, et al. Gut Microbiota Metabolites and Risk of Major Adverse Cardiovascular Disease Events and Death: A Systematic Review and Meta - Analysis of Prospective Studies. Journal of the American Heart Association Cardiovascular & Cerebrovascular Disease, 2017, 6(7): e4947.

[89] Zhu W, Buffa JA, Wang Z, et al. Flavin monooxygenase 3, the host hepatic enzyme in the metaorganismal trimethylamine N - oxide - generating pathway, modulates platelet responsiveness and thrombosis risk. Journal of Thrombosis and Haemostasis, 2018, 16(9): 1857-1872.

[90] Fujisaka S, Avila-Pacheco J, Soto MKA, et al. Diet, Genetics, and the Gut Microbiome Drive Dynamic Changes in Plasma Metabolites. Cell Reports, 2018, 22(11): 3072-3086.

[91] Romano KA, Martinez-Del Campo A, Kasahara K, et al. Metabolic, Epigenetic, and Transgenerational Effects of Gut Bacterial Choline Consumption. Cell Host & Microbe, 2017, 13;22(3): 279-290.

[92] Tofalo R, Cocchi S, Suzzi G. Polyamines and Gut Microbiota. Frontiers in Nutrition, 2019, 6: 16.

[93] Uda K, Tsujikawa T, Fujiyama Y, et al. Rapid absorption of luminal polyamines in a rat small intestine ex vivo model. Journal of Gastroenterology and Hepatology, 2003, 18(5): 554-559.

[94] Matsumoto M, Kibe R, Ooga T, et al. Impact of Intestinal Microbiota on Intestinal Luminal Metabolome. Scientific Reports, 2012, 2(1): 233.

[95] Di Martino ML, Campilongo R, Casalino M, et al. Polyamines: emerging players in bacteria-host interactions. International Journal of Medical Microbiology, 2013, 303(8): 484-491.

[96] Gong S, Richard H, JW F. YjdE (AdiC) Is the Arginine:Agmatine Antiporter Essential for Arginine-Dependent Acid Resistance in Escherichia coli. Journal of Bacteriology, 2003, 185(15): 4402-4409.

[97] Suarez C, Espariz M, Blancato VS, et al. Expression of the agmatine deiminase pathway in Enterococcus faecalis is activated by the AguR regulator and repressed by CcpA and PTS(Man) systems. Plos One, 2013, 8(10): e76170.

[98] Kitada Y, Muramatsu K, Toju H, et al. Bioactive polyamine production by a novel hybrid system comprising multiple indigenous gut bacterial strategies. Science Advances, 2018, 4(6): t62.

[99] Sugiyama Y, Nara M, Sakanaka M, et al. Comprehensive analysis of polyamine transport and biosynthesis in the dominant human gut bacteria: Potential presence of novel polyamine metabolism and transport genes. The International Journal of Biochemistry & Cell Biology, 2017, 93: 52-61.

[100] Sakanaka M, Sugiyama Y, Kitakata A, et al. Carboxyspermidine decarboxylase of the prominent intestinal microbiota species Bacteroides thetaiotaomicron is required for spermidine biosynthesis and contributes to normal growth. Amino Acids, 2016, 48(10): 2443-2451.

[101] Sturgill G, Rather PN. Evidence that putrescine acts as an extracellular signal required for swarming in Proteus mirabilis. Molecular Microbiology, 2004, 51(2): 437-446.

[102] Eisenberg T, Knauer H, Schauer A, et al. Induction of autophagy by spermidine promotes longevity. Nature Cell Biology, 2009, 11(11): 1305-1314.

[103] Zhang M, Wang H, Tracey KJ. Regulation of macrophage activation and inflammation by spermine: a new chapter in an old story. Critical Care Medicine, 2000, 28(4 Suppl): N60-N66.

[104] Kibe R, Kurihara S, Sakai Y, et al. Upregulation of colonic luminal polyamines produced by intestinal microbiota delays senescence in mice. Scientific Reports, 2014, 4: 4548.

[105] Matsumoto M, Kurihara S, Kibe R, et al. Longevity in Mice Is Promoted by Probiotic-Induced Suppression of Colonic Senescence Dependent on Upregulation of Gut Bacterial Polyamine Production. Plos One, 2011, 6(8): e23652.

[106] Minois N, Carmona-Gutierrez D, F. M. Polyamines in aging and disease. Aging, 2011, 3(8): 716-732.

[107] Pérez-Cano FJ, González-Castro A, Castellote C, et al. Influence of breast milk polyamines on suckling rat immune system maturation. Developmental &

Comparative Immunology, 2010, 34(2): 210-218.
[108] Miller-Fleming L, Olin-Sandoval V, Campbell K, et al. Remaining Mysteries of Molecular Biology: The Role of Polyamines in the Cell. Journal of Molecular Biology, 2015, 427(21): 3389-3406.
[109] Johnson CH, Dejea CM, Edler D, et al. Metabolism links bacterial biofilms and colon carcinogenesis. Cell Metabolism, 2015, 21(6): 891-897.
[110] Hayes CS, Shicora AC, Keough MP, et al. Polyamine-Blocking Therapy Reverses Immunosuppression in the Tumor Microenvironment. Cancer Immunology Research, 2014, 2(3): 274-285.
[111] Morita N, Umemoto E, Fujita S, et al. GPR31-dependent dendrite protrusion of intestinal CX3CR1(+) cells by bacterial metabolites. Nature, 2019, 566(7742): 110-114.
[112] Choi SY, Collins CC, Gout PW, et al. Cancer-generated lactic acid: a regulatory, immunosuppressive metabolite? Journal of Pathology, 2013, 230(4): 350-355.
[113] Liu C, Wu J, Zhu J, et al. Lactate Inhibits Lipolysis in Fat Cells through Activation of an Orphan G-protein-coupled Receptor, GPR81. Journal of Biological Chemistry, 2009, 284(5): 2811-2822.
[114] Ljungh A, T W. Lactic Acid Bacteria as Probiotics. Current issues in intestinal microbiology, 2006, 7(2): 73-89.
[115] Lee Y, Kim T, Kim Y, et al. Microbiota-Derived Lactate Accelerates Intestinal Stem-Cell-Mediated Epithelial Development. Cell Host & Microbe, 2018, 24(6): 833-846.
[116] Lee DC, Sohn HA, Park Z, et al. A Lactate-Induced Response to Hypoxia. Cell, 2015, 161(3): 595-609.
[117] Jakobsdottir G, Xu J, Molin G, et al. High-Fat Diet Reduces the Formation of Butyrate, but Increases Succinate, Inflammation, Liver Fat and Cholesterol in Rats, while Dietary Fibre Counteracts These Effects. Plos One, 2013, 8(11): e80476.
[118] Yang M, PJ. P. Succinate: a new epigenetic hacker. Cancer Cell, 2013, 23(6): 709-711.
[119] He W, Miao FJ, Lin DC, et al. Citric acid cycle intermediates asligands for orphan G-protein-coupled receptors. Nature, 2004, 429(6988): 188-193.
[120] Ferreyra JA, Wu KJ, Hryckowian AJ, et al. Gut Microbiota-Produced Succinate Promotes C. difficile Infection after Antibiotic Treatment or Motility Disturbance. Cell Host & Microbe, 2014, 16(6): 770-777.
[121] Curtis MM, Hu Z, Klimko C, et al. The Gut Commensal Bacteroides thetaiotaomicron Exacerbates Enteric Infection through Modification of the Metabolic Landscape. Cell Host & Microbe, 2014, 16(6): 759-769.
[122] Setchell KDR, Clerici C. Equol: History, Chemistry, and Formation. The Journal of Nutrition, 2010, 140(7): S1355-S1362.
[123] Rafii F. The Role of Colonic Bacteria in the Metabolism of the Natural Isoflavone Daidzin to Equol. Metabolites, 2015, 5(1): 56-73.
[124] Tanaka Y, Kimura S, Ishii Y, et al. Equol inhibits growth and spore formation of Clostridioides difficile. Journal of Applied Microbiology, 2019, 127(3): 932-940.
[125] Vázquez L, Flórez AB, Guadamuro L, et al. Effect of Soy Isoflavones on Growth of Representative Bacterial Species from the Human Gut. Nutrients, 2017, 9(7): 727.
[126] Feng W, Ao H, CP. Gut Microbiota, Short-Chain Fatty Acids, and Herbal Medicines. Frontiers in Pharmacology, 2018, 23(9): 1354.
[127] Postler TS, Ghosh S. Understanding the Holobiont: How Microbial Metabolites Affect Human Health and Shape the Immune System. Cell Metabolism, 2017, 26(1): 110-130.
[128] Kim EH, Kim W. An Insight into Ginsenoside Metabolite Compound K as a Potential Tool for Skin Disorder. Evidence-Based Complementary and Alternative Medicine, 2018, 2018: 8075870-8075878.
[129] Leung K, Wong A. Pharmacology of ginsenosides: a literature review. Chinese Medicine, 2010, 5(1): 20.
[130] Kim KA, Yoo HH, Gu W, et al. A prebiotic fiber increases the formation and subsequent absorption of compound K following oral administration of ginseng in rats. Journal of Ginseng Research, 2015, 39(2): 183-187.
[131] Hasebe T, Ueno N, Musch MW, et al. Daikenchuto (TU-100) shapes gut microbiota architecture and increases the production of ginsenoside metabolite compound K. Pharmacol Res Perspect, 2016, 4(1): e215.
[132] Mezzomo N, Ferreira SRS. Carotenoids Functionality, Sources, and Processing by Supercritical Technology: A Review. Journal of Chemistry, 2016, 2016: 1-16.
[133] Pandey KB, Rizvi SI. Plant Polyphenols as Dietary Antioxidants in Human Health and Disease. Oxidative Medicine and Cellular Longevity, 2009, 2(5): 270-278.

[134] Palafox-Carlos H, Ayala-Zavala JF, Gonzalez-Aguilar GA. The role of dietary fiber in the bioaccessibility and bioavailability of fruit and vegetable antioxidants. Journal of Food Science, 2011, 76(1): R6-R15.

[135] Klassen JL. Phylogenetic and Evolutionary Patterns in Microbial Carotenoid Biosynthesis Are Revealed by Comparative Genomics. Plos One, 2010, 5(6): e11257.

[136] Li X, Li S, Chen M, et al. (-)-Epigallocatechin-3-gallate (EGCG) inhibits starch digestion and improves glucose homeostasis through direct or indirect activation of PXR/CAR-mediated phase II metabolism in diabetic mice. Food Funct, 2018, 9(9): 4651-4663.

[137] Salem HA, Wadie W. Effect of Niacin on Inflammation and Angiogenesis in a Murine Model of Ulcerative Colitis. Scientific Reports, 2017, 7(1): 7139.

[138] Koh A, Molinaro A, Ståhlman M, et al. Microbially Produced Imidazole Propionate Impairs Insulin Signaling through mTORC1. Cell, 2018, 175(4): 947-961.

[139] Levy M, Thaiss CA, Zeevi D, et al. Microbiota-Modulated Metabolites Shape the Intestinal Microenvironment by Regulating NLRP6 Inflammasome Signaling. Cell, 2015, 163(6): 1428-1443.

[140] Levy M, Blacher E, E. E. Microbiome, metabolites and host immunity. Current Opinion in Microbiology, 2017, 37(4): 33.

[141] Zheng D, Liwinski T, Elinav E. Interaction between microbiota and immunity in health and disease. Cell Research, 2020, 30(6): 492-506.

[142] Donia MS, A F M. Small molecules from the human microbiota. Science, 2015, 349(6246): 1254766.

[143] Thorburn AN, Macia L, Mackay CR. Diet, metabolites, and "western-lifestyle" inflammatory diseases. Immunity, 2014, 40(6): 833-842.

[144] Thaiss CA, Zmora N, Levy M, et al. The microbiome and innate immunity. Nature, 2016, 535(7610): 65-74.

[145] Pott J, Hornef M. Innate immune signalling at the intestinal epithelium in homeostasis and disease. Embo Reports, 2012, 13(8): 684-698.

[146] Donohoe DR, Garge N, Zhang X, et al. The microbiome and butyrate regulate energy metabolism and autophagy in the mammalian colon. Cell Metabolism, 2011, 13(5): 517-526.

[147] Zhao Y, Chen F, Wu W, et al. GPR43 mediates microbiota metabolite SCFA regulation of antimicrobial peptide expression in intestinal epithelial cells via activation of mTOR and STAT3. Mucosal Immunology, 2018, 11(3): 752-762.

[148] Park J, Kotani T, Konno T, et al. Promotion of Intestinal Epithelial Cell Turnover by Commensal Bacteria: Role of Short-Chain Fatty Acids. Plos One, 2016, 11(5): e156334.

[149] Kim MH, Kang SG, Park JH, et al. Short-Chain Fatty Acids Activate GPR41 and GPR43 on Intestinal Epithelial Cells to Promote Inflammatory Responses in Mice. Gastroenterology, 2013, 145(2): 396-406.

[150] Zhao Y, Chen F, Wu W, et al. GPR43 mediates microbiota metabolite SCFA regulation of antimicrobial peptide expression in intestinal epithelial cells via activation of mTOR and STAT3. Mucosal Immunology, 2018, 11(3): 752-762.

[151] Ghorbani P, Santhakumar P, Hu Q, et al. Short-chain fatty acids affect cystic fibrosis airway inflammation and bacterial growth. European Respiratory Journal, 2015, 46(4): 1033-1045.

[152] Willemsen LE, Koetsier MA, van Deventer SJ, et al. Short chain fatty acids stimulate epithelial mucin 2 expression through differential effects on prostaglandin E(1) and E(2) production by intestinal myofibroblasts. Gut, 2003, 52(10): 1442-1447.

[153] Kaiko GE, Ryu SH, Koues OI, et al. The Colonic Crypt Protects Stem Cells from Microbiota-Derived Metabolites. Cell, 2016, 165(7): 1708-1720.

[154] Wlodarska M, Thaiss CA, Nowarski R, et al. NLRP6 inflammasome orchestrates the colonic host-microbial interface by regulating goblet cell mucus secretion. Cell, 2014, 156(5): 1045-1059.

[155] Zheng L, Kelly CJ, Battista KD, et al. Microbial-Derived Butyrate Promotes Epithelial Barrier Function through IL-10Receptor-Dependent Repression of Claudin-2. Journal of Immunology, 2017, 199(8): 2976-2984.

[156] Kelly CJ, Zheng L, Campbell EL, et al. Crosstalk between Microbiota-Derived Short-Chain Fatty Acids and Intestinal Epithelial HIF Augments Tissue Barrier Function. Cell Host & Microbe, 2015, 17(5): 662-671.

[157] Peng L, Li ZR, Green RS, et al. Butyrate enhances the intestinal barrier by facilitating tight junction assembly via activation of AMP-activated protein kinase in Caco-2 cell monolayers. Journal of Nutrition, 2009, 139(9): 1619-1625.

[158] Camille MG, Fabienne BC, Ludovica M, et al.

Butyrate produced by gut commensal bacteria activates TGF-beta1 expression through the transcription factor SP1 in human intestinal epithelial cells. Scientific Reports, 2018, 8(1): 9742.
[159] Feng Y, Wang Y, Wang P, et al. Short-Chain Fatty Acids Manifest Stimulative and Protective Effects on Intestinal Barrier Function Through the Inhibition of NLRP3 Inflammasome and Autophagy. Cellular Physiology and Biochemistry, 2018, 49(1): 190-205.
[160] Bansal T, Alaniz RC, Wood TK, et al. The bacterial signal indole increases epithelial-cell tight-junction resistance and attenuates indicators of inflammation. Proceedings of the National Academy of Sciences of the United States of America, 2010, 107(1): 228-233.
[161] Gerard P. Metabolism of cholesterol and bile acids by the gut microbiota. Pathogens, 2013, 3(1): 14-24.
[162] Hylemon PB, Zhou H, Pandak WM, et al. Bile acids as regulatory molecules. Journal of Lipid Research, 2009, 50(8): 1509-1520.
[163] Inagaki T, Moschetta A, Lee YK, et al. Regulation of antibacterial defense in the small intestine by the nuclear bile acid receptor. Proceedings of the National Academy of Sciences of the United States of America, 2006, 103(10): 3920-3925.
[164] Dossa AY, Escobar O, Golden J, et al. Bile acids regulate intestinal cell proliferation by modulating EGFR and FXR signaling. American journal of physiology. Gastrointestinal and Liver Physiology, 2016, 310(2): G81-G92.
[165] Zeng H, Umar S, Rust B, et al. Secondary Bile Acids and Short Chain Fatty Acids in the Colon: A Focus on Colonic Microbiome, Cell Proliferation, Inflammation, and Cancer. International Journal of Molecular Sciences, 2019, 20(5): 1214.
[166] Ma C, Han M, Heinrich B, et al. Gut microbiome-mediated bile acid metabolism regulates liver cancer via NKT cells. Science, 2018, 360(6391): n5931.
[167] Parker HE, Wallis K, le Roux CW, et al. Molecular mechanisms underlying bile acid-stimulated glucagon-like peptide-1 secretion. British Journal of Pharmacology, 2012, 165(2): 414-423.
[168] Carbonero F, Benefiel AC, Alizadeh-Ghamsari AH, et al. Microbial pathways in colonic sulfur metabolism and links with health and disease. Frontiers in Physiology, 2012, 3: 488.
[169] Rao JN, Liu SV, Zou T, et al. Rac1 promotes intestinal epithelial restitution by increasing Ca2+ influx through interaction with phospholipase C-(gamma)1 after wounding. American journal of physiology. Cell physiology, 2008, 295(6): C1499-C1509.
[170] Chen J, Rao JN, Zou T, et al. Polyamines are required for expression of Toll-like receptor 2 modulating intestinal epithelial barrier integrity. American journal of physiology. Gastrointestinal and liver physiology, 2007, 293(3): G568-G576.
[171] Liu L, Guo X, Rao JN, et al. Polyamines regulate E-cadherin transcription through c-Myc modulating intestinal epithelial barrier function. American journal of physiology. Cell physiology, 2009, 296(4): C801-C810.
[172] Lin X, Jiang S, Jiang Z, et al. Effects of equol on H_2O_2-induced oxidative stress in primary chicken intestinal epithelial cells. Poultry Science, 2016, 95(6): 1380-1386.
[173] Di Cagno R, Mazzacane F, Rizzello CG, et al. Synthesis of isoflavone aglycones and equol in soy milks fermented by food-related lactic acid bacteria and their effect on human intestinal Caco-2 cells. Journal of Agricultural and Food Chemistry, 2010, 58(19): 10338-10346.
[174] Harada K, Sada S, Sakaguchi H, et al. Bacterial metabolite S-equol modulates glucagon-like peptide-1 secretion from enteroendocrine L cell line GLUTag cells via actin polymerization. Biochemical and Biophysical Research Communications, 2018, 501(4): 1009-1015.
[175] Wang CW, Huang YC, Chan FN, et al. A gut microbial metabolite of ginsenosides, compound K, induces intestinal glucose absorption and Na(+) / glucose cotransporter 1 gene expression through activation of cAMP response element binding protein. Molecular Nutrition & Food Research, 2015, 59(4): 670-684.
[176] Yao H, Wan JY, Zeng J, et al. Effects of compound K, an enteric microbiome metabolite of ginseng, in the treatment of inflammation associated colon cancer. Oncology Letters, 2018, 15(6): 8339-8348.
[177] Lee IK, Kang KA, Lim CM, et al. Compound K, a metabolite of ginseng saponin, induces mitochondria-dependent and caspase-dependent apoptosis via the generation of reactive oxygen species in human colon cancer cells. International Journal of Molecular Sciences, 2010, 11(12): 4916-4931.
[178] Sonnenberg GF, Artis D. Innate lymphoid cell interactions with microbiota: implications for intestinal health and disease. Immunity, 2012, 37(4): 601-610.
[179] Walker JA, Barlow JL, McKenzie AN. Innate

lymphoid cells--how did we miss them? Nature Reviews Immunology, 2013, 13(2): 75-87.

[180] Lee JS, Cella M, McDonald KG, et al. AHR drives the development of gut ILC22 cells and postnatal lymphoid tissues via pathways dependent on and independent of Notch. Nature Immunology, 2011, 13(2): 144-151.

[181] Wang X, Ota N, Manzanillo P, et al. Interleukin-22 alleviates metabolic disorders and restores mucosal immunity in diabetes. Nature, 2014, 514(7521): 237-241.

[182] Sonnenberg GF, Artis D. Innate lymphoid cells in the initiation, regulation and resolution of inflammation. Nature Medicine, 2015, 21(7): 698-708.

[183] Goto Y, Obata T, Kunisawa J, et al. Innate lymphoid cells regulate intestinal epithelial cell glycosylation. Science, 2014, 345(6202): 1254009.

[184] Ota N, Wong K, Valdez PA, et al. IL-22 bridges the lymphotoxin pathway with the maintenance of colonic lymphoid structures during infection with Citrobacter rodentium. Nature Immunology, 2011, 12(10): 941-948.

[185] Wang Y, Koroleva EP, Kruglov AA, et al. Lymphotoxin beta receptor signaling in intestinal epithelial cells orchestrates innate immune responses against mucosal bacterial infection. Immunity, 2010, 32(3): 403-413.

[186] Akdis M, Burgler S, Crameri R, et al. Interleukins, from 1 to 37, and interferon-gamma: receptors, functions, and roles in diseases. The Journal of allergy and clinical immunology, 2011, 127(3): 701-721.

[187] O'Mahony L, Akdis M, Akdis CA. Regulation of the immune response and inflammation by histamine and histamine receptors. The Journal of allergy and clinical immunology, 2011, 128(6): 1153-1162.

[188] Claudia N, Marco C, Charlotte MB, et al. The effect of short-chain fatty acids on human monocyte-derived dendritic cells. Scientific Reports, 2015, 5: 16148.

[189] Singh N, Thangaraju M, Prasad PD, et al. Blockade of dendritic cell development by bacterial fermentation products butyrate and propionate through a transporter (Slc5a8)-dependent inhibition of histone deacetylases. Journal of Biological Chemistry, 2010, 285(36): 27601-27608.

[190] Millard AL, Mertes PM, Ittelet D, et al. Butyrate affects differentiation, maturation and function of human monocyte-derived dendritic cells and macrophages. Clinical and Experimental Immunology, 2002, 130(2): 245-255.

[191] Wang B, Morinobu A, Horiuchi M, et al. Butyrate inhibits functional differentiation of human monocyte-derived dendritic cells. Cellular Immunology, 2008, 253(1-2): 54-58.

[192] Liu L, Li L, Min J, et al. Butyrate interferes with the differentiation and function of human monocyte-derived dendritic cells. Cellular Immunology, 2012, 277(1-2): 66-73.

[193] Berndt BE, Zhang M, Owyang SY, et al. Butyrate increases IL-23 production by stimulated dendritic cells. American journal of physiology. Gastrointestinal and liver physiology, 2012, 303(12): G1384-G1392.

[194] Smolinska S, Jutel M, Crameri R, et al. Histamine and gut mucosal immune regulation. Allergy, 2014, 69(3): 273-281.

[195] Burris TP, Busby SA, Griffin PR. Targeting orphan nuclear receptors for treatment of metabolic diseases and autoimmunity. Chemistry & Biology, 2012, 19(1): 51-59.

[196] Fiorucci S, Mencarelli A, Palladino G, et al. Bile-acid-activated receptors: targeting TGR5 and farnesoid-X-receptor in lipid and glucose disorders. Trends in Pharmacological Sciences, 2009, 30(11): 570-580.

[197] Wang YD, Chen WD, Yu D, et al. The G-protein-coupled bile acid receptor, Gpbar1 (TGR5), negatively regulates hepatic inflammatory response through antagonizing nuclear factor kappa light-chain enhancer of activated B cells (NF-kappaB) in mice. Hepatology, 2011, 54(4): 1421-1432.

[198] Le Poul E, Loison C, Struyf S, et al. Functional characterization of human receptors for short chain fatty acids and their role in polymorphonuclear cell activation. Journal of Biological Chemistry, 2003, 278(28): 25481-25489.

[199] Wong JM, de Souza R, Kendall CW, et al. Colonic Health: Fermentation and Short Chain Fatty Acids. Journal of Clinical Gastroenterology, 2006, 40(3): 235-243.

[200] Maslowski KM, Vieira AT, Ng A, et al. Regulation of inflammatory responses by gut microbiota and chemoattractant receptor GPR43. Nature, 2009, 461(7268): 1282-1286.

[201] AW S. How neutrophils kill microbes. Annual Review of Immunology, 2005, 23: 197-223.

[202] Fialkow L, Wang Y, Downey GP. Reactive oxygen and nitrogen species as signaling molecules

regulating neutrophil function. Free radical biology & medicine, 2007, 42(2): 153-164.

[203] Smith K, McCoy KD, Macpherson AJ. Use of axenic animals in studying the adaptation of mammals to their commensal intestinal microbiota. Seminars in Immunology, 2007, 19(2): 59-69.

[204] Kim SV, Xiang WV, Kwak C, et al. GPR15-mediated homing controls immune homeostasis in the large intestine mucosa. Science, 2013, 340(6139): 1456-1459.

[205] Kinoshita M, Suzuki Y, Saito Y. Butyrate reduces colonic paracellular permeability by enhancing PPARgamma activation. Biochem Biophys Res Commun, 2002, 293(2): 827-831.

[206] Chen J, Wu H, Wang Q, et al. Ginsenoside metabolite compound K suppresses T-cell priming via modulation of dendritic cell trafficking and costimulatory signals, resulting in alleviation of collagen-induced arthritis. Journal of Pharmacology and Experimental Therapeutics, 2015, 353(1): 71-79.

[207] Kim J, Byeon H, Im K, et al. Effects of ginsenosides on regulatory T cell differentiation. Food Science and Biotechnology, 2018, 27(1): 227-232.

[208] Pols TWH, Puchner T, Korkmaz HI, et al. Lithocholic acid controls adaptive immune responses by inhibition of Th1 activation through the Vitamin D receptor. Plos One, 2017, 12(5): e176715.

[209] Kanai T, Kawamura T, Dohi T, et al. TH1/TH2-mediated colitis induced by adoptive transfer of CD4+CD45RBhigh T lymphocytes into nude mice. Inflammatory Bowel Diseases, 2006, 12(2): 89-99.

[210] Petra L, Georgina LH, Harry JF. The gut microbiota, bacterial metabolites and colorectal cancer. Nature Reviews Microbiology, 2014, 12(10): 661-672.

[211] Kimura A, Naka T, Nohara K, et al. Aryl hydrocarbon receptor regulates Stat1 activation and participates in the development of Th17 cells. Proceedings of the National Academy of Sciences of the United States of America, 2008, 105(28): 9721-9726.

[212] Shi Y, Mu L. An expanding stage for commensal microbes in host immune regulation. Cellular & Molecular Immunology, 2017, 14(4): 339-348.

[213] Brown EM, Kenny DJ, Xavier RJ. Gut Microbiota Regulation of T Cells During Inflammation and Autoimmunity. Annual Review of Immunology, 2019, 37(1): 599-624.

[214] Tanoue T, Atarashi K, Honda K. Development and maintenance of intestinal regulatory T cells. Nature Reviews Immunology, 2016, 16(5): 295-309.

[215] Smith PM, Howitt MR, Panikov N, et al. The Microbial Metabolites, Short-Chain Fatty Acids, Regulate Colonic T-reg Cell Homeostasis. Science, 2013, 341(6145): 569-573.

[216] Arpaia N, Campbell C, Fan X, et al. Metabolites produced by commensal bacteria promote peripheral regulatory T-cell generation. Nature, 2013, 504(7480): 451.

[217] Furusawa Y, Obata Y, Fukuda S, et al. Commensal microbe-derived butyrate induces the differentiation of colonic regulatory T cells. Nature, 2013, 504(7480): 446.

[218] Turroni S, Brigidi P, Cavalli A, et al. Microbiota–Host Transgenomic Metabolism, Bioactive Molecules from the Inside. Journal of Medicinal Chemistry, 2017, 61(1): 47-61.

[219] Park J, Kim M, Kang SG, et al. Short-chain fatty acids induce both effector and regulatory T cells by suppression of histone deacetylases and regulation of the mTOR-S6K pathway. Mucosal Immunology, 2015, 8(1): 80.

[220] Coombes JL, Siddiqui KRR, Arancibia-Cárcamo CV, et al. A functionally specialized population of mucosal CD103+ DCs induces Foxp3+ regulatory T cells via a TGF-beta and retinoic acid-dependent mechanism. The Journal of experimental medicine, 2007, 204(8): 1757-1764.

[221] Mucida D, Park Y, Kim G, et al. Reciprocal TH17 and Regulatory T Cell Differentiation Mediated by Retinoic Acid. Science, 2007, 317(5835): 256-260.

[222] Sun CM, Hall JA, Blank RB, et al. Small intestine lamina propria dendritic cells promote de novo generation of Foxp3 T reg cells via retinoic acid. Journal of Experimental Medicine, 2007, 204(8): 1775-1785.

[223] Chen W, Jin W, Hardegen N, et al. Conversion of peripheral CD4+CD25- naive T cells to CD4+CD25+ regulatory T cells by TGF-beta induction of transcription factor Foxp3. Journal of Experimental Medicine, 2003, 198(12): 1875-1886.

[224] Tone Y, Furuuchi K, Kojima Y, et al. Smad3 and NFAT cooperate to induce Foxp3 expression through its enhancer. Nature Immunology, 2008, 9(2): 194-202.

[225] Xu L, Kitani A, Stuelten C, et al. Positive and Negative Transcriptional Regulation of the Foxp3 Gene is Mediated by Access and Binding of the Smad3 Protein to Enhancer I. Immunity, 2010, 33(3): 313-325.

[226] Lochner M, Peduto L, Cherrier M, et al. In vivo equilibrium of proinflammatory IL-17+ and regulatory IL-10+ Foxp3+ RORgamma t+ T cells. Journal of Experimental Medicine, 2008, 205(6): 1381-1393.

[227] Ohnmacht C, Park JH, Cording S, et al. MUCOSAL IMMUNOLOGY. The microbiota regulates type 2 immunity through ROR γ (+) T cells. Science, 2015, 349(6251): 989-993.

[228] Kang SG, Lim HW, Andrisani OM, et al. Vitamin A metabolites induce gut-homing FoxP3+ regulatory T cells. Journal of Immunology, 2007, 179(6): 3724-3733.

[229] Huehn J, Siegmund K, Lehmann JC, et al. Developmental stage, phenotype, and migration distinguish naive- and effector/memory-like CD4+ regulatory T cells. Journal of Experimental Medicine, 2004, 199(3): 303-313.

[230] Strozzi GP, Mogna L. Quantification of folic acid in human feces after administration of Bifidobacterium probiotic strains. Journal of Clinical Gastroenterology, 2008, 42 Suppl 3 Pt 2: S179-S184.

[231] Kleerebezem M, Vaughan EE. Probiotic and gut lactobacilli and bifidobacteria: molecular approaches to study diversity and activity. Annual Review of Microbiology, 2009, 63: 269-290.

[232] Mezrich JD, Fechner JH, Zhang X, et al. An interaction between kynurenine and the aryl hydrocarbon receptor can generate regulatory T cells. Journal of Immunology, 2010, 185(6): 3190-3198.

[233] O'Mahony SM, Clarke G, Borre YE, et al. Serotonin, tryptophan metabolism and the brain-gut-microbiome axis. Behavioural Brain Research, 2015, 277: 32-48.

[234] Honda K, Littman DR. The microbiota in adaptive immune homeostasis and disease. Nature, 2016, 535(7610): 75-84.

[235] Ferstl R, Akdis CA, O'Mahony L. Histamine regulation of innate and adaptive immunity. Frontiers in bioscience (Landmark edition), 2012, 17: 40-53.

[236] Forward NA, Furlong SJ, Yang Y, et al. Mast cells down-regulate CD4+CD25+ T regulatory cell suppressor function via histamine H1 receptor interaction. Journal of Immunology, 2009, 183(5): 3014-3022.

[237] Kinoshita M, Kayama H, Kusu T, et al. Dietary folic acid promotes survival of Foxp3+ regulatory T cells in the colon. Journal of Immunology, 2012, 189(6): 2869-2878.

[238] Yamaguchi T, Hirota K, Nagahama K, et al. Control of immune responses by antigen-specific regulatory T cells expressing the folate receptor. Immunity, 2007, 27(1): 145-159.

[239] Macpherson AJ, Uhr T. Induction of protective IgA by intestinal dendritic cells carrying commensal bacteria. Science, 2004, 303(5664): 1662-1665.

[240] Kim M, Qie Y, Park J, et al. Gut Microbial Metabolites Fuel Host Antibody Responses. Cell Host & Microbe, 2016, 20(2): 202-214.

[241] Peterson DA, McNulty NP, Guruge JL, et al. IgA response to symbiotic bacteria as a mediator of gut homeostasis. Cell Host & Microbe, 2007, 2(5): 328-339.

[242] Caro-Maldonado A, Wang R, Nichols AG, et al. Metabolic reprogramming is required for antibody production that is suppressed in anergic but exaggerated in chronically BAFF-exposed B cells. Journal of Immunology, 2014, 192(8): 3626-3636.

[243] Agace WW, McCoy KD. Regionalized Development and Maintenance of the Intestinal Adaptive Immune Landscape. Immunity, 2017, 46(4): 532-548.

[244] Imhann F, Vich VA, Bonder MJ, et al. Interplay of host genetics and gut microbiota underlying the onset and clinical presentation of inflammatory bowel disease. Gut, 2018, 67(1): 108-119.

[245] Haberman Y, Tickle TL, Dexheimer PJ, et al. Pediatric Crohn disease patients exhibit specific ileal transcriptome and microbiome signature. Journal of Clinical Investigation, 2014, 124(8): 3617-3633.

[246] Forrest CM, Gould SR, Darlington LG, et al. Levels of purine, kynurenine and lipid peroxidation products in patients with inflammatory bowel disease. Advances in Experimental Medicine and Biology, 2003, 527: 395-400.

[247] Martin-Subero M, Anderson G, Kanchanatawan B, et al. Comorbidity between depression and inflammatory bowel disease explained by immune-inflammatory, oxidative, and nitrosative stress; tryptophan catabolite; and gut-brain pathways. Cns Spectrums, 2016, 21(2): 184-198.

[248] Furuzawa-Carballeda J, Fonseca-Camarillo G, Lima G, et al. Indoleamine 2,3-dioxygenase: expressing cells in inflammatory bowel disease-a cross-sectional study. Clinical & developmental immunology, 2013, 2013: 278035.

[249] Zhao L, Suolang Y, Zhou D, et al. Bifidobacteria alleviate experimentally induced colitis by upregulating indoleamine 2, 3-dioxygenase expression. Microbiology and Immunology, 2018,

62(2): 71-79.

[250] Wolf AM, Wolf D, Rumpold H, et al. Overexpression of indoleamine 2,3-dioxygenase in human inflammatory bowel disease. Clinical Immunology, 2004, 113(1): 47-55.

[251] Cervenka I, Agudelo LZ, Ruas JL. Kynurenines: Tryptophan's metabolites in exercise, inflammation, and mental health. Science, 2017, 357(6349): eaaf 9794.

[252] Sugimoto S, Naganuma M, Kanai T. Indole compounds may be promising medicines for ulcerative colitis. Journal of Gastroenterology, 2016, 51(9): 853-861.

[253] Levin AD, van den Brink GR. Selective inhibition of mucosal serotonin as treatment for IBD? Gut, 2014, 63(6): 866-867.

[254] Matsuoka K, Kanai T. The gut microbiota and inflammatory bowel disease. Seminars in Immunopathology, 2015, 37(1): 47-55.

[255] Gao J, Xu K, Liu H, et al. Impact of the Gut Microbiota on Intestinal Immunity Mediated by Tryptophan Metabolism. Frontiers in Cellular and Infection Microbiology, 2018, 8: 13.

[256] Kiss EA, Vonarbourg C, Kopfmann S, et al. Natural aryl hydrocarbon receptor ligands control organogenesis of intestinal lymphoid follicles. Science, 2011, 334(6062): 1561-1565.

[257] Zhu XM, Shi YZ, Cheng M, et al. Serum IL-6, IL-23 profile and Treg/Th17 peripheral cell populations in pediatric patients with inflammatory bowel disease. Pharmazie, 2017, 72(5): 283-287.

[258] Machiels K, Joossens M, Sabino J, et al. A decrease of the butyrate-producing species Roseburia hominis and Faecalibacterium prausnitzii defines dysbiosis in patients with ulcerative colitis. Gut, 2014, 63(8): 1275-1283.

[259] Tan JK, McKenzie C, Marino E, et al. Metabolite-Sensing G Protein-Coupled Receptors-Facilitators of Diet-Related Immune Regulation. Annual Review of Immunology, 2017, 35: 371-402.

[260] Huda-Faujan N, Abdulamir AS, Fatimah AB, et al. The impact of the level of the intestinal short chain Fatty acids in inflammatory bowel disease patients versus healthy subjects. The open biochemistry journal, 2010, 4: 53-58.

[261] Vernia P, Monteleone G, Grandinetti G, et al. Combined oral sodium butyrate and mesalazine treatment compared to oral mesalazine alone in ulcerative colitis: randomized, double-blind, placebo-controlled pilot study. Digestive Diseases and Sciences, 2000, 45(5): 976-981.

[262] Luhrs H, Gerke T, Boxberger F, et al. Butyrate inhibits interleukin-1-mediated nuclear factor-kappa B activation in human epithelial cells. Digestive Diseases and Sciences, 2001, 46(9): 1968-1973.

[263] Tan J, McKenzie C, Vuillermin PJ, et al. Dietary Fiber and Bacterial SCFA Enhance Oral Tolerance and Protect against Food Allergy through Diverse Cellular Pathways. Cell Reports, 2016, 15(12): 2809-2824.

[264] Bilotta AJ, Ma C, Huang X, et al. Microbiota metabolites SCFA stimulate epithelial migration to promote wound healing through MFGE8 and PAK1. Journal of Immunology, 2019, 202S(1).

[265] Duboc H, Rajca S, Rainteau D, et al. Connecting dysbiosis, bile-acid dysmetabolism and gut inflammation in inflammatory bowel diseases. Gut, 2013, 62(4): 531-539.

[266] Pavlidis P, Powell N, Vincent RP, et al. Systematic review: bile acids and intestinal inflammation-luminal aggressors or regulators of mucosal defence? Alimentary Pharmacology & Therapeutics, 2015, 42(7): 802-817.

[267] Wellman AS, Metukuri MR, Kazgan N, et al. Intestinal Epithelial Sirtuin 1 Regulates Intestinal Inflammation During Aging in Mice by Altering the Intestinal Microbiota. Gastroenterology, 2017, 153(3): 772-786.

[268] Li J, Zhong W, Wang W, et al. Ginsenoside metabolite compound K promotes recovery of dextran sulfate sodium-induced colitis and inhibits inflammatory responses by suppressing NF-kappaB activation. Plos One, 2014, 9(2): e87810.

[269] Soares RLS. Irritable bowel syndrome: a clinical review. World Journal of Gastroenterology, 2014, 20(34): 12144-12160.

[270] Jenkins TA, Nguyen JC, Polglaze KE, et al. Influence of Tryptophan and Serotonin on Mood and Cognition with a Possible Role of the Gut-Brain Axis. Nutrients, 2016, 8(1): 56.

[271] Fitzgerald P, Cassidy EM, Clarke G, et al. Tryptophan catabolism in females with irritable bowel syndrome: relationship to interferon-gamma, severity of symptoms and psychiatric co-morbidity. Neurogastroenterology and motility : the official journal of the European Gastrointestinal Motility Society, 2008, 20(12): 1291-1297.

[272] Reigstad CS, Salmonson CE, Rainey JR, et al. Gut

microbes promote colonic serotonin production through an effect of short-chain fatty acids on enterochromaffin cells. FASEB journal : official publication of the Federation of American Societies for Experimental Biology, 2015, 29(4): 1395-1403.

[273] Essien BE, Grasberger H, Romain RD, et al. ZBP-89 regulates expression of tryptophan hydroxylase I and mucosal defense against Salmonella typhimurium in mice. Gastroenterology, 2013, 144(7): 1466-1477.

[274] Peng Y, Nie Y, Yu J, et al. Microbial Metabolites in Colorectal Cancer: Basic and Clinical Implications. Metabolites, 2021, 11(3): 159.

[275] Li Q, Ren Y, Fu X. Inter-kingdom signaling between gut microbiota and their host. Cellular and Molecular Life Sciences, 2019, 76(12): 2383-2389.

[276] Noureldein MH, Eid AA. Gut microbiota and mTOR signaling: Insight on a new pathophysiological interaction. Microbial Pathogenesis, 2018, 118: 98-104.

[277] Lavoie S, Chun E, Bae S, et al. Expression of Free Fatty Acid Receptor 2 by Dendritic Cells Prevents Their Expression of Interleukin 27 and Is Required for Maintenance of Mucosal Barrier and Immune Response Against Colorectal Tumors in Mice. Gastroenterology, 2020, 158(5): 1359-1372.

[278] Mohseni AH, Taghinezhad-S S, Fu X. Gut microbiota-derived metabolites and colorectal cancer: New insights and updates. Microbial Pathogenesis, 2020, 149: 104569.

[279] Belcheva A, Irrazabal T, Robertson SJ, et al. Gut microbial metabolism drives transformation of MSH2-deficient colon epithelial cells. Cell, 2014, 158(2): 288-299.

[280] Scharlau D, Borowicki A, Habermann N, et al. Mechanisms of primary cancer prevention by butyrate and other products formed during gut flora-mediated fermentation of dietary fibre. Mutation research, 2009, 682(1): 39-53.

[281] Hovhannisyan G, Aroutiounian R, Glei M. Butyrate reduces the frequency of micronuclei in human colon carcinoma cells in vitro. Toxicology in Vitro, 2009, 23(6): 1028-1033.

[282] Chen G, Ran X, Li B, et al. Sodium Butyrate Inhibits Inflammation and Maintains Epithelium Barrier Integrity in a TNBS-induced Inflammatory Bowel Disease Mice Model. Ebiomedicine, 2018, 30: 317-325.

[283] Xiao T, Wu S, Yan C, et al. Butyrate upregulates the TLR4 expression and the phosphorylation of MAPKs and NK-κ B in colon cancer cell in vitro. Oncology Letters, 2018, 16(4): 4439-4447.

[284] Sun X, MJ Z. Butyrate Inhibits Indices of Colorectal Carcinogenesis via Enhancing α-Ketoglutarate-Dependent DNA Demethylation of Mismatch Repair Genes. Molecular Nutrition & Food Research, 2018, 62(10): e1700932.

[285] Li Q, Cao L, Tian Y, et al. Butyrate Suppresses the Proliferation of Colorectal Cancer Cells via Targeting Pyruvate Kinase M2 and Metabolic Reprogramming. Molecular & Cellular Proteomics, 2018, 17(8): 1531-1545.

[286] Liang S, Mao Y, Liao M, et al. Gut microbiome associated with APC gene mutation in patients with intestinal adenomatous polyps. International Journal of Biological Sciences, 2020, 16(1): 135-146.

[287] Louis P, Flint HJ. Formation of propionate and butyrate by the human colonic microbiota. Environmental Microbiology, 2017, 19(1): 29-41.

[288] Luu M, Weigand K, Wedi F, et al. Regulation of the effector function of CD8+ T cells by gut microbiota-derived metabolite butyrate. Scientific Reports, 2018, 8(1): 14430.

[289] Casanova MR, Azevedo-Silva J, Rodrigues LR, et al. Colorectal Cancer Cells Increase the Production of Short Chain Fatty Acids by Propionibacterium freudenreichii Impacting on Cancer Cells Survival. Frontiers in Nutrition, 2018, 5: 44.

[290] Marques C, Oliveira C, Alves S, et al. Acetate-induced apoptosis in colorectal carcinoma cells involves lysosomal membrane permeabilization and cathepsin D release. Cell Death & Disease, 2013, 4(2): e507.

[291] Hao N, Whitelaw ML. The emerging roles of AhR in physiology and immunity. Biochemical Pharmacology, 2013, 86(5): 561-570.

[292] Jin UH, Cheng Y, Park H, et al. Short Chain Fatty Acids Enhance Aryl Hydrocarbon (Ah) Responsiveness in Mouse Colonocytes and Caco-2 Human Colon Cancer Cells. Scientific Reports, 2017, 7(1): 10163.

[293] Théate I, van Baren N, Pilotte L, et al. Extensive profiling of the expression of the indoleamine 2,3-dioxygenase 1 protein in normal and tumoral human tissues. Cancer Immunology Research, 2015, 3(2): 161-172.

[294] Weinlich G, Murr C, Richardsen L, et al. Decreased serum tryptophan concentration predicts poor prognosis in malignant melanoma patients.

Dermatology, 2007, 214(1): 8-14.
[295] Engin AB, Karahalil B, Karakaya AE, et al. Helicobacter pylori and serum kynurenine-tryptophan ratio in patients with colorectal cancer. World Journal of Gastroenterology, 2015, 21(12): 3636-3643.
[296] Gargaro M, Manni G, Scalisi G, et al. Tryptophan Metabolites at the Crossroad of Immune-Cell Interaction via the Aryl Hydrocarbon Receptor: Implications for Tumor Immunotherapy. International Journal of Molecular Sciences, 2021, 22(9): 4644.
[297] Chan K, Kim JH, Jin SK, et al. A novel dual inhibitor of IDO and TDO, CMG017, potently suppresses the kynurenine pathway and overcomes resistance to immune checkpoint inhibitors. Journal of Clinical Oncology, 2019, 37(15_suppl): e14228.
[298] Romagnolo DF, Donovan MG, Doetschman TC, et al. n-6 Linoleic Acid Induces Epigenetics Alterations Associated with Colonic Inflammation and Cancer. Nutrients, 2019, 11(1): 171.
[299] Mahboob S, Ahn SB, Cheruku HR, et al. A novel multiplexed immunoassay identifies CEA, IL-8 and prolactin as prospective markers for Dukes'stages A-D colorectal cancers. Clinical Proteomics, 2015, 12(1): 10.
[300] Nguyen TT, Lian S, Ung TT, et al. Lithocholic Acid Stimulates ILExpression in Human Colorectal Cancer Cells Via Activation of Erk1/2 MAPK and Suppression of STAT3 Activity. Journal of Cellular Biochemistry, 2017, 118(9): 2958-2967.
[301] Nguyen TT, Ung TT, Li S, et al. Metformin inhibits lithocholic acid-induced interleukin 8 upregulation in colorectal cancer cells by suppressing ROS production and NF-κB activity. Scientific Reports, 2019, 9(1): 2003.
[302] Min KB, Park JS, Ji HP, et al. Lithocholic acid upregulates uPAR and cell invasiveness via MAPK and AP-1 signaling in colon cancer cells. Cancer Letters, 2010, 290(1): 123-128.
[303] Centuori SM, Gomes CJ, Trujillo J, et al. Deoxycholic acid mediates non-canonical EGFR-MAPK activation through the induction of calcium signaling in colon cancer cells. BBA - Molecular and Cell Biology of Lipids, 2016, 1861(7): 663-670.
[304] Zeng H, Claycombe KJ, Reindl KM. Butyrate and deoxycholic acid play common and distinct roles in HCT116 human colon cell proliferation. Journal of Nutritional Biochemistry, 2015, 26(10): 1022-1028.
[305] Liu L, Dong W, Wang S, et al. Deoxycholic acid disrupts the intestinal mucosal barrier and promotes intestinal tumorigenesis. Food & Function, 2018, 9(11): 5588-5597.
[306] Lee DG, Hori S, Kohmoto O, et al. Ingestion of difructose anhydride III partially suppresses the deconjugation and 7α-dehydroxylation of bile acids in rats fed with a cholic acid-supplemented diet. Bioscience Biotechnology & Biochemistry, 2019, 83(7): 1329-1335.
[307] Yan XL, Zhang XB, Ao R, et al. Effects of shRNA - Mediated Silencing of PKM2 Gene on Aerobic Glycolysis, Cell Migration, Cell Invasion, and Apoptosis in Colorectal Cancer Cells. Journal of Cellular Biochemistry, 2017, 118(12): 4792-4803.
[308] Graziano F, Ruzzo A, Giacomini E, et al. Glycolysis gene expression analysis and selective metabolic advantage in the clinical progression of colorectal cancer. Pharmacogenomics Journal, 2017, 17(3): 258-264.
[309] Haraguchi T, Kayashima T, Okazaki Y, et al. Cecal succinate elevated by some dietary polyphenols may inhibit colon cancer cell proliferation and angiogenesis. Journal of Agricultural & Food Chemistry, 2014, 62(24): 5589.
[310] Wu JY, Huang TW, Hsieh YT, et al. Cancer-Derived Succinate Promotes Macrophage Polarization and Cancer Metastasis via Succinate Receptor - ScienceDirect. Molecular Cell, 2020, 77(2): 213-227.
[311] Beaumont M, Andriamihaja M, Lan A, et al. Detrimental effects for colonocytes of an increased exposure to luminal hydrogen sulfide: The adaptive response. Free Radical Biology & Medicine, 2016, 93: 155-164.
[312] Libiad M, Vitvitsky V, Bostelaar T, et al. Hydrogen sulfide perturbs mitochondrial bioenergetics and triggers metabolic reprogramming in colon cells. Journal of Biological Chemistry, 2019, 294(32): A119-A9442.
[313] Wang W, Chen M, Xi J, et al. H 2 S induces Th1/Th2 imbalance with triggered NF-κ B pathway to exacerbate LPS-induce chicken pneumonia response. Chemosphere, 2018, 208: 241-246.
[314] Sakuma S, Minamino S, Takase M, et al. Hydrogen sulfide donor GYY4137 suppresses proliferation of human colorectal cancer Caco-2 cells by inducing both cell cycle arrest and cell death - ScienceDirect. Heliyon, 2019, 5(8): e2244.
[315] Armand L, Andriamihaja M, Gellenoncourt S, et al. In vitro impact of amino acid-derived bacterial metabolites on colonocyte mitochondrial activity,

oxidative stress response and DNA integrity. Biochimica et biophysica acta. General subjects, 2019, 1863(8): 1292-1301.

[316] Beaumont M, Portune KJ, Steuer N, et al. Quantity and source of dietary protein influence metabolite production by gut microbiota and rectal mucosa gene expression: a randomized, parallel, double-blind trial in overweight humans. The American journal of clinical nutrition, 2017, 106(4): 1005-1019.

[317] Haghi F, Goli E, Mirzaei B, et al. The association between fecal enterotoxigenic B. fragilis with colorectal cancer. Bmc Cancer, 2019, 19(1): 879.

[318] Thiele Orberg E, Fan H, Tam AJ, et al. The Myeloid Immune Signature of Enterotoxigenic Bacteroides Fragilis-Induced Murine Colon Tumorigenesis. Mucosal Immunology, 2017, 10(2): 421-433.

[319] Chung L, Orberg ET, Geis AL, et al. Bacteroides fragilis Toxin Coordinates a Pro-carcinogenic Inflammatory Cascade via Targeting of Colonic Epithelial Cells. Cell Host & Microbe, 2018, 23(3): 421.

[320] Allen J, Sears CL. Impact of the gut microbiome on the genome and epigenome of colon epithelial cells: contributions to colorectal cancer development. Genome Medicine, 2019, 11(1): 11.

[321] Wilson MR, Jiang Y, Villalta PW, et al. The human gut bacterial genotoxin colibactin alkylates DNA. Science, 2019, 363(6428): r7785.

[322] Ho M, Mettouchi A, Wilson BA, et al. CNF1-like deamidase domains: common Lego bricks among cancer-promoting immunomodulatory bacterial virulence factors. Pathogens & Disease, 2018, 76(5): y45.

[323] Zhang Z, Aung KM, Uhlin BE, et al. Reversible senescence of human colon cancer cells after blockage of mitosis/cytokinesis caused by the CNF1 cyclomodulin from Escherichia coli. Scientific Reports, 2018, 12;8(1): 17780.

[324] Gall GL, Guttula K, Kellingray L, et al. Correction: Metabolite quantification of faecal extracts from colorectal cancer patients and healthy controls. Oncotarget, 2019, 10(17): 1660.

[325] Wang Q, Li L, R. X. A systems biology approach to predict and characterize human gut microbial metabolites in colorectal cancer. Scientific Reports, 2018, 8(1): 6225.

[326] Y Achida S, Mizutani S, Shiroma H, et al. Metagenomic and metabolomic analyses reveal distinct stage-specific phenotypes of the gut microbiota in colorectal cancer. Nature Medicine, 2019, 25(6): 968-976.

[327] Yoshimoto S, Loo TM, Atarashi K, et al. Obesity-induced gut microbial metabolite promotes liver cancer through senescence secretome. Nature, 2013, 499(7456): 97-101.

[328] Ma C, Han M, Heinrich B, et al. Gut microbiome–mediated bile acid metabolism regulates liver cancer via NKT cells. Science, 2018, 360(6391): n5931.

[329] Younossi ZM, Koenig AB, Abdelatif D, et al. Global epidemiology of nonalcoholic fatty liver disease-Meta-analytic assessment of prevalence, incidence, and outcomes. Hepatology, 2016, 64(1): 73-84.

[330] Kim D, Touros A, Kim WR. Nonalcoholic Fatty Liver Disease and Metabolic Syndrome. Clinics in Liver Disease, 2018, 22(1): 133-140.

[331] Sahuri-Arisoylu M, Brody LP, Parkinson JR, et al. Reprogramming of hepatic fat accumulation and 'browning' of adipose tissue by the short-chain fatty acid acetate. International Journal of Obesity, 2016, 40(6): 955-963.

[332] Zhou D, Pan Q, Xin FZ, et al. Sodium butyrate attenuates high-fat diet-induced steatohepatitis in mice by improving gut microbiota and gastrointestinal barrier. World Journal of Gastroenterology, 2017, 23(1): 60-75.

[333] Zhou D, Chen YW, Zhao ZH, et al. Sodium butyrate reduces high-fat diet-induced non-alcoholic steatohepatitis through upregulation of hepatic GLP-1R expression. Experimental and Molecular Medicine, 2018, 50(12): 1-12.

[334] Jin CJ, Sellmann C, Engstler AJ, et al. Supplementation of sodium butyrate protects mice from the development of non-alcoholic steatohepatitis (NASH). The British journal of nutrition, 2015, 114(11): 1745-1755.

[335] Ye J, Lv L, Wu W, et al. Butyrate Protects Mice Against Methionine-Choline-Deficient Diet-Induced Non-alcoholic Steatohepatitis by Improving Gut Barrier Function, Attenuating Inflammation and Reducing Endotoxin Levels. Frontiers in Microbiology, 2018, 9: 1967.

[336] Mollica MP, Mattace RG, Cavaliere G, et al. Butyrate Regulates Liver Mitochondrial Function, Efficiency, and Dynamics in Insulin-Resistant Obese Mice. Diabetes, 2017, 66(5): 1405-1418.

[337] Fukunishi S, Sujishi T, Takeshita A, et al. Lipopolysaccharides accelerate hepatic steatosis in the development of nonalcoholic fatty liver disease

in Zucker rats. Journal of Clinical Biochemistry and Nutrition, 2014, 54(1): 39-44.

[338] Christiansen CB, Gabe M, Svendsen B, et al. The impact of short-chain fatty acids on GLP-1 and PYY secretion from the isolated perfused rat colon. American journal of physiology. Gastrointestinal and liver physiology, 2018, 315(1): G53-G65.

[339] Lee J, Hong S, Rhee E, et al. GLP-1 Receptor Agonist and Non-Alcoholic Fatty Liver Disease. Diabetes & Metabolism Journal, 2012, 36(4): 262-267.

[340] Rau M, Rehman A, Dittrich M, et al. Fecal SCFAs and SCFA-producing bacteria in gut microbiome of human NAFLD as a putative link to systemic T-cell activation and advanced disease. United European Gastroenterology Journal, 2018, 6(10): 1496-1507.

[341] Yang ZX, Shen W, Sun H. Effects of nuclear receptor FXR on the regulation of liver lipid metabolism in patients with non-alcoholic fatty liver disease. Hepatology International, 2010, 4(4): 741-748.

[342] Lou G, Ma X, Fu X, et al. GPBAR1/TGR5 mediates bile acid-induced cytokine expression in murine Kupffer cells. Plos One, 2014, 9(4): e93567.

[343] Fon TK, Rozman D. Nonalcoholic Fatty liver disease: focus on lipoprotein and lipid deregulation. Journal of lipids, 2011: 783976.

[344] Oellgaard J, Winther SA, Hansen TS, et al. Trimethylamine N-oxide (TMAO) as a New Potential Therapeutic Target for Insulin Resistance and Cancer. Current Pharmaceutical Design, 2017, 23(25): 3699-3712.

[345] Warrier M, Shih DM, Burrows AC, et al. The TMAO-Generating Enzyme Flavin Monooxygenase 3 Is a Central Regulator of Cholesterol Balance. Cell Reports, 2015, 10(3): 326-338.

[346] Duttaroy AK. Role of Gut Microbiota and Their Metabolites on Atherosclerosis, Hypertension and Human Blood Platelet Function: A Review. Nutrients, 2021, 13(1): 144.

[347] Moore KJ, Sheedy FJ, EA. F. Macrophages in atherosclerosis: a dynamic balance. Nature Reviews Immunology, 2013, 13(10): 709-721.

[348] Wang Z, Roberts AB, Buffa JA, et al. Nonlethal Inhibition of Gut Microbial Trimethylamine Production for the Treatment of Atherosclerosis. Cell, 2015, 163(7): 1585-1595.

[349] Maehre HK, Jensen IJ, Elvevoll EO, et al. omega-3 Fatty Acids and Cardiovascular Diseases: Effects, Mechanisms and Dietary Relevance. International Journal of Molecular Sciences, 2015, 16(9): 22636-22661.

[350] Ussher JR, Lopaschuk GD, Arduini A. Gut microbiota metabolism of L-carnitine and cardiovascular risk. Atherosclerosis, 2013, 231(2): 456-461.

[351] Nagata C, Wada K, Tamura T, et al. Choline and Betaine Intakes Are Not Associated with Cardiovascular Disease Mortality Risk in Japanese Men and Women. Journal of Nutrition, 2015, 145(8): 1787-1792.

[352] DiNicolantonio JJ, Lavie CJ, Fares H, et al. L-carnitine in the secondary prevention of cardiovascular disease: systematic review and meta-analysis. Mayo Clinic Proceedings, 2013, 88(6): 544-551.

[353] Collins HL, Drazul-Schrader D, Sulpizio AC, et al. L-Carnitine intake and high trimethylamine N-oxide plasma levels correlate with low aortic lesions in ApoE(-/-) transgenic mice expressing CETP. Atherosclerosis, 2016, 244: 29-37.

[354] Wang Z, Klipfell E, Bennett BJ, et al. Gut flora metabolism of phosphatidylcholine promotes cardiovascular disease. Nature, 2011, 472(7341): 57-82.

[355] Koeth R, Levison B, Culley M, et al. γ-Butyrobetaine Is a Proatherogenic Intermediate in Gut Microbial Metabolism of L-Carnitine to TMAO. Cell Metabolism, 2014, 20(5): 799-812.

[356] Erré MG, Hu FB, Anela MR, et al. Plasma Metabolites From Choline Pathway and Risk of Cardiovascular Disease in the PREDIMED (Prevention With Mediterranean Diet) Study. Journal of the American Heart Association, 2017, 6(11): e6524.

[357] Miao J, Ling AV, Manthena PV, et al. Flavin-containing monooxygenase 3 as a potential player in diabetes-associated atherosclerosis. Nature Communications, 2015, 7;6: 6498.

[358] Seldin MM, Meng Y, Qi H, et al. Trimethylamine N-Oxide Promotes Vascular Inflammation Through Signaling of Mitogen-Activated Protein Kinase and Nuclear Factor-kappaB. Journal of the American Heart Association, 2016, 5(2): e2767.

[359] Ma GH, Bing P, Chen Y, et al. Trimethylamine N-oxide in atherogenesis: impairing endothelial Self-repair capacity and enhancing monocyte adhesion. Bioscience Reports, 2017, 37(2): R20160244.

[360] Jaensson-Gyllenback E, Kotarsky K, Zapata F, et al. Bile retinoids imprint intestinal CD103+ dendritic cells with the ability to generate gut-tropic T cells. Mucosal Immunology, 2011, 4(4): 438-447.

[361] Ridlon JM, Harris SC, Bhowmik S, et al. Consequences of bile salt biotransformations by intestinal bacteria. Gut Microbes, 2016, 7(1): 22-39.

[362] Hsu C, Chan JYH, Wu KLH, et al. Altered Gut Microbiota and Its Metabolites in Hypertension of Developmental Origins: Exploring Differences between Fructose and Antibiotics Exposure. International Journal of Molecular Sciences, 2021, 22(5): 2674.

[363] van Nood E, Vrieze A, Nieuwdorp M, et al. Duodenal infusion of donor feces for recurrent Clostridium difficile. The New England journal of medicine, 2013, 368(5): 407-415.

[364] Tan JK, Mckenzie C, Mari OE, et al. Metabolite-Sensing G Protein–Coupled Receptors—Facilitators of Diet-Related Immune Regulation. Annual Review of Immunology, 2017, 35(1): 371-402.

[365] Natarajan N, Hori D, Flavahan S, et al. Microbial short chain fatty acid metabolites lower blood pressure via endothelial G-protein coupled receptor 41. Physiological Genomics, 2016, 48(11): 826-834.

[366] Yamashita T, Yoshida N, Emoto T, et al. Two Gut Microbiota-Derived Toxins Are Closely Associated with Cardiovascular Diseases: A Review. Toxins, 2021, 13(5): 297.

[367] Pastori D, Carnevale R, Nocella C, et al. Gut - Derived Serum Lipopolysaccharide is Associated With Enhanced Risk of Major Adverse Cardiovascular Events in Atrial Fibrillation: Effect of Adherence to Mediterranean Diet. Journal of the American Heart Association, 2017, 6(6): e5784.

[368] Neves AL, Coelho J, Couto L, et al. Metabolic endotoxemia: A molecular link between obesity and cardiovascular risk. Journal of Molecular Endocrinology, 2013, 51(2): R51-R64.

[369] Yin L, Li X, Ghosh S, et al. Role of gut microbiota - derived metabolites on vascular calcification in CKD. Journal of Cellular and Molecular Medicine, 2021, 25(3): 1332-1341.

[370] Strauss HW, Nakahara T, Narula N, et al. Vascular Calcification: The evolving relationship of vascular calcification to major acute coronary events. Journal of Nuclear Medicine, 2019, 60(9): 1207-1212.

[371] Yang T, Richards EM, Pepine CJ, et al. The gut microbiota and the brain–gut–kidney axis in hypertension and chronic kidney disease. Nature Reviews Nephrology, 2018, 14(7): 442-456.

[372] Chen YY, Chen DQ, Chen L, et al. Microbiome-metabolome reveals the contribution of gut-kidney axis on kidney disease. Journal of Translational Medicine, 2019, 17(1): 5.

[373] Ascher S, Reinhardt C. The gut microbiota: An emerging risk factor for cardiovascular and cerebrovascular disease. European Journal of Immunology, 2017, 48(4): 564-575.

[374] Hill E, Sapa H, Negrea L, et al. Effect of Oat β-Glucan Supplementation on Chronic Kidney Disease: A Feasibility Study. Journal of Renal Nutrition, 2019, 30(3): 208-215.

[375] Janakiraman M, G. K. Emerging role of diet and Microbiota interactions in neuroinflammation. Frontiers in Immunology, 2018, 9: 2067.

[376] Kocur M, Schneider R, Pulm AK, et al. IFNbeta secreted by microglia mediates clearance of myelin debris in CNS autoimmunity. Acta Neuropathol Commun, 2015, 3: 20.

[377] Reemst K, Noctor SC, Lucassen PJ, et al. The Indispensable Roles of Microglia and Astrocytes during Brain Development. Frontiers in Human Neuroscience, 2016, 10: 566.

[378] Sherry CL, Kim SS, Dilger RN, et al. Sickness behavior induced by endotoxin can be mitigated by the dietary soluble fiber, pectin, through up-regulation of IL-4 and Th2 polarization. Brain Behavior and Immunity, 2010, 24(4): 631-640.

[379] Braniste V, Al-Asmakh M, Kowal C, et al. The gut microbiota influences blood-brain barrier permeability in mice. Science Translational Medicine, 2014, 6(263): 158r-263r.

[380] Schilderink R, Verseijden C, Seppen J, et al. The SCFA butyrate stimulates the epithelial production of retinoic acid via inhibition of epithelial HDAC. American journal of physiology. Gastrointestinal and liver physiology, 2016, 310(11): G1138-G1146.

[381] Hill JA, Hall JA, Sun CM, et al. Retinoic acid enhances Foxp3 induction indirectly by relieving inhibition from CD4+CD44hi Cells. Immunity, 2008, 29(5): 758-770.

[382] Haase S, Haghikia A, Wilck N, et al. Impacts of microbiome metabolites on immune regulation and autoimmunity. Immunology, 2018, 154(2): 230-238.

[383] Colombo AV, Sadler RK, Llovera G, et al. Microbiota-derived short chain fatty acids modulate microglia and promote Aβ plaque deposition. Elife, 2021, 10: e59826.

[384] Marizzoni M, Cattaneo A, Mirabelli P, et al. Short-Chain Fatty Acids and Lipopolysaccharide as Mediators Between Gut Dysbiosis and Amyloid

Pathology in Alzheimer's Disease. Journal of Alzheimer's Disease, 2020, 78(2): 683-697.

[385] Sampson TR, Debelius JW, Thron T, et al. Gut Microbiota Regulate Motor Deficits and Neuroinflammation in a Model of Parkinson's Disease. Cell, 2016, 167(6): 1469-1480.

[386] Wakade C, Chong R, Bradley E, et al. Upregulation of GPR109A in Parkinson's disease. Plos One, 2014, 9(10): e109818.

[387] Fu SP, Wang JF, Xue WJ, et al. Anti-inflammatory effects of BHBA in both in vivo and in vitro Parkinson's disease models are mediated by GPR109A-dependent mechanisms. Journal of Neuroinflammation, 2015, 12: 9.

[388] Hsiao EY, McBride SW, Hsien S, et al. Microbiota modulate behavioral and physiological abnormalities associated with neurodevelopmental disorders. Cell, 2013, 155(7): 1451-1463.

[389] Rothhammer V, Mascanfroni ID, Bunse L, et al. Type I interferons and microbial metabolites of tryptophan modulate astrocyte activity and central nervous system inflammation via the aryl hydrocarbon receptor. Nature Medicine, 2016, 22(6): 586-597.

[390] Veldhoen M, Hirota K, Westendorf AM, et al. The aryl hydrocarbon receptor links TH17-cell-mediated autoimmunity to environmental toxins. Nature, 2008, 453(7191): 106-109.

[391] Yeste A, Nadeau M, Burns EJ, et al. Nanoparticle-mediated codelivery of myelin antigen and a tolerogenic small molecule suppresses experimental autoimmune encephalomyelitis. Proceedings of the National Academy of Sciences of the United States of America, 2012, 109(28): 11270-11275.

[392] Lawrence RC, Felson DT, Helmick CG, et al. Estimates of the prevalence of arthritis and other rheumatic conditions in the United States. Part II. Arthritis and rheumatism, 2008, 58(1): 26-35.

[393] Vereecke L, Elewaut D. Spondyloarthropathies: Ruminococcus on the horizon in arthritic disease. Nature reviews. Rheumatology, 2017, 13(10): 574-576.

[394] Kang KY, Lee SH, Jung SM, et al. Downregulation of Tryptophan-related Metabolomic Profile in Rheumatoid Arthritis Synovial Fluid. The Journal of rheumatology, 2015, 42(11): 2003-2011.

[395] Langan D, Perkins DJ, Vogel SN, et al. Microbiota-Derived Metabolites, Indole-3-aldehyde and Indole-3-acetic Acid, Differentially Modulate Innate Cytokines and Stromal Remodeling Processes Associated with Autoimmune Arthritis. International Journal of Molecular Sciences, 2021, 22(4): 2017.

[396] Pan H, Guo R, Ju Y, et al. A single bacterium restores the microbiome dysbiosis to protect bones from destruction in a rat model of rheumatoid arthritis. Microbiome, 2019, 7(1): 107.

[397] Mullaney JA, Stephens JE, Costello M, et al. Type 1 diabetes susceptibility alleles are associated with distinct alterations in the gut microbiota. Microbiome, 2018, 6(1): 35.

[398] Mariño E, Richards JL, McLeod KH, et al. Gut microbial metabolites limit the frequency of autoimmune T cells and protect against type 1 diabetes. Nature Immunology, 2017, 18(5): 552-562.

[399] Miani M, Le Naour J, Waeckel-Enee E, et al. Gut Microbiota-Stimulated Innate Lymphoid Cells Support beta-Defensin 14 Expression in Pancreatic Endocrine Cells, Preventing Autoimmune Diabetes. Cell Metabolism, 2018, 28(4): 557-572.

[400] Sun J, Furio L, Mecheri R, et al. Pancreatic beta-Cells Limit Autoimmune Diabetes via an Immunoregulatory Antimicrobial Peptide Expressed under the Influence of the Gut Microbiota. Immunity, 2015, 43(2): 304-317.

[401] Wen L, FS. W. Dietary short-chain fatty acids protect against type 1 diabetes. Nature Immunology, 2017, 18(5): 484-486.

[402] Vatanen T, Franzosa EA, Schwager R, et al. The human gut microbiome in early-onset type 1 diabetes from the TEDDY study. Nature, 2018, 562(7728): 589-594.

[403] Reilly SM, Saltiel AR. Adapting to obesity with adipose tissue inflammation. Nature reviews. Endocrinology, 2017, 13(11): 633-643.

[404] Canfora EE, Meex RCR, Venema K, et al. Gut microbial metabolites in obesity, NAFLD and T2DM. Nature Reviews Endocrinology, 2019, 15(5): 261-273.

[405] Chambers ES, Viardot A, Psichas A, et al. Effects of targeted delivery of propionate to the human colon on appetite regulation, body weight maintenance and adiposity in overweight adults. Gut, 2015, 64(11): 1744-1754.

[406] Gao Z, Yin J, Zhang J, et al. Butyrate improves insulin sensitivity and increases energy expenditure in mice. Diabetes, 2009, 58(7): 1509-1517.

[407] Chimerel C, Emery E, Summers DK, et al. Bacterial Metabolite Indole Modulates Incretin Secretion from Intestinal Enteroendocrine L Cells. Cell Reports,

2014, 9(4): 1202-1208.

[408] Ramos-Molina B, Queipo-Ortuño MI, Lambertos A, et al. Dietary and Gut Microbiota Polyamines in Obesity- and Age-Related Diseases. Frontiers in nutrition, 2019, 6: 24.

[409] Mandaliya DK, Seshadri S. Short Chain Fatty Acids, pancreatic dysfunction and type 2 diabetes. Pancreatology, 2019, 19(2): 280-284.

[410] Puddu A, Sanguineti R, Montecucco F, et al. Evidence for the gut microbiota short-chain fatty acids as key pathophysiological molecules improving diabetes. Mediators of Inflammation, 2014, 2014: 162021.

[411] Shen G, Wu J, Ye B, et al. Gut Microbiota-Derived Metabolites in the Development of Diseases. Canadian Journal of Infectious Diseases and Medical Microbiology, 2021, 2021: 6658674.

[412] Yan X, Li P, Tang Z, et al. The relationship between bile acid concentration, glucagon-like-peptide 1, fibroblast growth factor 15 and bile acid receptors in rats during progression of glucose intolerance. Bmc Endocrine Disorders, 2017, 17(1): 60.

[413] Tuomainen M, Lindström J, Lehtonen M, et al. Associations of serum indolepropionic acid, a gut microbiota metabolite, with type 2 diabetes and low-grade inflammation in high-risk individuals. Nutrition & Diabetes, 2018, 8(1): 35.

[414] Pedersen HK, Gudmundsdottir V, Nielsen HBR, et al. Human gut microbes impact host serum metabolome and insulin sensitivity. Nature, 2016, 535(7612): 376-381.

[415] Campbell C, Dikiy S, Bhattarai SK, et al. Extrathymically Generated Regulatory T Cells Establish a Niche for Intestinal Border-Dwelling Bacteria and Affect Physiologic Metabolite Balance. Immunity, 2018, 48(6): 1245-1257.

[416] Sanna S, van Zuydam NR, Mahajan A, et al. Causal relationships among the gut microbiome, short-chain fatty acids and metabolic diseases. Nature Genetics, 2019, 51(4): 600-605.

第 7 章
肠道菌群与肿瘤免疫

第一节　概　述

共生微生物是健康或病理条件（包括癌症）的关键决定因素[1]。肿瘤的发生是由宿主和微生物群共同发起的，微生物群是生活在身体外部和内部上皮表面的大量微生物的集合[2]。人类最大的微生物群落存在于胃肠道中，共生微生物和致病微生物与肠道屏障和肠道黏膜淋巴组织相互作用，形成肿瘤微环境，癌细胞在其中生长或死亡。肠道生态失调是共生菌和致病菌之间以及微生物抗原和代谢物产生之间的不平衡[2]。肠道微生物群的组成异常和功能失调可以通过诱导炎症、促进细胞生长和增殖、削弱免疫监视、改变宿主的食物和药物代谢或其他生化功能促进肿瘤的发生（图 7-1）[3]。肠道微生物群除了具有致癌作用外，还被认为会影响不同治疗方法的疗效，包括手术、化疗、放射治疗和免疫治疗[4]。

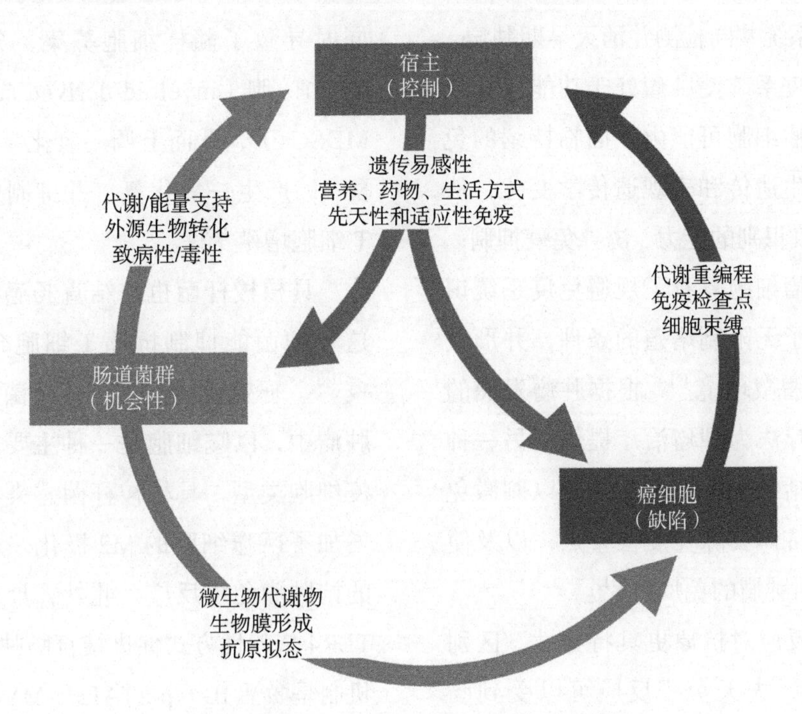

图 7-1　宿主 - 微生物 - 癌症相互作用示意图

每个多细胞群落在三角形关系中表现出不同的行为（如括号所示）。遗传和环境因素（如饮食和药物）可能决定宿主介导的微生物群和肿瘤发生。先天性免疫和适应性免疫在这两种情况下都是必不可少的，而额外的结构（例如肠道上皮屏障）和功能元素（例如抗菌肽）可调节微生物群并维持宿主体内平衡肿瘤免疫可分为先天性免疫和适应性免疫。先天性免疫包括各种类型的髓系细胞和先天淋巴细胞，包括它们产生的免疫因子[5]。作为第一道防御屏障，先天免疫的特点是其即时和广谱反应，这种反应是通过有限的种系编码受体的直接识别而启动的[6]。相反，适应性免疫可以更具体、更准确地执行目标，它始于将肿瘤抗原呈递给 T 细胞受体（T cell receptor，TCR）。肿瘤发生过程中产生的新抗原可以被肿瘤细胞或抗原呈递细胞（antigen presenting cell，APC）呈递，尤其是树突状细胞[7]。处理后的抗原肽以肽 - 主要组织相容性复合物（peptide-major histocompatibility complex，pMHC）的形式呈现给 T 细胞受体，TCR-pMHC 相互作用结合共刺激信号导致效应 T 细胞的启动，然后被激活的 T 细胞通过直接的细胞毒作用或产生细胞因子以募集更多的免疫细胞迁移到肿瘤，杀死癌细胞。此外，B 细胞还作为 APC，分泌细胞因子和抗体，在抗肿瘤免疫中发挥作用[6]。

第二节　肠道菌群与肿瘤免疫机制

免疫系统在肿瘤的发展过程中起着双重作用，一方面可以在称为免疫监视的过程中识别和控制新生肿瘤细胞，另一方面也可以通过多种机制抑制免疫从而促进肿瘤发展，这一双重机制也被称为肿瘤免疫编辑[8]。一般认为肿瘤免疫编辑是一个动态过程，包括 3 个阶段：消除、平衡和逃逸[9, 10]。在消除阶段，先天性免疫系统和适应性免疫系统共同检测并消灭早期肿瘤；在平衡阶段，免疫系统使肿瘤处于功能性休眠状态，但部分肿瘤细胞可能由于面临持续的免疫监视压力，发生遗传和表观遗传学变化，从而进化出抵抗免疫识别的能力，诱导免疫抑制；在逃逸阶段，肿瘤细胞具有了规避免疫系统识别的能力，获得了无限制增殖的条件，并形成了免疫抑制性肿瘤微环境[9]。根据肿瘤发展的这一特点，免疫疗法为肿瘤治疗提供了另一种可能，其策略包括给予促炎细胞因子以刺激免疫，使用单克隆抗体去除免疫检查点，以及使用癌症疫苗增强对肿瘤的免疫力[11]。

获得性免疫反应对抗原更具特异性，区别于先天免疫反应，先天免疫反应可以受到肠道微生物的有益或有害影响[10]。例如，来自 $Apc^{Min/+}$ 小鼠的粪便细菌，特别是脆弱拟杆菌门，与结直肠癌黏膜不典型增生、息肉数量增加以及辅助性 T 细胞（Th）17（$CD4^{+}IL\text{-}17^{+}$）和 Th1（$CD4^{+}IFN\text{-}\gamma^{+}$）细胞比例增加有关，从而触发信号转导和转录激活蛋白 3（signal transducer and activator of transcription 3，STAT3）的活化[12]。产肠毒素的脆弱拟杆菌门（enterotoxigenic bacteroides fragilis，ETBF）还通过其毒素和 IL-17 促进结肠上皮细胞肿瘤发生。同时导致了髓样细胞募集，分化为髓源性抑制细胞（myeloid-derived suppressor cells，MDSCs），从而上调一氧化氮合酶 2 和精氨酸酶 1，产生一氧化氮，并抑制肿瘤微环境中的 T 细胞增殖[13]。

具核梭杆菌也与结直肠癌有关，已经发现这种细菌能抑制抗癌 T 细胞介导的适应性免疫[14]。研究发现，具核梭杆菌感染的人结直肠肿瘤中，巨噬细胞是一种主要的肿瘤浸润性免疫细胞类型，具核梭杆菌感染在体外和体内都增加了巨噬细胞的 M2 极化，从而抑制了机体的抗肿瘤免疫反应。此外，具核梭杆菌感染以 TLR4 依赖性方式促进结直肠肿瘤生长，其分子机制是激活 IL-6/p-STAT3/c-MYC 信号通路[15]。但也有研究发现，脆弱拟杆菌门能促进巨噬细胞的吞噬作用，并向 M1 表型极化，从而发挥抗肿瘤效应[16]。具核梭杆菌的 Fap2 黏附素能与人 T 细胞免疫球蛋白以及自然杀伤细胞（nature killer cell，NK）上表达的 ITIM 结构域相互作用，从而阻断 NK 细胞活性，使具核梭杆菌逃避抗肿瘤免疫 [10]。具核梭杆菌还选择性地招募肿瘤浸润的髓样细胞，从而促进肿瘤微环境的炎症，这有利于结肠肿瘤的形成。在具核梭杆菌喂养的 $Apc^{Min/+}$ 小鼠中，MDSC 细胞富集，并且可以显著抑制 T 细胞的功能 [10]。

T 细胞能识别某些细菌，表明细菌可能与肿瘤细胞共享共同抗原 [17]。共生微生物群可以通过促进肿瘤免疫监测，从而增强肿瘤细胞的免疫清除。与健康对照组相比，乙型肝炎病毒相关肝细胞癌患者的循环 $CD8^+T$ 细胞的细菌反应性依赖于抗原呈递单核细胞。此外，肿瘤切除后的无病生存期与海氏肠球菌（*enterococcus hirae*）反应性和长双歧杆菌（*bifidobacterium longum*）反应性 $CD8^+T$ 细胞的数量呈正相关 [18]。在小鼠结肠癌模型中，广谱抗生素混合物诱导的肠道菌群失调可通过诱导 $CD8^+IFN\text{-}\gamma^+T$ 细胞，增加结肠癌易感性 [19]。在这项研究中，结肠固有层中 $CD8^+IFN\text{-}\gamma^+T$ 细胞的增加引起浆膜炎症，促进肿瘤发展。肿瘤形成后通过诱导功能性 $CD8^+T$ 细胞的耗竭，抑制了抗肿瘤免疫反应 [19]。

肠道黏膜或其他消化器官中富集的固有淋巴细胞有助于协调免疫平衡，并表达细胞因子以发挥免疫调节活性 [10]。在一组大肠癌患者中，细菌依赖的激活转录因子 6 的激活诱导了早期的肠道生物失调、上皮屏障损伤和促进肿瘤发生的先天免疫信号。先天免疫系统中存在数量有限的种系编码模式识别受体（pattern-recognition receptors，PRRs），这些受体识别微生物的病原体，称为病原体相关分子模式（pathogen-associated molecular pattern，PAMP）。TLR 家族一旦被微生物群衍生产物（如细菌脂多糖、脂磷壁酸、脂蛋白、脂肽、鞭毛蛋白、单链或双链 DNA 和 CpG DNA）激活，将在炎症扩散和肿瘤生长加速中发挥显著作用。除了检测保守的微生物相关分子模式外，TLR 还可以通过炎症或损伤相关分子模式激活。作为这些模式的下游靶点，核因子 κ B（nuclear factor-κ B，NF-κ B）和丝裂原活化蛋白激酶（mitogen-activated protein kinase，MAPK）信号通路被激活，最终启动细胞因子产生和促炎细胞的进一步募集，甚至有助于癌症的发生和发展 [20]。

细菌致癌的间接机制表现在细菌感染引起的慢性炎症过程。在这种情况下，微生物群持续产生几种炎症介质，如 TNF-α 和 IL-1，这些介质进一步介导 NF-κB 通路，并促进癌症的发生 [21]。此外，细菌的致癌作用也可能是通过微生物代谢产物或毒素的作用直接实现的。大量研究表明，几种肠道微生物与不同类型癌症的发生有关，如胃癌、结直肠癌和肝细胞癌（表 7-1）[21]。

表 7-1　可能导致胃肠道肿瘤的微生物

肿瘤	涉及的微生物
食管癌	幽门螺杆菌（*H.pylori*），厚壁菌门（Firmicutes），拟杆菌门（Bacteroidetes），变形菌门（Proteobacteria），放线菌门（Actinobacteria）和梭杆菌门（Fusobacteria phyla）
胃癌	幽门螺杆菌（*H.pylori*），卟啉单胞菌（*Porphyromonas*），奈瑟菌（*Neisseria*），苍白普雷沃菌（*Prevotella pallens*），中国链球菌（*Streptococcus sinensis*），人阴道乳杆菌（*Lactobacillus coleohominis*），肺炎克雷伯菌（*Klebsiella pneumoniae*）和鲍曼不动杆菌（*Acinetobacter baumannii*）
结直肠癌	普拉梭菌（*Faecalibacterium prausnitzii*），直肠真杆菌（*Eubacterium rectale*），变形菌门（Proteobacteria），拟杆菌门（Bacteroidetes），梭杆菌门（Fusobacteria）

续表

肿瘤	涉及的微生物
肝细胞癌 胆管癌	幽门螺杆菌（*H.pylori*），大肠埃希菌（*Escherichia coli*），假单胞菌（*Pseudomonadaceae*），草酸杆菌（*Oxalobacteraceae*），华支睾吸虫（Clonorchis sinensis）和泰国肝吸虫（Opisthorchis viverrini）
胰腺癌	幽门螺杆菌（*H. pylori*）

第三节 肠道菌群与肿瘤发生

肿瘤是世界范围内导致死亡的主要原因之一。据估计，微生物可能与 15% ～ 20% 的癌症有关[22]。在过去的几年里，人们越来越认识到肠道微生物群在致癌过程中的作用[23]。肠道中的微生物失调和某些细菌可以通过激活致瘤途径、诱导炎症和破坏宿主 DNA 来诱发癌症或促进癌症进程[24, 25]。一些细菌能产生促进 β-连环蛋白与 E- 钙黏蛋白分离的蛋白质，激活参与致癌作用的 β- 连环蛋白信号通路。肠道生态失调导致细菌衍生的保护性短链脂肪酸产生减少。肠道生态失调通过 Toll 样受体（Toll-like Receptors，TLRs）的微生物相关分子模式发挥促炎作用，促进细胞产生致炎因子，从而增加致癌作用。除了诱导炎症，许多细菌还具有释放特定代谢物破坏 DNA 的能力，这也会促进癌症的进展[26]。肠道微生物群可以通过以下几种机制促进肿瘤的发生：①微生物及其产物的直接致癌作用；②改变循环中的代谢物，使这些代谢产物成为致癌物质；③刺激宿主合成营养因子；④诱导促炎和免疫抑制途径，破坏宿主对癌症的免疫监视[27]。

一、肠道菌群与食管癌

食管癌是癌症相关死亡的第六大常见原因，也是全球第八大常见的癌症[28]。食管癌（esophageal carcinoma，EC）在组织学上可分为两大类：食管鳞状细胞癌和食管腺癌[29]。鉴于肠道微生物群在人类恶性肿瘤中的重要作用，更好地了解食管癌中的微生物群变得越来越重要。20 世纪 80 年代，基于培养的方法和手术切除的食管腺癌和鳞癌标本，在正常组织和癌组织中发现了相同的微生物群[28]。通过依赖培养和非培养方法，研究人员比较了对照组、胃食管反流病、Barrett 食管和食管腺癌患者的微生物群。弯曲杆菌在胃食管反流病和 Barrett 食管中的含量明显高于对照组和食管腺癌组[30]。此外，与癌变相关的细胞因子（如 IL-18）在弯曲杆菌定植的组织中高表达。鉴于弯曲杆菌对人类的潜在致病性[31]，弯曲杆菌在食管腺癌进展中的作用可能与幽门螺杆菌在胃癌中的作用相似[28]。

在大鼠食管空肠吻合术模型中，抗生素组与对照组的食管菌群存在差异，如抗生素组乳酸杆菌和梭状芽孢杆菌的比例分别降低和升高。然而，改变的微生物群并不影响食管腺癌的发病率[32]。在食管空肠吻合术的大鼠模型中，大肠埃希菌在Barrett食管和食管腺癌中普遍存在，而且 TLR1-3、TLR6、TLR7 和 TLR9 在食管腺癌中的表达明显高于正常上皮。这表明 TLR 信号通路与大肠埃希菌之间存在关联，提示在食管腺癌大鼠模型中，早期的分子变化是由微生物介导的[33]。目前，关于微生物组对食管腺癌影响的结论性信息较少。然而，微生物组状态的改变可能参与胃食管反流病和 Barrett 食管向腺癌的发展[28]。

与食管腺癌相比，食管鳞状细胞癌的微生物组特征较差[34]。研究发现，食管微生物丰度与食管鳞状上皮不典型增生（食管鳞状细胞癌的癌前病变）呈负相关，这提示食管微生物复杂性较低的个体更容易发生食管鳞状上皮不典型增生[35]。另有研究发现，在食管鳞状上皮不典型增生和食管鳞状细胞癌患者中，胃微生物群富含梭状芽孢杆菌和丹毒丝菌，这表明胃生物失调与食管鳞状上皮不典型增生向鳞状细胞癌的进展有关[36]。牙龈卟啉单胞菌能感染食管鳞癌患者的癌旁食管黏膜，但不感染对照组的正常黏膜，提示这种微生物在食管鳞状细胞癌发病机制中的作用。卟啉单胞菌还与食管鳞癌的严重程度（即癌细胞分化和转移）呈正相关，且与临床预后较差有关。因此，卟啉单胞菌可作为食管鳞癌的生物标志物[37]。

最近还发现，食管鳞状细胞癌的预后与具核梭杆菌的存在有关，具核梭杆菌主要寄生于口腔并引起牙周病[38]。具核梭杆菌是结肠癌组织中常见的一种细菌，可能影响结直肠癌的发生发展。由于食管靠近口腔，因此具核梭杆菌可能在食管癌中也起着重要作用。在 325 例食管癌手术切除癌组织中，近 23% 的癌组织中含有具核梭杆菌 DNA。重要的是，癌组织中具核梭杆菌的存在与显著缩短的生存时间相关。在具核梭杆菌阳性组织中，最重要的 KEGG 途径是“细胞因子 - 细胞因子受体相互作用”途径，并且特异性趋化因子（即 CCL20）基因数量增加，这表明具核梭杆菌通过激活 CCL20 等趋化因子，促进肿瘤获得了侵袭性生物学行为[38]。越来越多的证据表明，肠道微生物群在食管鳞状细胞癌的发生发展中起着至关重要的作用[28]。

近 20 年来，幽门螺杆菌（*H.pylori*）感染人群中食管腺癌的发病率呈下降趋势，尤其是东部人群。与此同时，食管鳞癌的发病率也有所下降[39]。胃食管反流病是 Barrett 食管的主要原因，Barrett 食管是食管腺癌的一种癌前状态[40]。慢性幽门螺杆菌感染可通过抑制壁细胞功能和（或）诱导萎缩性胃炎的发展，抑制壁细胞分泌盐酸，从而提高胃肠道的 pH，最终导致食管腺癌发病率降低。与正常人群相比，食管炎和 Barrett 食管患者胃中肠杆菌科细菌的相对丰度较高。已有研究表明，抗生素可能会改变胃食管反流病患者的食管微生物群[41]。经质子泵抑制剂（proton pump inhibitor，PPI）治疗后，定植于食管和胃的微生物群发生显著改变。然而，PPI 引起的变化是否有益尚不能确定[42]。最新的系统综述和 Meta 分析表明，质子泵抑制剂并不能减少异型增生和 Barrett 食管相关食管腺癌的发展[43]。

食管传统上被认为是一个无微生物的地方，只有少量来自吞咽和胃食管反流的微生物。然而，研究发现食管黏膜中存在一些特异微生物，包括厚壁菌门（Firmicutes）、拟杆菌门（Bacteroidetes）、变形菌（*Proteobacteria*）、放线菌（*Actinobacteria*）和具核梭杆菌（*Fusobacteria*）。此外，与正常食管相比，食管鳞状细胞癌（Ⅰ～Ⅱ期）和食管鳞状上皮不典型增生患者的食管中具有不同的微生物群落[36]。与正常胃黏膜微生物群一致，早期食管鳞状细胞癌和食管鳞状上皮不典型增生样本中最常见的菌群是变形菌（*Proteobacteria*）、厚壁菌门（Firmicutes）和拟杆菌门（Bacteroidetes），它们在食管微生物群失调时参与食管肿瘤的形成过程[44, 45]。研究还发现，人的远端食管有其特有的微生物群。革兰阳性菌，包括厚壁菌门（Firmicutes）和链球菌（*Streptococcus*），在正常食管中占主导地位，而革兰阴性厌氧 / 微需氧菌，如拟杆菌门（Bacteroidetes）、变形菌（*Proteobacteria*）、梭杆菌门（Fusobacteria）和螺旋体门（Spirochaetes），主要与食管炎和 Barrett 食管有关[29]。脂多糖是革兰阴性细菌细胞壁的重要组成部分，通过多种机制参与肿瘤发生过程。这些机制包括激活导致 NF-κ B 激活的先天免疫反应，促进炎症相关介质（包括 IL-1β、IL-6、IL-8 和 TNF-α）的释放，提高诱导型一氧化氮合成酶和一氧化氮的水平，通过松弛食管下括约肌和延迟胃排空，从而增加反流的风险[29]。

二、肠道菌群与胃癌

胃癌被认为是一种炎症相关性癌症。幽门螺杆菌感染是已知的Ⅰ类危险因子，可刺激免疫反应和炎症反应，调节多种信号通路，并导致贲门失弛缓症、上皮萎缩和异型增生。因此，有效根除幽门螺杆菌可以预防胃癌[46]。

癌蛋白细胞毒素相关基因A（cytotoxin associated gene A，CagA）和细胞空泡毒素基因A（vacuolating roxin A，VacA）是幽门螺杆菌的关键毒力因子[47]。Cag^+菌株感染可增加胃癌风险[48]。幽门螺杆菌感染者胃部炎性细胞因子积聚增加，包括IFN-γ、TNF-α、IL-1、IL-1β、IL-6、IL-7、IL-8、IL-10和IL-18。因此，各种类型的免疫细胞被刺激，包括淋巴细胞、外周血单核细胞、嗜酸性粒细胞、巨噬细胞、中性粒细胞、肥大细胞和树突状细胞。Cag^+幽门螺杆菌株感染后，ERK/MAPK、PI3K/Akt、NF-κ B、Wnt/β-catenin、Ras、sonic hedgehog、STAT3等致癌通路活性上调。另一方面，肿瘤抑制通路被诱导的P53突变所灭活[49-51]。

VacA可直接作用于线粒体，上调MAPK和细胞外信号调节激酶1/2（extracellular regulated protein kinases 1/2，ERK1/2）的表达，激活血管内皮生长因子，上调细胞生长和分化所必需的Wnt/β-catenin信号通路，通过PI3K/Akt信号通路抑制GSK3，从而导致人胃上皮细胞的空泡化和诱导自噬[29]。此外，幽门螺杆菌感染可导致E-钙黏蛋白和肿瘤抑制基因的CpG岛甲基化，包括编码TFF2和FOXD3的基因甲基化，导致胃腺癌的风险显著增加[52]。

幽门螺杆菌阳性患者的微生物群落特征是变形菌门（Proteobacteria）、螺旋体门（Spirochaetes）和酸杆菌门（Acidobacteria）的数量增加，而放线菌门（Actinobacteria）、拟杆菌门（Bacteroidetes）和厚壁菌门（Firmicutes）的数量减少[53]。相反，幽门螺杆菌阴性的个体携带更丰富的厚壁菌门（Firmicutes）、拟杆菌门（Bacteroidetes）和放线菌门（Actinobacteria）。微生物失调也与胃癌的发生有关[54]。定量PCR检测结果表明，胃癌患者体内微生物群组成多样化，表现为卟啉单胞菌（*Porphyromonas*）、奈瑟菌（*Neisseria*）、TM7群（the TM7 group）、苍白普雷沃菌（*Prevotella pallens*）、中华链球菌（*Streptococcus sinensis*）减少，而人阴道乳杆菌（*Lactobacillus coleohominis*）、肺炎克雷伯菌（*Klebsiella pneumoniae*）、鲍曼不动杆菌（*Acinetobacter baumannii*）和毛螺旋菌（*Lachnospiraceae*）增多[55-57]。来源于除幽门螺杆菌以外的其他幽门螺杆菌属的致病成分，如外膜蛋白磷脂酶C-γ2、BAK蛋白和镍结合蛋白，帮助微生物在胃黏膜层定植，进而促进胃炎的进程，最终增加了胃癌发生的可能性[58]。

三、肠道菌群与结直肠癌

目前存在两种模型来解释肠道微生物群在结直肠癌发病机制中的作用。在第一种模型中，肠道微生物群充当具有致癌特性的“驱动力”，它可能通过诱导上皮DNA损伤来启动结直肠癌（colorectal cancer，CRC）的发生，然后被能够促进或阻碍癌变的“乘客”细菌所取代，并在肿瘤微环境中形成生长优势。另一种模型考虑了宿主遗传学，这些遗传学使整个微生物群落失调，从而引起促炎性反应和上皮细胞转化，最终导致癌症（图7-2）[59]。肠道微生物群失调包括某些细菌种类的扩张和耗尽。与对照组相比，在大肠癌患者的粪便样本中也发现了特异性肠道病毒组。一些病毒标志物与大肠癌患者的生存率降低有关[60]。具核梭杆菌是一种革兰阴性厌氧菌，是结直肠癌中最常见的肠道细菌，在不同的种族中已得到验证[61, 62]。携带大肠癌异种移植物的小鼠给予抗生素不仅降低了具核梭杆菌的负荷，而且降低了癌细胞的增殖和肿瘤的总体生长[63]。具核梭杆菌被认为是结直肠癌的一个预后标志物，结直肠癌组织中高水平的具核梭杆菌与总生存率降低呈显著相关[64, 65]。

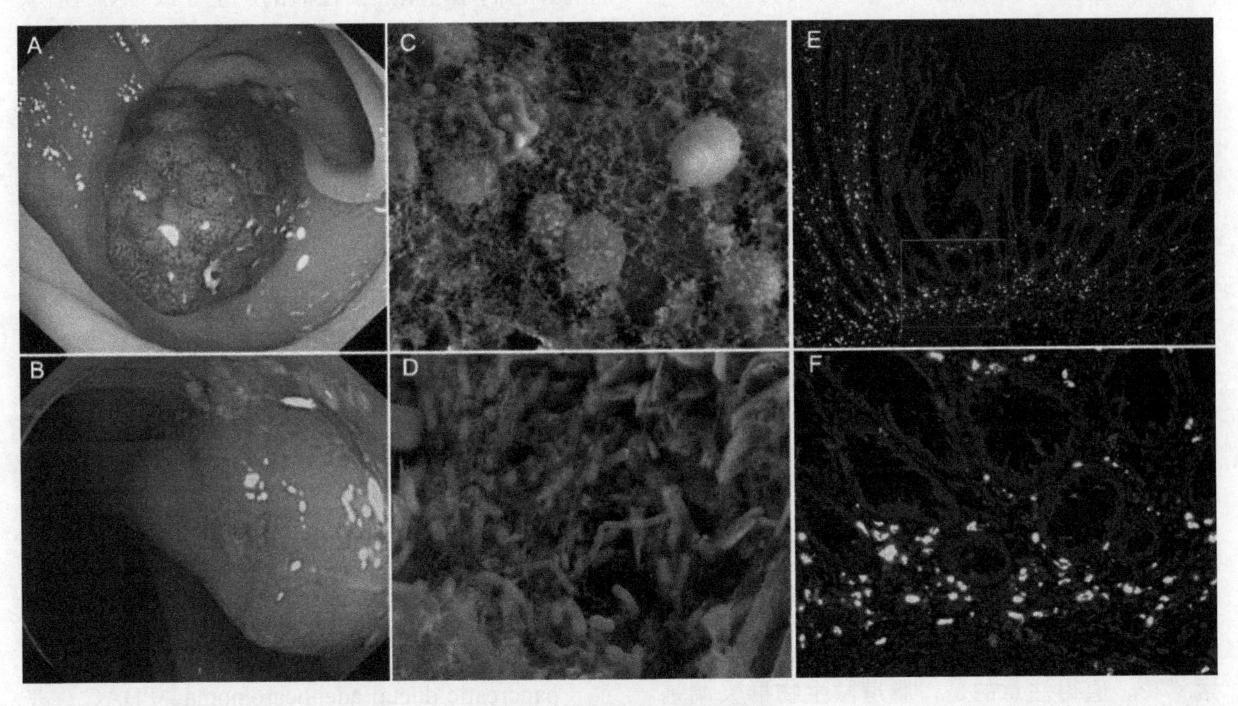

图 7-2　肠道菌群与肠道肿瘤

A、B. 肠镜检查发现人结直肠肿瘤；C、D. 扫描电镜发现肿瘤表面大量球菌及杆菌；E、F. 荧光原位杂交显示大量细菌侵入肠道肿瘤

表达在结肠细胞上的脂多糖（lipopolysaccharides，LPS）受体亚基抑制细胞死亡，通过 TLR2 激活细胞免疫反应，然后刺激下游的促炎细胞因子信号转导，导致肿瘤发生[66, 67]。脂磷壁酸是一种源于革兰阳性细菌细胞壁的成分，被认为是与革兰阴性细菌细胞壁成分 LPS 的对应物[68]。高脂饮食增加了硫酸盐还原菌的相对丰度，如普通脱硫弧菌（Desulfovibrio vulgaris），它能将初级胆汁酸转化为次级胆汁酸，如石胆酸和脱氧胆酸，具有潜在的致癌作用。相反，丁酸，一种重要的短链脂肪酸，由结肠细菌发酵饮食中的纤维产生，已被证明具有抗肿瘤作用。发酵过程中最重要的丁酸产生菌群是普拉梭菌（*Faecalibacterium prausnitzii*）和直肠真杆菌（*Eubacterium rectale*）。丁酸被结肠细胞线粒体利用，有助于维持健康的能量平衡，有利于结肠上皮细胞增殖[69]。免疫细胞上表达的单链脂肪酸受体 GPR109a 首先激活丁酸的配体，然后抑制炎性细胞因子，从而抑制炎症过程[70]。宿主免疫应答通过 IFNγ 抑制 DNA 甲基化介导的 GPR109a 沉默，从而促进抗癌作用[71, 72]。此外，丁酸还通过诱导 *P21* 基因表达、抑制活化蛋白 -1 信号通路、增加 c-Fos 和 ERK1/2 的磷酸化，从而发挥多种化学预防作用[73, 74]。

在与微生物及其基因产物相互作用时，树突状细胞被激活，启动肠道免疫反应。宿主天然免疫系统可以识别微生物分子，包括 LPS、鞭毛蛋白、肽聚糖和其他微生物相关分子模式。模式识别受体的激活，例如 NLRs 和 TLRs，调节炎症途径和多种细胞类型的增殖。NOD 样受体（NOD-like receptors，NLR）介导的炎症小体激活，以及 TLR2 的表达增强，通过抑制炎症环境来维持结肠上皮的完整结构和功能 [29]。

四、肠道菌群与肝癌

虽然肝脏通常被认为是无菌的，但肝脏环境很大程度上受到胃肠道微生物群通过肝门静脉系统带来的病原体或代谢物的影响。肝脏通过过滤血液、代谢和中和肠道微生物产生的毒素，对宿主微生物群落起着至关重要的作用。肠道微生物失调可导致肝癌的发生，因为肝脏内的免疫细胞可检测到微生物群和微生物代谢产物，并可改变肝脏代谢 [75]。

肝细胞癌和胆管癌是最常见的肝癌组织学类型。酒精性肝病、非酒精性脂肪性肝病以及食源性污染物黄曲霉毒素 B_1、乙型肝炎或丙型肝炎病毒感染被认为是肝癌的主要危险因素 [76-78]。值得注意的是，肠道微生物群失调是非酒精性脂肪性肝病的关键诱因之一 [79, 80]。肝细胞癌患者粪便中大肠埃希菌的丰度明显高于健康对照组，而胆管癌患者胆管标本中迪茨菌（*Dietziaceae*）、假单胞菌（*Pseudomonadaceae*）和草酸杆菌（*Oxalobacteraceae*）的含量明显高于非胆管癌患者。已有假说认为肠道内微生物过度生长可能促进肝癌的发生 [29]。

幽门螺杆菌通常生活在人的胃里。然而，来自肠道的幽门螺杆菌可以在吞噬细胞消除后通过门静脉血流到达肝组织，也可以通过十二指肠反向迁移到达肝组织。幽门螺杆菌产生的 VacA 和 CagA 已在肝细胞癌组织中被发现。幽门螺杆菌产生的 LPS 通过增加 IL-8 和 TGF-β1 的水平，直接促进肝癌的生长和迁移 [29]。微生物代谢产物能干扰肝脏的代谢途径和免疫反应，TLR4 识别细菌来源的 LPS 并通过 LPS 诱导的 TNF-β 和 IL-6 激活肝 Kupffer 细胞，它还可以通过表皮调节素等生长因子刺激星状细胞，启动各种炎症和致癌途径。LPS-TLR4 途径促进肝细胞癌的发生，而 LPS 的去除或 TLR4 的基因失活可能会抑制肝细胞癌的发展 [29]。

胆酸和鹅去氧胆酸是肝脏产生的主要胆汁酸。它们通过增加活性氧的产生导致 DNA 损伤，从而诱发肝癌。此外，胆汁酸也被证实对肠道微生物群有调节作用。胆汁酸的减少会导致肠道微生物群过度生长，加速炎症 [81]。梭状芽孢杆菌（*Clostridium*）产生的脱氧胆酸（deoxycholic acid，DCA）的肝肠循环导致肝星状细胞 DNA 损伤和衰老相关的分泌表型。这个过程涉及大量炎症细胞因子和生长因子，从而导致炎症和肥胖相关的肝癌 [82]，脱氧胆酸和石胆酸通过 DNA 损伤直接促进癌症 [83]。

五、肠道菌群与胰腺癌

胰腺是胃外的消化器官。胰腺导管腺癌（pancreatic ductal adenocarcinoma，PDAC）是全球最致命的癌症之一，也是胰腺癌中最常见的一种。越来越多的研究表明，肠道微生物群可能通过促进炎症、激活免疫反应和维持与癌症相关的炎症来影响胰腺癌的发生 [84-86]。胰腺癌的危险因素包括年龄、吸烟、肥胖、慢性胰腺炎和糖尿病。对数百个胰腺癌患者的 Meta 分析表明，幽门螺杆菌感染是胰腺导管腺癌的另一个重要危险因素。除胰腺导管腺癌外，幽门螺杆菌还参与急性和慢性胰腺炎以及自身免疫性胰腺炎 [87-89]。幽门螺杆菌的许多致病成分，包括氨、脂多糖以及由此产生的大量炎症细胞因子，都会损害胰腺。幽门螺杆菌感染同时激活 NF-κ B 和活化蛋白 -1，导致细胞周期的失调。IL-8 水平升高会加速炎症反应，最终导致胰腺癌变。KRAS 在正常组织信号转导中起重要作用，而 *KRAS* 基因突变存在于 90% 以上的胰腺癌中。幽门螺杆菌的脂多糖被证实能促进 *KRAS* 基因突变，并启动胰腺癌的发生过程。此外，幽门螺杆菌感染持续激活 STAT3，可通过上调抗凋亡和促增殖蛋白（包括 Bcl-xL、MCL-1、survivin、c-myc 和 cyclin D1）的表达来促进胰腺癌的进展 [29]。

微生物引起轻微和持续的免疫反应和炎症反应，导致胰腺癌的形成 [90] 。在各种免疫细胞上表达的 TLR 使免疫细胞能够识别多种微生物相关分子模式和非感染性炎症损伤相关分子模式，然后激活 NF-κ B 和 MAPK 信号通路 [91]。因此，这些过程引发并延续胰腺炎，最终促进胰腺癌的进展 [92, 93] 。NLR 是细胞质模式识别受体，参与 NF-κ B 的活化和炎症小体的形成。P38 MAPK 参与细胞分化、凋亡和自噬，从而加速胰腺导管腺癌的进程。因此，p38 抑制剂可能是治疗胰腺癌的药物 [94]。

T2R38 是一种苦味受体。有趣的是，T2R38 不仅在口腔细胞中表达，而且在胰腺癌细胞中也有表达，铜绿假单胞菌是 T2R38 的唯一配体，被认为可以激活 T2R38，诱导多药耐药转运蛋白 ATP 结合盒式跨膜转运蛋白超家族 B 亚群 1，参与肿瘤的侵袭和转移 [95] 。此外，8.8% 的胰腺癌组织中存在梭状芽孢杆菌，值得注意的是，梭状芽孢杆菌是胰腺癌独立的阴性预后生物标志物 [96]。也有研究表明，肠道微生物群失调诱导的肽酪氨酸表达与胰腺癌的发生有关 [2]。

六、肠道菌群与前列腺癌

长期以来细菌被认为是慢性、低度炎症的来源，这种炎症可能诱发前列腺癌 [97]。尽管人们普遍认为尿液是无菌的，但有几项研究证实，尿路中存在独特的微生物群 [98]。前列腺癌组织中的丙酸杆菌（*Propionibacterium*）总体上是最丰富的，而葡萄球菌（*Staphylococcus*）在肿瘤和瘤周组织中的比例更高 [99]。采用整合的宏基因组和转录组分析方法，从 65 例中国前列腺癌标本和邻近良性组织中，鉴定出 40 多个独特的细菌属，其中包括假单胞菌（*Pseudomonas*）、大肠埃希菌（*Escherichia*）、不动杆菌（*Acinetobacter*）和丙酸杆菌（*Propionibacterium*）。没有检测到与性病相关的微生物，也没有检测到病毒。此外，肿瘤和良性组织在总体（α）细菌多样性或群体（β）多样性方面没有差异，尽管来自同一患者的组织优先聚集在 β 多样性上 [100]。

除了独特的泌尿生殖道菌群，研究最多、最大的共生菌群来源是肠道（图 7-3）。肠道本身在产生各种激素方面具有动态和复杂的作用，这些激素可以与中枢神经系统、应激反应和其他各种身体系统相互作用 [101]。反过来，肠道内的微生物既可以感知机体产生的激素并对其做出反应，也可以分泌自身的分子 [102]。因此，肠道菌群组成的变化和（或）失调可能对全身产生影响。事实上，肠道微生物群变化与肥胖、药物代谢差异、能量消耗、对免疫系统的直接影响等有关 [103, 104]。

有研究首次展示了整个肠道菌群对癌症生长的影响，来自不同实验室相同小鼠的黑色素瘤生长速率不同，但通过共同寄居或粪便移植肠道菌群，则导致肿瘤以相似的速率生长，从而促进了生长较慢的肿瘤 [105]。最终，这个研究小组发现，双歧杆菌是减缓肿瘤生长的微生物。在无病原体的小鼠体内，由于 *Apc* 基因突变，容易发生癌症，肠道感染肝型螺杆菌（*Helicobacter hepaticus*）后，可诱导 TNF-α 依赖性细胞因子反应，导致乳腺癌的发生。该研究小组后来的研究发现，批量转移来自感染了肝螺杆菌的 $Apc^{min/+}$ 肠系膜淋巴结细胞，可以将前列腺癌以 TNF-α 依赖的方式转移到野生型小鼠体内。这些研究表明肠道微生物群的失调可能会导致异常的炎症信号，从而产生全身效应，其中一种炎症信号可能是 LPS[97]。

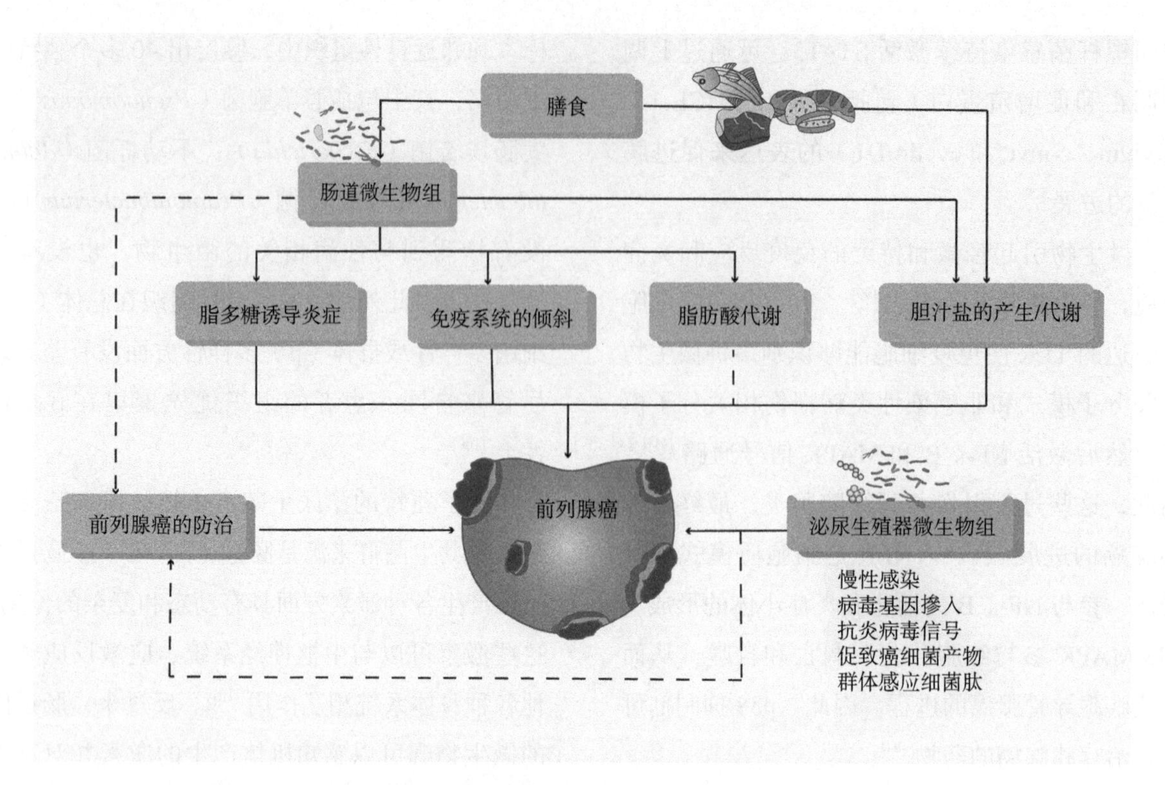

图 7-3　肠道和泌尿生殖道微生物群的相互作用与前列腺癌

示意图描述了肠道微生物群和泌尿生殖系微生物群在前列腺癌发生中的作用机制和相互作用

LPS 是细菌细胞壁中的一种化合物，通常被免疫系统识别为“危险信号”，能够引起免疫细胞活化，帮助清除体内的致病微生物。细菌性 LPS 被认为是糖尿病发生的触发因素，并与肥胖和前列腺癌有关。在一项关于高脂饮食与常规饮食小鼠的研究中，高脂饮食小鼠肠道中含有 LPS 的细菌增多；持续输注 LPS 可以恢复高血糖水平，高脂饮食喂养的小鼠体重增加。这种 LPS 诱导的代谢综合征依赖于 CD14，CD14 是 LPS 与其主要受体 TLR4 结合所必需的主要可溶性转运分子 [106]。人前列腺癌表达 TLR4，前列腺癌的实验室模型已经被用来证明 LPS 激活 TLR4 可以促进血清饥饿条件下的生存，并诱导 VEGF 和 CCL2 的产生 [107]。LPS 通过 TLR4 受体激活 NF-κ B，NF-κ B 介导多种应激相关化合物的转录，在侵袭性前列腺癌中上调。有趣的是，另一个小鼠模型发现，高脂饮食小鼠前列腺 NF-κ B 信号增加。这些结果都与以下假设一致：生理失调通过 LPS 诱导的炎症机制，导致肥胖和代谢综合征，对前列腺癌和其他癌症有显著影响 [97]。然而并非所有 LPS 都是一样的，大肠埃希菌 LPS 比其他细菌（如拟杆菌属）LPS 具有更大的致炎作用，而细菌会产生 LPS 的拮抗形式，从而抑制促炎性 LPS 信号传导 [108]。最终，来自健康肠道的总 LPS 负荷是免疫抑制的 [109]。占优势的细菌只是影响血液中 LPS 水平的一个成分。此外，肠道允许细菌移位或激素扩散（又称肠道“渗漏”）的倾向可由多种饮食、疾病和宿主相关因素决定 [110]。

七、肠道菌群与乳腺癌

2018 年的统计数据显示，全球女性乳腺癌（breast cancer，BC）的发病率和死亡率分别为 $46.3/10^5$ 和 $13.0/10^5$，且呈上升趋势，已成为当今世界许多地区女性的主要癌症。虽然已经确定遗传、表观遗传和环境因素在乳腺癌中的综合作用，但高达 70% 的乳腺癌患者发病机制仍然不明确。宿主内的细菌群落可能是与乳腺癌相关的重要环境因素。研究表明，人体不同环境中菌落各不相同，并且细菌和宿主之间建立了复杂的联系，这引起了研究者对不同健康状态下身体不同部位微生物群的浓厚兴趣 [111，112]。有研究探讨了肠道细菌基因组，发现其产物会分解雌激素及其代谢物。当微生物群 / 雌激素代谢紊乱时，会导致循环中雌激素及其代谢物水平升高，从而增加乳腺癌的风险 [113]。也有临床研究确定了肠道微生物群与尿雌激素和雌激素代谢物之间的联系 [114]。

除了通过调控不同信号途径来调节炎症和影响宿主细胞的基因稳定性外，肠道微生物群还通过肠肝循环影响雌激素的代谢途径来参与癌症的进展 [115]。某些肠道微生物可能通过促进抗肿瘤免疫、免疫监视和（或）通过调节全身雌激素水平，在乳腺癌的发生中发挥作用 [116]。病例对照试验对肠道微生物群失调与乳腺癌之间的联系进行了验证，并推测肠道微生物可能参与了雌激素的代谢 [117]。

雌激素在肝脏进行代谢，先在肝脏与胆汁结合，然后被细菌 β- 葡萄糖醛酸酶解离后排入肠道，再通过肠肝循环重新吸收为游离雌激素，到达乳腺等不同器官。梭状芽孢杆菌（*Clostridia*）和瘤胃球菌（*Ruminococcaceae*）科的几种细菌在雌激素的这一代谢过程中发挥了重要作用 [115]。此外，其他类似雌激素的代谢物也可通过肠道内的氧化还原反应，以及雌激素诱导的生长因子合成而产生，这可能具有致癌潜力。此外，细菌 β- 葡萄糖醛酸酶可以参与外源物质和（或）外源雌激素的解离，导致它们通过肠 - 肝途径重摄取，从而增加它们在体内的停留时间 [118]。

在厚壁菌门（Firmicutes phylum）的两个优势亚群中发现了许多 β- 葡萄糖醛酸酶细菌，即柔嫩梭菌（*Clostridium leptum*）和球形梭状芽孢杆菌（*Clostridium coccoides*）。大肠埃希菌 / 志贺氏菌群是变形菌门（Proteobacteria phylum）的一员，同时也具有 β- 葡萄糖醛酸酶 [119]。其他研究主要集中在肠道微生物群与乳腺癌风险之间的关系，这些关系为雌激素非依赖性途径 [120，121]。一项病例对照研究比较了乳腺癌患者和配对对照组之间的粪便微生物群，发现绝经后乳腺癌患者的肠道细菌种类减少，菌群组成也发生了显著变化 [120]。乳腺癌患者肠道微生物组中梭菌科（Clostridiaceae）、粪杆菌（*Faecalibacterium*）和瘤胃球菌（*Ruminococcaceae*）的含量较高，而多尔菌属（Dorea）和毛螺菌科（Lachnospiraceae）的含量较低。此外，癌症患者粪便微生物群的多样性（α- 多样性）较低。

出乎意料的是，乳腺癌患者体内雌激素水平较对照组高，虽然没有统计学意义。一个可能的原因是其他乳腺癌危险因素，如肥胖，在这种情况下，肠道微生物群的多样性较低。肠道微生物群改变也与肥胖症有关，众所周知，超重和肥胖的妇女比健康体重的妇女患乳腺癌的风险更高，尤其是在绝经后时期 [111]。在肠道内，厚壁菌门和拟杆菌门是两个主要的门类，参与肠道内营养物质的代谢，如膳食纤维和多酚 [122]。肥胖人群中厚壁菌门 / 拟杆菌门的比例更高，尽管结果存在争议。在早期乳腺癌妇女的粪便中观察到细菌总数的绝对数量和 3 种细菌群 [厚壁菌门、普氏粪杆菌（*Faecalibacterium prausnitzii*）和布劳特菌（*Blautia*）] 的绝对数量与早期乳腺癌患者的体重指数之间存在相关性，超重和肥胖患者的细菌数量较低，超重受试者中观察到的厚壁菌门数量也较低 [117]。

一项病例对照研究调查了免疫和炎症在乳腺癌发生风险中的作用，以及肠道微生物群在免疫识别微生物群组成中是否存在差异 [121]。病例组和对照组的 IgA^+ 菌群组成明显不同。将病例组分为 IgA^+ 和 IgA^- 微生物群时，在调整雌激素水平和其他变量后，乳腺癌患者 IgA^+ 的粪便微生物群组成发生改变，丰度降低，α- 多样性显著高于 IgA$^-$ 菌群的病例。雌激素代谢物与雌激素的水平在病例和对照组基本相同。与对照组相比，雌激素类物质（雌酮、雌二醇）和雌激素代谢物的含量也无显著性差异。与对照组相比，绝经后乳腺癌患者与 IgA$^+$ 和 IgA$^-$ 肠道菌群的雌激素依赖性显著不同，这表明肠道菌群可能通过改变代谢、雌激素循环和免疫途径影响乳腺癌患病风险 [121]。

有研究评估了不同类型和不同分期乳腺癌患者的肠道菌群组成，大多数患者是早期侵袭性导管癌，Ⅲ期癌症患者的布劳特菌（*Blautia spp*）绝对数高于Ⅰ期患者。此外，双歧杆菌（*Bifidobacterium*）和布劳特菌的绝对数量，以及粪便中普拉梭菌（*Prausnitzii*）和布劳特菌的比例随临床分期而变化，提示肠道微生物群可能与乳腺癌的发生和进展有关。在这项研究中，不同体重指数的患者，细菌总数量和一些特殊细菌群（普拉梭菌、厚壁菌门和布劳特菌）也存在显著差异 [111]。

人类的乳房不是无菌的，而是存在与身体其他部位菌群不同且独特的细菌群落，这与乳房内取样的位置、年龄、国籍、妊娠史、是否存在乳腺恶性肿瘤，以及 DNA 制备和测序技术方法无关 [123，124]。到目前为止，一些研究证明了部分乳房组织微生物群的潜在来源，即从胃肠道移位到皮肤、通过乳头乳晕开口、通过哺乳和（或）性接触进行乳头 - 口腔接触 [125]。有学者提出，乳房微生物群通过刺激常驻免疫细胞，来帮助维持健康的乳房组织，且细菌的种类及其代谢活动，如降解致癌物的能力，也可能起到维持健康乳房的作用 [123]。几项研究确定了乳腺肿瘤组织和正常癌旁组织存在的细菌，乳腺癌患者与对照组的乳腺组织微生物群存在差异。这些研究仍处于初级阶段，尚不清楚癌症患者肿瘤和邻近正常组织之间是否存在差异。一些研究报道，肿瘤组织和正常癌旁组织的细菌群落在种类上没有差异 [124，126]。但与健康对照组相比，女性癌症患者体内的大肠埃希菌（具有促癌活性）含量更高。然而，虽然乳腺微生物组成的变化可能通过多种途径促进疾病的发生与发展，但目前尚不清楚宿主微生物差异是该疾病的结果还是原因。也不清楚是否有特定的微生物群（是存在致病菌株，还是缺乏有益菌株）与乳腺癌相关 [120，126]。

健康妇女和乳腺癌患者的乳腺组织中存在不同的细菌谱 [125，126]。乳腺癌患者的正常癌旁组织与健康女性组织的比较显示，健康患者的普雷沃特菌（*Prevotella*）、乳球菌（*Lactococcus*）、链球菌（*Streptococcus*）、棒状杆菌（*Corynebacterium*）和微球菌（*Micrococcus*）的相对丰度显著高于乳腺癌患者，而乳腺癌患者正常癌旁组织的芽孢杆菌（*Bacillus*）、葡萄球菌（*Staphylococcus*）、肠杆菌（*Enterobacteriaceae*）、丛毛单胞菌（*Comamondaceae*）和拟杆菌门（Bacteroidetes）的相对丰度显著高于健康女性；这些细菌在体外会造成 DNA 损伤。此外，一些具有抗癌特性等对健康有益的乳酸菌在乳腺癌患者正常癌旁组织中也减少 [127]。

肿瘤周围的微环境包括多种细胞和微生物，细胞及微生物群发生的病理生理改变也可能对肿瘤生长产生重要影响 [128]。微生物对乳腺肿瘤微环境的作用研究较少。有研究从癌症基因组图谱中分析了 668 个乳腺肿瘤组织和 72 个乳腺癌旁组织的微生物群，并报道了疾病亚型之间的微生物组成差异 [129]。肿瘤部位最常见的细菌依次是变形菌门（Proteobacteria）（48.0%）、放线菌门（Actinobacteria）（26.3%）和厚壁菌门（Firmicutes）（16.2%），这些发现与先前的研究结果一致 [125，126]。偶发分枝杆菌（*Mycobacterium fortuitum*）和草分枝杆菌（*Mycobacterium phlei*）是两种在肿瘤标本中差异显著的优势菌。变形菌门在肿瘤组织中也显著增加，而放线菌门在癌旁组织样本中也存在 [129]。

除了正常乳腺和乳腺肿瘤组织外，通过分析乳腺癌妇女（导管癌）和健康对照妇女乳头抽吸液中存在的微生物，发现乳腺癌患者乳头抽吸液的群落组成与健康组相比存在差异（β-多样性）。有乳腺癌病史也显著影响乳头吸出液的微生物组成。最丰富的细菌门是厚壁菌门（42.1%）、蛋白菌门（32.9%）和拟杆菌门（14.5%）。在乳腺癌患者采集的乳头抽吸液中，另支菌属（*Alistipes*）的发现率相对较高，而鞘脂单胞菌科（Sphingomonadaceae）的一个未分类属在健康对照妇女的乳头抽吸液中则相对丰富[130]。

八、肠道菌群与人体其他肿瘤

最近的证据表明，肠道微生物群与黑色素瘤的进展和治疗效果有关。具有不同肠道微生物组成的两种小鼠（JAX和TAC），其黑色素瘤的生长及对抗程序性死亡配体-1（programmed death-ligand 1，PD-L1）免疫治疗的反应明显不同。通过对肠道微生物群的基因组分析，发现双歧杆菌有助于提高PD-L1的治疗效果[105]。

血液系统相关肿瘤如淋巴瘤、白血病恶性程度高，除一些早期、分型较好的类型外，大多预后差，病死率高，且患者最后通常由于免疫力低下引起感染死亡，而肠道菌群作为免疫系统中重要组成部分，与血液系统肿瘤的相关性也值得我们探索。研究显示：将血液恶性肿瘤患者与健康人群的肠道菌群进行对比，发现大肠埃希菌在血液恶性肿瘤患者肠道中数量增加，益生菌如柔嫩梭菌数量减少或缺失。此外，机会致病菌如粪肠球菌、约氏不动杆菌、硫黄肠球菌在个别患者的肠道中数量增加。这表明，血液系统恶性肿瘤的患者体内存在菌群失调，主要表现为机会致病菌的数量和种类增多，益生菌的数量和种类减少。非霍奇金淋巴瘤患者化疗后，粪便中的厚壁菌门数量明显下降，而变形菌门数量则显著增多，这可能与化疗后胃肠道黏膜炎症的发生相关[131]。

近年来的研究表明，肺癌的发生发展也与人体肠道菌群之间存在关联，这种相互作用可能与多种途径有关，如代谢、炎症或免疫途径[132-135]。炎症微环境在肿瘤的发生、发展，以及肿瘤治疗的敏感性中起重要作用，是癌症微环境的重要组成部分[136]。肠道微环境稳态的失衡可以通过部分循环炎性因子的增加，使发生在肠道的炎症影响远离肠道部位（如呼吸道）疾病的发展[9]。在白血病小鼠中，肠道屏障的破坏及促炎细菌的移位引起全身炎症反应增强，主要表现为血清脂多糖结合蛋白（serum lipopolysaccharide binding protein，sLBP）和IL-6水平升高，这也是几种恶病质症状（肌萎缩、厌食和体重减轻）的主要驱动因素[137]。在从不吸烟的人群中，膳食纤维（益生元的主要来源）或酸奶（益生菌食品）的摄入量与肺癌发生的风险成负相关：最高摄入量人群肺癌的患病风险相较于最低摄入量人群降低了30%以上[138]。这与益生元和益生菌在肠道中增加抗炎细胞因子IL-10、降低促炎症细胞因子IL-1β和IL-6的分泌有关[139]。

目前，肠道菌群发挥肺部免疫调节作用的机制尚未完全明确，但可能涉及以下两种途径。①“肺-肠轴”理论[140, 141]：一方面肠道菌群及其产物被抗原呈递细胞吞噬并转移到肠系膜淋巴结，刺激T和B细胞的活化。一旦激活，这些细胞就会表达某些趋化因子受体（如CCR4和CCR9）获得归巢特性，可以通过淋巴和血液循环迁移回到原始位置（肠黏膜）或远端位置，如气道。在那里，它们可以直接作用于目标或继续刺激其他免疫细胞；另一方面，来自肠道的菌群产物或活的菌群也可以直接通过血液或淋巴循环到达肺部，以刺激免疫系统。根据组织受到的刺激类型以及所处的免疫状态，结果可以是多样的，可能产生有效的抗炎或抗肿瘤活性，也可能进一步促进组织损伤、病原体定植和肿瘤进展。②“肠 - 骨髓”调节机制[142]：肠道菌群的代谢产物是微生物相关分子模式（microbe-associated molecular patterns，MAMPs）和 PAMP 的来源，MAMP 或 PAMP 通过结合诸如单核细胞、巨噬细胞和自然杀伤细胞等免疫细胞上的 PRRs 使其活化。同时，这些微生物来源的抗原通过血液循环到达骨髓，影响髓系免疫细胞的分化和功能，如诱导具有长期“记忆”特性细胞的产生。这些来自肠道菌群的抗原和肿瘤抗原之间具有相似性，从而通过抗原模拟或交叉反应来激发免疫细胞的活性，并形成了相应的 T 细胞库，从而提高了免疫系统在识别癌细胞时的反应性和抗肿瘤能力，即增强免疫监视能力[143]。肠道菌群产物中的短链脂肪酸可以通过促进骨髓造血前体的生成，发挥免疫调节的功能[144]。

第四节　肠道菌群与肿瘤治疗

肠道微生物可通过影响化疗药物疗效、消除抗癌作用和介导药物毒性，在抗癌治疗反应中发挥重要作用[22]。肠道菌群在适应性免疫和先天免疫的发育和调节中起着至关重要的作用，因此利用宿主免疫系统的免疫疗法在很多肿瘤的治疗中产生了巨大的效果。由于微生物群对炎症和免疫的显著调节作用，近年来大量研究报道了调控肠道菌群对肿瘤免疫治疗效果的影响。

一、肠道菌群在化疗中的作用

肠道微生物群系似乎通过多种机制影响化疗效果，包括异种代谢、免疫相互作用和改变菌群群落结构[145]。肠道微生物群能够直接修饰或代谢某些外源物质，如抗癌药物。这种微生物介导的异种代谢可能与化疗成分毒性的增加有关，从而导致治疗效果的降低[22]。

越来越多的证据支持肠道微生物群可以影响化疗效果。人类肠道微生物群是高度复杂的，包含许多细菌种类。这些细菌即使在很小的比例中也可能影响药物的疗效，并且可能相互影响，现有的研究进一步说明了潜在的细菌 - 药物相互作用的复杂性。当某些菌种的细菌生长受到化疗药物的影响时，其菌群组成可能发生变化，进而影响药物疗效和整体健康[146]。氟嘧啶是抗代谢药物，主要用于治疗癌症。原型氟嘧啶，5- 氟尿嘧啶（5-fluorouracil，5-FU）是用于结直肠癌的主要治疗方案。5- 氟尿嘧啶及其前体药物，如卡培他滨，通过抑制胸苷酸合成酶的活性，阻碍核苷酸的生物合成，从而阻碍细胞分裂。尽管以 5-FU 为基础的化疗被广泛使用，但还没有公认的剂量分类，而且患者之间存在显著的药动学差异。

最近，有研究报道氟嘧啶类药物与微生物群的代谢有关，这些药物由细菌的核糖核苷酸代谢激活或失活。抑制细菌核糖核酸代谢可显著对抗药物疗效，而抑制脱氧核糖核酸代谢可改善疗效[147]。例如，大肠埃希菌和丛毛单胞菌（*Comamonas*）可以通过调节细菌核苷酸代谢网络，以相反的方向影响秀丽隐杆线虫（Caenorhabditis elegans）对 5- 氟 -2′ - 脱氧尿苷（5-fluoro-2′ -deoxyuridine，FUDR）和喜树碱的反应。大肠埃希菌或丛毛单胞菌的突变可以使 5-FU 产生 5- 氟尿苷 -5′ - 单磷酸盐，但不能产生 5- 氟 -2′ - 脱氧尿嘧啶 -5′ - 单磷酸盐，因而降低 5-FU 和 FUDR 的疗效[146]。其他机制涉及细菌脱氧核苷酸库对宿主药物治疗反应的影响。细菌脱氧核苷酸库的改变放大了 5-FU 诱导的宿主细胞自噬和细胞死亡，从而提高了治疗的有效性，这一过程由核苷二磷酸激酶特异性地调节[147]。

最近研究发现，一种特殊的细菌——具核梭杆菌，可直接促进大肠癌对奥沙利铂和5-FU的耐药性[148]。2012年以来，具核梭杆菌作为一种潜在的促癌微生物受到了广泛关注，当时两个实验室同时发现，与健康人相比，该细菌在人结直肠癌中的含量更高。随后进一步研究表明，大肠癌组织中较高的具核梭杆菌丰度与转移、预后不良及特定的肿瘤分子特征有关，包括高CpG岛甲基化表型、微卫星不稳定性和基因突变（如BRFA、KRAS和TP53），可能导致腺瘤-癌变途径和锯齿状癌变途径[149-151]。这些现象背后的机制非常复杂，目前研究表明，具核梭杆菌可通过其毒力因子黏附素A（FadA）、梭杆菌自身转运蛋白2（Fap2）和梭杆菌外膜蛋白A（FomA）黏附和侵袭内皮细胞和上皮细胞，激活炎症反应，诱导肿瘤免疫逃逸，促进肿瘤的发展[152, 153]。具核梭杆菌感染后，miRNA-21的表达增加，这被认为是具核梭杆菌的另一种致癌作用[150]。这种细菌对化疗反应的影响最近才受到关注，提示肿瘤微环境促进了化疗耐药。具核梭杆菌通过激活TLR4/MYD88通路，下调miRNA18a和miRNA-4082，诱导大肠癌细胞从凋亡转为自噬，进而抵抗药物的治疗作用[151]。

环磷酰胺（cyclophosphamide，CTX）是一种重要的抗癌药物，用于乳腺癌、淋巴瘤和脑肿瘤等肿瘤的治疗。这种化合物诱导免疫原性癌细胞死亡，破坏免疫抑制性T细胞，并促进Th1和Th17细胞控制肿瘤的生长。CTX能改变小肠内微生物群的组成，并诱导某些革兰阳性菌转移到次级淋巴器官中[154]。最近确定了两种共栖菌，即海氏肠球菌（*Enterococcus hirae*）和肠道巴恩斯菌（*Barnesiella intestinihomi*），参与了CTX的治疗效果。革兰阳性海氏肠球菌存在于小肠，而革兰阴性肠道巴恩斯菌则存在于结肠。经CTX处理后，海氏肠球菌能从小肠转位到淋巴结和脾脏等次级淋巴器官，诱导Th17和Th1反应，提高细胞毒性$CD8^+T$/Treg比值。与海氏肠球菌相反，肠道巴恩斯菌不会在结肠中移位和积聚，在结肠中它会诱导系统性多功能Th1和Tc1反应，并诱导肿瘤内产生IFNγ的γδT细胞增加。

除环磷酰胺外，吉西他滨（gemcitabine，GEM）在微生物学领域引起了广泛关注。GEM是一种核苷类似物（2′，2′-二氟脱氧尿苷），用于治疗胰腺癌、肺癌、乳腺癌或膀胱癌。在大肠癌小鼠模型中，肿瘤内的γ-变形菌（*Gammaprotobacteria*）能够将GEM代谢成不活跃的形式，即2′，2′-二氟脱氧尿苷，从而使其失效。细菌胞苷脱氨酶分为长型和短型，长亚型的表达可能是这一代谢过程的原因。这种GEM耐药性可以通过使用抗生素，或从细菌基因组中删除长型胞苷脱氨酶来消除[155, 156]。铂是治疗肺癌最常用的治疗方案之一，在肺癌异种移植模型中，顺铂（cisplatin，DDP）和ABX（万古霉素、氨苄西林和新霉素）联合治疗后的肿瘤体积比单用DDP大，生存率显著降低。此外，基因表达分析显示，抗生素可通过上调VEGFA（癌基因）和下调$CD8^+T$细胞亚群中的BAX、CDKN1B（抑制基因）、IFN-γ、颗粒酶B和穿孔素1而部分减弱DDP的效果[157]。

特定类型的肠道细菌可以保护其他有益细菌免受癌症治疗的影响——减轻有害的、药物引起的肠道菌群变化。通过代谢化疗药物，这种保护性细菌可以缓解治疗的短期和长期副作用。尽管癌症治疗可以挽救生命，但它们也会引起极其严重和痛苦的副作用，包括胃肠道问题。尤其是化学疗法，会消灭人体肠道中健康的“有益”细菌。化疗药物不能区分是杀死癌细胞还是微生物，肠道中的微生物有助于消化食物，保持健康。杀死这些微生物对儿童尤其有害，因为有证据表明，生命早期肠道微生物组的破坏会导致日后潜在的健康问题。植生拉乌尔菌（*Raoultella planticola*）以低丰度自然存在于人体肠道，研究证实它能分解化疗药物多柔比星。通过降解多柔比星，这种细菌可以降低药物对肠道其他部位的毒性。因为儿童化疗相关的微生物组群变化与后期的健康并发症（包括肥胖症、哮喘和糖尿病）有关，所以发现保护肠道的新策略对儿童癌症患者尤其重要[158]。

二、肠道菌群在免疫治疗中的作用

免疫治疗是癌症治疗的支柱[159]。在过去10年间，癌症免疫治疗的兴起和成功彻底改变了一系列以前预后不良恶性肿瘤的临床治疗[160]。利用宿主免疫系统是一种很有前途的癌症控制策略，因为它可以特异性靶向肿瘤细胞并减少对正常组织的不良影响。由于微生物能显著调节炎症和免疫，所以微生物成分的改变可能会影响免疫治疗的反应[161]。免疫系统在抗癌中起着至关重要的作用。虽然致癌作用是由新的基因改变引起的，但它的持续发展取决于它逃避宿主免疫的能力[162, 163]。随着肿瘤免疫学的革命，现在在肿瘤持续和免疫监测失败之间建立了直接联系[164]。癌细胞通过免疫检查点途径直接抑制 $CD8^+$ 细胞毒性T细胞来逃避免疫，如程序性细胞死亡-1（programmed cell death–1，PD-1）和细胞毒性T淋巴细胞相关抗原-4（cytotoxic T lymphocyte associated antigen-4，CTLA-4）[162, 163]。PD-1、PD-L1和PD-L2配体在实体瘤和白血病/淋巴瘤中都有上调[7, 165]。阻断PD-1/PD-L1和CTLA-4/配体的相互作用，在多种实体肿瘤和恶性血液肿瘤中显示出有希望的治疗前景，可能会促进PD-1和CLTA-4抑制剂的“双免疫疗法”的组合策略[166, 167]。这些药物现在通常用于治疗晚期黑色素瘤、非小细胞肺癌、头颈癌、肾细胞癌、肝细胞癌、膀胱癌和霍奇金淋巴瘤等几种癌症[168, 169]。患者的肠道微生物群被认为是肿瘤免疫疗法疗效的决定因素之一[170]。

免疫治疗发展的最前沿是免疫检查点抑制剂（immune checkpoint blocker，ICB），由于其广泛的生物活性和反应的持久性，在癌症治疗中取得了巨大而无与伦比的成功，有时甚至涉及转移性和化疗耐药疾病的治疗成功案例[160]。近年来，许多免疫检查点抑制剂分子被开发出来并投入市场，包括针对CTLA4（伊匹单抗）、PD1（纳武利尤单抗）和PD1配体1或PDL1（帕博利珠单抗）的单克隆抗体，在一些难以治疗的癌症中被证明是高效的[171, 172]。富含CpG寡聚脱氧核苷酸（unmethylated cytosine-phosphate-guanine oligodeoxynucleotides，CpG-ODNs）的小寡核苷酸已被用作肿瘤免疫治疗的免疫佐剂。作为合成配体，CpG-ODNs可以模拟细菌感染并激活病原体相关分子模式受体TLR9[161, 173]。除了强大的免疫刺激特性外，肿瘤内注射CpG-ODNs可以诱导有效的抗肿瘤活性[173]。

另一种抗癌免疫治疗策略是阻断免疫检查点，特别是CTLA-4和PD-1及其配体（PD-L1）轴。美国食品药品监督管理局已经批准使用免疫疗法，包括检查点封锁治疗晚期黑色素瘤和肺癌[174]。通过靶向免疫检查点（如PD-1–PD-L1轴）来释放适应性免疫反应的能力，已成为治疗实体瘤的一种很有前途的癌症治疗方法[175]。检查点抑制剂在某些类型癌症的慢性治疗中显示出一致的疗效。然而，只有25%的患者对PD-1阻断剂有反应[176]。个体之间的可变免疫应答的潜在机制尚不清楚，有趣的是，肠道微生物组的组成与患者对抗PD-1治疗的反应有关，这已被众多研究证明[154]。

研究表明，对检查点抑制剂有反应的患者表现出更丰富的微生物群多样性，特别是一些特定的细菌[176]。据报道，瘤胃球菌科（Ruminococcaceae）、粪杆菌属（*Faecalibacterium*）、梭状芽孢杆菌（*Clostridales*）、长双歧杆菌（*Bifidobacterium longum*）、产气柯林斯菌（*Collinsella aerofaciens*）、粪肠球菌（*Enterococcus faecium*）和嗜黏蛋白阿克曼菌（*Akkermansia muciniphila*）在对治疗有反应的患者中富集[176-178]。这些物种被认为是“有利细菌”，而拟杆菌目在无反应者中富集，因此被认为是“不利细菌”。应答和非应答患者之间代谢途径的比较也表明了代谢作用的差异。前者以合成代谢为主，包括氨基酸的生物合成，促进宿主免疫，后者以分解代谢为主。

此外，有大量“有利细菌”的患者在全身循环中表现出更多的效应CD4^{+}和CD8^{+}T细胞，并且对抗PD-1治疗的细胞因子反应保持不变，而“不良细菌”数量较高的患者，其全身循环中调节性T细胞和MDSCs数量较高，细胞因子反应迟钝。这些发现与无菌受体小鼠粪便微生物群移植实验一致。这些数据表明，肠道菌群良好的患者通过增加抗原呈递，改善周边和肿瘤微环境中的效应T细胞功能，从而增强抗肿瘤免疫反应。相反，肠道菌群不良的患者表现出抗肿瘤免疫反应受损，这可能是由于肿瘤内淋巴和髓细胞浸润减少，以及抗原呈递能力减弱所致[179]。除黑色素瘤外，非小细胞肺癌、肾癌、尿路上皮癌患者在治疗各种感染时对抗PD-1治疗及抗生素暴露的反应均呈负相关[175, 178]。然而，这些发现与最近一项针对胰腺导管腺癌的研究相矛盾。在胰腺导管腺癌中，抗生素引起的细菌耗竭通过上调PD-1表达和重组肿瘤微环境（如减少MDSCs、增加M1巨噬细胞分化、促进CD4^{+}T和CD8^{+}T细胞TH1分化）来增强免疫治疗反应[180]。这些矛盾的发现表明，不同类型的癌症会引起微生物组的多种变化，并且不同的微生物组成对免疫检查点治疗产生正或负作用[181]。

抗PD-1主要诱导特异性肿瘤滤过衰竭型CD8细胞亚群的扩增。与这种类型的治疗不同，抗CTLA-4除了参与特定的衰竭型CD8细胞亚群外，还诱导ICOS^{+}Th1样CD4效应细胞群的扩增。对抗CTLA-4的最佳免疫介导反应也仅限于少数患者。例如，仅22%接受抗CTLA-4治疗的晚期黑色素瘤患者表现出超过10年的持久反应[159]，这种药物的抗肿瘤作用依赖于不同的拟杆菌[179]。无菌或经抗生素治疗的小鼠黑素瘤对CTLA-4抑制剂治疗无效[182]。

癌症治疗的最佳反应需要一个完整的共生菌群，肠道微生物群对免疫治疗效果有着重要作用，细菌和宿主免疫反应之间的复杂通信参与了抗肿瘤活性[183, 184]。研究表明，在黑色素瘤小鼠模型中，通过肠道细菌移位到肠系膜淋巴结，小鼠全身照射后抗肿瘤CD8^{+}T细胞过继转移的效果明显增强。辐射诱导微生物LPS释放，通过TLR4途径激活先天免疫反应，进而增强抗肿瘤CD8^{+}T细胞功能，而抗生素治疗或中和LPS（多黏菌素B）与抗肿瘤反应降低有关[22]。

肠道菌群不仅可以调节免疫系统发育及功能，而且能够诱导巨噬细胞、自然杀伤细胞等免疫细胞的固有免疫记忆，其机制一方面可能是菌群代谢产物（如丁酸盐、乙酸盐、丙酸盐）通过激活特异性组蛋白促进记忆反应，另一方面可能是肠道菌群来源的配体和固有免疫细胞受体相结合后，DNA甲基化、组蛋白修饰会再次快速启动进而激活如STAT1、c-Jun氨基末端激酶和MAPK等各种免疫记忆相关的信号通路[142]。正因为肠道菌群与机体免疫存在密切联系，越来越多的研究开始关注肠道菌群是否能够影响肿瘤免疫治疗的效果或毒副反应。通过分析接受PD-1抑制剂治疗的转移性黑色瘤患者粪便样本发现，长双歧杆菌、产气柯林斯菌及粪肠球菌显著富集于治疗敏感患者的粪便中，其机制为上述细菌通过减少调节性T细胞、增加树突状细胞，以及增强辅助性T细胞的作用来改善免疫治疗功效[177]。运用宏基因组学分析接受PD-1抗体治疗的非小细胞肺癌及肾细胞癌患者粪便样本发现，嗜黏蛋白阿克曼菌在治疗敏感的患者粪便样本中富集，进一步机制研究表明该细菌以依赖于 IL-12 途径募集 $CCR9^+CXCR3^+CD4^+$T 淋巴细胞至肿瘤，进而增强 PD-1 抑制剂的抗肿瘤效果[178]。

最近的一项研究发现，肿瘤突变负荷或肿瘤浸润淋巴细胞可能是接受免疫检查点抑制剂治疗患者的相关生物标志物[185, 186]，越来越多的证据支持这样的假设，即肠道微生物群对包括检查点抑制剂在内的免疫治疗有很大影响[29]。肠道微生物群在平衡炎症、感染和共生抗原方面起着至关重要的作用，这些抗原可以局部和系统地调节宿主免疫系统[187]。随着人们对微生物群对免疫治疗影响的兴趣不断增强，微生物群与癌症、特定的肠道微生物以及抗生素的使用引起了广泛关注[188]。

三、特异性菌群预测免疫治疗效果

越来越多的证据表明，肿瘤的发生、发展与肠道微生态变化有关。研究显示，九大癌症（急性淋巴细胞白血病、头颈部鳞状细胞癌、口腔癌、肺癌、乳腺癌、胰腺癌、胆管癌、人类乳头瘤病毒相关性宫颈癌、尿路上皮癌）均与特定的肠道微生态组成及丰度相关[27]。对来自 5 个国家的 526 例粪便标本进行宏基因组分析，发现结直肠癌患者肠道内富集脆弱拟杆菌、具核梭杆菌等 7 种细菌，另有 62 种细菌减少。因此，这些潜在的细菌标志物可用于结直肠癌的无创性诊断[189]。

现有的肿瘤检查方法包括肿瘤标志物检查、影像学检查、内镜检查及病理检查等。这些方法存在假阳性率或假阴性率高、价格昂贵、有创伤等缺点。因此急需一种同时具有高特异性及敏感性、无创、且价格适中的新方法。通过对比结直肠癌患者与对照组的粪便标本，发现前者的肠道菌群多样性降低，其中梭状芽孢杆菌减少最为显著[190]。这一研究结果证实，结直肠癌患者肠道菌群的多样性及菌群构成发生了改变，可考虑将其作为一种新的肿瘤检测指标。此外，与粪便隐血试验及相关临床危险因素评估相比，检测粪便中的肠道菌群可提高腺瘤性息肉及结直肠癌的检出率和区分度，为结直肠癌的早诊早治提供了可能性[191]。部分急性白血病患者在接受化疗后会出现肠道感染症状，对接受诱导化疗的成年急性髓系白血病患者进行粪便微生物分析发现，肠道菌群多样性降低的患者更容易并发化疗后感染[192]。根据这一结果，研究者们认为，肠道菌群可预测患者对治疗的反应及预后情况，对于治疗的评价提供了新的指标[193]。

免疫相关性结肠炎是最常见的与抗 CTLA-4 单抗相关的不良反应。一项前瞻性临床研究提示，拟杆菌门和涉及多胺转运、B 族维生素生物合成的肠道菌群在未发生结肠炎的患者中更为丰富。根据这些信息构建的预测结肠炎风险模型，其灵敏度为 70%，特异度为 83%[194]。研究发现，未发生结肠炎的患者，其基线全身炎症因子和拟杆菌门比例较高[195]。拟杆菌门和伯克菌（Burkholderiales）混合物可改善抗生素处理小鼠中抗 CTLA-4 单抗诱导的亚临床结肠炎和结肠炎症评分[189]。因此推测，高丰度的拟杆菌门可作为预测未发生免疫相关性结肠炎的标志物。目前抗 CTLA-4 单抗和抗 PD-1 单抗联合使用方案被批准用于黑色素瘤的治疗，可增加有效率（50% ～ 60%）和延长无进展生存期[196]。然而，联合用药会导致接近 60% 的 3 ～ 5 级不良事件，40% 的患者因毒性而中断治疗。因此，可将检测基线肠道菌群作为预测因子，用于鉴别联合用药与单药治疗的患者，以减少严重不良事件的发生。

综上所述，动物实验与临床试验研究得出了相似的结论：①肠道微生态组成差异决定机体免疫差异，从而决定肿瘤免疫治疗差异。肠道微生态能预测肿瘤免疫治疗效果，通过补充特定的有益肠道菌群及促使肠道微生态多样化，有助于获得更高的有效率。②肠道微生态能预测并改善肿瘤免疫治疗的不良反应。③多种抗生素可破坏免疫检查点抑制剂抗肿瘤效应，但万古霉素可增强其疗效。上述研究得出的影响免疫治疗的肠道菌群不尽相同，可能与肿瘤类型、人种和所选药物相关。此外，上述研究对肠道微生态改善肿瘤免疫治疗效果的机制进行了简单的探讨。多样化和高丰度的有益肠道菌群（如粪杆菌、疣微菌）可增强抗原呈递介导的抗肿瘤免疫应答，并改善效应性 T 细胞功能及肿瘤微环境。相反，有害肠道微生物群（如大肠埃希菌）使得抗原呈递能力减弱，从而影响免疫治疗效果 [179]。

肠道菌群除了与外部有着复杂的联系外，其内部也是动态平衡的，肠道菌群之间相互协同，共同维持人体肠道微生态的平衡。肠道菌群失调不仅仅是一些菌种相对丰度的变化，整体的菌群多样性同样能够对肠道环境的稳态产生重要影响。有研究指出，菌群多样性显著影响患者能否从免疫治疗中获益 [179]。但在这一问题上同样存在相反的研究结果，例如在接受免疫治疗的黑色素瘤患者中，疗效不同的患者间整体菌群的多样性并无显著差异，而仅有个别菌种丰度不同 [197]。因此，肠道菌群多样性能否决定免疫治疗的结局，仍然有待研究者们进一步探究。

特异性功能菌株的存在与鉴别对预测免疫治疗效果及并发症、提高有效率和减轻免疫相关炎症等具有重要意义。迄今报道的潜在功能细菌包括双歧杆菌（*Bifidobacterium*）、嗜黏蛋白阿克曼菌（*Akkermansia muciniphil*a）、瘤胃菌科（*Ruminococcaceae*）、产气柯林斯菌（*Collinsella aerofciens*）、屎肠球菌（*Enterococcus faecium*）、大肠埃希菌（*Escherichia coli*）和拟杆菌属（*Bacterodies*）等。小鼠模型中，双歧杆菌能够增强树突状细胞功能，增强抗肿瘤免疫，并提高 PD-L1 抑制剂治疗恶性黑色素瘤的效果，双歧杆菌的抗癌效果与单独使用 PD-L1 抗体相当，联合使用几乎可以完全阻止肿瘤生长 [105]。嗜黏蛋白阿克曼菌增强了 PD-1/PD-L1 抑制剂治疗上皮性肿瘤的疗效，该作用与 IL-12 释放以及肿瘤部位 $CCR9^{+}CXCR3^{+}CD4^{+}$ T 细胞募集有关 [178]。与 PD-1 抑制剂治疗恶性黑色素瘤效果有关的肠道菌群还包括瘤胃菌科、产气柯林斯菌和屎肠球菌等，相关菌群的作用在临床与动物实验中得到了反复验证 [177, 179]。

大肠埃希菌能够定向定植在前列腺癌病灶，并可增加浸润免疫细胞如 $CD8^{+}$T 细胞、Th17、DCs 巨噬细胞和 NK 细胞等，减少 Treg 细胞与血管内皮生长因子浓度，增强 PD-1 抑制剂的疗效 [198]。有文献报道，CTLA-4 抑制剂伊匹单抗对黑色素瘤患者的疗效与肠道菌群相关。拟杆菌属尤其是脆弱拟杆菌（*B.fragilis*）是起作用的关键，无论是脆弱拟杆菌本身，还是它分泌的脂多糖，或是针对它的特异性 T 细胞，均可强化 CTLA-4 抑制剂的抗肿瘤效应 [182]。研究还发现，双歧杆菌的使用能够减轻接受免疫治疗小鼠的结肠炎症，在黑色素瘤小鼠模型中，接受 CTLA-4 抑制剂治疗的同时用葡聚糖硫酸钠盐诱导结肠炎，发现给予双歧杆菌的小鼠结肠炎症较轻，其机制可能与 Treg 细胞有关 [199]。

越来越多的文献报道了特异性肠道细菌对免疫检查点抑制剂治疗效果的影响。因此，肠道菌群被视为预测免疫检查点抑制剂疗效的生物标记之一 [200-202]。通过分析接受伊匹单抗治疗的转移性黑色素瘤患者的临床资料、粪便菌群和血液，发现粪便菌群以厚壁菌门为主的患者临床疗效更好，但免疫相关结肠炎发生率较高，血液中 Treg 细胞占比减少。研究人员用 4 种细菌的组合（*Faecalibacterim* + *Gemmiger* + *Clostridium XIVa* + *Bacteroides*）预测伊匹单抗的疗效，绘制受试者工作特征曲线，曲线下面积达到了 0.895[195]。从健康人粪便中分离出 11 株细菌，联合使用可诱导分泌IFN-γ的CD8⁺T细胞，增强PD-1抑制剂、CTLA-4抑制剂的效果，并且不引起结肠炎，有望用于癌症免疫治疗的辅助治疗[203]。

除了细菌丰度，预测模型亦可为肠道菌群代谢通路的组合。对黑色素瘤患者接受免疫检查点抑制剂治疗前的粪便进行宏基因组检测，发现以4种肠道菌群代谢通路作为模型，包括聚胺转运通路和维生素B_1、维生素B_2、维生素B_5的合成通路，预测免疫治疗后结肠炎的发生率，其敏感度达到70%，特异度大于80%[194]。因此，肠道菌群是影响肿瘤免疫治疗的关键，特异性菌株的鉴定和生物模型构建对临床治疗有着深远的指导意义。

四、调节肠道菌群改善肿瘤治疗

在临床上，是否可以应用微生物来预防癌症或阻止其复发？根据最近的研究，通过饮食调节、粪菌移植（fecal microbiota transplantation，FMT）或其他方法来操纵微生物群落，可以在肠道微生物的治疗调节方面取得理想效果。对肠道菌群的调控可分为对有利肠道菌群的补充或激活和对不利肠道菌群的抑制[154]。

服用抗生素是导致微生物组分改变的一个重要因素，因此，可以引起一系列级联效应。如前所述，抗生素可以作为抗癌药物[161]。肿瘤微环境中的细菌可以产生一种胞苷脱氨酶，通过灭活吉西他滨，从而产生对这种药物的耐药性。与单独使用吉西他滨的小鼠相比，用吉西他滨加抗生素环丙沙星治疗的结肠癌小鼠，表现出更强的药物反应[155]。除环丙沙星外，梭状芽孢杆菌还导致原发性肝肿瘤和肝转移瘤的免疫应答降低，其机制包括改变宿主胆汁酸的产生，而不是调节吉西他滨代谢物的浓度[204]。此项研究分析了转基因小鼠的原发性肝肿瘤，并将淋巴或黑色素瘤细胞系的肝转移瘤植入肝外。所有小鼠通过饮用水接受ABX治疗（万古霉素、新霉素和亚胺培南），以减少肠道共生菌的定植[205]。结果发现，抗生素治疗可诱导肝脏选择性抗肿瘤作用，同时增加肝脏CXCR6⁺自然杀伤T细胞数量和抗原刺激后IFN-γ生成水平。

进一步的研究表明，肠道微生物组参与介导初级到次级胆汁酸的转化，控制肝窦内皮细胞CXCL16的表达水平。在ABX治疗后，肝窦内皮细胞产生了更高水平的CXCL16，将自然杀伤T细胞招募到肝脏，有效地抑制了肿瘤细胞的生长[204]。抗生素甲硝唑也可能有治疗癌症的作用。甲硝唑治疗结肠癌异种移植小鼠，可降低梭状芽孢杆菌负荷、癌细胞增殖和肿瘤生长。为了确定梭状芽孢杆菌阳性结肠癌异种移植瘤使用耐梭状芽孢杆菌抗生素或对梭杆菌敏感的抗生素是否会影响肿瘤生长，研究者对两种不同的小鼠模型分别使用红霉素和甲硝唑。口服甲硝唑后，含有梭状芽孢杆菌阳性患者来源的异种移植瘤的小鼠肿瘤生长轨迹显著降低。肿瘤组织中梭状芽孢杆菌载量和肿瘤细胞增殖也显著降低。然而，红霉素组在这些参数上无显著差异[63]。最近一项研究表明，梭状芽孢杆菌载量高的大肠肿瘤更容易复发[148]。这些发现表明，梭状芽孢杆菌阳性的肿瘤将从抗菌疗法中受益[63]。

活菌为人类带来健康益处的概念最初是由E´lie Metchnikov（1908年诺贝尔奖得主）提出的[206]。大多数益生菌是产乳酸的细菌，特别是那些属于乳杆菌和双歧杆菌属的细菌；也包括其他属，如链球菌、芽孢杆菌和肠球菌。益生菌除了在减轻化疗不良反应方面的作用外，也可以改善化疗和免疫治疗的效果[161]。在膀胱癌小鼠模型中，膀胱内灌注乳杆菌Shirota株（LC9018）显示出抗肿瘤作用。这种细菌的补充显著降低了肿瘤的进展速度，局部增加了IFN-γ和TNF-α水平，并诱导了巨噬细胞周围中性粒细胞的浸润。在一项前瞻性、随机、对照试验中，使用干酪乳杆菌（商业上称为养乐多）和表柔比星膀胱内灌注治疗的患者，其复发率比单纯接受化疗的患者低15%[154]。

此外，抗生素的使用削弱了顺铂在肺癌小鼠模型中的疗效。与顺铂单一治疗相比，顺铂和嗜酸乳杆菌治疗的小鼠肿瘤体积减小，生存率提高，提示益生菌联合治疗可促进顺铂的细胞抑制和促凋亡作用[157]。此外，根据以往的研究，几种乳酸菌在结直肠癌治疗中具有很大的潜力。小鼠经口给予干酪乳酸杆菌BL23和唾液乳杆菌（*lactobacillus salivarius Ren*）可显著保护动物免受1，2-二甲基肼（1，2-Dimethylhydrazine，DMH）诱导的大肠癌形成[207]。注射DMH会影响肠道菌群的组成，其特征是瘤胃球菌（*Ruminococcus spp*）和梭状芽孢杆菌（*Clostridiales*）的数量增加，而普雷沃特菌（*Prevotella spp*）的数量减少。与单用DMH组相比，补充唾液乳杆菌可使菌群失调恢复到健康状态，癌症发病率从87.5%下降到25%[208]。对于干酪乳酸杆菌BL23，进一步的研究表明其抗肿瘤作用可能与调节Treg和Th17细胞群有关[207]。其他属于乳酸菌属的细菌，如嗜酸乳酸杆菌NCFM，已被报道具有减少结肠癌发生的作用，肿瘤大小减少了50.3%。

口服双歧杆菌可达到与抗PD-L1治疗类似的抗肿瘤效果，而联合治疗几乎可以消除肿瘤的生长[105]。FMT已被频繁地用于无菌小鼠实验，给小鼠移植抗PD-1治疗（R-FMT）有效患者的粪便后，小鼠的肿瘤体积显著减小，T细胞反应增强，抗PD-L1的抗肿瘤效果更高[177，179]。小鼠模型中，口服一种特定的细菌嗜黏蛋白阿克曼菌（*Akkermansia muciniphila*），通过增加CCR9$^+$CXCR3$^+$CD4$^+$T淋巴细胞向小鼠肿瘤巢的募集，以IL-12依赖的方式恢复PD-1阻断的效果[178]。其他方法，包括用脆弱拟杆菌（*Bacteroides fragilis*）灌胃，用脆弱拟杆菌多糖免疫，或过继转移脆弱拟杆菌特异性T细胞，都能扭转荷瘤小鼠对抗生素治疗引起的抗CTLA-4治疗的不良反应[182]。

益生元是不易消化的食品成分，主要包括可溶性纤维、抗性淀粉和低聚糖，可由肠道微生物选择性发酵。这些大分子可以选择性地刺激结肠中一种或几种微生物的生长和（或）活性，从而促进宿主健康[209]。通过添加富含双歧杆菌（*Bifidobacteria*）的纤维，可以防止因低纤维摄入和高肉类摄入而导致的结肠黏液层病变[210，211]。已经证明，富含纤维的饮食与结直肠癌风险降低有关，特别是具核梭杆菌丰度高的结直肠癌[212，213]。然而，富含可溶性纤维的饮食，如菊粉，可在小鼠模型中诱发肝癌，而不溶性纤维则没有此作用。通过抑制发酵菌或其发酵过程降低短链脂肪酸水平，可以预防肝细胞癌的发生[214]。

抗性淀粉是目前研究最多的益生元之一，能够促进丁酸生产相关细菌的生长。后者是一种广为人知的具有抗炎和抗癌活性的后生元[161]。以工程抗性淀粉饲料喂养的小鼠，胰腺癌异种移植瘤显示出明显的生长迟缓。这表明工程抗性淀粉饮食干预可作为胰腺癌患者潜在的替代辅助治疗[215]。除抗性淀粉外，在移植性肝肿瘤小鼠中，菊粉和低聚果糖饮食治疗增强了6种细胞毒性药物[5-FU、多柔比星、长春新碱、环磷酰胺、甲氨蝶呤和阿糖胞苷]的效果，并能进一步延长动物的寿命。低聚果糖、β（1-4）低聚半乳糖和乳果糖已被证明与胃肠道中乳酸和短链脂肪酸水平的升高，以及粪便中次级胆汁酸水平的降低有关，这表明它们在大肠癌预防中具有重要作用[216]。

饮食中的其他食物成分也有预防癌症的作用，这可能与它们影响肠道微生物群的能力有关。例如，较高的海洋ω-3脂肪酸（MO3FA）摄入量与短链脂肪酸产生菌（主要是乳酸杆菌和双歧杆菌）的丰度增加有关，与具核梭杆菌、嗜黏蛋白阿克曼菌和脂多糖产生菌（例如大肠埃希菌）的丰度降低有关，进而使结直肠癌的病理进展减轻[217，218]。此外，运动还可以影响肠道微生物的组成和功能。职业运动员的粪便样本决定了他们微生物群的多样性更高，并且产生了更高的免疫友好的次级代谢物，如短链脂肪酸。这表明运动可能通过抑制免疫系统来提高结直肠癌患者的存活率[219]。

五、粪菌移植

粪菌移植是指将粪便悬浮液注入个体的消化道，目的是通过微生物群来治疗或预防疾病[220]。FMT对复发或难治性艰难梭菌感染（clostridium difficile infection，CDI）的疗效已经在多个随机对照试验中得到证实，并得到了医学指南的支持[220-222]。随着对肠道菌群的深入了解，再加上FMT在治疗CDI方面的成功，人们对FMT在各种疾病中的应用产生了广泛的热情，有几项试验证明了FMT在诱导溃疡性结肠炎缓解方面的有效性[223]。然而，FMT在肿瘤中的应用是一个新兴的研究方向，目前正在研究实验性FMT在肿瘤治疗中是否能够调节治疗效果和减轻治疗的严重并发症，包括脓毒症和移植物抗宿主病[224]。虽然有一些令人鼓舞的成功案例和临床研究，但FMT在癌症治疗中的证据质量仍然较低，仍然需要高质量的临床研究（图7-4）[26]。

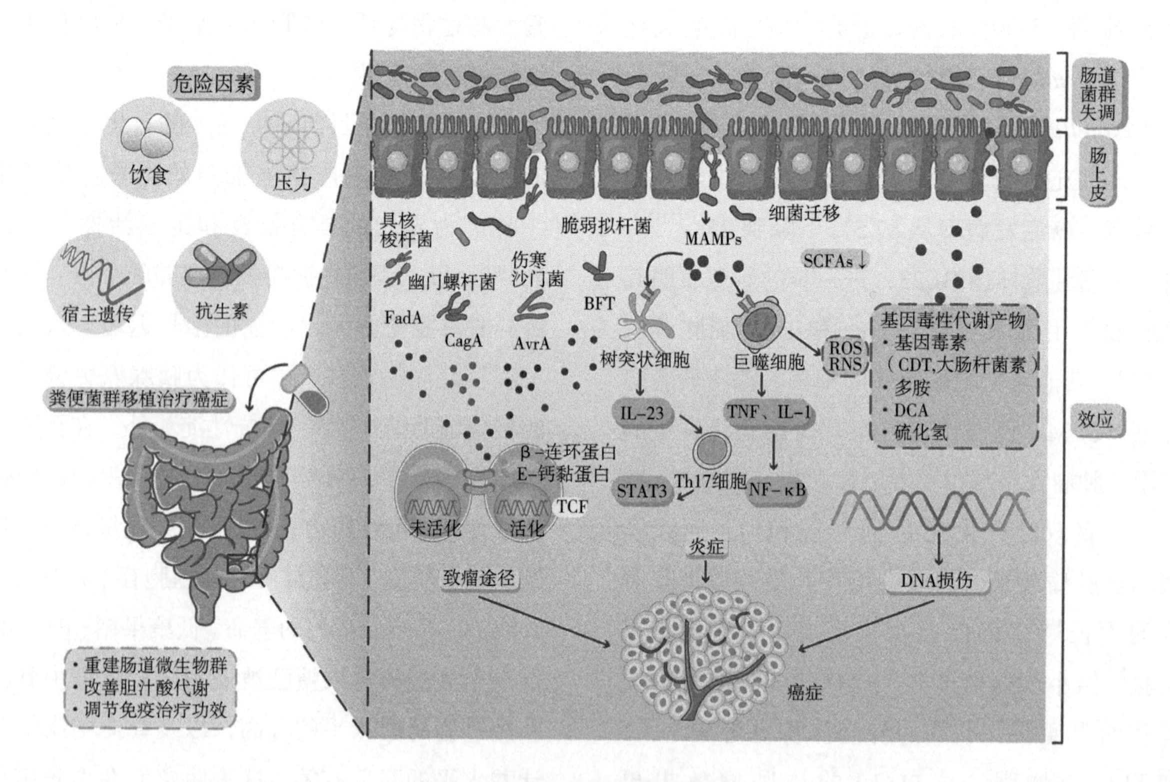

图7-4　粪便菌群移植治疗癌症

粪菌移植是一种通过重建肠道微生物群、改善胆汁酸代谢和调节免疫治疗功效来治疗癌症的潜在策略。肠道微生物失调和特殊细菌能够通过激活致瘤途径、诱导炎症和破坏宿主DNA来影响癌症的发展和进展。特殊的细菌产物，如来自具核梭杆菌的FadA毒素、来自幽门螺杆菌的CagA蛋白、来自伤寒沙门菌的AvrA蛋白和来自产肠毒素脆弱拟杆菌的BFT，可以促进β-连环蛋白与E-钙黏蛋白的分离，从而触发β-连环蛋白的活化并促进肿瘤的发生。BFT. 脆弱拟杆菌肠毒素；MAMPs. 微生物相关分子模式；IL-23. 白细胞介素23；TNF. 肿瘤坏死因子；IL-1. 白细胞介素1；STAT3. 信号转导和转录激活蛋白3；NF-κB. 核因子κB；ROS. 活性氧；RNS. 活性氮；CDT. 细胞致死膨胀毒素；DCA. 脱氧胆酸；CagA. 细胞毒素相关基因A；SCFAs. 短链脂肪酸；TCF. 转录因子

肠道菌群失调可通过多种途径促进结直肠肿瘤的发生，FMT通过恢复正常的肠道微生物

群发挥抗肿瘤作用。最近的临床研究证实，结直肠癌患者的粪便微生物群促进了动物模型肠道肿瘤的形成，降低了给予致癌物的无菌和常规小鼠体内的微生物丰度 [12]。此外，与对照组相比，移植了野生小鼠肠道微生物组的实验室小鼠对结直肠癌表现出更好的抵抗力和炎症的改善，提示FMT可能具有治疗大肠癌的潜力 [225]。

最近一项对 39 名接受免疫检查点治疗的转移性黑色素瘤患者的研究表明，微生物的含量与免疫治疗的反应之间存在显著的相关性 [197]。在癌症免疫治疗的应答者中，多形拟杆菌（*Bacteroides thetaiotaomicron*）、普氏栖粪杆菌（*Faecalibacterium prausnitzii*）和丝状霍尔德曼氏菌（*Holdemania filiformis*）在肠道中含量丰富 [197]。将黑色素瘤患者的粪便转移到小鼠体内，证实了 FMT 能够提高免疫疗法的有效性，从而优化目前的疗法 [179]。一项临床研究正在测试通过 FMT 将 PD-1 应答者的菌群移植入黑色素瘤无应答者肠道后抗肿瘤效果的变化 [226]。最近的研究报道了 FMT 可以克服黑色素瘤患者对免疫治疗的耐药性 [227, 228]。因此，FMT 似乎有希望通过转移有利的肠道微生物群，来增强黑色素瘤患者的抗肿瘤免疫 [26]。

在使用 FMT 之前有三个重要的注意事项。①必须准确地分离和筛选细菌种类，以确保它们能够提高宿主抗肿瘤免疫治疗的效果；②必须清除有害细菌、病毒和寄生虫；③应注意分离和培养数量较少但重要的微生物 [229]。

第五节　小结与展望

综上所述，肠道菌群已成为影响抗肿瘤免疫治疗效果和毒副反应的重要因素之一，而干预肠道菌群有望增强肿瘤对免疫治疗的敏感性，并减轻相关毒副反应。但是目前仍存在诸多问题，例如肠道菌群影响免疫治疗效果或毒副反应的具体分子机制尚未清楚；肠道菌群能否作为评估免疫治疗效果的生物标志物，其评估标准如何制订及验证；肿瘤患者个体肠道菌群差异较大，如何精准干预肠道菌群以达到预期治疗效果也有待解决；肠道菌群干预本身是否有相关的副反应也需要更多的临床研究来明确。肠道菌群在调控人体免疫过程中发挥着重要作用，相信随着肠道菌群相关研究的深入，肿瘤免疫治疗联合肠道菌群干预，将成为延长患者生存及改善患者生活质量的新策略。

参考文献

[1] Zhang YJ, Li S, Gan RY, et al. Impacts of gut bacteria on human health and diseases. International Journal of Molecular Sciences, 2015, 16(4): 7493-7519.

[2] Ge Y, Wang X, Guo Y, et al. Gut microbiota influence tumor development and Alter interactions with the human immune system. Journal of Experimental and Clinical Cancer Research, 2021, 40(1): 42.

[3] Baffy G. Gut Microbiota and Cancer of the Host: Colliding Interests. Advances in Experimental Medicine and Biology, 2020, 1219: 93-107.

[4] Scott AJ, Merrifield CA, Younes JA, et al. Pre-, pro- and synbiotics in cancer prevention and treatment-a review of basic and clinical research. Ecancermedicalscience, 2018, 12: 869.

[5] Demaria O, Cornen S, Daeron M, et al. Harnessing innate immunity in cancer therapy. Nature, 2019, 574(7776): 45-56.

[6] Liu X, Chen Y, Zhang S, et al. Gut microbiota-mediated immunomodulation in tumor. Journal of Experimental and Clinical Cancer Research, 2021, 40(1): 221.

[7] Chen DS, Mellman I. Oncology meets immunology: the cancer-immunity cycle. Immunity, 2013, 39(1): 1-10.

[8] O'Donnell JS, Teng MWL, Smyth MJ. Cancer immunoediting and resistance to T cell-based immunotherapy. Nature Reviews: Clinical Oncology, 2019, 16(3): 151-167.

[9] Teng J, Zhao Y, Jiang Y, et al. Correlation between Gut Microbiota and Lung Cancer. Chinese Journal of Lung Cancer, 2020, 23(10): 909-915.

[10] Zhou CB, Zhou YL, Fang JY. Gut Microbiota in Cancer

Immune Response and Immunotherapy. Trends Cancer, 2021, 7(7): 647-660.

[11] Chan CW, Tsui SK, Law BM, et al. The Utilization of the Immune System in Lung Cancer Treatment: Beyond Chemotherapy. International Journal of Molecular Sciences, 2016, 17(3): 286.

[12] Wong SH, Zhao L, Zhang X, et al. Gavage of Fecal Samples From Patients With Colorectal Cancer Promotes Intestinal Carcinogenesis in Germ-Free and Conventional Mice. Gastroenterology, 2017, 153(6): 1621-1633. e1626.

[13] Thiele Orberg E, Fan H, Tam AJ, et al. The myeloid immune signature of enterotoxigenic Bacteroides fragilis-induced murine colon tumorigenesis. Mucosal Immunology, 2017, 10(2): 421-433.

[14] Nosho K, Sukawa Y, Adachi Y, et al. Association of Fusobacterium nucleatum with immunity and molecular alterations in colorectal cancer. World Journal of Gastroenterology, 2016, 22(2): 557-566.

[15] Chen T, Li Q, Wu J, et al. Fusobacterium nucleatum promotes M2 polarization of macrophages in the microenvironment of colorectal tumours via a TLR4-dependent mechanism. Cancer Immunology, Immunotherapy, 2018, 67(10): 1635-1646.

[16] Deng H, Li Z, Tan Y, et al. A novel strain of Bacteroides fragilis enhances phagocytosis and polarises M1 macrophages. Scientific Reports, 2016, 6: 29401.

[17] Buchta Rosean CM, Rutkowski MR. The influence of the commensal microbiota on distal tumor-promoting inflammation. Seminars in Immunology, 2017, 32: 62-73.

[18] Rong Y, Dong Z, Hong Z, et al. Reactivity toward Bifidobacterium longum and Enterococcus hirae demonstrate robust CD8(+) T cell response and better prognosis in HBV-related hepatocellular carcinoma. Experimental Cell Research, 2017, 358(2): 352-359.

[19] Yu AI, Zhao L, Eaton KA, et al. Gut Microbiota Modulate CD8 T Cell Responses to Influence Colitis-Associated Tumorigenesis. Cell Reports, 2020, 31(1): 107471.

[20] Li Q, Jin M, Liu Y, et al. Gut Microbiota: Its Potential Roles in Pancreatic Cancer. Frontiers in Cellular and Infection Microbiology, 2020, 10: 572492.

[21] Li W, Deng X, Chen T. Exploring the Modulatory Effects of Gut Microbiota in Anti-Cancer Therapy. Frontiers in Oncology, 2021, 11: 644454.

[22] Villeger R, Lopes A, Carrier G, et al. Intestinal Microbiota: A Novel Target to Improve Anti-Tumor Treatment? International Journal of Molecular Sciences, 2019, 20(18):4584.

[23] Shahanavaj K, Gil-Bazo I, Castiglia M, et al. Cancer and the microbiome: potential applications as new tumor biomarker. Expert Review of Anticancer Therapy, 2015, 15(3): 317-330.

[24] Yu LX, Schwabe RF. The gut microbiome and liver cancer: mechanisms and clinical translation. Nature Reviews: Gastroenterology & Hepatology, 2017, 14(9): 527-539.

[25] Lam SY, Yu J, Wong SH, et al. The gastrointestinal microbiota and its role in oncogenesis. Best Practice & Research: Clinical Gastroenterology, 2017, 31(6): 607-618.

[26] Chen D, Wu J, Jin D, et al. Fecal microbiota transplantation in cancer management: Current status and perspectives. International Journal of Cancer, 2019, 145(8): 2021-2031.

[27] Zitvogel L, Daillere R, Roberti MP, et al. Anticancer effects of the microbiome and its products. Nature Reviews: Microbiology, 2017, 15(8): 465-478.

[28] Baba Y, Iwatsuki M, Yoshida N, et al. Review of the gut microbiome and esophageal cancer: Pathogenesis and potential clinical implications. Annals of Gastroenterological Surgery, 2017, 1(2): 99-104.

[29] Meng C, Bai C, Brown TD, et al. Human Gut Microbiota and Gastrointestinal Cancer. Genomics Proteomics Bioinformatics, 2018, 16(1): 33-49.

[30] Blackett KL, Siddhi SS, Cleary S, et al. Oesophageal bacterial biofilm changes in gastro-oesophageal reflux disease, Barrett's and oesophageal carcinoma: association or causality? Alimentary Pharmacology and Therapeutics, 2013, 37(11): 1084-1092.

[31] Man SM. The clinical importance of emerging Campylobacter species. Nature Reviews: Gastroenterology & Hepatology, 2011, 8(12): 669-685.

[32] Sawada A, Fujiwara Y, Nagami Y, et al. Alteration of Esophageal Microbiome by Antibiotic Treatment Does Not Affect Incidence of Rat Esophageal Adenocarcinoma. Digestive Diseases and Sciences, 2016, 61(11): 3161-3168.

[33] Zaidi AH, Kelly LA, Kreft RE, et al. Associations of microbiota and toll-like receptor signaling pathway in esophageal adenocarcinoma. BMC Cancer, 2016, 16: 52.

[34] Di Pilato V, Freschi G, Ringressi MN, et al. The esophageal microbiota in health and disease. Annals of the New York Academy of Sciences, 2016, 1381(1): 21-33.

[35] Yu G, Gail MH, Shi J, et al. Association between upper

digestive tract microbiota and cancerpredisposing states in the esophagus and stomach. Cancer Epidemiology, Biomarkers & Prevention, 2014, 23(5): 735-741.

[36] Nasrollahzadeh D, Malekzadeh R, Ploner A, et al. Variations of gastric corpus microbiota are associated with early esophageal squamous cell carcinoma and squamous dysplasia. Scientific Reports, 2015, 5: 8820.

[37] Gao S, Li S, Ma Z, et al. Presence of Porphyromonas gingivalis in esophagus and its association with the clinicopathological characteristics and survival in patients with esophageal cancer. Infectious Agents and Cancer, 2016, 11: 3.

[38] Yamamura K, Baba Y, Nakagawa S, et al. Human Microbiome Fusobacterium Nucleatum in Esophageal Cancer Tissue Is Associated with Prognosis. Clinical Cancer Research, 2016, 22(22): 5574-5581.

[39] Xie FJ, Zhang YP, Zheng QQ, et al. Helicobacter pylori infection and esophageal cancer risk: an updated meta-analysis. World Journal of Gastroenterology, 2013, 19(36): 6098-6107.

[40] Runge TM, Abrams JA, Shaheen NJ. Epidemiology of Barrett's Esophagus and Esophageal Adenocarcinoma. Gastroenterology Clinics of North America, 2015, 44(2): 203-231.

[41] Neto AG, Whitaker A, Pei Z. Microbiome and potential targets for chemoprevention of esophageal adenocarcinoma. Seminars in Oncology, 2016, 43(1): 86-96.

[42] Amir I, Konikoff FM, Oppenheim M, et al. Gastric microbiota is altered in oesophagitis and Barrett's oesophagus and further modified by proton pump inhibitors. Environmental Microbiology, 2014, 16(9): 2905-2914.

[43] Hu Q, Sun TT, Hong J, et al. Proton Pump Inhibitors Do Not Reduce the Risk of Esophageal Adenocarcinoma in Patients with Barrett's Esophagus: A Systematic Review and Meta-Analysis. PloS One, 2017, 12(1): e0169691.

[44] Xu W, Liu Z, Bao Q, et al. Viruses, Other Pathogenic Microorganisms and Esophageal Cancer. Gastrointest Tumors, 2015, 2(1): 2-13.

[45] Patel T, Bhattacharya P, Das S. Gut microbiota: an Indicator to Gastrointestinal Tract Diseases. J Gastrointest Cancer, 2016, 47(3): 232-238.

[46] Doorakkers E, Lagergren J, Engstrand L, et al. Eradication of Helicobacter pylori and Gastric Cancer: A Systematic Review and Meta-analysis of Cohort Studies. Journal of the National Cancer Institute, 2016, 108(9):djw132.

[47] Khatoon J, Rai RP, Prasad KN. Role of Helicobacter pylori in gastric cancer: Updates. World Journal of Gastrointestinal Oncology, 2016, 8(2): 147-158.

[48] Kwok T, Zabler D, Urman S, et al. Helicobacter exploits integrin for type IV secretion and kinase activation. Nature, 2007, 449(7164): 862-866.

[49] Moyat M, Velin D. Immune responses to Helicobacter pylori infection. World Journal of Gastroenterology, 2014, 20(19): 5583-5593.

[50] Udhayakumar G, Jayanthi V, Devaraj N, et al. Interaction of MUC1 with beta-catenin modulates the Wnt target gene cyclinD1 in H. pylori-induced gastric cancer. Molecular Carcinogenesis, 2007, 46(9): 807-817.

[51] Yong X, Tang B, Li BS, et al. Helicobacter pylori virulence factor CagA promotes tumorigenesis of gastric cancer via multiple signaling pathways. Cell Commun Signal, 2015, 13: 30.

[52] Sitaraman R. Helicobacter pylori DNA methyltransferases and the epigenetic field effect in cancerization. Front Microbiol, 2014, 5: 115.

[53] Maldonado-Contreras A, Goldfarb KC, Godoy-Vitorino F, et al. Structure of the human gastric bacterial community in relation to Helicobacter pylori status. Multidisciplinary Journal of Microbial Ecology, 2011, 5(4): 574-579.

[54] Iizasa H, Ishihara S, Richardo T, et al. Dysbiotic infection in the stomach. World Journal of Gastroenterology, 2015, 21(40): 11450-11457.

[55] Aviles-Jimenez F, Vazquez-Jimenez F, Medrano-Guzman R, et al. Stomach microbiota composition varies between patients with non-atrophic gastritis and patients with intestinal type of gastric cancer. Scientific Reports, 2014, 4: 4202.

[56] Wang J, Zhao L, Yan H, et al. A Meta-Analysis and Systematic Review on the Association between Human Papillomavirus (Types 16 and 18) Infection and Esophageal Cancer Worldwide. PloS One, 2016, 11(7): e0159140.

[57] Dias-Jacome E, Libanio D, Borges-Canha M, et al. Gastric microbiota and carcinogenesis: the role of non-Helicobacter pylori bacteria - A systematic review. Revista Española de Enfermedades Digestivas, 2016, 108(9): 530-540.

[58] De Witte C, Schulz C, Smet A, et al. Other Helicobacters and gastric microbiota. Helicobacter, 2016, 21 Suppl 1: 62-68.

[59] Temraz S, Nassar F, Nasr R, et al. Gut Microbiome: A Promising Biomarker for Immunotherapy in Colorectal

Cancer. International Journal of Molecular Sciences, 2019, 20(17):4155.

[60] Nakatsu G, Zhou H, Wu WKK, et al. Alterations in Enteric Virome Are Associated With Colorectal Cancer and Survival Outcomes. Gastroenterology, 2018, 155(2): 529-541. e525.

[61] Ito M, Kanno S, Nosho K, et al. Association of Fusobacterium nucleatum with clinical and molecular features in colorectal serrated pathway. International Journal of Cancer, 2015, 137(6): 1258-1268.

[62] Li YY, Ge QX, Cao J, et al. Association of Fusobacterium nucleatum infection with colorectal cancer in Chinese patients. World Journal of Gastroenterology, 2016, 22(11): 3227-3233.

[63] Bullman S, Pedamallu CS, Sicinska E, et al. Analysis of Fusobacterium persistence and antibiotic response in colorectal cancer. Science, 2017, 358(6369): 1443-1448.

[64] Yamaoka Y, Suehiro Y, Hashimoto S, et al. Fusobacterium nucleatum as a prognostic marker of colorectal cancer in a Japanese population. Journal of Gastroenterology, 2018, 53(4): 517-524.

[65] Oh HJ, Kim JH, Bae JM, et al. Prognostic Impact of Fusobacterium nucleatum Depends on Combined Tumor Location and Microsatellite Instability Status in Stage II/III Colorectal Cancers Treated with Adjuvant Chemotherapy.Journal of Pathology and Translational Medicine, 2019, 53(1): 40-49.

[66] Kuo WT, Lee TC, Yang HY, et al. LPS receptor subunits have antagonistic roles in epithelial apoptosis and colonic carcinogenesis. Cell Death and Differentiation, 2015, 22(10): 1590-1604.

[67] Chiu WT, Lin YL, Chou CW, et al. Propofol inhibits lipoteichoic acid-induced iNOS gene expression in macrophages possibly through downregulation of toll-like receptor 2-mediated activation of Raf-MEK1/2-ERK1/2-IKK-NFkappaB. Chemico-Biological Interactions, 2009, 181(3): 430-439.

[68] Su SC, Hua KF, Lee H, et al. LTA and LPS mediated activation of protein kinases in the regulation of inflammatory cytokines expression in macrophages. Clinica Chimica Acta, 2006, 374(1-2): 106-115.

[69] Richards JL, Yap YA, McLeod KH, et al. Dietary metabolites and the gut microbiota: an alternative approach to control inflammatory and autoimmune diseases. Clinical Translational Immunology, 2016, 5(5): e82.

[70] Elangovan S, Pathania R, Ramachandran S, et al. The niacin/butyrate receptor GPR109A suppresses mammary tumorigenesis by inhibiting cell survival. Cancer Research, 2014, 74(4): 1166-1178.

[71] Singh N, Gurav A, Sivaprakasam S, et al. Activation of Gpr109a, receptor for niacin and the commensal metabolite butyrate, suppresses colonic inflammation and carcinogenesis. Immunity, 2014, 40(1): 128-139.

[72] Bardhan K, Paschall AV, Yang D, et al. IFNgamma Induces DNA Methylation-Silenced GPR109A Expression via pSTAT1/p300 and H3K18 Acetylation in Colon Cancer. Cancer Immunol Res, 2015, 3(7): 795-805.

[73] Pudlo NA, Urs K, Kumar SS, et al. Symbiotic Human Gut Bacteria with Variable Metabolic Priorities for Host Mucosal Glycans. mBio, 2015, 6(6): e01282-01215.

[74] Nepelska M, Cultrone A, Beguet-Crespel F, et al. Butyrate produced by commensal bacteria potentiates phorbol esters induced AP-1 response in human intestinal epithelial cells. PloS One, 2012, 7(12): e52869.

[75] Ohtani N. Microbiome and cancer. Seminars in Immunopathology, 2015, 37(1): 65-72.

[76] Schinzari V, Barnaba V, Piconese S. Chronic hepatitis B virus and hepatitis C virus infections and cancer: synergy between viral and host factors. Clinical Microbiology and Infection, 2015, 21(11): 969-974.

[77] Wong CR, Nguyen MH, Lim JK. Hepatocellular carcinoma in patients with non-alcoholic fatty liver disease. World Journal of Gastroenterology, 2016, 22(37): 8294-8303.

[78] French SW. Epigenetic events in liver cancer resulting from alcoholic liver disease. Alcohol Research, 2013, 35(1): 57-67.

[79] Yu J, Marsh S, Hu J, et al. The Pathogenesis of Nonalcoholic Fatty Liver Disease: Interplay between Diet, Gut Microbiota, and Genetic Background. Gastroenterology Research and Practice, 2016, 2016: 2862173.

[80] Aqel B, DiBaise JK. Role of the Gut Microbiome in Nonalcoholic Fatty Liver Disease. Nutrition in Clinical Practice, 2015, 30(6): 780-786.

[81] Ridlon JM, Kang DJ, Hylemon PB, et al. Bile acids and the gut microbiome. Current Opinion in Gastroenterology, 2014, 30(3): 332-338.

[82] Hara E. Relationship between Obesity, Gut Microbiome and Hepatocellular Carcinoma Development. Digestive Diseases, 2015, 33(3): 346-350.

[83] Louis P, Hold GL, Flint HJ. The gut microbiota, bacterial metabolites and colorectal cancer. Nature Reviews: Microbiology, 2014, 12(10): 661-672.

[84] Wang C, Li J. Pathogenic Microorganisms and Pancreatic Cancer. Gastrointest Tumors, 2015, 2(1): 41-47.

[85] Michaud DS, Izard J. Microbiota, oral microbiome, and pancreatic cancer. Cancer Journal, 2014, 20(3): 203-206.

[86] Zambirinis CP, Pushalkar S, Saxena D, et al. Pancreatic cancer, inflammation, and microbiome. Cancer Journal, 2014, 20(3): 195-202.

[87] Rabelo-Goncalves EM, Roesler BM, Zeitune JM. Extragastric manifestations of Helicobacter pylori infection: Possible role of bacterium in liver and pancreas diseases. World Journal of Hepatology, 2015, 7(30): 2968-2979.

[88] Bulajic M, Panic N, Lohr JM. Helicobacter pylori and pancreatic diseases. World Journal of Gastrointestinal Pathophysiology, 2014, 5(4): 380-383.

[89] Goni E, Franceschi F. Helicobacter pylori and extragastric diseases. Helicobacter, 2016, 21 Suppl 1: 45-48.

[90] Wormann SM, Diakopoulos KN, Lesina M, et al. The immune network in pancreatic cancer development and progression. Oncogene, 2014, 33(23): 2956-2967.

[91] Daniluk J, Liu Y, Deng D, et al. An NF-kappaB pathway-mediated positive feedback loop amplifies Ras activity to pathological levels in mice. Journal of Clinical Investigation, 2012, 122(4): 1519-1528.

[92] Ochi A, Graffeo CS, Zambirinis CP, et al. Toll-like receptor 7 regulates pancreatic carcinogenesis in mice and humans. Journal of Clinical Investigation, 2012, 122(11): 4118-4129.

[93] Ochi A, Nguyen AH, Bedrosian AS, et al. MyD88 inhibition amplifies dendritic cell capacity to promote pancreatic carcinogenesis via Th2 cells. Journal of Experimental Medicine, 2012, 209(9): 1671-1687.

[94] Alam MS, Gaida MM, Bergmann F, et al. Selective inhibition of the p38 alternative activation pathway in infiltrating T cells inhibits pancreatic cancer progression. Nature Medicine, 2015, 21(11): 1337-1343.

[95] Gaida MM, Mayer C, Dapunt U, et al. Expression of the bitter receptor T2R38 in pancreatic cancer: localization in lipid droplets and activation by a bacteria-derived quorum-sensing molecule. Oncotarget, 2016, 7(11): 12623-12632.

[96] Mitsuhashi K, Nosho K, Sukawa Y, et al. Association of Fusobacterium species in pancreatic cancer tissues with molecular features and prognosis. Oncotarget, 2015, 6(9): 7209-7220.

[97] Wheeler KM, Liss MA. The Microbiome and Prostate Cancer Risk. Current Urology Reports, 2019, 20(10): 66.

[98] Aragon IM, Herrera-Imbroda B, Queipo-Ortuno MI, et al. The Urinary Tract Microbiome in Health and Disease. European Urology Focus, 2018, 4(1): 128-138.

[99] Cavarretta I, Ferrarese R, Cazzaniga W, et al. The Microbiome of the Prostate Tumor Microenvironment. European Urology, 2017, 72(4): 625-631.

[100] Feng Y, Ramnarine VR, Bell R, et al. Metagenomic and metatranscriptomic analysis of human prostate microbiota from patients with prostate cancer. BMC Genomics, 2019, 20(1): 146.

[101] Sandrini S, Aldriwesh M, Alruways M, et al. Microbial endocrinology: host-bacteria communication within the gut microbiome. Journal of Endocrinology, 2015, 225(2): R21-R34.

[102] Carabotti M, Scirocco A, Maselli MA, et al. The gut-brain axis: interactions between enteric microbiota, central and enteric nervous systems. Annals of Gastroenterology, 2015, 28(2): 203-209.

[103] Vivarelli S, Salemi R, Candido S, et al. Gut Microbiota and Cancer: From Pathogenesis to Therapy. Cancers, 2019, 11(1):38.

[104] Lazar V, Ditu LM, Pircalabioru GG, et al. Aspects of Gut Microbiota and Immune System Interactions in Infectious Diseases, Immunopathology, and Cancer. Frontiers in Immunology, 2018, 9: 1830.

[105] Sivan A, Corrales L, Hubert N, et al. Commensal Bifidobacterium promotes antitumor immunity and facilitates anti-PD-L1 efficacy. Science, 2015, 350(6264): 1084-1089.

[106] Gnauck A, Lentle RG, Kruger MC. The Characteristics and Function of Bacterial Lipopolysaccharides and Their Endotoxic Potential in Humans. International Reviews of Immunology, 2016, 35(3): 189-218.

[107] Jain S, Suklabaidya S, Das B, et al. TLR4 activation by lipopolysaccharide confers survival advantage to growth factor deprived prostate cancer cells. Prostate, 2015, 75(10): 1020-1033.

[108] d'Hennezel E, Abubucker S, Murphy LO, et al. Total Lipopolysaccharide from the Human Gut Microbiome Silences Toll-Like Receptor Signaling. mSystems, 2017, 2(6):e00046-17.

[109] Vatanen T, Kostic AD, d'Hennezel E, et al. Variation in Microbiome LPS Immunogenicity Contributes to Autoimmunity in Humans. Cell, 2016, 165(4): 842-853.

[110] Quigley EM. Leaky gut - concept or clinical entity? Current Opinion in Gastroenterology, 2016, 32(2): 74-79.

[111] Fernandez MF, Reina-Perez I, Astorga JM, et al. Breast Cancer and Its Relationship with the Microbiota. International Journal of Environmental Research and Public Health, 2018, 15(8):1747.

[112] Laborda-Illanes A, Sanchez-Alcoholado L, Dominguez-Recio ME, et al. Breast and Gut Microbiota Action Mechanisms in Breast Cancer Pathogenesis and Treatment. Cancers, 2020, 12(9): 246.

[113] Plottel CS, Blaser MJ. Microbiome and malignancy. Cell Host & Microbe, 2011, 10(4): 324-335.

[114] Fuhrman BJ, Feigelson HS, Flores R, et al. Associations of the fecal microbiome with urinary estrogens and estrogen metabolites in postmenopausal women. Journal of Clinical Endocrinology and Metabolism, 2014, 99(12): 4632-4640.

[115] Rea D, Coppola G, Palma G, et al. Microbiota effects on cancer: from risks to therapies. Oncotarget, 2018, 9(25): 17915-17927.

[116] Wang H, Altemus J, Niazi F, et al. Breast tissue, oral and urinary microbiomes in breast cancer. Oncotarget, 2017, 8(50): 88122-88138.

[117] Luu TH, Michel C, Bard JM, et al. Intestinal Proportion of Blautia sp. is Associated with Clinical Stage and Histoprognostic Grade in Patients with Early-Stage Breast Cancer. Nutrition and Cancer, 2017, 69(2): 267-275.

[118] Yang J, Tan Q, Fu Q, et al. Gastrointestinal microbiome and breast cancer: correlations, mechanisms and potential clinical implications. Breast Cancer, 2017, 24(2): 220-228.

[119] Dabek M, McCrae SI, Stevens VJ, et al. Distribution of beta-glucosidase and beta-glucuronidase activity and of beta-glucuronidase gene gus in human colonic bacteria. FEMS Microbiology Ecology, 2008, 66(3): 487-495.

[120] Goedert JJ, Jones G, Hua X, et al. Investigation of the association between the fecal microbiota and breast cancer in postmenopausal women: a population-based case-control pilot study. Journal of the National Cancer Institute, 2015, 107(8):djv147.

[121] Goedert JJ, Hua X, Bielecka A, et al. Postmenopausal breast cancer and oestrogen associations with the IgA-coated and IgA-noncoated faecal microbiota. British Journal of Cancer, 2018, 118(4): 471-479.

[122] Jandhyala SM, Talukdar R, Subramanyam C, et al. Role of the normal gut microbiota. World Journal of Gastroenterology, 2015, 21(29): 8787-8803.

[123] Xuan C, Shamonki JM, Chung A, et al. Microbial dysbiosis is associated with human breast cancer. PloS One, 2014, 9(1): e83744.

[124] Urbaniak C, Cummins J, Brackstone M, et al. Microbiota of human breast tissue. Applied and Environmental Microbiology, 2014, 80(10): 3007-3014.

[125] Hieken TJ, Chen J, Hoskin TL, et al. The Microbiome of Aseptically Collected Human Breast Tissue in Benign and Malignant Disease. Scientific Reports, 2016, 6: 30751.

[126] Urbaniak C, Gloor GB, Brackstone M, et al. The Microbiota of Breast Tissue and Its Association with Breast Cancer. Applied and Environmental Microbiology, 2016, 82(16): 5039-5048.

[127] Koller VJ, Marian B, Stidl R, et al. Impact of lactic acid bacteria on oxidative DNA damage in human derived colon cells. Food and Chemical Toxicology, 2008, 46(4): 1221-1229.

[128] Schwabe RF, Jobin C. The microbiome and cancer. Nature Reviews: Cancer, 2013, 13(11): 800-812.

[129] Thompson KJ, Ingle JN, Tang X, et al. A comprehensive analysis of breast cancer microbiota and host gene expression. PloS One, 2017, 12(11): e0188873.

[130] Chan AA, Bashir M, Rivas MN, et al. Characterization of the microbiome of nipple aspirate fluid of breast cancer survivors. Scientific Reports, 2016, 6: 28061.

[131] Montassier E, Gastinne T, Vangay P, et al. Chemotherapy-driven dysbiosis in the intestinal microbiome. Alimentary Pharmacology and Therapeutics, 2015, 42(5): 515-528.

[132] Chen J, Domingue JC, Sears CL. Microbiota dysbiosis in select human cancers: Evidence of association and causality. Seminars in Immunology, 2017, 32: 25-34.

[133] Mao Q, Jiang F, Yin R, et al. Interplay between the lung microbiome and lung cancer. Cancer Letters, 2018, 415: 40-48.

[134] Dzutsev A, Goldszmid RS, Viaud S, et al. The role of the microbiota in inflammation, carcinogenesis, and cancer therapy. European Journal of Immunology, 2015, 45(1): 17-31.

[135] Zhao Y, Liu Y, Li S, et al. Role of lung and gut microbiota on lung cancer pathogenesis. Journal of Cancer Research and Clinical Oncology, 2021,

147(8): 2177-2186.

[136] Crusz SM, Balkwill FR. Inflammation and cancer: advances and new agents. Nature Reviews: Clinical Oncology, 2015, 12(10): 584-596.

[137] Bindels LB, Neyrinck AM, Loumaye A, et al. Increased gut permeability in cancer cachexia: mechanisms and clinical relevance. Oncotarget, 2018, 9(26): 18224-18238.

[138] Yang JJ, Yu D, Xiang YB, et al. Association of Dietary Fiber and Yogurt Consumption With Lung Cancer Risk: A Pooled Analysis. JAMA Oncology, 2020, 6(2): e194107.

[139] Sichetti M, De Marco S, Pagiotti R, et al. Anti-inflammatory effect of multistrain probiotic formulation (L. rhamnosus, B. lactis, and B. longum). Nutrition, 2018, 53: 95-102.

[140] Budden KF, Gellatly SL, Wood DL, et al. Emerging pathogenic links between microbiota and the gut-lung axis. Nature Reviews: Microbiology, 2017, 15(1): 55-63.

[141] Bingula R, Filaire M, Radosevic-Robin N, et al. Desired Turbulence? Gut-Lung Axis, Immunity, and Lung Cancer. Journal of Oncology, 2017, 2017: 5035371.

[142] Negi S, Das DK, Pahari S, et al. Potential Role of Gut Microbiota in Induction and Regulation of Innate Immune Memory. Frontiers in Immunology, 2019, 10: 2441.

[143] Zitvogel L, Ayyoub M, Routy B, et al. Microbiome and Anticancer Immunosurveillance. Cell, 2016, 165(2): 276-287.

[144] Dang AT, Marsland BJ. Microbes, metabolites, and the gut-lung axis. Mucosal Immunology, 2019, 12(4): 843-850.

[145] Alexander JL, Wilson ID, Teare J, et al. Gut microbiota modulation of chemotherapy efficacy and toxicity. Nature Reviews Gastroenterology Hepatology, 2017, 14(6): 356-365.

[146] Garcia-Gonzalez AP, Ritter AD, Shrestha S, et al. Bacterial Metabolism Affects the C. elegans Response to Cancer Chemotherapeutics. Cell, 2017, 169(3): 431-441. e438.

[147] Scott TA, Quintaneiro LM, Norvaisas P, et al. Host-Microbe Co-metabolism Dictates Cancer Drug Efficacy in C. elegans. Cell, 2017, 169(3): 442-456. e418.

[148] Yu T, Guo F, Yu Y, et al. Fusobacterium nucleatum Promotes Chemoresistance to Colorectal Cancer by Modulating Autophagy. Cell, 2017, 170(3): 548-563. e516.

[149] Mima K, Nishihara R, Qian ZR, et al. Fusobacterium nucleatum in colorectal carcinoma tissue and patient prognosis. Gut, 2016, 65(12): 1973-1980.

[150] Yang Y, Weng W, Peng J, et al. Fusobacterium nucleatum Increases Proliferation of Colorectal Cancer Cells and Tumor Development in Mice by Activating Toll-Like Receptor 4 Signaling to Nuclear Factor-kappaB, and Up-regulating Expression of MicroRNA-21. Gastroenterology, 2017, 152(4): 851-866. e824.

[151] Ramos A, Hemann MT. Drugs, Bugs, and Cancer: Fusobacterium nucleatum Promotes Chemoresistance in Colorectal Cancer. Cell, 2017, 170(3): 411-413.

[152] Abed J, Emgard JE, Zamir G, et al. Fap2 Mediates Fusobacterium nucleatum Colorectal Adenocarcinoma Enrichment by Binding to Tumor-Expressed Gal-GalNAc. Cell Host & Microbe, 2016, 20(2): 215-225.

[153] Ye X, Wang R, Bhattacharya R, et al. Fusobacterium Nucleatum Subspecies Animalis Influences Proinflammatory Cytokine Expression and Monocyte Activation in Human Colorectal Tumors. Cancer Prevention Research (Philadelphia, Pa), 2017, 10(7): 398-409.

[154] Chen B, Du G, Guo J, et al. Bugs, drugs, and cancer: can the microbiome be a potential therapeutic target for cancer management? Drug Discovery Today, 2019, 24(4): 1000-1009.

[155] Geller LT, Barzily-Rokni M, Danino T, et al. Potential role of intratumor bacteria in mediating tumor resistance to the chemotherapeutic drug gemcitabine. Science, 2017, 357(6356): 1156-1160.

[156] Jobin C. Cancer treatment: Bacterial snack attack deactivates a drug. Nature, 2017, 550(7676): 337-339.

[157] Gui QF, Lu HF, Zhang CX, et al. Well-balanced commensal microbiota contributes to anti-cancer response in a lung cancer mouse model. Genetics and Molecular Research, 2015, 14(2): 5642-5651.

[158] Blaustein RA, Seed PC, Hartmann EM. Biotransformation of Doxorubicin Promotes Resilience in Simplified Intestinal Microbial Communities. mSphere, 2021, 6(3): e0006821.

[159] Wei SC, Levine JH, Cogdill AP, et al. Distinct Cellular Mechanisms Underlie Anti-CTLA-4 and Anti-PD-1 Checkpoint Blockade. Cell, 2017, 170(6): 1120-1133. e1117.

[160] Daillere R, Derosa L, Bonvalet M, et al. Trial

watch : the gut microbiota as a tool to boost the clinical efficacy of anticancer immunotherapy. Oncoimmunology, 2020, 9(1): 1774298.

[161] Panebianco C, Andriulli A, Pazienza V. Pharmacomicrobiomics: exploiting the drug-microbiota interactions in anticancer therapies. Microbiome, 2018, 6(1): 92.

[162] Montanari F, Diefenbach CS. Hodgkin lymphoma: targeting the tumor microenvironment as a therapeutic strategy. Clinical Advances in Hematology & Oncology, 2015, 13(8): 518-524.

[163] Upadhyay R, Hammerich L, Peng P, et al. Lymphoma: immune evasion strategies. Cancers, 2015, 7(2): 736-762.

[164] Schreiber RD, Old LJ, Smyth MJ. Cancer immunoediting: integrating immunity's roles in cancer suppression and promotion. Science, 2011, 331(6024): 1565-1570.

[165] Topalian SL, Hodi FS, Brahmer JR, et al. Safety, activity, and immune correlates of anti-PD-1 antibody in cancer. New England Journal of Medicine, 2012, 366(26): 2443-2454.

[166] Weber JS, D'Angelo SP, Minor D, et al. Nivolumab versus chemotherapy in patients with advanced melanoma who progressed after anti-CTLA-4 treatment (CheckMate 037): a randomised, controlled, open-label, phase 3 trial. Lancet Oncology, 2015, 16(4): 375-384.

[167] Hellmann MD, Ciuleanu TE, Pluzanski A, et al. Nivolumab plus Ipilimumab in Lung Cancer with a High Tumor Mutational Burden. New England Journal of Medicine, 2018, 378(22): 2093-2104.

[168] Motzer RJ, Tannir NM, McDermott DF, et al. Nivolumab plus Ipilimumab versus Sunitinib in Advanced Renal-Cell Carcinoma. New England Journal of Medicine, 2018, 378(14): 1277-1290.

[169] Ansell SM, Lesokhin AM, Borrello I, et al. PD-1 blockade with nivolumab in relapsed or refractory Hodgkin's lymphoma. New England Journal of Medicine, 2015, 372(4): 311-319.

[170] Abid MB. Could the menagerie of the gut microbiome really cure cancer? Hope or hype. Journal for Immunotherapy Cancer, 2019, 7(1): 92.

[171] Fan CA, Reader J, Roque DM. Review of Immune Therapies Targeting Ovarian Cancer. Current Treatment Options in Oncology, 2018, 19(12): 74.

[172] Kudo M. Systemic Therapy for Hepatocellular Carcinoma: Latest Advances. Cancers, 2018, 10(11): 412.

[173] Appelbe OK, Moynihan KD, Flor A, et al. Radiation-enhanced delivery of systemically administered amphiphilic-CpG oligodeoxynucleotide. Journal of Controlled Release, 2017, 266: 248-255.

[174] Snyder A, Pamer E, Wolchok J. IMMUNOTHERAPY. Could microbial therapy boost cancer immunotherapy? Science, 2015, 350(6264): 1031-1032.

[175] Jobin C. Precision medicine using microbiota. Science, 2018, 359(6371): 32-34.

[176] Kaiser J. Gut microbes shape response to cancer immunotherapy. Science, 2017, 358(6363): 573.

[177] Matson V, Fessler J, Bao R, et al. The commensal microbiome is associated with anti-PD-1 efficacy in metastatic melanoma patients. Science, 2018, 359(6371): 104-108.

[178] Routy B, Le Chatelier E, Derosa L, et al. Gut microbiome influences efficacy of PD-1-based immunotherapy against epithelial tumors. Science, 2018, 359(6371): 91-97.

[179] Gopalakrishnan V, Spencer CN, Nezi L, et al. Gut microbiome modulates response to anti-PD-1 immunotherapy in melanoma patients. Science, 2018, 359(6371): 97-103.

[180] Pushalkar S, Hundeyin M, Daley D, et al. The Pancreatic Cancer Microbiome Promotes Oncogenesis by Induction of Innate and Adaptive Immune Suppression. Cancer Discovery, 2018, 8(4): 403-416.

[181] Riquelme E, Maitra A, McAllister F. Immunotherapy for Pancreatic Cancer: More Than Just a Gut Feeling. Cancer Discovery, 2018, 8(4): 386-388.

[182] Vetizou M, Pitt JM, Daillere R, et al. Anticancer immunotherapy by CTLA-4 blockade relies on the gut microbiota. Science, 2015, 350(6264): 1079-1084.

[183] Iida N, Dzutsev A, Stewart CA, et al. Commensal bacteria control cancer response to therapy by modulating the tumor microenvironment. Science, 2013, 342(6161): 967-970.

[184] Paulos CM, Wrzesinski C, Kaiser A, et al. Microbial translocation augments the function of adoptively transferred self/tumor-specific $CD8^+$ T cells via TLR4 signaling. Journal of Clinical Investigation, 2007, 117(8): 2197-2204.

[185] Biton J, Ouakrim H, Dechartres A, et al. Impaired Tumor-Infiltrating T Cells in Patients with Chronic Obstructive Pulmonary Disease Impact Lung Cancer Response to PD-1 Blockade. American Journal

of Respiratory and Critical Care Medicine, 2018, 198(7): 928-940.

[186] Sacher AG, Gandhi L. Biomarkers for the Clinical Use of PD-1/PD-L1 Inhibitors in Non-Small-Cell Lung Cancer: A Review. JAMA Oncology, 2016, 2(9): 1217-1222.

[187] Belkaid Y, Naik S. Compartmentalized and systemic control of tissue immunity by commensals. Nature Immunology, 2013, 14(7): 646-653.

[188] Yan C, Tu XX, Wu W, et al. Antibiotics and immunotherapy in gastrointestinal tumors: Friend or foe? World Journal of Clinical Cases, 2019, 7(11): 1253-1261.

[189] Dai Z, Coker OO, Nakatsu G, et al. Multi-cohort analysis of colorectal cancer metagenome identified altered bacteria across populations and universal bacterial markers. Microbiome, 2018, 6(1): 70.

[190] Ahn J, Sinha R, Pei Z, et al. Human gut microbiome and risk for colorectal cancer. Journal of the National Cancer Institute, 2013, 105(24): 1907-1911.

[191] Zackular JP, Rogers MA, Ruffin MT, et al. The human gut microbiome as a screening tool for colorectal cancer. Cancer Prevention Research (Philadelphia, Pa), 2014, 7(11): 1112-1121.

[192] Galloway-Pena JR, Smith DP, Sahasrabhojane P, et al. The role of the gastrointestinal microbiome in infectious complications during induction chemotherapy for acute myeloid leukemia. Cancer, 2016, 122(14): 2186-2196.

[193] 孟珅，朱婉琦，邢力刚．人体肠道菌群对肿瘤发生和治疗效果的影响．肿瘤研究与临床．2018, 30(07): 493-497.

[194] Dubin K, Callahan MK, Ren B, et al. Intestinal microbiome analyses identify melanoma patients at risk for checkpoint-blockade-induced colitis. Nature Communications, 2016, 7: 10391.

[195] Chaput N, Lepage P, Coutzac C, et al. Baseline gut microbiota predicts clinical response and colitis in metastatic melanoma patients treated with ipilimumab. Annals of Oncology, 2017, 28(6): 1368-1379.

[196] Hodi FS, Chesney J, Pavlick AC, et al. Combined nivolumab and ipilimumab versus ipilimumab alone in patients with advanced melanoma: 2-year overall survival outcomes in a multicentre, randomised, controlled, phase 2 trial. Lancet Oncology, 2016, 17(11): 1558-1568.

[197] Frankel AE, Coughlin LA, Kim J, et al. Metagenomic Shotgun Sequencing and Unbiased Metabolomic Profiling Identify Specific Human Gut Microbiota and Metabolites Associated with Immune Checkpoint Therapy Efficacy in Melanoma Patients. Neoplasia, 2017, 19(10): 848-855.

[198] Anker JF, Naseem AF, Mok H, et al. Multi-faceted immunomodulatory and tissue-tropic clinical bacterial isolate potentiates prostate cancer immunotherapy. Nature Communications, 2018, 9(1): 1591.

[199] Wang F, Yin Q, Chen L, et al. Bifidobacterium can mitigate intestinal immunopathology in the context of CTLA-4 blockade. Proceedings of the National Academy of Sciences of the United States of America, 2018, 115(1): 157-161.

[200] Adachi K, Tamada K. Microbial biomarkers for immune checkpoint blockade therapy against cancer. Journal of Gastroenterology, 2018, 53(9): 999-1005.

[201] Buder-Bakhaya K, Hassel JC. Biomarkers for Clinical Benefit of Immune Checkpoint Inhibitor Treatment-A Review From the Melanoma Perspective and Beyond. Frontiers in Immunology, 2018, 9: 1474.

[202] Galluzzi L, Chan TA, Kroemer G, et al. The hallmarks of successful anticancer immunotherapy. Science Translational Medicine, 2018, 10(459):eaat 7807.

[203] Tanoue T, Morita S, Plichta DR, et al. A defined commensal consortium elicits CD8 T cells and anti-cancer immunity. Nature, 2019, 565(7741): 600-605.

[204] Ma C, Han M, Heinrich B, et al. Gut microbiome-mediated bile acid metabolism regulates liver cancer via NKT cells. Science, 2018, 360(6391): eaan 5931.

[205] Hartmann N, Kronenberg M. Cancer immunity thwarted by the microbiome. Science, 2018, 360(6391): 858-859.

[206] Pamer EG. Resurrecting the intestinal microbiota to combat antibiotic-resistant pathogens. Science, 2016, 352(6285): 535-538.

[207] Lenoir M, Del Carmen S, Cortes-Perez NG, et al. Lactobacillus casei BL23 regulates Treg and Th17 T-cell populations and reduces DMH-associated colorectal cancer. Journal of Gastroenterology, 2016, 51(9): 862-873.

[208] Zhang M, Fan X, Fang B, et al. Effects of Lactobacillus salivarius Ren on cancer prevention and intestinal microbiota in 1, 2-dimethylhydrazine-induced rat model. Journal of Microbiology, 2015, 53(6): 398-405.

[209] Pandey KR, Naik SR, Vakil BV. Probiotics, prebiotics and synbiotics- a review. Journal of Food Science and Technology, 2015, 52(12): 7577-7587.

[210] Song M, Chan AT. The Potential Role of Exercise and Nutrition in Harnessing the Immune System to Improve Colorectal Cancer Survival. Gastroenterology, 2018, 155(3): 596-600.

[211] Schroeder BO, Birchenough GMH, Stahlman M, et al. Bifidobacteria or Fiber Protects against Diet-Induced Microbiota-Mediated Colonic Mucus Deterioration. Cell Host & Microbe, 2018, 23(1): 27-40. e27.

[212] Mehta RS, Nishihara R, Cao Y, et al. Association of Dietary Patterns With Risk of Colorectal Cancer Subtypes Classified by Fusobacterium nucleatum in Tumor Tissue. JAMA Oncology, 2017, 3(7): 921-927.

[213] Sebastian C, Mostoslavsky R. Untangling the fiber yarn: butyrate feeds Warburg to suppress colorectal cancer. Cancer Discovery, 2014, 4(12): 1368-1370.

[214] Singh V, Yeoh BS, Chassaing B, et al. Dysregulated Microbial Fermentation of Soluble Fiber Induces Cholestatic Liver Cancer. Cell, 2018, 175(3): 679-694. e622.

[215] Panebianco C, Adamberg K, Adamberg S, et al. Engineered Resistant-Starch (ERS) Diet Shapes Colon Microbiota Profile in Parallel with the Retardation of Tumor Growth in In Vitro and In Vivo Pancreatic Cancer Models. Nutrients, 2017, 9(4):331.

[216] Bruno-Barcena JM, Azcarate-Peril MA. Galacto-oligosaccharides and Colorectal Cancer: Feeding our Intestinal Probiome. Journal of Functional Foods, 2015, 12: 92-108.

[217] Song M, Chan AT. Diet, Gut Microbiota, and Colorectal Cancer Prevention: A Review of Potential Mechanisms and Promising Targets for Future Research. Current Colorectal Cancer Reports, 2017, 13(6): 429-439.

[218] Song M, Zhang X, Meyerhardt JA, et al. Marine omega-3 polyunsaturated fatty acid intake and survival after colorectal cancer diagnosis. Gut, 2017, 66(10): 1790-1796.

[219] Barton W, Penney NC, Cronin O, et al. The microbiome of professional athletes differs from that of more sedentary subjects in composition and particularly at the functional metabolic level. Gut, 2018, 67(4): 625-633.

[220] Cammarota G, Ianiro G, Tilg H, et al. European consensus conference on faecal microbiota transplantation in clinical practice. Gut, 2017, 66(4): 569-580.

[221] Trubiano JA, Cheng AC, Korman TM, et al. Australasian Society of Infectious Diseases updated guidelines for the management of Clostridium difficile infection in adults and children in Australia and New Zealand. Internal Medicine Journal, 2016, 46(4): 479-493.

[222] Quraishi MN, Widlak M, Bhala N, et al. Systematic review with meta-analysis: the efficacy of faecal microbiota transplantation for the treatment of recurrent and refractory Clostridium difficile infection. Alimentary Pharmacology and Therapeutics, 2017, 46(5): 479-493.

[223] Costello SP, Soo W, Bryant RV, et al. Systematic review with meta-analysis: faecal microbiota transplantation for the induction of remission for active ulcerative colitis. Alimentary Pharmacology and Therapeutics, 2017, 46(3): 213-224.

[224] Wardill HR, Secombe KR, Bryant RV, et al. Adjunctive fecal microbiota transplantation in supportive oncology: Emerging indications and considerations in immunocompromised patients. EBioMedicine, 2019, 44: 730-740.

[225] Rosshart P, Vassallo BG, Angeletti D, et al. Wild Mouse Gut Microbiota Promotes Host Fitness and Improves Disease Resistance. Cell, 2017, 171(5): 1015-1028. e1013.

[226] Mullard A. Oncologists tap the microbiome in bid to improve immunotherapy outcomes. Nature Reviews: Drug Discovery, 2018, 17(3): 153-155.

[227] Baruch EN, Youngster I, Ben-Betzalel G, et al. Fecal microbiota transplant promotes response in immunotherapy-refractory melanoma patients. Science, 2021, 371(6529): 602-609.

[228] Davar D, Dzutsev AK, McCulloch JA, et al. Fecal microbiota transplant overcomes resistance to anti-PD-1 therapy in melanoma patients. Science, 2021, 371(6529): 595-602.

[229] Li W, Deng Y, Chu Q, et al. Gut microbiome and cancer immunotherapy. Cancer Letters, 2019, 447: 41-47.

第 8 章
肠道菌群与免疫性疾病

肠道菌群具有复杂的蛋白质编码基因，除参与消化、合成维生素、抵抗病原菌定植等作用外，也具有潜在的致病性。研究显示，当肠道微环境发生改变，正常的肠道菌群构成发生改变，细菌及其代谢产物易位，可导致肠道免疫及全身免疫失衡，从而与多种免疫相关性疾病的发生发展相关，例如自身免疫性疾病与过敏性疾病。

自身免疫性肝病（autoimmune liver disease，AILD）是由持续、过度的自身免疫反应导致的以肝组织损伤、肝功能异常为特征的慢性进展性炎症性肝脏疾病，主要包括自身免疫性肝炎（autoimmune hepatitis，AIH）、原发性胆汁性胆管炎（primary biliary cholangitis，PBC）、原发性硬化性胆管炎（primary sclerosing cholangitis，PSC）、IgG4 相关性胆管炎及重叠综合征[1]。肠道菌群可参与调节 AILD 发病机制中的多个环节，主要涉及免疫失衡、代谢紊乱、炎性细胞因子激增、肠黏膜屏障破坏和肠道菌群组成及数量异常等方面。

炎性肠病（inflammatory bowel disease，IBD）包括溃疡性结肠炎（ulcerative colitis，UC）和克罗恩病（Crohn's disease，CD），是一类慢性、复发性肠道炎症性疾病。肠道微生物在营养、免疫系统和机体防御方面发挥重要的生理作用。近年研究发现，IBD 患者存在肠道微生物组成和功能的改变，即菌群失调。临床和相关基础研究均提示，菌群失调在 IBD 的病理生理中发挥着关键作用[2]。

此外，肠道菌群失衡在日益高发的过敏性疾病（主要包括过敏性哮喘、过敏性紫癜、过敏性湿疹等变态反应性疾病）[3]、类风湿关节炎[4]、系统性红斑狼疮[5]等多种免疫性疾病中的作用受到了众多学者的广泛关注，成为当下重点研究方向之一。本章将对肠道菌群在免疫性疾病中的最新研究进展进行总结分析，以期为后续寻求新的治疗方法、改善患者预后奠定基础。

第一节　肠道菌群与自身免疫性肝病

一、肠道菌群与自身免疫性肝炎

AIH 是一种由免疫介导的慢性炎症性肝脏疾病，其特征是肝细胞破坏、循环系统中自身抗体及血清 IgG 水平升高。AIH 在儿童及成年女性中发病率较高，且近年来呈现上升趋势[6]。新近研究发现，AIH 患者肠道菌群失调与疾病严重程度密切相关[7]，提示肠道菌群的研究将有助于提高对AIH发病机制的认识，为合理有效的个体化治疗提供新的理论依据。

（一）自身免疫性肝炎患者肠道菌群特征

近年研究发现，AIH患者和健康志愿者肠道菌群在门水平上并无显著差异，其中最主要的菌群为拟杆菌门和硬壁菌门。然而，在属水平上，AIH患者与健康人群的肠道菌群存在较大差异。AIH患者的肠道菌群特征主要表现为韦荣球菌属、克雷伯菌属、链球菌属和乳酸杆菌属显著增多，而梭菌属、瘤胃球菌属、理研菌属、螺杆菌属、狄氏副拟杆菌属和粪球菌属减少，其中韦荣球菌属的增多与AIH密切相关[7]。

另外，肠道菌群的构成还与AIH疾病活动度密切相关。在天冬氨酸转氨酶（aspartate aminotransferase，AST）升高或晚期活动性AIH患者中，肠道菌群中的韦荣球菌显著增多。韦荣球菌能分泌脂多糖（lipopolysaccharide，LPS），可解释AIH患者肠道中LPS的生物合成水平更高的原因。

（二）自身免疫性肝炎发病机制中肠道菌群的调控作用

AIH的病因尚不清楚，目前认为主要受遗传及环境因素两方面的影响。由外源或易位的病原体等因素引起的、以肝脏抗原为靶点的免疫耐受缺失被认为是AIH的始动因素。越来越多的证据表明，肠道菌群所携带的基因远多于人类基因组，肠道菌群可通过“肠-肝轴”在AIH的发生发展中发挥重要的调控作用。近期在小鼠模型上的研究结果已为肠道菌群参与AIH的发病机制提供了初步证据[5]。

肠道菌群失调时，微生物相关分子模式（microbe-associated molecular patterns，MAMPs）激活肠内Toll样受体（toll-like receptors，TLRs）。活化的TLRs可激活巨噬细胞中的核因子κB（nuclear factor-κB，NF-κB）信号通路，使巨噬细胞产生促炎性细胞因子[8]。IL-1β、IL-18等促炎症细胞因子可以增加抗原呈递细胞上主要组织相容性复合物的表达，并使$CD4^+$ T淋巴细胞识别细菌配体。另外，肠道菌群失调还可以产生内毒素、LPS和细菌成分，它们可以作为抗原引起过度的免疫应答反应。此时，肠上皮间的紧密连接受损，肠道中活化的淋巴细胞、细菌配体和内毒素等发生移位而进入血液。当这些肠源性成分通过门静脉到达肝脏，细菌配体在肝脏内可激活肝细胞、肝星状细胞、Kupffer细胞和窦状上皮细胞的TLRs，并产生促炎症细胞因子和活性氧。这些细胞因子和活性氧可激活损伤相关分子模式（damage-associated molecular patterns，DAMPs），并进一步发生正反馈，放大激活自身环路中的TLRs（图8-1右上）[9]。肝脏内的TLRs也可以促进$CD4^+$T淋巴细胞对细菌配体或细菌配体类似物进行自身抗原的结合致敏。另外，细菌配体和肠源性内毒素可以激活肝细胞和肝星状细胞内炎症小体的非肥胖型糖尿病样受体（non-obese diabetes-like receptor，NLR）；Caspase 1的释放可产生IL-1β和IL-18等促炎症细胞因子，进而促进肝组织损伤和过激的免疫应答[10]（图8-1）。综上，菌群失调总效应为增加肝脏炎症和肝损伤，并导致自身免疫反应和肝纤维化。

（三）调节肠道菌群在自身免疫性肝炎治疗中的应用

皮质类固醇及免疫抑制剂可用于治疗AIH，它们对AIH患者肠道菌群的影响目前研究报道相对较少。研究显示，糖皮质激素治疗可以引起肠道菌群的改变。给无菌雄性C57BL/6、Muc2杂合子（+/-）或Muc2基因敲除（-/-）小鼠注射地塞米松4周后，发现肠道菌群中梭菌属和乳酸杆菌显著下降[11]。同时，在小鼠的研究中还显示，应用免疫抑制剂会导致大肠埃希菌过度生长，引起肠道菌群失调。

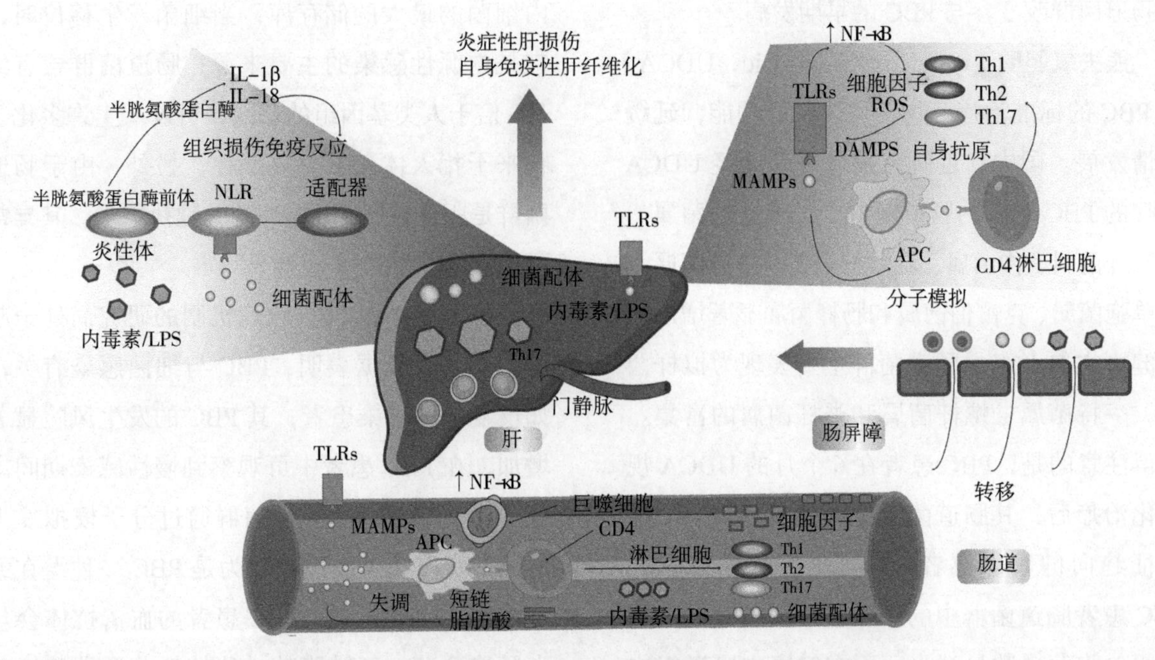

图 8-1　肠道菌群失调对自身免疫性肝炎发生发展的影响

肠道菌群失调产生病原相关分子模式，激活肠道内 Toll 样受体，进而刺激巨噬细胞分泌促炎细胞因子。它们还可以增加抗原呈递细胞上主要组织相容性复合体的表达，使 CD4 淋巴细胞增敏。肠道菌群失调还可产生短链脂肪酸、内毒素、脂多糖和细菌成分等，其可发生易位，进入门静脉并被输送到肝脏，进而激活肝细胞、肝星状细胞等，并产生促炎症细胞因子和活性氧，这些细胞可以产生损伤相关分子模式，在自放大环路中激活 Toll 样受体（右上角放大图）；同时，细菌配体和肠源性内毒素可以激活肝细胞和肝星状细胞内炎症小体的非肥胖糖尿病样受体（左上角放大图）。最终的效应是增加肝脏炎症和肝损伤，易发生自身免疫性炎症和肝纤维化。APC. 抗原呈递细胞；DAMPS. 损伤相关分子模式；ROS. 活性氧；LPS. 脂多糖；MAMPs. 微生物相关分子模式；NF-κ B. 核因子 κ B；NLR. 非肥胖性糖尿病样病变受体；TLRs.Toll 样受体

当肠道菌群失调时，可通过调整饮食、补充益生菌、维生素 A、维 A 酸和使用抗生素等调节肠道微生态，同时可通过调节肠道屏障、降低肠道通透性的药物控制病情。另外，还可通过阻断 TLR 信号和抑制促炎细胞因子的产生等手段干预、调节 AIH 患者的肠道菌群。明确肠道菌群在 AIH 中的作用具有重要意义，有助于制订相应的干预措施，以期改善 AIH 患者预后[12]。

二、肠道菌群与原发性胆汁性胆管炎

PBC 是肝内中、小胆管慢性进行性非化脓性炎症，小胆管进行性减少，进而发生肝内胆汁淤积、肝脏纤维化，最终导致肝硬化。其病因和发病机制尚存盲点，可能与遗传背景、环境因素相互作用导致异常的自身免疫反应相关。该病多见于中老年女性，初期无明显症状，病程进展缓慢，其中乏力、瘙痒是其最常见的症状，随着疾病进展会合并其他自身免疫性疾病[13]。碱性磷酸酶的持续升高是该病最突出的生化异常表现，此外，超过 90% 的 PBC 患者血清中抗线粒体抗体呈阳性，约 50% 患者抗核抗体呈阳性。肠道菌群在 PBC 的病理生理过程中也发挥了重要作用。

（一）原发性胆汁性胆管炎患者肠道菌群谱特征

近期，通过对 42 例早期 PBC 患者的肠道菌群 16S rRNA 基因组测序分析发现，患者肠道中有益菌，如乳酸杆菌、产酸拟杆菌、埃氏拟杆菌和布氏瘤胃球菌数量减少，而一些机会致病菌显著增加，包括肠杆菌科、链球菌、奈瑟球菌、γ 变形杆菌、幽门螺杆菌、韦荣球菌、克雷伯菌、阿氏肠杆菌和副流感嗜血杆菌等，提

示肠道菌群改变参与 PBC 的早期发病。

熊去氧胆酸（ursodeoxycholic acid，UDCA）是 PBC 的标准治疗药物，可改善肝功能，延缓病情发展。国内新近研究发现[14]，未经 UDCA 治疗的 PBC 患者的肠道菌群中，嗜血杆菌属、细脉杆菌属、梭菌属、乳酸杆菌属、链球菌属、假单胞菌属、克雷伯菌属和肠杆菌属显著增加，而健康志愿者组的肠道菌群主要表现为拟杆菌属、萨特菌属、螺杆菌属和粪杆菌属的富集。值得注意的是，PBC 患者在 6 个月的 UDCA 规范化治疗后，其肠道菌群发生改变，其菌群谱特征趋向健康志愿者。即经 UDCA 治疗后，PBC 患者肠道菌群中的嗜血杆菌属、链球菌属和假单胞菌属数量减少，而在健康志愿者组中富集的 4 个菌属中的 3 个菌属（包括拟杆菌属、萨特菌属、螺杆菌属）显著升高。在治疗后的 PBC 患者中，韦荣球菌属的数量并未下降至健康志愿者水平，但对 UDCA 应答较差的 PBC 患者与预后较好的患者相比，其韦荣球菌属的数量显著增高。因此，UDCA 治疗可部分改善 PBC 患者肠道菌群失调，并有助于评价治疗效果。

（二）原发性胆汁性胆管炎发病机制中肠道菌群的调控作用

PBC 是一种慢性胆汁淤积性肝病，目前认为 PBC 是一种自身免疫性疾病，发病主要受个体遗传及环境因素的触发。近期研究发现，肠道菌群失调可能诱发或加重 PBC。肠道作为体内细菌的最大的储存库，当细菌发生移位时，它是内源性感染的主要来源；肠道菌群含有约 100 倍于人类基因组的基因，可以产生许多化合物来干扰人体正常免疫调控；另外，由于肠道菌群是胆汁酸肠 - 肝循环的关键执行者，其异常可能有助于 PBC 的发生[15, 16]。

肠道菌群在 PBC 致病机制的研究尚处于起步阶段。有证据表明，PBC 与细菌感染有关，如反复尿路感染患者，其 PBC 的发生风险显著增加。在 PBC 患者中可观察到慢性感染期间出现的多克隆 IgM，肠道菌群通过分子模拟发生的异常免疫交叉反应被认为是 PBC 一种潜在致病机制。研究发现，PBC 患者的血清抗体会与大肠埃希菌、新鞘脂菌、分枝杆菌及乳酸杆菌来源的保守的丙酮酸脱氢酶复合物 E2 同系物发生交叉反应，提示肠道菌群失调可能在遗传易感宿主中启动 PBC 的分子模拟和疾病进展[17]。

此外，PBC 可通过引起肠道运动障碍、免疫紊乱、胆汁分泌缺陷和门静脉高压而引起肠道菌群失调[18]，当肠道菌群的稳态受到破坏后，上述病因所导致的恶性循环均可影响 PBC 患者的预后。

（三）肠道菌群在原发性胆汁性胆管炎研究中的混淆因素分析

在研究肠道菌群与 PBC 发生发展的相互关系中，必须面对某些干扰因素（表 8-1）。只有考虑并排除这些混杂因素，才能进一步准确地把握肠道菌群在 PBC 致病中发挥的作用。

表 8-1　肠道菌群在 PBC 相关研究中的混淆因素[19]

混淆因素
1. 饮食：是影响肠道菌群组分的主要决定因素之一，其在任何慢性疾病中都可能影响菌群的构成。目前对 PBC 患者是否有必要进行饮食调整尚不清楚
2. 药物：① PBC 患者若长期使用抗生素等，会对肠道菌群产生长远的影响
②由于不同菌株对胆汁酸的敏感性及耐受性存在差异，PBC 患者在使用熊去氧胆酸过程中也会影响到肠道菌群的组成
③其他常用的药物，如质子泵抑制剂也会对肠道菌群有所影响
3. 检测技术：目前对 PBC 患者肠道菌群的检测研究大多是基于单一时间点的粪便微生物组学评估。该方法存在一定缺陷，无法有针对性的呈现那些与 PBC 免疫反应更相关的菌群

另外，PBC 患者的并发症与其肠道菌群也存在相互影响。当 PBC 患者肠道生理发生改变时，易导致其肠道机会致病菌等过度生长，近年得益于高通量测序技术，对 PBC 患者肠道菌群进行了更详细的研究。尽管菌群偏离正常值的确切意义尚待确定，但肠道菌群组分的改变与 PBC 的严重程度和并发症（包括肝性脑病、自发性细菌性腹膜炎和菌血症等）的发生密切相关[20]。虽然 PBC 患者肝癌的发生率不高，但在肠道菌群失调的 PBC 患者中，其肝癌的发生率相对增高。在腹腔感染、炎性肠病和类风湿关节炎等疾病中亦可伴肠道菌群失调，此现象在 PBC 患者中更为常见[21]。关于 PBC 患者肠道菌群的研究，需要在更大的人群中进行探索，以提高结论的客观性和准确性。

三、肠道菌群与原发性硬化性胆管炎

PSC 是一种慢性进展性胆汁淤积性肝病，主要特征为胆管弥漫性炎症和纤维化，进而导致了肝内、外胆管多灶性狭窄。PSC 起病隐匿，患者早期无明显症状，若不及时有效干预，部分患者会出现进行性胆道梗阻及炎症反应，进而导致肝硬化、肝衰竭。迄今为止，PSC 的病因发病机制尚不明确，目前研究认为，肠道菌群的紊乱是参与调控该病发生发展的重要因素之一[22, 23]。

（一）原发性硬化性胆管炎肠道菌群的横断面研究概况

在关于 PSC 患者肠道菌群的首次研究报道中，Rossen 等分别从 PSC 患者（12 例）、UC 患者（11 例）和非炎症健康对照者（9 例）的结肠及回肠活检标本中提取细菌 DNA，利用 16S rRNA 基因进行 PCR 扩增，然后与包括了 5000 个以上的探针、靶向 1000 种以上细菌的人肠道芯片进行杂交，发现与 UC 组和非炎症健康对照组相比，PSC 患者肠道菌群的多样性和丰富性降低，伴Ⅱ型梭状芽孢杆菌减少。另外，根据 Torres 等对 20 例 PSC 患者、15 例 IBD 患者和 9 例健康对照者肠道菌群的 16S rRNA 基因测序发现，在菌门水平上观察到肠道菌群的整体差异不显著，但在菌属水平上，PSC 患者布劳特菌属丰度增加[24]。亦有研究显示，在 UC 患者中，无论其是否伴发 PSC，肠道菌群特征均无显著改变。以上研究结论不一致性的主要原因可能与取样差异、不同检测方法及统计方法的缺陷等有关。

一项涉及 85 名 PSC 患者、36 名 UC 患者与 263 名健康对照者肠道菌群的研究发现，PSC 患者肠道菌群多样性降低，相对于健康对照组及 UC 患者，PSC 患者韦荣球菌属的数量显著增多[25]。与此结果相似，另一项包括 66 名 PSC 患者、43 名 IBD 患者及 66 名健康对照者的研究发现，PSC 患者肠道菌群多样性降低，与 IBD 组和健康对照组相比，PSC 组患者肠球菌属、乳酸杆菌属和梭菌属均显著增加；另有报道对 PSC 患者肠道菌群的进一步研究中也观察到韦荣球菌属、链球菌属和肠球菌属的增多[26]。此外，虽然 PSC 患者肠道韦荣球菌属和链球菌属也会增加，但是当患者出现肝硬化或近期经抗生素治疗后，这两种菌属未增加。

总之，关于 PSC 患者肠道菌群的研究显示，与 IBD 和健康对照组相比，PSC 患者肠道菌群结构有显著的变化。更重要的是，PSC 患者肠道菌群组分的改变与 IBD 发病无关，这表明 PSC 与肠道菌群可能存在更为直接的联系。

（二）原发性硬化性胆管炎中肠道菌群的潜在致病机制

宿主不同的生理情况及其与肠道菌群之间的相互作用导致了 PSC 患者疾病进展及预后的多样性。目前认为，肠道菌群主要通过 3 个方面影响 PSC 的进程，即宿主免疫途径、内源途径和外源途径，PSC 的疾病发展过程可能也涉及其他尚未阐明的途径。

1. 宿主免疫途径　宿主免疫途径是指肠道菌群与患者宿主免疫系统之间的相互作用，主要通过保守的 TLR 或可变的 T 细胞受体等免疫受体发挥调控作用，进而影响 PSC 的病理生理反应。

近期研究发现，PSC 与 IBD 之间具有密切联系，肠道菌群可能参与 PSC 的进程。关于 PSC 发病机制的经典假设之一便是细菌及其产物从肠道易位并引起胆道炎症，该假设已在动物模型中得到了验证。此外，肠道细菌作为微生物抗原，与 PSC 的适应性免疫反应密切相关，其主要可能有 3 种方式导致 PSC 患者组织损伤：①直接引发免疫反应；②通过分子模拟，即机体免疫系统识别内源性多肽发生交叉免疫反应；③调节适应免疫反应的应答性或耐受性[27]。

在 26% ～ 94% 的 PSC 患者体内存在异常抗中性粒细胞胞质抗体（anti-neutrophil cytoplasmic antibody，ANCA），ANCA 可与细菌蛋白 FtsZ 发生交叉反应[28]。一项对 IBD 患者的前瞻性研究显示，诊断 IBD 10 年后的患者 pANCA 的阳性率与诊断 IBD 20 年后患者磁共振胰胆管造影出现 PSC 样病理改变的发生率呈正相关，提示 ANCA 与 PSC 相关[29]。但 ANCA 的增多是否由肠道菌群抗原引起的适应性免疫应答所致仍需进一步研究。

新近的研究发现，黏膜相关恒定 T 细胞（mucosal-associated invariant T cell，MAIT）和自然杀伤 T（natural killer T，NKT）细胞可能参与介导了 PSC 的胆道病理改变[30]，而且这些细胞主要富集于与细菌接触的部位，例如肠道。这些细胞的重要性在小鼠结肠炎模型中也得到了印证。

2. 内源途径　内源途径是指肠道菌群可通过自身合成关键的分子或与宿主产生共代谢分子调控宿主生理反应的途径，进而影响 PSC 患者疾病进程。

肠道菌群参与调控宿主代谢。肠道菌群在提供短链脂肪酸（丁酸、丙酸）、介导氨基酸修饰（色氨酸）、维持维生素稳态（维生素 K 生成）和参与胆汁酸转化等方面均发挥了重要作用。肠道菌群已成为一个极其重要的“代谢器官”，而且上述许多反应都与肝脏疾病的发生密切相关。

胆汁淤积是 PSC 的主要临床表现之一，其确切机制尚不清楚。研究发现，肠道菌群可参与调节胆汁酸代谢，在肠道菌群失调的 PSC 患者中，终末期肝脏胆汁酸的积聚更为显著。

由于缺乏足够有力的临床试验，UDCA 在 PSC 中的临床收益需要进一步评估。有报道发现，每天给予 28 ～ 30mg/kg UDCA 的 PSC 患者发生临床事件风险（尤其是与门静脉高压相关的事件）更高，该研究因此而停止[31]。此外，石胆酸是疏水性和毒性的，在小鼠模型中已被证明可引起胆管炎。服用高剂量 UDCA 的 PSC 患者石胆酸水平显著升高，可能是由于未被吸收的 UDCA 在结肠部位被肠道微生物转化所致[32]，由此推测，通过调节肠道菌群改善 PSC 患者胆汁酸代谢可能成为 PSC 治疗的新方向。

3. 外源途径　外源途径是指肠道菌群通过参与外源化合物（包括营养素、药物和暴露于某些环境因素）的生物转化，进而影响宿主 PSC 进程。

饮食是微生物群和宿主的重要营养来源，肠道菌群对基本营养物质，包括糖、脂肪、蛋白质等的代谢发挥了重要作用[33]。研究发现，长期红肉饮食可加速 PSC 进展。目前认为其潜在机制可能是肠道菌群和肠上皮细胞共同代谢产生半胱胺，而半胱胺是血管黏附蛋白 -1（vascular adhesion protein-1，VAP-1）活性最有效的诱导剂，循环中可溶性 VAP-1 水平与 PSC 严重程度相关。另外，红肉饮食会导致肠道菌群代谢产物——氧化三甲胺（trimethylamine-N-oxide，TMAO）增多，其与胆管细胞活化增生、纤维化有关，进而促进了 PSC 的进展[34]。以上研究中的 TMAO 和可溶性 VAP-1 为肠道菌群在 PSC 中的致病作用提供了间接证据，值得进一步研究。

另外，肠道菌群在药物及其他外源性化合物的生物转化、代谢方面也发挥着重要作用。当肠道菌群失调时，可能会导致药物失效甚至具有毒性，例如口服柳氮磺胺吡啶后，其代谢与肠道菌群有关，当菌群失调，其代谢产生的 5-氨基水杨酸几乎不被吸收，大部分以原型从粪便排出，导致用药失效[35]。在 PSC 中，以外源性的生物活性物质为底物，在异常的肠道菌群作用下，对疾病的进展会产生复杂的影响。随着我们对肠道菌群代谢机制认识的深入，进一步阐明其在 PSC 发病机制中的潜在作用将成为可能。

（三）肠道菌群与原发性硬化性胆管炎严重程度及预后的关系

PSC 患者肠道菌群的改变是复杂的，但具有一定特征，寻求其菌群特征与疾病严重程度之间的关系，为我们判断 PSC 患者临床预后提供了一定的帮助。在比利时的一项研究中，通过检测肠道肠球菌、乳酸杆菌和梭杆菌丰度的改变，可预测并区分 PSC 患者和对照组，其准确度达 71%～95%。

目前基于肠道菌群改变预测患者临床预后的报道相对较少，尚无定论，但在克罗恩病的长期研究中，已证实肠道菌群对该病的临床活动以及预后具有重要的提示作用[36]。PSC 肠道菌群分布的横断面研究表明，疾病严重程度与肠道菌群之间存在关联，如挪威的一项研究中，已在 PSC 的梅奥风险评分修订版中加入了肠道韦荣球菌属这一检测项目。

肠道某些生物活性分子可为肠道菌群与 PSC 疾病进展和预后提供直接证据。肉碱和胆碱经肠道细菌代谢生成三甲胺（trimethylamine，TMA），而在肝脏中 TMA 进一步代谢产生 TMAO，高水平的 TMAO 与 PSC 患者不良预后相关，提示肠道菌群可能影响 PSC 疾病进程[37]。在将来的研究中，有必要进一步增加 PSC 患者的前瞻性研究，评估肠道菌群对其病程进展的影响，以便能够更好地理解肠道菌群与 PSC 预后的关系。

第二节　肠道菌群与炎性肠病

一、概述

炎性肠病包括溃疡性结肠炎和克罗恩病，是一种慢性、复发性肠道炎症性疾病，其病因及发病机制尚未完全阐明。普遍认为，IBD 是遗传、环境和机体免疫系统之间复杂的相互作用而导致的异常免疫反应和慢性肠道炎症。人体肠道存在复杂而多样的微生物，统称为肠道微生物。肠道菌群可产生丰富的健康效应，主要与病原防御、营养、代谢和免疫反应有关。肠道微生物与人类共同进化，人体宿主与微生物之间的共生作用对维持人体健康是不可或缺的。菌群失调，即肠道细菌组成和功能的变化，会改变宿主 - 细菌之间的相互作用和宿主的免疫系统。研究证实，菌群失调与人体多种疾病密切相关，如 IBD、肠易激综合征、过敏、哮喘、代谢综合征和心血管疾病等。业已证实，IBD 患者肠道细菌的组成发生了明显的改变[2]，然而，菌群失调与 IBD 之间是否存在因果关系，目前仍无定论。

二、肠道微生物的生理作用

肠道微生物对宿主的生理作用大致分为 3

类：营养、免疫发生和宿主防御。

（一）肠道微生物与营养

肠道微生物可以为宿主提供能量和营养素。人体的共生菌，如双歧杆菌，可以合成和提供维生素K和水溶性维生素B。肠道细菌还可以通过发酵残留的淀粉或碳水化合物而提供短链脂肪酸（short-chain fatty acid，SCFA）。如厚壁菌门和拟杆菌门细菌通过与发酵寡糖的双歧杆菌共同利用碳水化合物而合成SCFAs。SCFAs是结肠内主要的阴离子，其主要形式为乙酸盐、丁酸盐和丙酸盐，其中丁酸盐是肠道上皮细胞最主要的能量来源。研究发现，IBD患者结肠内SCFAs水平显著下降，这是肠道免疫平衡发生紊乱的重要因素[38]。

（二）肠道微生物与免疫发生

肠道微生物是宿主免疫系统发育的基石，同时，宿主的免疫系统会塑造肠道微生物的组成和功能。无菌小鼠的免疫系统无法正常发育，主要表现为不成熟的淋巴组织、肠道淋巴细胞数量减少、抗菌肽和IgA水平的降低[39]，而重建无菌小鼠的肠道微生物后，则能够改善上述免疫缺陷，其中发挥特异性作用的细菌为分节丝状菌（*segmented filamentous bacteria*，SFB）。SFB可以促进黏膜淋巴系统的成熟。宿主免疫系统的成熟依赖于宿主特异性的微生物，因此，无菌小鼠定植人类的细菌也会导致免疫系统的发育不充分。

（三）肠道微生物与宿主防御

肠道微生物有助于宿主抵御病原体的感染。无菌动物模型对肠道病原体感染明显易感，可能与肠黏膜免疫系统的异常有关。机体另一个病原体的防御机制是共生细菌对外来微生物在肠道的定植或增殖的抑制，被称为竞争排斥或定植抗力，这种定植抗力可表现为直接或间接的方式。部分菌群可通过竞争营养或者产生抑制性物质而直接抑制肠道病原菌感染，如肠道的厌氧菌-多形拟杆菌能够消耗大量的碳水化合物而竞争性抑制艰难梭菌的感染。共生菌群还可以通过激活免疫反应而间接保护肠道免受病原体感染，如肠道菌群的脂多糖和鞭毛蛋白可以刺激$TLR4^+$间质细胞和$TLR5^+CD103^+$树突状细胞，进而促进上皮细胞表达抗菌肽和RegⅢγ[40]。分节丝状菌可以促进B细胞分泌IgA，促进肠黏膜抗菌肽的产生和Th17细胞的发育。

三、肠道微生物对炎性肠病的影响

研究已证实，肠道菌群失调与IBD密切相关。过去10年间，深度测序技术的大规模应用逐步揭示了微生物在IBD发生中的作用，也进一步加深了我们对微生物在IBD发病机制中的认识。肠道微生物包括细菌、真菌、病毒和其他微生物，作为一个次级器官系统，对宿主发挥着关键作用。IBD是目前研究最为广泛的，也是与肠道微生物关系最为密切的炎症性疾病。

（一）细菌

细菌是研究最为充分的肠道微生物。肠道细菌在IBD患者的肠道炎症过程中扮演重要的角色。研究发现，IBD患者存在较为广泛的细菌变化，包括多样性的降低、厚壁菌门和拟杆菌门丰度的下降以及变形菌门丰度的增加[41]。与健康人相比，CD患者粪便微生物的α多样性显著降低[42]。即使同一个CD患者，肠道炎症部位细菌的多样性和载量是降低的[43]。一项多中心研究纳入超过1000例初始治疗的儿童CD患者，其肠道菌群中韦荣球菌科、肠杆菌科和梭杆菌科明显增加，而拟杆菌目、丹毒丝菌目和梭菌目显著减少，且与疾病状态显著相关[36]。同时，该研究还对直肠黏膜相关的微生物进行测序分析，发现其可以作为诊断CD的生物学标志物。肠杆菌科在IBD患者和小鼠模型中均明显增加[36]。在CD和UC患者肠黏膜组织中，可以分离到大肠埃希菌，特别是黏附侵袭性大肠埃希菌（adherent-invasive Escherichia coli，AIEC），而且黏膜样本较粪便样本的改变更加明显，提示 IBD 患者的肠道炎症环境有助于肠杆菌科细菌的生长，抗炎药物美沙拉嗪可通过减轻肠道炎症而降低 IBD 患者大肠埃希菌 / 志贺杆菌的丰度。

梭杆菌门是另一种具有黏附和侵袭能力的细菌。梭杆菌主要定植在口腔和肠道，UC 患者结肠黏膜中梭杆菌明显增加[44]。在小鼠模型中，以梭杆菌灌肠也可导致结肠黏膜炎症反应[45]。梭杆菌的侵袭能力与 IBD 患者的疾病严重程度呈正相关，提示具有侵袭能力的梭杆菌能够促进 IBD 的病理变化。同时，肠道也存在具有保护作用的细菌，如乳杆菌、双歧杆菌和粪杆菌等，这些细菌可促进抗炎细胞因子 IL-10 和抑制促炎因子的产生而发挥保护性作用。在 CD 患者的回肠黏膜中，粪杆菌的数量显著下降，而大肠埃希菌的数量明显增加，且具有低丰度粪杆菌的 CD 患者术后更易复发[46]。反之，在 UC 患者中恢复粪杆菌的丰度，可以促进临床缓解[47]。在化学药物诱导的结肠炎小鼠中，无菌小鼠的结肠炎表型更加严重[48]。所以，IBD 患者存在广泛的细菌变化，不仅表现为多样性的降低，还表现为益生菌数量和丰度的下降及侵袭性细菌数量和丰度的增加。

（二）真菌

真菌仅占肠道微生物群的小部分，其所占比例＜0.1%。因没有完备的真菌基因组数据库，真菌的作用可能被低估。尽管如此，通过标记基因的靶向区域测序技术，如内转录间隔区和 18S rRNA，我们对肠道真菌微生物有了一定的认识。真菌种类组成在人体不同部位不尽一致。胃肠道、泌尿生殖道和口腔中存在大量的假丝酵母菌，共包含 160 余种。在哺乳动物体内，假丝酵母菌的定植存在种属特异性，白念珠菌和近平滑念珠菌最常定植于人体，而热带念珠菌主要定植在小鼠体内。与细菌不同的是，生活环境对真菌的影响较为明显。肠道内细菌和真菌存在竞争关系，长期服用抗生素可以促进人类和小鼠真菌的过度生长和感染[49]，而抗生素导致的肠道真菌过度生长会使机体对孢子虫更加易感，这些结果提示了肠道真菌在免疫性疾病中的作用[50]。

肠道真菌在 IBD 发病中发挥着重要作用。在健康人体内，酵母菌、念珠菌和枝孢菌是最主要的种属，而在 IBD 患者体内，担子菌、子囊菌和白念珠菌明显增加。真菌的细胞壁糖蛋白、壳质、β- 葡聚糖和甘露聚糖能够通过受体，包括 dectin-1、TLR4、补体系统和清道夫受体（CD5、SCARF1 和 CD36），触发固有免疫反应，进而激活下游免疫级联效应分子，如 CARD9、IL-17、IL-22、ITAM、NFAT 和 NF-κ B 等[51]。研究发现，真菌可通过破坏肠黏膜屏障，进而加重结肠炎小鼠的炎症反应[52]。在 IBD 患者中，肠黏膜炎症反应导致黏膜紧密连接破坏，出现肠道上皮细胞的完整性受损，真菌可以渗透到固有肌层并激活 TLRs、Dectin-1 和 CARD9，导致更加严重的炎症反应[53]。因此，IBD 患者存在真菌种属的变化，这些变化了的微生物通过破坏肠道黏膜屏障功能和激活下游炎症因子，进一步加重结肠的炎症反应。

（三）病毒

病毒微生物包含真核病毒和原核噬菌体。病毒在 IBD 病理过程中的作用仍未完全阐明，而噬菌体在其中发挥的重要作用已得到认可。研究发现，CD 患者肠道病毒微生物的多样性下降、变异性更高[54]。儿童 CD 患者的肠道组织和洗涤标本中噬菌体明显增加[55]，且电镜下 CD 肠道标本中可观察到噬菌体病毒颗粒的存在[56]。

噬菌体可能是通过与宿主直接相互作用而发挥作用。在啮齿动物、CD 患者和健康人中，噬菌体能够从胃肠道转位进入体循环。它们还可以诱发体液免疫反应。因此，噬菌体可以作为免疫配体或抗原而促进宿主的免疫和炎症反应。

某些病毒，如诺如病毒，可以在功能上替代共生细菌的保护作用，缓解无菌小鼠的肠道损伤，降低化学物和细菌感染导致的肠道炎症[57]。此外，黏附在黏膜上的病毒，可以保护上皮细胞免受细菌的感染。另有研究发现，肠道病毒会导致慢性肠道炎症反应[58]。当小鼠存在*ATG16L1*基因突变或诺如病毒感染时，可出现肠道CD表现，但无临床症状，而当二者同时存在时，则会导致显著的疾病进展和表型[59]。在化学药物诱导的结肠炎小鼠模型中，肠道病毒微生物发挥着重要的保护作用。给予抗病毒药物后，葡聚糖硫酸钠（dextran sulfate sodium，DSS）可以诱导小鼠严重的结肠炎症状，结肠长度和体重显著降低[59]。这些保护作用可能与TLR3和TLR7之间的协同作用有关，因为同时存在TLR3和TLR7突变的IBD患者，其住院率显著高于无突变的IBD患者[59]。所以，IBD患者肠道也存在病毒种类的变化，而这些肠道病毒微生物可能发挥着重要的黏膜保护和维持黏膜免疫平衡的作用。

（四）寄生虫

肠道寄生虫是肠道与细菌、病毒和真菌共存的重要微生物组成部分。如果儿童早期缺乏寄生虫感染，成年后则对免疫性疾病明显易感。在发达国家，自身免疫性疾病和其他免疫相关疾病的发病率较高，随着城市化和卫生条件的改善，IBD发病率的上升可能与宿主寄生虫感染的减少相关。

研究证实，鞭虫和多形螺旋线虫可通过抑制拟杆菌而保护NOD2-/-小鼠免于肠道炎症[60]，这种保护作用与寄生虫对宿主免疫调控作用有关，比如旋毛虫和血吸虫可促进Treg增殖，鞭虫可上调Th2反应，多形螺旋线虫可抑制Th17炎症反应[61]。寄生虫定植的个体，其肠道细菌的多样性明显高于无寄生虫定植的个体。寄生虫感染可促进肠道黏液和水分分泌而发挥抗炎作用。临床研究证实，猪鞭虫对CD和UC患者具有一定的治疗作用。所以，IBD的发生与缺乏寄生虫感染有关，寄生虫可通过增加肠道细菌多样性和提高免疫调控作用为IBD患者带来新的治疗手段。

四、微生物代谢产物与炎性肠病

（一）胆汁酸

胆汁酸是胆固醇在肝脏的代谢产物，属于小分子物质，其中胆酸（cholic acid，CA）和鹅脱氧胆酸（chenodeoxycholic acid，CDCA）可以与牛磺酸或甘氨酸发生耦合作用。初级胆汁酸具有亲水和亲脂两种特性，有助于小肠对脂肪的消化与吸收。在远端回肠，95%的初级胆汁酸被重吸收（肠-肝循环）。除调控自身合成外，胆汁酸还具代谢和免疫调控作用，主要是通过结合胆汁酸受体实现的，包括法尼醇X受体（farnesoid X Receptor，FXR）、胆汁酸G蛋白偶联受体5（takeda G protein-coupled receptor 5，TGR5）、孕烷X受体（pregnane X receptor，PXR）、维生素D受体（vitamin D receptor，VDR）和AR。TGR5能够促进胰岛素敏感性[通过刺激胰高血糖素样肽-1（glucagon-like peptide-1，GLP-1）]、肌肉和脂肪组织的能量消耗（通过活化Ⅱ型脱碘酶）以及胆囊松弛。TGR5还通过抑制NF-κB而抑制Kupffer细胞对LPS的反应，抑制外周血单核细胞释放IL-1、IL-6、TNF等细胞因子[62]。胆汁酸激活FXR后，对宿主代谢可产生多种效应，包括减少脂肪生成、降低肝脏糖原异生、促进肝脏再生和增加抗菌肽的合成。

胆汁酸与肠道微生物之间存在着相互作用。很多古细菌和细菌都可以通过胆盐水解酶解离初级胆汁酸的氨基酸残基。小鼠定植表达胆盐水解酶的大肠埃希菌后，会发挥对代谢、节律和上皮细胞完整性的调控作用（特别是增加抗菌肽 Reg Ⅲ γ 的表达）[63]。通过分析 MetaHit 和 Human Microbiome Project 数据库的宏基因组样本发现，在 IBD 患者中与厚壁菌门相关的胆盐水解酶基因簇丰度显著降低 [64]。在结肠中，胆汁酸由初级胆汁酸转换为次级胆汁酸，此过程主要由梭菌属细菌和 7α/β 脱羟基酶介导，最终 CA 和 CDCA 分别转换为脱氧胆酸（deoxycholic acid，DCA）、石胆酸及其他衍生物。

胆汁酸也可以影响肠道微生物的组成和密度。小肠 FXR 活化后，能够抑制细菌过度生长和转位。胆汁酸还具有直接的抗菌作用，比如 CA 和 DCA 可抑制双歧杆菌和乳酸杆菌，同时胆汁酸还可以刺激宿主产生抗菌肽、血管生成素 I 和 iNOS 等，发挥间接抗菌作用 [65]。CA 能够改变大鼠的肠道微生物组成，如增加梭状芽孢杆菌和产芽孢菌的数量，抑制拟杆菌和放线菌的数量 [66]。次级胆汁酸，如 DCA 能够促进对艰难梭菌的定植抗力，以及对产芽孢梭菌和艰难梭菌产生抑制作用。

在 IBD 患者中，肠道菌群失调与胆汁酸吸收障碍之间的关系较为复杂。研究发现，进食高脂饮食的小鼠体内牛磺酸结合胆汁酸比例显著增加，进而导致沃氏嗜胆菌异常活跃 [67]，后者可增加牛磺酸的合成，产生具有毒性的代谢产物硫化氢，此改变与 IL-10 敲除小鼠结肠炎密切相关，以上研究提示，在遗传易感的背景下，高脂饮食、宿主代谢产物改变、菌群失调与炎症之间存在相互作用 [67]。进一步研究发现，FXR 激动剂能够减轻化学药物诱导的结肠炎症状，而 FXR 敲除小鼠则会对化学药物诱导的结肠炎更加易感 [65]。CD 患者结肠组织中 FXR 的表达显著下降。体外以 FXR 激动剂诱导 IBD 患者固有层的单核细胞，能抑制促炎因子 IL-17、IFN-γ 和 TNF 的释放。新近研究发现，胆汁酸在小鼠结肠 $ROR\gamma^+$ 调节性 T 细胞的发育中发挥重要作用 [68]。敲除小鼠胆汁酸代谢相关的基因，这种调节性 T 细胞的数量减少，胆汁酸可以通过 VDR 维持 $ROR\gamma^+$T 细胞活性和数量，进而减轻 DSS 诱导的结肠炎症 [68]。

IBD 患者粪便中胆汁酸谱发生显著的改变，粪便中结合初级胆汁酸明显增加，血清和粪便中次级胆汁酸明显降低 [69]，活动性 IBD 患者粪便中 3-OH 硫酸化的胆汁酸水平升高，提示 IBD 患者体内胆汁酸解离、转化和脱硫酸化作用显著减弱。体外研究显示，缺乏脱硫酸化的胆汁酸具有促进炎症反应的作用。未来仍需要大规模的人群研究和模型系统，进一步揭示 IBD 患者肠道菌群与胆汁酸之间的相互作用。

（二）短链脂肪酸

SCFAs 是一类对机体有益的细菌代谢产物（图 8-2），主要来自于碳水化合物。SCFAs 主要包括乙酸盐、丙酸盐和丁酸盐，此外，还有甲烷、硫化氢和其他中介产物。SCFAs 的构成比例随饮食差异而有所不同，乙酸盐的比例为 50% ～ 70%，丙酸盐占 10% ～ 20%，其余为丁酸盐。通过饮食干预研究显示，由基础饮食过渡至动物性饮食后，体内蛋白质代谢会出现显著的改变，进而导致菌群失调，主要表现为胆汁耐受和诱导结肠炎的嗜胆菌和分解黏液的瘤胃球菌数量显著增加，这两种细菌均与 IBD 的发病密切相关 [70, 71]。

SCFAs 能够增加肠道调节性 T 细胞数量，进而抑制肿瘤细胞的增殖，在结肠炎和结直肠癌动物模型中发挥保护作用。SCFAs 还可以调节肠道巨噬细胞活性、脂肪积聚、肠道动力以及通过调控肠道内分泌 L 细胞分泌多肽 YY 和 GLP-1 而影响能量代谢。丁酸盐是结肠上皮细胞的主要能量来源，它可以抑制上皮干细胞，通过活化炎症小体而促进 IL-18 的产生，从而维持肠道上皮稳态。此外，SCFAs 可以促进肠道 B 细胞反应和黏膜免疫成熟。SCFAs 的这些生物学作用主要是通过与游离脂肪酸受体，如 GPR43 和 GPR109A 结合、抑制组蛋白去乙酰化酶活性有关。

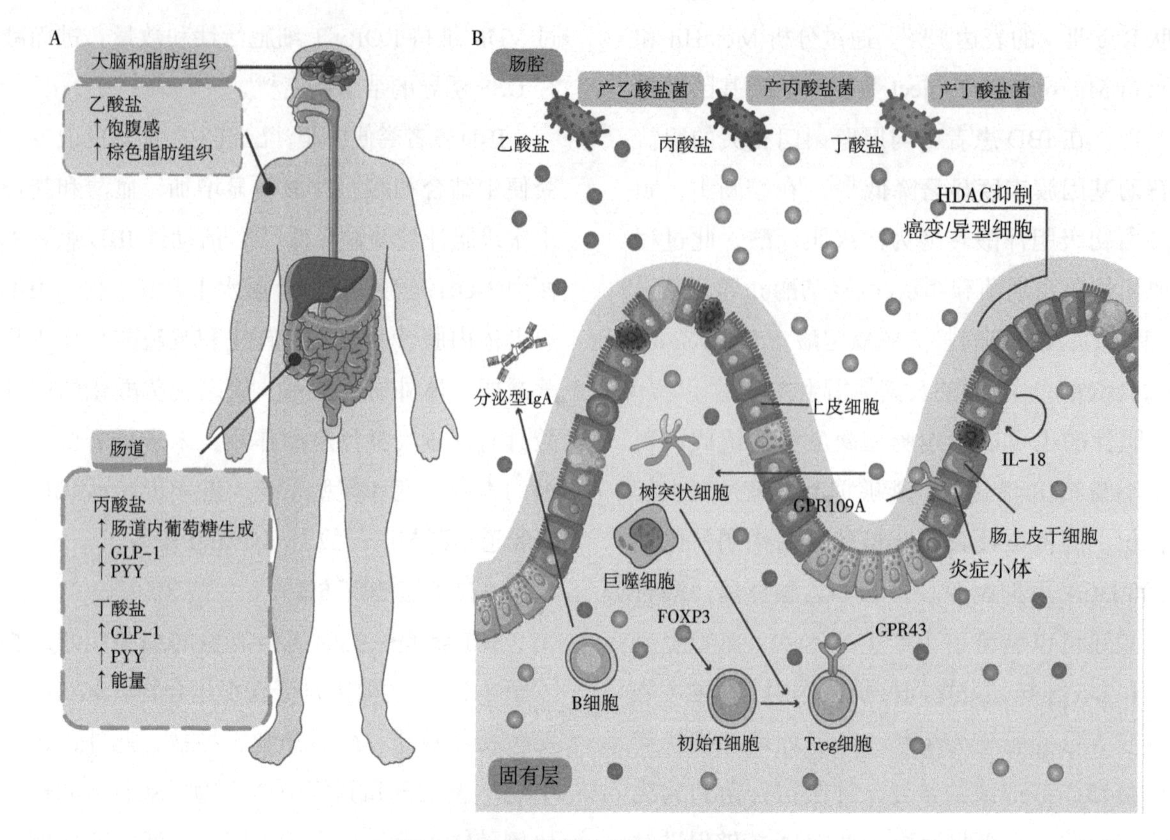

图 8-2　短链脂肪酸对宿主的生理作用

A. 短链脂肪酸（SCFA）介导宿主脂肪代谢，包括增加饱腹感和白色脂肪组织的褐变。丙酸盐是肠道糖异生的前体，丙酸盐和丁酸盐可增加多肽 YY（PYY）和胰高血糖素样肽 1（GLP1）的分泌。B. SCFA 介导黏膜免疫，包括促进 B 细胞发育、调节性 T（Treg）细胞的分化和增殖，通过抑制组蛋白去乙酰化酶（HDAC）的抗增殖作用，通过活化炎性体促进 IL-18 的产生而维持黏膜完整性，以及通过结肠树突状细胞和巨噬细胞促进 Treg 细胞的分化作用。GLP-1. 胰高血糖素样肽 -1；PYY. 多肽 YY；GPR.G 蛋白偶联受体；FOXP3. 叉头状家族转录因子 3；HDAC. 组蛋白去乙酰化酶

在 IBD 患者体内观察到的菌群失调，主要表现为产丁酸盐细菌数量的下降，如粪杆菌和人罗斯拜瑞弧菌，与 IBD 患者粪便代谢组学所显示的 SCFAs 下降的结果一致 [72]。结肠造口术后导致的转流性结肠炎可以通过 SCFAs 灌肠治疗得以缓解。在 UC 患者中，利用丁酸盐灌肠同样具有一定的治疗作用。产丁酸盐的膳食纤维添加剂有助于维持缓解，其效果不劣于 5- 氨基水杨酸（5-ASA）。

IBD 患者肠道内产丁酸盐的细菌明显减少，且丁酸盐氧化过程显著受损，此结果都在基因水平得到了验证。IBD 患者粪便中 SCFAs 水平明显下降，乙酸辅酶 A 转移酶（丁酸盐合成的主要酶）表达明显降低 [73]。相反，中间产物乳酸在活动性 UC 和 CD 患者中明显升高 [74]。在小鼠模型中，低水平的膳食纤维会导致 DSS 诱导的结肠炎更加易感，其机制与膳食来源的 SCFAs 及其受体 GPR43 和 GPR109A，以及下游 NOD、LRP 和 NLRP3 炎症小体 -IL-18 轴的活化密切相关。GPR43 敲除小鼠会出现 DSS 诱导的难治性结肠炎 [75]。这些改变直接与肠道微生物的组成改变密切相关。在小鼠体内过表达 NLRP1A 会引起产丁酸盐的梭菌数量下降，导致 IL-18 和 IFN-γ 生成增加 [76]。

（三）色氨酸

色氨酸是人体必需的芳香族氨基酸。家禽、鱼类、燕麦和乳制品是常见的色氨酸来源。色氨酸是体内合成重要生物活性分子的前体，如 5- 羟色胺、褪黑素、烟酰胺和维生素 B_3。消化道是色氨酸代谢的主要部位。通过饮食进入体内

的色氨酸，主要以 3 种途径代谢：两种宿主代谢途径，包括犬尿氨酸和 5- 羟色胺途径；第一种是微生物代谢途径（吲哚代谢途径）。膳食中大部分色氨酸通过犬尿氨酸途径代谢，其关键的限速酶包括黏膜和免疫细胞中的吲哚胺 2，3- 二加氧酶 1 和肝脏中的色氨酸 2，3- 二加氧酶。第二种宿主代谢途径是 5- 羟色胺代谢途径，由肠道嗜铬细胞中的关键限速酶色氨酸羟化酶 1 调控，是体内 5- 羟色胺的主要来源。最后，色氨酸还可以在肠道微生物作用下代谢为多种吲哚类代谢产物，其中部分物质可以作为芳香烃受体（aryl hydrocarbon receptor，AhR）的配体。重要的是，细菌及其代谢产物对宿主色氨酸代谢具有重要的调控作用。

吲哚作为肠道微生物的代谢产物，可以刺激 GLP-1 的释放，吲哚的衍生物，包括吲哚乙酸、吲哚 -3- 乙醛、吲哚丙烯酸等可作为 AhR 的激动剂。AhR 是体内调控肠道 T 细胞免疫和 IL-22 表达的重要转录因子，具有重要的抗炎作用。在近端小肠，膳食来源的 AhR 受体激动剂可以通过调控上皮内的淋巴细胞而维持细菌载量、组成和免疫耐受。反之，细菌来源的 AhR 受体激动剂则主要作用于远端小肠和结肠。消化道链球菌和乳酸杆菌可以产生 AhR 受体激动剂，产芽孢梭菌（*C. sporogenes*）可以产生吲哚 -3- 丙酸（Indolyl-3-propionic acid，IPA）。IPA 有助于维持肠道的黏膜屏障功能，还可以通过结合 PXR 而抑制黏膜 TNF 的产生。

研究发现，IBD 患者色氨酸代谢明显增加，且疾病活动度与色氨酸水平呈负相关，主要表现为犬尿氨酸代谢途径的增加 [77]。在 CD 患者黏膜中，AhR 的表达显著下降，在小鼠模型中，膳食色氨酸缺乏与结肠炎的严重程度密切相关 [78]。DSS 诱导的 CARD9 敲除小鼠结肠炎模型证实，吲哚衍生物吲哚乙酸的水平显著下降，同时细菌激活 AhR 的能力显著降低 [79]。IL-22 敲除小鼠来源的细菌无法激活 AhR，而给予 IL-22 则可以显著缓解 CARD9 敲除小鼠的结肠炎症状 [79]。进一步研究发现，AhR 激动剂可以引起 IBD 患者固有层单核细胞分泌 IFN-γ 减少和 IL-22 分泌增加，同时还可以减轻 DSS 和 T 细胞转移的小鼠结肠炎，这种效应在阻断 IL-22 后消失。具有 AhR 激动活性的乳酸杆菌同样可以减轻 DSS 诱导的小鼠结肠炎 [79]。

细菌来源的吲哚物质可以作为 AhR 的配体，调控 IL-10 信号通路。口服 IPA 可以减轻 DSS 诱导的小鼠结肠炎症状。在 UC 患者的血清中，IPA 水平明显降低 [80]。此外，人体共生的拉塞尔消化链球菌（*P. russellii*）可以通过代谢色氨酸成为 AhR 激动剂吲哚丙烯酸（indoleacrylic acid，IA）而减轻结肠炎症状、促进杯状细胞分化和抑制炎症信号通路 [80]。拉塞尔消化链球菌代谢产生 IA 的途径与产芽孢梭菌代谢产生 IPA 的途径相同，在拉塞尔消化链球菌中同样发现 fldAIBC 苯基乳酸基因簇的表达 [80]。以上研究提示，具有代谢黏液和色氨酸的细菌，可以促进上皮的完整性，而在 IBD 患者中，fldAIBC 基因簇的表达显著下调 [80]。

五、肠道微生物与炎性肠病的分型及诊治

由于 CD 和 UC 的疾病特征具有较大的异质性，IBD 的诊断、治疗和预防仍存在诸多挑战。如 CD 的疾病严重程度和分期主要是基于病变的部位和病理特征，如黏膜炎症、狭窄导致的梗阻和穿透性病变导致的内、外瘘或肠道穿孔。CD 的临床表型还包括肛周病变，如肛瘘和肛周脓肿，以及各种肠外表现等。部分 UC 患者在手术后会出现克罗恩样病变。近年来，内镜下的治疗终点和黏膜愈合备受关注，新近，又提出了将 UC 的组织学愈合和 CD 的透壁愈合作为可考虑的治疗目标。然而，目前可用于预测疾病进展与疗效评估的方法仍然有限，影像学和组织学的预测指标仍在探索中，尚需进一步临

床验证，故亟须开发无创的预后检测指标。基于以上的研究结果，微生物组学在预测IBD疾病进展、并发症和治疗反应等方面，具有一定的优势和潜力。

肠道微生物在IBD发病过程中发挥着重要作用，部分研究提示细菌特征与疾病表型密切相关[81]。通过分析粪便中包被IgA的不同种类细菌的丰度发现，CD相关的脊柱关节炎和肠道Th17细胞相关的患者中，大肠埃希菌较为富集[82]。通过IgA-seq技术可以重点聚焦免疫系统识别的细菌成员[83]。CD患者的菌群紊乱程度明显高于UC患者，粪便中菌群的组成主要依赖于疾病的部位和吸烟史[84]。通过利用8种细菌组成的粪便菌群特征，可以区分健康人群与IBD患者，其精确度可以达到85%，同时还可以区分CD与UC患者，精确度达到82%。

UC患者实施回肠储袋肛管吻合术后，炎症相关的并发症较为常见，通过分析患者术后组织相关的细菌组成发现，出现炎症并发症的患者（结肠袋炎或克罗恩样病变），较无并发症的患者，其肠道内拟杆菌数量相对较少，而变形菌门的细菌数量则相对较多[85]。在CD患者中，回肠黏膜中部分细菌种类组成的改变，包括厚壁菌门和拟杆菌门数量的减少与变形菌门数量的增加，提示更高的儿童克罗恩病活动指数（pediatric Crohn’s disease activity index，PCDAI）评分[86]。胃球菌属数量的增加提示狭窄并发症的存在，而韦荣氏球菌属则提示穿孔相关的并发症较为多见。

在儿童CD患者中，初始通过配方饮食或抗TNF单抗治疗的患者，整个治疗过程中粪便菌群失调的比例明显增加[87]。肠道部分细菌丰度的改变，包括埃希杆菌属、克雷伯菌和韦永球菌属的增加，真杆菌属、普氏菌属和瘤胃球菌属的减少，都与肠道炎症密切相关，而炎症、抗生素暴露、饮食都可以独立地影响不同细菌种类的丰度改变。同时，抗生素的使用提高了肠道真菌的水平，而饮食治疗则降低了真菌水平，提示肠道菌群的波动依赖于环境因素的改变[87]。CD和UC患者相比，结肠黏膜细菌的多样性并未发生明显的变化，但是与健康对照相比，CD或UC患者的关键细菌微生物的丰度发生了改变，产丁酸盐的细菌丰度下降而产硫化氢的细菌丰度则出现增加[88]，提示细菌的组成与疾病程度密切相关。

六、调整肠道微生物对炎性肠病的治疗作用

（一）膳食结构调整

膳食成分对IBD患者肠道微生物具有显著的影响。正常情况下，黏附侵袭性大肠埃希菌无法诱导炎症反应，当肠道上皮细胞和M细胞表达配体CEACAM6，AIEC可以形成小肠生物膜，通过M细胞进行黏附和转移，进而侵袭相关的上皮细胞，导致炎症反应。高脂和高果糖饮食可以促进IBD患者病原微生物的定植。CEABAC10小鼠表达CEACAM6，当给予高脂和高果糖饮食后，出现黏膜相关AIEC的定植，并产生严重的隐窝脓肿[89]。给无菌小鼠移植高脂/高果糖喂养小鼠的粪便后，会增加对AIEC的易感性[90]。高脂饮食可以导致小鼠肠道菌群失调，主要表现为变形菌门的增加和厚壁菌门的下降，与CD患者表现一致[91]。高脂饮食会导致次级胆汁酸DCA的积聚，进而抑制拟杆菌门和厚壁菌门的增殖，类似于CD患者的菌群失调特征。麦芽糖可以在无CECAM6的情况下促进AIEC生物膜的形成，当培养基中添加多糖，如麦芽糖和黄原胶等后，AIEC的增殖会显著增加[92]。

膳食中的脂肪可促进致结肠炎的细菌生长。给予IL-10敲除小鼠高脂饮食后，结肠炎症状会加重，而等热量的多不饱和脂肪酸饮食或低脂饮食却不会加重结肠炎。结肠炎的发生与沃氏嗜胆菌密切相关，此细菌主要依赖于牛奶脂肪，

并可促进牛磺酸结合胆汁酸的合成。

饮食治疗可能在 IBD 患者的管理中发挥关键作用。全肠内营养可用于早期或新发 CD 患者的诱导缓解治疗[93]。新的膳食策略可以用于维持缓解，也可以用于复杂病例或生物制剂治疗失败患者的诱导缓解治疗。

（二）抗生素、益生菌和益生元

抗生素、益生菌和益生元已被用于 IBD 患者的治疗。抗生素的应用在 CD 的治疗中显示了一定的价值，而益生菌和益生元的作用并不十分确定。单一抗生素治疗能够缓解 CD 患者的肠炎和并发症，并可预防术后复发，但对 UC 患者并未显示明显的效果。抗生素联用可能提高治疗效果，但长期抗生素的应用会导致肠道细菌对抗生素的抵抗作用。

益生菌在 UC 治疗中显示了一定的疗效。复合益生菌 VSL#3 联合大肠埃希菌的应用可以降低炎症活动度和维持缓解，然而，它们对 CD 患者无效[94]。普拉梭菌可以通过产生 SCFAs、刺激 Treg 细胞产生 IL-10，进而抑制 IBD 患者过度的免疫反应。多个小鼠模型研究证实，普拉梭菌、梭状芽孢杆菌和脆弱拟杆菌可以减轻结肠炎程度。以寡聚糖和纤维素作为益生元，通过增加产 SCFAs 细菌丰度的思路虽然极具诱惑力，但是研究结果并不令人满意。利用针对鞭毛蛋白的抗生素、甘油聚合物或 FimH 的拮抗剂来抑制 AIEC 的上皮黏附、侵袭和转位更具有现实意义[95]。抑制粪肠球菌的蛋白酶活性或蛋白酶受体结合能力可以降低黏膜的通透性[96]。因此，选择性地阻断病原微生物的致病产物或细菌活性，或许能够减轻 IBD 患者的菌群失调。最近的研究发现，钨酸盐可以选择性抑制细菌呼吸链途径阻断肠道炎症和肠杆菌的增殖，通过钨酸盐调控肠道细菌可以减轻小鼠的结肠炎[97]。

（三）粪菌移植

粪菌移植（fecal microbiota transplantation，FMT）已被指南推荐用于治疗复发性艰难梭菌感染，其机制是通过恢复肠道微生物的平衡和纠正菌群失调。随后 FMT 逐步扩展至其他疾病的临床研究，包括 IBD 和代谢综合征。目前针对 FMT 治疗 IBD 的有效性仍存在不一致性。一项系统分析纳入 18 项研究共 122 例 IBD 患者，发现仅 36% ～ 45% 的患者达到临床缓解，亚组分析显示 FMT 在 UC 患者中的有效率约 22%，而在 CD 中的有效率约 61%[98]。

两项关于 FMT 治疗 UC 的安慰剂对照研究出现相反结果，其中一项研究提示，粪菌供体的作用较为重要，部分 UC 患者 FMT 术后出现发热和 CRP 升高。在 FMT 治疗 CDI 的过程中，发现 UC 和 CD 患者本身的病情也会得到部分缓解[99]。最近一项多中心的 RCT 结果显示，疗程密集且来自多个供体的 FMT 可以诱导 UC 的临床缓解和内镜下黏膜愈合，与肠道菌群改变有关。FMT 组约 27% 的患者有效，而对照组仅有 8%[100]。FMT 术后患者肠道菌群的多样性有所增加并得以维持。这些研究证实，多个供体的 FMT 能够有效地治疗活动期 UC。

研究证实，30% ～ 50% 的供体细菌能够在受体体内持续生存[101]。两项研究显示，噬菌体在 FMT 过程中出现转移[102]，且其中一项研究发现，FMT 的有效性与供体的病毒丰度相关。如果供体的有尾噬菌体丰度高于受体，则 FMT 的治疗效果较好。这些研究进一步强调了供体的重要性，提示具有较高病毒丰度或多个供体的 FMT 具有更好的临床疗效。

七、未来研究方向

随着微生物与宿主之间相互作用研究的深入，基于微生物的治疗策略逐渐多样化，其中部分已显示了一定的临床疗效。遗传与环境因素对肠道微生物及其产物的研究备受关注。在不同个体之间存在微生物的多样性，而系统的研究方法能够精确地分析这一多样性问题，并进一步探索针对特异性疾病的治疗措施。目前仍然存在诸多的问题亟须解决。如，不同分子通路作用于不同遗传易感者可能促进微生物组成的改变，微生物可能异常地调节宿主的生理功能，直至超越特定的阈值而导致病理性的炎症反应，哪些膳食因素和微生物代谢产物与机体健康和疾病密切相关？它们又是通过作用于哪些分子受体和通路而发挥作用的？这些研究将进一步阐明宿主 - 微生物之间的相互作用，并有助于最终恢复肠道屏障、免疫系统和微生物平衡，最终为临床提供可行性的治疗策略，对诱导 IBD 患者长期缓解具有重要作用。

第三节　肠道菌群与过敏性疾病

一、概述

过敏性疾病被世界卫生组织列为 21 世纪需要重点防治的三大疾病之一。过敏是机体对过敏原的一种变态反应，是体液或细胞免疫机制紊乱介导的超敏反应。过敏反应累及某些特定器官及组织引发的一组疾病称为过敏性疾病 [103]。近年来，由于经济发展和生活环境的不断变化，过敏性疾病的发生率不断增加。世界变态反应组织（World Allergy Organization，WAO）对全球 30 个国家过敏性疾病的流行病学调查结果显示，约 22% 的人口患有不同种类的过敏性疾病，由于其表现不尽相同，过敏性疾病的患病率可能更高。过敏性疾病主要包括过敏性哮喘、过敏性紫癜、过敏性湿疹等变态反应性疾病，这些慢性炎症性疾病的特征是 $CD4^{+}$ T 辅助型 2（Th 2）细胞产生的细胞因子（如 IL–4、IL–5、IL–9 和 IL–13 等）、IgE 和效应细胞对组织炎症部位的影响 [104]。

在正常情况下，人体肠道微生物形成了一个相对平衡状态。一些因素，如服用抗生素、放化疗、免疫力下降等，可导致肠道菌群失调。研究证实，肠道群落组成的改变与过敏性疾病的发病具有一定的相关性。例如，患过敏症的婴儿表现出体内共生体的改变，接受广谱抗生素治疗的儿童患过敏性疾病的风险增加 [105]。过敏儿童肠道菌群拟杆菌属的数量增高，嗜黏蛋白阿克曼菌、普氏粪杆菌和梭状芽孢杆菌的数量降低，青春双歧杆菌（*B. adolescentis*）数量较高，链双歧杆菌（*B.catenulatum*）和金黄色葡萄球菌数量较低，并呈现较低的细菌多样性 [106]。动物模型研究发现，口服广谱抗生素有助于 Th2 细胞因子依赖性过敏性炎症的发生 [107]。益生菌有望成为预防和管理过敏性疾病的措施，然而其调节免疫的具体机制仍需进一步阐明。

二、 肠道菌群与过敏性哮喘

过敏性哮喘是一种由过敏原引起的气道慢性炎症性疾病，其病理生理特征是黏液堵塞、支气管黏膜增厚和支气管狭窄，最终可导致气道高反应性增高、气道结构重塑及支气管阻塞 [3, 108]。据统计，全世界约有 3 亿人患过敏性哮喘，预计到 2025 年将增加至 4 亿人，其中每年约 25 万人死于哮喘 [109]。过敏性哮喘的发病机制仍不完全明确，可能与遗传、环境、细菌感染和营养因素有关 [110]。

微生物组学研究显示，呼吸道和胃肠道微生物菌落对维持人体健康至关重要，肠道菌群在免疫稳态的建立、调节和维持方面发挥着关键作用。生活方式、环境暴露和儿童早期接触的微生物多样性等因素均会影响肠道菌群 [111]。肠道菌群失调及其导致的免疫失调会参与多种疾病的发生、临床表现及疗效 [112]。哮喘与早期肠道菌群失调密切相关，肠道菌群失调可通过“肠 - 肺轴”引起免疫反应的改变和气道疾病的发生 [113]。

（一）过敏性哮喘肠道菌群改变

近年来，关于过敏性哮喘与肠道菌群的研究备受关注。有对哮喘高风险儿童的肠道菌群研究发现，患儿 3 月龄时毛螺菌属（*Lachnospira*）、韦荣球菌属（*Veillonella*）、粪杆菌属（*Faecalibacterium*）和罗氏菌属（*Rothia genera*）细菌丰度显著降低 [114]。给予哮喘小鼠移植毛螺菌属、韦荣球菌属、粪杆菌属和罗氏菌属粪菌可减少气道炎症 [114]。该团队还发现，5 岁时过敏性哮喘易感性显著增加的患儿，其 3 月龄时粪便中链球菌属和拟杆菌属丰度增加，而双歧杆菌（*Bifidobacterium*）和瘤胃球菌（*Ruminococcus*）丰度减低 [115]。关于美国新生儿哮喘的研究发现，念珠菌属（*Candida*）和红酵母属（*Rhodotorula*）丰度增高时，过敏性哮喘发生风险最高，嗜黏蛋白阿克曼菌属（*Akkermansia*）、粪杆菌属、双歧杆菌属（*Bifidobacteria*）丰度降低时，哮喘发生风险最低 [116]。代谢组学的研究发现，与健康儿童相比，4 ～ 7 岁哮喘儿童粪便中粪脂杆菌属和罗氏菌属（厚壁菌门）的丰度显著降低，而肠球菌属和梭状芽孢杆菌属丰度增高。因此，厚壁菌门丰度降低可能会增加哮喘风险 [117, 118]。综上，肠道细菌丰度的变化，可能与过敏性哮喘的发病相关（图 8-3）。

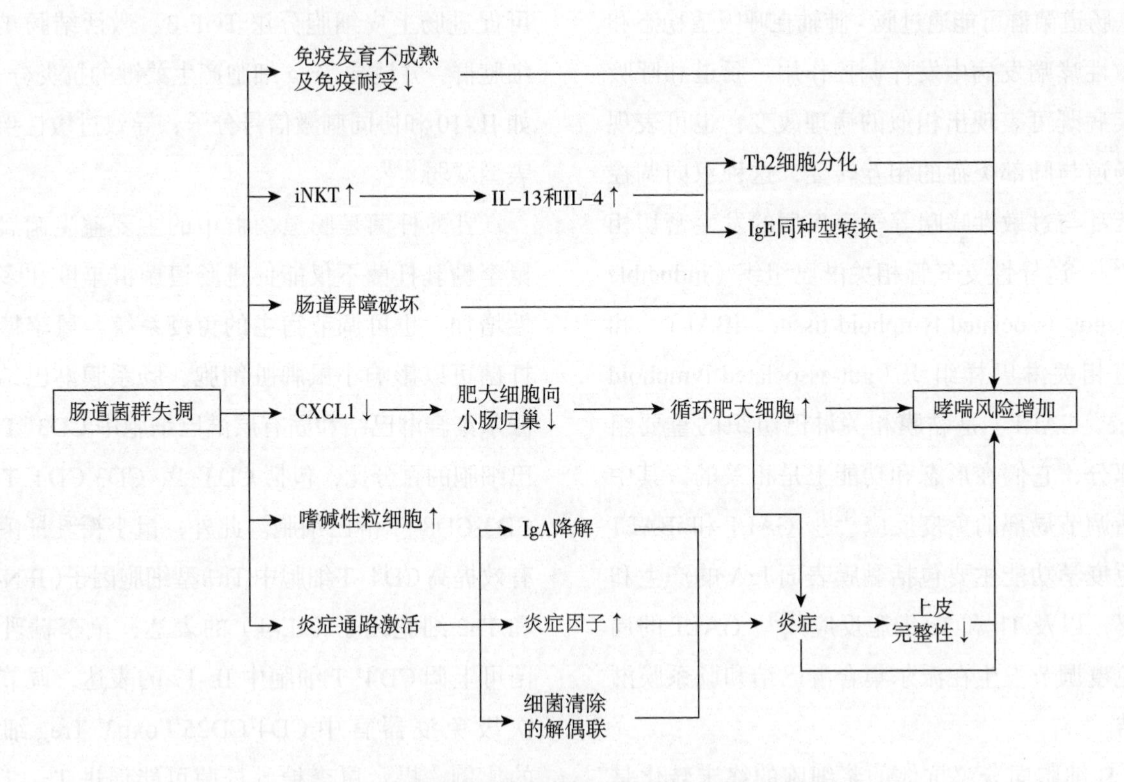

图 8-3　肠道菌群失调可能通过多种机制增加哮喘风险

菌群失调可能导致 TLR4 上调、IgA 降解和共刺激增加，从而导致炎症。此外，由于微生态失调，免疫系统可能无法在感染清除后终止炎症。炎症引起上皮完整性的丧失，导致哮喘风险增加。趋化因子 CXCL1 下调减少了肥大细胞向肠道的归巢，从而增加了循环肥大细胞。通过循环自然杀伤恒定 T 细胞（iNKT）产生 IL-13 和 IL-4，诱导 IgE 同种型转换和 Th2 细胞分化，这可能增加哮喘风险。iNKT. 恒定自然杀伤 T 细胞；CXCL1. 趋化因子配体 L1

人类肠道微生物组成在生命的最初几年逐渐成熟。据推测，这一时期的微生物组成异常可能导致机体免疫功能失调甚至引发哮喘。关于690名儿童出生后第1年肠道细菌定植模式与后期哮喘风险之间的相关性研究发现，1岁时肠道微生物不成熟的儿童，在5岁时患哮喘的风险显著增加，这种相关性在哮喘母亲所生的儿童中更明显。而未患哮喘的儿童肠道菌群的改变主要体现在韦荣球菌属丰度增高，罗氏菌属、另枝菌属（*Alistipes*）、粪杆菌属、双歧杆菌、瘤胃球菌、毛螺菌属（*Lachnospiraceaeincertaesedis*）及小杆菌属（*Dialister*）丰度降低[119]。这一研究表明，在出生后的第1年肠道菌群丰度的降低可增加儿童的哮喘发生风险，相反，这一时期肠道菌群丰度的成熟可以预防哮喘的发生。

（二）肠道菌群在过敏性哮喘发病机制中的作用

肠道菌群可能通过肠 - 肺轴在呼吸道稳态和过敏性哮喘发病中发挥调控作用。肠道和呼吸道疾病既可表现出相似的病理改变，也可表现为肠道与肺部炎症的相互转变，这种双向调控的紊乱与过敏性哮喘等气道疾病的发生密切相关[120]。诱导性支气管相关淋巴组织（inducible bronchus-associated lymphoid tissue，iBALT）和肠道相关淋巴样组织（gut-associated lymphoid tissue，GALT）是黏膜相关淋巴组织的重要组成部分，它们在形态和功能上是相关的，其中包括调节局部的免疫反应[121]。GALT和iBALT的免疫学功能主要包括黏膜表面IgA的产生和分泌，以及Th和Tc细胞反应[122]。GALT的局部免疫调节发生在派尔集合淋巴结和肠系膜淋巴结。

B细胞向分泌抗体的浆细胞的终末转化是在固有层中完成，而记忆和效应T细胞和B细胞也驻留在固有层中[123]。肠道中被激活的幼稚免疫细胞通过淋巴和血液系统到达呼吸道发挥免疫反应[124]。肠 - 肺轴是双向调控的，如用脂多糖刺激小鼠呼吸道可引起肠道细菌数量显著增加[125]，且肺炎可引发肠道损伤[126]并抑制肠上皮再生。呼吸道和肠道之间的相互作用已被反复证实，而肠道菌群如何参与哮喘的发病尚不十分明确[127]，其可能的机制如下所述。

1. *肠道免疫发育不成熟* 免疫成熟需要特异性微生物群的定植，微生物定植促进Tregs细胞的扩增，Tregs细胞对于免疫耐受和适应性免疫的调节至关重要。肠道菌群可表达外膜菌毛样结构蛋白，该蛋白可抑制与特应性疾病（包括哮喘）相关的Th2细胞的过度活化，从而影响Tregs细胞的发育和功能[128]。肠道菌群可参与$CD4^+$T细胞的发育和分化，而无菌小鼠肠道固有层中存在$CD4^+$T细胞数量减少和Th1/Th2细胞失衡[129]。宿主特异性微生物群的某些组分可能参与免疫成熟的某些环节。对过敏小鼠的研究发现，口服某些特异性梭状芽孢杆菌菌株可促进肠上皮细胞分泌TGF-β、激活结肠Treg细胞群，并诱导Treg细胞产生关键的抗炎分子，如IL-10和协同刺激信号分子，导致过敏性疾病表型减弱[130]。

乳酸杆菌是肠道菌群中的主要益生菌属，鼠李糖乳杆菌不仅能促进肠道菌群丰度和多样性增加，也可调节宿主的免疫系统。鼠李糖乳杆菌可以影响小鼠脾脏细胞、肠系膜淋巴结、派尔集合淋巴结和固有层淋巴细胞中$CD3^+$ T淋巴细胞的百分比，包括$CD3^+$ T、$CD3^+CD4^+$ T和$CD3^+CD8^+$ T淋巴细胞。此外，鼠李糖乳杆菌可有效提高$CD4^+$ T细胞中Th1型细胞因子（IFN-γ）和Th2细胞因子（IL-4）的表达，鼠李糖乳杆菌可上调$CD4^+$ T细胞中IL-17的表达，调节各次级免疫器官中$CD4^+CD25^+Foxp3^+$ Treg细胞的比例，提示鼠李糖乳杆菌可能促进Th-17和Treg细胞的平衡[131]。这些特定肠道菌群对过敏性宿主具有整体保护作用，菌群失调诱发的免疫不成熟可能是哮喘发生的重要机制。

2. *肠漏* 肠道上皮的通透性依赖于紧密连接，紧密连接破坏可通过增加肠道通透性而导致肠漏。紧密连接受损导致抗原物质进入血流进而诱发全身免疫反应。目前认为，肠漏与哮喘等多种慢性疾病有关[132]。紧密连接受细胞骨架肌动蛋白与肌球蛋白相互作用的调节，如肌球蛋白轻链激酶、Rho 激酶等激酶参与调节紧密连接[133]。来自肠道菌群的短链脂肪酸信号可改变肌球蛋白轻链激酶和 Rho 激酶的活性，可能诱导新紧密连接的组装。调节紧密连接最重要的信号是促炎和抗炎细胞因子，包括干扰素 -γ、TNF-α、IL-1、IL-6 和 IL-10。通过改变以上信号分子，可能影响胃肠道上皮的完整性，导致过敏性哮喘的发生[134]。而事实上，部分哮喘可以通过修复胃肠道黏膜屏障和改善肠道通透性，以防止免疫级联反应而得以控制。

3. 循环嗜碱性粒细胞增加　嗜酸性粒细胞可促进 Th2 免疫反应从而加重过敏。过敏性疾病炎症部位存在嗜酸性粒细胞和嗜碱性粒细胞浸润，而嗜碱性粒细胞被证明是“守门者”，可通过激活内皮细胞而促使嗜酸性粒细胞进入炎症部位[135]。最近的研究提示，肺晚期过敏反应中嗜碱性粒细胞浸润是疾病严重程度的特征。活化的嗜碱性粒细胞释放的组胺和促炎症介质（如 LTC4）可增加血管通透性和黏液生成、促进支气管周围平滑肌细胞收缩，从而诱发哮喘[136]。肠道微生物可能通过调节嗜碱细胞数量而调控过敏反应。使用广谱抗生素治疗的小鼠循环嗜碱性粒细胞数量增加，Th2 细胞反应加剧，IgE 浓度升高，嗜碱性粒细胞的消耗减弱了 Th2 细胞反应，表明嗜碱性粒细胞参与了过敏性炎症的发生[137]。

4. 循环自然杀伤恒定 T 细胞增加　由于微生态失调，自然杀伤恒定 T 细胞（invariant natural killer T cells，iNKT）可能通过释放 IL-4 和 IL-13 而诱导 Th-2 分化，这些细胞聚集于无菌小鼠的肺部并促进产生 IgE，导致哮喘易感性增加。此外，部分微生物群通过基因的表观遗传修饰，对肺中趋化因子配体 CXCL16 编码，使循环 iNKT 细胞迁移并与之结合[138]，诱发哮喘。

5. CXCL1 减少　由于微生态失调，感染后免疫系统可能无法终止炎症反应，持续存在的炎症与其引起的上皮完整性的丧失可导致哮喘的风险增加。同时，肠道菌群失调可导致趋化因子 CXCL1 下调，从而减少了肥大细胞向肠道的归巢，增加了血液循环中肥大细胞数量[139]。肥大细胞的增加可激活全身炎症反应，并促使循环 iNKT 产生 IL-13 和 IL-4，诱导 IgE 同种型转换和 Th2 细胞分化，增加哮喘的风险。

6. 肠道菌群失调通过激活多种炎症通路增加哮喘的发生

（1）炎症因子增加：肠道菌群失调可通过微生物产物，如脂多糖[140]，激活肠道巨噬细胞产生大量炎症因子，包括 IL-1、IL-6、IL-12、趋化因子配体 5（CCL5）、TGF-β 和 TNF-α 等，结合 TLRs，激活下游 NF-κ B 信号通路进一步诱导炎症因子表达[141]。

（2）IgA 降解增加：肠道菌群产生 SCFAs 等代谢产物，如丁酸盐、丙酸盐和乙酸盐，可能通过同种型转换刺激 IgA 产生，导致肠道的免疫耐受形成[142]。来自健康肠道微生物的某些信号可限制 IgE 同型转换，并增加 IgA 的水平，而肠道某些致病菌，如萨特菌属（*Sutterella*），可引起 IgA 降解[143]。肠黏膜表面低水平的 IgA 可能诱发免疫反应，导致过敏和哮喘的发生。肠道菌群还能促使 IgA 降解，从而导致哮喘易感性增加[144]。因此，IgE 水平升高和 IgA 水平降低与哮喘有关。

（3）细菌根除的解偶联：共生群落中某些细菌可破坏宿主的免疫系统，同时又在炎症等机体不健康的环境中繁殖，导致细菌根除与其引起的炎症解偶联。即，某些细菌可通过某种机制使自身能够安全地免受它们引起的炎症清除作用。有研究发现，牙龈假单胞菌可利用补体将口腔微生物群重塑为一种微生态失调状态，在中性粒细胞中，牙龈假单胞菌通过降解MyD88蛋白酶体解除保护宿主的TLR2-MyD88通路，并激活另一条TLR2-Mal-PI3K通路。这种替代的TLR2-Mal-PI3K通路可阻断吞噬作用，为易感细菌提供“旁观者”的保护作用，并诱发体内的非生物性炎症[145]。这种将细菌从炎症中分离出来的机制需要TLR2和补体受体C5aR之间密切的交互作用，并有助于微生物群落的持续存在，这些微生物群落驱动了菌群失调性疾病的发生。这种解偶联机制为肠道细菌提供了生存和致炎的能力。

综上所述，肠道菌群失调可通过影响免疫耐受、破坏肠上皮完整性、增加循环iNKT细胞和嗜碱性粒细胞、加重炎症反应等机制诱导或加重哮喘（图8-3）。

（三）益生菌及细菌裂解物与过敏性哮喘

1. 益生菌　益生菌在哮喘防治中的确切疗效尚不明确。益生菌包括乳杆菌属、链球菌属、肠球菌属、双歧杆菌属和非致病性大肠埃希菌等[146]。最近的一项Meta分析显示，在长期随访中，给予鼠李糖乳杆菌可预防哮喘发生[147]。益生菌可维护肠黏膜屏障的成熟、促进树突状细胞的发育、调节肠道及全身免疫反应[148]。给予哮喘小鼠口服鼠李糖乳杆菌2周（作为致敏前或致敏后治疗），结果显示，小鼠气道阻力、炎症细胞数量和肺组织中Th2细胞因子明显降低，哮喘症状显著改善[149]。给哮喘模型小鼠补充罗伊乳杆菌，可增加脾脏Tregs细胞数量、减轻气道嗜酸性粒细胞增多和气道高反应性，有效缓解哮喘[150]。以上研究结果提示，口服益生菌可能是过敏性哮喘的补充治疗手段[134]。

关于益生菌的最佳应用时间，动物实验进行了相关研究。BALB/c小鼠模型中，在围生期以鼠李糖乳杆菌灌胃，则母鼠胃内细菌定植，子代脾脏细胞中TNF-α、IFN-γ、IL-5和IL-1表达减少，肺变应性气道和支气管周围炎症减轻[151]。在围生期给予哮喘小鼠口服含双歧杆菌的益生菌，可减轻过敏性气道疾病。给予幼鼠含厚壁菌门和放线菌门益生菌时，肺泡灌洗液中$CD4^+$T细胞和盲肠丁酸盐的数量也随之增加。而当益生菌应用于成年动物时，仅有放线菌门显著升高，且未能显著改善哮喘[152]。因此，在动物的围生期及幼鼠期应用益生菌，可改善哮喘症状。

关于益生菌应用时间点的临床研究，一项Meta分析研究提示，在产前或婴儿期给予益生菌可降低儿童特应性致敏的风险，并可降低IgE水平[153]。给予4～10岁特应性哮喘患儿口服益生菌（嗜酸乳杆菌、双歧杆菌、德氏乳杆菌保加利亚种）12周后，与安慰剂组相比，益生菌治疗的儿童肺功能显著改善，哮喘急性发作次数及支气管扩张药使用量均减少[134]。给予6～18岁哮喘儿童副干酪乳杆菌、发酵乳杆菌或两者联合胶囊治疗3个月，可增加呼气峰流速，降低IgE水平，降低哮喘严重程度，改善哮喘症状，其中以两种菌株联合治疗时效果最为显著[154]。一项在哮喘儿童中开展的为期8周的随机试验结果显示，加用副干酪乳杆菌后，哮喘症状显著减轻，肺功能明显改善。最近的一项Meta分析显示，通过对特定菌株进行亚组分析，提示出生后补充鼠李糖乳杆菌可能对哮喘的预防更有益。综上，在不同时期补充益生菌会改善哮喘的症状，但并不会降低哮喘/喘息发作的风险[147]。

目前，国际健康组织和过敏组织指南指出，迄今为止的研究尚未明确“正确的益生菌、正确的剂量、正确的时间或持续时间和（或）人群”[155, 156]，需要对益生菌预防哮喘/喘息的有效性进行更多的探索。

2. 细菌裂解物　幼儿呼吸道感染可能是诱发哮喘的前提条件。除疫苗接种预防呼吸道感染外，使用非特异性免疫调节剂也得到了关注[157]。与益生菌相比，细菌裂解物发挥作用时间短暂，且不需要活制剂。细菌裂解物主要来自于儿童微生物群，其中包括嗜血杆菌、卡他莫拉菌和肺炎链球菌等[158]。一项回顾性研究中，200 名小于 6 岁有反复呼吸道感染病史的儿童在 2 年内接受了 OM-85（Broncho-Vaxom®）治疗，该药物由 8 种呼吸道病原体的 21 种细菌菌株碱裂解而成，结果显示，接受治疗者复发性呼吸道感染、喘息发作、抗生素使用以及新发感染均明显降低，其主要机制之一可能是通过肠 - 肺轴发挥作用[159]。

关于过敏性气道疾病小鼠模型的研究发现，口服 OM-85 可激活肠道树突状细胞，诱导 Tregs 细胞向肺转运[160]。口服 OM-85 可缓解复发型喘息儿童的呼吸道感染，并减少哮喘急性发作[161, 162]。同样，通过舌下途径给予 6 ～ 16 岁过敏性哮喘儿童多价细菌溶解物片剂（PMBL®）可显著减少急性哮喘发作[162, 163]。因此，细菌裂解产物可减少呼吸道感染，并可激活肠道树突状细胞，诱导 Tregs 细胞向肺转运，减少肺部炎症和高反应性[164]。然而，细菌裂解物对过敏性气道常驻菌群的影响，尤其在儿童早期的作用，仍需进一步研究。

三、肠道菌群与过敏性紫癜

过敏性紫癜（henoch-schonlein purpura，HSP）是一种 IgA 介导的血管过敏性出血性疾病，也称为 IgA 血管炎，是儿童时期最常见的血管炎，其病理特征是 IgA 免疫沉积物影响小血管，故以全身小血管炎症为主要病变而累及多器官及多系统，最常见的临床表现包括皮肤紫癜或瘀斑、（多）关节痛、胃肠道疾病和肾小球肾炎等[165, 166]。该病由于报道不足而低估了发病率，报道的年发病率为（3 ～ 26.7）/100 000[167]，儿童的发病率显著高于成人，年发病率为（10 ～ 20）/100 000，并在 4 ～ 6 岁时出现高峰[168]。仅 10% 的过敏性紫癜为成人，年发病率为（0.8 ～ 1.8）/100 000[167]。据报道，在青春期和成年患者中，男性过敏性紫癜的发病率均高于女性[169]。该病发病率也有一定地域差异，东南亚最高，欧洲和北美较低，在非洲很少发生[170]。

过敏性紫癜患者通常具有两种以上的临床表现，症状出现的顺序可能不同，紫癜通常是首发症状，胃肠道受累或关节炎可能是部分患者最突出的表现[171]。目前尚无特异性实验室检查用于诊断过敏性紫癜，其诊断主要依赖于各种临床表现[172]。皮肤出血是诊断的先决条件，由红细胞渗入皮肤或黏膜引起，可能是由真皮小血管坏死性血管炎引起的。通常，疾病急性期的症状是自限性的，无须干预即可缓解。一些患者会出现肾小球肾炎，部分可发展为终末期肾病[165]。成年人发病通常更为严重，胃肠道和肾脏受累是成年人发病和死亡的主要原因[173]。

肠道菌群是塑造和调节免疫反应的重要因素，它们通过刺激机体局部或全身免疫应答，促进肠黏膜相关淋巴组织的成熟，增强肠黏膜的屏障功能[174]。研究证实，以乳酸杆菌、双歧杆菌为代表的专性厌氧菌，通过与肠黏膜表面特异性受体结合，构成肠黏膜上皮重要的生物屏障，协助机体抵御外源性致病菌的入侵，在维持机体正常的免疫功能中发挥举足轻重的作用[175]。以往研究证实，过敏性疾病患者与健康人群的肠道菌群组成存在一定差异，肠道菌群失调与过敏性哮喘、过敏性紫癜及过敏性皮炎等多种过敏性疾病密切相关[176]。过敏性紫癜可能与感染、疫苗接种、食物和药物等因素有关，其发病机制仍未完全阐明。越来越多的证据表明，肠道菌群对过敏性紫癜的发病至关重要。

（一）过敏性紫癜中肠道菌群的变化

微生物的多样性与多种过敏性疾病的发病机制密切相关，过敏性疾病中肠道微生物群落的丰度和多样性显著降低。随着基因高通量测序方法的广泛应用，通过收集粪便标本测定不同人群肠道菌群结构和比例的临床研究不断成熟。研究发现，出生后第1个月内肠道微生物多样性低与随后的过敏性湿疹有关[177]。对早期食物过敏儿童的大便菌群分析发现，食物过敏患儿的肠道菌群多样性也较低[178]。与上述过敏性疾病的发现一致，过敏性紫癜患者的微生物群种属显著低于健康对照组，包括较低的多样性和较低的丰度[179]。大部分临床研究皆是通过收集粪便标本测定肠道菌群来代表肠道菌群的变化，通过收集口腔咽拭子标本进行基因测序发现，健康儿童和过敏性紫癜患儿的口腔微生物群存在差异，且过敏性紫癜患儿表现出更高的微生物多样性和丰度[180]。过敏性紫癜患者免疫功能紊乱，共生菌生存环境受到破坏，进一步加重了肠道微生态紊乱。

研究证实，过敏患儿肠道菌群中双歧杆菌和乳酸杆菌数量等益生菌显著减少，而大肠埃希菌等致病菌显著增多[176]。研究发现，过敏性紫癜患者肠道菌群中含有较少罗氏菌属和小杆菌属的种群[179]。与此相反，狄氏副拟杆菌和肠球菌属数量显著增加[179]。过敏性紫癜患者肠道中，益生菌的数量明显减少，致病菌或条件致病菌显著增加。过敏性紫癜患者IgA水平显著升高，肠道菌群双歧杆菌属的数量显著降低，IgA水平与双歧杆菌属的种群水平呈负相关[179]。

过敏性紫癜肠道菌群的影响因素很多，如年龄、药物的使用及饮食等。有研究发现，长期口服类固醇激素糖皮质激素会导致大鼠肠道菌群中微生物群落的丰度和多样性显著降低[181]。目前尚无临床研究证实糖皮质激素与肠道菌群的相关性。对于重症和难治性过敏性紫癜的治疗，尤其是合并肾脏损害的患儿，可能需要长期激素或利妥昔单抗等药物治疗，对于使用与未使用激素治疗的过敏性紫癜患儿肠道菌群是否存在差异性，目前尚无相关研究。

（二）肠道菌群对过敏性紫癜的调节机制

过敏性紫癜患者肠上皮发生损伤后，肠道菌群赖以生存的环境发生了改变，共生菌的保护机制被破坏，尤其是腹型过敏性紫癜患者，肠黏膜上皮损伤更严重，可造成肠道菌群失调[182]。国内外研究显示，过敏性紫癜患儿急性期双歧杆菌数量减少，大肠埃希菌数量增高，肠道菌群失调，肠黏膜生物屏障功能受到损伤[183]。罗氏菌属是健康人群中最常见的可移动的肠道细菌种属，而且它是已知的产丁酸盐厚壁菌门细菌[184]。丁酸盐通过诱导结肠Tregs细胞分化、抑制炎症反应、维持黏膜完整性来促进结肠健康和抗击肿瘤，在维持结肠黏膜健康方面发挥重要作用[185，186]。当罗氏菌属的细菌丰度降低时，肠道丁酸盐的生成也会减少[187]，这可能是过敏性紫癜发病的一个重要的危险因素。

罗氏菌属被认为是梭状芽孢杆菌类的重要一员，可以像梭状芽孢杆菌一样发挥抗炎作用。在过敏性紫癜患者的肠道菌群中，罗氏菌属的种群也有所减少[179]。对大量食物过敏病例的统计分析发现，食物过敏组罗氏菌属含量低于对照组[188]。罗氏菌属的丰度与湿疹也呈负相关[189]。因此，过敏性紫癜患者罗氏菌属的减少，导致其肠道抗炎作用及免疫力下降，加重了过敏性紫癜的进展。

过敏性紫癜患者的肠道菌群中，狄氏副拟杆菌属和肠球菌属数量显著增加[179]。在健康婴儿中，狄氏副拟杆菌属的数量减少，而在患有湿疹的婴儿中，狄氏副拟杆菌属的数量增加[190]。狄氏副拟杆菌属在感染性疾病中通常是条件致病菌，能够产生耐药性[191]。已有研究表明，狄氏副拟杆菌属的丰度与结肠紧密连接蛋白occludin和ZO-1的mRNA水平呈负相关，这两种主要的紧密连接蛋白与肠道通透性密切相关[192]，这与之前报道的过敏性紫癜肾炎患者肠道通透性增加一致。另外，在食物过敏或湿疹患儿中，肠球菌属的丰度也显著增加[193]。肠球菌属也是一种重要的机会性病原体，当肠道微生物群处于失调状态时，它可以导致机会感染[194]。因此，在过敏性紫癜患者菌群失调的情况下，肠球菌属可能引起肠道感染，而狄氏副拟杆菌属会通过增加肠道通透性而导致腹腔感染及菌血症的发生。

双歧杆菌属的数量在过敏性紫癜患者中下降。据报道，在不同的屏障损伤模型中，双歧杆菌能够通过恢复肠道通透性、结肠杯状细胞群和细胞因子水平来维护屏障功能[195]。双歧杆菌可能在保护过敏性紫癜患者肠道通透性、降低 IgA 水平方面发挥作用。

（三）益生菌在过敏性紫癜中的作用

益生菌可改善微生物的平衡，尤其是调节胃肠菌群平衡[196]。肠道菌群可能会调节免疫系统和全身炎症反应，从而影响致敏和过敏反应的发展，2015 年世界卫生组织（World Health Organization，WHO）已推荐使用乳酸杆菌和双歧杆菌等益生菌预防过敏性疾病[197]。益生菌可通过抑制肠道病原菌生长、改善肠道屏障功能，调节肠道免疫反应来协助肠道发挥正常免疫功能。益生菌的疗效已在随机对照试验中进行了评估，研究发现牛奶过敏的患儿在接受乳酸杆菌治疗后，肠道内的罗氏菌属特异性菌株有所增加[198]。益生菌可能有助于增加过敏性紫癜患者肠道菌群中有益的罗氏菌属的数量。

益生菌在预防和治疗过敏性疾病中得到了广泛关注，补充益生菌可预防婴儿湿疹，这为妊娠期和婴儿期使用益生菌提供了新的潜在适应证[199]。也有大量研究证实，补充益生菌有助于减轻特应性皮炎的严重程度[200]。过敏性疾病肠道菌群的紊乱常早于疾病，因此早期干预可以预防和治疗疾病，但是益生菌干预的种类、剂量及时间仍不确定，需要进一步大规模、多点临床试验。利用益生菌改善肠道菌群的策略可能会改善过敏性紫癜患者的临床预后，今后还需要设计药物临床试验，来进一步验证益生菌对过敏性紫癜的治疗效果。

第四节　肠道菌群与类风湿关节炎

一、概述

类风湿关节炎（rheumatoid arthritis，RA）是一种由免疫介导的慢性炎症性疾病，以慢性关节炎为特征，可并发肺部疾病、心血管疾病、恶性肿瘤及抑郁症等。RA 典型病理过程为关节滑膜炎、血管翳形成，关节软骨和骨渐进性破坏，随疾病进展，最终导致关节畸形和功能丧失，残疾率较高。流行病学调查显示，RA 的全球发病率为 0.5% ～ 1%，中国大陆地区发病率为 0.42%，总患病人群约 500 万。RA 任何年龄均可发病，男女患病比率约为 1 ∶ 4。RA 发病机制尚不明确，目前普遍认为，在遗传、感染、环境等多因素共同作用下，自身免疫反应引起的免疫损伤和修复是 RA 发生发展的基础[201, 202]。

RA 虽有家族聚集倾向，但单卵双生子 RA 的符合率约为 15%，提示除外遗传易感性，环境因素对 RA 的发生同样重要。吸烟、激素、微生物群和感染等多种环境因素可能与 RA 的发病有关[203-205]，其中肠道菌群在模型小鼠关节炎的发生发展中起着重要作用[206-209]。在 RA 发病的前几年，可以检测到类风湿因子和抗瓜氨酸蛋白抗体[210, 211]，表明 RA 与黏膜组织高度相关，如肠道和口腔。米诺环素或柳氮磺吡啶等抗菌药物对某些类风湿关节炎患者有效，也提示肠道和口腔菌群与 RA 相关[212]。

二、肠道菌群与类风湿关节炎的发生发展

RA 是遗传、环境和激素因素以及免疫系

统之间复杂相互作用的结果[206]，其发病机制复杂，其中包括先天性和适应性免疫反应之间的相互作用，如抗原呈递细胞、自身反应性T细胞形成、针对自身抗原的自身抗体、类风湿因子和抗环瓜氨酸肽抗体（anticitrullinated protein antibody，ACPA）。RA抗体常出现在临床症状之前，表明自身免疫反应的触发可能发生在关节之外，如消化道或气道[213]。RA的遗传易感性表现为家族聚集倾向，单卵双胞胎之间的一致性及与某些组织相容性抗原的关联性[214, 215]。目前，发现约有100个基因与RA相关[216]，人类白细胞抗原（human leucocyte antigen，HLA）多态性是最重要的遗传危险因素。

主要组织相容性复合物由抗原呈递细胞表达，抗原呈递细胞将抗原呈递给$CD4^+$T细胞，从而刺激B细胞产生抗体。翻译后修饰对蛋白质的功能和抗原性至关重要。3种主要参与类风湿关节炎的翻译后修饰是糖基化、氨甲酰化和瓜氨酸化。瓜氨酸化指精氨酸通过精氨酸脱氨酶转化为瓜氨酸的过程，是RA自我抗原识别的关键转录后修饰[4]。牙龈卟啉单胞菌及放线菌可以通过不同的精氨酸脱氨酶完成瓜氨酸化修饰[217]。瓜氨酸化可改变蛋白质结构，形成与ACPA相关的新抗原表位。随着时间的推移，尤其是在关节炎发病前，ACPA的滴度和表位多样性增加。ACPA可以是具有改变糖基化状态的同种IgG、IgA或IgM，而增强Fc受体和瓜氨酸抗原的亲和性[218]。ACPA本身可通过免疫复合物形成和Fc受体结合激活巨噬细胞或破骨细胞，也可能通过结合瓜氨酸化波形蛋白，从而促进骨丢失而致病。

（一）环境与遗传因素相互作用促进对含有瓜氨酸残基自身蛋白质的免疫应答

ACPA是一类针对含有瓜氨酸化表位自身抗原的抗体统称，其中抗环瓜氨酸抗体表达率较高，75%的患者呈阳性。ACPA在疾病早期出现，甚至在RA临床症状发作前几年就出现[219]。早期ACPA的靶点较少，随着关节炎病情进展，其靶点不断增加。瓜氨酸化蛋白修饰可以产生新抗原，激活T细胞，进而为B细胞产生ACPA呈递更多抗原。随着自身抗体进一步增加，RA患者关节滑膜中促炎症细胞因子的水平升高。ACPA靶向性扩增与炎性介质的产生和临床炎症的出现相关。事实上，瓜氨酸化增加了抗原与HLA-DRB1共有表位的亲和力[220]。

RA与某些HLA-DR等位基因之间存在着强烈的相关性，即所谓的“共有表位”假说[221]。此外，也有研究表明遗传因子HLA-DRB1共有表位与ACPA相关，而不是RA[222]。从RA患者外周血单个核细胞中分离到一种HLA-DR呈递肽Pc-p27，该肽源于普雷沃菌。在42%的受试者中，Pc-p27可刺激Th1反应，也对整个普雷沃菌呈现免疫反应[223]，进一步分析发现，RA患者滑膜中鉴定出两种由HLA-DR呈递的自身抗原，N-乙酰氨基葡萄糖-6-硫酸酯酶（N-acetylglucosamine-6-sulfatase，GNS）和细丝蛋白A（filamin A，FLNA）。RA患者T细胞和B细胞对GNS和FLNA的应答率分别为52%和56%。GNS呈现瓜氨酸化，其抗体值与ACPA水平相关。GNS和FLNA均与普雷沃菌抗原表位具有序列同源性[224]。此外，对GNS和FLNA产生T细胞反应的RA患者，也对普雷沃菌蛋白产生相同的反应，这表明RA对普雷沃菌的黏膜免疫与滑膜自身免疫反应之间存在着联系。

另一方面，RA与牙周炎存在显著相关性[225]。研究人员从牙周炎患者的牙龈细胞液检测出广泛的蛋白瓜氨酸化[226]，与RA高瓜氨酸化相似，而在没有牙周炎的对照组中显示出很少量的瓜氨酸化。牙龈沟中的放线菌通过宿主中性粒细胞诱导了高瓜氨酸化，且HLA-DRB1共有表位等位基因对自身抗体阳性的影响仅限于接触放线菌的RA患者。对牙周炎患者的龈沟液和牙龈组织进一步分析[205]，鉴定出一种新型的瓜氨酸化细胞角蛋白 13（cCK13），24% 的类风湿关节炎患者存在抗 cCK13 的 ACPA，并与普雷沃菌相关，而抗 α- 烯醇化酶、波形蛋白和纤维蛋白原的 ACPA 与吸烟和 HLA-DRB1 共有表位相关。

牙龈卟啉单胞菌是牙周炎的主要致病菌，其特殊的精氨酸脱氨酶可使游离 L- 精氨酸和 C- 末端精氨酸残基瓜氨酸化。在实验小鼠模型中，牙龈卟啉单胞菌可致关节炎恶化 [217]。吸烟和其他刺激也可能通过精氨酸脱氨酶激活启动瓜氨酸化，形成可增强抗原呈递的淋巴结构以及 T 和 B 细胞的产生 [227]。在 RA 患者和对照组的结肠组织中发现了 21 种瓜氨酸肽 [228]，RA 患者的肺组织和滑液中亦发现这些肽类的存在。3 种瓜氨酸蛋白（瓜氨酸波形蛋白、纤维蛋白原 α 和肌动蛋白）是已知的 ACPA 靶点，提示结肠黏膜可能是瓜氨酸表位免疫耐受的潜在破坏位点。综上所述，消化道菌群与多种因素共同促进了 RA 的发生。

（二）模式识别受体介导的免疫反应

肠黏膜调节机制较为复杂，不仅要维护屏障功能以免受外来抗原和正常肠道微生物抗原所致的免疫损伤，又要对病原体产生有效的免疫应答以抵御致病微生物的入侵 [229]。模式识别受体（pattern-recognition receptors，PRRs）是能识别病原微生物的重要成分即病原体相关分子模式（pathogen-associated molecular pattern，PAMP），可激活一系列信号通路，是引发先天免疫反应的受体，对免疫系统识别自我和非我成分有重要作用。PAMP 是许多病原微生物所共有，而人体宿主所没有的结构恒定且进化保守的分子结构。PAMP 包括多种微生物组成成分，如蛋白质、脂类、糖和核酸。PRRs 包括多种受体，如同样表达在膜上的 Toll 样受体、C 型凝集素受体、表达在胞质的 NLR、维 A 酸诱导基因 I 样受体和胞质多种 DNA 感应分子等。肠上皮中的所有细胞都表达 PRRs，是第一批与肠道微生物群接触的细胞。人类和小鼠模型的几项研究表明，肠道中的 PRR 信号传导适应于防止常驻菌群持续过度激活，上皮细胞中的 PRR 信号在维持上皮细胞的稳态和屏障功能方面具有重要作用。

Toll 样受体具有保守的跨膜糖蛋白结构域，胞外区域有配体识别结构域，胞质尾部有 Toll/IL-1 受体同源性结构域，其对于 TLR 亚基之间的相互作用以及适配器蛋白（如 MyD88）的募集以启动下游信号级联是必不可少的。TLR 分布在细胞表面及胞内囊泡（如内质网、溶酶体等）上，可以诱导炎症反应和产生促炎症细胞因子如 TNF-α、IL-6、IL-1β。人体中发现 10 余种 TLR，每个 TLR 都有自己的配体，如脂多糖、脂蛋白、鞭毛蛋白和细菌核酸。TLR2 参与了多种革兰阳性细菌化合物和钩端螺旋体脂多糖的识别，而 TLR4 和 TLR9 分别参与了脂多糖和非甲基化 CpG DNA 的固有反应。此外，热休克蛋白 60 是 TLR4 的内源性配体，在 RA 的诱导中起一定作用 [230]。研究表明，细菌合成脂肽或脂多糖刺激 RA 滑膜成纤维细胞后，TLR2 的表达显著增加；用促炎细胞因子 IL-1β 和 TNF-α 刺激成纤维细胞也导致 TLR2 mRNA 的增加；原位杂交显示，类风湿关节炎患者的滑膜和软骨或骨组织 TLR2 mRNA 表达明显增加。早期类风湿关节炎患者的微生物组成与对照组不同，拟杆菌科细菌减少 [231, 232]，而抗风湿药物治疗后，改变的微生物群部分恢复，临床症状改善 [214, 232, 233]。

NOD 样受体是 PRRs 的另一个主要家族，其共同特点是都具有核酸结合域，这些受体位于细胞质中。NOD1 和 NOD2 是这个家族在肠道中表达最多的成员，NOD2 基本上局限于小肠的潘氏细胞，两者都能识别肽聚糖。细菌肽聚糖中的一种保守的分子结构——胞壁酰二肽（N-acetylmuramyl-L-alanyl-D-isoglutamine or Muramyldipeptide，MDP）可以促进小鼠胶原诱导关节炎的发展[234]。利用单克隆抗体2E7中和MDP后，小鼠关节炎显著改善。而恢复正常的小鼠注射小剂量脂多糖病情再次恶化，后再次注射2E7依然可以有效控制关节炎。此外，发病前注射2E7抗体的小鼠几乎不会发生关节症状。研究还发现，2E7可以保护关节软骨，减少关节中的中性粒细胞和巨噬细胞，抑制Th17细胞的分化而削弱适应性免疫；给NOD2缺陷的小鼠注射2E7却没有保护效应，表明2E7是通过抑制NOD2对循环肽聚糖激活的炎症反应来发挥抗关节炎作用。

（三）诱导T细胞亚群分化调控免疫系统

黏膜中树突状细胞与抗原相互作用后，可促进幼稚CD4$^+$淋巴细胞分化为Th1、Th2、Th17、Treg和Tfh等亚群。菌群组成的多样性和丰富度水平的改变，即菌群失调，可通过诱导T细胞亚群分化失衡而引发多种自身免疫性疾病[235]。RA患者存在循环Treg细胞功能缺陷，血浆和滑液中Th17增多[236]。部分机制是巨噬细胞源性和树突状细胞源性TGF-β、IL-1β、IL-6、IL-21和IL-23促进Th17细胞分化，同时抑制Treg细胞分化，从而将内环境稳态转化为炎症状态[209]。一个或多个黏膜部位的菌群失调可导致免疫系统改变和对瓜氨酸化自身抗原的免疫耐受破坏[237]。已经证实，肠道共生菌群可以调节Treg细胞，这有助于宿主防御病原体的入侵，同时可避免自身免疫反应被过度激活[228]。

Treg细胞亚群可调节T细胞免疫反应。肠道Treg细胞在维持对饮食抗原和肠道微生物群的免疫耐受方面发挥着重要作用[238]。CD4$^+$CD25$^+$Treg细胞是表达转录因子Foxp3的抑制性细胞，通过抑制异常或过度免疫应答，维持机体的免疫耐受和稳态。乳酸杆菌和双歧杆菌通过诱导CD4$^+$CD25$^+$FoxP3$^+$Treg细胞分化发挥抗过敏作用[239]。脆弱类杆菌多糖A作为免疫调节剂，通过IL-2刺激CD4$^+$Treg细胞分泌IL-10，抑制T细胞增殖，减少炎性介质，缓解滑膜炎[239]。

肠道菌群失调可通过影响T细胞亚群的分化而诱发关节炎，也可影响抗原呈递细胞Toll样受体的表达水平，导致Th17/Treg细胞比例失衡。随着Th17细胞的发育，组织损伤和易感个体激活局部免疫级联反应，这种局部免疫反应可发展为系统性自身免疫反应，波及多个器官。研究表明，分节丝状菌可以诱导Th17分化，促进K/BxN小鼠模型关节炎进展[240]。Tfh细胞主要功能是辅助B细胞增殖和产生抗体，参与体液免疫，分节丝状菌通过诱导Tfh细胞向全身淋巴部位迁移，导致自身抗体产生增加，触发自身免疫性关节炎，导致疾病进展[241]。肠道菌群可通过Tfh细胞促进关节炎的发展，而不依赖于Th17细胞[240，242]。

三、不同菌株在类风湿关节炎中扮演不同角色

乳酸杆菌是肠道重要的益生菌属和微生物之一，在维持和调节肠道菌群稳态中发挥重要作用[243]。动物模型研究发现，口服干酪乳杆菌和德氏乳杆菌后，关节肿胀和炎症减轻，炎症因子分泌减少，说明乳酸杆菌可以影响RA的病情，抑制疾病进展[244-246]。然而，在随机、对照、双盲临床试验中，口服鼠李糖乳杆菌或罗伊氏乳杆菌后，RA患者的症状无明显缓解[247，248]。另外，有学者利用干酪乳杆菌细胞壁成功诱导出了动物类风湿关节炎[249]，证明乳酸杆菌细胞壁是促使类风湿关节炎发生的因素之一[250]。因此，乳杆菌属的不同菌株在RA中的作用不尽一致。

与健康受试者相比，新发未经治疗的RA患者的肠道菌群谱显示，普雷沃菌数量增加而拟杆菌减少。普雷沃菌能够增加IL-17的产生以及ACPA反应，导致RA进展[223]。而另有研究发现，普雷沃菌可通过调节肠道树突状细

胞和 $CD4^+CD25^+FoxP3^+$Treg 细胞的生成，抑制 Th17 应答，减少促炎细胞因子 IL-2、IL-17 和 TNF-α，同时增加 IL-10 的转录，从而抑制关节炎的发展[251]。因此，这取决于环境因素。这解释了普雷沃菌在健康人菌群中大量存在的原因，并表明只有某些菌株可能具有致病性。

同菌种的不同菌株在 RA 病程中发挥不同作用，提示肠道微环境对 RA 患者菌群失衡有重要作用。综上，肠道环境多因素共同作用，特定菌种的丰度明显增加，促进 RA 病情进展。

四、类风湿关节炎治疗药物对肠道菌群的影响

抗类风湿药物具有免疫调节作用，通过多种途径抑制免疫应答，在控制 RA 症状和疾病进展方面发挥着重要作用。肠道菌群也直接或间接地调节宿主免疫系统，RA 与肠道菌群的失调有关，用抗类风湿药物治疗 RA 患者，其肠道益生菌群可部分恢复[252]。微生物群具有多样的代谢潜能，可以代谢药物等外源性化合物，并对药物稳定性和活性有重大影响[253, 254]。研究表明，RA 患者肠道菌群失调与革兰阴性菌的减少和革兰阳性菌的富集有关；RA 患者与健康受试者相比有大量的柯林斯菌群，而抗类风湿药物治疗的患者肠道菌群部分正常[214, 255]。

传统抗类风湿药物，包括甲氨蝶呤、柳氮磺吡啶、羟氯喹和来氟米特，可以减轻炎症，控制疾病活动，减少关节破坏，改善生活质量，通常作为一线药物。其中部分药物还具有抗菌活性，它们可能直接或间接地引起患者菌群的改变，而菌群组分的改变可能对类风湿关节炎患者有益[214, 255]。

经甲氨蝶呤治疗后，RA 患者口腔菌群丰度与抗环瓜氨酸肽抗体、类风湿因子和 C 反应蛋白水平呈负相关，表明甲氨蝶呤可改善患者口腔菌群[214]。在另一项研究中也观察到类似的结果，甲氨蝶呤治疗的患者与接受其他治疗的患者相比，菌群丰度和多样性增加，这表明通过治疗菌群可能恢复正常[255]。

柳氮磺吡啶含有磺胺，具有抗菌活性，可能对肠道微生物有直接影响[206]。另一方面，宿主的肠道菌群可以影响柳氮磺吡啶的激活及其反应，因为柳氮磺吡啶需要在细菌酶水解下才能乙酰化为活性代谢物 5- 氨基水杨酸[236]。研究显示，柳氮磺吡啶可以改变粪便菌群，导致 RA 患者的大肠埃希菌、拟杆菌、变形菌减少，而芽孢杆菌增加，促进短链脂肪酸产生，对肠道产生有益作用[256]。

羟氯喹是一种抗疟药物，兼具抗炎活性，也用于治疗 RA。它能减轻关节炎疼痛、关节肿胀，并可降低致残风险。羟氯喹通过增加溶酶体 pH 和减少 MHC Ⅱ类蛋白中的抗原呈递及肽载量来抑制自身抗原的生成，从而导致 $CD4^+$T 细胞增殖减少和细胞因子释放减少[257]。此外，它还可以干扰 TLR 依赖的信号传导。羟氯喹治疗的患者与未接受治疗的患者相比，微生物多样性有所增加[255]。

对传统的抗类风湿药物治疗无反应时，可选用生物制剂。生物制剂作用于特异性的免疫介质，如 TNF-α、IL-1 和 IL-6，在减轻症状、减缓疾病进展和改善生活质量方面有效。生物制剂根据作用靶点可分为多个亚组：TNF-α 阻滞剂通过结合 TNF-α 阻断其下游效应，包括依那西普、英夫利昔单抗、阿达木单抗和高利木单抗等；T 细胞共刺激调节剂，靶向 T 细胞和抗原呈递细胞刺激受体 CTLA4；靶向 $CD20^+$ B 细胞和细胞因子，如 IL-1、IL-1R 拮抗剂和 IL-6、IL-6R 拮抗剂。肿瘤坏死因子阻滞剂依那西普临床运用最为广泛。TNF-α 是一种促炎症细胞因子，参与 B 细胞途径的炎性核因子 κ 轻链增强剂的激活，还激活 c-Jun 氨基末端激酶（JNK）通路，诱导细胞分化和增殖。依那西普通过抑制 TNF-α 及其相关下游通路抑制宿主免

疫反应。蓝细菌门产生的某些次级代谢产物具有抗炎、抑制细胞分裂、抑制蛋白酶等免疫抑制活性，可能对RA患者有益[233]。与未治疗的RA患者相比，依那西普治疗组蓝藻的丰度显著增加。

五、展望

研究表明，肠道菌群与RA的发生及发展密切相关，但这些结果主要是基于模型小鼠的研究数据，啮齿类与人体复杂的肠道环境存在较大差异，将来仍需进一步研究菌群与菌群之间、菌群与RA患者之间的复杂关系。与宿主共同进化的微生物谱系的改变与人类代谢性疾病和自身免疫疾病的发生有关，因此，将细菌作为RA患者治疗手段备受重视，其中益生菌的研究已成为一大热点。

另一方面，作为肠道生态系统稳定的主导力量，肠道菌群和上皮细胞之间的相互作用是新型益生菌或治疗药物的基础，益生菌有望作为肠道修复、平衡肠道菌群、调节免疫反应的重要手段而用于RA治疗[248, 258]。此外，一些抗类风湿药物通过诱导宿主菌群改变、产生肠道优化作用而产生“益生菌”效应，或可为抗风湿药物的使用与研发拓展新方向。ACPA的出现早于RA的临床表现，这段空窗期是否可作为预防窗口，肠道菌群调节作为干预手段是否可以阻止或延缓疾病的进展，这有待进一步研究，也值得期待。

第五节　肠道菌群与系统性红斑狼疮

一、概述

系统性红斑狼疮（systemic lupus erythematosus，SLE）是一种侵犯全身多系统、多器官的慢性自身免疫性结缔组织病。SLE的特征是先天和适应性免疫系统失调，自身抗原抗体复合物沉积导致的全身多脏器和多组织受损。SLE好发于育龄女性，以15～45岁年龄段最多见，男女比例是1∶9，除性别、年龄外，还有人种差异，黑种人最高，我国汉族次之，全球平均患病率为（12～39）/10万，黑种人为100/10万，我国发病率为（30.31～70.41）/10万。目前SLE仍然无法彻底治愈，常用的治疗药物是类固醇皮质激素联合免疫抑制剂，其他治疗药物包括非甾体抗炎药和羟氯喹等。

SLE最突出的免疫异常表现是T淋巴细胞功能异常、B淋巴细胞的高度活化和产生多种自身抗体、免疫复合物沉积于多种组织器官。SLE表现为对各种自身抗原的耐受性丧失。生理情况下，凋亡过程中释放的核抗原通常会在不激活免疫系统的情况下被清除。然而在SLE中，外部刺激（例如紫外线、感染或毒素）可能会增加凋亡细胞的负荷[259]，导致含有核酸的细胞碎片水平增高，通过TLR9或TLR7等核酸识别受体的作用激活Ⅰ型干扰素途径[260, 261]。Ⅰ型干扰素和其他细胞因子促进B细胞成熟和活化[262, 263]。B细胞过度活跃是SLE的标志，B细胞产生针对自身抗原的高亲和力自身抗体，自身抗体和免疫复合物通过激活补体，并结合炎性细胞上的Fcγ受体来介导炎症和组织损伤。研究表明，T细胞参与了B细胞的活化和自身抗体的产生。辅助性T细胞的增加和调节性T（Treg）细胞的缺陷与SLE发病有关[264, 265]。SLE患者还表现出双阴性（$CD4^-CD8^-$）T细胞的扩增，从而促进致病性自身抗体的产生[266, 267]。因此，B细胞和T细胞的激活都会增强针对自身的先天性和适应性免疫反应[268, 269]。

SLE 的病因尚不清楚，目前认为与环境、雌激素和遗传等因素有关。单卵双胎之间 SLE 发病率的差异提示，除外遗传易感性，环境因素也发挥重要作用。紫外线、激素、感染（如埃伯斯坦 - 巴尔病毒和巨细胞病毒）、吸烟、毒素（如硅尘）和药物都是 SLE 的潜在诱因[270, 271]。紫外线会诱导 DNA 损伤和角质形成细胞凋亡，从而增加自身抗原的暴露[272]。紫外线可通过抑制狼疮小鼠 I 型 IFN 依赖性的 Treg 细胞，诱导 T 细胞活化[273]。雌激素和催乳素可激活免疫系统，与活动性 SLE 密切相关[274, 275]。口服避孕药和绝经后激素治疗可增加 SLE 的风险[276, 277]。

有学者分析 SLE 与肠道微生物菌群之间的关系，发现肠道菌群可能是被忽略了的重要因素[5, 278, 279]。这一发现也再次证实了肠道微生物在人体健康和疾病中扮演重要角色，肠道菌群对于宿主的新陈代谢、激素分泌、消化功能和免疫系统的发育和维持免疫稳态至关重要[280, 281]。研究发现，SLE 患者的肠道菌群与 Th17 和 Treg 细胞比例失衡有关[282]。此外，早期肠道菌群的组成决定了抗核抗体的发展[283]。对 SLE 肠道微生物的组成及作用的深入研究有助于进一步阐明 SLE 病因、发病机制和发现新的诊断标志物。

二、系统性红斑狼疮患者肠道菌群组成特点

迄今为止，SLE 的病因、各种危险因素之间的联系和信号通路仍不完全明确。不少学者尝试破译 SLE 相关的肠道微生物特征，期望对 SLE 有进一步的了解。新近的研究对 SLE 缓解期患者与健康对照组人群的肠道菌群进行了分析。其中，一项研究以 20 例西班牙女性 SLE 患者为研究对象，发现其肠道菌群中硬壁菌门 / 拟杆菌门比例降低，厚壁菌门的丰度降低[284]，在 2 型糖尿病和克罗恩病中也发现了同样的改变[285]，因此，这并不是 SLE 的特征性变化[286]。第二项研究分析了 45 例 SLE 患者的肠道菌群，发现肠道厚壁菌门数量和丰度较低，而拟杆菌门数量和丰度较高[287]。此外，SLE 患者与对照组相比，红球菌、伊格尔兹菌、克雷伯菌、普雷沃菌、真杆菌和梭状芽孢杆菌显著富集，而侵肺拟杆菌和假丁酸弧菌明显降低。第三项研究比较了 83 例 SLE 患者和 16 例健康成人的肠道菌群，发现 SLE 患者肠菌多样性降低，表现为变形杆菌的丰度增高，而沙门菌的丰度降低。在未经治疗的或正在接受治疗的 SLE 患者均可见到这些肠菌的改变。此外，普雷沃菌在 SLE 组明显增高。粪便代谢组学研究发现，SLE 患者嘌呤、嘧啶和氨基酸代谢均减低[288]。

此外，在 SLE 患者中，厚壁菌门的相对丰度降低，而拟杆菌的相对丰度增加，硬毛虫与拟杆菌的比例降低[289-291]。有研究首次报道 61 例女性 SLE 患者肠道菌群中存在过多的鲁米诺球菌[292]。活泼瘤胃球菌（*ruminococcus gnavus*，RG）RG2 菌株与抗双链 DNA（double-stranded DNA，dsDNA）抗体存在交叉反应，而 RG1 菌株与该抗体无交叉反应[292]。两个独立的队列研究（包括男性患者和多样化的种族）发现，活动性狼疮肾炎患者表现出对 RG2 的免疫反应[292]。因此，血清抗 RG2 IgG 的水平可作为 SLE 疾病活动性和狼疮性肾炎的指标。然而，血清抗 RG2 IgG 应答水平升高对 SLE 是否具有特异性尚不十分清楚。

迄今为止，尚无研究分析 SLE 患者活动期与缓解期之间的微生物谱变化，因多数患者在活动期均接受药物治疗，可能会改变其真实的肠道微生物群。

三、系统性红斑狼疮小鼠模型肠道菌群组成特点

自发性和诱发性 SLE 动物模型已被用于研究肠道微生物群对 SLE 的影响。在 Fas 蛋白编码基因中具有 lpr 突变的 MRL/lpr 小鼠可以自发地出现多种狼疮自身抗体和狼疮表现[293]。MRL/lpr 狼疮小鼠的肠道菌群表现为乳杆菌科比例降低，毛螺菌科比例增高[293]。乳酸杆菌作为人类胃肠道和泌尿生殖道的常驻微生物群，其中一些乳杆菌属因具有抗炎作用而被作为益生菌用于临床[294]。毛螺菌科（属于梭状芽孢杆菌）是人类肠道菌群的主要成分，包括许多产生丁酸盐的细菌[295]，可预防结肠癌并可能减轻肥胖。SLE 疾病活动指数（systemic lupus erythematosus disease activity index，SLEDAI）的严重程度（淋巴结病和肾小球肾炎）与乳酸杆菌科的相对丰度成反比，与毛螺菌科的相对丰度成正比。

乳酸杆菌属的相对缺乏在疾病发作之前最为突出[296]。因此，乳酸杆菌可能在狼疮的发病机制中起到预防作用。乳酸杆菌的干预可降低 MRL/lpr 小鼠的蛋白尿和狼疮自身抗体水平，并改善肾脏病理学评分[296]。然而，乳酸杆菌在不同的狼疮小鼠模型中的作用不尽一致。NZB/WF1 小鼠肠道菌群的动力学研究发现[297]，狼疮发作前后，乳酸杆菌水平变化尤为明显。狼疮的发展过程中，乳酸杆菌的相对丰度急剧增加，地塞米松可逆转其上升趋势。乳酸杆菌的相对丰度与较差的肾功能和较高水平的全身免疫反应呈正相关。

在 TLR7 依赖的小鼠狼疮模型中，如 TLR7.1 Tg 小鼠和咪喹莫特（一种 TLR7 激动剂）诱导的小鼠模型，发现罗伊乳杆菌会加重狼疮[298]。在这类模型小鼠粪便中，罗伊乳杆菌、脱硫弧菌属和里肯菌科较为丰富。但是，仅乳杆菌属（罗伊乳杆菌和约翰逊乳杆菌）会转移到内部器官，从而诱导 IFN 基因表达和介导全身免疫反应。抗性淀粉（发酵成短链脂肪酸）在体外和体内均抑制罗伊氏乳杆菌的生长，降低肠道上皮的通透性，降低 I 型 IFN 的表达并改善狼疮性肾炎。在各种狼疮小鼠模型中，乳酸杆菌的相对丰度和由乳酸杆菌介导的全身免疫如何影响狼疮发展的潜在机制并不完全相同，还可能与肠道菌群组成及其与宿主间的相互作用有关，如 IFN-α 在 TLR7 依赖性小鼠模型中起主要作用，而 IFN-γ 在 MRL/lpr 小鼠中显得更为重要[299]。

研究发现，在 NZB 无菌鼠自发性红斑狼疮模型中，肾脏疾病发生率和严重程度均较低，其血清抗核因子水平较高，循环血中可产生免疫复合物，然而肾脏中免疫复合物的沉积减少。相反，MRL/lpr 模型的无菌小鼠和常规小鼠之间并未观察到此差异，当无菌 MRL/lpr 小鼠接受无抗原超滤饮食后，发现肾炎减轻和淋巴结缩小，而自身抗体水平无明显差异。研究发现，MRL/lpr 模型中肠道乳酸杆菌丰度减少，鞭毛科丰度增加，雌性小鼠的菌群多样性增加，在狼疮进展中毛螺菌和梭菌水平增高[293]。此外，在自发性红斑狼疮鼠模型中，雌性和雄性模型之间的细菌存在差异，表现为雌性小鼠模型发病较早和症状较重，推测与其鞭毛藻科数量升高有关。给予雌性小鼠维 A 酸（维生素 A）后，乳酸杆菌数量恢复正常水平，其临床症状也相应改善[293]。既往研究已证实，维生素 A 可调节宿主免疫反应并改变肠道微生物组成[300]。此外，罗伊乳杆菌可抑制 B6Sle.123 和 NZBx NZW F1 两种自发模型小鼠狼疮的发展，出现 Tregs 水平升高，生存率提高[301]。而另一种 SNF1 小鼠模型中，疾病严重程度与其拟杆菌水平有关[302]。

四、肠道菌群与系统性红斑狼疮的发病机制

（一）肠漏和肠道菌群易位

健康人肠黏膜屏障始终保持完整性，肠道

细菌不会转移到内部器官。如果肠道屏障功能破坏，肠道共生细菌便会暴露于免疫系统，“泄漏的肠道”会引发全身性自身免疫反应。在小鼠模型中已充分证实，在狼疮发病过程中，肠道屏障受损 [5，296，298]。易患狼疮的 MRL/lpr 和（NZW BXSB）F1 小鼠的肠上皮紧密连接蛋白水平降低 [5]。在肠漏的情况下，细菌会转移到全身组织进而激活抗原呈递细胞，抗原呈递细胞又会迁移到肠系膜淋巴结 [5]，继之，激活 $CD4^+T$ 细胞，后者活化后释放炎症细胞因子，如 IL-6。IL-6 通过诱导 B 细胞产生自身抗体并抑制 Treg 细胞的活性，在 MRL/lpr 小鼠的狼疮进展中起关键作用 [303]。

抗生素治疗或无菌狼疮小鼠表现出肠道完整性增强和全身自身免疫反应性降低 [5，298]。因此，肠道菌群影响肠上皮的完整性和狼疮的发展。Manfredo-Vieira 等最早报道了鸡肠球菌在肠漏和细菌易位诱导狼疮自身抗体发展中的作用 [5]。通过组织培养和原位检测，在（NZW BXSB）F1 小鼠的肠系膜淋巴结和肝脏中均检测到肠道内共生的鸡肠球菌 [5]。鸡肠球菌可增加浆细胞样树突细胞的数量，此外，鸡肠球菌诱导的自身抗原可激活 AhR-CYP1A1 途径，触发 Th17 细胞活化和抗 dsDNA 抗体的产生 [5]。对 SLE 患者的研究显示，肠道屏障受损后，肠腔内容物不仅渗透至全身组织，而且全身免疫球蛋白也会渗入肠腔 [292]。

（二）分子模拟

研究发现，狼疮自身抗原 Ro60 与 EB 病毒核抗原 -1（epstein-Barr virus nuclear antigen-1，EBNA-1）可发生交叉反应，表明狼疮体液免疫是通过 EBNA-1 和 Ro60 抗原的分子模拟而启动的 [304]。

共生细菌可在易感人群中触发自身反应性 T 和 B 细胞反应。研究发现，细菌可表达人类 Ro60 自身抗原的直系同源物 [279]。在人体（包括 SLE 患者）的口腔、皮肤和肠道中普遍存在表达 Ro60 直系同源物的共生体，其可向免疫细胞传递抗原 [279]。口腔、肠、皮肤衍生的肽和阴道菌群可激活 Ro60 反应性 T 细胞 [305]。来自 SLE 患者的反应性 T 细胞克隆与人 Ro60 自身抗原可发生交叉反应 [279]。研究发现，抗 Ro60 阳性患者的血清中含有 SLE 免疫沉淀的细菌核糖核蛋白复合物，后者含有 Ro60 直系同源物 [279]。此外，C57BL/6 小鼠定植了含直系同源物的 Ro60 肠道拟杆菌，后者在血液中表达人抗 Ro60 抗体 [279]。总之，抗体和 T 细胞交叉反应都会触发 SLE 患者的抗 Ro 反应。

活泼瘤胃球菌的过度生长与狼疮疾病活动和狼疮肾炎有关 [292]。IgG 抗 RG2 抗体滴度高的患者表现出活动性狼疮性肾炎。在各种活泼瘤胃球菌菌株中，RG2 的提取物（而非 RG1）与狼疮抗 dsDNA 抗体发生交叉反应，这表明交叉反应性受到菌株限制。RG2 菌株的细胞壁脂多糖含有抗 dsDNA 抗体反应的抗原特性。RG 细胞壁部分与天然 DNA 分子之间的分子模拟可能触发或加剧 SLE 和狼疮性肾炎。

（三）肠道菌群的性别二态性

微生物种群可通过影响宿主激素水平而参与 SLE 的发病机制。SLE 在育龄妇女中比在男性中更为普遍，表明性激素在该疾病中可能起一定作用。以往研究发现，雌激素可加剧 SLE[306]。睾丸激素可能在 1 型糖尿病中具有保护作用，这种保护效应可能是由肠道微生物群所介导的 [307]。在 WF1 狼疮模型中，雌性和雄性小鼠肠道微生物组成具有明显的差异，雌性小鼠在干预早期即进展为 SLE，而雄性小鼠不易发展为 SLE 或进展较为缓慢，阉割后的雄性小鼠发病则与雌性小鼠相似。给雌性 BWF1 小鼠喂食雄性 BWF1 小鼠的粪便则可显著提高其生存率，减轻肾脏症状和减少抗 dsDNA 抗体的产生。提示，肠道微生物可能在 SLE 的性别差异中起作用。

狼疮小鼠中，雌性小鼠比雄性小鼠疾病程度更为严重。雌性和雄性 MRL/lpr 小鼠的肠道菌群研究发现[293]，发现雌性 MRL/lp 小鼠比雄性 MRL/lpr 小鼠含有更多的鞭毛纲科细菌和双歧杆菌，而乳杆菌科的水平却相当。雌性 MRL/lpr 小鼠的鞭毛纲科细菌数量比雌性 MRL 对照小鼠多，但雄性 MRL/lpr 小鼠和 MRL 对照小鼠则无明显差异。因此，雌性狼疮小鼠的疾病更严重，可能是肠道微生物群落的差异所致。

肠道菌群的性别差异在成年后出现，这表明肠道菌群受到性激素的影响[307]。雄性小鼠去势后逆转了雄性和雌性小鼠肠道菌群的差异，表明肠道菌群的变化是雄激素依赖性的[308]。非肥胖糖尿病小鼠是 1 型糖尿病的小鼠模型，在小鼠 1 型糖尿病发生率中表现出以雌性为主的性别差异。将雄性非肥胖糖尿病小鼠的肠道微生物群移植给雌性非肥胖糖尿病小鼠，改变了受体的肠道微生物组成，导致睾丸激素水平升高、胰腺胰岛炎症抑制和自身抗体生成，并降低了雌性小鼠糖尿病的发生率[307]。研究发现，雄性小鼠中某些肠道菌群谱系可以诱导对糖尿病的保护。因此，性激素和肠道微生物相互作用，可以保护雄性非肥胖糖尿病小鼠免于糖尿病（这被称为“双信号模型”）。有研究认为，增强的 IFN 信号传导是支持非肥胖糖尿病小鼠性别差异的微生物 - 激素轴保护途径之一。尽管确切机制尚未明确，但性激素和肠道菌群可能共同参与了小鼠模型和人类狼疮的性别二态性。

性激素可能会影响狼疮小鼠的肠道菌群。乳杆菌治疗可逆转雌性 MRL/lpr 小鼠的蛋白尿并改善肾脏病理评分[296]。在给予乳杆菌的雄性 MRL/lpr 小鼠中，阉割去势小鼠和模拟阉割小鼠均表现出乳杆菌的肠道定植增加。然而，只有阉割去势的小鼠才出现蛋白尿减少，肾脏病理评分升高，模拟阉割去势小鼠则无改善。总之，乳杆菌在雌性狼疮小鼠和阉割去势的雄性狼疮小鼠中均具有治疗作用，但在完整的雄性小鼠中则不能。因此，雄激素可以抑制肠道菌群的治疗效果。

五、肠道菌群与系统性红斑狼疮的免疫应答

研究发现，在未经抗生素、类固醇或免疫抑制治疗的非活动性 SLE 患者中，肠道分离的培养物在体外可促进淋巴细胞激活和 Th17 分化[282]。SLE 患者肠道菌群中含有诱导 Treg 的丰富的微生物群，其中两种梭菌菌株的混合物可显著降低 Th17/Th1 的平衡。此外，沙门菌与血清促炎症细胞因子，包括 IFN-α 水平密切相关。小肠互养菌门水平与硬壁菌门呈负相关，与拟杆菌属呈正相关。另外，互养菌门浓度与 dsDNA 抗体效价和 IL-6 水平之间存在负相关，而与天然保护抗体，如抗磷酰胆碱 IgM 抗体水平呈正相关，抗磷酰胆碱 IgM 抗体可促进吞噬作用并抑制炎症反应。在 SLE 患者中，较高滴度的抗 PC 抗体与较低的疾病活动度相关。双歧杆菌可以防止 $CD4^+$ 淋巴细胞过度活化，表明益生菌干预可能有益于改善 SLE 的发展。

肠腔内容物会影响多物种（包括鸡、兔、猪和羔羊）的 B 细胞多样化。研究发现，在小鼠体内，最早的 B 细胞出现在小肠固有层内并引发肠道细菌定植[278]。定植于肠道的微生物抗原与 T 细胞的基础水平、共生反应性 IgA 应答水平，以及细菌免疫反应的 IgG 水平有关[278]。因此，肠道微生物共生体可以丰富免疫前 B 细胞库中的抗菌特异性。

六、展望

肠道菌群失调与疾病发病机制的相关研究结果，使调整肠道微生物群作为 SLE 等疾病治疗新靶点的设想成为可能。目前，纠正肠道微生物失衡治疗 SLE 的一些研究虽已初显成效，然而仍需要设计严格的多中心前瞻性研究予以证实。例如，已经发现肠道微生物可以通过饮食模式调节，即饮食中增加特定微生物种群，

这可能是 SLE 的一种治疗策略[284]。饮食习惯、益生菌或靶向抗生素已被证实可用于治疗和预防自身免疫性疾病[309]及恶性肿瘤[310]。抗生素能够清除或抑制有害的微生物群，益生菌可补充宿主缺少的有益微生物，益生元可增强益生菌。此外，尚需结合宿主的遗传背景以确定最佳的个体化治疗策略。FMT 可能通过移植健康供体粪菌从而纠正与SLE相关的肠道菌群失调，但目前该研究尚未在动物模型中开展。新近研究发现，性激素与肠道菌群的相互作用可能导致了 SLE 的性别差异，这促使我们进一步深入探索导致肠道菌群失调的特定性激素，这可能又是 SLE 发病机制研究的一个新领域。

参考文献

[1] Czaja AJ. Transitioning from Idiopathic to Explainable Autoimmune Hepatitis. Digestive Diseases and Sciences, 2015, 60(10): 2881-2900.

[2] Nishino K, Nishida A, Inoue R, et al. Analysis of endoscopic brush samples identified mucosa-associated dysbiosis in inflammatory bowel disease. Journal of Gastroenterology, 2018, 53(1): 95-106.

[3] Salameh M, Burney Z, Mhaimeed N, et al. The role of gut microbiota in atopic asthma and allergy, implications in the understanding of disease pathogenesis. The Scandinavian Foundation for Immunology, 2020, 91(3): e12855.

[4] Lourido L, Blanco FJ, Ruiz-Romero C. Defining the proteomic landscape of rheumatoid arthritis: progress and prospective clinical applications. Expert Review of Proteomics, 2017, 14(5): 431-444.

[5] Manfredo VS, Hiltensperger M, Kumar V, et al. Translocation of a gut pathobiont drives autoimmunity in mice and humans. Science, 2018, 359(6380): 1156-1161.

[6] Mieli-Vergani G, Vergani D, Czaja AJ, et al. Autoimmune hepatitis. Nature Reviews Disease Primers, 2018, 4: 18017.

[7] Wei Y, Li Y, Yan L, et al. Alterations of gut microbiome in autoimmune hepatitis. Gut, 2020, 69(3): 569-577.

[8] Rogier R, Koenders MI, Abdollahi-Roodsaz S. Toll-like receptor mediated modulation of T cell response by commensal intestinal microbiota as a trigger for autoimmune arthritis. Journal of Immunology Research, 2015, 2015: 527696.

[9] Brenner C, Galluzzi L, Kepp O, et al. Decoding cell death signals in liver inflammation. Journal of Hepatology, 2013, 59(3): 583-594.

[10] Liao J, Yang F, Tang Z, et al. Inhibition of Caspase-1-dependent pyroptosis attenuates copper-induced apoptosis in chicken hepatocytes. Ecotoxicology and Environmental Safety, 2019, 174: 110-119.

[11] Huang EY, Inoue T, Leone VA, et al. Using corticosteroids to reshape the gut microbiome: implications for inflammatory bowel diseases. Inflammatory Bowel Diseases, 2015, 21(5): 963-972.

[12] Tabibian JH, Weeding E, Jorgensen RA, et al. Randomised clinical trial: vancomycin or metronidazole in patients with primary sclerosing cholangitis - a pilot study. Alimentary Pharmacology and Therapeutics, 2013, 37(6): 604-612.

[13] Carey EJ, Ali AH, Lindor KD. Primary biliary cirrhosis. Lancet, 2015, 386(10003): 1565-1575.

[14] Tang R, Wei Y, Li Y, et al. Gut microbial profile is altered in primary biliary cholangitis and partially restored after UDCA therapy. Gut, 2018, 67(3): 534-541.

[15] Donia MS, Fischbach MA. HUMAN MICROBIOTA. Small molecules from the human microbiota. Science, 2015, 349(6246): 1254766.

[16] Visschers RG, Luyer MD, Schaap FG, et al. The gut-liver axis. Current Opinion in Clinical Nutrition and Metabolic Care, 2013, 16(5): 576-581.

[17] Mattner J, Savage PB, Leung P, et al. Liver autoimmunity triggered by microbial activation of natural killer T cells. Cell Host Microbe, 2008, 3(5): 304-315.

[18] Dyson JK, Hirschfield GM, Adams DH, et al. Novel therapeutic targets in primary biliary cirrhosis. Nature Reviews Gastroenterology & Hepatology, 2015, 12(3): 147-158.

[19] Lv LX, Fang DQ, Shi D, et al. Alterations and correlations of the gut microbiome, metabolism and immunity in patients with primary biliary cirrhosis. Environmental Microbiology, 2016, 18(7): 2272-2286.

[20] Chen Y, Yang F, Lu H, et al. Characterization of fecal microbial communities in patients with liver cirrhosis. Hepatology, 2011, 54(2): 562-572.

[21] Iqbal S, Quigley EM. Progress in Our Understanding of the Gut Microbiome: Implications for the Clinician. Current Gastroenterology Reports, 2016, 18(9): 49.

[22] Ozdirik B, Muller T, Wree A, et al. The Role of

Microbiota in Primary Sclerosing Cholangitis and Related Biliary Malignancies. International Journal of Molecular Sciences, 2021, 22(13) : 6975.

[23] Kummen M, Thingholm LB, Ruhlemann MC, et al. Altered Gut Microbial Metabolism of Essential Nutrients in Primary Sclerosing Cholangitis. Gastroenterology, 2021, 160(5): 1784-1798. e0.

[24] Torres J, Bao X, Goel A, et al. The features of mucosa-associated microbiota in primary sclerosing cholangitis. Alimentary Pharmacology and Therapeutics, 2016, 43(7): 790-801.

[25] Kummen M, Holm K, Anmarkrud JA, et al. The gut microbial profile in patients with primary sclerosing cholangitis is distinct from patients with ulcerative colitis without biliary disease and healthy controls. Gut, 2017, 66(4): 611-619.

[26] Iwasawa K, Suda W, Tsunoda T, et al. Characterisation of the faecal microbiota in Japanese patients with paediatric-onset primary sclerosing cholangitis. Gut, 2017, 66(7): 1344-1346.

[27] Kain R, Exner M, Brandes R, et al. Molecular mimicry in pauci-immune focal necrotizing glomerulonephritis. Nature Medicine, 2008, 14(10): 1088-1096.

[28] Terjung B, Sohne J, Lechtenberg B, et al. p-ANCAs in autoimmune liver disorders recognise human beta-tubulin isotype 5 and cross-react with microbial protein FtsZ. Gut, 2010, 59(6): 808-816.

[29] Lunder AK, Hov JR, Borthne A, et al. Prevalence of Sclerosing Cholangitis Detected by Magnetic Resonance Cholangiography in Patients With Long-term Inflammatory Bowel Disease. Gastroenterology, 2016, 151(4): 660-669.e664.

[30] Legoux F, Salou M, Lantz O. Unconventional or Preset alphabeta T Cells: Evolutionarily Conserved Tissue-Resident T Cells Recognizing Nonpeptidic Ligands. Annual Review of Cell and Developmental Biology, 2017, 33: 511-535.

[31] Olsson R, Boberg KM, De Muckadell OS, et al. High-dose ursodeoxycholic acid in primary sclerosing cholangitis: a 5-year multicenter, randomized, controlled study. Gastroenterology, 2005, 129(5): 1464-1472.

[32] Sinakos E, Marschall HU, Kowdley KV, et al. Bile acid changes after high-dose ursodeoxycholic acid treatment in primary sclerosing cholangitis: Relation to disease progression. Hepatology, 2010, 52(1): 197-203.

[33] Sender R, Fuchs S, Milo R. Revised Estimates for the Number of Human and Bacteria Cells in the Body. PLOS Biology, 2016, 14(8): e1002533.

[34] Lin JK, Ho YS. Hepatotoxicity and hepatocarcinogenicity in rats fed squid with or without exogenous nitrite. Food and Chemical Toxicology, 1992, 30(8): 695-702.

[35] Spanogiannopoulos P, Bess EN, Carmody RN, et al. The microbial pharmacists within us: a metagenomic view of xenobiotic metabolism. Nature reviews Microbiology, 2016, 14(5): 273-287.

[36] Gevers D, Kugathasan S, Denson LA, et al. The treatment-naive microbiome in new-onset Crohn's disease. Cell Host Microbe, 2014, 15(3): 382-392.

[37] Kummen M, Vesterhus M, Troseid M, et al. Elevated trimethylamine-N-oxide (TMAO) is associated with poor prognosis in primary sclerosing cholangitis patients with normal liver function. United European Gastroenterology Journal, 2017, 5(4): 532-541.

[38] Machiels K, Joossens M, Sabino J, et al. A decrease of the butyrate-producing species Roseburia hominis and Faecalibacterium prausnitzii defines dysbiosis in patients with ulcerative colitis. Gut, 2014, 63(8): 1275-1283.

[39] Bouskra D, Brezillon C, Berard M, et al. Lymphoid tissue genesis induced by commensals through NOD1 regulates intestinal homeostasis. Nature, 2008, 456(7221): 507-510.

[40] Kinnebrew MA, Ubeda C, Zenewicz LA, et al. Bacterial flagellin stimulates Toll-like receptor 5-dependent defense against vancomycin-resistant Enterococcus infection. The Journal of Infectious Diseases, 2010, 201(4): 534-543.

[41] Morgan XC, Tickle TL, Sokol H, et al. Dysfunction of the intestinal microbiome in inflammatory bowel disease and treatment. Genome Biology, 2012, 13(9): R79.

[42] Manichanh C, Rigottier-Gois L, Bonnaud E, et al. Reduced diversity of faecal microbiota in Crohn's disease revealed by a metagenomic approach. Gut, 2006, 55(2): 205-211.

[43] Sepehri S, Kotlowski R, Bernstein CN, et al. Microbial diversity of inflamed and noninflamed gut biopsy tissues in inflammatory bowel disease. Inflammatory Bowel Diseases, 2007, 13(6): 675-683.

[44] Ohkusa T, Yoshida T, Sato N, et al. Commensal bacteria can enter colonic epithelial cells and induce proinflammatory cytokine secretion: a possible pathogenic mechanism of ulcerative colitis. Journal of Medical Microbiology, 2009, 58(Pt 5): 535-545.

[45] Ohkusa T, Okayasu I, Ogihara T, et al. Induction of experimental ulcerative colitis by Fusobacterium

varium isolated from colonic mucosa of patients with ulcerative colitis. Gut, 2003, 52(1): 79-83.

[46] Willing B, Halfvarson J, Dicksved J, et al. Twin studies reveal specific imbalances in the mucosa-associated microbiota of patients with ileal Crohn's disease. Inflammatory Bowel Diseases, 2009, 15(5): 653-660.

[47] Varela E, Manichanh C, Gallart M, et al. Colonisation by Faecalibacterium prausnitzii and maintenance of clinical remission in patients with ulcerative colitis. Alimentary Pharmacology and Therapeutics, 2013, 38(2): 151-161.

[48] Kitajima S, Morimoto M, Sagara E, et al. Dextran sodium sulfate-induced colitis in germ-free IQI/Jic mice. Experimental Animals, 2001, 50(5): 387-395.

[49] Dollive S, Chen YY, Grunberg S, et al. Fungi of the murine gut: episodic variation and proliferation during antibiotic treatment. PLoS One, 2013, 8(8): e71806.

[50] Noverr MC, Noggle RM, Toews GB, et al. Role of antibiotics and fungal microbiota in driving pulmonary allergic responses. Infection and Immunity, 2004, 72(9): 4996-5003.

[51] Sartor RB, Wu GD. Roles for Intestinal Bacteria, Viruses, and Fungi in Pathogenesis of Inflammatory Bowel Diseases and Therapeutic Approaches. Gastroenterology, 2017, 152(2): 327-339.e4.

[52] Underhill DM, Iliev ID. The mycobiota: interactions between commensal fungi and the host immune system. Nature Reviews Immunology, 2014, 14(6): 405-416.

[53] Iliev ID, Funari VA, Taylor KD, et al. Interactions between commensal fungi and the C-type lectin receptor Dectin-1 influence colitis. Science, 2012, 336(6086): 1314-1317.

[54] Perez-Brocal V, Garcia-Lopez R, Vazquez-Castellanos JF, et al. Study of the viral and microbial communities associated with Crohn's disease: a metagenomic approach. Clinical and Translational Gastroenterology, 2013, 4(6): e36.

[55] Wagner J, Coupland P, Browne HP, et al. Evaluation of PacBio sequencing for full-length bacterial 16S rRNA gene classification. BMC Microbiology, 2016, 16(1): 274.

[56] Lepage P, Colombet J, Marteau P, et al. Dysbiosis in inflammatory bowel disease: a role for bacteriophages? Gut, 2008, 57(3): 424-425.

[57] Kernbauer E, Ding Y, Cadwell K. An enteric virus can replace the beneficial function of commensal bacteria. Nature, 2014, 516(7529): 94-98.

[58] Cadwell K, Patel KK, Maloney NS, et al. Virus-plus-susceptibility gene interaction determines Crohn's disease gene Atg16L1 phenotypes in intestine. Cell, 2010, 141(7): 1135-1145.

[59] Yang JY, Kim MS, Kim E, et al. Enteric Viruses Ameliorate Gut Inflammation via Toll-like Receptor 3 and Toll-like Receptor 7-Mediated Interferon-beta Production. Immunity, 2016, 44(4): 889-900.

[60] Yang X, Yang Y, Wang Y, et al. Excretory/secretory products from Trichinella spiralis adult worms ameliorate DSS-induced colitis in mice. PLoS One, 2014, 9(5): e96454.

[61] Broadhurst MJ, Leung JM, Kashyap V, et al. IL-22+ $CD4^+$ T cells are associated with therapeutic trichuris trichiura infection in an ulcerative colitis patient. Science Translational Medicine, 2010, 2(60): 60ra88.

[62] Wang YD, Chen WD, Yu D, et al. The G-protein-coupled bile acid receptor, Gpbar1 (TGR5), negatively regulates hepatic inflammatory response through antagonizing nuclear factor kappa light-chain enhancer of activated B cells (NF-kappaB) in mice. Hepatology, 2011, 54(4): 1421-1432.

[63] Joyce SA, Macsharry J, Casey PG, et al. Regulation of host weight gain and lipid metabolism by bacterial bile acid modification in the gut. Proceedings of the National Academy of Sciences of the United States of America, 2014, 111(20): 7421-7426.

[64] Labbe A, Ganopolsky JG, Martoni CJ, et al. Bacterial bile metabolising gene abundance in Crohn's, ulcerative colitis and type 2 diabetes metagenomes. PLoS One, 2014, 9(12): e115175.

[65] Gadaleta RM, Van Erpecum KJ, Oldenburg B, et al. Farnesoid X receptor activation inhibits inflammation and preserves the intestinal barrier in inflammatory bowel disease. Gut, 2011, 60(4): 463-472.

[66] Islam KB, Fukiya S, Hagio M, et al. Bile acid is a host factor that regulates the composition of the cecal microbiota in rats. Gastroenterology, 2011, 141(5): 1773-1781.

[67] Devkota S, Wang Y, Musch MW, et al. Dietary-fat-induced taurocholic acid promotes pathobiont expansion and colitis in Il10-/- mice. Nature, 2012, 487(7405): 104-108.

[68] Song X, Sun X, Oh SF, et al. Microbial bile acid metabolites modulate gut RORgamma(+) regulatory T cell homeostasis. Nature, 2020, 577(7790): 410-415.

[69] Duboc H, Rajca S, Rainteau D, et al. Connecting dysbiosis, bile-acid dysmetabolism and gut inflammation in inflammatory bowel diseases. Gut, 2013, 62(4): 531-539.

[70] Hall AB, Yassour M, Sauk J, et al. A novel

Ruminococcus gnavus clade enriched in inflammatory bowel disease patients. Genome Medicine, 2017, 9(1): 103.

[71] David LA, Maurice CF, Carmody RN, et al. Diet rapidly and reproducibly alters the human gut microbiome. Nature, 2014, 505(7484): 559-563.

[72] Machiels K, Sabino J, Vandermosten L, et al. Specific members of the predominant gut microbiota predict pouchitis following colectomy and IPAA in UC. Gut, 2017, 66(1): 79-88.

[73] Laserna-Mendieta EJ, Clooney AG, Carretero-Gomez JF, et al. Determinants of Reduced Genetic Capacity for Butyrate Synthesis by the Gut Microbiome in Crohn's Disease and Ulcerative Colitis. Journal of Crohn's and Colitis, 2018, 12(2): 204-216.

[74] Hove H, Mortensen PB. Influence of intestinal inflammation (IBD) and small and large bowel length on fecal short-chain fatty acids and lactate. Digestive Diseases and Sciences, 1995, 40(6): 1372-1380.

[75] Maslowski KM, Vieira AT, Ng A, et al. Regulation of inflammatory responses by gut microbiota and chemoattractant receptor GPR43. Nature, 2009, 461(7268): 1282-1286.

[76] Tye H, Yu CH, Simms LA, et al. NLRP1 restricts butyrate producing commensals to exacerbate inflammatory bowel disease. Nature Communications, 2018, 9(1): 3728.

[77] Nikolaus S, Schulte B, Al-Massad N, et al. Increased Tryptophan Metabolism Is Associated With Activity of Inflammatory Bowel Diseases. Gastroenterology, 2017, 153(6): 1504-1516. e1502.

[78] Hashimoto T, Perlot T, Rehman A, et al. ACE2 links amino acid malnutrition to microbial ecology and intestinal inflammation. Nature, 2012, 487(7408): 477-481.

[79] Lamas B, Richard ML, Leducq V, et al. CARD9 impacts colitis by altering gut microbiota metabolism of tryptophan into aryl hydrocarbon receptor ligands. Nature Medicine, 2016, 22(6): 598-605.

[80] Alexeev EE, Lanis JM, Kao DJ, et al. Microbiota-Derived Indole Metabolites Promote Human and Murine Intestinal Homeostasis through Regulation of Interleukin-10 Receptor. The American Journal of Pathology, 2018, 188(5): 1183-1194.

[81] Lynch SV, Pedersen O. The Human Intestinal Microbiome in Health and Disease. The New England Journal of Medicine, 2016, 375(24): 2369-2379.

[82] Viladomiu M, Kivolowitz C, Abdulhamid A, et al. IgA-coated E. coli enriched in Crohn's disease spondyloarthritis promote TH17-dependent inflammation. Science Translational Medicine, 2017, 9(376): e9655.

[83] Palm NW, De Zoete MR, Cullen TW, et al. Immunoglobulin A coating identifies colitogenic bacteria in inflammatory bowel disease. Cell, 2014, 158(5): 1000-1010.

[84] Pascal V, Pozuelo M, Borruel N, et al. A microbial signature for Crohn's disease. Gut, 2017, 66(5): 813-822.

[85] Tyler AD, Knox N, Kabakchiev B, et al. Characterization of the gut-associated microbiome in inflammatory pouch complications following ileal pouch-anal anastomosis. PLoS One, 2013, 8(9): e66934.

[86] Haberman Y, Tickle TL, Dexheimer PJ, et al. Pediatric Crohn disease patients exhibit specific ileal transcriptome and microbiome signature. The Journal of Clinical Investigation, 2014, 124(8): 3617-3633.

[87] Lewis JD, Chen EZ, Baldassano RN, et al. Inflammation, Antibiotics, and Diet as Environmental Stressors of the Gut Microbiome in Pediatric Crohn's Disease. Cell Host Microbe, 2015, 18(4): 489-500.

[88] Mottawea W, Chiang CK, Muhlbauer M, et al. Altered intestinal microbiota-host mitochondria crosstalk in new onset Crohn's disease. Nature Communications, 2016, 7: 13419.

[89] Martinez-Medina M, Denizot J, Dreux N, et al. Western diet induces dysbiosis with increased E coli in CEABAC10 mice, alters host barrier function favouring AIEC colonisation. Gut, 2014, 63(1): 116-124.

[90] Agus A, Denizot J, Thevenot J, et al. Western diet induces a shift in microbiota composition enhancing susceptibility to Adherent-Invasive E. coli infection and intestinal inflammation. Scientific Reports, 2016, 6: 19032.

[91] Lai KP, Chung YT, Li R, et al. Bisphenol A alters gut microbiome: Comparative metagenomics analysis. Environmental Pollution, 2016, 218: 923-930.

[92] Nickerson KP, Mcdonald C. Crohn's disease-associated adherent-invasive Escherichia coli adhesion is enhanced by exposure to the ubiquitous dietary polysaccharide maltodextrin. PLoS One, 2012, 7(12): e52132.

[93] Ruemmele FM, Veres G, Kolho KL, et al. Consensus guidelines of ECCO/ESPGHAN on the medical management of pediatric Crohn's disease. Journal of Crohn's and Colitis, 2014, 8(10): 1179-1207.

[94] Bibiloni R, Fedorak RN, Tannock GW, et al. VSL#3 probiotic-mixture induces remission in patients with active ulcerative colitis. The American Journal of

Gastroenterology, 2005, 100(7): 1539-1546.

[95] Yan X, Sivignon A, Yamakawa N, et al. Glycopolymers as Antiadhesives of E. coli Strains Inducing Inflammatory Bowel Diseases. Biomacromolecules, 2015, 16(6): 1827-1836.

[96] Maharshak N, Huh EY, Paiboonrungruang C, et al. Enterococcus faecalis Gelatinase Mediates Intestinal Permeability via Protease-Activated Receptor 2. Infection and Immunity, 2015, 83(7): 2762-2770.

[97] Zhu W, Winter MG, Byndloss MX, et al. Precision editing of the gut microbiota ameliorates colitis. Nature, 2018, 553(7687): 208-211.

[98] Paramsothy S, Paramsothy R, Rubin DT, et al. Faecal Microbiota Transplantation for Inflammatory Bowel Disease: A Systematic Review and Meta-analysis. Journal of Crohn's and Colitis, 2017, 11(10): 1180-1199.

[99] Angelberger S, Reinisch W, Makristathis A, et al. Temporal bacterial community dynamics vary among ulcerative colitis patients after fecal microbiota transplantation. The American Journal of Gastroenterology, 2013, 108(10): 1620-1630.

[100] Paramsothy S, Kamm MA, Kaakoush NO, et al. Multidonor intensive faecal microbiota transplantation for active ulcerative colitis: a randomised placebo-controlled trial. Lancet, 2017, 389(10075): 1218-1228.

[101] Li SS, Zhu A, Benes V, et al. Durable coexistence of donor and recipient strains after fecal microbiota transplantation. Science, 2016, 352(6285): 586-589.

[102] Chehoud C, Dryga A, Hwang Y, et al. Transfer of Viral Communities between Human Individuals during Fecal Microbiota Transplantation. MBio, 2016, 7(2): e00322.

[103] The role of gut microbiota in the pathogenesis and management of allergic diseases. European Review for Medical and Pharmacological Sciences, 2014, 18(17): 2400.

[104] Kalliomaki M, Antoine JM, Herz U, et al. Guidance for substantiating the evidence for beneficial effects of probiotics: prevention and management of allergic diseases by probiotics. The Journal of Nutrition, 2010, 140(3): S713-S721.

[105] Kalliomaki M, Kirjavainen P, Eerola E, et al. Distinct patterns of neonatal gut microflora in infants in whom atopy was and was not developing. The Journal of Allergy and Clinical Immunology, 2001, 107(1): 129-134.

[106] Melli LC, Do Carmo-Rodrigues MS, Araujo-Filho HB, et al. Intestinal microbiota and allergic diseases: A systematic review. Allergologia et Immunopathologia, 2016, 44(2): 177-188.

[107] Bashir ME, Louie S, Shi HN, et al. Toll-like receptor 4 signaling by intestinal microbes influences susceptibility to food allergy. The Journal of Immunology, 2004, 172(11): 6978-6987.

[108] Borbet TC, Zhang X, Müller A, et al. The role of the changing human microbiome in the asthma pandemic. The Journal of Allergy and Clinical Immunology, 2019, 144(6): 1457-1466.

[109] Ch L. Treatment of Hypertension in Patients with Asthma. The New England Journal of Medicine, 2019, 381(23): 2278-2279.

[110] Desai M, Oppenheimer J. Elucidating asthma phenotypes and endotypes: progress towards personalized medicine. Annals of allergy, Asthma & Immunology, 2016, 116(5): 394-401.

[111] Hm W, Bl C, Nv F, et al. Siblings Promote a Type 1/Type 17-oriented immune response in the airways of asymptomatic neonates. Allergy, 2016, 71(6): 820-828.

[112] Sl T, Lex L, Jm C, et al. Inflammatory phenotypes in patients with severe asthma are associated with distinct airway microbiology. The Journal of Allergy and Clinical Immunology, 2018, 141(1): 94-103. e15.

[113] At D, Bj M. Microbes, metabolites, and the gut-lung axis. Mucosal Immunology, 2019, 12(4): 843-850.

[114] Mc A, Lt S, Pa D, et al. Early infancy microbial and metabolic alterations affect risk of childhood asthma. Science Translational Medicine, 2015, 7(307): 307. ra152.

[115] Arrieta MC, Arévalo A, Stiemsma L, et al. Associations between infant fungal and bacterial dysbiosis and childhood atopic wheeze in a nonindustrialized setting. The Journal of Allergy and Clinical Immunology, 2018, 142(2): 424-434.e410.

[116] Fujimura KE, Sitarik AR, Havstad S, et al. Neonatal gut microbiota associates with childhood multisensitized atopy and T cell differentiation. Nature Medicine, 2016, 22(10): 1187-1191.

[117] Van Tilburg Bernardes E, Arrieta MC. Hygiene Hypothesis in Asthma Development: Is Hygiene to Blame? Archives of Medical Research, 2017, 48(8): 717-726.

[118] Cy C, Ml C, Mh C, et al. Gut microbial-derived butyrate is inversely associated with IgE responses to allergens in childhood asthma. Pediatric Allergy and Immunology, 2019, 30(7): 689-697.

[119] Stokholm J, Blaser MJ, Thorsen J, et al. Maturation of the gut microbiome and risk of asthma in childhood. Nature Communications, 2018, 9(1): 141.

[120] Barcik W, Boutin RCT, Sokolowska M, et al. The Role of Lung and Gut Microbiota in the Pathology of Asthma. Immunity, 2020, 52(2): 241-255.

[121] Sa E. Enhanced histopathology of mucosa-associated lymphoid tissue. Toxicologic Pathology, 2006, 34(5): 687-696.

[122] Willard-Mack CL. Normal structure, function, and histology of lymph nodes. Toxicologic Pathology, 2006, 34(5): 409-424.

[123] Brandtzaeg P. The mucosal immune system and its integration with the mammary glands. The Journal of Pediatrics, 2010, 156(2 Suppl): S8-S15.

[124] Hillman ET, Lu H, Yao T, et al. Microbial Ecology along the Gastrointestinal Tract. Microbes and Environments, 2017, 32(4): 300-313.

[125] Sze MA, Tsuruta M, Yang SW, et al. Changes in the bacterial microbiota in gut, blood, and lungs following acute LPS instillation into mice lungs. PloS one, 2014, 9(10): e111228.

[126] Perrone EE, Jung E, Breed E, et al. Mechanisms of methicillin-resistant Staphylococcus aureus pneumonia-induced intestinal epithelial apoptosis. Shock, 2012, 38(1): 68-75.

[127] Loverdos K, Bellos G, Kokolatou L, et al. Lung Microbiome in Asthma: Current Perspectives. Journal of Clinical Medicine, 2019, 8(11) : 1967.

[128] Abell GC, Cooke CM, Bennett CN, et al. Phylotypes related to Ruminococcus bromii are abundant in the large bowel of humans and increase in response to a diet high in resistant starch. FEMS Microbiology Ecology, 2008, 66(3): 505-515.

[129] Huang YJ, Boushey HA. The microbiome in asthma. The Journal of Allergy and Clinical Immunology, 2015, 135(1): 25-30.

[130] Atarashi K, Tanoue T, Oshima K, et al. Treg induction by a rationally selected mixture of Clostridia strains from the human microbiota. Nature, 2013, 500(7461): 232-236.

[131] Shi CW, Cheng MY, Yang X, et al. Probiotic Lactobacillus rhamnosus GG Promotes Mouse Gut Microbiota Diversity and T Cell Differentiation. Frontiers in Microbiology, 2020, 11: 607735.

[132] Peng L, Li ZR, Green RS, et al. Butyrate enhances the intestinal barrier by facilitating tight junction assembly via activation of AMP-activated protein kinase in Caco-2 cell monolayers. The Journal of Nutrition, 2009, 139(9): 1619-1625.

[133] Capaldo CT, Nusrat A. Cytokine regulation of tight junctions. Biochimica et Biophysica Acta, 2009, 1788(4): 864-871.

[134] Hufnagl K, Pali-Schöll I, Roth-Walter F, et al. Dysbiosis of the gut and lung microbiome has a role in asthma. Seminars in Immunopathology, 2020, 42(1): 75-93.

[135] Sokol CL, Chu NQ, Yu S, et al. Basophils function as antigen-presenting cells for an allergen-induced T helper type 2 response. Nature Immunology, 2009, 10(7): 713-720.

[136] Siracusa MC, Kim BS, Spergel JM, et al. Basophils and allergic inflammation. The Journal of Allergy and Clinical Immunology, 2013, 132(4): 789-801.

[137] Zhao W, Ho HE, Bunyavanich S. The gut microbiome in food allergy. Annals of Allergy Asthma & Immunology, 2019, 122(3): 276-282.

[138] Akbari O, Stock P, Meyer E, et al. Essential role of NKT cells producing IL-4 and IL-13 in the development of allergen-induced airway hyperreactivity. Nature Medicine, 2003, 9(5): 582-588.

[139] Kunii J, Takahashi K, Kasakura K, et al. Commensal bacteria promote migration of mast cells into the intestine. Immunobiology, 2011, 216(6): 692-697.

[140] Gordon S. Alternative activation of macrophages. Nature Reviews Immunology, 2003, 3(1): 23-35.

[141] Demirci M, Tokman HB, Uysal HK, et al. Reduced Akkermansia muciniphila and Faecalibacterium prausnitzii levels in the gut microbiota of children with allergic asthma. Allergologia et Immunopathologia, 2019, 47(4): 365-371.

[142] Young RP, Hopkins RJ, Marsland B. The Gut-Liver-Lung Axis. Modulation of the Innate Immune Response and Its Possible Role in Chronic Obstructive Pulmonary Disease. American Journal of Respiratory Cell and Molecular Biology, 2016, 54(2): 161-169.

[143] Moon C, Baldridge MT, Wallace MA, et al. Vertically transmitted faecal IgA levels determine extra-chromosomal phenotypic variation. Nature, 2015, 521(7550): 90-93.

[144] Pascal M, Perez-Gordo M, Caballero T, et al. Microbiome and Allergic Diseases. Frontiers in Immunology, 2018, 9: 1584.

[145] Maekawa T, Krauss JL, Abe T, et al. Porphyromonas gingivalis manipulates complement and TLR signaling to uncouple bacterial clearance from

inflammation and promote dysbiosis. Cell Host Microbe, 2014, 15(6): 768-778.

[146] Dargahi N, Johnson J, Donkor O, et al. Immunomodulatory effects of probiotics: Can they be used to treat allergies and autoimmune diseases? Maturitas, 2019, 119: 25-38.

[147] Du X, Wang L, Wu S, et al. Efficacy of probiotic supplementary therapy for asthma, allergic rhinitis, and wheeze: a meta-analysis of randomized controlled trials. Allergy and Asthma Proceedings, 2019, 40(4): 250-260.

[148] Arpaia N, Campbell C, Fan X, et al. Metabolites produced by commensal bacteria promote peripheral regulatory T-cell generation. Nature, 2013, 504(7480): 451-455.

[149] Wu CT, Chen PJ, Lee YT, et al. Effects of immunomodulatory supplementation with Lactobacillus rhamnosus on airway inflammation in a mouse asthma model. Journal of Microbiology, Immunology and Infection, 2016, 49(5): 625-635.

[150] Forsythe P, Inman MD, Bienenstock J. Oral treatment with live Lactobacillus reuteri inhibits the allergic airway response in mice. American Journal of Respiratory and Critical Care Medicine, 2007, 175(6): 561-569.

[151] Blümer N, Sel S, Virna S, et al. Perinatal maternal application of Lactobacillus rhamnosus GG suppresses allergic airway inflammation in mouse offspring. Clinical and Experimental Allergy, 2007, 37(3): 348-357.

[152] Feleszko W, Jaworska J, Rha RD, et al. Probiotic-induced suppression of allergic sensitization and airway inflammation is associated with an increase of T regulatory-dependent mechanisms in a murine model of asthma. Clinical and Experimental Allergy, 2007, 37(4): 498-505.

[153] Elazab N, Mendy A, Gasana J, et al. Probiotic administration in early life, atopy, and asthma: a meta-analysis of clinical trials. Pediatrics, 2013, 132(3): e666-e676.

[154] Huang CF, Chie WC, Wang IJ. Efficacy of Lactobacillus Administration in School-Age Children with Asthma: A Randomized, Placebo-Controlled Trial. Nutrients, 2018, 10(11). 1678.

[155] Cuello-Garcia CA, Brożek JL, Fiocchi A, et al. Probiotics for the prevention of allergy: A systematic review and meta-analysis of randomized controlled trials. The Journal of Allergy and Clinical Immunology, 2015, 136(4): 952-961.

[156] Edwards MR, Walton RP, Jackson DJ, et al. The potential of anti-infectives and immunomodulators as therapies for asthma and asthma exacerbations. Allergy, 2018, 73(1): 50-63.

[157] Esposito S, Soto-Martinez ME, Feleszko W, et al. Nonspecific immunomodulators for recurrent respiratory tract infections, wheezing and asthma in children: a systematic review of mechanistic and clinical evidence. Current Opinion in Allergy and Clinical Immunology, 2018, 18(3): 198-209.

[158] Teo SM, Tang HHF, Mok D, et al. Airway Microbiota Dynamics Uncover a Critical Window for Interplay of Pathogenic Bacteria and Allergy in Childhood Respiratory Disease. Cell Host & Microbe, 2018, 24(3): 341-352.

[159] Esposito S, Bianchini S, Polinori I, et al. Impact of OM-85 Given during Two Consecutive Years to Children with a History of Recurrent Respiratory Tract Infections: A Retrospective Study. International Journal of Environmental Research and Public Health, 2019, 16(6): 1065.

[160] Navarro S, Cossalter G, Chiavaroli C, et al. The oral administration of bacterial extracts prevents asthma via the recruitment of regulatory T cells to the airways. Mucosal Immunology, 2011, 4(1): 53-65.

[161] Lu Y, Li Y, Xu L, et al. Bacterial lysate increases the percentage of natural killer T cells in peripheral blood and alleviates asthma in children. Pharmacology, 2015, 95(3-4): 139-144.

[162] Razi CH, Harmancı K, Abacı A, et al. The immunostimulant OM-85 BV prevents wheezing attacks in preschool children. The Journal of Allergy and Clinical Immunology, 2010, 126(4): 763-769.

[163] Emeryk A, Bartkowiak-Emeryk M, Raus Z, et al. Mechanical bacterial lysate administration prevents exacerbation in allergic asthmatic children-The EOLIA study. Pediatric Allergy and Immunology, 2018, 29(4): 394-401.

[164] Beigelman A, Rosas-Salazar C, Hartert TV. Childhood Asthma: Is It All About Bacteria and Not About Viruses? A Pro/Con Debate. The Journal of Allergy and Clinical Immunology Practice, 2018, 6(3): 719-725.

[165] Sunderkotter CH, Zelger B, Chen KR, et al. Nomenclature of Cutaneous Vasculitis: Dermatologic Addendum to the 2012 Revised International Chapel Hill Consensus Conference Nomenclature of Vasculitides. Arthritis Rheumatology, 2018, 70(2): 171-184.

[166] Heineke MH, Ballering AV, Jamin A, et al. New insights in the pathogenesis of immunoglobulin A vasculitis (Henoch-Schonlein purpura). Autoimmunity Reviews, 2017, 16(12): 1246-1253.

[167] Piram M, Mahr A. Epidemiology of immunoglobulin A vasculitis (Henoch-Schonlein): current state of knowledge. Current Opinion in Rheumatology, 2013, 25(2): 171-178.

[168] Reid-Adam J. Henoch-Schonlein purpura. Pediatrics in Review, 2014, 35(10): 447-449.

[169] Brom M, Gandino IJ, Scolnik M. IgA Vasculitis (Henoch-Schonlein Purpura) in Argentina: Comparison Between Pediatric and Adult Populations. Mayo Clinic Proceedings, 2020, 95(2): 422-424.

[170] Gardner-Medwin JM, Dolezalova P, Cummins C, et al. Incidence of Henoch-Schonlein purpura, Kawasaki disease, and rare vasculitides in children of different ethnic origins. Lancet, 2002, 360(9341): 1197-1202.

[171] Ronkainen J, Koskimies O, Ala-Houhala M, et al. Early prednisone therapy in Henoch-Schonlein purpura: a randomized, double-blind, placebo-controlled trial. The Journal of Pediatrics, 2006, 149(2): 241-247.

[172] Purevdorj N, Mu Y, Gu Y, et al. Clinical significance of the serum biomarker index detection in children with Henoch-Schonlein purpura. Clinical Biochemistry, 2018, 52: 167-170.

[173] Audemard-Verger A, Pillebout E, Guillevin L, et al. IgA vasculitis (Henoch-Shonlein purpura) in adults: Diagnostic and therapeutic aspects. Autoimmunity Reviews, 2015, 14(7): 579-585.

[174] Zheng D, Liwinski T, Elinav E. Interaction between microbiota and immunity in health and disease. Cell Research, 2020, 30(6): 492-506.

[175] Hooper LV, Littman DR, Macpherson AJ. Interactions between the microbiota and the immune system. Science, 2012, 336(6086): 1268-1273.

[176] Chua HH, Chou HC, Tung YL, et al. Intestinal Dysbiosis Featuring Abundance of Ruminococcus gnavus Associates With Allergic Diseases in Infants. Gastroenterology, 2018, 154(1): 154-167.

[177] Abrahamsson TR, Jakobsson HE, Andersson AF, et al. Low diversity of the gut microbiota in infants with atopic eczema. The Journal of Allergy and Clinical Immunology, 2012, 129(2): 434-440, 440. e431-e432.

[178] Chen CC, Chen KJ, Kong MS, et al. Alterations in the gut microbiotas of children with food sensitization in early life. Pediatric Allergy and Immunology, 2016, 27(3): 254-262.

[179] Wang X, Zhang L, Wang Y, et al. Gut microbiota dysbiosis is associated with Henoch-Schonlein Purpura in children. International Immunopharmacology, 2018, 58: 1-8.

[180] Chen B, Wang J, Wang Y, et al. Oral microbiota dysbiosis and its association with Henoch-Schonlein Purpura in children. International Immunopharmacology, 2018, 65: 295-302.

[181] Wu T, Yang L, Jiang J, et al. Chronic glucocorticoid treatment induced circadian clock disorder leads to lipid metabolism and gut microbiota alterations in rats. Life Sciences, 2018, 192: 173-182.

[182] Crayne CB, Eloseily E, Mannion ML, et al. Rituximab treatment for chronic steroid-dependent Henoch-Schonlein purpura: 8 cases and a review of the literature. Pediatric Rheumatology, 2018, 16(1): 71.

[183] Shimizu J, Kubota T, Takada E, et al. Bifidobacteria Abundance-Featured Gut Microbiota Compositional Change in Patients with Behcet's Disease. PLoS One, 2016, 11(4): e0153746.

[184] Tamanai-Shacoori Z, Smida I, Bousarghin L, et al. Roseburia spp.: a marker of health? Future Microbiology, 2017, 12: 157-170.

[185] Furusawa Y, Obata Y, Fukuda S, et al. Commensal microbe-derived butyrate induces the differentiation of colonic regulatory T cells. Nature, 2013, 504(7480): 446-450.

[186] O'Keefe SJ. Diet, microorganisms and their metabolites, and colon cancer. Nature Reviews Gastroenterology & Hepatology, 2016, 13(12): 691-706.

[187] Duncan SH, Belenguer A, Holtrop G, et al. Reduced dietary intake of carbohydrates by obese subjects results in decreased concentrations of butyrate and butyrate-producing bacteria in feces. Applied and Environmental Microbiology, 2007, 73(4): 1073-1078.

[188] Savage JH, Lee-Sarwar KA, Sordillo J, et al. A prospective microbiome-wide association study of food sensitization and food allergy in early childhood. Allergy, 2018, 73(1): 145-152.

[189] Zheng H, Liang H, Wang Y, et al. Altered Gut Microbiota Composition Associated with Eczema in Infants. PLoS One, 2016, 11(11): e0166026.

[190] Huang YJ, Marsland BJ, Bunyavanich S, et al. The microbiome in allergic disease: Current understanding and future opportunities-2017 PRACTALL document

of the American Academy of Allergy, Asthma & Immunology and the European Academy of Allergy and Clinical Immunology. The Journal of Allergy and Clinical Immunology, 2017, 139(4): 1099-1110.

[191] Awadel-Kariem FM, Patel P, Kapoor J, et al. First report of Parabacteroides goldsteinii bacteraemia in a patient with complicated intra-abdominal infection. Anaerobe, 2010, 16(3): 223-225.

[192] Lee SM, Han HW, Yim SY. Beneficial effects of soy milk and fiber on high cholesterol diet-induced alteration of gut microbiota and inflammatory gene expression in rats. Food & Function, 2015, 6(2): 492-500.

[193] Hong PY, Lee BW, Aw M, et al. Comparative analysis of fecal microbiota in infants with and without eczema. PLoS One, 2010, 5(4): e9964.

[194] Taur Y, Pamer EG. Microbiome mediation of infections in the cancer setting. Genome Medicine, 2016, 8(1): 40.

[195] Ling X, Linglong P, Weixia D, et al. Protective Effects of Bifidobacterium on Intestinal Barrier Function in LPS-Induced Enterocyte Barrier Injury of Caco-2 Monolayers and in a Rat NEC Model. PLoS One, 2016, 11(8): e0161635.

[196] Sanders ME. Probiotics: definition, sources, selection, and uses. Clinical Infectious Diseases, 2008, 46 Suppl 2: S58-61; discussion S144-S151.

[197] Fiocchi A, Pawankar R, Cuello-Garcia C, et al. World Allergy Organization-McMaster University Guidelines for Allergic Disease Prevention (GLAD-P): Probiotics. The World Allergy Organization, 2015, 8(1): 4.

[198] Berni Canani R, Sangwan N, Stefka AT, et al. Lactobacillus rhamnosus GG-supplemented formula expands butyrate-producing bacterial strains in food allergic infants. The ISME Journal, 2016, 10(3): 742-750.

[199] Zuccotti G, Meneghin F, Aceti A, et al. Probiotics for prevention of atopic diseases in infants: systematic review and meta-analysis. Allergy, 2015, 70(11): 1356-1371.

[200] Foolad N, Armstrong AW. Prebiotics and probiotics: the prevention and reduction in severity of atopic dermatitis in children. Beneficial Microbes, 2014, 5(2): 151-160.

[201] Chinese Rheumatology A. [2018 Chinese guideline for the diagnosis and treatment of rheumatoid arthritis]. Zhonghua Nei Ke Za Zhi, 2018, 57(4): 242-251.

[202] Parisi S, Bortoluzzi A, Sebastiani GD, et al. The Italian Society for Rheumatology clinical practice guidelines for rheumatoid arthritis. Reumatismo, 2019, 71(S1): 22-49.

[203] Caminer AC, Haberman R, Scher JU. Human microbiome, infections, and rheumatic disease. Clinical Rheumatology, 2017, 36(12): 2645-2653.

[204] Chang K, Yang SM, Kim SH, et al. Smoking and rheumatoid arthritis. International Journal of Molecular Sciences, 2014, 15(12): 22279-22295.

[205] Schwenzer A, Quirke AM, Marzeda AM, et al. Association of Distinct Fine Specificities of Anti-Citrullinated Peptide Antibodies With Elevated Immune Responses to Prevotella intermedia in a Subgroup of Patients With Rheumatoid Arthritis and Periodontitis. Arthritis Rheumatology, 2017, 69(12): 2303-2313.

[206] Horta-Baas G, Romero-Figueroa MDS, Montiel-Jarquin AJ, et al. Intestinal Dysbiosis and Rheumatoid Arthritis: A Link between Gut Microbiota and the Pathogenesis of Rheumatoid Arthritis. Journal of Immunology Research, 2017, 2017: 4835189.

[207] Garrett WS, Gordon JI, Glimcher LH. Homeostasis and inflammation in the intestine. Cell, 2010, 140(6): 859-870.

[208] Maeda Y, Takeda K. Host-microbiota interactions in rheumatoid arthritis. Experimental & Molecular Medicine, 2019, 51(12): 150.

[209] Mcinnes IB, Schett G. The pathogenesis of rheumatoid arthritis. The New England Journal of Medicine, 2011, 365(23): 2205-2219.

[210] Nielen MM, Van Schaardenburg D, Reesink HW, et al. Specific autoantibodies precede the symptoms of rheumatoid arthritis: a study of serial measurements in blood donors. Arthritis Rheumatology, 2004, 50(2): 380-386.

[211] Rantapaa-Dahlqvist S, De Jong BA, Berglin E, et al. Antibodies against cyclic citrullinated peptide and IgA rheumatoid factor predict the development of rheumatoid arthritis. Arthritis Rheumatology, 2003, 48(10): 2741-2749.

[212] O'dell JR, Blakely KW, Mallek JA, et al. Treatment of early seropositive rheumatoid arthritis: a two-year, double-blind comparison of minocycline and hydroxychloroquine. Arthritis Rheumatology, 2001, 44(10): 2235-2241.

[213] Malmstrom V, Catrina AI, Klareskog L. The immunopathogenesis of seropositive rheumatoid arthritis: from triggering to targeting. Nature Reviews

Immunology, 2017, 17(1): 60-75.
[214] Zhang X, Zhang D, Jia H, et al. The oral and gut microbiomes are perturbed in rheumatoid arthritis and partly normalized after treatment. Nature Medicine, 2015, 21(8): 895-905.
[215] Fung I, Garrett JP, Shahane A, et al. Do bugs control our fate? The influence of the microbiome on autoimmunity. Current Allergy and Asthma Reports, 2012, 12(6): 511-519.
[216] Rodriguez-Elias AK, Maldonado-Murillo K, Lopez-Mendoza LF, et al. [Genetics and genomics in rheumatoid arthritis (RA): An update]. Gaceta Medica de Mexico, 2016, 152(2): 218-227.
[217] Konig MF. The microbiome in autoimmune rheumatic disease. Best Practice & Research Clinical Rheumatology, 2020, 34(1): 101473.
[218] Smolen JS, Aletaha D, Mcinnes IB. Rheumatoid arthritis. The Lancet, 2016, 388(10055): 2023-2038.
[219] Van Der Woude D, Rantapaa-Dahlqvist S, Ioan-Facsinay A, et al. Epitope spreading of the anti-citrullinated protein antibody response occurs before disease onset and is associated with the disease course of early arthritis. Annals of the Rheumatic Diseases, 2010, 69(8): 1554-1561.
[220] Sakkas LI, Bogdanos DP. Multiple hit infection and autoimmunity: the dysbiotic microbiota-ACPA connection in rheumatoid arthritis. Current Opinion in Rheumatology, 2018, 30(4): 403-409.
[221] Highlander SK, Gomez A, Luckey D, et al. Loss of Sex and Age Driven Differences in the Gut Microbiome Characterize Arthritis-Susceptible *0401 Mice but Not Arthritis-Resistant *0402 Mice. PLoS ONE, 2012, 7(4): e36095.
[222] Van Der Helm-Van Mil AH, Verpoort KN, Breedveld FC, et al. The HLA-DRB1 shared epitope alleles are primarily a risk factor for anti-cyclic citrullinated peptide antibodies and are not an independent risk factor for development of rheumatoid arthritis. Arthritis and Rheumatism, 2006, 54(4): 1117-1121.
[223] Pianta A, Arvikar S, Strle K, et al. Evidence of the Immune Relevance of Prevotella copri, a Gut Microbe, in Patients With Rheumatoid Arthritis. Arthritis & Rheumatology, 2017, 69(5): 964-975.
[224] Pianta A, Arvikar SL, Strle K, et al. Two rheumatoid arthritis-specific autoantigens correlate microbial immunity with autoimmune responses in joints. The Journal of Clinical Investigation, 2017, 127(8): 2946-2956.
[225] Fuggle NR, Smith TO, Kaul A, et al. Hand to Mouth: A Systematic Review and Meta-Analysis of the Association between Rheumatoid Arthritis and Periodontitis. Frontiers in Immunology, 2016, 7: 80.
[226] Konig MF, Abusleme L, Reinholdt J, et al. Aggregatibacter actinomycetemcomitans-induced hypercitrullination links periodontal infection to autoimmunity in rheumatoid arthritis. Science Translational Medicine, 2016, 8(369): 369. ra176.
[227] Catrina AI, Joshua V, Klareskog L, et al. Mechanisms involved in triggering rheumatoid arthritis. Immunological Reviews, 2016, 269(1): 162-174.
[228] Bennike TB, Ellingsen T, Glerup H, et al. Proteome Analysis of Rheumatoid Arthritis Gut Mucosa. Gut Mucosa. Journal of Proteome Research, 2017, 16(1): 346-354.
[229] Torow N, Marsland BJ, Hornef MW, et al. Neonatal mucosal immunology. Mucosal Immunology, 2017, 10(1): 5-17.
[230] Ohashi K, Burkart V, Flohe S, et al. Cutting edge: heat shock protein 60 is a putative endogenous ligand of the toll-like receptor-4 complex. Journal of Immunology, 2000, 164(2): 558-561.
[231] Vaahtovuo J, Munukka E, Korkeamaki M, et al. Fecal microbiota in early rheumatoid arthritis. The Journal of Rheumatology, 2008, 35(8): 1500-1505.
[232] Scher JU, Sczesnak A, Longman RS, et al. Expansion of intestinal Prevotella copri correlates with enhanced susceptibility to arthritis. Elife, 2013, 2: e01202.
[233] Picchianti-Diamanti A, Panebianco C, Salemi S, et al. Analysis of Gut Microbiota in Rheumatoid Arthritis Patients: Disease-Related Dysbiosis and Modifications Induced by Etanercept. International Journal of Molecular Sciences, 2018, 19(10): 2938.
[234] Huang Z, Wang J, Xu X, et al. Antibody neutralization of microbiota-derived circulating peptidoglycan dampens inflammation and ameliorates autoimmunity. Nature Microbiology, 2019, 4(5): 766-773.
[235] Lee N, Kim WU. Microbiota in T-cell homeostasis and inflammatory diseases. Experimental & Molecular Medicine, 2017, 49(5): e340.
[236] Scher JU, Abramson SB. The microbiome and rheumatoid arthritis. Nature Reviews Rheumatology, 2011, 7(10): 569-578.
[237] Holers VM. Autoimmunity to citrullinated proteins and the initiation of rheumatoid arthritis. Current Opinion in Immunology, 2013, 25(6): 728-735.
[238] Honda K, Littman DR. The microbiota in adaptive immune homeostasis and disease. Nature, 2016,

535(7610): 75-84.

[239] Luckey D, Gomez A, Murray J, et al. Bugs & us: the role of the gut in autoimmunity. The Indian Journal of Medical Research, 2013, 138(5): 732-743.

[240] Block KE, Zheng Z, Dent AL, et al. Gut Microbiota Regulates K/BxN Autoimmune Arthritis through Follicular Helper T but Not Th17 Cells. Journal of Immunology, 2016, 196(4): 1550-1557.

[241] Teng F, Klinger CN, Felix KM, et al. Gut Microbiota Drive Autoimmune Arthritis by Promoting Differentiation and Migration of Peyer's Patch T Follicular Helper Cells. Immunity, 2016, 44(4): 875-888.

[242] Onuora S. Autoimmunity: TFH cells link gut microbiota and arthritis. Nature Reviews Rheumatology, 2016, 12(3): 133.

[243] Lorca GL, Wadstrom T, Valdez GF, et al. Lactobacillus acidophilus autolysins inhibit Helicobacter pylori in vitro. Current Microbiology, 2001, 42(1): 39-44.

[244] Amdekar S, Singh V, Singh R, et al. Lactobacillus casei reduces the inflammatory joint damage associated with collagen-induced arthritis (CIA) by reducing the pro-inflammatory cytokines: Lactobacillus casei: COX-2 inhibitor. Journal of Clinical Immunology, 2011, 31(2): 147-154.

[245] So JS, Lee CG, Kwon HK, et al. Lactobacillus casei potentiates induction of oral tolerance in experimental arthritis. Molecular Immunology, 2008, 46(1): 172-180.

[246] Tarantilis CD, Kiranoudis CT. Distribution of fresh meat. Journal of Food Engineering, 2002, 51(1): 85-91.

[247] Hatakka K, Martio J, Korpela M, et al. Effects of probiotic therapy on the activity and activation of mild rheumatoid arthritis--a pilot study. Scandinavian Journal of Rheumatology, 2003, 32(4): 211-215.

[248] Pineda Mde L, Thompson SF, Summers K, et al. A randomized, double-blinded, placebo-controlled pilot study of probiotics in active rheumatoid arthritis. Medical Science Monitor, 2011, 17(6): CR347-CR354.

[249] Simelyte E, Rimpilainen M, Lehtonen L, et al. Bacterial cell wall-induced arthritis: chemical composition and tissue distribution of four Lactobacillus strains. Infection and Immunity, 2000, 68(6): 3535-3540.

[250] Abdollahi-Roodsaz S, Joosten LA, Koenders MI, et al. Stimulation of TLR2 and TLR4 differentially skews the balance of T cells in a mouse model of arthritis. The Journal of Clinical Investigation, 2008, 118(1): 205-216.

[251] Marietta EV, Murray JA, Luckey DH, et al. Suppression of Inflammatory Arthritis by Human Gut-Derived Prevotella histicola in Humanized Mice. Arthritis & Rheumatology, 2016, 68(12): 2878-2888.

[252] Magnusdottir S, Ravcheev D, De Crecy-Lagard V, et al. Systematic genome assessment of B-vitamin biosynthesis suggests co-operation among gut microbes. Frontiers in Genetics, 2015, 6: 148.

[253] Geller LT, Barzily-Rokni M, Danino T, et al. Potential role of intratumor bacteria in mediating tumor resistance to the chemotherapeutic drug gemcitabine. Science, 2017, 357(6356): 1156-1160.

[254] Lehouritis P, Cummins J, Stanton M, et al. Local bacteria affect the efficacy of chemotherapeutic drugs. Scientific Reports, 2015, 5: 14554.

[255] Chen J, Wright K, Davis JM, et al. An expansion of rare lineage intestinal microbes characterizes rheumatoid arthritis. Genome Medicine, 2016, 8(1): 43.

[256] Zheng H, Chen M, Li Y, et al. Modulation of Gut Microbiome Composition and Function in Experimental Colitis Treated with Sulfasalazine. Frontiers in Microbiology, 2017, 8: 1703.

[257] Kyburz D, Brentano F, Gay S. Mode of action of hydroxychloroquine in RA-evidence of an inhibitory effect on toll-like receptor signaling. Nature Clinical Practice Rheumatology, 2006, 2(9): 458-459.

[258] Zamani B, Golkar HR, Farshbaf S, et al. Clinical and metabolic response to probiotic supplementation in patients with rheumatoid arthritis: a randomized, double-blind, placebo-controlled trial. International Journal of Rheumatic Diseases, 2016, 19(9): 869-879.

[259] Tsokos GC, Lo MS, Reis PC, et al. New insights into the immunopathogenesis of systemic lupus erythematosus. Nature Reviews Rheumatology, 2016, 12(12): 716-730.

[260] Barrat FJ, Meeker T, Gregorio J, et al. Nucleic acids of mammalian origin can act as endogenous ligands for toll-like receptors and may promote systemic lupus erythematosus. Journal of Experimental Medicine, 2005, 202(8): 1131-1139.

[261] Lovgren T, Eloranta ML, Bave U, et al. Induction of interferon-a production in plasmacytoid dendritic cells by immune complexes containing nucleic acid released by necrotic or late anoptotic cells and lupus IgG. Arthritis and Rheumatism, 2004, 50(6): 1861-

1872.

[262] Jego G, Palucka AK, Blanck JP, et al. Plasmacytoid dendritic cells induce plasma cell differentiation through type I interferon and interleukin 6. Immunity, 2003, 19(2): 225-234.

[263] Le Bon A, Schiavoni G, D'agostino G, et al. Type I interferons potently enhance humoral immunity and can promote isotype switching by stimulating dendritic cells in vivo. Immunity, 2001, 14(4): 461-470.

[264] Choi J-Y, Ho John H-E, Pasoto SG, et al. Circulating Follicular Helper-Like T Cells in Systemic Lupus Erythematosus. Arthritis & Rheumatology, 2015, 67(4): 988-999.

[265] Von Spee-Mayer C, Siegert E, Abdirama D, et al. Low-dose interleukin-2 selectively corrects regulatory T cell defects in patients with systemic lupus erythematosus. Annals of the Rheumatic Diseases, 2016, 75(7): 1407-1415.

[266] Crispin JC, Oukka M, Bayliss G, et al. Expanded Double Negative T Cells in Patients with Systemic Lupus Erythematosus Produce IL-17 and Infiltrate the Kidneys. Journal of Immunology, 2008, 181(12): 8761-8766.

[267] Sieling PA, Porcelli SA, Duong BT, et al. Human double-negative T cells in systemic lupus erythematosus provide help for IgG and are restricted by CD1c. Journal of Immunology, 2000, 165(9): 5338-5344.

[268] Berggren O, Hagberg N, Weber G, et al. B lymphocytes enhance interferon-a production by plasmacytoid dendritic cells. Arthritis and Rheumatism, 2012, 64(10): 3409-3419.

[269] Leonard D, Eloranta M-L, Hagberg N, et al. Activated T cells enhance interferon- production by plasmacytoid dendritic cells stimulated with RNA-containing immune complexes. Annals of the Rheumatic Diseases, 2016, 75(9): 1728-1734.

[270] Barbhaiya M, Costenbader KH. Environmental exposures and the development of systemic lupus erythematosus. Current Opinion in Rheumatology, 2016, 28(5): 497-505.

[271] James JA, Neas BR, Moser KL, et al. Systemic lupus erythematosus in adults is associated with previous Epstein-Barr virus exposure. Arthritis and Rheumatism, 2001, 44(5): 1122-1126.

[272] Wolf SJ, Estadt SN, Gudjonsson JE, et al. Human and Murine Evidence for Mechanisms Driving Autoimmune Photosensitivity. Frontiers in Immunology, 2018, 9: 2430.

[273] Wolf SJ, Estadt SN, Theros J, et al. Ultraviolet light induces increased T cell activation in lupus-prone mice via type I IFN-dependent inhibition of T regulatory cells. Journal of Autoimmunity, 2019, 103: 102291.

[274] Grimaldi CM. Sex and systemic lupus erythematosus: the role of the sex hormones estrogen and prolactin on the regulation of autoreactive B cells. Current Opinion in Rheumatology, 2006, 18(5): 456-461.

[275] Vera-Lastra O, Jara LJ, Espinoza LR. Prolactin and autoimmunity. Autoimmunity Reviews, 2002, 1(6): 360-364.

[276] Bernier M-O, Mikaeloff Y, Hudson M, et al. Combined Oral Contraceptive Use and the Risk of Systemic Lupus Erythematosus. Arthritis Care & Research, 2009, 61(4): 476-481.

[277] Costenbader KH, Feskanich D, Stampfer MJ, et al. Reproductive and menopausal factors and risk of systemic lupus erythematosus in women. Arthritis and Rheumatism, 2007, 56(4): 1251-1262.

[278] Chen Y, Chaudhary N, Yang N, et al. Microbial symbionts regulate the primary Ig repertoire. The Journal of Experimental Medicine, 2018, 215(5): 1397-1415.

[279] Greiling TM, Dehner C, Chen X, et al. Commensal orthologs of the human autoantigen Ro60 as triggers of autoimmunity in lupus. Science Translational Medicine, 2018, 10(434): eaan2306.

[280] Shamriz O, Mizrahi H, Werbner M, et al. Microbiota at the crossroads of autoimmunity. Autoimmunity Reviews, 2016, 15(9): 859-869.

[281] Rosenbaum JT, Silverman GJ. The Microbiome and Systemic Lupus Erythematosus. The New England Journal of Medicine, 2018, 378(23): 2236-2237.

[282] Lopez P, De Paz B, Rodriguez-Carrio J, et al. Th17 responses and natural IgM antibodies are related to gut microbiota composition in systemic lupus erythematosus patients. Scientific Reports, 2016, 6: 24072.

[283] Van Praet JT, Donovan E, Vanassche I, et al. Commensal microbiota influence systemic autoimmune responses. The Embo Journal, 2015, 34(4): 466-474.

[284] Hevia A, Milani C, Lopez P, et al. Intestinal dysbiosis associated with systemic lupus erythematosus. MBio, 2014, 5(5): e01548-e01514.

[285] Man SM, Kaakoush NO, Mitchell HM. The role of bacteria and pattern-recognition receptors in

Crohn's disease. Nature Reviews Gastroenterology & Hepatology, 2011, 8(3): 152-168.

[286] Larsen N, Vogensen FK, Van Den Berg FW, et al. Gut microbiota in human adults with type 2 diabetes differs from non-diabetic adults. PLoS One, 2010, 5(2): e9085.

[287] He Z, Shao T, Li H, et al. Alterations of the gut microbiome in Chinese patients with systemic lupus erythematosus. Gut Pathogens, 2016, 8: 64.

[288] David R, Arancha H, Rafael B, et al. Ranking the impact of human health disorders on gut metabolism: systemic lupus erythematosus and obesity as study cases. Scientific Reports, 2015, 5: 8310.

[289] Hevia A, Milani C, Lopez P, et al. Intestinal Dysbiosis Associated with Systemic Lupus Erythematosus. Mbio, 2014, 5(5): e01548-14.

[290] He Z, Shao T, Li H, et al. Alterations of the gut microbiome in Chinese patients with systemic lupus erythematosus. Gut Pathogens, 2016, 8: 64.

[291] Van Der Meulen TA, Harrnsen HJM, Vila AV, et al. Shared gut, but distinct oral microbiota composition in primary Sjogren's syndrome and systemic lupus erythematosus. Journal of Autoimmunity, 2019, 97: 77-87.

[292] Azzouz D, Omarbekova A, Heguy A, et al. Lupus nephritis is linked to disease-activity associated expansions and immunity to a gut commensal. Annals of the Rheumatic Diseases, 2019, 78(7): 947-956.

[293] Zhang H, Liao X, Sparks JB, et al. Dynamics of Gut Microbiota in Autoimmune Lupus. Applied and Environmental Microbiology, 2014, 80(24): 7551-7560.

[294] Klaenhammer TR, Kleerebezem M, Kopp MV, et al. The impact of probiotics and prebiotics on the immune system. Nature Reviews Immunology, 2012, 12(10): 728-734.

[295] Meehan CJ, Beiko RG. A Phylogenomic View of Ecological Specialization in the Lachnospiraceae, a Family of Digestive Tract-Associated Bacteria. Genome Biology and Evolution, 2014, 6(3): 703-713.

[296] Mu Q, Zhang H, Liao X, et al. Control of lupus nephritis by changes of gut microbiota. Microbiome, 2017, 5(1): 73.

[297] Luo XM, Edwards MR, Mu Q, et al. Gut Microbiota in Human Systemic Lupus Erythematosus and a Mouse Model of Lupus. Applied and Environmental Microbiology, 2018, 84(4): e02288-17.

[298] Zegarra-Ruiz DF, El Beidaq A, Iniguez AJ, et al. A Diet-Sensitive Commensal Lactobacillus Strain Mediates TLR7-Dependent Systemic Autoimmunity. Cell Host & Microbe, 2019, 25(1): 113-127.e6.

[299] Richard ML, Gilkeson G. Mouse models of lupus: what they tell us and what they don't. Lupus Science & Medicine, 2018, 5(1): e000199.

[300] Kau AL, Ahern PP, Griffin NW, et al. Human nutrition, the gut microbiome and the immune system. Nature, 2011, 474(7351): 327-336.

[301] Kosiewicz MM, Dryden GW, Chhabra A, et al. Relationship between gut microbiota and development of T cell associated disease. FEBS Letters, 2014, 588(22): 4195-4206.

[302] Johnson BM, Gaudreau MC, Al-Gadban MM, et al. Impact of dietary deviation on disease progression and gut microbiome composition in lupus-prone SNF1 mice. Clinical and Experimental Immunology, 2015, 181(2): 323-337.

[303] Kimura A, Kishimoto T. IL-6: Regulator of Treg/Th17 balance. European Journal of Immunology, 2010, 40(7): 1830-1835.

[304] Mcclain MT, Heinlen LD, Dennis GJ, et al. Early events in lupus humoral autoimmunity suggest initiation through molecular mimicry. Nature Medicine, 2005, 11(1): 85-89.

[305] Szymula A, Rosenthal J, Szczerba BM, et al. T cell epitope mimicry between Sjogren's syndrome Antigen A (SSA)/Ro60 and oral, gut, skin and vaginal bacteria. Clinical Immunology, 2014, 152(1-2): 1-9.

[306] Khan D, Ansar Ahmed S. The Immune System Is a Natural Target for Estrogen Action: Opposing Effects of Estrogen in Two Prototypical Autoimmune Diseases. Front Immunology, 2015, 6: 635.

[307] Markle JG, Frank DN, Mortin-Toth S, et al. Sex differences in the gut microbiome drive hormone-dependent regulation of autoimmunity. Science, 2013, 339(6123): 1084-1088.

[308] Yurkovetskiy L, Burrows M, Khan AA, et al. Gender Bias in Autoimmunity Is Influenced by Microbiota. Immunity, 2013, 39(2): 400-412.

[309] Preidis GA, Versalovic J. Targeting the human microbiome with antibiotics, probiotics, and prebiotics: gastroenterology enters the metagenomics era. Gastroenterology, 2009, 136(6): 2015-2031.

[310] Bultman SJ. Emerging roles of the microbiome in cancer. Carcinogenesis, 2014, 35(2): 249-255.

第 9 章
益生菌与机体免疫

益生菌是指一类具有生物活性的菌群，当人体此类菌群含量适当时，可以起到调节肠道微生态环境、改善健康状况、调节人体免疫、协同过敏治疗等对健康有益的作用[1]。一些常用的益生菌种如双歧杆菌等凭借其安全可靠、性能优良的特点而备受青睐，并早已被广泛应用于食品、医药和饲料工业之中。

越来越多的研究发现，益生菌与肠道屏障的生理状态及病理变化存在着密不可分的联系。本章在介绍益生菌相关定义的同时，也将为读者呈现益生菌在肠道免疫方面的前沿内容，以供参考。

第一节　益生菌相关产品概述

一、益生菌

益生菌是一类可以在肠道定植、参与维护肠道菌群平衡、刺激和调节肠黏膜免疫的菌群[1, 2]。在 2001 年联合国粮食及农业组织（Food and Agriculture Organization of the United Nations，FAO）和世界卫生组织（World Health Organization，WHO）联合报告上，益生菌的定义首次出现在公众的视野中。然而，由于肠道菌群领域的飞速发展与各种观点的冲突，益生菌的定义不断地被质疑、推翻和重新确立。2014 年，益生菌的定义被更新为："适量摄入后，可以对宿主产生有益作用的活菌群"[3]，这一新的定义得到了大多数相关研究人员的肯定。

根据新定义，一种细菌在被确定为益生菌前，其有益作用必须要经过科学有效的论证证实。其间不仅要确定益生菌的菌种，更要弄清楚它有没有发挥有益的效应。因此，一些一直以来被称作"益生菌"的传统菌群实际上并不合格。例如传统发酵食品中的菌群，这些菌群成分复杂多样，难以一一分类。而且发酵食物及其内含的菌群对机体的影响混杂难辨，很难确定两者各自的作用到底如何。因此发酵食物中的细菌并不属于益生菌。同理，用于粪菌移植的粪便菌群彼此差异很大，成分也很复杂，且其各自的作用效果不能确定，因此也不符合益生菌的定义。

一般来说，评价一种合格的益生菌要包含以下 4 个步骤：①属 / 种 / 株的鉴定；②体外实验，如对胃酸、胆汁的抵抗力、对人肠上皮细胞和细胞系和（或）黏液的黏附力、对潜在致病菌的抗菌活性等；③安全性评估；④成熟的动物和人体试验证据。目前已经确认的常见益生菌有双歧杆菌、嗜热链球菌、布拉氏酵母菌等。这些益生菌被加工成各种类型的产品，例如双歧杆菌三联活菌肠溶胶囊等，在防治感染、调节机体正常菌群平衡、提高机体免疫力和儿童生长发育等方面均有应用。

益生菌在人体中的作用大致可以分为 4 类：①通过其本身成分，如磷壁酸、脂多糖等，刺激宿主免疫细胞活化，强化宿主免疫功能，参与肠黏膜免疫系统的调控；②肠道内某些益生菌可诱发肠道低度炎症，可以增强肠道免疫系统对外来抗原的防御性；③部分肠道益生菌通过与致病菌竞争生存空间和营养物质来减少致病菌的定植，并通过产生短链脂肪酸、调节机体免疫杀灭致病菌；④某些益生菌通过代谢肠上皮细胞产生的黏液及肠道内的食物残渣，产生营养物质（如维生素 K、维生素 D），为宿主及自身供给营养。尽管与其他菌群相比，益生菌具有更高的安全性，但对于人体来说，它们始终是“非己”的外来物种，可能会引起全身感染、排放有毒代谢物、引发敏感个体的免疫不良事件以及导致耐药基因的转移。虽然这些副作用仅偶见报道，但对于免疫功能受损的患者而言，使用益生菌应非常谨慎。

当前，益生菌的研究还面临着重要挑战：益生菌就像一个功能繁多却机制不明的“黑箱”，研究者们所能观测到的仅仅是“输入”与“输出”之间的变化，而鲜少能窥明其中过程。这就使得我们即使明知应用益生菌的确能改善个体的健康状况，但仍然应该时刻担忧益生菌可能对机体带来的隐患。

益生菌的作用机制是近年来肠道菌群研究的热门之一。以往研究表明，定植于肠道的益生菌可通过激活肠道免疫，抑制肠上皮细胞异常凋亡、调节紧密连接蛋白等多种途径增强肠道屏障功能，从而有利于宿主健康[4]。新近报道提示，非蛋白编码小分子 RNA（microRNA 或 miRNA）作为调控宿主基因表达的重要因子之一，不仅是肠道屏障稳态的监督者，也是肠道益生菌与宿主相互作用的重要纽带[5]。

二、益生元

益生元是一类可以调节肠道菌群组成的化合物。它们可以为肠道菌群提供生长所需的基本营养物质，也可以通过这一方式改变菌群的组成。1995 年，Gibson 和 Roberfioid 率先提出了益生元的概念：一种在肠道中不被消化、通过选择性地促进消化道中某些细菌物种的生长和（或）增加其活性，从而对宿主健康产生有益影响的物质。此后，有关专家围绕该概念进行了多次讨论，大多数认为：益生元必须作用于有益宿主健康的菌群（主要是双歧杆菌和乳酸菌属）或有助于人体代谢活动，但由于其中的机制不明朗，研究者们始终难以达成一致。直到 2017 年，业界才发布了益生元的共识定义：“一种有益于健康的，被宿主菌群选择性利用的物质”。该定义确定，益生元必须是被肠道菌群利用的化合物，并且其发挥的有益作用必须来自利用它的菌群，而不是益生元本身。新版的定义也扩大了可以利用益生元的菌群范围，只要是对机体有益的、能够明确种类的菌群都可被纳入其中（图 9-1）。

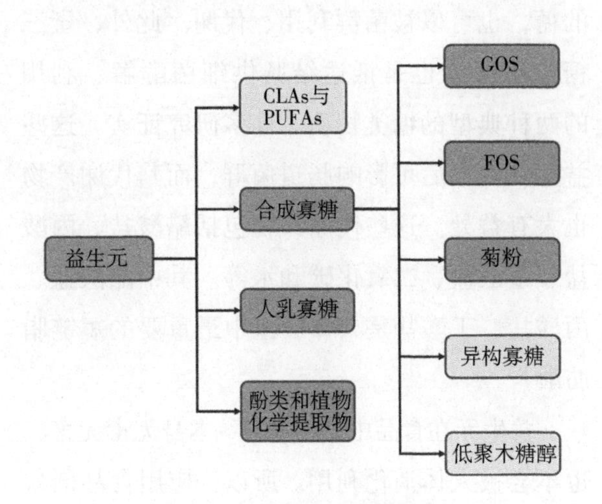

图 9-1　益生元的简单分类

CLAs. 亚油酸；PUFAs. 不饱和脂肪酸；GOS. 益生元；FOS. 低聚果糖

益生元安全可靠，可以直观且显著的影响

肠道菌群的功能。例如，低聚糖类益生元可促进肠道有益菌乳杆菌的增殖、抑制有害菌的生长，还可促进淋巴细胞增殖，提高免疫水平。益生元被肠道菌群利用，产生的有机酸可以调节结肠菌群的生长，还可以刺激细胞免疫的建立，从而提高机体免疫能力[3]。益生元还对机体养分的吸收具有调节作用。例如，益生元对矿物质代谢具有调控作用，还可以提高结肠钙离子浓度，促进肠道蠕动和激素的肠肝循环[6]。

传统意义上的益生元包括菊粉、低聚果糖和低聚半乳糖等[7]。菊粉降解产生低聚果糖，作为果聚糖，它们天然存在于某些容易获得的食物中，例如洋葱、芹菜、芦笋等。低聚果糖一般用作低度甜味剂进行调味，或者用来代替食品加工中的脂肪。低聚半乳糖来自乳糖，是一种包含 2 ～ 8 个糖基单元的混合物，最常用于婴儿的营养喂养，对胃肠道的消化过程具有保护作用，可以促进肠道微生物群中对机体有益细菌的选择性发育；维持消化系统的结构和功能；通过激活免疫细胞、淋巴组织的发育来维持免疫平衡；维持肠道菌群动态平衡[8]。

乳果糖则是一种由半乳糖和果糖单元合成的糖，也可以被菌群利用、代谢。此外，蜜三糖与水苏糖也是抵达结肠供细菌酵解、利用的两种典型的糖类物质。许多研究证实，这些益生元代谢后可影响肠道菌群，而其代谢产物也大有益处。这些代谢产物包括醋酸盐、丙酸盐、丁酸盐、二氧化碳和水等。其中醋酸盐、丙酸盐、丁酸盐是人体肠道中最重要的短链脂肪酸[9]。

益生元在食品中广泛存在，本身无毒无害，也不会被人体消化利用。所以，应用食品剂量的益生元是不会出现副作用的。尽管北欧食品毒理学和风险评估工作组对低聚果糖进行安全评估后，指出一些人在食用低聚果糖后可能会产生腹痛、腹胀和腹泻等消化道不良症状，但对于大多数成人来说，只要每日摄入的低聚果糖不多于 20g，这些不良反应便不会发生。此外还有一些证据显示[10-12]，在非母乳喂养的婴幼儿食品中添加益生元成分（低聚果糖 / 低聚半乳糖），这些婴幼儿的粪便与各项指标都会与母乳喂养的婴幼儿相接近，而且这种效应与剂量呈现正相关[13, 14]，这为益生菌制品在婴幼儿早期哺育中的使用提供了参考依据。

目前，所有已发现的益生元都是碳水化合物（膳食纤维），但根据新定义，未来益生元将不再局限于此，其概念可能也将适用于消化系统以外的其他部位。

三、合生元

合生元是至少一种益生菌与至少一种益生元的组合。鉴于益生菌和益生元的确切效果尚不十分清楚，合生元的作用机制显然更加复杂。但不可否认的是，目前已经投入使用的一些合生元的确发挥了“1+1 ＞ 2”的生物作用，其疗效与成果是真实可见的。研究显示，在肝硬化患者中进行合生元的临床对照试验，实验组患者肠道菌群组成明显优于安慰剂组，同时消化系统的功能也得到了显著改善。合生元主要是通过作用于调节性 T 细胞、效应性 T 细胞、自然杀伤细胞和 B 细胞来改善免疫功能的[15]。当前合生元的应用尚不普及，有关研究也多处于起步阶段。但由于其低廉的成本与特殊的效果，合生元有望在未来的临床治疗中大展身手，造福万千患者。

四、后生元

益生菌虽然已日趋成熟，但仍有不耐高温、胃酸和胆碱的先天劣势，商品应用的形态备受限制。“后生元”即灭活型的有益菌应运而生，具体来说，后生元制剂是益生菌通过自然发酵过程产生的有益代谢物或细胞壁成分，通过提供额外的生物活性为宿主提供生理益处[16]。这些代谢物包括胆汁酸、短链脂肪酸、酶、肽、磷壁酸、肽聚糖衍生的多肽、内多糖和外多糖、细胞表面蛋白、维生素和有机酸[17-19]。后生元稳定，不仅耐高温、耐胃酸、耐胆碱，而且后期加工不受限制，并且根据成分不同，具有不同的功能特性，包括抗菌、抗氧化和免疫调节。这些特性可能会影响微生物群稳态、宿主代谢、信号通路、生理、免疫系统的调节、神经激素调节和代谢反应[20, 21]。后生元在骨骼矿物吸收和生物利用、生长因子的调节、炎症细胞因子的分泌和抗萎缩作用、调节血脂、慢性胃肠道疾病（如胃炎、克罗恩病、炎性肠病）和2型糖尿病的防治中均发挥有益作用[22]。

五、益生菌产品的发展

过去由于不同的文化与地域，不同地方的人们发展了属于各自的“传统益生菌产品”。益生菌最早的应用并非治疗，而是储存。例如，在亚热带和热带国家的高温环境下，通常会用发酵的手段建立优势有益菌群、抵抗腐败菌，从而预防食源性疾病[23]。

而现代益生菌产品主要集中在乳制品行业，它们以食物的形式进入人体后，通过定植占位、代谢产物等途径，发挥调节菌群组成、促进肠道吸收、调节肠道免疫、缓解肝脏负荷、治疗便秘等作用。国外很早便展开了益生菌发酵产品工业化生产的探索，例如1977年，Morinage开发了家庭型双歧杆菌乳；1978年，日本Yakul公司开发了双歧杆菌液体酸奶MilMilTM；法国市场上含有双歧杆菌和嗜酸乳杆菌的产品已占酸奶销售量的11%；丹麦医师已经开始推荐通过摄食双歧杆菌制品治疗肠道紊乱；英国更是开发了一种可用来治疗肠炎的益生菌混合产品，这种包含了8种细菌的混合物可以通过刺激IL-10增加，减少炎症介质IL-12的产生；美国人则相信含有益生菌的酸奶能够增进免疫系统的功能，并认为这种食品不仅安全，更是药品的良好补充（图9-2）。

我国有关益生菌产品的开发起步则较晚。在中国，非食品益生菌制品最早主要以胶囊类药物形式生产，经常被临床医师用来调节肠道菌群平衡，仅有少量的商家生产含有益生菌的奶制品。随着我国科研工作者向肠道菌群领域的不断进军，以及大众认知水平的不断提升，益生菌制品开始走出实验室和医院，并走入千家万户，不断地在功能性食品方面取得发展。尽管行业发展形势喜人，但目前益生菌产品所涵盖的范围还比较局限，其应用多在保健食品方面，而临床应用尚处于探索阶段[24]。

随着更好的培养方法、更先进的编辑和修改细菌基因工具的出现，以及基因测序费用的不断降低，极大地扩大了具有潜在健康益处的生物体的范围，一些具有强大调节人体免疫，改善健康状况的菌株出现在大众视野（图9-2）。如可逆转肥胖和胰岛素抵抗的普雷沃特菌、克里斯滕森菌、嗜黏蛋白阿克曼菌和多形拟杆菌；能够保护小鼠抵抗肠道疾病的普拉梭菌，以及减少炎症和显示抗癌作用的脆弱拟杆菌。这些菌株被称为二代益生菌（next-generation probiotics，NGP）。它们在美国新的监管体系下也被称为活生物治疗产品（Live biotherapeutic products，LBP）[25]。美国食品药品监督管理局指出，LBP是一类专门为药物应用而开发的活体生物，包含3个方面内容，即含有活生物体，如细菌；适用于人类疾病的预防及治疗；不是疫苗。

在美国，LBP是一个正式认可的概念，有一些研究人员对NGP一词是否有必要单独存在提出了质疑。随着研究的进展，NGPs 逐渐开发出了除药物以外的其他应用，如在食品运输工具中作为补充剂，在市面作为功能性食品等。但又区别于传统益生菌，NGP 的可能上市途径将遵循一条以临床前作用模式，即需要以安全性、药动学、药效学和 1 ～ 3 期临床试验研究为标志的路径，并需要更严格的监管与批准条件，因此将这些生物体简单地称为益生菌会引起混淆。从开发方面来说，NGP 倾向于将既往关于益生菌的实验数据进一步深入研究，从而发现新的益生菌，并将其用于实际应用；LBP 往往由生物技术公司或制药公司进行开发，具有明确的目的，寻找治疗改善疾病的具有药品潜能的益生菌。NGP 是一种合理的尝试，它标志着从具有长期安全使用历史的传统肠道菌群中，不断发掘新的对健康有益的菌群。随着时间的推移，NGP 一词将消失，其成员将要么与现有益生菌合并，要么开发为 LBP，通过制药途径进入市场（图 9-3）[26]。

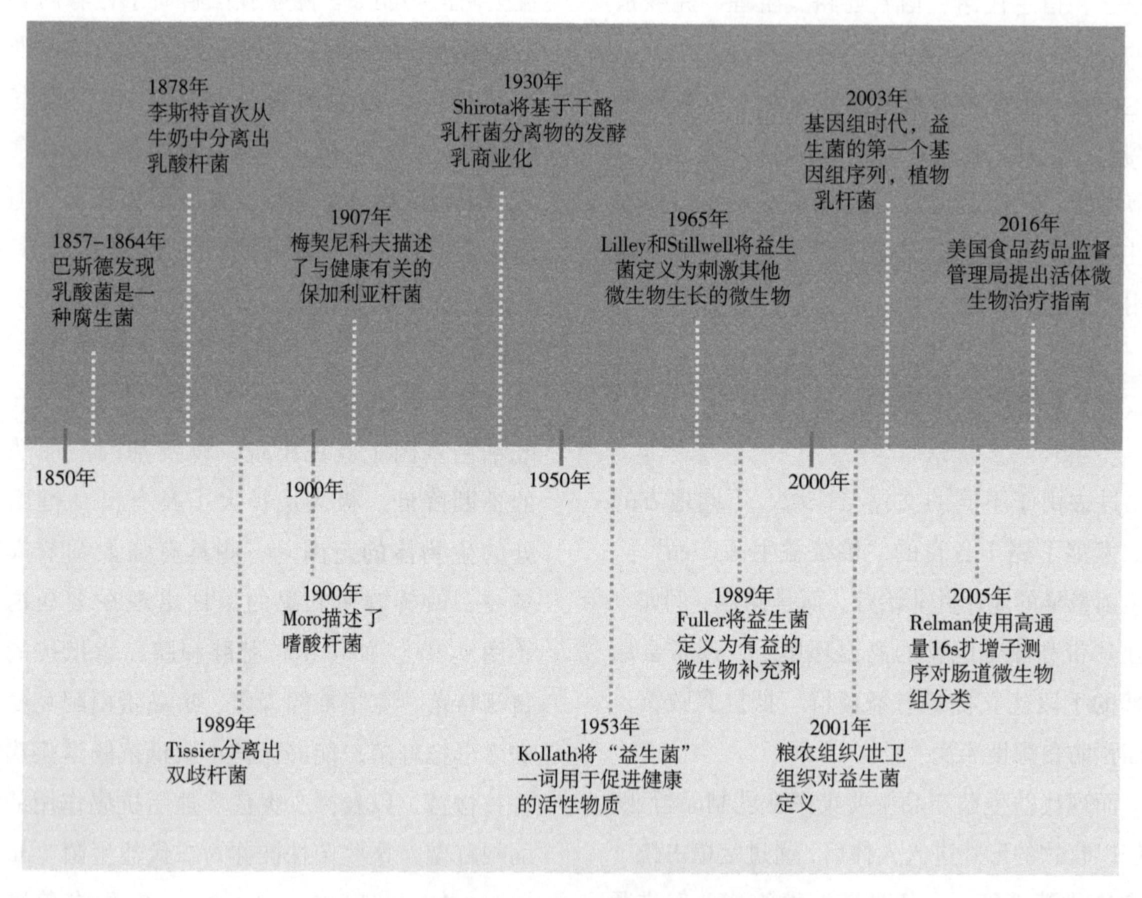

图 9-2　益生菌及二代益生菌历史里程碑时间轴

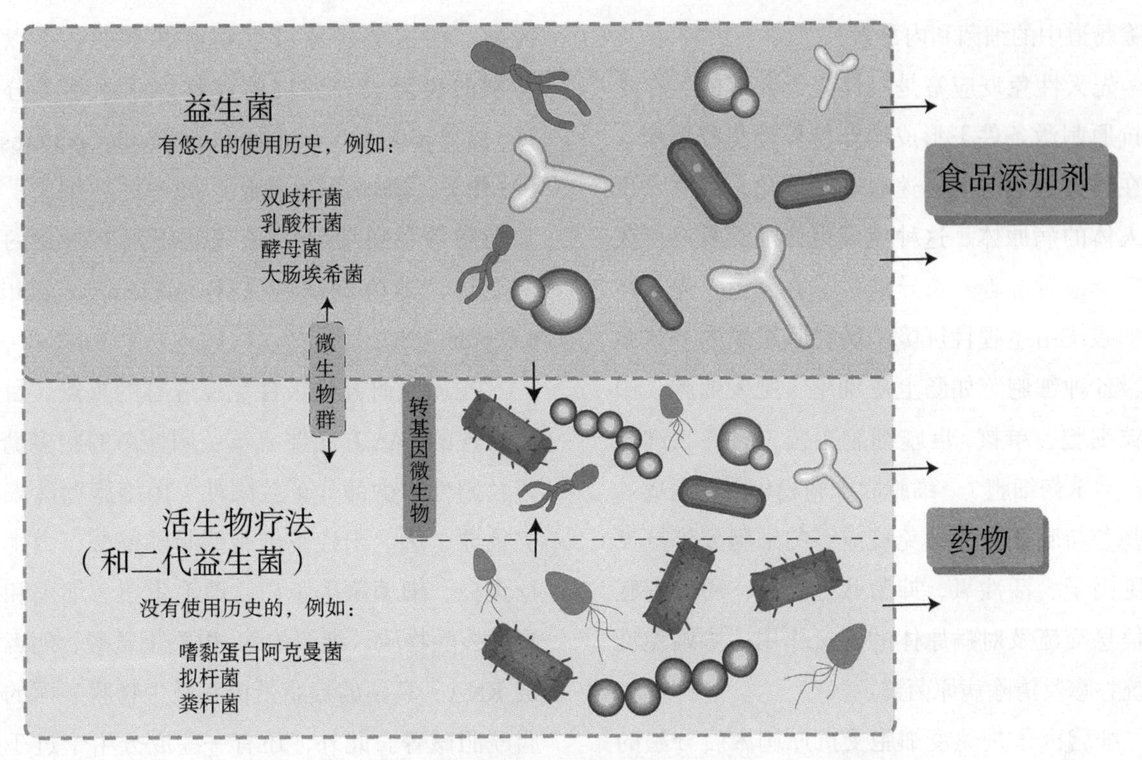

图 9-3　益生菌、二代益生菌和活生物治疗产品的历史和市场途径

第二节　益生菌与肠黏膜免疫

一、什么是肠黏膜免疫

消化道是外界抗原或病原菌群进入体内的门户，同时也充当着机体的机械、化学、生物及免疫屏障。胃肠黏膜机械屏障主要由黏膜上皮细胞、细胞间的紧密连接和黏液层构成。肠黏膜杯状细胞可以参与分泌糖蛋白，覆盖于肠黏膜表面，具有保护肠黏膜、润滑黏膜表面和阻止条件致病菌黏附的作用。现已发现小肠黏膜下集合淋巴结（Peyer's patches，派尔淋巴结）表面的微折叠细胞（M 细胞）是肠道上皮唯一允许物质通透的上皮细胞，食物抗原、细菌、病毒和其他病原菌群可通过微折叠细胞进入人体，引发机体免疫反应。

胃肠黏膜化学屏障主要由胃酸、胆汁、溶菌酶、蛋白分解酶、防御素、三叶因子和前列腺素 E2 等构成[27, 28]。胃肠黏膜的生物屏障则主要由肠道黏液、黏膜内的大量免疫细胞和固有层供血血管构成，其作用是抑制侵入性菌群的增殖并阻止细菌穿透。食物或细菌抗原接触胃肠道后，大部分抗原成分主要由派尔淋巴结表面的微折叠细胞吸收[28]，少部分则经肠黏膜固有层内树突状细胞的突触摄取，或通过上皮细胞间的缝隙直接吸收，引起肠黏膜局部免疫反应，并诱导肠黏膜免疫耐受。若打破此免疫耐受状态，可引起食物过敏反应或炎症损伤[29]。肠道内病理性抗原被吸收进入派尔淋巴结和肠黏膜固有层后，在 TGF-β、IL-4、IL-5、IFN-γ 和抗原的刺激下，B 淋巴细胞增殖并转化成细胞表面的浆细胞，T 淋巴细胞激活分化成免疫效应细胞，对肠黏膜起免疫保护作用[28, 30]。分泌型免疫球蛋白 A（secretory immunoglobulin A，sIgA）是胃肠道黏膜表面主要的免疫球蛋白，对消化道黏膜的防御起重要保护作用。肠腔内 sIgA 通过结合细菌，形成抗原抗体复合物，刺激肠道黏液的分泌并加速黏液的更新，有助于排除肠道中的细菌和内毒素[31-33]。

先天性免疫应答是机体先天固有的、在外界抗原刺激条件下形成的非特异性免疫反应，可在感染数小时后迅速启动和活化，以清除侵入人体的病原体。这种免疫力伴随终身，持久而广泛地对抗着外来异物[34]。在肠道，先天性免疫系统由生理性屏障、肠黏膜组织内补体系统、各种细胞（如肠上皮细胞、肥大细胞、中性粒细胞、单核 - 巨噬细胞系统、树突状细胞和自然杀伤细胞）、细胞因子和趋化因子组成。这些参与肠道先天性免疫应答的细胞分泌抗炎细胞因子、活性氧、抑菌肽等物质，通过细胞间信息交流及对病原体的直接作用，协调免疫系统吞噬及清除病原菌群。

细胞因子是免疫细胞受抗原刺激后分泌的小分子可溶性蛋白，是调节先天性和获得性免疫应答的化学信使。目前，人们认为参与先天性免疫应答的细胞因子主要包括 TNF-α 和 IL-1。TNF-α 能刺激其他促炎症细胞因子的生成和释放，激活中性粒细胞并诱导其表达分化抗原 CD11/CD18 复合物。此外，TNF-α 还可以作用于血管内皮细胞，使其表达细胞间黏附分子 -1 和内皮细胞黏附分子 -1，从而导致白细胞和血管内皮细胞之间的相互作用，介导白细胞的黏附、游出，对入侵的病原体针对性地诱发趋化作用，进而导致大量活性氧和弹性蛋白酶的释放与聚集，杀灭病原体的同时也损害血管内皮细胞和器官组织细胞[35]。

巨噬细胞是一种功能强大的细胞，它不仅可以直接吞噬外来抗原，还具有分泌功能，如 IL-1 和趋化因子。IL-1 可增强 NK 细胞的杀伤活性，刺激内皮细胞和中性粒细胞合成黏附分子，激活巨噬细胞分泌表达 IL-6、IL-8、TNF-α 等，提高免疫系统活性，促使炎症反应。

先天性免疫系统好比是人类免疫系统代代相传的“记忆”，伴随着人类进化而不断更新、传递，使之拥有了识别病原微生物的能力。而这类病原微生物所共有的、可以被识别的表面成分和抗原被称为病原相关分子模式（pathogen-associated molecular patterns，PAMP），机体免疫细胞（如中性粒细胞、巨噬细胞等吞噬细胞）能表达识别 PAMP 的相应受体，即模式识别受体（pattern recognition receptor，PRR）[36, 37]。

PRR 是研究先天性免疫应答的重点。目前已发现的 PAMP 主要有革兰阴性菌的脂多糖、革兰阳性（少部分革兰阴性）菌细胞壁肽聚糖类抗原、革兰阳性菌细胞壁脂磷壁酸、微生物糖蛋白、细菌鞭毛蛋白、菌毛蛋白、细菌和病毒的核酸物质、细菌的 N- 甲酰蛋氨酸、病毒双链 RNA、真菌的糖脂蛋白和微生物膜表面的磷脂酰胆碱等。此外，还有一些成分由于过于复杂而暂时不能被精准识别。PAMP 具有以下特点[37]：①它们仅为病原微生物所具有，宿主的细胞并不存在，借此宿主细胞可以通过 PRR 识别非自身的抗原成分；②它们的分子组成和构型相对保守，是微生物生存所必需的共有物质，基因突变率较低；③稳定性较好，即便宿主的 PRR 有限，也足以用来识别大多数病原体。

Toll 样受体（Toll-like receptor，TLR）是一类重要的 PRR。前发现 TLR 主要表达于具有免疫功能的细胞（如单核 - 巨噬细胞、B 细胞、T 细胞、树突状细胞和中性粒细胞），此外，在脂肪细胞和肠黏膜上皮细胞中也有表达。TLR 有许多亚型，每种可以识别不同的物质。从来源上看，TLR 配体可分为内源性和外源性两大类[38]。目前，研究较为成熟的配体是外源性配体，其中较重要的外源性配体有内毒素、肽聚糖、鞭毛蛋白、dsRNA、含非甲基化 CpG 的 DNA 等。而对于内源性配体，目前我们仍知之甚少。已有研究发现细胞由于感染或者其他应激情况而坏死后释放的热休克蛋白 60 是 TLR4 的配体之一，这为内源性配体的研究方向打开新的门户，也使得研究者们对肠道黏膜免疫有更进一步的认知[39, 40]。

二、益生菌对肠黏膜免疫的作用

人的消化道中生存着大量的菌群，其数量甚至多于人体本身的细胞数目。这些菌群不断地增殖、更新，利用肠道中的营养物质进行着活跃的代谢活动。这些菌群中不乏有具致病能力的菌种，但是为什么人体很少因为它们而患病呢？这与益生菌的定植和肠黏膜免疫系统有关[2，41-43]。

消化道中的益生菌作为一种活的菌群，对人体发挥着双重的保护效应[44]。一方面它可以在肠道内定植，通过生物学行为维护肠道菌群的平衡[45]；另一方面益生菌还可作为抗原直接作用于宿主的免疫系统。其作用方式包括：①诱发肠道免疫并刺激胸腺、脾脏或法氏囊等免疫器官的发育，促进巨噬细胞活力或发挥佐剂作用（如联用益生菌与霍乱疫苗可增强效果）[46]；②通过增强 T、B 细胞对抗原刺激的反应性，发挥特异性免疫作用；③活化肠黏膜内的相关淋巴组织，使 sIgA 生物合成增加，增强肠道黏膜免疫功能；④诱导淋巴细胞和巨噬细胞产生细胞因子，发挥免疫调节作用[47-49]，从而增强机体免疫功能。此外，肠黏膜上皮的通透性也具有免疫保护活性，并受紧密连接蛋白的调控，这些蛋白的表达已被证明是由特定的益生菌调节的[50]。益生菌进入消化道后被肠黏膜上特有的微折叠细胞摄取，传递给抗原呈递细胞，通过产生最适剂量的抗原促进辅助性 T 细胞（Th）0 向 Th1 的分化，进一步刺激 B 细胞分泌 sIgA，抑制免疫球蛋白 E 的分泌，它产生的一些细胞因（（IL-2、IFN-α、IFN-β 和 IFN-γ）可抑制肿瘤和病毒的生长。

sIgA 经肠黏膜上皮细胞释放到肠腔内，溶解到黏液中，与肠黏膜表面的正常菌群共同存在，共同抑制致病菌的侵入、定植，维持肠道内的正常菌群平衡。益生菌在肠道黏附定植，增强 T、B 细胞对抗原刺激的反应性，发挥特异性免疫作用活化肠黏膜内的相关淋巴组织，使 sIgA 生物合成增加、提高消化道黏膜免疫功能[51]。用益生菌鲜奶酪灌喂食小鼠后发现，在小肠和大肠中产生 sIgA 的浆细胞数量明显增多。用乳杆菌灌喂小鼠，发现小肠产生 sIgA 的浆细胞数量有所增加，这说明益生菌饲喂的动物肠道产生了更多的 sIgA，产生 sIgA 的细胞可以在益生菌的作用下逐渐增加，机体免疫功能因此得到加强。由此佐证了益生菌能够通过竞争机制在肠道内定植，并通过提供抗原等方式刺激肠道免疫细胞，发挥免疫保护作用。

益生菌还有诱导细胞 NF-κ B 介导反应的能力。NF-κ B 是一种可以活化肠腔内 Toll 样受体的物质，而 Toll 样受体又是免疫启动的关键因子[52]。这些细胞因子通过合成和分泌的相互调节、受体表达的相互调控、生物学效应的相互影响而共同组成了一个纷繁而高效的细胞因子网络。用储存不同时期的酸奶饲喂小白鼠，结果发现乳酸菌的浓度与小鼠肠道淋巴细胞数目以及 IL-2 的水平呈正相关。有研究对一种含双歧杆菌的益生菌产品进行测评，结果发现来源于婴儿双歧杆菌和短双歧杆菌的 DNA 可以明显促进 IL-10 和 IL-1 的产生[53]。此外，双歧杆菌 DNA 中鸟嘌呤和胞嘧啶含量相对较高，可以更好地与 TLR9 结合。用嗜酸乳杆菌与副干酪乳杆菌灌喂小鼠后，发现小鼠肠道中 sIgA、IL-10 和 IFN-γ 的分泌明显增多，由此证实摄入这两种菌都可以激活小鼠免疫系统的功能[54]。

肠黏膜免疫反应不仅取决于机体自身的免疫调节系统，同时也受控于肠道菌群的代谢活性和肠道菌群的情况，如嗜酸乳杆菌中提取的肽聚糖可以对诱导型一氧化氮酶和环氧合酶 -2 有明显的抑制作用[55，56]，而植物乳杆菌可以明显提高肠道内 sIgA 的含量[57，58]。

早期研究发现，喂食含双歧杆菌酵奶的小鼠肠黏膜下淋巴结明显增生。此后的进一步研究表明，短双歧杆菌不仅可以直接促进小肠黏膜下淋巴结细胞增殖，还可促进B细胞的增殖[59]。益生菌对肠上皮内淋巴细胞的增殖也具有刺激作用，具体来说，就是益生菌对肠道上皮层内的T细胞和自然杀伤细胞有刺激作用，其中T细胞以$CD8^+$T细胞为主。有研究证明，对母猪和仔猪灌喂益生菌，能够增强肠道上皮细胞的功能及活性，并且促进$CD8^+$T细胞数量增多[60]。巨噬细胞在益生菌的刺激下活化并产生细胞因子，后者可以促进免疫细胞的增殖、分化或增强免疫反应。同时，巨噬细胞也可以通过自身或分泌物抑制免疫应答[61-63]。

三、益生菌与肠道菌群的关系

肠道菌群是一个“鱼龙混杂”的大家庭。人们在动物研究中发现，无菌小鼠要比正常条件饲喂的小鼠对病原菌更加易感。除此以外，人或者小鼠在接受抗生素处理后，消化道内优势菌群若发生变化，也更容易被病原菌感染。这些现象说明肠道内的正常菌群尤其是益生菌在保护宿主抵抗病原菌感染方面发挥着重要作用。目前的研究表明，正常菌群可能通过直接和间接两种途径来帮助宿主对抗病原菌感染。

一些研究表明，益生菌与正常菌群可以通过占位作用[64]和限制营养的方式来阻止病原菌在肠道定植。例如，非致病性大肠埃希菌能够和肠出血性大肠埃希菌竞争有机酸、氨基酸和其他营养成分来抑制对方的生长，从而发挥抗菌的作用[65, 66]。虽然正常菌群与病原菌进行竞争的具体机制还不很清楚，但就目前的研究来看，肠道菌群在与代谢方面类似的病原菌对抗时，能表现出更强大的竞争能力。例如有研究表明，大肠埃希菌比与其代谢类似的病原菌-柠檬酸埃希菌更具竞争力[66]。除了直接竞争营养外，正常菌群也能够通过其他方面的代谢活动来阻止病原菌在肠道的定植。例如，多形拟埃希菌通过利用宿主肠道里的多糖产生多种岩藻糖苷酶，这些酶能够激活肠出血性大肠埃希菌的甘露糖受体，干扰肠出血性大肠埃希菌毒性因子的表达，从而来阻止其定植到肠道内。

在益生菌与正常菌群长期处于优势地位的情况下，病原菌也进化出一些策略来对抗这种竞争。其中一种策略是利用正常菌群并不需要的一些营养成分来满足其营养需要。例如，肠出血性大肠埃希菌能够利用非致病性大肠埃希菌不常利用的一些营养成分（如半乳糖、甘露糖、核糖和氨基乙醇）来作为其代谢的原料来源，避免了与优势菌群的竞争。病原菌的另外一个策略是通过表达毒性因子来诱导炎症反应，造成组织损伤并为自己提供适宜的生存环境。例如，鼠伤寒沙门菌感染会导致中性粒细胞活性氧的过度释放，促进内源性的硫代硫酸盐向连四硫酸盐转变[67]，这些硫代硫酸盐可以被鼠伤寒沙门菌利用，从而促进其生长；另外，肠出血性大肠埃希菌和肠致病性大肠埃希菌还可以侵入到接近肠上皮的一些部位，远离优势菌群的定植部位或者使用一些特定的营养成分来进行生长，从而避免直接与共生菌竞争营养物质[66, 68]。病原菌也可以通过直接的杀伤作用来对抗正常菌群，如一些革兰阴性病原菌能够通过分泌一些毒素，直接对与其有竞争性的共生菌进行杀伤[69]。

正常菌群与益生菌对抗菌物质的分泌也具有显著刺激作用。研究发现，潘氏细胞能够分泌REG Ⅲ g和REG Ⅲ b来调节宿主和细菌之间的相互作用[70]，进一步实验发现，REG Ⅲ g和REG Ⅲ b的含量在敲除了上皮细胞MyD88的小鼠体内显著下降，这说明TLR分子能够通过识别正常菌群来调控抗菌分子的表达。而REG Ⅲ g和REG Ⅲ b表达量低的小鼠也更容易被一些在正常个体体内不易致病的肠道病原菌（如单核李斯特菌、假结核耶尔森菌和肠道沙门菌等）所感染[71]。正常菌群与益生菌还通过其代谢产物来间接增强肠上皮的屏障功能。例如，双歧杆菌能够通过产生醋酸盐作用于肠上皮细胞，从而抑制大肠埃希菌 O157: H7 产生的志贺毒素的移位[72]。

位于肠黏膜固有层的单核 - 吞噬细胞如巨噬细胞和树突状细胞是机体建立对正常菌免疫耐受的基础[73]，免疫耐受的建立有助于维持肠道的稳态，防止炎症的过度激活。肠道吞噬细胞对于正常菌群和它们的微生物模式分子受体往往表现为低反应，受到刺激后不会产生很高的促炎性分子[74]。正常菌群的存在可以提高单核 - 吞噬细胞中 IL-1β 的前体水平。当肠上皮比较完整时，正常菌群并不能诱导 IL-1β 的前体成熟为有活性的 IL-1β，因此能够维持肠道的这种低免疫反应。但当肠道被一些病原菌（如铜绿假单胞菌）感染时，这些病原菌能够通过Ⅲ型分泌系统将 NLRC4 受体注入到细胞质内来激活炎性小体，从而诱导 IL-1β 的前体为成熟的 IL-1β。

肠道的正常菌群还可以通过诱导 ILC3s 分泌 IL-22 来间接地提高肠道的免疫反应能力[75]。研究发现，无菌小鼠肠道中产生的 IL-22 要显著低于正常小鼠，表明肠道正常菌群或者其代谢产物能够调控 IL-22 的表达，进一步研究发现，与野生型小鼠相比，IL-22 缺失的小鼠对于柠檬酸杆菌的感染更加敏感，表明正常菌群诱导 ILC3s 产生的 IL-22 对于肠道防止病原菌感染发挥着重要作用[76]。

某些共生菌还可以促进肠道不同的 T 细胞亚群的产生，而这些 T 细胞亚群在对抗病原菌感染过程中发挥着独特的作用，可以对特定的病原体进行杀伤，颇有些“借刀杀人”的意味。例如，分节丝状菌在肠道里的定植能够诱导 Th17 细胞分化，从而加强宿主抵抗柠檬酸杆菌感染的能力[77]。共生菌也能够诱导调节性 T 细胞的生成[78]，减轻由于病原菌感染诱导的过强免疫反应造成的炎性损伤。例如，脆弱类杆菌能够诱导调节性 T 细胞分泌 IL-10 来保护宿主免受肝螺杆菌感染引起的机体损伤，而婴儿双歧杆菌则能够通过增加调节性 T 细胞的数量来降低沙门菌感染造成的炎症反应。共生菌诱导的针对病原菌特异性的获得性免疫反应不仅可以限制其感染，而且能够通过调理作用或者其他的免疫机制来消灭病原菌，阻止感染的蔓延和全身化[79]。但其作用机制究竟如何，还有待于后续的研究。

第三节　益生菌与特殊人群免疫

一、益生菌与婴幼儿和儿童免疫

（一）婴幼儿和儿童免疫概述

作为刚刚问世的新生命体，婴儿在呱呱坠地那一刻，其体内的免疫系统便开始发育并在此后不断地行使着它重要的职能。普遍观点认为，婴儿免疫系统的初次抗原接触是在通过产道时接触的生殖道菌群[80]。这些以乳酸杆菌为主的共生菌保护了母亲，也将通过诱导免疫系统发育的方式保护她的孩子。

婴儿的免疫系统组成包括免疫器官、组织和免疫细胞。免疫器官和组织包括脾、胸腺、骨髓及淋巴结，免疫细胞则包括尚不成熟的 T、B 细胞，白细胞等[81, 82]。婴儿的肠黏膜免疫系统与成人相比存在着很大的不同。首先，由于缺乏固定的肠道菌群定植[83]，婴儿的肠黏膜免疫系统很不成熟，对于外来的致病菌抵抗力很弱，此时有益菌与抗原接触的缺乏也很容易导致成年后的肠道甚至身体其他部位的过敏性疾病[84]。不过，婴儿的免疫系统虽然较弱，却也并不是手无寸铁。来自母乳的 sIgA 代替了婴儿尚不成熟的 B 细胞免疫功能，而透过胎盘获得的一些母体抗体（如抗 -HAV）也将在一定时间内保护婴儿不受相应病原体的感染[84]。此外婴儿的 T 细胞可以分泌 IL-8 对抗致病菌。总之，婴儿的免疫系统较为弱小，但对于某些疾病具有独特的抵抗力，也具有特殊的免疫特性。

儿童的免疫系统并不是单纯的“婴儿的放大”或“成人的缩小”，它也具有自己的一些特点。与婴儿时期相比，儿童的免疫系统通过接触抗原，得到了一定的发育。同时，此时儿童的消化道内也开始有菌群定植。但另一方面，婴儿曾经从母亲那里获得的一些免疫能力荡然无存，这就导致尽管免疫系统相对成熟，儿童对于某些疾病的抵抗力却并不如婴儿。而与成人相比，儿童的免疫系统虽然总体而言更加脆弱，但由于胸腺功能的活跃，儿童对于某些病毒感染性疾病的抵抗能力会更强。

（二）益生菌对婴幼儿和儿童免疫的作用

新生儿的肠道一开始是无菌的，其后所有定植的菌群都来自后天饮食或非饮食的接触[85]。对于那些剖宫产的新生儿而言，由于缺少了来自母亲菌群的定植，他们的免疫能力明显要弱于自然分娩的新生儿。菌群的定植有一定时间顺序，这也间接导致了每个人的肠道菌群组成不同。对于新生儿与儿童而言，如果能尽快尽好地培植益生菌作为其肠道菌群的优势菌群，不仅有利于其免疫系统的成熟，也可以预防成年后的一些肠道菌群失调相关疾病[86, 87]。

婴幼儿的天然益生菌来源是母乳[88]，目前已有证明母乳是婴儿肠道中共生菌和潜在益生菌的良好和连续来源。初乳是婴儿肠道免疫系统的启动环节，母乳中丰富的 sIgA 和益生菌不仅可以为脆弱的婴儿消化系统提供保护，还可以作为初次的抗原接触，开启它的“成长之路”。益生菌在进入婴幼儿体内后，首先是作为抗原被肠黏膜固有层的 M 细胞摄取、识别、呈递，直接刺激肠黏膜免疫系统的发育。此后，随着益生菌优势菌群的建立，它们开始发挥占位作用，抑制外来致病菌的定植或大量繁殖。例如，肠道内益生菌的缺乏可导致肉毒杆菌感染；婴儿可以通过补充双歧杆菌刺激调节性 T 细胞的数量，维持免疫稳定，以此来降低沙门菌感染造成的肠道炎症性损伤[89]。

益生菌不仅刺激免疫系统的应答，也协助建立起免疫耐受。据俄罗斯的一项研究表明，相对于那些从小就因疾病原因而远离有菌环境的孩子而言，在正常环境里长大的孩子成年后发生炎性肠病等免疫性疾病的概率要更低。总的来说，对于婴幼儿和儿童的免疫系统而言，益生菌就像一条缰绳、一根皮鞭和一面盾牌，它不仅刺激着免疫系统的发育，促进免疫耐受的建立[90-93]，也在免疫系统尚且脆弱的时候担当保护者。

（三）益生菌产品在婴幼儿和儿童免疫中的应用

除天然的益生菌以外，人工益生菌产品也是一种值得关注的来源。一般来说，推荐给儿童与婴幼儿使用的益生菌制品主要是类似酸奶这样的乳制品。这些乳制品经过长期的食用实践，不仅安全性相对较高，而且口味良好，易于被儿童及婴幼儿接受并主动摄入。近年来除乳制品以外，一些合生元性质的胶囊等补充剂也争相问世。但是这些补充剂由于其单纯的用途与性质，更多是应用于临床的补充治疗，鲜少推荐用于日常的补充。随着人们对益生菌认识的进一步加深，益生菌产品将以多种多样的形式参与儿童的免疫活动。

二、益生菌与老年人健康和免疫

随着时间的推移，人体将会经历由幼稚到成熟再到衰老的自然过程。在衰老的过程中，人体各个系统功能将发生不可避免的退化，一些疾病和不良状态也随之而来。虽然衰老的过程不可逆转，但是如何避免或者减轻由各个系统尤其是免疫系统衰退所引发的生活质量下降，是老年医学领域经久不衰的研究热点。当前一些研究显示，益生菌制剂的应用可以从多个方面改善老年人的身体健康状况，且与传统的医疗保健手段相比在某些方面具有更好的效果和更小的风险。探索益生菌制剂在老年人群中的应用，或许会为部分长久以来困扰老年群体的问题找到独特的解决方案。

（一）人体免疫衰老概述

免疫衰老是指机体免疫系统组分和功能随着年龄增长而发生退行性变化的过程，可以影响免疫系统的各个层面，是老年人健康问题的核心影响因素。当免疫系统发生衰老时，机体对体内免疫稳态的调控能力逐渐减弱，使得机体长期处于炎症环境中。慢性炎症的状态又加速机体衰老进程，加重本就存在的健康问题。

研究发现，老年人体内持续慢性的炎症状态可能是衰弱发生发展的重要原因之一，主要表现为IL-6、C反应蛋白、TNF-α的明显增高[94]。IL-6是一种急性炎症反应细胞因子，参与免疫调节、造血、炎症及肿瘤发生等多个生物学过程。老年人的IL-6水平显著增加，推测这与多种衰弱表型的发生相关，包括肌少症、握力减弱和骨质疏松症等[95]。C反应蛋白是一种急性炎症反应产物，在老年衰弱人群中呈现高表达状态，并与衰弱患者死亡率的增加相关[96, 97]。TNF-α在衰弱患者外周血中也明显升高，现有研究提示TNF-α与老年人心脑血管事件发生之间存在重要联系[98, 99]。此外，衰弱与CXC趋化因子配体10、新蝶呤和IL-1的表达上调之间存在着关联，还与抑制性细胞因子IL-10以及IL-1Rα的低水平表达相关[100-102]。

免疫衰老还会导致免疫系统的重构，主要表现为固有免疫和适应性免疫细胞的表型和功能发生变化。在老年人群体之中，固有免疫细胞数量增加，且多呈促炎症介质高表达的表型[103]。相较于固有免疫，适应性免疫受年龄和衰弱的影响更为严重。当衰弱发生时，适应性免疫系统表现出明显的衰弱相关变化。衰弱女性相较于衰弱前期和非衰弱女性，$CD8^+$和$CD28^+$T细胞显著增加[104]。另有研究发现，衰弱个体的T细胞总数无显著性变化，但$CD8^+$ T细胞呈现显著性升高，$CD4^+$ T细胞则显著降低[105]。此后有研究验证了衰弱老年人出现CD4/CD8比值下降这一现象[106]。同样，B淋巴细胞也受到衰弱的影响。一项基于瑞典NONA的免疫纵向研究显示，B细胞多样性在衰弱的老年个体中随着年龄增长而降低，具体来说，老年人外周血和淋巴结中的B细胞多样性低于青壮年组，而老年人脾脏中的B细胞多样性则高于青壮年组[107]。这些结果表明，B淋巴细胞多样性的改变和T淋巴细胞比例的演替很可能是老年人免疫衰弱时免疫重构作用的主要表现形式。

（二）益生菌制剂在老年群体中的应用

许多研究证实，适当地应用益生菌制剂可以调节老年人肠道菌群，通过生成短链脂肪酸、胆汁酸、色氨酸等代谢产物，影响机体糖脂代谢和神经系统的功能，改善肠道屏障、增强机体免疫力，防止病原易位，降低慢性炎症因子对骨骼肌和神经系统的不良影响。以老年人慢性腹泻为例，老年慢性腹泻是一种在老年人中较为常见的消化系统功能紊乱，许多学者猜测老年慢性腹泻的发生很可能与肠道菌群的变化有关。

shannon指数和simpson指数可以评价人肠道菌群的情况，它们不仅能够反映群落中总的物种数量，还综合考量了群落中每个物种的丰度情况。研究发现[108]，shannon指数从出生时开始增长，其高峰出现在青春期，到老年期时出现下降，且年龄越高，shannon指数的减少就越明显。结合老年人普遍降低的免疫力水平，我们可以推测，肠道菌群多样性的减少可能影响宿主的免疫功能。有研究对比了功能性便秘组和健康对照组的粪便菌群后发现，功能性便秘患者的simpson指数明显小于健康对照组，这说明同健康对照组相比，便秘组肠道菌群的数量或菌群丰度更少[109]。通过PCR-DGGE技术分析处理后也发现，慢性传输型便秘组和健康对照组的粪便细菌均以拟杆菌门和厚壁菌门最为丰富，但慢性传输型便秘组中放线菌门相对丰度升高，梭杆菌门相对丰度降低。老年人肠道菌群的多样性下降和种类丰度变化而引起的免疫功能紊乱，很可能是老年慢性便秘的发病机制之一，因而以菌群多样性和菌群丰度正常的微生物制剂治疗老年慢性便秘在理论上是可行的。

对老年便秘人群和老年非便秘人群粪便中菌群进行对比后发现，老年便秘组中的菌群丰度由多至少排序依次为肠杆菌、乳酸杆菌、双歧杆菌、葡萄球菌和酵母菌，而老年非便秘组的排序则是双歧杆菌、肠杆菌、乳酸杆菌、酵母菌和葡萄球菌。进一步对比分析发现，老年便秘组中的双歧杆菌、乳酸杆菌及酵母菌等益生菌的含量显著减少，而葡萄球菌和肠杆菌等潜在致病菌则较非便秘组多。也有学者发现，肠道菌群失调在老年人群体中存在性别差异，普雷沃菌只在男性组明显减少，双歧杆菌则只在女性组中显著减少[110]，提示在采用益生菌制剂治疗不仅要结合患者的疾病特点，更要考虑到性别等其他因素带来的影响，以期取得最佳的治疗效果。

老年人的免疫力下降与肠道菌群发生的改变有密切联系，适度补充益生菌，可在一定程度上帮助老年人重建免疫力，还可以预防胰岛素抵抗、提高机体抗氧化活性，进而抑制慢性炎症反应，缓解肌萎缩、情绪和认知功能障碍，对促进老年人机体健康，提高生活独立性，改善生活质量具有积极有益的长远效应。采用益生菌制剂治疗不仅要结合患者的疾病特异性，更要考虑到其他可能影响因素，如性别等，以获得取得最好的预期。

第四节　益生菌与人类疾病

近年来，随着研究的不断进展，肠道菌群不仅在基础理论方面有了很大的发展，同时与人体疾病的密切联系也逐步被揭示。目前已有许多证据证明，肠道菌群的组成差异会影响个体对高血压、肥胖、自身免疫性疾病、神经系统疾病、癌症的易感性及疫苗的效果[111-116]，那么能够调节菌群组成、维持菌群稳态的益生菌，显然也可以对相应系统发挥作用。下文将举例叙述益生菌对各个系统疾病的治疗作用。

一、益生菌与消化系统疾病

（一）益生菌与儿童感染性腹泻

感染性腹泻是儿科的常见病，根据菌种的不同，感染性腹泻可分为侵袭性细菌性感染性腹泻和非侵袭性细菌性感染性腹泻。儿童时期，肠道免疫功能较弱，尤其是体内的菌群，其肠道定植力等生理功能也处于相对不稳定状态。儿童的年龄越小，肠道微生态平衡就越脆弱，更易受到各种因素如病毒、细菌、食物毒素、药物作用、变态反应、全身性疾病等的影响。腹泻时，肠道菌群紊乱，以双歧杆菌为主的益生菌数量明显减少，使得病原菌易于侵袭和定植，大量有毒物质产生，导致腹泻。而腹泻本身又加重了肠道菌群紊乱，造成恶性循环。

目前临床实践表明，联合、早期应用益生菌制剂效果最好，例如酪酸梭菌、粪肠球菌、糖化菌三联剂Bio-Three治疗小儿腹泻可提高疗效，缩短病程[117]。补充合生元制剂也对小儿腹泻有较好的治疗作用[118]。益生菌一方面补充了人体肠道正常菌群，在肠道内定植、增殖，形成一层有保护作用的生物屏障。肠道益生菌菌群发酵糖产生的醋酸和乳酸，连同其本身产生的一些抗菌活性物质一起形成有广泛抗菌作用的化学屏障。另一方面，双歧杆菌等还可以刺激机体免疫系统产生 sIgA 等免疫活性物质，形成免疫屏障。

研究发现，布拉酵母菌能分泌一种 54kD 的蛋白，用以水解艰难梭菌毒素 A 和 B，对于艰难梭菌感染性腹泻的患者，补充布拉酵母菌可以显著缓解症状。布拉氏酵母菌还可以产生一种作用于肠道黏膜的蛋白，可以抑制霍乱毒素激活腺苷酸环化酶，从而降低氯的分泌，缓解腹泻带来的水及电解质流失。此外，布拉氏酵母菌本身也能黏附霍乱毒素而将其排出体外，对治疗霍乱弧菌性腹泻有一定疗效。不过近来也有一些证据表明，对于 4 岁以下的儿童而言，应用益生菌并不能显著改变急性感染性胃肠炎的结局[119, 120]，这提示我们或许还需要进一步分析，比较益生菌在各型腹泻中发挥的作用。

（二）益生菌与肠易激综合征

肠易激综合征是一种功能性疾病，其临床上症状如腹痛、排便习惯的改变、胃肠胀气等明显，但肠道并无显著病理变化。肠易激综合征的发生机制极为复杂，有报道称这是一种与精神状态和心理健康联系紧密的心身疾病，也有学者认为肠道菌群的失调在肠易激综合征的发生中也占有一席之地。目前还没有根治肠易激综合征的方法，临床上一般采取对症和支持治疗，有时会辅以心理疏导来缓解患者的焦虑情绪，但效果差异大，且往往不尽如人意。

使用特定益生菌是改善症状的有效方法[121]。例如双歧杆菌可使肠易激综合征患者腹泻症状缓解率显著增高[122]；植物乳杆菌能有效减轻肠易激综合征患者腹痛[123]。最近在一项随机对照人体试验中发现，使用婴儿双歧杆菌株能有效缓解肠易激综合征的症状，提示了婴幼儿应用益生菌制剂预防肠易激综合征的可能。

随着对益生菌治疗肠易激综合征的不断探索，目前临床倾向于使用联合制剂进行治疗，例如将植物乳杆菌和乳酸片球菌配伍使用[124]。相比单用以上任意一种益生菌，联合制剂在改善肠易激综合征患者的生活质量方面更为有效。

（三）益生菌与非酒精性脂肪肝

肝脏作为消化系统的重要组成部分，不仅在功能上与消化道息息相关，更通过门静脉系统与肠道有着紧密的联系。肠道吸收的一些毒素和营养物质可以通过门静脉系统直接运送至肝脏，肝脏则通过强大的单核 - 巨噬细胞系统和肝细胞的解毒功能净化血液。鉴于肝脏与肠道之间密切的联系，“肝 - 肠轴”这一新兴名词应运而生。具体来说，微生物可能是慢性肝脏损害的一个辅助因素。这些慢性损害可通过酒精或引起的肝性脑病（氨、乙醇、乙醛、酚类、内毒素、苯二氮䓬类等产物导致）等并发症而发生。益生菌通过增加肠道的屏障功能，阻止细菌易位和毒素的释放来阻止肝病的进展。

革兰阴性菌能增加肠黏膜的渗透性，破坏免疫功能，从而导致细菌易位，细菌易位的速率与肝硬化的严重程度有很大关联。益生菌可以加强肠道保护机制，与革兰阴性菌对抗。随着肥胖和 2 型糖尿病的患病率增加，非酒精性脂肪肝的患病率显著上升。脂肪肝临床上一般无症状，但却使肝脏的解毒功能受损，进而易遭受如脂多糖或肝毒素的损害，导致迁延不愈的慢性炎症，进而导致肝硬化的发生。非酒精性脂肪性肝炎是非酒精性脂肪肝的中间状态，有肝小叶的炎症损害，而肝硬化是最严重的结局。非酒精性肝炎的组织病理学变化与酒精性肝病相似，发展途径也可能相同。目前，研究者对于其发生过程提出了“多次打击”假设：首先，肥胖和亚临床胰岛素抵抗促进脂肪肝的发生，从而导致肝脏易受乙醇、脂多糖等因子的损害，引起促炎症因子如 TNF-α 的产生；进而导致胰岛素抵抗，产生氧化应激，促进肝细胞和库普弗细胞（一种位于肝脏的特殊巨噬细胞）产生活性氧，导致细胞器损害甚至肝细胞死亡，促进肝脏炎症的发展。

在非酒精性脂肪肝发展中，肠腔内的微生物通过产生乙醇和脂多糖，使肝脏发生氧化应激反应，使肠上皮和肝巨噬细胞产生致炎症细胞因子，加重肝脏解毒代谢负担并导致肝脏炎症发生。这些产物还可以导致肠上皮损伤，损坏肠屏障功能，加重肝脏暴露于肠道衍生的毒素中。

研究发现，肠道内非正常菌群的过度生长能促进鼠和人脂肪肝的发展，非酒精性脂肪肝的肥胖患者肠蠕动功能降低，又会促进细菌过度增殖，形成恶性循环。最近的临床试验发现，非酒精性脂肪肝患者肠黏膜屏障的渗透性增加，这也可以刺激小肠内非正常菌群过度生长。鉴于以上发现，控制非酒精性脂肪肝的关键环节应当是对肠道菌群的控制。虽然抗生素疗效肯定，应用广泛，但由于其存在的副作用，临床上对于抗生素在非酒精性脂肪肝的使用仍然十分谨慎。而采用益生菌治疗能从不同方面阻止非酒精性脂肪肝的发展。益生菌不仅竞争性抑制病原体，改变由于肠道细菌过度生长引起的炎症影响，还能提高肠上皮的功能，减少细菌易位和内毒素血症。小鼠实验发现，分别采用益生菌和 TNF-α 抗体来治疗非酒精性脂肪肝，两种方法都能有效改善肝脏功能，降低肝脏脂肪酸的水平，干扰 NF-κ B 信号，减少肝脏脂肪酸的 β 氧化，使实验组小鼠肝功能接近于正常小鼠水平，改善胰岛素抵抗[125]。

（四）益生菌与肝硬化

肝硬化是一种严重的消化系统疾病，我国最常见的肝硬化病因是慢性乙型病毒性肝炎。目前，肝硬化的治疗主要是预防并发症和保护残存的肝功能，唯一的根治手段是进行肝脏移植。

肝硬化时，由于肝功能严重受损及门静脉高压，机体出现激素灭活障碍、顽固性腹水等临床表现。肝硬化患者肠道菌群失衡，如肠杆菌、肠球菌）和梭菌增加而双歧杆菌减少，出现消化道功能失调、屏障能力下降、非正常细菌过度生长甚至出现细菌易位。细菌的易位又会增加肠黏膜通透性，降低宿主的防御功能，加重肝硬化的症状。肝硬化的严重程度与细菌易位和菌群平衡失调有很大关联。利用乳酸杆菌和双歧杆菌治疗急性肝损伤的大鼠，其肝损伤程度明显下降，细菌易位感染的情况得到改善，肝 TNF-α 和谷胱甘肽水平趋向正常。这说明乳酸杆菌有抑制细菌易位的作用，其机制可能如下：①促进厌氧菌和革兰阳性菌的生长，同时抑制革兰阴性菌的增殖；②增加短链脂肪酸的生成，同时降低肠道 pH，诱导正常菌群的生长，抑制病原体的侵袭和黏附[126]。

在益生菌治疗的肝硬化患者（主要是乙型和丙型肝炎引起的肝炎后肝硬化）的干预实验中，患者被给予两种不同的益生菌胶囊（双歧杆菌和嗜酸菌制剂 + 肠球菌，或枯草芽孢杆菌 + 屎肠球菌）。这两组益生菌治疗后，双歧杆菌数量增加，粪便的 pH 和氨水平及血氨都降低。而且，在用枯草芽孢杆菌 + 屎肠球菌治疗组中，艰难梭菌的数量减少，血中的内毒素水平也降低。目前的观点倾向于肝硬化时菌群失调可能是由于肠道运动功能降低、sIgA、溶菌酶、黏液、酸的分泌减少、pH 升高、胆汁酸缺乏、乙醇吸收增加等原因所致，而补充益生菌可以大大改善上述情况[127]。

（五）益生菌与炎性肠病

炎性肠病是一组以肠道异常免疫反应为核心，以肠道慢性长期性炎症损伤为表现的一组疾病，主要包括溃疡性结肠炎和克罗恩病。炎性肠病的发病机制目前还不清楚，但是许多研究证明，肠道菌群的失衡可能是导致疾病的关键因素。研究者发现，溃疡性结肠炎患者体内的菌群组成发生了较大变化，可以激活炎症发生的细菌如肠杆菌、脆弱类杆菌等数量增加，而起保护作用的细菌如乳酸杆菌和双歧杆菌的数量明显减少。这些效应协同患者自身不正常的免疫功能，导致了肠黏膜炎症的发生。

益生菌制剂通过与肠上皮细胞之间的相互作用，可以在一定程度上缓解溃疡性结肠炎患者的症状，这可能与短链脂肪酸增加，促炎症细胞因子分泌减少和 Th2 细胞的升高有关。一项研究发现，给炎性肠病模型的小鼠模型喂食乳酸杆菌，能减少细菌易位，降低结肠黏膜渗透性。除了直接补充益生菌外，研究者发现植物乳杆菌培养的上清液也可以抑制肠道的炎症反应，例如抑制小鼠结肠细胞和巨噬细胞的 NF-κ B 和蛋白酶体的活化，这说明益生菌可以通过直接和间接两种途径改善炎性肠病患者的病情。

二、益生菌与呼吸系统疾病

肺部健康和肠道菌群分属两个系统，看似风马牛不相及，实际上却借助复杂的分子机制而密切相关。研究发现，在 133 例存在肠道菌群异常的肠应激综合征患者中，约有 33% 患有一定程度的呼吸系统问题[128]。除此之外，慢性阻塞性肺疾病患者发生溃疡性结肠炎和克罗恩病的可能性是健康对照组的近 3 倍。肠道菌群可以产生多种短链脂肪酸，其中丁酸盐与炎症的发生密切相关，它可以刺激单核 - 巨噬系统增生，这些增生的单核细胞迁移到肺部变成树突状细胞后，可以抑制幼稚 T 细胞向 Th2 分化，促进其分化为调节性 T 细胞，进而抑制炎症反应，维持免疫平衡[129]。这不仅仅证实了肺 - 肠轴的存在，也提示益生菌可以通过自身及代谢产物影响肠道菌群，进而通过肺 - 肠轴的作用调节肺部免疫稳态。目前的研究发现，益生菌联合使用的效果往往更好，故强调治疗时需考虑微生物 - 微生物和微生物 - 宿主之间的关系[128]。

（一）益生菌与哮喘

哮喘是一种呈急性发作的慢性气道炎症性疾病。目前针对哮喘的药物治疗已经较为成熟。在药物治疗以外，研究者们也一直在探索其他有效而安全的治疗方案。研究表明，50% ～ 60% 哮喘的成人和儿童中存在特应性反应，在激活共同因子的同时，树突状细胞也参与特异性反应，并刺激适应性 T 细胞 2 产生 IL-5、IL-4 和 IL-13[130，131]。一些研究表明，应用益生菌可以显著地缓解哮喘的症状、延缓哮喘引发的呼吸道和血管重塑。在治疗和改善哮喘症状方面，益生菌辅助传统药物治疗能够改善支气管哮喘患儿肺功能，提高哮喘控制效果，减轻炎性反应，这为益生菌在哮喘治疗中的应用提供了一条可行的思路[132，133]。此外。也有研究应用随机抽样问卷调查法对较大样本的先天性支气管哮喘患儿的家长进行跟踪调查，详细记录患儿出生后呼吸道疾病、抗生素应用史、维生素 AD 制剂及益生菌等相关药物应用情况，结果表明早期补充益生菌是防止哮喘的保护性因素。尽管对于应用益生菌能否预防哮喘的发生仍然存在争议[134，135]，但益生菌制剂安全性较好，依然不失为一种值得参考的新辅助治疗方案。

（二）益生菌与过敏性鼻炎

过敏性鼻炎同支气管哮喘一样，是一种气道的变态反应性疾病。其临床表现主要是接触变应原后反复发作的打喷嚏、流涕和鼻黏膜供血改变，长期的过敏性鼻炎会导致嗅觉功能的周围性损伤。

流行病学资料显示，过敏性鼻炎与支气管哮喘的发病率有较大的相关性。据报道，约 40% 的过敏性鼻炎患者可合并哮喘[136]。一些专家提出了“同一个气道，同一种疾病”的观点，认为过敏性鼻炎的存在是哮喘发病的危险因素之一。过敏性鼻炎的症状和治疗存在很大的异质性，不同个体之间的表现可以存在较大差异。目前，已经有不少研究者开始着手应用益生菌来调节过敏性鼻炎患者的免疫功能[137-139]，以图能缓解或预防过敏性鼻炎的发生。而就他们的实验结果来看，早期应用益生菌的确有助于预防成年后过敏性鼻炎的发生，同时应用益生菌也可以缓解过敏性鼻炎发作期时的症状。但由于过敏性鼻炎发病机制尚不十分明朗，益生菌制剂究竟通过何种途径来缓解过敏性鼻炎也暂时不得而知，亟待后续的研究。

（三）益生菌与儿童反复性呼吸道感染

反复呼吸道感染是指患儿在 1 年内呼吸道感染次数频繁、超出正常范围，是一种儿科临床常见病，发病率高达 20%。由于婴幼儿免疫系统发育并不成熟，病毒性呼吸道感染疾病的发病率和病死率都比较高，而且到目前为止也没有特效的抗病毒药物治疗。研究人员发现，益生菌或益生元对预防和治疗病毒性呼吸道感染有一定的作用，并可减少抗生素的使用剂量。目前研究均认为益生菌预防和协助治疗儿童反复性呼吸道感染具有较高的安全性，并且有研究证实凝结芽孢杆菌活菌片可有效降低婴幼儿及儿童反复性呼吸道感染的次数。以上实验均证明了益生菌在减少呼吸道感染次数以及抗生素使用方面卓有成效，且有较高的安全性。

（四）益生菌与机械通气相关肺炎

机械通气（mechanical ventilation，MV）作为支持基本生命系统的核心手段，目前已在临床广泛应用于危重症患者的生命维持，尤其用于呼吸衰竭和危重患者的呼吸支持。但随着机械通气技术的不断普及，呼吸机相关性肺炎的发生率也逐年升高。呼吸机相关性肺炎指气管插管或气管切开患者接受机械通气 48 小时后发生的肺炎，是机械通气患者中最常见的并发症。临床工作者普遍认为口咽部和胃肠内细菌的定植和吸入是呼吸机相关性肺炎的主要发病机制之一。研究发现，益生菌可以减少晚期呼吸机相关性肺炎的发生率，有助于维持消化道适宜 pH 的相对稳定，减少胃 - 肺途径的呼吸机相关性肺炎。使用益生菌的重症监护室患者呼吸机相关性肺炎发生率和平均重症监护室住院时间明显低于对照组。考虑到重症监护室患者病情的个体差异巨大，益生菌如何作用于重症患者及应用方式还需进一步研究。

三、益生菌与循环系统疾病

心脑血管疾病是目前城市人口死亡的主要原因。这与心血管疾病的常见危险因素（如肥胖、2 型糖尿病和代谢综合征）在人群中的分布日益增加密不可分。当人体内菌群失调时，一些有毒有害的代谢产物会经过肠黏膜进入血液，对血管内皮造成损伤。当肠黏膜屏障功能极度低下时，正常菌群甚至可能进入血液，引起严重的感染。益生菌作为可以调节人体代谢、维持肠道菌群稳态的“新星”，其在循环系统中的应用也越发受到关注。

（一）益生菌与高血压

原发性高血压是一种以体循环动脉压升高为主，晚期可伴有多系统损害的常见慢性疾病。高血压是我国的常见病，我国自 20 世纪 50 年代以来进行了 4 次（1959 年、1979 年、1991 年、2002 年）大规模的成人血压普查，结果显示高血压病患病率分别为 5.11%、7.73%、13.58%、18.80%，呈现明显上升的趋势。高血压病的发病机制至今尚不清楚，不同个体之间也存在着很强的异质性。目前学者倾向于其是一种多因素共同导致的多环节、多阶段、多差异的疾病。

当前，高血压的治疗主要是药物治疗。但由于各类降压药物都有难以避免的影响生活质量的副作用，而患者也往往缺乏相关知识和预防意识，药物治疗的依从性并不佳。因此，医务工作者们也在积极探索化学药物治疗以外的有效控制措施。

肠道菌群与高血压有着密切的关联。肠道菌群代谢产物可以影响高血压的进程，高血压本身也可以造成肠道菌群变化，例如高血压患者，以及自发性高血压和输注血管紧张素诱发的高血压大鼠模型，均显示肠道菌群数量和多样性存在失调，厚壁菌 / 拟杆菌比例增加 [136, 140, 141]。大量体内外研究证实，益生菌及其发酵食品对高血压有确切控制效果。根据一项对 702 例受试者服用益生菌发酵乳前后变化的研究，益生菌可降低高血压前期和轻度高血压患者收缩压和舒张压 [142]；另有一项连续 12 周对 39 例轻度高血压患者进行跟踪记录的研究发现，患者在服用益生菌制品后，其舒张压和收缩压出现不同程度的降低 [143]。

益生菌对高血压的影响主要是通过代谢产物短链脂肪酸来实现。短链脂肪酸是肠上皮组织的能量来源之一，可以影响细胞的分化、发育及肠道运动，还可以维持黏膜上皮屏障完整性，直接影响免疫细胞活性和减少交感神经兴奋性；短链脂肪酸进入循环后，可与肾脏、心脏、交感神经节和血管分布相关受体结合，发挥调节血压的作用 [144]；短链脂肪酸还可以刺激 G 蛋白结合受体途径，影响肾素分泌和血压调节 [145]。短链脂肪酸与芳香基烃受体结合后，可以增加 IL-10R 的转录，调节肠道免疫应答，减轻回肠上皮发生的炎症，发挥降低血压的作用 [146]。肥胖的孕妇肠道丁酸盐产生菌减少，这会造成丁酸盐产生量降低，导致血压升高 [147]。还有一些研究发现，益生菌不仅可以通过在胃肠道发挥的作用来间接促进循环系统健康，还可以通过对其他系统的影响直接改善高血压患者的健康状况。

（二）益生菌与冠状动脉粥样硬化

一些研究发现，动脉粥样硬化（尤其颈动脉与冠状动脉）者的肠道菌群常表现为拟杆菌门减少、厚壁菌与拟杆菌门之比增加。病情也往往与普拉梭菌等益生菌的数量密切 [148] 相关。目前一些研究认为，当益生菌相对减少，造成肠道菌群异常时，免疫系统将过度激活，导致非正常优势菌群大量繁殖后又大量地被杀死，释放多量内毒素经由肠黏膜进入血液循环，造成内皮的直接损伤，参与粥样斑块的形成与破溃 [149]。肠道菌群还通过其独有的三甲胺代谢胆碱等生成氧化三甲胺（TMAO）促进粥样斑块的进展 [149, 150]。这也是红肉提高动脉粥样硬化的风险机制之一 [116]。

心力衰竭是指心脏泵血功能绝对或相对不足导致肺循环或全身循环衰竭的一种病理状态。心力衰竭的成因很多，包括感染、先天畸形等。当前的一些研究显示，心力衰竭患者肠道功能常受损，其粪便内可分离出多种致病菌 [151, 152]。右心心力衰竭会导致体循环淤血，尤其是静脉系统。这会导致消化道水肿、淤血，进而影响到肠黏膜的屏障作用 [152]。在这种情况下，病原菌不仅有穿越肠黏膜导致感染的可能，其代谢产物如氧化三甲胺、硫酸吲哚酚等更加会直接损伤心血管系统，加重心力衰竭的症状 [153-155]。若合理应用益生菌，调整肠道菌群平衡，便可以有效地防止肠道菌群失调引发的继发性心血管系统损害，对于心力衰竭患者的生活质量与转归大有裨益。

四、益生菌与内分泌代谢疾病

（一）益生菌与肥胖

随着社会生活水平的不断发展，肥胖逐渐成为影响人们健康的一大问题 [156]。肥胖是高血压、冠心病、糖尿病等多种疾病的危险因素 [129]，因此它的预防和治疗也备受关注。由于各种显著的副作用，大多数减肥药物在昙花一现后便纷纷被撤市，仅有奥利司他仍允许销售。在这种情况下，开发安全可靠的减肥药物成为各个相关领域研究者竞相参与的项目。许多动物和人类的研究阐明 [157, 158]，食用益生菌可通过纠正肠道菌群失调来改善肥胖。例如乳酸菌和双歧杆菌属最有潜力，可以起到减轻胰岛素抵抗、降低低密度脂蛋白等作用 [159, 160]；植物乳杆菌可以降低脂肪细胞分化早期阶段相关的脂肪酸合酶、CCAAT/增强结合蛋白-α和过氧化物酶体增殖物激活受体-γ的表达水平，从而降低三酰甘油水平[161]。不过这一领域还有待开发，在益生菌被广泛用于超重患者的预防和治疗之前，我们还需要进一步的研究来确定合适的剂量、给药时间以及副作用。

（二）肠道菌群与2型糖尿病

众所周知，肠道菌群失调或称为“肠道营养不良”，在2型糖尿病的病情发展中起着关键作用[162]。肠道菌群及其代谢物可能会影响葡萄糖稳态、能量消耗和肠道屏障完整性，并导致系统性轻度炎症，这都是胰岛素抵抗和2型糖尿病的主要发病因素[163]。

糖尿病患者的肠道菌群可能出现机会致病菌的增多，产丁酸细菌的减少[164]。普拉梭菌是一种重要的丁酸盐产生菌，糖尿病患者肠道普拉梭菌、嗜黏蛋白阿克曼菌丰度明显减少，这似乎说明了普拉梭菌和嗜黏蛋白阿克曼菌对糖尿病的治疗潜力[165]。研究证明，植物乳杆菌显著降低血糖和糖化血红蛋白水平并改善葡萄糖耐量。此外，它还增加了胰岛中胰岛素阳性B细胞的面积，并降低了与糖异生相关的磷酸烯醇式丙酮酸羧激酶和葡萄糖6-磷酸酶的mRNA表达水平，增加了短链脂肪酸含量。简单来说，植物乳杆菌可以通过调节肝脏中的葡萄糖代谢、保护胰岛B细胞群、恢复肠道微生物群和短链脂肪酸来缓解高血糖和2型糖尿病[166, 167]。因此，植物乳杆菌可能是2型糖尿病的有效治疗剂。越来越多的研究提出益生菌将作为糖尿病及其并发症的新治疗方式，以在获得最大疗效的同时改善不良副作用[168]。

五、益生菌与神经系统疾病

肠道菌群通过脑-肠轴与神经系统发生双向联系，一方面肠道菌群可以通过神经、内分泌、免疫以及代谢等途径参与调节肠道和中枢神经系统的功能，另一方面机体可以通过这些途径监控和调控肠道菌群的变化，从而保持肠道微生态的平衡状态[169]。目前较为普遍承认的一种途径是：肠道菌群通过影响迷走神经来影响中枢神经系统[170]。在此理论基础上，针对慢性神经系统疾病的益生菌疗法开始受到越来越多的关注。

（一）益生菌与阿尔茨海默病

阿尔茨海默病是一种进行性中枢神经系统退行性病变，主要病理学特征为β淀粉样蛋白在脑内沉积形成老年斑，导致大量神经纤维缠结并伴随神经元数量减少，临床上主要表现为认知功能的退变和行为异常。越来越多的研究认为，肠道菌群可以通过微生物-肠-脑轴影响大脑的免疫稳态，并且可以在包括阿尔茨海默病在内的神经退行性疾病的发病机制中发挥关键作用。肠道菌群失调可以增强脂多糖和淀粉样蛋白的分泌，从而干扰肠道通透性和血脑屏障。此外，它还可以促进氧化应激、神经炎症、β淀粉样蛋白的形成、胰岛素抵抗，最终导致神经死亡。因此，通过益生菌或粪便微生物群移植进行的转基因调节可能成为阿尔茨海默病的潜在治疗方法[171]。

早期研究表明，应用益生菌可以预防阿尔茨海默病的发生，其可能的机制是益生菌通过激活沉默信息调节因子2同源蛋白1（SIRT1）依赖性机制，显著降低脑中的氧化应激，进而发挥保护作用[58]。此后进一步研究发现，对阿尔茨海默病模型的大鼠喂食罗伊乳杆菌、鼠李糖乳杆菌和婴儿双歧杆菌三联益生菌后，小鼠脑细胞和血清中β淀粉样蛋白的水平明显下降，此外，炎症指标IL-1β和TNF-α也明显降低，临床症状也得到了一定改善[172]。而嗜黏蛋白阿克曼菌可以有效改善阿尔茨海默病模型大鼠的糖耐量，肠屏障功能障碍和血脂异常，并可以延缓其大脑的病理变化，减轻其对空间学习和记忆的损害，这为阿尔茨海默病的防治提供了新的策略[173]。尽管临床试验中益生菌的表现不总是尽如人意，效果也不尽相同[56, 174]，但作为一种安全的治疗方法，益生菌疗法的后续发展仍然值得期待。

（二）益生菌与帕金森病

帕金森病也是一种慢性神经系统退行性疾病，其临床表现主要是运动能力的不断下降和静止性震颤等。随着药物的推陈出新，帕金森的治疗方法也不断迭代更新，但目前这些药物只能改善帕金森病的临床表现和延缓疾病的进展。由于复杂的发病机制，我们始终找不到根本的治疗手段。随着疾病的进展，多巴胺补充药物的作用也越发有限，怎样从根本上阻止帕金森病的恶化，以及寻找到更有效的替代治疗方法是当前研究者们共同关注的热点。

近来有部分研究者认为，帕金森病的发病原因之一可能是胃肠道的肠道菌群失调，因为胃肠道症状可能先于运动症状出现并且帕金森病患者的肠道菌群显示出独特的变化。研究认为益生菌及其代谢物可以调节神经炎症、屏障功能和神经递质活性，而这正是帕金森病的发病基础[175, 176]。一项帕金森病的随机双盲对照研究表明，使用益生菌实验组患者的运动、行为等观察指标明显好于对照组[177]。此外，帕金森模型小鼠在喂食含有嗜酸乳杆菌、双歧杆菌、罗伊乳杆菌和发酵乳杆菌后，会出现旋转行为、认知功能、脂质过氧化和神经元损伤等指标的明显改善。益生菌醋酸梭菌可以改善肠道菌群 - 肠 - 脑轴，逆转帕金森小鼠模型肠道菌群的营养不良和结肠、脑胰高血糖素样肽 1 及结肠 G 蛋白偶联受体 41 / 43 的水平下降，并可以改善小鼠的运动功能障碍、多巴胺能神经元丢失、突触功能障碍和小胶质细胞活化[178]。这表明益生菌制剂的确可以改善帕金森病的临床表现，为后续的治疗措施更新提供了一种良好的选择。

（三）益生菌与焦虑（抑郁）

焦虑和抑郁是常见的神经精神功能紊乱表现，它们不仅会带给患者心理上的痛苦，还会引发生理上的不适，它的发病机制往往与促炎刺激有关。而益生菌则可以通过降低促炎细胞因子水平和氧化应激，增加抗炎症细胞因子水平，发挥免疫调节作用，抑制炎症反应。因此，补充益生菌有助于改善抑郁和焦虑症状，从而全面改善患者的生活质量。长期暴露于应激环境诱导的抑郁小鼠使用益生菌干预后，小鼠体内的皮质醇、IL-6 和 TNF-α 水平较安慰剂组降低，小鼠的抑郁症也得到改善[179]。存在运动障碍和焦虑抑郁行为的无菌小鼠应用植物乳杆菌后，抑郁小鼠的状态明显改善，运动量显著增加[180]。尽管目前从益生菌的临床应用来看，其效果并非一直如人所愿，但作为一种新兴的、实惠可靠的治疗方法，益生菌制剂在抗焦虑抑郁中的后续应用仍然值得期待[181, 182]。

（四）益生菌与自闭症

自闭症又名自闭谱系障碍，是一种与遗传因素有关的疾病，其特征是早期不同程度的神经功能障碍及行为异常等综合症状，会严重破坏人的社会功能。研究发现，自闭症儿童粪便中的有益菌群（主要是双歧杆菌）较正常儿童少，而在补充益生菌治疗后，患儿的语言沟通能力、社交能力、感知 / 认知能力和健康行为均得到一定的改善[183]。另一项对 131 名自闭症儿童和青少年展开的实验表明：服用了益生菌植物乳杆菌的患者，行为可以得到显著的改善，这些积极影响在年幼的儿童中更为明显[184]。而在动物实验中发现，给自闭症模型小鼠喂食瑞士乳杆菌，可以改善 5- 羟色胺合成分解代谢、平衡神经兴奋抑制，并可以上调丁酸水平来改善神经递质稳态，从而改善自闭行为[185]。这些发现将有助于开发用于治疗自闭症的益生菌产品，并有助于建立模仿自闭症患者肠道环境的治疗模型。

六、益生菌与肿瘤辅助治疗

恶性肿瘤具有发生部位广、表现多样等特

点，因而不仅难以被早期发现诊断，同时也缺乏针对性治疗措施，给人类的生命健康造成了不容忽视的威胁。当前晚期恶性肿瘤的治疗多采用放化疗等姑息疗法，这些方法虽然可以在一定程度上控制疾病的进展，但是其本身同样可以对人体造成损伤，引起放射性结直肠炎等疾病，进一步降低患者的生存质量。目前一些研究表明，益生菌制剂的合理应用可以在一定程度上辅助治疗恶性肿瘤，并可有效改善由肿瘤治疗措施所引起的继发病变，这为临床治疗恶性肿瘤提供了新的思路与工具[186]。下面以常见的鼻咽癌和结直肠癌为例，介绍益生菌在肿瘤辅助治疗中的作用。

（一）益生菌与鼻咽癌

鼻咽癌是常见的高发恶性肿瘤之一，常发生于鼻咽腔顶部和侧壁，主要表现为头痛、鼻塞、面麻、耳鸣、听力下降、嗅觉下降或消失及颈部淋巴结肿大等[187]。发生在中国的鼻咽癌病例占据全球总病例数目的80%[188]。目前，放射治疗为鼻咽癌的主要治疗方式[189]。然而，放射治疗常导致一些副作用的发生，口腔黏膜炎是其中最常见的并发症之一，约80%采用放射治疗的鼻咽癌患者会发生口腔黏膜炎[190]。口腔黏膜炎会影响患者的饮食和休息，严重降低患者治疗后的生活品质，而且还会降低患者依从性。根据统计，约有19%的鼻咽癌患者因无法耐受口腔黏膜炎的不良反应而中断放射治疗。

尽管当前已有大量改善口腔黏膜炎的药物（如帕利夫明、氯己定、康复新）投入临床应用，然而这些药物疗效并不显著，且往往在缓解了口腔黏膜炎的痛苦后又增添了另外的副作用。因此，探究一种能在癌症治疗期间有效安全地预防口腔黏膜炎的药物显得尤其重要。当前一些人体和实验动物的证据表明，益生菌可以调节抗癌症免疫反应并减轻与癌症治疗相关的毒性副作用，这为苦于放射疗法副作用的鼻咽癌患者带来了福音。研究发现，采用特定的益生菌组合对于放、化疗引起的口腔黏膜炎的发生率及严重程度有明显的改善作用。研究认为，益生菌制剂通过增加肠紧密连接蛋白的表达，降低促炎因子来发挥抗炎作用。因此，益生菌有望成为缓解放化疗引起的口腔黏膜炎的新型药物。

（二）益生菌与结直肠癌

结直肠癌是常见的胃肠道恶性肿瘤。近30年来，我国结直肠癌发病率及死亡率呈上升趋势。化疗是结直肠癌后期治疗中重要的方法之一，但化疗常会导致患者肠道菌群紊乱，引起食欲降低、恶心、呕吐、腹泻等并发症，从而导致营养不良，影响患者生存质量。益生菌具有维持肠道菌群平衡、调节肠道功能及免疫应答等作用，近年来大量研究表明，益生菌在调节免疫功能、肿瘤防治方面具有积极作用。此外，一些研究显示，结直肠癌患者治疗过程中常出现肠道屏障功能受损，肠道菌群失衡，从而引起细菌移位，增加发生感染性并发症的风险。化疗引起的营养不良也会导致患者免疫力下降，感染率升高，使患者死亡率、住院费用及住院时间上升。

研究显示，肠黏膜的厚度可在益生菌的定植下相对增加，其完整性亦被保护而免受辐射损害。肠道内初级胆汁酸与双歧杆菌、乳酸杆菌等结合，经系列反应生成次级胆汁酸，后者可抑制肠道内细菌的过度增生，保持肠道内细菌的制约平衡，保障生物屏障的稳定性，进而显著改善胃癌等消化道肿瘤患者术后营养状态。尽管益生菌的使用可以减轻患者化疗期间胃肠道反应、降低炎性反应、改善患者营养状况等，但由于机制尚不明朗，并不能作为一种通用的治疗措施加以推广。当前的有关实验亦存在着许多不足之处，未来仍需大样本的随机对照研究加以论证，并探讨益生菌制剂的具体作用机制。

七、益生菌与人体自身免疫性疾病

自身免疫性疾病具有广谱性、难治性的特点。目前已发现超过百种以自身免疫异常为基础的疾病，其中包括类风湿关节炎、系统性红斑狼疮、多发性硬化等，另有其他至少 40 种疾病与自身免疫异常有着密切的联系 [191]。自身免疫性疾病是一种常见病，据统计，全球自身免疫性疾病的总发病率约为 0.09%，整体发病率呈逐年上升趋势 [192]。自身免疫性疾病多见于女性，提示了性激素在其发生发展中可能发挥着不可或缺的作用。另外也有证据表明其发生与遗传、环境、肠道菌群等因素密切相关 [193]。目前，医疗工作者已经在尝试通过干预肠道菌群以治疗自身免疫性疾病的措施，例如“粪菌移植”技术因可改变肠道菌群的组成和多样性，已广泛应用于溃疡性结肠炎等自身免疫性疾病的治疗。但肠道菌群组成复杂，异质性也很强，因此对粪菌移植患者前后菌群相对变化的研究甚少，也就无法客观地评估疗效。

已有研究表明益生菌具有调节免疫系统功能的能力，包括促进非特异性免疫、上调体液免疫和细胞免疫。乳酸菌能诱导健康人巨噬细胞 NF-κ B 的信号传导及转录激活因子活化，激活巨噬细胞，增强 T 细胞和 B 细胞介导的特异性免疫，促进促炎因子和趋化因子调节免疫应答 [194]，这为治疗自身免疫性疾病带来了新的思路。目前大部分相关研究关注的是益生菌对免疫系统产生的间接作用，而对益生菌产生的直接影响研究甚少。某些特定种类的益生菌可以刺激抗体的分泌增加，提高体内 $CD4^+$T 细胞水平，上调抑炎因子（如 IL-10、TNF-β）的分泌和抑制促炎症因子（如 IL-1β、IL-2、IL-6、IL-12、IL-17、IFN-γ 和 TNF-α 等）的分泌，还可以调节髓源性树突状细胞表面抗原的表达 [195]。综上所述，益生菌可以在一定程度上参与自身免疫性疾病的调控，因此合理应用益生菌制剂来改善甚至治疗自身免疫性疾病并非天方夜谭。

（一）益生菌与系统性红斑狼疮

系统性红斑狼疮是一种发病机制不明的异质性自身免疫病。其主要病因是自身反应性 B 细胞产生自身抗体，导致免疫复合物的大量形成与沉积，诱发超敏反应，造成全身非器官特异性的组织损伤。系统性红斑狼疮患者体内，参与超敏反应的除了 Th1/Th2 细胞免疫应答模式偏移外，还包括 Th17 和调节性 T 细胞的改变。Th17 细胞作为系统性红斑狼疮的主要驱动因素之一，通过分泌促炎因子（包括 IL-17、IL-22 和 IL-23 等）造成局部组织炎性破坏。

研究显示，具有诱导调节性 T 细胞能力的益生菌株可以调节系统性红斑狼疮患者过度的炎症反应，恢复其免疫稳态。通过给系统性红斑狼疮患者补充短双歧杆菌，可显著降低血清内 IL-17/IFNγ 水平，从而恢复 Th1 的偏倚，短双歧杆菌的富集还可防止 $CD4^+$ 淋巴细胞过度活化 [196]。给系统性红斑狼疮小鼠模型喂食德氏乳杆菌和鼠李糖乳杆菌后，小鼠体内 Th17、Th1 及其相关细胞因子水平降低，减轻了小鼠的炎症相关症状 [197]。喂食德氏乳杆菌和鼠李糖乳杆菌可以通过促进调节性 T 细胞的产生，减少 IL-6 的表达来抑制系统性红斑狼疮的进展，且鼠李糖乳杆菌疗效更好，提示在治疗系统性红斑狼疮中应该考虑菌株特异性带来的不同收益。

此外，在体外条件下，德氏乳杆菌和鼠李糖乳杆菌可诱导产生调节性细胞因子和转录因子，可诱导免疫耐受，改善炎症反应。这为攻克系统性红斑狼疮提供了新的思路。但是值得注意的是，如果益生菌使用不当，则可能致系统性红斑狼疮患者体内色氨酸代谢过度增强，这可能会引发患者产生继发性的精神障碍，因此采用这种方式治疗红斑狼疮尚存在一定的风险，如何合理有效地利用益生菌治疗系统性红斑狼疮，值得进一步研究 [198]。

（二）益生菌与类风湿关节炎

类风湿关节炎是一种以侵蚀性、对称性多关节炎为主要临床表现的慢性全身性自身免疫疾病。类风湿关节炎在全球均有分布，是目前人类丧失劳动力和致残的主要原因之一，严重威胁着人类的生命健康。当前，治疗类风湿关节炎的主要措施是药物，但一直以来其毒副作用所带来的继发性生活质量下降都得不到解决。近年来，一些有关益生菌制剂在类风湿关节炎患者中的应用研究为这种疾病的治疗开拓了新的思路。

类风湿关节炎患者多关节软骨和骨质受累破坏，导致关节畸形，其病因可能是促炎因子的过度产生和抑炎因子的减少[199]，胞内 p38 MAPK 信号通路的持续激活很大程度上维持了类风湿关节炎的症状[200, 201]。在类风湿关节炎小鼠模型中，连续多日给予有效剂量的益生菌后，其痛觉过敏和水肿症状明显减轻，血清 IL-1β 水平、脊髓 p38-MAPK 通路活性有所下降，而 μ- 阿片受体表达增加。其中 IL-1β 在诱导痛觉过敏中起着重要作用，同时还能刺激组织间质细胞释放蛋白酶，造成滑膜的破坏。p38-MAPK 在促炎因子生成中起关键作用，在关节炎中 p38-MAPK 的激活可增加促炎因子 TNF-α、IL-1β 的产生[202]，p38-MAPK 活性下降可使其介导的促炎作用减弱。

在一项随机双盲临床试验中，将 46 例类风湿关节炎患者平均分成益生菌组和安慰剂组，分别服用干酪乳杆菌和安慰剂 8 周，试验结束后对患者进行评估，发现益生菌组血清促炎因子 TNF-α、IL-6 和 IL-12 水平显著降低，血清抑炎因子 IL-10 水平明显升高，IL-10/IL-12 也显著增加。而干酪乳杆菌对患者血清 IL-1β 水平无明显影响，推测可能与菌株特异性有关。益生菌影响免疫系统的机制可能是特定的菌群相关分子模式结合了抗原呈递细胞上相应的模式识别受体，最终上调或下调炎症相关分子的分泌水平[199]。但具体效应不仅与人体有关，还受到益生菌株的种类和益生菌制剂剂型等的影响。

（三）益生菌与银屑病

银屑病是一种遗传因素参与介入、环境因素诱发、免疫异常介导的、具有慢性复发性特点的全身炎症性皮肤疾病。流行病学调查显示，银屑病可发生于任何年龄，好发于中青年，男女患病率并无明显差异。银屑病分布广泛，影响着世界 2% ～ 3% 的人口。由于其皮损具有多形性，而典型的银屑病皮损酷似一张张银白色鳞片，故民间有人称银屑病为“龙鳞身”[203]。尽管有此美名，银屑病带给患者的却是身心的双重折磨。因其缠绵难愈、反复发作，银屑病为患者带来巨大的经济负担，严重影响患者的生存质量。银屑病自发现以来便被医学家们视作一种顽固的难治性疾病，其有效治疗措施的开发也一直是当前皮肤病学和风湿免疫学的热点和难点。虽然随着现代医学高速发展，越来越多的治疗措施被应用到银屑病的治疗中，但目前常用的治疗方法的效果往往不尽如人意，更不必说它们难以避免的毒副作用。

皮肤和肠道作为人体最大的两套器官，有着血管密集、功能丰富的共同特点。两者均具有重要的免疫调节、防御、神经内分泌功能，且都是人类与外部环境联系的主要接口，对维持生理平衡至关重要。越来越多的证据表明，肠道和皮肤之间存在密切的双向联系，胃肠健康与皮肤的平衡稳态有关。胃肠道疾病常伴有皮肤症状，胃肠道系统，尤其是肠道菌群也参与了许多炎症性疾病的生理病理过程。近年来的大量科学研究表明，银屑病患者存在明显的皮肤和肠道生态失调。在银屑病患者中肠道菌群 α 多样性和 β 多样性显著下降。这种差异具体表现在银屑病患者角质杆菌、伯克霍尔德菌和乳杆菌的数量减少，而克氏棒状杆菌、模仿棒状杆菌、奈瑟菌属和细粒杆菌属增多[204]。

重塑肠道菌群能改善银屑病症状，如中草

药土奎银可通过重塑肠道菌群，下调 IL-17 的表达量，显著减轻银屑病小鼠皮肤角质细胞的过度增殖和炎症细胞浸润[205]。可见，肠道与皮肤有着紧密的关联性。肠 - 皮肤轴学说为肠道微生态与皮肤疾病相关性研究提供了依据，肠道菌群紊乱在银屑病的发生、发展中有较大影响[206]。研究证明，益生菌、益生元对皮肤具有免疫调节作用，可以通过降低皮肤细菌负荷和拮抗侵袭性共生菌，增强皮肤的屏障和修复功能。

乳酸杆菌可以改善小鼠银屑病模型的皮损严重程度，还可以降低小鼠体内银屑病相关的促炎症细胞因子如 IL-17、IL-19、IL-23 的表达水平。由此可认为乳酸杆菌或许可以缓解临床症状，甚至能够在发病环节层面控制银屑病的发生。此外，给银屑病小鼠模型喂食戊糖乳杆菌后，发现小鼠红斑、鳞屑和表皮增厚等皮肤损害明显较少，且发生的皮损也相对较轻。目前发现有至少 3 种不同的益生菌都能改善银屑病病情[207]。但总的来说，当前针对银屑病患者的益生菌制剂应用研究并不多，且各项实验质量之间也存在差距，因此尚且不足以制订银屑病益生菌制剂疗法的统一评价标准和提供建设性的医疗建议。

第五节　小结与展望

一、目前益生菌研究中存在的问题

虽然近年来在益生菌对肠道免疫调节方面的研究众多，但是研究的焦点仍主要集中在高质量益生菌和新型菌种的研究和工业应用方面，对于机制的研究相较之下就显得不那么出彩。国外对于益生菌和肠道关系的基础理论研究较为领先，我国的研究则更多地关注益生菌的应用和工业生产，对于基础理论的研究还比较落后。基础领域方面如群落的菌种组成结构，关键功能菌类型，群落的基因组成，群落结构与生态功能在不同环境下变化规律的研究鲜少见到，即便有也较为浅显。在科研领域，人们往往对益生菌的作用表现得过于乐观[208]。并且作为一种菌群补充剂，由于每个人肠道菌群组成的不同，显然益生菌的作用也不会是一致的。

另外，实验室目前使用的各种菌群群落多是自行培养，往往质控水平低、质量不稳定，缺乏干预、调控手段，更缺少统一的标准，不能实现对群落结构和功能的优化，对于实验的重复性也有一定影响。如果想要更好地了解益生菌和肠道免疫以及肠道菌群群落与人类健康的关系，我们必须首先为益生菌机制研究提供一个可以广泛参考的基准，以此为基础进行深入研究，为前沿的产品生产提供坚实可靠的理论基础。

二、益生菌制剂的现状与展望

在全球，随着市场需求增加和生物技术发展，益生菌及其制剂的市场规模以每年 15% ～ 20% 的速度增长，我国的增速甚至高达 25 %。目前，用于保健食品的益生菌多为乳酸菌，我国常用于食品生产加工的益生菌主要是双歧杆菌属和乳杆菌属两类。其中，5 种常见的双歧杆菌属分别是两歧双歧杆菌、长双歧杆菌、短双歧杆菌、婴儿双歧杆菌和青春双歧杆菌；乳杆菌属常见有 10 种，包括嗜酸乳杆菌、植物乳杆菌和干酪乳杆菌等。还有一类为兼性厌氧球菌，包括肠球菌属的粪肠球菌、乳球菌属的乳酸乳球菌等。

益生菌制剂与其他药物的最大区别在于其

本质是一种具有生命活性的复合物质，因此无论是应用还是储存，都与单纯的化学药品有很大区别。因益生菌对温度、氧气和酸碱等极为敏感，而其在肠道的定植情况又被认为是发挥功能的前提，因此如何确保益生菌在食品加工、运输和贮藏等过程中保持良好活性，提升其经过机体胆汁、胃酸等环境后的存活率，是益生菌制剂行业重点关注的话题。当前益生菌制剂的剂型很多，包括胶囊、粉剂、口服液和片剂等。以目前已经被普遍应用于食品行业的微胶囊包埋技术为例：微胶囊包埋技术是指用天然或合成的高分子材料，将分散的固体颗粒、液滴甚至气体进行包裹，得到粒径微小、有半透性或密封性囊膜粒子的技术。经微胶囊包裹的益生菌可更长时间地维持其活性，有利于益生菌制剂发挥应有的作用。而内容物能否在肠道中完全释放，是判断微胶囊品质的一个重要标准[209]，如何得到粒径足够小的益生菌微胶囊，则是目前食品级益生菌制剂研究的主要课题之一。缩小粒径一方面可以减少其与空气等外界不良环境的接触，另一方面可以更大程度地保持食品本身的感官性质免受不良影响。传统微胶囊粒子直径大小一般为 5 ～ 200μm，新型微胶囊（纳米微胶囊）粒径可小至 1 ～ 1000nm，极好地保证了益生菌制剂有效成分的充分利用。微胶囊技术的不断成熟得益于食品加工行业与益生菌基础研究的不断发展交流，随着益生菌基础机制研究的不断更新完善，以后将会有更多类似于微胶囊保存技术这样的微生态制剂领域技术进步出现。

当前，发酵食品在新生儿哺育领域的应用是一个全新的热点领域。不少研究显示，剖宫产儿相较于顺产儿抵抗力更加低下，更易发生免疫系统的紊乱。这与剖宫产儿自然菌群的定植异常有密切联系。应用益生菌制剂和益生菌发酵食品可以有效改善新生儿菌群定植，为“生命早期关键一千天”保驾护航。

总的来说，益生菌制剂的开发与应用方兴未艾，亟待更多科研工作者投入其中，从食品加工和作用机制等各个研究领域展开广泛的合作与研究。

参考文献

[1] Da Silva TF, Casarotti SN, de Oliveira G, et al. The impact of probiotics, prebiotics, and synbiotics on the biochemical, clinical, and immunological markers, as well as on the gut microbiota of obese hosts. Critical Reviews in Food Science and Nutrition, 2021, 61(2): 337-355.

[2] Gibson GR, Hutkins R, Sanders ME, et al. Expert consensus document: The International Scientific Association for Probiotics and Prebiotics (ISAPP) consensus statement on the definition and scope of prebiotics. Nature Reviews. Gastroenterology & Hepatology, 2017, 14(8): 491-502.

[3] Miqdady M, Al Mistarihi J, Azaz A, et al. Prebiotics in the Infant Microbiome: The Past, Present, and Future. Pediatric Gastroenterology, Hepatology & Nutrition, 2020, 23(1): 1-14.

[4] Qiu Y, Yang Y, Yang H. The unique surface molecules on intestinal intraepithelial lymphocytes: from tethering to recognizing. Digestive Diseases and Sciences, 2014, 59(3): 520-529.

[5] Moein S, Vaghari-Tabari M, Qujeq D, et al. MiRNAs and inflammatory bowel disease: An interesting new story. Journal of Cellular Physiology, 2019, 234(4): 3277-3293.

[6] Ismail AS, Hooper LV. Epithelial cells and their neighbors. IV. Bacterial contributions to intestinal epithelial barrier integrity. American Journal of Physiology. Gastrointestinal and Liver Physiology, 2005, 289(5): G779-G784.

[7] Paulo AFS, Baú TR, Ida EI, et al. Edible coatings and films with incorporation of prebiotics -A review. Food Research International. 2021, 148:110629.

[8] Pusceddu MM, Murray K, Gareau MG. Targeting the Microbiota, from Irritable Bowel Syndrome to Mood Disorders: Focus on Probiotics and Prebiotics. Current Pathobiology Reports, 2018, 6(1): 1-13.

[9] Mao Y, Nobaek S, Kasravi B, et al. The effects of Lactobacillus strains and oat fiber on methotrexate-induced enterocolitis in rats. Gastroenterology, 1996, 111(2): 334-344.

[10] Boehm G, Lidestri M, Casetta P, et al. Supplementation of a bovine milk formula with an oligosaccharide mixture increases counts of faecal bifidobacteria in preterm infants. Archives of Disease in Childhood. Fetal and Neonatal Edition, 2002, 86(3): F178-F181.

[11] Boehm G, Jelinek J, Stahl B, et al. Prebiotics in Infant Formulas. Journal of Clinical Gastroenterology, 2004, 38: S76-S79.

[12] Knol J, Scholtens P, Kafka C, et al. Colon microflora in infants fed formula with galacto- and fructo-oligosaccharides: more like breast-fed infants. Journal of Pediatric Gastroenterology and Nutrition, 2005, 40(1): 36-42.

[13] Moreno Villares JM . Probiotics in infant formulae. Could we modify the immune response? Anales de Pediatria (Barcelona, Spain : 2003), 2008, 68(3): 286-294.

[14] Moro G, Minoli I, Mosca M, et al. Dosage-related bifidogenic effects of galacto- and fructooligosaccharides in formula-fed term infants. Journal of Pediatric Gastroenterology and Nutrition, 2002, 34(3): 291-295.

[15] Frei R, Akdis M, O'Mahony L. Prebiotics, probiotics, synbiotics, and the immune system: experimental data and clinical evidence. Current Opinion in Gastroenterology, 2015, 31(2): 153-158.

[16] Jäger R, Mohr AE, Carpenter KC, et al. International Society of Sports Nutrition Position Stand: Probiotics. Journal of The International Society of Sports Nutrition, 2019, 16(1): 62.

[17] Tsilingiri K, Rescigno M. Postbiotics: what else? Beneficial Microbes, 2013, 4(1): 101-107.

[18] Oberg TS, Steele JL, Ingham SC, et al. Intrinsic and inducible resistance to hydrogen peroxide in Bifidobacterium species. Journal of Industrial Microbiology & Biotechnology, 2011, 38(12): 1947-1953.

[19] Konstantinov SR, Kuipers EJ, Peppelenbosch MP. Functional genomic analyses of the gut microbiota for CRC screening.Nature reviews. Gastroenterology & Hepatology, 2013, 10(12): 741-745.

[20] Shenderov BA. Metabiotics: novel idea or natural development of probiotic conception. Microbial Ecology in Health and Disease, 2013, 24:10.3402/mehd.v24i0.20399.

[21] Canfora EE, Jocken JW, Blaak EE. Short-chain fatty acids in control of body weight and insulin sensitivity. Nature Reviews. Endocrinology, 2015, 11(10): 577-591.

[22] Ilesanmi Oyelere BL, Kruger MC. The Role of Milk Components, Pro-, Pre-, and Synbiotic Foods in Calcium Absorption and Bone Health Maintenance. Frontiers in Nutrition, 2020, 75:78702.

[23] Mahomoodally MF. Traditional medicines in Africa: an appraisal of ten potent african medicinal plants. Evidence-based Complementary and Alternative Medicine : eCAM, 2013, 2013:617459.

[24] Mugula JK, Sørhaug T, Stepaniak L. Proteolytic activities in togwa, a Tanzanian fermented food. International journal of Food Microbiology, 2003, 84(1): 1-12.

[25] Chang CJ, Lin TL, Tsai YL, et al. Next generation probiotics in disease amelioration. Journal of Food and Drug Analysis, 2019, 27(3): 615-622.

[26] O'Toole PW, Marchesi JR, Hill C. Next-generation probiotics: the spectrum from probiotics to live biotherapeutics. Nature Microbiology, 2017, 2:17057.

[27] Sartor RB. Therapeutic manipulation of the enteric microflora in inflammatory bowel diseases: antibiotics, probiotics, and prebiotics. Gastroenterology, 2004, 126(6): 1620-1633.

[28] Sartor RB. Mechanisms of disease: pathogenesis of Crohn's disease and ulcerative colitis. Gastroenterology & Hepatology, 2006, 3(7): 390-407.

[29] Strobel S, Mowat AM. Oral tolerance and allergic responses to food proteins. Current Opinion in Allergy and Clinical Immunology, 2006, 6(3): 207-213.

[30] Lefrançois L, Puddington L. Intestinal and pulmonary mucosal T cells: local heroes fight to maintain the status quo. Annual Review of Immunology, 2006, 24:681-704.

[31] Suzuki K, Maruya M, Kawamoto S, et al. The sensing of environmental stimuli by follicular dendritic cells promotes immunoglobulin A generation in the gut. Immunity, 2010, 33(1): 71-83.

[32] Tezuka H, Abe Y, Asano J, et al. Prominent role for plasmacytoid dendritic cells in mucosal T cell-independent IgA induction. Immunity, 2011, 34(2): 247-257.

[33] He P, Wang H, Huang C, He L. Hematuria was a high risk for renal progression and ESRD in immunoglobulin a nephropathy: a systematic review and meta-analysis. Renal Failure, 2021;43(1):488-499.

[34] Fritz JH, Rojas OL, Simard N, et al. Acquisition of a multifunctional IgA+ plasma cell phenotype in the gut. Nature, 2011, 481(7380): 199-203.

[35] Akira S, Uematsu S, Takeuchi O. Pathogen recognition and innate immunity. Cell, 2006, 124(4): 783-801.

[36] Dalpke A, Heeg K. Signal integration following Toll-like receptor triggering. Critical Reviews in Immunology, 2002, 22(3): 217-250.

[37] Medzhitov R, Janeway C. Innate immune recognition: mechanisms and pathways. Immunological Reviews, 2000, 173:89-97.

[38] Wickelgren I. Immunology Targeting The Tolls. Science (New York, N.Y.), 2006, 312(5771): 184-187.

[39] Cario E, Podolsky DK. Toll-like receptor signaling and its relevance to intestinal inflammation. Annals of The New York Academy of Sciences, 2006, 1072:332-338.

[40] Takeda K, Kaisho T, Akira S. Toll-like receptors.Annual review of immunology, 2003, 21:335-376.

[41] Shokryazdan P, Faseleh Jahromi M, Navidshad B, et al. Effects of prebiotics on immune system and cytokine expression.Medical microbiology and immunology, 2017, 206(1): 1-9.

[42] Klampfer L, Huang J, Sasazuki T, et al. Inhibition of interferon gamma signaling by the short chain fatty acid butyrate. Molecular Cancer Research: MCR, 2003, 1(11): 855-862.

[43] Gibson GR, McCartney AL, Rastall RA. Prebiotics and resistance to gastrointestinal infections. The British Journal of Nutrition, 2005, 93:S31-S34.

[44] Maslowski KM, Vieira AT, Ng A, et al. Regulation of inflammatory responses by gut microbiota and chemoattractant receptor GPR43. Nature, 2009, 461(7268): 1282-1286.

[45] Abdo Z, LeCureux J, LaVoy A, et al. Impact of oral probiotic Lactobacillus acidophilus vaccine strains on the immune response and gut microbiome of mice. PloS One, 2019, 14(12): e225842.

[46] Di Luccia B, Ahern PP, Griffin NW, et al. Combined Prebiotic and Microbial Intervention Improves Oral Cholera Vaccination Responses in a Mouse Model of Childhood Undernutrition. Cell Host & Microbe, 2020, 27(6): 899-908.

[47] Sartor RB. Therapeutic manipulation of the enteric microflora in inflammatory bowel diseases: antibiotics, probiotics, and prebiotics. Gastroenterology, 2004, 126(6): 1620-1633.

[48] Sartor RB. Mechanisms of disease: pathogenesis of Crohn's disease and ulcerative colitis.Nature clinical practice. Gastroenterology & Hepatology, 2006, 3(7): 390-407.

[49] Gibson GR. Dietary modulation of the human gut microflora using the prebiotics oligofructose and inulin. The Journal of Nutrition, 1999, 129: S1438-S1441.

[50] Bischoff SC, Barbara G, Buurman W, et al. Intestinal permeability—a new target for disease prevention and therapy. BMC Gastroenterology, 2014, 14:189.

[51] Saxelin M, Tynkkynen S, Mattila-Sandholm T, et al. Probiotic and other functional microbes: from markets to mechanisms. Current Opinion in Biotechnology, 2005, 16(2): 204-211.

[52] Galdeano CM, Perdigón G. The probiotic bacterium Lactobacillus casei induces activation of the gut mucosal immune system through innate immunity. Clinical and Vaccine Immunology : CVI, 2006, 13(2): 219-226.

[53] Lammers KM, Brigidi P, Vitali B, et al. Immunomodulatory effects of probiotic bacteria DNA: IL-1 and IL-10 response in human peripheral blood mononuclear cells. FEMS Immunology and Medical Microbiology. 2003,38(2):165-172.

[54] Paturi G, Phillips M, Jones M, et al. Immune enhancing effects of Lactobacillus acidophilus LAFTI L10 and Lactobacillus paracasei LAFTI L26 in mice. International Journal of Food Microbiology, 2007, 115(1): 115-118.

[55] Wu Z, Pan DD, Guo Y, et al. Structure and anti-inflammatory capacity of peptidoglycan from Lactobacillus acidophilus in RAW-264.7 cells. Carbohydrate Polymers, 2013, 96(2): 466-473.

[56] Leblhuber F, Steiner K, Schuetz B, et al. Probiotic Supplementation in Patients with Alzheimer's Dementia - An Explorative Intervention Study. Current Alzheimer Research, 2018, 15(12): 1106-1113.

[57] Wu Z, Pan D, Guo Y, et al. Peptidoglycan diversity and anti-inflammatory capacity in Lactobacillus strains. Carbohydr Polym, 2015,128:130-137.

[58] Bonfili L, Cecarini V, Cuccioloni M, et al. SLAB51 Probiotic Formulation Activates SIRT1 Pathway Promoting Antioxidant and Neuroprotective Effects in an AD Mouse Model. Molecular Neurobiology, 2018, 55(10): 7987-8000.

[59] Yasui H, Mike A, Ohwaki M. Immunogenicity of Bifidobacterium breve and change in antibody production in Peyer's patches after oral administration. Journal of Dairy Science, 1989, 72(1): 30-35.

[60] Scharek L, Guth J, Reiter K, et al. Influence of a probiotic Enterococcus faecium strain on development of the immune system of sows and piglets.Veterinary Immunology and Immunopathology, 2005, 105:151-161.

[61] Gentek R, Molawi K, Sieweke MH. Tissue macrophage identity and self-renewal. Immunological Reviews, 2014, 262(1): 56-73.

[62] Shapouri-Moghaddam A, Mohammadian S, Vazini H, et al. Macrophage plasticity, polarization, and function in health and disease. Journal of Cellular Physiology, 2018, 233(9): 6425-6440.

[63] Essandoh K, Li Y, Huo J, et al. MiRNA-Mediated Macrophage Polarization and its Potential Role in the Regulation of Inflammatory Response. Shock (Augusta, Ga.), 2016, 46(2): 122-131.

[64] Shoaf K, Mulvey GL, Armstrong GD, et al. Prebiotic galactooligosaccharides reduce adherence of enteropathogenic Escherichia coli to tissue culture cells. Infection and Immunity, 2006, 74(12): 6920-6928.

[65] Pacheco AR, Curtis MM, Ritchie JM, et al. Fucose sensing regulates bacterial intestinal colonization. Nature, 2012, 492(7427): 113-117.

[66] Kamada N, Kim YG, Sham HP, et al. Regulated virulence controls the ability of a pathogen to compete with the gut microbiota. Science (New York, N.Y.), 2012, 336(6086): 1325-1329.

[67] Winter SE, Thiennimitr P, Winter MG, et al. Gut inflammation provides a respiratory electron acceptor for Salmonella. Nature, 2010, 467(7314): 426-429.

[68] Bertin Y, Girardeau JP, Chaucheyras Durand F, et al. Enterohaemorrhagic Escherichia coli gains a competitive advantage by using ethanolamine as a nitrogen source in the bovine intestinal content. Environmental Microbiology, 2011, 13(2): 365-377.

[69] Murdoch SL, Trunk K, English G, et al. The opportunistic pathogen Serratia marcescens utilizes type VI secretion to target bacterial competitors. Journal of Bacteriology, 2011, 193(21): 6057-6069.

[70] Cash HL, Whitham CV, Behrendt CL, et al. Symbiotic bacteria direct expression of an intestinal bactericidal lectin. Science (New York, N.Y.), 2006, 313(5790): 1126-1130.

[71] Brandl K, Plitas G, Schnabl B, et al. MyD88-mediated signals induce the bactericidal lectin RegIII gamma and protect mice against intestinal Listeria monocytogenes infection. The Journal of Experimental Medicine, 2007, 204(8): 1891-1900.

[72] Fukuda S, Toh H, Hase K, et al. Bifidobacteria can protect from enteropathogenic infection through production of acetate. Nature, 2011, 469(7331): 543-547.

[73] Denning TL, Wang YC, Patel SR, et al. Lamina propria macrophages and dendritic cells differentially induce regulatory and interleukin 17-producing T cell responses. Nature Immunology, 2007, 8(10): 1086-1094.

[74] Franchi L, Kamada N, Nakamura Y, et al. NLRC4-driven production of IL-1β discriminates between pathogenic and commensal bacteria and promotes host intestinal defense. Nature Immunology, 2012, 13(5): 449-456.

[75] Mazmanian SK, Round JL, Kasper DL. A microbial symbiosis factor prevents intestinal inflammatory disease. Nature, 2008, 453(7195): 620-625.

[76] Satoh Takayama N, Vosshenrich CA, Lesjean Pottier S, et al. Microbial flora drives interleukin 22 production in intestinal NKp46+ cells that provide innate mucosal immune defense. Immunity, 2008, 29(6): 958-970.

[77] Ivanov II, Atarashi K, Manel N, et al. Induction of intestinal Th17 cells by segmented filamentous bacteria. Cell, 2009, 139(3): 485-498.

[78] Round JL, Lee SM, Li J, et al. The Toll-like receptor 2 pathway establishes colonization by a commensal of the human microbiota. Science (New York, N.Y.), 2011, 332(6032): 974-977.

[79] Fagundes CT, Amaral FA, Vieira AT, et al. Transient TLR activation restores inflammatory response and ability to control pulmonary bacterial infection in germfree mice. Journal of Immunology (Baltimore, Md. 1950), 2012, 188(3): 1411-1420.

[80] Martín R, Heilig GH, Zoetendal EG, et al. Diversity of the Lactobacillus group in breast milk and vagina of healthy women and potential role in the colonization of the infant gut. Journal of Applied Microbiology, 2007, 103(6): 2638-2644.

[81] Koleva PT, Kim JS, Scott JA, et al. Microbial programming of health and disease starts during fetal life. Birth Defects Research. Part C, Embryo Today : Reviews, 2015, 105(4): 265-277.

[82] Gomez Gallego C, Garcia Mantrana I, Salminen S, et al. The human milk microbiome and factors influencing its composition and activity. Seminars in Fetal & Neonatal Medicine, 2016, 21(6): 400-405.

[83] Orrhage K, Nord CE. Factors controlling the bacterial colonization of the intestine in breastfed infants. Acta Paediatrica (Oslo, Norway: 1992). Supplement, 1999, 88(430): 47-57.

[84] Rutayisire E, Huang K, Liu Y, et al. The mode of delivery affects the diversity and colonization pattern of the gut microbiota during the first year of infants' life: a systematic review. BMC Gastroenterology, 2016, 16(1): 86.

[85] Grönlund MM, Lehtonen OP, Eerola E, et al. Fecal microflora in healthy infants born by different methods of delivery: permanent changes in intestinal flora after

cesarean delivery. Journal of Pediatric Gastroenterology and Nutrition, 1999, 28(1): 19-25.

[86] Salminen S, Gibson GR, McCartney AL, et al. Influence of mode of delivery on gut microbiota composition in seven year old children. Gut, 2004, 53(9): 1388-1389.

[87] Negele K, Heinrich J, Borte M, et al. Mode of delivery and development of atopic disease during the first 2 years of life. Pediatric Allergy and Immunology: Official Publication of The European Society of Pediatric Allergy and Immunology, 2004, 15(1): 48-54.

[88] McGuire MK, McGuire MA. Human milk: mother nature's prototypical probiotic food? Advances in Nutrition (Bethesda, Md.), 2015, 6(1): 112-123.

[89] O'Mahony C, Scully P, O'Mahony D, et al. Commensal-induced regulatory T cells mediate protection against pathogen-stimulated NF-kappaB activation. PLoS Pathogens, 2008, 4(8): e1000112.

[90] Aitoro R, Paparo L, Amoroso A, et al. Gut Microbiota as a Target for Preventive and Therapeutic Intervention against Food Allergy. Nutrients, 2017, 9(7):672.

[91] Stiemsma LT, Michels KB. The Role of the Microbiome in the Developmental Origins of Health and Disease. Pediatrics, 2018, 141(4):e20172437.

[92] West CE, Renz H, Jenmalm MC, et al. The gut microbiota and inflammatory noncommunicable diseases: associations and potentials for gut microbiota therapies. The Journal of Allergy and Clinical Immunology, 2015, 135(1): 3-13, 14.

[93] Kalliomäki M, Kirjavainen P, Eerola E, et al. Distinct patterns of neonatal gut microflora in infants in whom atopy was and was not developing. The Journal of Allergy and Clinical Immunology, 2001, 107(1): 129-134.

[94] Ferrucci L, Fabbri E. Inflammageing: chronic inflammation in ageing, cardiovascular disease, and frailty. Nature Reviews. Cardiology, 2018, 15(9): 505-522.

[95] Ma L, Sha G, Zhang Y, et al. Elevated serum IL-6 and adiponectin levels are associated with frailty and physical function in Chinese older adults. Clinical Interventions in Aging, 2018, 13: 2013-2020.

[96] Puzianowska Kuźnicka M, Owczarz M, Wieczorowska-Tobis K, et al. Interleukin-6 and C-reactive protein, successful aging, and mortality: the PolSenior study. Immunity & Ageing, 2016: 13-21.

[97] Giovannini S, Onder G, Liperoti R, et al. Interleukin-6, C-reactive protein, and tumor necrosis factor-alpha as predictors of mortality in frail, community-living elderly individuals. Journal of The American Geriatrics Society, 2011, 59(9): 1679-1685.

[98] Collerton J, Martin-Ruiz C, Davies K, et al. Frailty and the role of inflammation, immunosenescence and cellular ageing in the very old: cross-sectional findings from the Newcastle 85+ Study. Mechanisms of Ageing and Development, 2012, 133(6): 456-466.

[99] Bruunsgaard H, Andersen Ranberg K, Hjelmborg J, et al. Elevated levels of tumor necrosis factor alpha and mortality in centenarians. The American Journal of Medicine, 2003, 115(4): 278-283.

[100] Wilson D, Jackson T, Sapey E, et al. Frailty and sarcopenia: The potential role of an aged immune system. Ageing Research Reviews, 2017, 36:1-10.

[101] Qu T, Yang H, Walston JD, et al. Upregulated monocytic expression of CXC chemokine ligand 10 (CXCL-10) and its relationship with serum interleukin-6 levels in the syndrome of frailty. Cytokine, 2009, 46(3): 319-324.

[102] Spencer ME, Jain A, Matteini A, et al. Serum levels of the immune activation marker neopterin change with age and gender and are modified by race, BMI, and percentage of body fat.The journals of gerontology. Series A, Biological Sciences and Medical Sciences, 2010, 65(8): 858-865.

[103] Pansarasa O, Pistono C, Davin A, et al. Altered immune system in frailty: Genetics and diet may influence inflammation. Ageing Research Reviews, 2019, 54:100935.

[104] Semba RD, Margolick JB, Leng S, et al. T cell subsets and mortality in older community-dwelling women. Experimental Gerontology, 2005, 40:81-87.

[105] De Fanis U, Wang GC, Fedarko NS, et al. T-lymphocytes expressing CC chemokine receptor-5 are increased in frail older adults. Journal of The American Geriatrics Society, 2008, 56(5): 904-908.

[106] Yao X, Li H, Leng SX. Inflammation and immune system alterations in frailty. Clinics in Geriatric Medicine, 2011, 27(1): 79-87.

[107] Tabibian Keissar H, Hazanov L, Schiby G, et al. Aging affects B-cell antigen receptor repertoire diversity in primary and secondary lymphoid tissues. European Journal of Immunology, 2016, 46(2): 480-492.

[108] Odamaki T, Kato K, Sugahara H, et al. Age-related changes in gut microbiota composition from newborn to centenarian: a cross-sectional study. BMC Microbiology, 2016,16:90.

[109] Huang LS, Kong C, Gao RY, et al. Analysis of fecal microbiota in patients with functional constipation

undergoing treatment with synbiotics. European Journal of Clinical Microbiology & Infectious Diseases: Official Publication of The European Society of Clinical Microbiology, 2018, 37(3): 555-563.

[110] Nourrisson C, Scanzi J, Pereira B, et al. Blastocystis is associated with decrease of fecal microbiota protective bacteria: comparative analysis between patients with irritable bowel syndrome and control subjects. PloS One, 2014, 9(11): e111868.

[111] Wang Y, Kasper LH. The role of microbiome in central nervous system disorders. Brain, Behavior, and Immunity, 2014, 38:1-12.

[112] Yacoub R, Jacob A, Wlaschin J, et al. Lupus: The microbiome angle. Immunobiology, 2018, 223:460-465.

[113] Karbach SH, Schönfelder T, Brandão I, et al. Gut Microbiota Promote Angiotensin II-Induced Arterial Hypertension and Vascular Dysfunction. Journal of The American Heart Association, 2016, 5(9):e003698.

[114] Zhao L. The gut microbiota and obesity: from correlation to causality. Nature Reviews. Microbiology, 2013, 11(9): 639-647.

[115] Huda MN, Lewis Z, Kalanetra KM, et al. Stool microbiota and vaccine responses of infants. Pediatrics, 2014, 134(2): e362-e372.

[116] Tang WH, Hazen SL. The contributory role of gut microbiota in cardiovascular disease. The Journal of Clinical Investigation, 2014, 124(10): 4204-4211.

[117] Braegger C, Chmielewska A, Decsi T, et al. Supplementation of infant formula with probiotics and/or prebiotics: a systematic review and comment by the ESPGHAN committee on nutrition. Journal of Pediatric Gastroenterology and Nutrition, 2011, 52(2): 238-250.

[118] Chen CC, Kong MS, Lai MW, et al. Probiotics have clinical, microbiologic, and immunologic efficacy in acute infectious diarrhea. The Pediatric Infectious Disease Journal, 2010, 29(2): 135-138.

[119] Freedman SB, Williamson Urquhart S, Farion KJ, et al. Multicenter Trial of a Combination Probiotic for Children with Gastroenteritis. The New England Journal of Medicine, 2018, 379(21): 2015-2026.

[120] Schnadower D, Tarr PI, Casper TC, et al. Lactobacillus rhamnosus GG versus Placebo for Acute Gastroenteritis in Children. The New England Journal of Medicine, 2018, 379(21): 2002-2014.

[121] Simrén M, Barbara G, Flint HJ, et al. Intestinal microbiota in functional bowel disorders: a Rome foundation report. Gut, 2013, 62(1): 159-176.

[122] Ducrotté P, Sawant P, Jayanthi V. Clinical trial: Lactobacillus plantarum 299v (DSM 9843) improves symptoms of irritable bowel syndrome. World Journal of Gastroenterology, 2012, 18(30): 4012-4018.

[123] Lorenzo Zúñiga V, Llop E, Suárez C, et al. I.31, a new combination of probiotics, improves irritable bowel syndrome-related quality of life. World Journal of Gastroenterology, 2014, 20(26): 8709-8716.

[124] Guglielmetti S, Mora D, Gschwender M, et al. Randomised clinical trial: Bifidobacterium bifidum MIMBb75 significantly alleviates irritable bowel syndrome and improves quality of life--a double-blind, placebo-controlled study. Alimentary Pharmacology & Therapeutics, 2011, 33(10): 1123-1132.

[125] Li Z, Yang S, Lin H, et al. Probiotics and antibodies to TNF inhibit inflammatory activity and improve nonalcoholic fatty liver disease. Hepatology (Baltimore, Md.), 2003, 37(2): 343-350.

[126] Adawi D, Ahrné S, Molin G. Effects of different probiotic strains of Lactobacillus and Bifidobacterium on bacterial translocation and liver injury in an acute liver injury model. International Journal of Food Microbiology, 2001, 70(3): 213-220.

[127] Loguercio C, De Simone T, Federico A, et al. Gut-liver axis: a new point of attack to treat chronic liver damage? The American Journal of Gastroenterology, 2002, 97(8): 2144-2146.

[128] Chakradhar S. A curious connection: Teasing apart the link between gut microbes and lung disease. Nature Medicine, 2017, 23(4): 402-404.

[129] Daniali M, Nikfar S, Abdollahi M. A brief overview on the use of probiotics to treat overweight and obese patients. Expert Review of Endocrinology & Metabolism, 2020, 15(1): 1-4.

[130] Papi A, Brightling C, Pedersen SE, et al. Asthma. Lancet (London, England), 2018, 391(10122): 783-800.

[131] Kulkarni NS, Hollins F, Sutcliffe A, et al. Eosinophil protein in airway macrophages: a novel biomarker of eosinophilic inflammation in patients with asthma. The Journal of Allergy and Clinical Immunology, 2010, 126(1): 61-69.

[132] Du X, Wang L, Wu S, et al. Efficacy of probiotic supplementary therapy for asthma, allergic rhinitis, and wheeze: a meta-analysis of randomized controlled trials. Allergy and Asthma Proceedings, 2019, 40(4): 250-260.

[133] Jamalkandi SA, Ahmadi A, Ahrari I, et al. Oral and nasal probiotic administration for the prevention and alleviation of allergic diseases, asthma and chronic obstructive pulmonary disease. Nutrition Research Reviews, 2021, 34(1): 1-16.

[134] Kallio S, Kukkonen AK, Savilahti E, et al. Perinatal probiotic intervention prevented allergic disease in a Caesarean-delivered subgroup at 13-year follow-up. Clinical and Experimental Allergy : Journal of The British Society for Allergy and Clinical Immunology, 2019, 49(4): 506-515.

[135] Bertelsen RJ, Brantsæter AL, Magnus MC, et al. Probiotic milk consumption in pregnancy and infancy and subsequent childhood allergic diseases. The Journal of Allergy and Clinical Immunology, 2014, 133(1): 165-171.

[136] Julia V, Macia L, Dombrowicz D. The impact of diet on asthma and allergic diseases. Nature Reviews Immunology, 2015, 15(5): 308-322.

[137] Schaefer M, Enck P. Enterococcus faecalisEffects of a probiotic treatment and open-label placebo on symptoms of allergic rhinitis: study protocol for a randomised controlled trial. BMJ Open, 2019, 9(10): e31339.

[138] West NP, Watts AM, Smith PK, et al. Digital Immune Gene Expression Profiling Discriminates Allergic Rhinitis Responders from Non-Responders to Probiotic Supplementation. Genes, 2019, 10(11):889.

[139] Cao X, Zhong P, Li G, et al. Application of probiotics in adjuvant treatment of infant allergic rhinitis: A randomized controlled study. Medicine, 2020, 99(18): e20095.

[140] Qi Y, Aranda JM, Rodriguez V, et al. Impact of antibiotics on arterial blood pressure in a patient with resistant hypertension - A case report. International Journal of Cardiology, 2015, 201: 157-158.

[141] Li J, Zhao F, Wang Y, et al. Gut microbiota dysbiosis contributes to the development of hypertension. Microbiome, 2017, 5(1): 14.

[142] Dong JY, Szeto IM, Makinen K, et al. Effect of probiotic fermented milk on blood pressure: a meta-analysis of randomised controlled trials. The British Journal of Nutrition, 2013, 110(7): 1188-1194.

[143] Inoue K, Shirai T, Ochiai H, et al. Blood-pressure-lowering effect of a novel fermented milk containing gamma-aminobutyric acid (GABA) in mild hypertensives .European Journal of Clinical Nutrition, 2003, 57(3): 490-495.

[144] Brown AJ, Goldsworthy SM, Barnes AA, et al. The Orphan G protein-coupled receptors GPR41 and GPR43 are activated by propionate and other short chain carboxylic acids. The Journal of Biological Chemistry, 2003, 278(13): 11312-11319.

[145] Pluznick JL, Protzko RJ, Gevorgyan H, et al. Olfactory receptor responding to gut microbiota-derived signals plays a role in renin secretion and blood pressure regulation. Proceedings of The National Academy of Sciences of The United States of America, 2013, 110(11): 4410-4415.

[146] Sanduzzi Zamparelli M, Compare D, Coccoli P, et al. The Metabolic Role of Gut Microbiota in the Development of Nonalcoholic Fatty Liver Disease and Cardiovascular Disease. International Journal of Molecular Sciences, 2016, 17(8):395.

[147] Gomez-Arango LF, Barrett HL, McIntyre HD, et al. Increased Systolic and Diastolic Blood Pressure Is Associated With Altered Gut Microbiota Composition and Butyrate Production in Early Pregnancy. Hypertension (Dallas, Tex: 1979), 2016, 68(4): 974-981.

[148] Li J, Zhao F, Wang Y, et al. Gut microbiota dysbiosis contributes to the development of hypertension. Microbiome, 2017, 5(1): 14.

[149] Curtiss LK , Tobias PS. Emerging role of Toll-like receptors in atherosclerosis. Journal of Lipid Research, 2009, 50: S340-S345.

[150] Wang Z, Tang WH, Buffa JA, et al. Prognostic value of choline and betaine depends on intestinal microbiota-generated metabolite trimethylamine-N-oxide. European Heart Journal, 2014, 35(14): 904-910.

[151] Pasini E, Aquilani R, Testa C, et al. Pathogenic Gut Flora in Patients With Chronic Heart Failure.JACC. Heart Failure, 2016, 4(3): 220-227.

[152] Kamo T, Akazawa H, Suzuki JI, et al. Novel Concept of a Heart-Gut Axis in the Pathophysiology of Heart Failure. Korean Ccirculation Journal, 2017, 47(5): 663-669.

[153] Zabell A, Tang WH. Targeting the Microbiome in Heart Failure.Current treatment options in cardiovascular. Medicine, 2017, 19(4): 27.

[154] Yang K, Wang C, Nie L, et al. Klotho Protects Against Indoxyl Sulphate-Induced Myocardial Hypertrophy. Journal of The American Society of Nephrology: JASN, 2015, 26(10): 2434-2446.

[155] Tang WH, Wang Z, Levison BS, et al. Intestinal microbial metabolism of phosphatidylcholine and cardiovascular risk. The New England Journal of

Medicine, 2013, 368(17): 1575-1584.

[156] Michels N, Zouiouich S, Vanderbauwhede B, et al. Human microbiome and metabolic health: An overview of systematic reviews. Obesity Reviews, 2022,23(4):e13409.

[157] Sotoudegan F, Daniali M, Hassani S, et al. Reappraisal of probiotics' safety in human. Food and Chemical Toxicology: An International Journal Published for The British Industrial Biological Research Association, 2019, 129:22-29.

[158] Tchernof A, Després JP. Pathophysiology of human visceral obesity: an update. Physiological Reviews, 2013, 93(1): 359-404.

[159] Agerbaek M, Gerdes LU, Richelsen B. Hypocholesterolaemic effect of a new fermented milk product in healthy middle-aged men. European Journal of Clinical Nutrition, 1995, 49(5): 346-352.

[160] Rajkumar H, Mahmood N, Kumar M, et al. Effect of probiotic (VSL#3) and omega-3 on lipid profile, insulin sensitivity, inflammatory markers, and gut colonization in overweight adults: a randomized, controlled trial. Mediators of Inflammation, 2014, 2014:348959.

[161] Han KJ, Lee NK, Yu HS, et al. Anti-adipogenic Effects of the Probiotic Lactiplantibacillus plantarum KU15117 on 3T3-L1 Adipocytes. Probiotics and Antimicrobial Proteins, 2021,10.1007/s12602-021-09818-z.

[162] Allin KH, Tremaroli V, Caesar R, et al. Aberrant intestinal microbiota in individuals with prediabetes. Diabetologia, 2018, 61(4): 810-820.

[163] Ghorbani Y, Schwenger K, Allard JP. Manipulation of intestinal microbiome as potential treatment for insulin resistance and type 2 diabetes. European Journal of Nutrition, 2021, 60(5): 2361-2379.

[164] Rodriguez J, Hiel S, Delzenne NM. Metformin: old friend, new ways of action-implication of the gut microbiome? Current opinion in clinical nutrition and metabolic care, 2018, 21(4): 294-301.

[165] Hippe B, Remely M, Aumueller E, et al. Faecalibacterium prausnitzii phylotypes in type two diabetic, obese, and lean control subjects. Beneficial Microbes, 2016, 7(4): 511-517.

[166] Lee YS, Lee D, Park GS, et al. Lactobacillus plantarum HAC01 ameliorates type 2 diabetes in high-fat diet and streptozotocin-induced diabetic mice in association with modulating the gut microbiota. Food & function, 2021, 12(14): 6363-6373.

[167] Won G, Choi SI, Park N, et al. In Vitro Antidiabetic, Antioxidant Activity, and Probiotic Activities of Lactiplantibacillus plantarum and Lacticaseibacillus paracasei Strains. Current Microbiology, 2021, 78(8): 3181-3191.

[168] Bauer PV, Duca FA. Targeting the gastrointestinal tract to treat type 2 diabetes. The Journal of Endocrinology, 2016, 230(3): R95-R113.

[169] Dalal N, Jalandra R, Bayal N, et al. Gut microbiota-derived metabolites in CRC progression and causation. Journal of Cancer Research and Clinical Oncology, 2021;147(11):3141-3155.

[170] Burokas A, Moloney RD, Dinan TG, et al. Microbiota regulation of the Mammalian gut-brain axis. Advances in Applied Microbiology, 2015: 911-962.

[171] Shabbir U, Arshad MS, Sameen A, et al. Crosstalk between Gut and Brain in Alzheimer's Disease: The Role of Gut Microbiota Modulation Strategies. Nutrients, 2021, 13(2):690.

[172] Mehrabadi S, Sadr SS. Assessment of Probiotics Mixture on Memory Function, Inflammation Markers, and Oxidative Stress in an Alzheimer's Disease Model of Rats. Iranian Biomedical Journal, 2020, 24(4): 220-228.

[173] Ou Z, Deng L, Lu Z, et al. Protective effects of Akkermansia muciniphila on cognitive deficits and amyloid pathology in a mouse model of Alzheimer's disease. Nutrition & Diabetes, 2020, 10(1): 12.

[174] Akbari E, Asemi Z, Daneshvar Kakhaki R, et al. Effect of Probiotic Supplementation on Cognitive Function and Metabolic Status in Alzheimer's Disease: A Randomized, Double-Blind and Controlled Trial. Frontiers in Aging Neuroscience, 2016, 8:256.

[175] Klingelhoefer L, Reichmann H. Pathogenesis of Parkinson disease--the gut-brain axis and environmental factors.Nature reviews. Neurology, 2015, 11(11): 625-636.

[176] Wang Q, Luo Y, Chaudhuri KR, et al. The role of gut dysbiosis in Parkinson's disease: mechanistic insights andtherapeutic options. Brain: A Journal of Neurology, 2021:144(9):2571-2593.

[177] Barichella M, Pacchetti C, Bolliri C, et al. Probiotics and prebiotic fiber for constipation associated with Parkinson disease: An RCT. Neurology, 2016, 87(12): 1274-1280.

[178] Sun J, Li H, Jin Y, et al. Probiotic Clostridium butyricum ameliorated motor deficits in a mouse model of Parkinson's disease via gut microbiota-GLP-1 pathway. Brain, Behavior, and Immunity, 2021, 91:703-715.

[179] Burokas A, Arboleya S, Moloney RD, et al. Targeting the Microbiota-Gut-Brain Axis: Prebiotics Have Anxiolytic and Antidepressant-like Effects and Reverse the Impact of Chronic Stress in Mice. Biological Psychiatry, 2017, 82(7): 472-487.

[180] Liu WH, Chuang HL, Huang YT, et al. Alteration of behavior and monoamine levels attributable to Lactobacillus plantarum PS128 in germ-free mice. Behavioural Brain Research, 2016, 298:202-209.

[181] Pirbaglou M, Katz J, de Souza R-J, et al. Probiotic supplementation can positively affect anxiety and depressive symptoms: a systematic review of randomized controlled trials. Nutrition Research (New York, N.Y.), 2016, 36(9): 889-898.

[182] Messaoudi M, Lalonde R, Violle N, et al. Assessment of psychotropic-like properties of a probiotic formulation (Lactobacillus helveticus R0052 and Bifidobacterium longum R0175) in rats and human subjects.The British Journal of Nutrition, 2011, 105(5): 755-764.

[183] Shaaban SY, El Gendy YG, Mehanna NS, et al. The role of probiotics in children with autism spectrum disorder: A prospective, open-label study. Nutritional Neuroscience, 2018, 21(9): 676-681.

[184] Mensi MM, Rogantini C, Marchesi M, et al. Lactobacillus plantarum PS128 and Other Probiotics in Children and Adolescents with Autism Spectrum Disorder: A Real-World Experience. Nutrients, 2021, 13(6):2036.

[185] Kong Q, Wang B, Tian P, et al. Daily intake of Lactobacillus alleviates autistic-like behaviors by ameliorating the 5-hydroxytryptamine metabolic disorder in VPA-treated rats during weaning and sexual maturation. Food & Function, 2021, 12(6): 2591-2604.

[186] Hassan H, Rompola M, Glaser AW, et al. Systematic review and meta-analysis investigating the efficacy and safety of probiotics in people with cancer. Supportive Care in Cancer : Official Journal of The Multinational Association of Supportive Care in Cancer, 2018, 26(8): 2503-2509.

[187] Ben Haj Ayed A, Moussa A, Ghedira R, et al. Prognostic value of indoleamine 2,3-dioxygenase activity and expression in nasopharyngeal carcinoma. Immunology Letters, 2016, 169:23-32.

[188] Chan K, Woo J, King A, et al. Analysis of Plasma Epstein-Barr Virus DNA to Screen for Nasopharyngeal Cancer. The New England Journal of Medicine, 2017, 377(6): 513-522.

[189] Guo R, Tang LL, Mao YP, et al. Clinical Outcomes of Volume-Modulated Arc Therapy in 205 Patients with Nasopharyngeal Carcinoma: An Analysis of Survival and Treatment Toxicities. PloS One, 2015, 10(7): e129679.

[190] Nishimura N, Nakano K, Ueda K, et al. Prospective evaluation of incidence and severity of oral mucositis induced by conventional chemotherapy in solid tumors and malignant lymphomas. Supportive Care in Cancer : Official Journal of The Multinational Association of Supportive Care in Cancer, 2012, 20(9): 2053-2059.

[191] Luo X, Miller SD, Shea LD. Immune Tolerance for Autoimmune Disease and Cell Transplantation. Annual Review of Biomedical Engineering, 2016, 18:181-205.

[192] Ortona E, Pierdominici M, Maselli A, et al. Sex-based differences in autoimmune diseases.Annali dell'Istituto superiore di sanita, 2016, 52(2): 205-212.

[193] Lin CS, Chang CJ, Lu CC, et al. Impact of the gut microbiota, prebiotics, and probiotics on human health and disease. Biomedical Journal, 2014, 37(5): 259-268.

[194] Miettinen M, Lehtonen A, Julkunen I, et al. Lactobacilli and Streptococci activate NF-kappa B and STAT signaling pathways in human macrophages. Journal of Immunology (Baltimore, Md: 1950), 2000, 164(7): 3733-3740.

[195] Drakes M, Blanchard T, Czinn S. Bacterial probiotic modulation of dendritic cells. Infection and Immunity, 2004, 72(6): 3299-3309.

[196] López P, de Paz B, Rodríguez-Carrio J, et al. Th17 responses and natural IgM antibodies are related to gut microbiota composition in systemic lupus erythematosus patients. Scientific Reports, 2016, 6:24072.

[197] Mardani F, Mahmoudi M, Esmaeili SA, et al. In vivo study: Th1-Th17 reduction in pristane-induced systemic lupus erythematosus mice after treatment with tolerogenic Lactobacillus probiotics. Journal of Cellular Physiology, 2018, 234(1): 642-649.

[198] Khorasani S, Mahmoudi M, Kalantari MR, et al. Amelioration of regulatory T cells by Lactobacillus delbrueckii and Lactobacillus rhamnosus in pristane-induced lupus mice model. Journal of Cellular Physiology, 2019, 234(6): 9778-9786.

[199] Vaghef-Mehrabany E, Alipour B, Homayouni-Rad A, et al. Probiotic supplementation improves inflammatory status in patients with rheumatoid

arthritis. Nutrition (Burbank, Los Angeles County, Calif.), 2014, 30(4): 430-435.

[200] Shadnoush M, Nazemian V, Manaheji H, et al. The Effect of Orally Administered Probiotics on the Behavioral, Cellular, and Molecular Aspects of Adjuvant-Induced Arthritis. Basic and Clinical Neuroscience, 2018, 9(5): 325-336.

[201] Mohammed AT, Khattab M, Ahmed AM, et al. The therapeutic effect of probiotics on rheumatoid arthritis: a systematic review and meta-analysis of randomized control trials. Clinical Rheumatology, 2017, 36(12): 2697-2707.

[202] Joos H, Albrecht W, Laufer S, et al. Differential effects of p38MAP kinase inhibitors on the expression of inflammation-associated genes in primary, interleukin-1beta-stimulated human chondrocytes. British Journal of Pharmacology, 2010, 160(5): 1252-1262.

[203] Svendsen MT, Feldman SR, Tiedemann SN, et al. Psoriasis patient preferences for topical drugs: a systematic review. The Journal of Dermatological Treatment, 2021, 32(5): 478-483.

[204] Olejniczak-Staruch I, Ciążyń ska M, Sobolewska Sztychny D, et al. Alterations of the Skin and Gut Microbiome in Psoriasis and Psoriatic Arthritis. International Journal of Molecular Sciences, 2021, 22(8):3998.

[205] Di T, Zhao J, Wang Y, et al. Tuhuaiyin alleviates imiquimod-induced psoriasis via inhibiting the properties of IL-17-producing cells and remodels the gut microbiota. Biomedicine & Pharmacotherapy , 2021, 141:111884.

[206] Nayak RR. Western Diet and Psoriatic-Like Skin and Joint Diseases: A Potential Role for the Gut Microbiota. The Journal of Investigative Dermatology, 2021, 141(7): 1630-1632.

[207] Rigon RB, de Freitas A, Bicas JL, et al. Skin microbiota as a therapeutic target for psoriasis treatment: Trends and perspectives. Journal of Cosmetic Dermatology, 2021, 20(4): 1066-1072.

[208] Piewngam P, Zheng Y, Nguyen TH, et al. Pathogen elimination by probiotic Bacillus via signalling interference. Nature, 2018, 562(7728): 532-537.

[209] Naidu A-S, Bidlack WR, Clemens RA. Probiotic spectra of lactic acid bacteria (LAB). Critical Reviews in Food Science and Nutrition, 1999, 39(1): 13-126.

第 10 章
膳食，功能食品与机体免疫

第一节 概 述

机体免疫系统受遗传、生理和生活方式等多方面因素的影响，例如年龄、性别、压力、激素、运动、饮酒、吸烟、健康状况和营养状况等[1]。在这些因素中，营养是一个可控的因素，对维持健康免疫系统的稳定至关重要[2, 3]。在免疫系统中发挥较大作用的营养成分可被视为免疫调节剂，包括免疫佐剂、免疫刺激剂和免疫抑制剂。免疫佐剂是能增强疫苗效力的特异性免疫刺激剂。免疫刺激剂激活免疫系统的介质或成分，其作用是增强机体对感染、自身免疫、癌症和过敏的抵抗力。免疫抑制剂可以抑制免疫系统，从而控制器官移植后的病理免疫反应[4]。例如，氨基酸是免疫细胞增殖和免疫效应分子合成的基石，蛋白质缺乏会导致抗体和免疫细胞数量减少，而增加感染风险[5]。通常由营养过剩引起的肥胖与白细胞发育、表型和活性的改变，以及先天性和适应性免疫反应的改变有关[6]。此外，即使在健康个体中，潜在的饥饿状态对最佳免疫反应的影响也很明显[4]。

营养对不同免疫系统的影响程度并不总是遵循线性剂量 - 反应关系，有些免疫系统对营养相对不敏感[7]。此外，个人的健康状况和营养状况会影响免疫系统对营养摄入的反应。例如，已经处于最佳营养状态的健康个体，可能无法从补充某些营养素中受益，除非这种状态存在缺陷[7, 8]。另一方面，摄入增强免疫系统的相关营养物质，可以为营养不足的人群带来免疫益处，例如住院患者。由于这些患者通常年龄较大，并且随着年龄的增长，其免疫系统会发生各种结构和功能的变化，例如循环促炎症细胞因子水平增加，以及幼稚免疫细胞产生不足。此外，记忆免疫细胞的单克隆扩增，导致先天性和适应性免疫反应的有效性降低，自身免疫反应的风险增加，感染的易感性增加（免疫衰老）[1]。同时，饮食还可以通过影响肿瘤监视和塑造宿主免疫反应来影响癌症的发展[9]。所有这些变化都表明：机体最佳营养状态对抵抗传染病的功能性免疫非常重要[4]。不仅如此，饮食对肠道微血管炎症有明显影响。研究表明，高脂饮食喂养可增强 T 淋巴细胞在小肠微血管中的黏附，导致肠道血管屏障的破坏[10]。

一、膳食与肠黏膜屏障

肠道免疫细胞网络位于上皮细胞内，由 3 个不同的淋巴样结构组成。这种对共生微生物和外源性病原微生物的物理和生化屏障是由上皮层、固有层和肠相关淋巴组织（gut-associated lymphoid tissue，GALT）形成的。GALT 包括肠系膜淋巴结（mesenteric lymph nodes，MLN）、派尔集合淋巴结（peyer patch）和孤立淋巴滤泡（isolated lymphoid follicle，ILF）[11]。肠上皮细胞（intestinal epithelial cell，IEC）具有基本的免疫调节功能。例如，肠上皮细胞过度表达 IL-15 后驱动了免疫调节性 CD8α 肠道上皮内淋巴细胞（intraepithelial lymphocytes，IELs）的增殖和分化[12]。此外，上皮细胞 IL-15 的过表达对于保护性 γ/δ IEL 的运动和在小肠上皮中的定位至关重要[13]。肠上皮细胞还表达抗炎症细胞因子，例如 IL-10，它可能通过上皮 - 巨噬细胞作用增强对共生细菌的耐受性和保护上皮完整性[14]。除此之外，肠上皮细胞通过紧密连接形成相互连接的网络，该网络作为肠黏膜表面的主要机械屏障。紧密连接，例如黏蛋白、闭合蛋白和密封蛋白，对于肠黏膜屏障至关重要，通过严格响应各种信号的调节，保持细胞极性[14]。特殊的肠上皮细胞（例如杯状细胞和帕内特细胞）通过分泌抗菌肽（antimicrobial peptides，AMPs）来增强屏障功能[15]。肠道先天性免疫系统介导宿主与肠道菌群之间的共生关系。髓样分化初级应答基因 88（myeloid differentiation factor 88，MyD88）是涉及病原体识别的大多数 Toll 样受体（Toll-like receptor，TLR）的中央衔接子分子。为了保持共生关系，肠上皮细胞需要识别病原体分子模式。研究认为，MyD88 通过塑造肠道菌群，在调节免疫系统中起关键作用[15]。此外，肠上皮细胞 MyD88 可以作为主要传感器，参与饮食诱发肥胖症期间营养素、肠道菌群和宿主之间的相互作用[16]。特定敲除肠上皮细胞 MyD88，可保护机体免受饮食诱发的肥胖症、糖尿病和炎症。此外，肠上皮细胞还参与了由 IgA 介导的肠道稳态的适应性免疫调节[17]。

另一类特殊的肠上皮细胞是滤泡相关上皮细胞，该上皮细胞覆盖在包括派尔集合淋巴结和孤立淋巴滤泡在内的淋巴结构表面。这些微折叠细胞（microfold cells，也称 M 细胞）有助于对腔内抗原和微生物进行收集，以呈递给肠道免疫系统[17, 18]。尽管 M 细胞和杯状细胞都参与了向固有层的抗原传递，但它们对病原体反应或免疫耐受调节的重要性和影响仍不十分清楚。肠道菌群与黏膜免疫之间的联系在很大程度上受到膳食纤维和微生物短链脂肪酸的影响。肠道微生物对不同食物的发酵，例如难消化的多糖，会产生乙酸，丙酸和丁酸等有益的短链脂肪酸[19, 20]。

肠内稳态可防止产生过量的黏液，这对于健康肠道上皮的结构和组成是必不可少的[21]。研究发现，慢性或间歇性膳食纤维缺乏，会导致肠道菌群把宿主分泌的黏液糖蛋白作为营养源，从而导致结肠黏液屏障的侵蚀[22]。这些黏蛋白中含量最多的是黏蛋白 2，它在形成肠上皮表面的黏液层中起着关键作用。缺乏黏蛋白 2 的小鼠在感染啮齿类柠檬酸杆菌后，会发生致命性结肠炎[23]。此外，膳食纤维的长期缺乏与黏蛋白降解细菌如嗜黏蛋白阿克曼菌的丰度增加有关[22]。与此一致的是，研究证实，给小鼠喂食纤维含量极低的西方饮食，会严重影响肠道菌群的组成，降低黏液的生成速率并增加结肠黏液层的渗透性[24]。鉴于纤维是肠道菌群的主要能源，上述研究结果证实了微生物短链脂肪酸的关键作用。因此，双歧埃希菌菌株长期以来被用作益生菌，其部分有益作用归因于乙酸盐产量的增加[25]。例如，用双歧杆菌定植的无菌小鼠阻止了肠致病性大肠埃希菌的移位[25]。尽管补充 1% 的菊粉（具有促进双歧杆菌增殖的益生元）或长双歧杆菌可预防黏液缺陷，但不足以改善肥胖动物的代谢指标。但是，大量摄入 20% 的菊粉可减少微生物进入黏膜，防止结肠萎缩并促进肠上皮增殖，从而保护小鼠免受肥胖的危害[26]。

研究表明，环境因素在形成人类肠道菌群中占主导地位[27]，其中饮食是影响肠道菌群和免疫系统的最重要的环境因素之一[28]。例如，主要由植物来源的碳水化合物组成的饮食有助于形成以普雷沃菌为主的肠道微生物群，而动物蛋白和富含饱和脂肪的饮食与拟杆菌为主的肠道微生物群有关[29]。高膳食脂肪摄入与促炎细胞因子水平升高[30,31]和肠道通透性增加有关，从而增加对革兰阴性细菌感染的易感性，尤其是肠侵袭性大肠埃希菌感染[32]。

二、膳食，细菌代谢物与肠道免疫

饮食由不同的宏量和微量营养素组成，天然纤维对于短链脂肪酸的产生是必不可少的[18]。饮食习惯的改变，如摄入高度加工和精制食品、添加剂（如乳化剂和防腐剂），会对肠道菌群产生负面影响，从而对机体健康产生有害作用。而乙酸、丙酸和丁酸有益于健康的肠道菌群和成熟的免疫系统[19]。小鼠模型研究证实，葡聚糖硫酸钠（dextran sodium sulfate，DSS）诱导的结肠炎中，短链脂肪酸如乙酸盐和丁酸盐可防止肠道上皮损伤[33]。高纤维饮食促进短链脂肪酸的产生，从而通过 GPR43 和 GPR109A 激活炎症小体[33]，由此导致 IL-18 的产生和分泌。IL-18 是一种参与修复肠上皮完整性和体内稳态的细胞因子，其高水平状态及其对 DSS 结肠炎的保护机制与拟杆菌的显著增加有关，而拟杆菌的增加是在喂食高纤维饮食的小鼠中发现的[33]。在饲喂零纤维饮食的小鼠中，普雷沃菌属的细菌相当低或不存在，已知该细菌含有一组纤维素和木聚糖水解的基因。这与来自西非国家布基纳法索农村地区儿童的研究结果一致，该地区儿童的微生物多样性与从富含植物纤维的食物中摄取有价值的营养素有关[29, 30]。

短链脂肪酸还能调节驻留在固有层中的免疫细胞。因此，缺乏 GPR43 或 GPR109A（乙酸盐和丁酸盐的受体）的小鼠表现出明显的食物过敏和 $CD103^+$ 树突状细胞减少。共生细菌产生的丁酸盐通过抑制组蛋白脱乙酰基酶（histone deacetylase，HDAC）[34]或诱导耐受性树突状细胞[35]，从而促进外周 Treg 细胞的生成。体外研究表明，丁酸盐通过抑制 HDAC 和通过 GPR109A 传递信号来诱导 1 型调节性 T 细胞（Tr1）分化，从而成为人类耐受性树突状细胞的有效诱导剂[36]。肠道菌群分析表明，炎性肠病[37]和大肠癌患者[38]结肠中产生丁酸盐的细菌数量显著减少，例如厚壁菌门和毛螺菌科。通过 GPR109A 信号传递的肠道菌群，能促进结肠巨噬细胞和树突状细胞诱导 Treg 细胞和 T 细胞的分化，从而发挥抗炎特性[39]。GPR109A 也能通过结肠上皮中的丁酸途径诱导 IL-18 的产生，这对于维持肠道完整性是必不可少的。除丁酸盐外，GPR109A 还充当烟酸（或维生素 B_3）的受体，烟酸也由肠道菌群产生。因此，烟酸缺乏的小鼠表现出更明显的 DSS 诱导结肠炎。肠道菌群的完全消失，大大增加了 DSS 诱发小鼠结肠炎和结肠癌的风险[39]。研究表明，高乙酸盐饮食以 GPR43 依赖性方式减少了感染柠檬酸杆菌小鼠的结肠萎缩和炎症[40]。GPCR 介导的短链脂肪酸促使肠黏膜上皮细胞产生 IL-6、CXCL1 和 CXCL10，从而及时抵抗细菌感染。这些发现也再次证实了饮食、肠道菌群和肠道免疫之间的相互作用关系[40]。

三、膳食，细菌代谢物与全身免疫

饮食通过介导肠道微生物群的组成和功能变化，通过影响肠道微生物群衍生的代谢物，以及直接通过特定食物成分的抗炎特性来影响机体免疫[41]。最初认为，肠道是膳食代谢产物通过肠道上皮或黏膜免疫作用于机体的主要部位。的确，远端结肠的共生细菌发酵纤维，可产生大量的乙酸盐、丙酸盐和丁酸盐（分别为 40mmol/L，20mmol/L 和 20mmol/L）[42]。但是最近的研究表明，细菌代谢物（尤其是乙酸盐）会全身分布。实际上，在无菌小鼠疾病模型中观察到的炎症反应加剧现象[43, 44]，可能部分与肠道、血液或组织中缺乏短链脂肪酸有关。一项最新的研究表明，丙酸酯会影响骨髓中的树突状细胞和巨噬细胞的功能，并影响气道中 Th2

细胞的反应。循环中的短链脂肪酸可以对全身的巨噬细胞和树突状细胞的功能产生显著影响，这表明在树突状细胞或巨噬细胞的调控下，膳食纤维摄入与多种类型免疫反应之间存在密切联系。

代谢产物与免疫系统相互作用的另一个主要方面是它们可以穿过胎盘转运至发育中的胎儿，尽管该过程的具体作用尚待确定。有趣的是，最近发现胎盘中含有共生微生物[44]。这提示母亲饮食和共生微生物在胎儿发育过程中指导免疫系统的潜在作用。同样，母乳中也存在诸如短链脂肪酸之类的代谢物，这可能是代谢物与免疫系统之间相互作用的重要途径。迄今为止，母乳在保护（或促进）过敏和哮喘方面的作用尚无定论。但是动物实验发现，高脂喂养的哺乳期小鼠，其幼鼠在成年后更容易罹患肥胖症和代谢综合征[45]。膳食、细菌代谢物与免疫系统之间的相互作用对炎症性疾病的影响，尚需要进一步研究确定（图 10-1）。

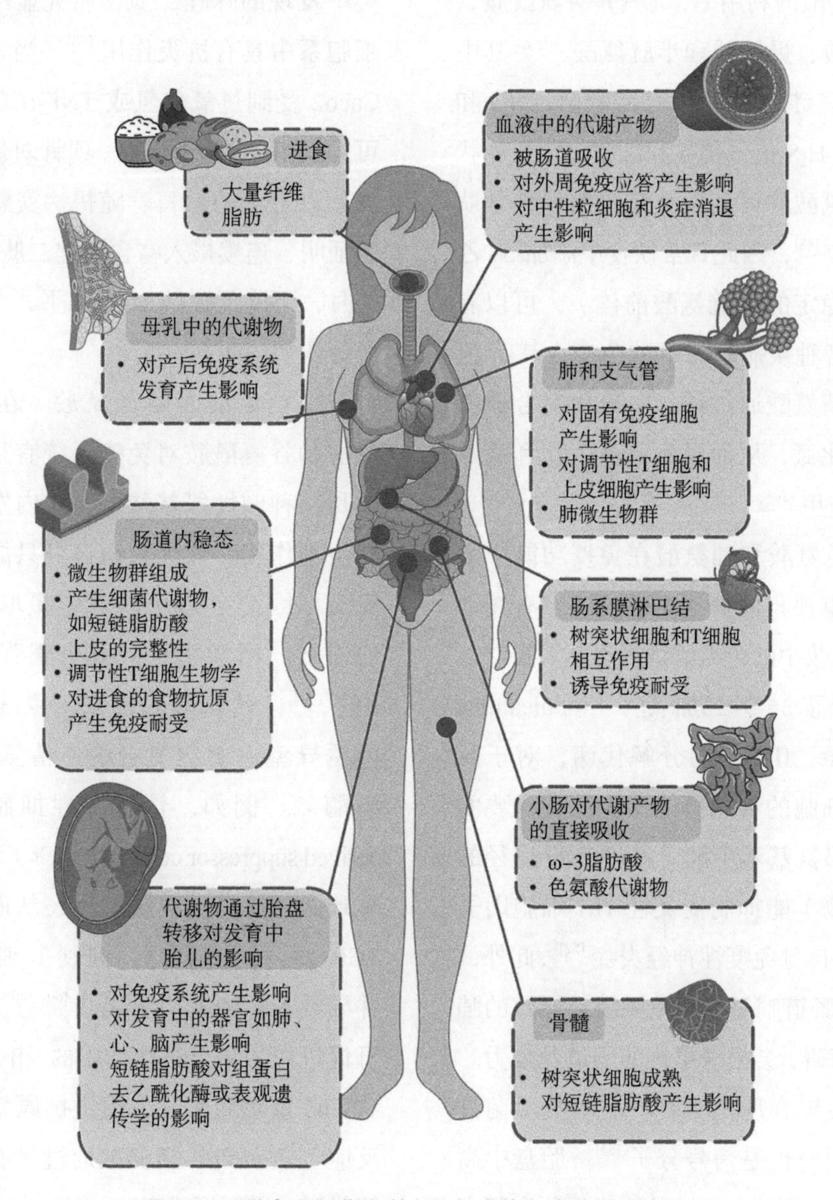

图 10-1　膳食或细菌代谢产物与人体免疫的相互作用

在胃肠道中，膳食纤维主要被结肠中的共生细菌消化，这会产生高浓度的短链脂肪酸，例如乙酸盐、丙酸盐和丁酸盐。其他代谢物（例如 ω-3 脂肪酸、琥珀酸或犬尿酸）直接在整个胃肠道中消耗和吸收。另外，代谢物可以直接在小肠中吸收。短链脂肪酸（主要是乙酸盐）从肠道转移到血液，借此影响整个人体的骨髓和许多细胞类型。另一个重要的相互作用是代谢产物向发育中的胎儿转移。短链脂肪酸能够穿过胎盘或通过母乳传递，影响基因表达和免疫系统的发育。改自：Thorburn AN，Macia L，Mackay CR. Diet，metabolites，and “western-lifestyle” inflammatory diseases. Immunity, 2014,40（6）:833-842[46]

第二节 膳食及其成分对免疫系统的影响

一、蛋白质／氨基酸

长期以来，人们都知道蛋白质缺乏会削弱免疫功能，并增加动物对疾病的易感性。然而，其中潜在的细胞和分子机制在最近20年才开始被人们所发现。饮食中蛋白质的缺乏会降低血浆中大多数氨基酸的利用率，尤其是谷氨酰胺、精氨酸、色氨酸、蛋氨酸和半胱氨酸[47]。其中潜在的机制可能涉及mTOR通路激活，NO和谷胱甘肽合成，H_2S信号传导和细胞氧化还原状态。由于半胱氨酸的可利用性是限制谷胱甘肽合成的主要因素[48]，因此日常饮食中添加N-乙酰半胱氨酸（稳定的半胱氨酸前体），可以有效增强机体在各种疾病状态下的免疫力[49]。值得注意的是，精氨酸通过诱导一氧化氮合成酶产生大量一氧化氮，从而对病原微生物和病毒产生细胞毒性作用[50]。

近年来，色氨酸和脯氨酸在免疫功能中的作用逐渐引起重视。值得注意的是，患有慢性肺部炎症的猪血浆中的色氨酸浓度逐渐下降[51]。色氨酸通过吲哚胺2，3-二加氧酶（indoleamine 2，3-dioxygenase，IDO）的分解代谢，对于巨噬细胞和淋巴细胞的功能可能产生至关重要的作用。例如，邻氨基苯甲酸（通过IDO途径的色氨酸代谢产物）能抑制促炎性Th1细胞因子的产生，并防止自身免疫性神经炎症[52]。此外，Ha等发现[53]，肠道脯氨酸氧化酶缺乏导致的脯氨酸分解代谢障碍，会损害果蝇的肠道免疫力。脯氨酸氧化的主要介质是H_2O_2，它对病原菌具有细胞毒性，同时也是信号分子[54]。胎盘中高活性的脯氨酸氧化酶[55]以及哺乳动物的小肠中的脯氨酸氧化酶在胎儿和新生儿发育的关键时期可能在保护这些器官免受感染中起着至关重要的作用。此外，脯氨酸氧化酶存在于母乳中，并可能在保护新生儿肠道免受感染中发挥作用[56]。这些发现解释了为什么与母乳喂养的婴儿相比，非母乳喂养的新生儿出现肠道功能障碍的风险更高[57]。

1. *肌肽* 已知某些肽类具有抗炎功能。肌肽（β-丙氨酰-l-组氨酸）是一种在肉类和鱼类中发现的咪唑二肽，研究显示肌肽在肠上皮细胞系中具有抗炎作用[58]。当人类肠道细胞系Caco2受到过氧化氢或TNF-α的刺激时，肌肽可抑制IL-8的分泌。一项针对健康志愿者（60岁或以上）的双盲、随机、安慰剂对照临床试验证明，连续摄入富含咪唑二肽的鸡肉提取物3个月，可降低血清IL-8水平，并可以改善受试者的认知功能[59]。

2. *精氨酸和谷氨酰胺* 在氨基酸中，精氨酸和谷氨酰胺对免疫系统有很大影响。精氨酸是一种碱性氨基酸，在体内发挥着非常重要的生理作用。哺乳动物自身只能合成少量的精氨酸，因此必须从饮食中摄取才能满足机体的需要。已知精氨酸具有很强的免疫增强作用[60, 61]，精氨酸代谢主要受3种酶的调控，即诱导型一氧化氮合酶、精氨酸酶-1和精氨酸酶-2。例如，骨髓源性抑制细胞（myeloid derived suppressor cells，MDSCs）释放精氨酸酶-1导致微环境精氨酸缺乏[62]，从而降低细胞表面TCR复合物的表达，抑制T细胞增殖[61, 63]，并导致NK细胞功能降低[60, 64]。此外，精氨酸可促进派尔集合淋巴结中的αβ T细胞产生Th1和Th2型细胞因子，促进抗原特异性黏膜免疫反应[65]。另外，精氨酸通过鸟氨酸循环完成分解代谢，进而形成多胺，例如精胺和亚精胺，多胺在增强肠道屏障功能方面发挥重要作用[66, 67]。

谷氨酰胺是氨基酸、蛋白质、核酸和许多其他生物分子的前体物质，也是快速分裂细胞（如淋巴细胞和肠上皮细胞）的重要能量来源[68, 69]。另外，谷氨酰胺是一种对免疫细胞如淋巴细胞、自然杀伤细胞和巨噬细胞增殖很重要的氨基酸，并且对于谷胱甘肽的合成也很重要[70, 71]。当巨噬细胞进行吞噬和胞饮作用时，葡萄糖不能通过磷酸戊糖途径合成还原型辅酶，只能通过谷氨酰胺生成，而还原型辅酶是脂质合成的必需辅酶，在这种情况下，谷氨酰胺可作为还原型辅酶的来源发挥重要作用。由于感染、伤口或手术期间谷氨酰胺需求增加，导致血清谷氨酰胺水平降低，这促使骨骼肌的谷氨酰胺产生量增加，以补偿血清的谷氨酰胺水平。如果血清谷氨酰胺没有达到需要的水平，就会抑制 B 淋巴细胞和 T 淋巴细胞的增殖和分化，使细胞因子的产生减少、抗原呈递和吞噬活性降低。研究发现，口服谷氨酰胺可改善肠道免疫功能并抑制细菌易位[72]。同时，在炎性肠病模型中，研究者发现谷氨酰胺具有抗炎特性[73-83]。

不管是在健康还是患病个体中，免疫细胞对谷氨酰胺的消耗速率与葡萄糖相似或更高。体外和体内研究已经确定，谷氨酰胺是淋巴细胞增殖、产生细胞因子的必需营养素，也是巨噬细胞吞噬、分泌活性物质以及嗜中性粒细胞杀死细菌的必需营养素。细胞因子在炎性肠病的发病机制中起主要作用，炎性肠病（克罗恩病或溃疡性结肠炎）患者的肠道黏膜会产生大量促炎症细胞因子，例如 IL-1β、IL-6、IL-8 和 TNF -α，而抗炎性细胞因子（例如 IL-10）增加不明显。Coëffier 等证实，谷氨酰胺可以减少人肠道黏膜促炎性细胞因子的产生，这种作用可能是通过调控转录后途径实现的。因此，谷氨酰胺可通过调节细胞因子的产生，抑制肠道炎症反应[70]。

3. 鸡蛋蛋白质　蛋白质对于免疫细胞和效应分子（如抗体）的产生至关重要，据估计，机体感染期间每天消耗多达 60g 蛋白质。并且，与蛋白质含量充足的人相比，蛋白质含量低的人更容易患感染性疾病，在接种疫苗后抗体滴度反应较低。因此，维持蛋白质稳态对于维持最佳免疫力非常重要，并且在感染期间摄入足量的高质量蛋白质对于预防症状加重和并发症至关重要。

鸡蛋是具有免疫调节作用的蛋白质成分的最佳例子[26]。研究表明，鸡蛋蛋白质及其衍生肽可作用于多种免疫调节途径[84]。例如，二羟基甲基乙醛酸修饰的卵清蛋白会刺激巨噬细胞分泌 TNF-α[85]。此外，Rupa 等报道，热变性卵清蛋白可调节 $CD4^+$ T 细胞产生细胞因子。热变性卵白蛋白还通过下调 IL-4 和上调 IL-10、IL-12 和 IL-17 来调节免疫[86]。由胃蛋白酶和胰凝乳蛋白酶消化卵清蛋白产生的卵清蛋白衍生肽 OA77-84 和 OA126-134，能增强巨噬细胞吞噬活性[87]，卵清蛋白还可增强癌症治疗中机体的免疫应答[88]。研究还报道，卵转铁蛋白通过 MAPK 信号通路刺激小鼠巨噬细胞产生促炎细胞因子[89]，如 IL-6 等。而 IL-6 是适应性免疫的成分之一，在慢性炎症期间，IL-6 激活 T 细胞，促进 B 细胞增殖并上调抗体产生。

卵黏蛋白被碱性蛋白酶裂解时产生的肽，也可通过抑制 TNF 介导的 NF-κ B 途径表现出抗炎活性[90]。谷崎等报道，卵黏蛋白糖蛋白可以激活巨噬细胞，诱导巨噬细胞发生形态学改变，并增加 H_2O_2 和 IL-1 的生成[91]。此外，在 IFN-γ 激活的巨噬细胞中，胱抑素 C 可促进 IL-10 和 TNF-α 这两种细胞因子产生，而这两种细胞因子进一步促进巨噬细胞产生一氧化氮(Nitric Oxide，NO）[92]。胱抑素 C 也会影响成纤维细胞产生 IL-6 等细胞因子[93]。

蛋黄活性蛋白（α、β 和 γ 活性蛋白）及其水解物可抑制巨噬细胞中 NO 和促炎细胞因子的产生，例如 IL-1β、IL-6、IL-10、TNF-α，以及诱导型一氧化氮合酶，从而抑制炎症。并且，蛋黄活性蛋白及其水解物可显著增强巨噬细胞的吞噬活性[94]。据报道，蛋黄中的蛋黄蛋白还能抑制自由基的产生，从而抑制氧化应激和促炎症细胞因子，如巨噬细胞中的 IL-1β、IL-6、IL-10 及 TNF-α[95]。

4. 乳清蛋白　乳清蛋白是一种重要的免疫调节剂，具有抗菌和抗氧化作用。动物实验和人体试验表明，口服乳清蛋白可有效对抗感染、癌症和炎症，从而使其能够作为食品添加剂使用[96]。实验证据表明，乳清蛋白具有增强机体免疫反应的特性。例如，与正常饮食的对照组相比，喂食 12 周乳清蛋白后的小鼠对霍乱毒素和卵清蛋白的黏膜抗体反应显著增强[97]。与喂食等热量酪蛋白饮食的对照组相比，接受未变性乳清蛋白（25g）4 周的小鼠的辅助 T 细胞（$CD4^+$）数量也有所增加。不仅 $CD4^+$ 辅助 T 细胞群升高，而且 $CD4^+$ 辅助 T 细胞群与抑制性 T 细胞（$CD8^+$）的比值也升高。与富含酪蛋白和大豆蛋白的饮食相比，乳清蛋白饮食与 $CD4^+$ 和 $CD8^+$ 淋巴细胞、总白细胞计数增加有关[98]。乳清蛋白还会增加血浆中的谷胱甘肽水平，并增强慢性乙型肝炎患者体内的NK细胞活性[84]。然而，乳清蛋白也可能导致过强的免疫反应，一项研究发现，给小鼠喂食乳清蛋白（5～8 周）导致足垫发生延迟型超敏反应[97]。

二、脂肪

西方饮食可能因高脂肪含量而加剧机体炎症。脂肪酸通过多种机制，如对免疫细胞、TLR 和细胞因子信号的直接作用，以及通过影响肠道黏膜通透性促进炎症[99-101]。即使是健康受试者，高脂肪西方饮食也可能导致其处于全身性轻度炎症状态[102]。高脂饮食诱导肠道炎症，增加回肠肿瘤坏死因子和脂肪的产生，值得注意的是，这种现象仅存在于常规饲养的无特定病原体的小鼠，而在无菌小鼠中则无此现象[103]。高脂饮食可能导致小鼠严重肺损伤和死亡，同时减少其体内 TLR2 和 TLR4 的表达。这种现象可通过粪便移植传播给同舍的野生型小鼠；通过抗生素可以预防其肺部疾病，这表明有某些病原体参与此过程[104]。

ω-3 脂肪酸是地中海饮食的关键成分，可以抑制机体的炎症反应[105-109]。膳食 ω-3 脂肪酸可能对多种炎症性疾病产生有益影响，例如动脉粥样硬化和心血管疾病[108]、炎性肠病[106]和过敏性疾病[110, 111]。母亲在妊娠期间摄入 ω-3 脂肪酸可防止婴儿和儿童发生过敏性和炎症性疾病[110, 112]。此外，ω-3 脂肪酸直接与转录因子（如 NF-κB 和 PPAR -γ）相互作用，从而下调促炎基因的表达[106, 108]，抑制 TLR4 的激活[100]。ω-3 脂肪酸可通过分解素和保护素（由二十碳五烯酸和二十二碳六烯酸合成的抗炎介质）进一步调节免疫反应[107, 113-115]。这些介质减少炎症诱导的中性粒细胞浸润，促进炎症趋化因子的清除[114]，并增强巨噬细胞吞噬作用以清除凋亡细胞[107]。动物模型还表明，ω-3 脂肪酸可通过调节 TLR2 和 TLR4 依赖性炎症，实现动物体内的抗炎平衡[116-118]。

三、碳水化合物

大多数具有免疫调节功能的多糖是不可消化的[119]。其中 β- 葡聚糖是一种典型的不易消化的多糖。从蘑菇到酵母的 β- 葡聚糖，包括提取自香菇的香菇多糖，都是抗癌物质，具有 β-1, 3- 葡聚糖骨架和 β-1, 6- 葡聚糖侧链[120]。据报道，这些 β- 葡聚糖可增强机体对细菌和病毒感染的抵抗力[121-123]。源自燕麦的 β- 葡聚糖通过 β-1, 3- 和 β-1, 4- 糖苷键连接，据报道口服这种 β- 葡聚糖可以保护小鼠免受疱疹病毒[124]和细菌的感染[125]。具有 β-1, 3- 键的 β- 葡聚糖被 dectin-1 识别，dectin-1 是一种 C 型凝集素，在巨噬细胞、单核细胞、树突状细胞和 B 细胞上表达，是一种模式识别受体[126]。β- 葡聚糖激活这些细胞产生促炎细胞因子，包括 IL-6、TNF-α 和 Th1 诱导细胞因子 IL-12。IL-12 通过上调 Th1 细胞产生 IFN-γ 来促进 NK 细胞和吞噬细胞的活性 [119]。

据报道，酶促合成糖原具有免疫调节作用 [127-129]。在一项体外研究中，酶促合成糖原能刺激小鼠巨噬细胞系 RAW264.7，产生 NO 和促炎症细胞因子，如 IL-6 和 TNF-α[127]。该糖原引起 TLR2 介导的效应，刺激源自野生型小鼠的腹膜巨噬细胞，但不刺激源自 TLR2 缺陷小鼠的腹膜巨噬细胞，并且发现酶促合成糖原可以和 TLR2 直接结合 [128]。口服酶促合成糖原可以增强小鼠的抗肿瘤活性和 NK 细胞活性 [129]。

褐藻糖胶是一组存在于食用褐藻中的硫酸化多糖。褐藻糖胶含有硫酸化 L- 岩藻糖和其他几种糖类，包括甘露糖、半乳糖和木糖。褐藻糖胶具有免疫调节功能，包括抗癌 [130] 和抗病毒活性 [131, 132]。研究表明，口服来自裙带菜的褐藻糖胶可上调 NK 细胞和细胞毒性 T 淋巴细胞的活性，从而保护小鼠免受 1 型单纯疱疹病毒或流感病毒感染 [131, 132]。

甲壳素是一种含有 β-1-4- 键的 N- 乙酰基 -D- 氨基葡萄糖聚合物，存在于甲壳类动物（如螃蟹和虾）的外骨骼以及真菌的细胞壁中。甲壳素的脱乙酰衍生物是壳聚糖。据报道，壳聚糖具有免疫调节作用，包括抗癌作用 [133] 和巨噬细胞活化作用 [134]。尽管尚未鉴定出壳聚糖的细胞受体，但研究表明，不同分子大小的壳聚糖能被不同受体识别，壳聚糖的分子大小决定了其免疫调节功能 [133, 134]。在一项临床研究中，与对照组相比，肺癌放疗患者服用水溶性低分子量壳聚糖，导致 NK 细胞数量和 IL-6 和 TNF-α 水平升高 [135]。

四、纤维

膳食纤维大多是复合碳水化合物，是影响炎症的重要因素 [136, 137]。随着纤维消耗量的增加，超敏 C 反应蛋白浓度显著降低 [138]。膳食纤维并不是影响肠道健康的唯一因素，但它们确实具有重要的影响 [3]。许多对膳食纤维采取干预措施的研究认为，纤维通过增强上皮屏障功能、抑制病原体诱导的细胞毒性和防止病原菌定植来维持肠道稳态 [139]。研究显示，美国成年人的膳食纤维摄入量与呼吸道和传染病的死亡风险呈负相关 [140]。在这项研究中，每天膳食纤维每增加 10g，男性感染和呼吸系统疾病的死亡率相对风险分别降低 34% 和 18%，女性分别降低 39% 和 34%。

如前所述，高纤维饮食可促进短链脂肪酸的产生，短链脂肪酸通过抑制组蛋白脱乙酰基酶发挥抗炎作用。给小鼠喂食高纤维饮食可以提高 Treg 细胞的数量和功能，从而保护小鼠发生气道过敏性疾病 [141]。该研究证明，膳食乙酸盐通过增加 Foxp3 启动子处的乙酰化作用来抑制组蛋白脱乙酰基酶 9，从而导致 Treg 细胞增多 [141]。同样，另一项研究报道，膳食纤维和丙酸盐以 GPCR41 依赖性方式保护小鼠免受气道过敏性疾病的侵害 [142]。小鼠和人类的研究表明，妊娠期间膳食纤维摄入量增加与血清乙酸盐浓度升高以及对后代气道疾病的保护作用增强有关 [141]。

五、无麸质饮食

无麸质饮食的起源可以追溯到 1941 年，当时在儿科医师关于饮食治疗乳糜泻的报告中首次出现 [143]。如今，无麸质饮食继续用于各种其他健康目的，包括用于控制肠易激综合征、糖尿病、炎症和肥胖症等 [144]。

乳糜泻（celiac disease，CD）是一种由麸质引起的自身免疫介导的肠道炎症性疾病，由易感个体对麸质蛋白的永久性不耐受引起。在这些患者中，谷蛋白肽触发异常的免疫反应，导致典型的乳糜泻组织病变，其特征是绒毛萎缩、隐窝增生以及上皮内和固有层淋巴细胞数量增加[145，146]。先天免疫细胞（单核细胞、巨噬细胞和树突状细胞）促炎症细胞因子的产生增加，并通过介导淋巴细胞向固有层和上皮细胞积聚，从而导致该病的发生[147]。20世纪60年代，无麸质饮食被认为是乳糜泻潜在的治疗方法，可以恢复患者正常的肠黏膜[148]。使用无麸质饮食进行治疗，可促进患者临床症状缓解，黏膜组织学恢复正常。然而，这种饮食疗法的依从性非常复杂，患者经常产生更高的健康风险和营养缺乏[149，150]。

目前，无麸质饮食仍被认为是乳糜泻的唯一治疗方法，也是世界上最常用的一种特殊饮食[151]。De Palma等[152]研究了无麸质饮食1个月以上健康受试者的肠道菌群，发现双歧杆菌、象牙海岸梭菌、普拉梭菌、普雷沃菌的比例下降，肠杆菌科和大肠埃希菌的数量增加。此外，导致肠道菌群变异最强的无麸质饮食是禾本科植物，无麸质饮食肠道菌群的丰度显著下降[153]。

最近的研究表明，未经无麸质饮食治疗和无麸质饮食治疗的乳糜泻患者具有不同的微生物群组成[154，155]。然而，部分检测到的微生物群变化可能不仅是由于潜在的疾病状态，还因为无麸质饮食对乳糜泻患者的饮食干预。无麸质饮食可以影响健康人肠道菌群的组成和免疫功能，而不受任何潜在疾病的影响。因此，无麸质饮食可能有益于麸质相关性疾病的重要方式是通过调节肠道菌群。不含麸质的饮食能够对乳糜泻患者肠道细菌的组成和功能产生有益的影响[156]。

随着无麸质饮食的临床应用和普及程度的提高，消费者的需求在更大程度上继续影响着无麸质产品的标准。无麸质饮食包括自然不含麸质的食物，例如新鲜水果、蔬菜、海鲜、肉、家禽、豆类、坚果和大多数乳制品[157]。但是，其中一些产品可能还包含“隐藏”的麸质。因此，消费者需要仔细检查产品标签和成分列表。目前已经有各种各样的无麸质食品可供选择，这些食品使用无麸质谷物和假谷物，例如大米、玉米、藜麦、小米和苋菜作为基本成分[158]。

六、素食饮食

长期以来，富含植物的饮食一直是重要的饮食建议，素食饮食可促进健康和降低疾病风险[159]。这些有益的影响可能与肠道菌群有关。植物性食物构成了微生物可用性碳水化合物（microbial accessible carbohydrate，MAC）的主要来源，食用素食或以植物性饮食为主的宿主微生物群显示出更大的MAC发酵能力。然而，一些干预和横断面研究发现，杂食动物和素食动物之间的微生物群差异并不大，表明饮食模式主要在属和种水平影响微生物群，而在更广泛的组成特征（如多样性和丰富度）上影响相对较小[160，161]。尽管没有微生物群组成的总体变化，但物种水平的变化似乎足以改变代谢产物，因为在素食者中，短链脂肪酸的产量通常会增加。这些微生物代谢产物在多大程度上调节素食的有益效果尚不清楚。

除了提供MAC，植物性食品还提供多种来源的植物化学物质，即可能影响人类健康的生物活性小分子。在植物内部，许多植物化学物质是糖基化的，食用时降低了它们的生物利用度和生物活性。因此，植物化学物质通常会到达远端肠道，在肠道内发挥直接的抗菌和抗炎作用。此外，植物性化学物质可以被微生物酶修饰成具有更高生物利用度和不同生物活性的代谢物[162]。一个突出的例子是将天然大豆异黄酮生物转化为马醇[163]，与母代化合物相比，马醇显示出更高的生物利用度。因此，植物性化学物质生物利用度的改变，可能代表了植物性饮食有益效果的另一种机制。

杂食者、素食者和纯素食者的肠道菌群具有明显的差异。研究表明，纯素食者和素食者的肠道双歧杆菌和拟杆菌数量都较少。但对粪便短链脂肪酸水平和呼吸产生的甲烷的量化检测显示，纯素食者和杂食者之间没有区别，表明纯素食和素食可以降低肠道菌群多样性，但不会降低短链脂肪酸和甲烷水平 [164]。事实上，由于不同的微生物群鉴定方法，不同的样本量，以及地理来源、年龄、性别和体重的影响，素食对肠道菌群的影响结果都应该谨慎地分析 [165]。此外，应该考虑多酚对肠道菌群调节的影响，多酚在植物性食物中含量丰富，因此在素食和素食饮食中也是如此。事实上，这些成分增加了有益细菌的丰度，如双歧杆菌和乳酸杆菌。然而，需要进一步的研究来阐明素食饮食和肠道菌群之间的复杂机制和相互关系。

第三节　功能食品与免疫调控

众所周知，营养在慢性疾病中起着至关重要的作用。含有特定营养素或食物成分的功能性食品对身体的一种或多种功能有着有益的影响。那么功能性食品是如何对机体产生影响的呢？近 20 年关于功能性食品与机体免疫的研究给出了答案，即功能性食品通过免疫调节，从而对机体产生有益影响 [166]。

饮食除了营养作用之外，还可以调节机体免疫。这些发现正在将营养的概念从“均衡”饮食转变为“优化”营养。“优化”营养的结果是通过优化饮食来最大限度地延长预期寿命和质量。后者提供了功能食品的概念 [166]。功能食品的一般定义指出，如果一种食品除了足够的营养作用外，被证明对身体中的一个或多个功能产生有益的影响，还与提高健康和（或）降低疾病风险相关，则可以将其视为功能食品 [167]。

使食物具有“功能性”的成分可以是摄入量高于每日推荐量的具有特定生理作用的必需常量营养素（如 ω-3 脂肪酸）或必需微量营养素。也可以是是非必需的食物成分（如益生元）或不提供营养功能的食物成分（如益生菌或植物化学物质）[166]。

功能性食品对免疫系统有两个主要作用：①克服或预防营养不良对机体的影响；②作为临床治疗慢性疾病的辅助手段。能量和常量营养素摄入不足或特定微量营养素缺乏导致的营养不良会损害免疫系统，抑制免疫功能，而这些功能对于宿主抵抗病原微生物至关重要 [7]。营养不良的影响在发展中国家最大，但在发达国家也很常见，特别是对老年人、饮食失调者、酗酒者、某些疾病患者及早产儿。动物和人体研究表明，免疫系统有效运作所需的营养素包括必需氨基酸、必需脂肪酸、亚油酸、维生素 A、叶酸、维生素 B_6、维生素 B_{12}、维生素 C、维生素 E、Zn、Cu、Fe 和 Se[168-172]。如果免疫反应要有效运作，这些营养素在适当水平上的可用性是必不可少的。实际上，任何免疫反应都可能受到这些营养素中的一种或多种的影响，缺乏这些营养素的动物或个体会更容易受到感染。动物和人体研究表明，将缺乏的营养素重新添加到饮食中，可以恢复免疫功能和对感染的抵抗力 [169, 171, 172]。

一、多酚类

在各种植物和食品中，如水果、蔬菜、茶叶、药用植物、微藻、草药、种子和谷类，以及咖啡、茶、可可和葡萄酒等饮料中，已鉴定出超过 10 000 种多酚化合物 [165]。一些食用和野生水果，如葡萄、橄榄、蓝莓、甜瓜、芒果和柑橘类水果多酚含量非常丰富 [173]。研究表明，多酚可以预防与氧化应激有关的疾病，例如心血管疾病、癌症、炎症、神经退行性疾病和糖尿病等 [174]。此外有学者提出，多酚的有益作用取决于它们的抗氧化、抗微生物、抗癌、心脏和神经保护活性[175]。

最近的研究表明，多酚可以干扰异种代谢酶[176]，并且能够与药物[177]、食物毒素[178]和其他保健食品（例如益生菌）相互作用[179]。例如，柚皮苷（葡萄柚中的主要黄烷酮）可以通过抑制肠内细胞色素P4503A4（一种解毒酶）来增加药物（例如非洛地平和维拉帕米）的毒性；槲皮素（一种洋葱富含的黄酮）通过竞争性抑制与多药耐药性相关的蛋白质外排泵，诱导Caco-2细胞中曲毒素A（一种食物传播的霉菌毒素）的蓄积。但是，所有这些发现都强调需要进一步研究多酚与其他摄入物质，例如药物、营养素、异种生物、食物污染物等之间的相互作用。

人类缺乏特异性的多酚消化酶，饮食中超过90%的多酚没有吸收就通过上消化道。未吸收的多酚在结肠中积聚，其中大多数被肠道菌群广泛代谢。代谢途径随不同的多酚和微生物群组成而变化，但通常首先从解偶联或水解（糖基或其他部分从酚类骨架上裂解）开始，以产生糖苷配基或单体。然后将产物进行进一步转化，例如裂变（环或内酯）、还原、脱羟基、脱甲基化、脱羧和（或）异构化，产生广泛的微生物多酚代谢产物，主要包括芳族酸[180, 181]。从而使多酚的微生物群代谢物更易于在肠道中吸收，与它们的母体化合物相比，肠肝循环使它们在血浆中的停留时间延长，并最终从尿液中排泄[182]。

现在，我们已经越来越多地认识到，多酚对机体产生的益处可能与肠道菌群组成的改变有关。多酚可以促进益生菌（主要是乳酸杆菌）的生长、增殖或存活，从而发挥类似益生元的作用。多酚还可能抑制某些病原细菌（如沙门菌和幽门螺杆菌）的生长[183]。因此，多酚可能通过调节肠道微生态为宿主带来益处。然而，多酚与特定肠道菌群功能之间的相互作用尚待进一步探索。

研究人员一致认为，肠道菌群和酚类化合物的相互作用对酚类化合物的生物利用度有重要影响[184]。肠道菌群在调节原花青素的生物利用度方面起着关键作用[185]。原花青素可以对结肠上皮细胞发挥局部有益的生物学作用，从而对包括结直肠癌在内的炎症介导的疾病发挥保护作用[185]。此外，肠道菌群对酚类化合物的生物转化也有影响[184]。

多酚的免疫抑制作用包括调节T细胞功能、抑制肥大细胞脱颗粒和下调炎症细胞因子反应。在小鼠模型中，膳食补充绿茶多酚，结果可以使Th1和Th17细胞种群减少[186]。黄芩苷是一种在中药黄芩中发现的黄酮苷，目前已被证明对Th17细胞分化具有抑制作用[187]。绿茶多酚和黄芩苷，已被证明可以诱导小鼠Treg细胞的分化[188, 189]。

在抗过敏反应方面，从“红风”绿茶中提取出来的多酚物质已被证明可以抑制人类肥大细胞的脱颗粒作用[190]，从而在人体临床试验中用于缓解过敏症状[191]。从番茄皮提取物中分离出的柚皮素查耳酮也被证明具有抗过敏作用，其作用是通过抑制肥大细胞的脱颗粒作用[192]。

多酚还具有抗炎作用。例如，槲皮素是一种黄酮醇，广泛存在于水果和蔬菜中，具有多种生物学效应。除了茶多酚外，槲皮素是研究最广泛的多酚之一，动物模型和人体临床试验结果都证明了其抗炎活性[193]。口服给药时，槲皮素糖苷的体内生物活性远高于其他形式的槲皮素[194]。据报道，绿原酸可抑制人肠上皮衍生的Caco-2细胞分泌趋化因子IL-8，并减少DSS诱导的结肠炎小鼠肠道中的趋化因子MIP-2水平[194]。尽管目前的临床研究尚未完全证实多酚对人体的免疫调节作用，但一项系统评价报告称，富含类黄酮的食物有降低感染性炎症患者血清TNF-α水平的作用[195]。

二、生酮饮食

生酮饮食是一种高脂肪、极低碳水化合物的等热量饮食，其特点是碳水化合物消耗量非常低（占总卡路里摄入量的 5% ～ 10%），但酮的产量增高。它最初被用于治疗难治性儿童癫痫，肠道菌群对生酮饮食的反应似乎对癫痫儿童的治疗起到了作用[196]。近年来，生酮饮食的好处已经超出了控制癫痫发作的范围，这种饮食通常被用于减肥，并已被证明可以延长实验动物的寿命，减少疾病的发生[197]。相反，一些生酮饮食的临床研究表明，生酮饮食对微生态和肠道健康有负面影响。然而，这些研究是在具有特定代谢条件的小队列中进行的[198，199]，限制了对较大人群的推广。由于生酮饮食正在迅速普及，因此我们有必要确定它们的长期安全性以及对肠道菌群和肠道环境的影响[200]。

在肥胖患者中，生酮饮食似乎是有效的减肥饮食疗法。然而，减肥后保持体重通常是一个主要问题。在接受生酮饮食治疗的患者中，比较他们在饮食 3 个月前和 3 个月后的粪便微生物群组成，发现脱硫弧菌增加，这与肠道炎症加剧有关[199]。另一项研究分析了生酮饮食开始前和 3 个月后儿童微生物菌群的分类变化，发现 α- 多样性在饮食过程中没有显著变化。然而，在分类学和功能成分上都检测到了差异，双歧杆菌、大肠埃希菌和直肠真杆菌的丰度下降，而大肠埃希菌的丰度增加[201]。Lindefeldt 等[201]在生酮饮食后 1 周的难治性癫痫儿童中发现，生酮饮食后肠道菌群的丰富度降低，拟杆菌增加，变形杆菌减少。在属水平上，生酮饮食后拟杆菌、双歧杆菌和普雷沃菌增加，而迟缓芽孢杆菌减少[202]。Olson 等的研究[203]更进一步显示了生酮饮食在小鼠体内的抗癫痫作用，其作用由嗜黏蛋白阿克曼菌和副拟杆菌属介导，涉及全身 γ- 谷氨酰化氨基酸的变化和海马 γ 氨基丁酸 / 谷氨酸水平的升高。生酮饮食导致碳水化合物摄入量减少，并导致多糖含量减少，同时有益的肠道菌群（如双歧杆菌）减少。此外，所有这些研究都表明，在生酮饮食过程中，嗜黏蛋白阿克曼菌和大肠埃希菌等微生物的数量都有所增加。因此，虽然生酮饮食对多种疾病都有积极影响[201]，但我们应该考虑生酮饮食对肠道菌群组成的长期影响，进而对黏液层动态平衡和免疫功能的长期影响。尤其是在采用生酮饮食减肥的健康受试者中，有必要进行进一步的研究，以了解肠道菌群变化在生酮饮食过程中的作用及其治疗效果[204]。

最近的一项研究揭示了短期生酮饮食对运动员在休息和剧烈运动后的适应性和黏膜免疫的影响[205]。生酮饮食初始约 4 周的适应期期间，似乎在基因和蛋白质水平上改变了一些免疫标志物的表达，在体外实验中促进了对抗原的促炎和抗炎 T 细胞相关的细胞反应，同时在体内发挥黏膜保护作用。并且，短期生酮饮食不会降低运动员在休息状态时对皮质醇的反应，因而不会影响运动员的专业能力[206]。此外，最新的临床前数据表明，在多发性硬化症小鼠模型中，生酮饮食可以调节免疫、降低疾病严重程度并促进髓鞘再生[207]。

三、微量营养素

大量研究表明，锌、硒、铁、铜、β 胡萝卜素、维生素 A、维生素 C、维生素 E 和叶酸等微量营养素可以影响免疫系统的多个组成部分，并在疾病预防和促进健康中发挥作用。某些微量营养素如维生素 C、维生素 E、锌和铁已被证明对治疗幽门螺杆菌感染有效，并有助于调节免疫反应，同时降低致癌风险[208]。

1. 锌　锌的主要膳食来源是动物产品，如肉、鱼、蛋和奶制品，全谷物、坚果和豆类中也含有锌[209]。与植物来源的锌相比，动物来源的锌具有更高的生物利用度[210]。

锌具有调节炎症活动、抗病毒和抗氧化功能[211]。在大鼠模型中，锌缺乏会增加氧化应激、促炎性 TNF-α 和血管细胞黏附分子 VCAM-1 的表达，并导致肺组织重塑，而锌补充剂部分逆转了这些损害[212]。缺锌还可以调节呼吸道病毒的侵入、融合、复制、病毒蛋白翻译和病毒出芽[211, 213]。研究表明，口服补充锌可使急性呼吸道感染的发生率降低 35%，由于其抗炎、抗氧化和直接抗病毒作用，锌被认为对新型冠状病毒 COVID-19 感染具有潜在的支持治疗作用[214, 215]。

锌对免疫系统的影响是多方面的，固有免疫和获得免疫都受到锌的影响。锌可以诱导骨髓单核细胞黏附到内皮[216]，缺锌对骨髓有显著影响，缺锌会减少有核细胞的数量以及淋巴前体细胞的数量和比例[217, 218]。因此，即使在免疫反应的最早期，锌也是必不可少的。缺锌会影响中性粒细胞的募集，降低中性粒细胞的趋化性。缺锌还会导致自然杀伤细胞的活性降低，巨噬细胞和中性粒细胞的吞噬作用减弱，以及氧化爆发的受损[219, 220]。研究发现，由于肠道对锌的吸收减少导致的肠病性肢端皮炎患者，会出现胸腺萎缩、淋巴细胞发育受损、CD4 细胞数量减少，以及淋巴细胞反应性和迟发型超敏反应降低[221, 222]。

轻度或中度锌缺乏或实验性锌缺乏导致胸腺素活性降低、自然杀伤细胞活性降低、 CD4/CD8 比值降低、淋巴细胞增殖减少、IL-2 产生减少、迟发型超敏反应降低；所有这些反应都可以通过补充锌来纠正[223]。人体试验性锌缺乏会减少有丝分裂原刺激的淋巴细胞 IL-2、IFN-γ 和 TNF-α 产生[224]。同时，给早产低出生体重婴儿食用锌（每天 1mg/kg，连续 30 天）会增加其体内循环 T 淋巴细胞的数量和淋巴细胞增殖[169]。另一项研究发现，为早产低出生体重婴儿提供锌 5mg/d，持续 6 个月后其体内细胞介导的免疫功能增强，并且其胃肠道和上呼吸道感染的发生率降低[225]，而低剂量的锌（1 mg/d）则没有此效果。

在分子水平上，NK 细胞上的 p58 杀伤细胞抑制受体与靶细胞上的主要组织相容性复合物 I 类分子（主要是人类白细胞抗原 C）之间的相互作用需要锌[226]，缺锌会导致 NK 细胞杀伤活性受到抑制。在与镰状细胞病相关的缺锌患者中，自然杀伤细胞的活性降低，但可以通过补锌恢复正常[227]。因此，维持自然杀伤细胞的正常功能需要锌，缺锌可能导致非特异性杀伤活性和功能丧失。依赖锌的不仅是免疫系统的增殖，病原体的增殖也依赖于锌。因此，血浆中的锌减少是应对感染的急性期反应。此外，中性粒细胞降解释放的 S-100 钙离子结合蛋白——钙网蛋白，通过锌螯合来抑制细菌和白念珠菌的繁殖[228-230]。

2. 膳食抗氧化剂——维生素 A　吞噬细胞会产生活性氧（自由基），这是人体防御感染的一部分。这些物质可能会损坏免疫细胞，削弱细胞之间的通信，从而削弱免疫反应性。除了内源性氧化应激外，暴露于环境中的氧化剂和自由基（如香烟烟雾、紫外线和臭氧）也会导致体内氧化剂的水平升高。从饮食中可以获得许多抗氧化剂，但是需要足够量的中和抗氧化剂，以防止吞噬细胞产生的活性氧损害免疫细胞。既往研究表明，富含抗氧化剂的饮食与减少癌症发病率之间存在联系，至少部分原因是由于抗氧化剂可以增强人体的免疫系统，抵抗吞噬细胞产生的有毒产物（即活性氧）[231]。

维生素 A 影响许多不同类型的免疫细胞。维生素 A 缺乏会导致吞噬功能缺陷（如趋化性、黏附性及中性粒细胞中产生活性氧代谢产物的能力不足）、T 细胞和 B 细胞功能受损、NK 细胞活性降低、减少 IFN 的产生[231]。维生素 A 缺乏还会削弱肠上皮的完整性，影响免疫应答反应，从而导致细菌易位（如肠道细菌向肠外器官的移动），甚至可能引起全身感染[232]。

维生素 A 在先天免疫细胞的分化、成熟和功能调节中起着至关重要的作用。先天性免疫细胞由巨噬细胞和中性粒细胞组成，它们通过吞噬作用和对自然杀伤性 T 细胞的激活，从而引发对病原体入侵的即时反应，而自然杀伤性 T 细胞则通过细胞毒性活性发挥免疫调节功能。有报告显示，维生素 A 对结肠 $CD169^+$ 巨噬细胞的正常发育和分化至关重要 [233]。巨噬细胞主要包括分泌促炎细胞因子的 M1 巨噬细胞和表达抗炎因子的 M2 巨噬细胞。维 A 酸是体内维生素 A 的代谢中间产物，全反式维 A 酸通过诱导单核细胞向巨噬细胞谱系分化而抑制炎症反应，同时抑制巨噬细胞释放炎症因子，从而诱导骨髓中的 M1 巨噬细胞转化为 M2 巨噬细胞 [234, 235]。全反式维 A 酸作用于中性粒细胞核中的维 A 酸受体，通过激活 mTOR 信号通路诱导中性粒细胞分化和异质性。该途径增强了中性粒细胞的细胞外捕获能力和细胞毒性，从而可以有效杀死肿瘤细胞 [236]。通过下调 IFN-γ 的表达水平和上调 IL-5 的分泌，维 A 酸在天然杀伤性 T 细胞的早期分化阶段起调节作用。

树突状细胞是强大而通用的抗原呈递细胞，是免疫系统的专职哨兵，能够协调先天性和适应性免疫反应 [237]。骨髓驻留前树突状细胞具有分化为黏膜前树突状细胞的潜力，其特征是肠归巢受体的表达。全反式维 A 酸在肠道固有的树突状细胞发育中以细胞内方式起作用，以实现肠道 $CD103^+$ 细胞的定向分化和迁移 [238]。前树突状细胞可以迁移到脾脏，全反式维 A 酸使其分化趋向于 $CD11b^+CD8^-$，而不是 $CD11b^-CD8\alpha^+$ [239]。目前的普遍共识是，全反式维 A 酸促进肠道树突状细胞的抗炎表型特征 [240, 241]。然而，在存在 IL-15 的情况下，全反式维 A 酸是促进树突状细胞促炎性细胞因子 IL-12 和 IL-23 分泌的佐剂 [242]，并且具有不可预见的辅佐剂特性，可诱导 Th1 细胞对饮食抗原的免疫反应。这表明在与肠黏膜中 IL-15 和 IL-6 诱导相关的感染条件下，全反式维 A 酸也会促进 Th17 免疫反应 [243]。这些观察结果提示我们，不要将维生素 A 和全反式维 A 酸用于高水平 IL-15 相关的自身免疫性疾病和炎性肠道疾病的治疗 [244]。

先天性淋巴细胞（innate lymphoid cell，ILC）是不同于 T 细胞和 B 细胞的淋巴细胞亚群。ILC 位于肠黏膜表面，可增强免疫反应，维持黏膜完整性并促进淋巴器官形成。ILC 可以分为 3 组：ILC1、ILC2 和 ILC3。ILC3 的特征是转录因子 RORγt 和细胞因子 IL-22 和 IL-17 的表达 [245]。在胎儿期，继发性淋巴器官的形成取决于 ILC3 的一个亚群，称为淋巴组织诱导物（lymphoid tissue inducer，LTi）[245, 246]。胎儿 ILC3 受子宫内细胞自主性维甲酸信号控制，从而预设了成年后的免疫适应性。胚胎淋巴器官包含 ILC 祖细胞，可局部分化为成熟的 LTi 细胞。IL-22 和 IL-17 都介导抗菌免疫反应，并防止细菌跨壁转运。ILC3 的异常调节，尤其是 IL-17 的表达异常是慢性胃肠道炎症的潜在驱动因素 [247, 248]。与饲喂维生素 A 的小鼠相比，缺乏维生素 A 的动物 ILC3 数量减少。ILC3 减少会影响肠道免疫功能，这些小鼠更容易受到病原菌啮齿类柠檬酸杆菌的感染 [249]。维甲酸通过促进 γδT 细胞和 ILC 合成 IL-22，从而塑造早期肠道免疫反应 [250]。

3. 膳食抗氧化剂——维生素 E　维生素 E 是一种脂溶性抗氧化剂，可以保护细胞膜中的多不饱和脂肪酸免受氧化，调节活性氧（reactive oxygen species，ROS）和活性氮（reactive nitrogen species，RNS）的产生，并调节信号转导。在正常和疾病条件下的动物和人体模型中均已观察到维生素 E 的免疫调节作用。

维生素 E 缺乏会减少动物中性粒细胞的脾淋巴细胞的增殖，降低 NK 细胞活性和吞噬作用。维生素 E 缺乏症还会增加动物对传染性病原体的敏感性。实际上，对鸡、火鸡、小鼠、绵羊、猪和牛的研究表明，维生素 E 摄入量的增加会增强动物对病原体的抵抗力。值得注意的是，如果给动物饲喂含有大量多不饱和脂肪酸的饮食，则维生素 E 缺乏症的影响更为明显。此外，发挥作用所需维生素 E 的最适剂量取决于年龄（即随着年龄的增长而增加，部分原因是长期与自由基接触）。饮食中过量的维生素 E 会损害免疫功能：每天 300mg 的维生素 E 会降低中性粒细胞的吞噬能力，减少单核细胞呼吸暴发和 IL-13 的产生 [3]。膳食补充维生素 E 还可以减轻大鼠低氧引起的肠道损伤 [251]。

NK 细胞活性与维生素 E 有关。生育酚为维生素 E 的水解产物，患有严重维生素 E 缺乏症的 Shwachman 综合征男孩的 NK 细胞活性较低，补充 α- 生育酚 8 周后（100mg/d），其 NK 细胞活性有所改善。但停止补充 α- 生育酚后，NK 细胞活性和 $CD16^+CD56^+$ 细胞减少。恢复补充 α- 生育酚 8 周后，NK 细胞活性和 $CD16^+CD56^+$ 细胞数量恢复 [252]。在 37 位 90 ～ 106 岁的女性中，NK 细胞的细胞毒性与血浆维生素 E 浓度呈正相关 [253]。大肠癌患者补充 2 周维生素 E，可使大部分 NK 细胞活性增加，但尚不明确维生素 E 改善 NK 细胞活性的机制 [254]。

NO 似乎与 NK 细胞功能受损有关，NK 细胞和 MDSCs 的共培养显示，NK 细胞的细胞毒性和分泌 IFN-γ 的能力受到 MDSCs 的损伤，而诱导型一氧化氮合酶的抑制可以挽救 MDSCs 导致的损伤。使 NK 细胞暴露于 NO，可导致 $CD16^+$NK 细胞酪氨酸残基硝化。这些结果表明，MDSCs 通过产生 NO 和酪氨酸残基的硝化作用损害 NK 细胞的功能 [255]。维生素 E 可能通过调节 NO 水平，发挥其对 NK 细胞功能的调控作用 [256]。

感染流感病毒的小鼠和结直肠癌患者中，补充维生素 E 可增强 Th1 型免疫反应 [257, 258]。结直肠癌患者补充 2 周维生素 E，可使产生 IL-2 的 $CD4^+$ T 细胞数量增加，并且使 IFN-γ 的分泌增加 [257]。感染流感病毒的小鼠中，在感染前 8 周补充维生素 E 可降低肺中的病毒滴度，而维生素 E 的这种保护作用与 Th1 型免疫反应增强有关。IFN-γ 的产生水平与病毒滴度呈负相关，饲喂维生素 E 饮食的小鼠产生的 IFN-γ 和 IL-2 水平明显增高 [258]。

4. 维生素 C　维生素 C（抗坏血酸）是一种水溶性维生素，是人体必需的微量营养素 [259]。维生素 C 的主要来源是柑橘类水果、西红柿、马铃薯和绿叶蔬菜。膳食维生素 C 的摄取依赖于食物，很容易因食物的长期储存、过度烹饪和加工而被破坏。母乳是新生儿和婴儿维生素 C 的主要来源 [210]。维生素 C 是一种抗氧化剂，参与多种生物过程：胶原蛋白的合成、神经递质代谢、胆固醇代谢、脂肪酸转运（肉碱的合成）、作为金属酶的辅助因子。此外，维生素 C 会影响造血系统的细胞和免疫功能，因为它可以增强非血红素铁的吸收、将铁从转铁蛋白转移到铁蛋白以及形成四氢叶酸 [260]。

维生素 C 可通过抑制 3- 磷酸甘油醛脱氢酶（glyceraldehyde 3-phosphate dehydrogenase，GAPDH）来防止免疫细胞过度活化。糖酵解酶 GAPDH 可以调节活化的髓样和淋巴细胞中的糖酵解速率 [261]。维生素 C 可以在细胞内和细胞外氧化成无活性的脱氢抗坏血酸 [262]。在细胞内，脱氢抗坏血酸被还原为抗坏血酸，而还原型谷胱甘肽被氧化 [263]。氧化型谷胱甘肽（二硫化谷胱甘肽）可以被烟酰胺腺嘌呤二核苷酸磷酸（nicotinamide adenine dinucleotide phosphate，NADPH）还原为谷胱甘肽 [263]。维生素 C 具有抗氧化能力；然而，高剂量维生素 C 可以通过减少包括还原型谷胱甘肽和 NADPH 在内的 ROS 清除系统来促进氧化活性 [264]，ROS 增高可诱导 DNA 损伤 [38]。因此，高剂量维生素 C 对 GAPDH 的抑制可能会通过减少细胞中三磷酸腺苷的产生，降低免疫细胞的活化 [261]。

维生素 C 有助于增强体液和细胞免疫反应 [259]。维生素 C 在吞噬细胞（如中性粒细胞）

中积累，可以增强趋化性、吞噬作用和 ROS 的产生，并最终杀死微生物[265]。维生素 C 在淋巴细胞中的作用尚不清楚，但已经证明维生素 C 可以增强 B 细胞和 T 细胞的分化和增殖，这可能是由于其基因调节作用。维生素 C 对细胞因子产生的影响似乎取决于细胞类型和（或）炎症刺激因子。最近的研究表明，维生素 C 治疗会减弱促炎细胞因子 TNF、IL-6 和 IL-1β 的合成[265]。体外研究表明，维生素 C 可作为人体抗体产生（IgM 和 IgG）的有效免疫刺激剂，细胞内维生素 C 的含量是建立外周血淋巴细胞免疫反应的关键因素[266]。维生素 C 具有促进 T 细胞成熟的作用[267]，最近的研究表明，维生素 C 可能通过组蛋白去甲基化的表观遗传等机制，在调节 T 细胞成熟中发挥作用[268]。

最近国内武汉大学一项研究表明，静脉注射维生素 C 可以抑制 COVID-19 引起的细胞因子风暴，改善肺功能，并降低 COVID-19 急性呼吸窘迫综合征的风险[269]。

5. 维生素 D　维生素 D 是一种脂溶性维生素，主要通过暴露于阳光中的紫外线后在皮肤中合成（以维生素 D_3 的形式），少部分来源于膳食摄入的维生素 D_2 或维生素 D_3（维生素 D 的主要来源是富含脂肪的鱼、鱼油、蛋黄、奶酪和维生素 D 强化食品）。维生素 D 在皮肤中产生或通过胃肠道吸收后，通过维生素 D 结合蛋白转运至肝脏。在肝脏中，维生素 D 转化为 25- 羟基维生素 D[25 hydroxy vitamin D，25（OH）D]。然后，25- 羟基维生素 D 被输送到肾脏，在那里最终转化为其活性形式：1，25- 二羟基维生素 D[1, 25 dihydroxyvitamin D; 1，25（OH）2D][210]。

活性维生素 D[1，25（OH）2 D] 是一种类固醇激素，对人体免疫力具有重要的影响[270]。作为免疫系统调节剂，1，25- 二羟维生素 D 能防止炎症细胞因子的过度表达，并增加巨噬细胞“氧化爆发”的潜力[271]。更重要的是，它刺激了中性粒细胞、单核细胞、自然杀伤细胞和呼吸道上皮细胞中强效抗菌肽的表达，这在保护肺部免受病原菌感染方面发挥着重要作用[272]。维生素 D 还能降低 1 型促炎症细胞因子的表达，例如 IL-12、IL-16、IL-8、TNF-α、IFN-γ，同时增加 2 型细胞因子的水平，例如 IL-4、IL-5、IL -10 和调节性 T 细胞[209]。由于这些原因，维生素 D 缺乏会使增加呼吸道感染的风险[209, 273]。维生素 D 还可以调节 T 细胞中 FOXP3 的表达，从而诱导该细胞分化为具有免疫抑制活性的 FOXP3 调节性 T 细胞，促进抗炎细胞因子 IL-10 的分泌[259]。

近几十年来，研究证实了维生素 D 与先天性和获得性免疫系统之间的重要相互作用[213, 274]。数据表明，包括免疫细胞在内的大多数组织细胞都表达维生素 D 代谢酶，为维生素 D 的天然形式向活性形式的转化提供了一种生物学上的可信机制。这一过程对正常的免疫功能至关重要，因此维生素 D 水平不足可能导致免疫反应失调。关于维生素 D 水平是否与自身免疫性疾病的发生风险有关，以及补充维生素 D 是否可以改变自身免疫性疾病的病程问题，最近的一项系统综述认为：过去 40 年的几项研究支持维生素 D 在预防自身免疫性疾病中的作用，但这一领域仍然缺乏随机对照临床试验[275]。综合所有现有的证据，维生素 D 是一种有前途的、相对安全的营养物质，可用于预防和辅助治疗免疫稳态受损引起的疾病[276]。

维生素 D 代谢酶和维生素 D 受体存在于许多细胞类型中，包括各种免疫细胞，例如抗原呈递细胞、T 细胞、B 细胞和单核细胞[277]。体外研究显示，除了调节先天性免疫细胞外，维生素 D 还可以增强免疫原性。

四、核苷酸

和许多氨基酸一样，核苷酸被认为是对人

体有益的。多年来，核苷酸被添加到婴儿配方食品中以改善免疫功能。在快速生长或营养摄入有限的时期，或在某些胃肠道疾病状态下，经膳食摄入可以使机体无须从头合成核苷酸，从而使组织代谢水平达到充分的工作条件[278]。

核苷酸及其代谢产物是人体几个生物学过程中的关键成分。从头合成是健康人合成的主要途径，主要发生在肝脏，仅需饮食核苷酸的5%，几乎不依赖于外源核苷酸。但是，人体对外源核苷酸的需求可能会相差很大，并且在某些情况下可能会增加。例如，当机体从严重组织损伤中恢复时，例如大手术、全身感染或广泛烧伤，或者当身体快速生长时，例如婴儿早期或青春期快速生长时，可能需要外源核苷酸。当肝功能受到抑制时，人体对饮食中核苷酸的需求也可能会大大增加。喂食无核苷酸饮食的大鼠中，小肠中的RNA和蛋白质水平显著降低，表明在核苷酸相对缺乏的状态下，DNA含量的维持会以肠道RNA和蛋白质池的损失为代价[279]。

众所周知，人母乳中含有大量核苷酸，这与母乳喂养婴儿胃肠道的生长发育有关[280]。这些发现进一步表明，饮食核苷酸在某些情况下对维持人体最佳功能非常重要。饮食核苷酸对免疫系统的重要性尚未完全了解。但是，淋巴细胞是免疫系统的关键组成部分，淋巴细胞处于细胞周期的某些时期时，不能有效地合成核苷酸，它们依赖于其他器官（主要是肝脏）从头合成的核苷酸[281]。因此，在如上所述的压力条件下，当人体对核苷酸的需求量超过了从头合成获得的水平时，饮食中的核苷酸可能在维持淋巴细胞的最佳功能方面发挥了重要作用。

免疫系统通常分为两个主要部分：细胞免疫和体液免疫。重要的是，这两个系统都必须正常工作，才能有效抵抗病原体入侵并抑制肿瘤发生。在一系列研究中，鲁道夫及其同事表明，长期给动物饲喂无核苷酸饮食可以显著抑制其细胞免疫[281，282]。使用鼠类脾细胞的体外研究证明，酵母RNA制备物显著提高了T细胞依赖性抗原的抗体生成能力[283]。这一发现进一步证实了核苷酸对动物和人类体液免疫反应的影响[284]。体内研究结果表明，在核苷酸相对缺乏的状态下，单核苷酸和核苷可能会迅速整合到组织核苷酸库中，从而恢复机体的免疫功能。喂食无核苷酸饮食的小鼠中，人体免疫应答在较短的时间内恢复是令人鼓舞的，因为这表明在某些临床情况下，外源核苷酸可快速恢复免疫功能[284]。

五、植物化学物质

所有植物性食品中都含有植物化学物质，它们是植物中含有的生物活性小分子，例如姜黄素、生物黄酮类、胡萝卜素类、芥子油苷、有机硫化物、黄酮类、吲哚类等。这些有益的物质在下列食物中含量特别丰富：全谷类、绿菜花、甘蓝、菜花、柑橘类水果、深绿色叶蔬菜、蒜、茶、洋葱、西红柿等。

姜黄素具有广泛的生物作用，包括抗细菌、抗病毒、抗真菌、抗氧化和抗炎活性。姜黄素对多种病毒具有抗病毒作用，包括流感病毒、腺病毒、肝炎病毒、人乳头瘤病毒、人类免疫缺陷病毒、单纯疱疹病毒-2和寨卡病毒。姜黄素通过各种机制发挥抗病毒作用，例如抑制病毒进入细胞、抑制病毒和病毒蛋白酶的包裹、抑制病毒复制以及调节多种信号通路[285]。最近的研究表明，姜黄素可抑制ACE2，调节脂质双层的特性，以及抑制病毒进入细胞的病毒S245蛋白[285，286]，抑制病毒蛋白酶[287]，刺激宿主干扰素的产生并激活宿主先天免疫等[286]。这些研究揭示了姜黄素潜在的增强免疫、抗氧化和抗病毒作用。因此，姜黄素可能是治疗COVID-19的潜在药物[209]。

类黄酮是一种天然化合物，存在于水果、

蔬菜和植物性食物中，如茶、巧克力和葡萄酒。黄酮类化合物会被人体肠道微生物分解，且与心血管疾病之间存在联系。研究发现，摄入富含类黄酮食物最多的参与者（包括浆果、红酒、苹果和梨），比摄入富含类黄酮食物最少的参与者的收缩压水平更低，肠道微生物群的多样性更丰富。摄入富含类黄酮的食物和收缩压水平密切相关，高达 15.2% 的关联可以用肠道微生物群的多样性来解释 [288]。

六、奶酪

奶酪是一种牛奶衍生产品，富含脂肪、蛋白质、必需矿物质和维生素等营养物质，通过使牛奶变酸或与肾素凝结而成 [289]。奶酪加工形成具有多种风味、质地和营养价值的多功能食品 [84]。除了半胱氨酸和蛋氨酸外，奶酪富含所有必需氨基酸，足以满足推荐的人类摄入量。富含必需氨基酸的蛋白质膳食可以调节餐后血糖并增加饱腹感，因为它们在胃部的停留时间和胃肠转运时间更长 [290]。此外，奶酪还可以减少餐后脂肪生成和炎症活动 [91]。这表明高质量奶酪衍生的营养物质的重要性，一方面提供热量和饱和脂肪，另一方面可以纳入具有免疫调节作用的抗炎饮食 [291]。斯坦福大学医学院的研究团队利用组学分析，包括最先进的免疫分析，研究两种饮食（高发酵或高纤维食物）对人类免疫系统的影响。结果发现，食用发酵食品（例如奶酪、酸奶、酸乳酒、康普茶、蔬菜盐水饮料和泡菜等）的参与者许多炎症指标水平下降，同时微生物群的多样性增加。这些结果表明，发酵食品可能是人类微生物 - 免疫系统轴的强大调节器，并为人类提供了一个对抗非传染性疾病的可能途径 [292]。

第四节 食品添加剂与肠道菌群

随着以西式饮食为主的加工食品的发展，近几十年来，业界批准的非营养性甜味剂、乳化剂等食品添加剂数量激增 [293]。食品添加剂对人类健康具有潜在的影响，但是，食品添加剂对肠道菌群和肠道稳态的影响是目前研究不足的领域。虽然西方饮食对微生物和健康的影响通常归因于大量营养素的含量和组成，例如三大营养物质，但一些研究表明，西方饮食对机体的有害影响可能是由食品添加剂造成的。例如，食品添加剂通过影响肠道菌群，进而增加机体患炎症性肠病的风险 [294]。此外，在没有其他饮食影响的情况下，给予饮食乳化剂聚山梨酸酯 80 和羧甲基纤维素可以诱导小鼠肥胖、肠道炎症和代谢功能障碍 [295]。炎性肠病患者食用乳化剂会增加肠道屏障通透性和肠道炎症 [294]。

食品乳化剂，如卵磷脂、脂肪酸单甘酯和双甘酯，可以增加细菌跨上皮细胞移位，促进全身炎症，改变肠道菌群的定位和组成 [295]。乳化剂的摄入降低了肠道微生物多样性，降低了拟杆菌的丰度，增加了嗜黏蛋白阿克曼菌、变形杆菌和包括瘤胃球菌在内的黏液溶解菌的丰度。这些肠道菌群的改变导致了微生态失调和慢性肠道炎症，增加了结肠炎和代谢综合征风险 [295]。除了乳化剂，非营养甜味剂也与肠道相关的代谢改变有关。在一系列动物实验和人体试验中，食用非营养甜味剂通过改变肠道菌群，诱导受试者葡萄糖不耐受 [296]。然而，非营养甜味剂对肠道菌群和代谢功能的确切影响尚不清楚 [200]。

人工甜味剂被加入到几乎所有加工食品中，其目的通常是为了提高食品的稳定性和延长其保质期，并改善其口味和质地。人造甜味剂最初是作为一种对健康无害的食品销售的，可以用来代替天然糖。然而最新证据表明，各种人造甜味剂实际上比葡萄糖和蔗糖更容易引起葡萄糖不耐受。目前认为，人造甜味剂是通过改变肠道菌群从而介导这种不良作用。例如，糖精喂养的小鼠患有肠道营养不良，拟杆菌的相对丰度增加，罗伊氏乳杆菌的丰度相对减少[296]。这些微生物变化与摄入天然糖（葡萄糖、果糖和蔗糖）引起的变化相反。

上述证据表明，与大众普遍的看法相反，人造甜味剂实际上可能比天然糖更加不健康。许多研究已经表明，食用人工甜味剂会改变肠道菌群，并在宿主体内诱导肠道菌群介导的各种不良反应（如葡萄糖耐量异常）[296, 297]。无热量人工甜味剂改变了微生物的代谢途径，这些变化与宿主对代谢性疾病的易感性有关[296]。在这项研究中，健康志愿者服用了1周的糖精，剂量为5mg/kg体重，他们出现了较差的葡萄糖耐量。无热量人工甜味剂的摄入增加了拟杆菌和某些梭菌的丰度，而降低了某些梭菌、双歧杆菌和乳杆菌的丰度[297, 298]。另一项研究报道，食用低剂量（每天5～7mg）阿斯巴甜的大鼠中，肠道菌群发生了显著变化，例如肠杆菌科细菌和亮色梭菌数量增加，空腹血糖水平升高，胰岛素反应受损[299]。

体外研究表明，加工过的单糖会减弱白细胞的吞噬作用，并可能增加血液中的炎症细胞因子水平[300, 301]。也有研究持相反观点，认为甜味剂具有抗炎作用[302, 303]。对新型甜味剂的研究很少，但有限的细胞培养证据表明，甜菊糖具有抗炎特性，同时可改善T细胞和B细胞的吞噬作用和有丝分裂原反应[304-306]。

第五节　小结与展望

营养不足会对免疫系统产生负面影响，从而增加机体患病风险，并会加重疾病的严重程度、增加并发症。因此，保持最佳营养状态对于人类预防各种疾病（包括流感和COVID-19）至关重要。通过这一章节内容，我们了解到微量营养素，包括锌、维生素A、维生素C、维生素D、维生素E，以及三大营养物质糖、脂肪、蛋白质，还有膳食纤维和植物化学物质，对于维持我们机体的正常免疫功能至关重要。虽然这些分子可以直接调节免疫系统，但也可以通过间接机制提高和维持免疫力，例如改变肠道微生物群的组成和功能，以及增加短链脂肪酸的产生。

我们仍然需要进一步的研究，评估各种营养素在对抗病毒感染的免疫反应中的协同作用。这些营养物质应与健康饮食相辅相成，并应在营养专家和相关机构推荐的安全范围内食用。保持最佳营养状态，补充微量营养素（例如锌）和ω-3脂肪酸，可能是一种有效且低成本的方式，并可能有助于减轻全球性传染病（包括COVID-19）的负担。尽管如此，我们依然要认识到，营养补充不一定能预防感染或治愈疾病，可能只是有助于减轻症状并促进康复[210]。

目前已经证明饮食中的不同成分可以作用于免疫系统，对机体的免疫功能产生有益（抑制炎症）或有害（促进炎症）的影响，从而与多种疾病的发生发展和转归相关联，因此，饮食可能是影响患者对免疫疗法反应的重要因素。尽管一些饮食因素（例如维生素D）对免疫疗法有效性的影响是众所周知的，但仍有许多饮食因素（例如碳水化合物摄入量）对免疫疗法的影响仍不确定，需要进一步深入研究。目前人类掌握的信息不足以完全确定不同的饮食成分如何影响免疫疗法的有效性，阐明膳食成分对机体免疫功能的影响及其机制，将成为未来有趣的研究方向和重要的研究领域[307]。

参考文献

[1] Wintergerst ES, Maggini S, Hornig DH. Contribution of selected vitamins and trace elements to immune function. Annals of Nutrition and Metabolism, 2007, 51(4): 301-323.

[2] Iddir M, Brito A, Dingeo G, et al. Strengthening the immune system and reducing inflammation and oxidative stress through diet and nutrition: considerations during the covid-19 crisis. Nutrients, 2020, 12(6): 1562.

[3] Childs CE, Calder PC, Miles EA. Diet and immune function. Nutrients, 2019, 11(8): 1933.

[4] Chen O, Mah E, Dioum E, et al. The role of oat nutrients in the immune system: a narrative review. Nutrients, 2021, 13(4): 1048.

[5] Rodríguez L, Cervantes E, Ortiz R. Malnutrition and gastrointestinal and respiratory infections in children: a public health problem. International Journal of Environmental Research and Public Health, 2011, 8(4): 1174-1205.

[6] Andersen CJ, Murphy KE, Fernandez ML. Impact of obesity and metabolic syndrome on immunity. Advances in Nutrition, 2016, 7(1): 66-75.

[7] Calder PC, Kew S. The immune system: a target for functional foods? British Journal of Nutrition, 2002, 88(S2): S165-S176.

[8] Calder PC, Carr AC, Gombart AF, et al. Optimal nutritional status for a well-functioning immune system is an important factor to protect against viral infections. Nutrients, 2020, 12(4): 1181.

[9] Del Cornò M, Varì R, Scazzocchio B, et al. Dietary fatty acids at the crossroad between obesity and colorectal cancer: fine regulators of adipose tissue homeostasis and immune response. Cells, 2021, 10(7): 1738.

[10] Bayer F, Dremova O, Khuu MP, et al. The interplay between nutrition, innate immunity, and the commensal microbiota in adaptive intestinal morphogenesis. Nutrients, 2021, 13(7): 2198.

[11] Flach M, Diefenbach A. Development of gut-associated lymphoid tissues. Mucosal Immunology. Elsevier, 2015: 31-42.

[12] Ma LJ, Acero LF, Zal T, et al. Trans-presentation of il-15 by intestinal epithelial cells drives development of cd8$\alpha\alpha$ iels. The Journal of Immunology, 2009, 183(2): 1044-1054.

[13] Hu MD, Ethridge AD, Lipstein R, et al. Epithelial il-15 is a critical regulator of $\gamma\delta$ intraepithelial lymphocyte motility within the intestinal mucosa. The Journal of Immunology, 2018, 201(2): 747-756.

[14] Hyun J, Romero L, Riveron R, et al. Human intestinal epithelial cells express interleukin-10 through toll-like receptor 4-mediated epithelial-macrophage crosstalk. Journal of Innate Immunity, 2015, 7(1): 87-101.

[15] Peterson LW, Artis D. Intestinal epithelial cells: regulators of barrier function and immune homeostasis. Nature Reviews Immunology, 2014, 14(3): 141-153.

[16] Everard A, Geurts L, Caesar R, et al. Intestinal epithelial myd88 is a sensor switching host metabolism towards obesity according to nutritional status. Nature Communications, 2014, 5(1): 1-12.

[17] Shulzhenko N, Morgun A, Hsiao W, et al. Crosstalk between b lymphocytes, microbiota and the intestinal epithelium governs immunity versus metabolism in the gut. Nature Medicine, 2011, 17(12): 1585-1593.

[18] Mowat AM. Anatomical basis of tolerance and immunity to intestinal antigens. Nature Reviews Immunology, 2003, 3(4): 331-341.

[19] Rescigno M. Intestinal microbiota and its effects on the immune system. Cellular Microbiology, 2014, 16(7): 1004-1013.

[20] Makki K, Deehan EC, Walter J, et al. The impact of dietary fiber on gut microbiota in host health and disease. Cell Host & Microbe, 2018, 23(6): 705-715.

[21] Wrzosek L, Miquel S, Noordine M-L, et al. Bacteroides thetaiotaomicron and faecalibacterium prausnitzii influence the production of mucus glycans and the development of goblet cells in the colonic epithelium of a gnotobiotic model rodent. BMC Biology, 2013, 11(1): 1-13.

[22] Desai MS, Seekatz AM, Koropatkin NM, et al. A dietary fiber-deprived gut microbiota degrades the colonic mucus barrier and enhances pathogen susceptibility. Cell, 2016, 167(5): 1339-1353. e1321.

[23] Bergstrom KS, Kissoon-Singh V, Gibson DL, et al. Muc2 protects against lethal infectious colitis by disassociating pathogenic and commensal bacteria from the colonic mucosa. Plos Pathogens, 2010, 6(5): e1000902.

[24] Schroeder BO, Birchenough GM, Ståhlman M, et al. Bifidobacteria or fiber protects against diet-induced microbiota-mediated colonic mucus deterioration. Cell Host & Microbe, 2018, 23(1): 27-40. e27.

[25] Fukuda S, Toh H, Hase K, et al. Bifidobacteria can protect from enteropathogenic infection through production of acetate. Nature, 2011, 469(7331): 543-547.

[26] Zou J, Chassaing B, Singh V, et al. Fiber-mediated nourishment of gut microbiota protects against diet-induced obesity by restoring il-22-mediated colonic health. Cell Host & Microbe, 2018, 23(1): 41-53. e44.

[27] Rothschild D, Weissbrod O, Barkan E, et al. Environment dominates over host genetics in shaping human gut microbiota. Nature, 2018, 555(7695): 210-215.

[28] Isibor PO, Akinduti P, Aworunse OS, et al. Significance of african diets in biotherapeutic modulation of the gut microbiome. Bioinformatics and Biology Insights, 2021, 15: 11779322211012697.

[29] Brodin P, Davis MM. Human immune system variation. Nature Reviews Immunology, 2017, 17(1): 21-29.

[30] Reichardt F, Chassaing B, Nezami BG, et al. Western diet induces colonic nitrergic myenteric neuropathy and dysmotility in mice via saturated fatty acid - and lipopolysaccharide - induced tlr4 signalling. The Journal of Physiology, 2017, 595(5): 1831-1846.

[31] Duan Y, Zeng L, Zheng C, et al. Inflammatory links between high fat diets and diseases. Frontiers in Immunology, 2018, 9: 2649.

[32] Pickard JM, Zeng MY, Caruso R, et al. Gut microbiota: role in pathogen colonization, immune responses, and inflammatory disease. Immunological Reviews, 2017, 279(1): 70-89.

[33] Macia L, Tan J, Vieira AT, et al. Metabolite-sensing receptors gpr43 and gpr109a facilitate dietary fibre-induced gut homeostasis through regulation of the inflammasome. Nature Communications, 2015, 6(1): 1-15.

[34] Arpaia N, Campbell C, Fan X, et al. Metabolites produced by commensal bacteria promote peripheral regulatory t-cell generation. Nature, 2013, 504(7480): 451-455.

[35] Tan J, McKenzie C, Vuillermin PJ, et al. Dietary fiber and bacterial scfa enhance oral tolerance and protect against food allergy through diverse cellular pathways. Cell Reports, 2016, 15(12): 2809-2824.

[36] Kaisar MM, Pelgrom LR, van der Ham AJ, et al. Butyrate conditions human dendritic cells to prime type 1 regulatory t cells via both histone deacetylase inhibition and g protein-coupled receptor 109a signaling. Frontiers in Immunology, 2017, 8: 1429.

[37] Frank DN, Amand ALS, Feldman RA, et al. Molecular-phylogenetic characterization of microbial community imbalances in human inflammatory bowel diseases. Proceedings of the National Academy of Sciences, 2007, 104(34): 13780-13785.

[38] Wang T, Cai G, Qiu Y, et al. Structural segregation of gut microbiota between colorectal cancer patients and healthy volunteers. The ISME Journal, 2012, 6(2): 320-329.

[39] Singh N, Gurav A, Sivaprakasam S, et al. Activation of gpr109a, receptor for niacin and the commensal metabolite butyrate, suppresses colonic inflammation and carcinogenesis. Immunity, 2014, 40(1): 128-139.

[40] Kim MH, Kang SG, Park JH, et al. Short-chain fatty acids activate gpr41 and gpr43 on intestinal epithelial cells to promote inflammatory responses in mice. Gastroenterology, 2013, 145(2): 396-406. e310.

[41] Perez-Muñoz ME, Sugden S, Harmsen HJ, et al. Nutritional and ecological perspectives of the interrelationships between diet and the gut microbiome in multiple sclerosis: insights from marmosets. Iscience, 2021, 24(7): 102709.

[42] Tan J, McKenzie C, Potamitis M, et al. The role of short-chain fatty acids in health and disease. Advances in Immunology, 2014, 121: 91-119.

[43] Herbst T, Sichelstiel A, Schär C, et al. Dysregulation of allergic airway inflammation in the absence of microbial colonization. American Journal of Respiratory and Critical Care Medicine, 2011, 184(2): 198-205.

[44] Maslowski KM, Vieira AT, Ng A, et al. Regulation of inflammatory responses by gut microbiota and chemoattractant receptor gpr43. Nature, 2009, 461(7268): 1282-1286.

[45] Vogt MC, Paeger L, Hess S, et al. Neonatal insulin action impairs hypothalamic neurocircuit formation in response to maternal high-fat feeding. Cell, 2014, 156(3): 495-509.

[46] Thorburn AN, Macia L, Mackay CR. Diet, metabolites, and "western-lifestyle" inflammatory diseases. Immunity, 2014, 40(6): 833-842.

[47] Li P, Yin Y-L, Li D, et al. Amino acids and immune function. British Journal of Nutrition, 2007, 98(2): 237-252.

[48] Wu G, Fang Y-Z, Yang S, et al. Glutathione metabolism and its implications for health. The Journal of Nutrition, 2004, 134(3): 489-492.

[49] Grimble RF. The effects of sulfur amino acid intake on immune function in humans. The Journal of Nutrition, 2006, 136(6): S1660-S1665.

[50] Bronte V, Zanovello P. Regulation of immune responses by l-arginine metabolism. Nature Reviews Immunology, 2005, 5(8): 641-654.

[51] Melchior D, Le Melchior N, Sève B. Effects of chronic

lung inflammation on tryptophan metabolism in piglets. Developments in tryptophan and serotonin metabolism. Springer, 2003: 359-362.

[52] Platten M, Ho PP, Youssef S, et al. Treatment of autoimmune neuroinflammation with a synthetic tryptophan metabolite. Science, 2005, 310(5749): 850-855.

[53] Ha E-M, Oh C-T, Bae YS, et al. A direct role for dual oxidase in drosophila gut immunity. Science, 2005, 310(5749): 847-850.

[54] Shi W, Meininger CJ, Haynes TE, et al. Regulation of tetrahydrobiopterin synthesis and bioavailability in endothelial cells. Cell Biochemistry and Biophysics, 2004, 41(3): 415-433.

[55] Wu G, Bazer F, Datta S, et al. Proline metabolism in the conceptus: implications for fetal growth and development. Amino Acids, 2008, 35(4): 691-702.

[56] Sun Y, Nonobe E, Kobayashi Y, et al. Characterization and expression of l-amino acid oxidase of mouse milk. Journal of Biological Chemistry, 2002, 277(21): 19080-19086.

[57] Wu G, Knabe DA, Flynn NE, et al. Arginine degradation in developing porcine enterocytes. American Journal of Physiology-Gastrointestinal and Liver Physiology, 1996, 271(5): G913-G919.

[58] Son DO, Satsu H, Kiso Y, et al. Inhibitory effect of carnosine on interleukin-8 production in intestinal epithelial cells through translational regulation. Cytokine, 2008, 42(2): 265-276.

[59] Hisatsune T, Kaneko J, Kurashige H, et al. Effect of anserine/carnosine supplementation on verbal episodic memory in elderly people. Journal of Alzheimer's Disease, 2016, 50(1): 149-159.

[60] Lamas B, Vergnaud-Gauduchon J, Goncalves-Mendes N, et al. Altered functions of natural killer cells in response to l-arginine availability. Cellular Immunology, 2012, 280(2): 182-190.

[61] Rodriguez PC, Quiceno DG, Ochoa AC. L-arginine availability regulates t-lymphocyte cell-cycle progression. Blood, 2007, 109(4): 1568-1573.

[62] Rodríguez PC, Ochoa AC. Arginine regulation by myeloid derived suppressor cells and tolerance in cancer: mechanisms and therapeutic perspectives. Immunological Reviews, 2008, 222(1): 180-191.

[63] Norian LA, Rodriguez PC, O'Mara LA, et al. Tumor-infiltrating regulatory dendritic cells inhibit cd8+ t cell function via l-arginine metabolism. Cancer Research, 2009, 69(7): 3086-3094.

[64] Oberlies J, Watzl C, Giese T, et al. Regulation of nk cell function by human granulocyte arginase. The Journal of Immunology, 2009, 182(9): 5259-5267.

[65] Kobayashi T, Yamamoto M, Hiroi T, et al. Arginine enhances induction of t helper 1 and t helper 2 cytokine synthesis by peyer's patch alpha beta t cells and antigen-specific mucosal immune response. Biosci Biotechnol Biochem, 1998, 62(12): 2334-2340.

[66] Dufour C, Dandrifosse G, Forget P, et al. Spermine and spermidine induce intestinal maturation in the rat. Gastroenterology, 1988, 95(1): 112-116.

[67] Deloyer P, Peulen O, Dandrifosse G. Dietary polyamines and non-neoplastic growth and disease. European Journal of Gastroenterology & Hepatology, 2001, 13(9): 1027-1032.

[68] Van der Hulst R, Von Meyenfeldt M, Soeters P. Glutamine: an essential amino acid for the gut. Nutrition (Burbank, Los Angeles County, Calif), 1996, 12(11-12 Suppl): S78-S81.

[69] Calder P, Yaqoob P. Glutamine and the immune system. Amino Acids, 1999, 17(3): 227-241.

[70] Cruzat V, Macedo Rogero M, Noel Keane K, et al. Glutamine: metabolism and immune function, supplementation and clinical translation. Nutrients, 2018, 10(11): 1564.

[71] McCarthy MS, Martindale RG. Immunonutrition in critical illness: what is the role? Nutrition in Clinical Practice, 2018, 33(3): 348-358.

[72] Santos RG, Quirino IE, Viana ML, et al. Effects of nitric oxide synthase inhibition on glutamine action in a bacterial translocation model. British Journal of Nutrition, 2014, 111(1): 93-100.

[73] Bertrand J, Goichon A, Chan P, et al. Enteral glutamine infusion modulates ubiquitination of heat shock proteins, grp-75 and apg-2, in the human duodenal mucosa. Amino Acids, 2014, 46(4): 1059-1067.

[74] Coëffier M, Le Pessot F, Leplingard A, et al. Acute enteral glutamine infusion enhances heme oxygenase-1 expression in human duodenal mucosa. The Journal of Nutrition, 2002, 132(9): 2570-2573.

[75] Coëffier M, Marion R, Ducrotte P, et al. Modulating effect of glutamine on il-1β-induced cytokine production by human gut. Clinical Nutrition, 2003, 22(4): 407-413.

[76] Coëffier M, Marion R, Leplingard A, et al. Glutamine decreases interleukin-8 and interleukin-6 but not nitric oxide and prostaglandins e2 production by human gut in-vitro. Cytokine, 2002, 18(2): 92-97.

[77] Coëffier M, Marion-Letellier R, Déchelotte P. Potential for amino acids supplementation during inflammatory

bowel diseases. Inflammatory Bowel Diseases, 2010, 16(3): 518-524.

[78] Deniel N, Marion-Letellier R, Charlionet R, et al. Glutamine regulates the human epithelial intestinal hct-8 cell proteome under apoptotic conditions. Molecular & Cellular Proteomics, 2007, 6(10): 1671-1679.

[79] Lecleire S, Hassan A, Marion-Letellier R, et al. Combined glutamine and arginine decrease proinflammatory cytokine production by biopsies from Crohn's patients in association with changes in nuclear factor-κ b and p38 mitogen-activated protein kinase pathways. The Journal of Nutrition, 2008, 138(12): 2481-2486.

[80] Marion R, Coeffier M, Leplingard A, et al. Cytokine-stimulated nitric oxide production and inducible no-synthase mrna level in human intestinal cells: lack of modulation by glutamine. Clinical Nutrition, 2003, 22(6): 523-528.

[81] Marion R, Coëffier M s, Gargala G, et al. Glutamine and cxc chemokines il-8, mig, ip-10 and i-tac in human intestinal epithelial cells. Clinical Nutrition, 2004, 23(4): 579-585.

[82] Mbodji K, Torre S, Haas V, et al. Alanyl-glutamine restores maternal deprivation-induced tlr4 levels in a rat neonatal model. Clinical Nutrition, 2011, 30(5): 672-677.

[83] Thébault S, Deniel N, Marion R, et al. Proteomic analysis of glutamine - treated human intestinal epithelial hct - 8 cells under basal and inflammatory conditions. Proteomics, 2006, 6(13): 3926-3937.

[84] Batiha GE, Alqarni M, Awad DAB, et al. Dairy-derived and egg white proteins in enhancing immune system against covid-19. Frontiers in Nutrition, 2021, 8: 629440.

[85] Fan X, Subramaniam R, Weiss M, et al. Methylglyoxal–bovine serum albumin stimulates tumor necrosis factor alpha secretion in raw 264.7 cells through activation of mitogen-activating protein kinase, nuclear factor κ b and intracellular reactive oxygen species formation. Archives of Biochemistry and Biophysics, 2003, 409(2): 274-286.

[86] Rupa P, Schnarr L, Mine Y. Effect of heat denaturation of egg white proteins ovalbumin and ovomucoid on cd4+ t cell cytokine production and human mast cell histamine production. Journal of Functional Foods, 2015, 18: 28-34.

[87] Mine Y. Egg bioscience and biotechnology. John Wiley & Sons, Inc., 2008.

[88] He X, Tsang TC, Luo P, et al. Enhanced tumor immunogenicity through coupling cytokine expression with antigen presentation. Cancer Gene Therapy, 2003, 10(9): 669-677.

[89] Xie H, Huff GR, Huff WE, et al. Effects of ovotransferrin on chicken macrophages and heterophil-granulocytes. Developmental & Comparative Immunology, 2002, 26(9): 805-815.

[90] Sun X, Chakrabarti S, Fang J, et al. Low-molecular-weight fractions of alcalase hydrolyzed egg ovomucin extract exert anti-inflammatory activity in human dermal fibroblasts through the inhibition of tumor necrosis factor–mediated nuclear factor κ b pathway. Nutrition Research, 2016, 36(7): 648-657.

[91] Tanizaki H, Tanaka H, Iwata H, et al. Activation of macrophages by sulfated glycopeptides in ovomucin, yolk membrane, and chalazae in chicken eggs. Bioscience, Biotechnology, and Biochemistry, 1997, 61(11): 1883-1889.

[92] Verdot L, Lalmanach G, Vercruysse V, et al. Chicken cystatin stimulates nitric oxide release from interferon - γ - activated mouse peritoneal macrophages via cytokine synthesis. European Journal of Biochemistry, 1999, 266(3): 1111-1117.

[93] Kato T, Imatani T, Miura T, et al. Cytokine-inducing activity of family 2 cystatins. Biological Chemistry, 2000, 381(11): 1143-1147.

[94] Meram C, Wu J. Anti-inflammatory effects of egg yolk livetins (α, β, and γ -livetin) fraction and its enzymatic hydrolysates in lipopolysaccharide-induced raw 264.7 macrophages. Food Research International, 2017, 100: 449-459.

[95] Sava G. Pharmacological aspects and therapeutic applications of lysozymes. Exs, 1996, 75: 433-449.

[96] El-Loly MM, Mahfouz MB. Lactoferrin in relation to biological functions and applications: a review. International Journal of Dairy Science, 2011, 6(2): 79-111.

[97] Wong CW, Watson DL. Immunomodulatory effects of dietary whey proteins in mice. Journal of Dairy Research, 1995, 62(2): 359-368.

[98] Ford JT, Wong CW, Colditz IG. Effects of dietary protein types on immune responses and levels of infection with eimeria vermiformis in mice. Immunology and Cell Biology, 2001, 79(1): 23-28.

[99] Huang S, Rutkowsky JM, Snodgrass RG, et al. Saturated fatty acids activate tlr-mediated proinflammatory signaling pathways. Journal of Lipid Research, 2012, 53(9): 2002-2013.

[100] Lee JY, Sohn KH, Rhee SH, et al. Saturated fatty

acids, but not unsaturated fatty acids, induce the expression of cyclooxygenase-2 mediated through toll-like receptor 4. Journal of Biological Chemistry, 2001, 276(20): 16683-16689.

[101] Calder PC. Fatty acids and inflammation: the cutting edge between food and pharma. European Journal of Pharmacology, 2011, 668: S50-S58.

[102] Galli C, Calder PC. Effects of fat and fatty acid intake on inflammatory and immune responses. Annals of Nutrition & Metabolism, 2009, 55(1/3): 123-139.

[103] Ding S, Chi MM, Scull BP, et al. High-fat diet: bacteria interactions promote intestinal inflammation which precedes and correlates with obesity and insulin resistance in mouse. Plos One, 2010, 5(8): e12191.

[104] Ji Y, Sun S, Goodrich JK, et al. Diet-induced alterations in gut microflora contribute to lethal pulmonary damage in tlr2/tlr4-deficient mice. Cell Reports, 2014, 8(1): 137-149.

[105] Endres S, Ghorbani R, Kelley VE, et al. The effect of dietary supplementation with n-3 polyunsaturated fatty acids on the synthesis of interleukin-1 and tumor necrosis factor by mononuclear cells. New England Journal of Medicine, 1989, 320(5): 265-271.

[106] Calder PC. Fatty acids and immune function: relevance to inflammatory bowel diseases. International Reviews of Immunology, 2009, 28(6): 506-534.

[107] Weylandt KH, Chiu C-Y, Gomolka B, et al. Omega-3 fatty acids and their lipid mediators: towards an understanding of resolvin and protectin formation. Prostaglandins & Other Lipid Mediators, 2012, 97(3-4): 73-82.

[108] Calder PC. Omega-3 fatty acids and inflammatory processes. Nutrients, 2010, 2(3): 355-374.

[109] Fredman G, Oh SF, Ayilavarapu S, et al. Impaired phagocytosis in localized aggressive periodontitis: rescue by resolvin e1. Plos One, 2011, 6(9): e24422.

[110] Kremmyda L-S, Vlachava M, Noakes PS, et al. Atopy risk in infants and children in relation to early exposure to fish, oily fish, or long-chain omega-3 fatty acids: a systematic review. Clinical Reviews in Allergy & Immunology, 2011, 41(1): 36-66.

[111] Shek LP, Chong MF, Lim JY, et al. Role of dietary long-chain polyunsaturated fatty acids in infant allergies and respiratory diseases. Clinical and Developmental Immunology, 2012, 2012: 730568.

[112] Nwaru BI, Erkkola M, Lumia M, et al. Maternal intake of fatty acids during pregnancy and allergies in the offspring. British Journal of Nutrition, 2012, 108(4): 720-732.

[113] Levy BD. Resolvins and protectins: natural pharmacophores for resolution biology. Prostaglandins, Leukotrienes and Essential Fatty Acids (PLEFA), 2010, 82(4-6): 327-332.

[114] Ariel A, Serhan CN. Resolvins and protectins in the termination program of acute inflammation. Trends in Immunology, 2007, 28(4): 176-183.

[115] Merched AJ, Ko K, Gotlinger KH, et al. Atherosclerosis: evidence for impairment of resolution of vascular inflammation governed by specific lipid mediators. The FASEB Journal, 2008, 22(10): 3595-3606.

[116] Kim K, Jung N, Lee K, et al. Dietary omega-3 polyunsaturated fatty acids attenuate hepatic ischemia/reperfusion injury in rats by modulating toll-like receptor recruitment into lipid rafts. Clinical Nutrition, 2013, 32(5): 855-862.

[117] Lee JY, Plakidas A, Lee WH, et al. Differential modulation of toll-like receptors by fatty acids: preferential inhibition by n-3 polyunsaturated fatty acids. Journal of Lipid Research, 2003, 44(3): 479-486.

[118] Lee JY, Zhao L, Hwang DH. Modulation of pattern recognition receptor-mediated inflammation and risk of chronic diseases by dietary fatty acids. Nutrition Reviews, 2010, 68(1): 38-61.

[119] Hachimura S, Totsuka M, Hosono A. Immunomodulation by food: Impact on gut immunity and immune cell function. Bioscience, Biotechnology, and Biochemistry, 2018, 82(4): 584-599.

[120] Dalonso N, Goldman GH, Gern RMM. B-(1 → 3),(1 → 6)-glucans: medicinal activities, characterization, biosynthesis and new horizons. Applied Microbiology and Biotechnology, 2015, 99(19): 7893-7906.

[121] Brown GD, Gordon S. Immune recognition of fungal β - glucans. Cellular Microbiology, 2005, 7(4): 471-479.

[122] Zeković DB, Kwiatkowski S, Vrvić MM, et al. Natural and modified (1 → 3)-β-d-glucans in health promotion and disease alleviation. Critical Reviews in Biotechnology, 2005, 25(4): 205-230.

[123] Shi S-H, Yang W-T, Huang K-Y, et al. B-glucans from coriolus versicolor protect mice againsts. Typhimurium challenge by activation of macrophages. International Journal of Biological Macromolecules, 2016, 86: 352-361.

[124] Davis JM, Murphy EA, Brown AS, et al. Effects of moderate exercise and oat β-glucan on innate immune function and susceptibility to respiratory infection. American Journal of Physiology-Regulatory, Integrative and Comparative Physiology, 2004, 286(2): R366-R372.

[125] Yun C-H, Estrada A, Van Kessel A, et al. B-glucan, extracted from oat, enhances disease resistance against bacterial and parasitic infections. FEMS Immunology & Medical Microbiology, 2003, 35(1): 67-75.

[126] Dambuza IM, Brown GD. C-type lectins in immunity: recent developments. Current Opinion in Immunology, 2015, 32: 21-27.

[127] Kakutani R, Adachi Y, Kajiura H, et al. Relationship between structure and immunostimulating activity of enzymatically synthesized glycogen. Carbohydrate Research, 2007, 342(16): 2371-2379.

[128] Kakutani R, Adachi Y, Takata H, et al. Essential role of toll-like receptor 2 in macrophage activation by glycogen. Glycobiology, 2012, 22(1): 146-159.

[129] Kakutani R, Adachi Y, Kajiura H, et al. The effect of orally administered glycogen on anti-tumor activity and natural killer cell activity in mice. International Immunopharmacology, 2012, 12(1): 80-87.

[130] Maruyama H, Tamauchi H, Iizuka M, et al. The role of nk cells in antitumor activity of dietary fucoidan from undaria pinnatifida sporophylls (mekabu). Planta Medica, 2006, 72(15): 1415-1417.

[131] Hayashi K, Nakano T, Hashimoto M, et al. Defensive effects of a fucoidan from brown alga undaria pinnatifida against herpes simplex virus infection. International Immunopharmacology, 2008, 8(1): 109-116.

[132] Hayashi K, Lee J-B, Nakano T, et al. Anti-influenza a virus characteristics of a fucoidan from sporophyll of undaria pinnatifida in mice with normal and compromised immunity. Microbes and Infection, 2013, 15(4): 302-309.

[133] Zou P, Yang X, Wang J, et al. Advances in characterisation and biological activities of chitosan and chitosan oligosaccharides. Food Chemistry, 2016, 190: 1174-1181.

[134] Lundahl ML, Scanlan EM, Lavelle EC. Therapeutic potential of carbohydrates as regulators of macrophage activation. Biochemical Pharmacology, 2017, 146: 23-41.

[135] Ma J-X, Qian L, Zhou Y. Stimulation effect of chitosan on the immunity of radiotherapy patients suffered from lung cancer. International Journal of Biological Macromolecules, 2015, 72: 195-198.

[136] Bo S, Ciccone G, Guidi S, et al. Diet or exercise: What is more effective in preventing or reducing metabolic alterations? European Journal of Endocrinology, 2008, 159(6): 685.

[137] Galland L. Diet and inflammation. Nutrition in Clinical Practice, 2010, 25(6): 634-640.

[138] North C, Venter C, Jerling J. The effects of dietary fibre on c-reactive protein, an inflammation marker predicting cardiovascular disease. European Journal of Clinical Nutrition, 2009, 63(8): 921-933.

[139] Venter C, Eyerich S, Sarin T, et al. Nutrition and the immune system: a complicated tango. Nutrients, 2020, 12(3): 818.

[140] Park Y, Subar AF, Hollenbeck A, et al. Dietary fiber intake and mortality in the nih-aarp diet and health study. Archives of Internal Medicine, 2011, 171(12): 1061-1068.

[141] Thorburn AN, McKenzie CI, Shen S, et al. Evidence that asthma is a developmental origin disease influenced by maternal diet and bacterial metabolites. Nature Communications, 2015, 6(1): 1-13.

[142] Trompette A, Gollwitzer ES, Yadava K, et al. Gut microbiota metabolism of dietary fiber influences allergic airway disease and hematopoiesis. Nature Medicine, 2014, 20(2): 159-166.

[143] van Berge-Henegouwen G, Mulder C. Pioneer in the gluten free diet: willem-karel dicke 1905-1962, over 50 years of gluten free diet. Gut, 1993, 34(11): 1473.

[144] Di Liberto D, Carlisi D, D'Anneo A, et al. Gluten free diet for the management of non celiac diseases: the two sides of the coin. Healthcare (Basel), 2020, 8(4):400.

[145] Marsh MN. Gluten, major histocompatibility complex, and the small intestine: a molecular and immunobiologic approach to the spectrum of gluten sensitivity ('celiac sprue'). Gastroenterology, 1992, 102(1): 330-354.

[146] Sollid LM. Coeliac disease: dissecting a complex inflammatory disorder. Nature Reviews Immunology, 2002, 2(9): 647-655.

[147] Cinova J, Palová-Jelínková L, Smythies LE, et al. Gliadin peptides activate blood monocytes from patients with celiac disease. Journal of Clinical Immunology, 2007, 27(2): 201-209.

[148] Newnham ED. Coeliac disease in the 21st century: paradigm shifts in the modern age. Journal of Gastroenterology and Hepatology, 2017, 32: 82-85.

[149] Cosnes J, Cellier C, Viola S, et al. Incidence of autoimmune diseases in celiac disease: protective effect of the gluten-free diet. Clinical Gastroenterology and Hepatology, 2008, 6(7): 753-758.

[150] Malandrino N, Capristo E, Farnetti S, et al. Metabolic and nutritional features in adult celiac patients. Digestive Diseases, 2008, 26(2): 128-133.

[151] McAllister BP, Williams E, Clarke K. A comprehensive review of celiac disease/gluten-sensitive enteropathies. Clinical Reviews in Allergy & Immunology, 2019, 57(2): 226-243.

[152] De Palma G, Nadal I, Collado MC, et al. Effects of a gluten-free diet on gut microbiota and immune function in healthy adult human subjects. British Journal of Nutrition, 2009, 102(8): 1154-1160.

[153] Bonder MJ, Tigchelaar EF, Cai X, et al. The influence of a short-term gluten-free diet on the human gut microbiome. Genome Medicine, 2016, 8(1): 1-11.

[154] Nadal I, Donant E, Ribes-Koninckx C, et al. Imbalance in the composition of the duodenal microbiota of children with coeliac disease. Journal of Medical Microbiology, 2007, 56(12): 1669-1674.

[155] Sanz Y, Sánchez E, De Palma G, et al. Indigenous gut microbiota, probiotics, and coeliac disease. Child Nutrition & Physiology, 2008: 211-224.

[156] Nistal E, Caminero A, Herrán AR, et al. Differences of small intestinal bacteria populations in adults and children with/without celiac disease: effect of age, gluten diet, and disease. Inflammatory Bowel Diseases, 2012, 18(4): 649-656.

[157] Theethira TG, Dennis M. Celiac disease and the gluten-free diet: consequences and recommendations for improvement. Digestive Diseases, 2015, 33(2): 175-182.

[158] Bascuñán KA, Vespa MC, Araya M. Celiac disease: understanding the gluten-free diet. European Journal of Nutrition, 2017, 56(2): 449-459.

[159] Salas-Salvadó J, Bulló M, Estruch R, et al. Prevention of diabetes with mediterranean diets: a subgroup analysis of a randomized trial. Annals of Internal Medicine, 2014, 160(1): 1-10.

[160] Losasso C, Eckert EM, Mastrorilli E, et al. Assessing the influence of vegan, vegetarian and omnivore oriented westernized dietary styles on human gut microbiota: a cross sectional study. Frontiers in Microbiology, 2018, 9: 317.

[161] Wu GD, Compher C, Chen EZ, et al. Comparative metabolomics in vegans and omnivores reveal constraints on diet-dependent gut microbiota metabolite production. Gut, 2016, 65(1): 63-72.

[162] van Duynhoven J, Vaughan EE, Jacobs DM, et al. Metabolic fate of polyphenols in the human superorganism. Proceedings of the National Academy of Sciences, 2011, 108(Supplement 1): 4531-4538.

[163] Wang X-L, Hur H-G, Lee JH, et al. Enantioselective synthesis of s-equol from dihydrodaidzein by a newly isolated anaerobic human intestinal bacterium. Applied and Environmental Microbiology, 2005, 71(1): 214-219.

[164] Zimmer J, Lange B, Frick J-S, et al. A vegan or vegetarian diet substantially alters the human colonic faecal microbiota. European Journal of Clinical Nutrition, 2012, 66(1): 53-60.

[165] Wong M-W, Yi C-H, Liu T-T, et al. Impact of vegan diets on gut microbiota: an update on the clinical implications. Tzu-Chi Medical Journal, 2018, 30(4): 200.

[166] Hoyles L, Vulevic J. Diet, immunity and functional foods. GI Microbiota and Regulation of the Immune System, 2008: 79-92.

[167] Scientific concepts of functional foods in europe. Consensus Document. British Journal of Nutrition, 1999, 81 Suppl 1: S1-S27.

[168] Gross RL, Newberne PM. Role of nutrition in immunologic function. Physiological Reviews, 1980, 60(1): 188-302.

[169] Chandra R. Nutrition and immunity: lesson from the past and new insights into the future system: an introduction. The American Journal of Clinical Nutrition, 1991, 53: 1087-1101.

[170] Kuvibidila S, Yu L, Ode D, et al. The immune response in protein-energy malnutrition and single nutrient deficiencies. Nutrition and Immunology. Springer, 1993: 121-155.

[171] Scrimshaw NS, SanGiovanni JP. Synergism of nutrition, infection, and immunity: an overview. The American Journal of Clinical Nutrition, 1997, 66(2): S464-S477.

[172] Calder PC, Jackson AA. Undernutrition, infection and immune function. Nutrition Research Reviews, 2000, 13(1): 3-29.

[173] Li A-N, Li S, Zhang Y-J, et al. Resources and biological activities of natural polyphenols. Nutrients, 2014, 6(12): 6020-6047.

[174] Scalbert A, Manach C, Morand C, et al. Dietary polyphenols and the prevention of diseases. Critical Reviews in Food Science and Nutrition, 2005, 45(4):

287-306.

[175] Middleton E, Kandaswami C, Theoharides TC. The effects of plant flavonoids on mammalian cells: Implications for inflammation, heart disease, and cancer. Pharmacological Reviews, 2000, 52(4): 673-751.

[176] Sergent T, Dupont I, Van Der Heiden E, et al. Cyp1a1 and cyp3a4 modulation by dietary flavonoids in human intestinal caco-2 cells. Toxicology Letters, 2009, 191(2-3): 216-222.

[177] Yang CS, Pan E. The effects of green tea polyphenols on drug metabolism. Expert Opinion on Drug Metabolism & Toxicology, 2012, 8(6): 677-689.

[178] Sergent T, Garsou S, Schaut A, et al. Differential modulation of ochratoxin a absorption across caco-2 cells by dietary polyphenols, used at realistic intestinal concentrations. Toxicology Letters, 2005, 159(1): 60-70.

[179] Bustos I, Garcia-Cayuela T, Hernandez-Ledesma B, et al. Effect of flavan-3-ols on the adhesion of potential probiotic lactobacilli to intestinal cells. Journal of Agricultural and Food Chemistry, 2012, 60(36): 9082-9088.

[180] Aura A-M. Microbial metabolism of dietary phenolic compounds in the colon. Phytochemistry Reviews, 2008, 7(3): 407-429.

[181] Selma MV, Espin JC, Tomas-Barberan FA. Interaction between phenolics and gut microbiota: role in human health. Journal of Agricultural and Food Chemistry, 2009, 57(15): 6485-6501.

[182] Laparra JM, Sanz Y. Interactions of gut microbiota with functional food components and nutraceuticals. Pharmacological Research, 2010, 61(3): 219-225.

[183] Hervert-Hernández D, Goñi I. Dietary polyphenols and human gut microbiota: a review. Food Reviews International, 2011, 27(2): 154-169.

[184] Ozdal T, Sela DA, Xiao J, et al. The reciprocal interactions between polyphenols and gut microbiota and effects on bioaccessibility. Nutrients, 2016, 8(2): 78.

[185] He J, Magnuson BA, Giusti MM. Analysis of anthocyanins in rat intestinal contents impact of anthocyanin chemical structure on fecal excretion. Journal of Agricultural and Food Chemistry, 2005, 53(8): 2859-2866.

[186] Wang J, Pae M, Meydani SN, et al. Green tea epigallocatechin-3-gallate modulates differentiation of naïve cd4+ t cells into specific lineage effector cells. Journal of Molecular Medicine, 2013, 91(4): 485-495.

[187] Yang J, Yang X, Chu Y, et al. Identification of baicalin as an immunoregulatory compound by controlling th17 cell differentiation. Plos One, 2011, 6(2): e17164.

[188] Wong CP, Nguyen LP, Noh SK, et al. Induction of regulatory t cells by green tea polyphenol egcg. Immunology Letters, 2011, 139(1-2): 7-13.

[189] Yang J, Yang X, Li M. Baicalin, a natural compound, promotes regulatory t cell differentiation. BMC Complementary and Alternative Medicine, 2012, 12(1): 1-7.

[190] Tachibana H, Sunada Y, Miyase T, et al. Identification of a methylated tea catechin as an inhibitor of degranulation in human basophilic ku812 cells. Bioscience, Biotechnology, and Biochemistry, 2000, 64(2): 452-454.

[191] Maeda-Yamamoto M. Human clinical studies of tea polyphenols in allergy or life style-related diseases. Current Pharmaceutical Design, 2013, 19(34): 6148-6155.

[192] Yamamoto T, Yoshimura M, Yamaguchi F, et al. Anti-allergic activity of naringenin chalcone from a tomato skin extract. Bioscience, Biotechnology, and Biochemistry, 2004, 68(8): 1706-1711.

[193] Li Y, Yao J, Han C, et al. Quercetin, inflammation and immunity. Nutrients, 2016, 8(3): 167.

[194] Shin HS, Satsu H, Bae M-J, et al. Anti-inflammatory effect of chlorogenic acid on the il-8 production in caco-2 cells and the dextran sulphate sodium-induced colitis symptoms in c57bl/6 mice. Food Chemistry, 2015, 168: 167-175.

[195] Peluso I, Miglio C, Morabito G, et al. Flavonoids and immune function in human: a systematic review. Critical Reviews in Food Science and Nutrition, 2015, 55(3): 383-395.

[196] Zhang Y, Zhou S, Zhou Y, et al. Altered gut microbiome composition in children with refractory epilepsy after ketogenic diet. Epilepsy Research, 2018, 145: 163-168.

[197] Roberts MN, Wallace MA, Tomilov AA, et al. A ketogenic diet extends longevity and healthspan in adult mice. Cell Metabolism, 2017, 26(3): 539-546. e535.

[198] Swidsinski A, Dörffel Y, Loening-Baucke V, et al. Reduced mass and diversity of the colonic microbiome in patients with multiple sclerosis and their improvement with ketogenic diet. Frontiers in Microbiology, 2017, 8: 1141.

[199] Tagliabue A, Ferraris C, Uggeri F, et al. Short-term impact of a classical ketogenic diet on gut microbiota in glut1 deficiency syndrome: a 3-month prospective observational study. Clinical Nutrition ESPEN, 2017, 17: 33-37.

[200] Gentile CL, Weir TL. The gut microbiota at the intersection of diet and human health. Science, 2018, 362(6416): 776-780.

[201] Lindefeldt M, Eng A, Darban H, et al. The ketogenic diet influences taxonomic and functional composition of the gut microbiota in children with severe epilepsy. NPJ Biofilms and Microbiomes, 2019, 5(1): 1-13.

[202] Xie G, Zhou Q, Qiu C-Z, et al. Ketogenic diet poses a significant effect on imbalanced gut microbiota in infants with refractory epilepsy. World Journal of Gastroenterology, 2017, 23(33): 6164.

[203] Olson CA, Vuong HE, Yano JM, et al. The gut microbiota mediates the anti-seizure effects of the ketogenic diet. Cell, 2018, 173(7): 1728-1741. e1713.

[204] Rinninella E, Cintoni M, Raoul P, et al. Food components and dietary habits: keys for a healthy gut microbiota composition. Nutrients, 2019, 11(10): 2393.

[205] Shaw DM, Merien F, Braakhuis A, et al. Adaptation to a ketogenic diet modulates adaptive and mucosal immune markers in trained male endurance athletes. Scandinavian Journal of Medicine & Science in Sports, 2021, 31(1): 140-152.

[206] Terink R, Witkamp RF, Hopman MT, et al. A 2 week cross-over intervention with a low carbohydrate, high fat diet compared to a high carbohydrate diet attenuates exercise-induced cortisol response, but not the reduction of exercise capacity, in recreational athletes. Nutrients, 2021, 13(1): 157.

[207] Bahr LS, Bock M, Liebscher D, et al. Ketogenic diet and fasting diet as nutritional approaches in multiple sclerosis (nams): protocol of a randomized controlled study. Trials, 2020, 21(1): 1-9.

[208] Alam W, Ullah H, Santarcangelo C, et al. Micronutrient food supplements in patients with gastro-intestinal and hepatic cancers. International Journal of Molecular Sciences, 2021, 22(15): 8014.

[209] Mrityunjaya M, Pavithra V, Neelam R, et al. Immune-boosting, antioxidant and anti-inflammatory food supplements targeting pathogenesis of covid-19. Frontiers in Immunology, 2020, 11: 570122.

[210] Pecora F, Persico F, Argentiero A, et al. The role of micronutrients in support of the immune response against viral infections. Nutrients, 2020, 12(10): 3198.

[211] Read SA, Obeid S, Ahlenstiel C, et al. The role of zinc in antiviral immunity. Advances in Nutrition, 2019, 10(4): 696-710.

[212] Biaggio VS, Pérez Chaca MV, Valdéz SR, et al. Alteration in the expression of inflammatory parameters as a result of oxidative stress produced by moderate zinc deficiency in rat lung. Experimental Lung Research, 2010, 36(1): 31-44.

[213] Shakoor H, Feehan J, Al Dhaheri AS, et al. Immune-boosting role of vitamins d, c, e, zinc, selenium and omega-3 fatty acids: could they help against covid-19? Maturitas, 2021, 143: 1-9.

[214] Zhang L, Liu Y. Potential interventions for novel coronavirus in china: a systematic review. Journal of Medical Virology, 2020, 92(5): 479-490.

[215] Souza ACR, Vasconcelos AR, Prado PS, et al. Zinc, vitamin d and vitamin c: perspectives for covid-19 with a focus on physical tissue barrier integrity. Frontiers in Nutrition, 2020, 7: 295.

[216] Chavakis T, May AE, Preissner KT, et al. Molecular mechanisms of zinc-dependent leukocyte adhesion involving the urokinase receptor and β2-integrins. Blood, The Journal of the American Society of Hematology, 1999, 93(9): 2976-2983.

[217] Fraker P, King L. Changes in regulation of lymphopoiesis and myelopoiesis in the zinc-deficient mouse. Nutrition Reviews, 1998, 56(1): S65.

[218] Fraker PJ, King LE, Garvy BA, et al. The immunopathology of zinc deficiency in humans and rodents. Nutrition and Immunology. Springer, 1993: 267-283.

[219] Allen JI, Perri RT, McClain CJ, et al. Alterations in human natural killer cell activity and monocyte cytotoxicity induced by zinc deficiency. The Journal of Laboratory and Clinical Medicine, 1983, 102(4): 577-589.

[220] Keen CL, Gershwin ME. Zinc deficiency and immune function. Annual Review of Nutrition, 1990, 10(1): 415-431.

[221] Chandra RK, Dayton DH. Trace element regulation of immunity and infection. Nutrition Research, 1982, 2(6): 721-733.

[222] Fraker PJ, Gershwin ME, Good RA, et al. Interrelationships between zinc and immune function. Fed Proc, 1986, 45(5): 1474-1479.

[223] Shankar AH, Prasad AS. Zinc and immune function: the biological basis of altered resistance to infection. The American Journal of Clinical Nutrition, 1998,

68(2): S447-S463.

[224] Beck F, Prasad A, Kaplan J, et al. Changes in cytokine production and t cell subpopulations in experimentally induced zinc-deficient humans. American Journal of Physiology-Endocrinology and Metabolism, 1997, 272(6): E1002-E1007.

[225] Lira P, Ashworth A, Morris SS. Effect of zinc supplementation on the morbidity, immune function, and growth of low-birth-weight, full-term infants in northeast brazil. The American Journal of Clinical Nutrition, 1998, 68(2): S418-S424.

[226] Rajagopalan S, Winter CC, Wagtmann N, et al. The ig-related killer cell inhibitory receptor binds zinc and requires zinc for recognition of hla-c on target cells. The Journal of Immunology, 1995, 155(9): 4143-4146.

[227] Tapazoglou E, Prasad AS, Hill G, et al. Decreased natural killer cell activity in patients with zinc deficiency with sickle cell disease. The Journal of Laboratory and Clinical Medicine, 1985, 105(1): 19-22.

[228] Sohnle PG, Collins-Lech C, Wiessner JH. The zinc-reversible antimicrobial activity of neutrophil lysates and abscess fluid supernatants. Journal of Infectious Diseases, 1991, 164(1): 137-142.

[229] Murthy A, Lehrer R, Harwig S, et al. In vitro candidastatic properties of the human neutrophil calprotectin complex. The Journal of Immunology, 1993, 151(11): 6291-6301.

[230] Clohessy P, Golden B. Calprotectin - mediated zinc chelation as a biostatic mechanism in host defence. Scandinavian Journal of Immunology, 1995, 42(5): 551-556.

[231] López-Varela S, González-Gross M, Marcos A. Functional foods and the immune system: a review. European Journal of Clinical Nutrition, 2002, 56(3): S29-S33.

[232] Wiedermann U, Hanson L, Bremell T, et al. Increased translocation of escherichia coli and development of arthritis in vitamin a-deficient rats. Infection and Immunity, 1995, 63(8): 3062-3068.

[233] Hiemstra IH, Beijer MR, Veninga H, et al. The identification and developmental requirements of colonic cd 169+ macrophages. Immunology, 2014, 142(2): 269-278.

[234] Pereira WF, Ribeiro - Gomes FL, Guillermo LVC, et al. Myeloid - derived suppressor cells help protective immunity to leishmania major infection despite suppressed t cell responses. Journal of Leukocyte Biology, 2011, 90(6): 1191-1197.

[235] Vellozo NS, Pereira-Marques ST, Cabral-Piccin MP, et al. All-trans retinoic acid promotes an m1- to m2-phenotype shift and inhibits macrophage-mediated immunity to leishmania major. Frontiers in Immunology, 2017, 8: 1560.

[236] Shrestha S, Kim S-Y, Yun Y-J, et al. Retinoic acid induces hypersegmentation and enhances cytotoxicity of neutrophils against cancer cells. Immunology Letters, 2017, 182: 24-29.

[237] Worbs T, Hammerschmidt SI, Foerster R. Dendritic cell migration in health and disease. Nature Reviews Immunology, 2017, 17(1): 30-48.

[238] Zeng R, Bscheider M, Lahl K, et al. Generation and transcriptional programming of intestinal dendritic cells: essential role of retinoic acid. Mucosal Immunology, 2016, 9(1): 183-193.

[239] Klebanoff CA, Spencer SP, Torabi-Parizi P, et al. Retinoic acid controls the homeostasis of pre-cdc–derived splenic and intestinal dendritic cells. Journal of Experimental Medicine, 2013, 210(10): 1961-1976.

[240] Sun C-M, Hall JA, Blank RB, et al. Small intestine lamina propria dendritic cells promote de novo generation of foxp3 t reg cells via retinoic acid. The Journal of Experimental Medicine, 2007, 204(8): 1775-1785.

[241] Coombes JL, Siddiqui KR, Arancibia-Cárcamo CV, et al. A functionally specialized population of mucosal cd103+ dcs induces foxp3+ regulatory t cells via a tgf-β–and retinoic acid–dependent mechanism. Journal of Experimental Medicine, 2007, 204(8): 1757-1764.

[242] Scott CL, Aumeunier AM, Mowat AM. Intestinal cd103+ dendritic cells: master regulators of tolerance? Trends in Immunology, 2011, 32(9): 412-419.

[243] DePaolo R, Abadie V, Tang F, et al. Co-adjuvant effects of retinoic acid and il-15 induce inflammatory immunity to dietary antigens. Nature, 2011, 471(7337): 220-224.

[244] Huang Z, Liu Y, Qi G, et al. Role of vitamin a in the immune system. Journal of Clinical Medicine, 2018, 7(9): 258.

[245] Van de Pavert SA, Mebius RE. New insights into the development of lymphoid tissues. Nature Reviews Immunology, 2010, 10(9): 664-674.

[246] Eberl G, Marmon S, Sunshine M-J, et al. An essential function for the nuclear receptor rorγt in the genera-

tion of fetal lymphoid tissue inducer cells. Nature Immunology, 2004, 5(1): 64-73.

[247] Buonocore S, Ahern PP, Uhlig HH, et al. Innate lymphoid cells drive interleukin-23-dependent innate intestinal pathology. Nature, 2010, 464(7293): 1371-1375.

[248] Geremia A, Arancibia-Cárcamo CV, Fleming MP, et al. Il-23–responsive innate lymphoid cells are increased in inflammatory bowel disease. Journal of Experimental Medicine, 2011, 208(6): 1127-1133.

[249] Spencer S, Wilhelm C, Yang Q, et al. Adaptation of innate lymphoid cells to a micronutrient deficiency promotes type 2 barrier immunity. Science, 2014, 343(6169): 432-437.

[250] Mielke LA, Jones SA, Raverdeau M, et al. Retinoic acid expression associates with enhanced il-22 production by γδ t cells and innate lymphoid cells and attenuation of intestinal inflammation. Journal of Experimental Medicine, 2013, 210(6): 1117-1124.

[251] Xu C, Sun R, Qiao X, et al. Effect of vitamin e supplementation on intestinal barrier function in rats exposed to high altitude hypoxia environment. The Korean Journal of Physiology & Pharmacology, 2014, 18(4): 313-320.

[252] Adachi N, Migita M, Ohta T, et al. Depressed natural killer cell activity due to decreased natural killer cell population in a vitamin e-deficient patient with shwachman syndrome: reversible natural killer cell abnormality by α-tocopherol supplementation. European Journal of Pediatrics, 1997, 156(6): 444-448.

[253] Ravaglia G, Forti P, Maioli F, et al. Effect of micronutrient status on natural killer cell immune function in healthy free-living subjects aged ≥ 90y. The American Journal of Clinical Nutrition, 2000, 71(2): 590-598.

[254] Hanson MG, Özenci V, Carlsten MC, et al. A short-term dietary supplementation with high doses of vitamin e increases nk cell cytolytic activity in advanced colorectal cancer patients. Cancer Immunology, Immunotherapy, 2007, 56(7): 973-984.

[255] Stiff A, Trikha P, Mundy-Bosse B, et al. Nitric oxide production by myeloid-derived suppressor cells plays a role in impairing fc receptor–mediated natural killer cell function. Clinical Cancer Research, 2018, 24(8): 1891-1904.

[256] Lee GY, Han SN. The role of vitamin e in immunity. Nutrients, 2018, 10(11): 1614.

[257] Malmberg K-J, Lenkei R, Petersson M, et al. A short-term dietary supplementation of high doses of vitamin e increases t helper 1 cytokine production in patients with advanced colorectal cancer. Clinical Cancer Research, 2002, 8(6): 1772-1778.

[258] Han S, Wu D, Ha W, et al. Vitamin e supplementation increases t helper 1 cytokine production in old mice infected with influenza virus. Immunology, 2000, 100(4): 487-493.

[259] Bae M, Kim H. The role of vitamin c, vitamin d, and selenium in immune system against covid-19. Molecules, 2020, 25(22): 5346.

[260] Abdullah M, Jamil RT, Attia FN. Vitamin c (ascorbic acid). StatPearls Publishing, 2021.

[261] Kornberg MD, Bhargava P, Kim PM, et al. Dimethyl fumarate targets gapdh and aerobic glycolysis to modulate immunity. Science, 2018, 360(6387): 449-453.

[262] Ngo B, Van Riper JM, Cantley LC, et al. Targeting cancer vulnerabilities with high-dose vitamin c. Nature Reviews Cancer, 2019, 19(5): 271-282.

[263] Wilson JX. The physiological role of dehydroascorbic acid. FEBS Letters, 2002, 527(1-3): 5-9.

[264] Yun J, Mullarky E, Lu C, et al. Vitamin c selectively kills kras and braf mutant colorectal cancer cells by targeting gapdh. Science, 2015, 350(6266): 1391-1396.

[265] Raj PA, Dentino AR. Current status of defensins and their role in innate and adaptive immunity. FEMS Microbiology Letters, 2002, 206(1): 9-18.

[266] Barlow PG, Svoboda P, Mackellar A, et al. Antiviral activity and increased host defense against influenza infection elicited by the human cathelicidin ll-37. Plos One, 2011, 6(10): e25333.

[267] Currie SM, Findlay EG, McHugh BJ, et al. The human cathelicidin ll-37 has antiviral activity against respiratory syncytial virus. Plos One, 2013, 8(8): e73659.

[268] Kota S, Sabbah A, Harnack R, et al. Role of human β-defensin-2 during tumor necrosis factor-α/nf-κb-mediated innate antiviral response against human respiratory syncytial virus. Journal of Biological Chemistry, 2008, 283(33): 22417-22429.

[269] Liu F, Zhu Y, Zhang J, et al. Intravenous high-dose vitamin c for the treatment of severe covid-19: study protocol for a multicentre randomised controlled trial. BMJ Open, 2020, 10(7): e039519.

[270] Suardi C, Cazzaniga E, Graci S, et al. Link between viral infections, immune system, inflammation and diet. International Journal of Environmental Research

and Public Health, 2021, 18(5): 2455.

[271] Martens P-J, Gysemans C, Verstuyf A, et al. Vitamin d's effect on immune function. Nutrients, 2020, 12(5): 1248.

[272] Cannell J, Vieth R, Umhau J, et al. Epidemic influenza and vitamin d. Epidemiology & Infection, 2006, 134(6): 1129-1140.

[273] Goncalves-Mendes N, Talvas J, Dualé C, et al. Impact of vitamin d supplementation on influenza vaccine response and immune functions in deficient elderly persons: a randomized placebo-controlled trial. Frontiers in Immunology, 2019, 10: 65.

[274] Siddiqui M, Manansala JS, Abdulrahman HA, et al. Immune modulatory effects of vitamin d on viral infections. Nutrients, 2020, 12(9): 2879.

[275] Antico A, Tampoia M, Tozzoli R, et al. Can supplementation with vitamin d reduce the risk or modify the course of autoimmune diseases? A systematic review of the literature. Autoimmunity Reviews, 2012, 12(2): 127-136.

[276] Prietl B, Treiber G, Pieber TR, et al. Vitamin d and immune function. Nutrients, 2013, 5(7): 2502-2521.

[277] Boulkrane MS, Ilina V, Melchakov R, et al. Covid-19 disease and vitamin d: a mini-review. Frontiers in Pharmacology, 2020, 11: 2107.

[278] Domeneghini C, Di Giancamillo A, Arrighi S, et al. Gut-trophic feed additives and their effects upon the gut structure and intestinal metabolism. State of the art in the pig, and perspectives towards humans. Histology and Histopathology, 2006, 21(3): 273-283.

[279] Leleiko NS, Martin BA, Walsh M, et al. Tissue-specific gene expression results from a purine-and pyrimidine-free diet and 6-mercaptopurine in the rat small intestine and colon. Gastroenterology, 1987, 93(5): 1014-1020.

[280] Uauy R. Dietary nucleotides and requirements in early life. Textbook of Gastroenterology and Nutrition in Infancy, 1989: 265-280.

[281] Rudolph FB, Kulkarni AD, Fanslow W, et al. Role of rna as a dietary source of pyrimidines and purines in immune function. Nutrition (Burbank), 1990, 6(1): 45-52.

[282] Van Buren CT, Kim E, Kulkarni AD, et al. Nucleotide-free diet and suppression of immune response. Transplantation Proceedings, 1987, 19(4 Suppl 5): 57-59.

[283] Jyonouchi H, Hill RJ, Good RA. Rna/nucleotide enhances antibody production in vitro and is moderately mitogenic to murine spleen lymphocytes. Proceedings of the Society for Experimental Biology and Medicine, 1992, 200(1): 101-108.

[284] Jyonouchi H. Nucleotide actions on humoral immune responses. The Journal of Nutrition, 1994, 124(suppl_1): 138S-143S.

[285] Zahedipour F, Hosseini SA, Sathyapalan T, et al. Potential effects of curcumin in the treatment of covid - 19 infection. Phytotherapy Research, 2020, 34(11): 2911-2920.

[286] Ting D, Dong N, Fang L, et al. Multisite inhibitors for enteric coronavirus: antiviral cationic carbon dots based on curcumin. American Chemical Society Applied Nano Materials, 2018, 1(10): 5451-5459.

[287] Khaerunnisa S, Kurniawan H, Awaluddin R, et al. Potential inhibitor of covid-19 main protease (mpro) from several medicinal plant compounds by molecular docking study. Preprints, 2020, 2020: 2020030226.

[288] Jennings A, Koch M, Bang C, et al. Microbial diversity and abundance of parabacteroides mediate the associations between higher intake of flavonoid-rich foods and lower blood pressure. Hypertension, 2021, 78(4): 1016-1026.

[289] Kress-Rogers E, Brimelow CJ. Instrumentation and sensors for the food industry. Woodhead Publishing, 2001.

[290] Arora SK, McFarlane SI. The case for low carbohydrate diets in diabetes management. Nutrition & Metabolism, 2005, 2(1): 1-9.

[291] O'Keefe JH, Gheewala NM, O'Keefe JO. Dietary strategies for improving post-prandial glucose, lipids, inflammation, and cardiovascular health. Journal of the American College of Cardiology, 2008, 51(3): 249-255.

[292] Wastyk HC, Fragiadakis GK, Perelman D, et al. Gut-microbiota-targeted diets modulate human immune status. Cell, 2021, 184(16): 4137-4153.e4114.

[293] Carocho M, Barreiro MF, Morales P, et al. Adding molecules to food, pros and cons: a review on synthetic and natural food additives. Comprehensive Reviews in Food Science and Food Safety, 2014, 13(4): 377-399.

[294] Rinninella E, Cintoni M, Raoul P, et al. Food additives, gut microbiota, and irritable bowel syndrome: a hidden track. International Journal of Environmental Research and Public Health, 2020, 17(23): 8816.

[295] Chassaing B, Koren O, Goodrich JK, et al. Dietary emulsifiers impact the mouse gut microbiota

promoting colitis and metabolic syndrome. Nature, 2015, 519(7541): 92-96.

[296] Suez J, Korem T, Zeevi D, et al. Artificial sweeteners induce glucose intolerance by altering the gut microbiota. Nature, 2014, 514(7521): 181-186.

[297] Spencer M, Gupta A, Van Dam L, et al. Artificial sweeteners: a systematic review and primer for gastroenterologists. Journal of Neurogastroenterology and Motility, 2016, 22(2): 168.

[298] Roca-Saavedra P, Mendez-Vilabrille V, Miranda JM, et al. Food additives, contaminants and other minor components: effects on human gut microbiota—a review. Journal of Physiology and Biochemistry, 2018, 74(1): 69-83.

[299] Palmnäs MS, Cowan TE, Bomhof MR, et al. Low-dose aspartame consumption differentially affects gut microbiota-host metabolic interactions in the diet-induced obese rat. Plos One, 2014, 9(10): e109841.

[300] Sanchez A, Reeser J, Lau H, et al. Role of sugars in human neutrophilic phagocytosis. The American Journal of Clinical Nutrition, 1973, 26(11): 1180-1184.

[301] Sørensen LB, Raben A, Stender S, et al. Effect of sucrose on inflammatory markers in overweight humans. The American Journal of Clinical Nutrition, 2005, 82(2): 421-427.

[302] Szucs E, Barrett K, Metcalfe D. The effects of aspartame on mast cells and basophils. Food and Chemical Toxicology, 1986, 24(2): 171-174.

[303] Rahiman F, Pool EJ. The in vitro effects of artificial and natural sweeteners on the immune system using whole blood culture assays. Journal of Immunoassay and Immunochemistry, 2014, 35(1): 26-36.

[304] Sehar I, Kaul A, Bani S, et al. Immune up regulatory response of a non-caloric natural sweetener, stevioside. Chemico-Biological Interactions, 2008, 173(2): 115-121.

[305] Boonkaewwan C, Toskulkao C, Vongsakul M. Anti-inflammatory and immunomodulatory activities of stevioside and its metabolite steviol on thp-1 cells. Journal of Agricultural and Food Chemistry, 2006, 54(3): 785-789.

[306] Yingkun N, Zhenyu W, Jing L, et al. Stevioside protects lps-induced acute lung injury in mice. Inflammation, 2013, 36(1): 242-250.

[307] Szczyrek M, Bitkowska P, Chunowski P, et al. Diet, microbiome, and cancer immunotherapy—a comprehensive review. Nutrients, 2021, 13(7): 2217.

彩 图

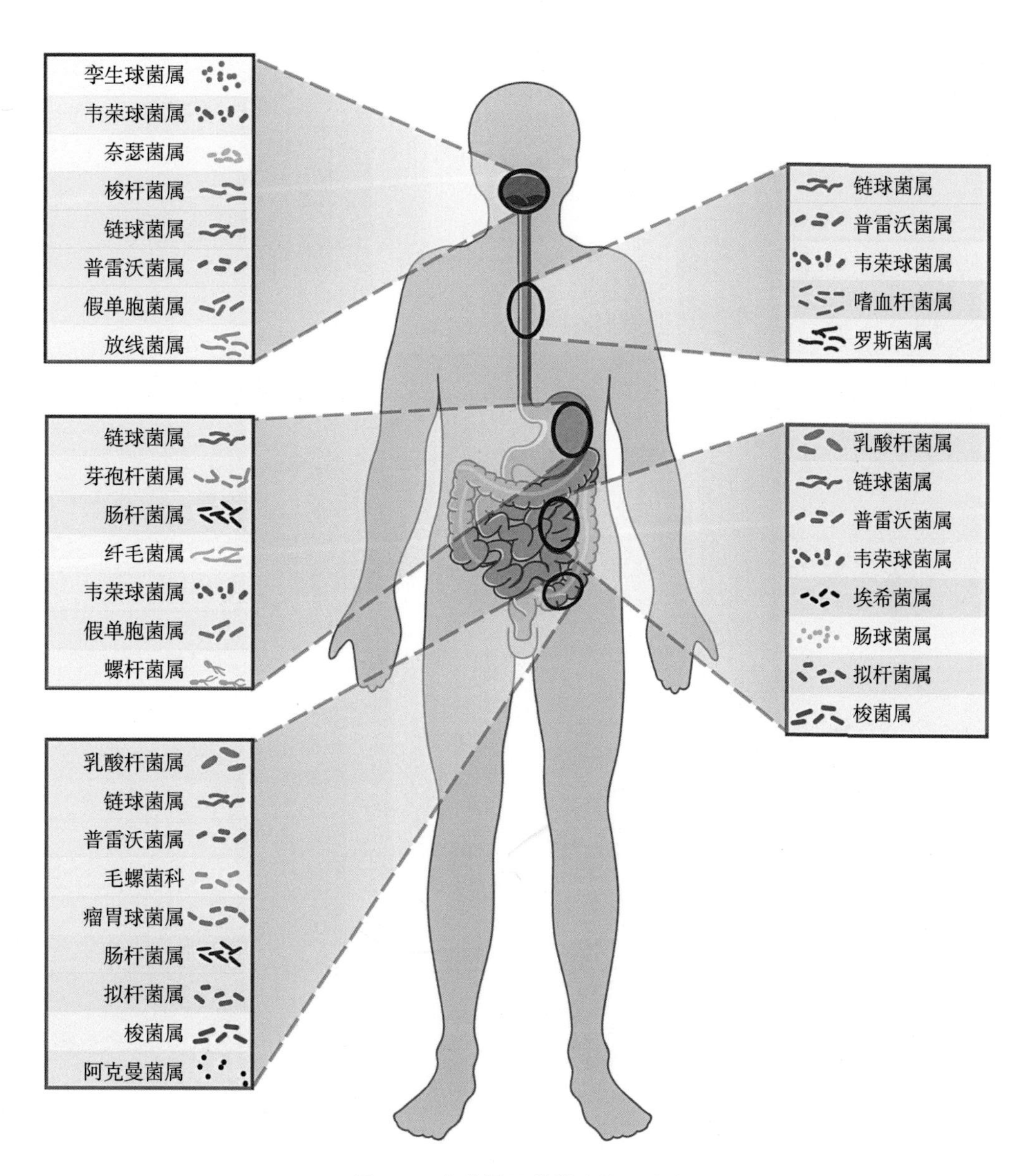

图 1-1　人类胃肠道微生物群分布

人类胃肠道不同部位具有不同的微生物群组成。口腔中的优势菌群为孪生球菌属、韦荣球菌属、奈瑟菌属、梭杆菌属、链球菌属、普雷沃菌属、假单胞菌属、放线菌属；食管的优势菌群为链球菌属、普雷沃菌属、韦荣球菌属、嗜血杆菌属、罗斯菌属；胃内优势菌群为链球菌属、芽孢杆菌属、肠杆菌属、纤毛菌属、韦荣球菌属、假单胞菌属、螺杆菌属；小肠中的优势菌群为乳酸杆菌属、链球菌属、普雷沃菌属、韦荣球菌属、埃希菌属、肠球菌属、拟杆菌属、梭菌属；大肠中优势菌群为乳酸杆菌属、链球菌属、普雷沃菌属、毛螺菌科、瘤胃球菌属、肠杆菌属、拟杆菌属、梭菌属、阿克曼菌属

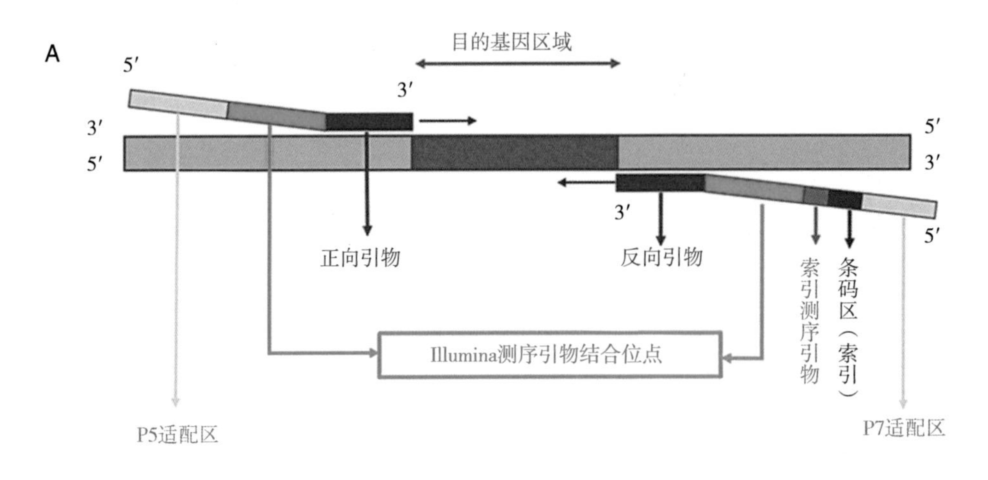

B

引物名称	引物核苷酸序列
V3F 5′	–aatgatacggcgaccaccgagatct acactctttccctacacgacgctcttccgatct NNNNCCTACGGGAGGCAGCAG–3′
V3R 5′	–caagcagaagacggcatacgagat xxxxx gtgactggagttcagacgtgtgctcttccgatct ATTACCGCGGCTGCTGG–3′

图 1-2　用于扩增 16S rRNA 基因区域，以生成 Illumina DNA 测序文库的引物中包含的各种功能元素

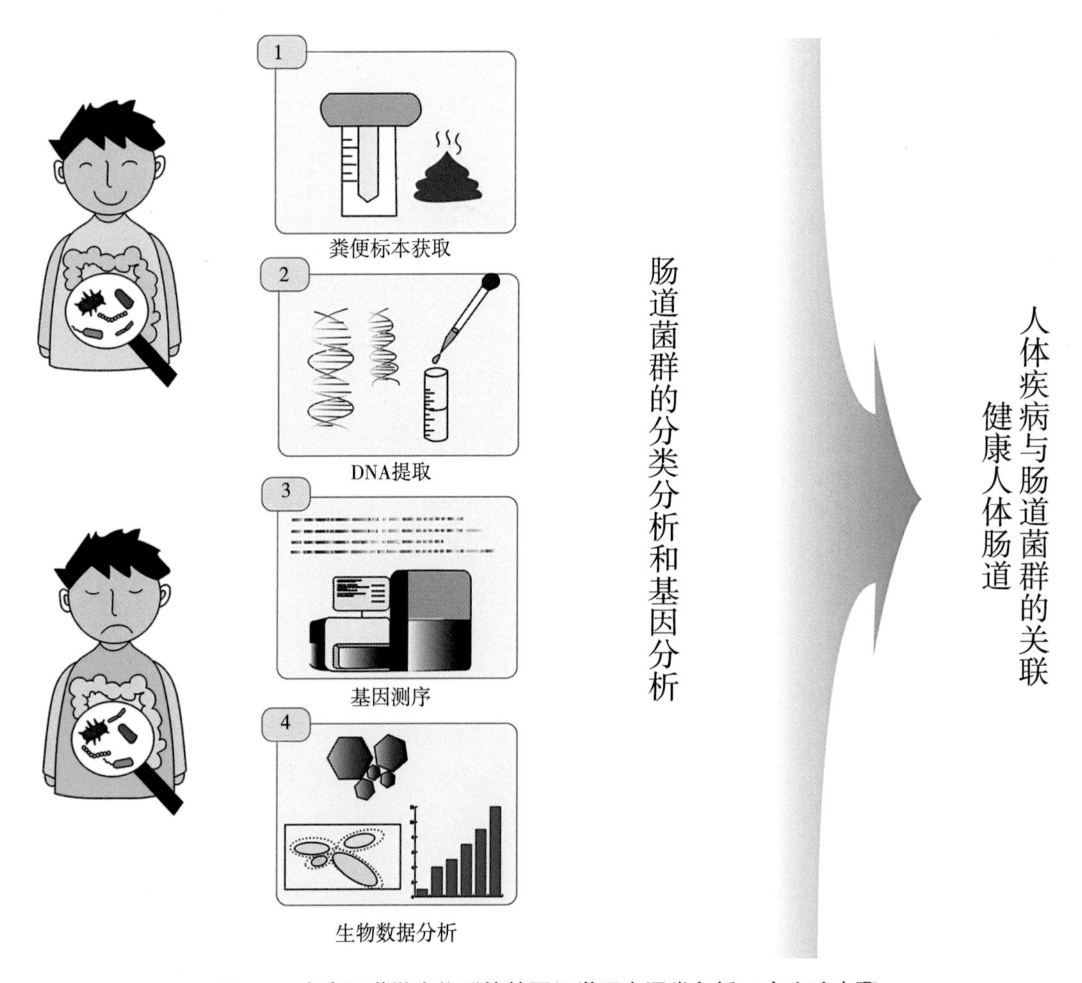

图 1-3　人类肠道微生物群的基因组学研究通常包括 4 个实验步骤

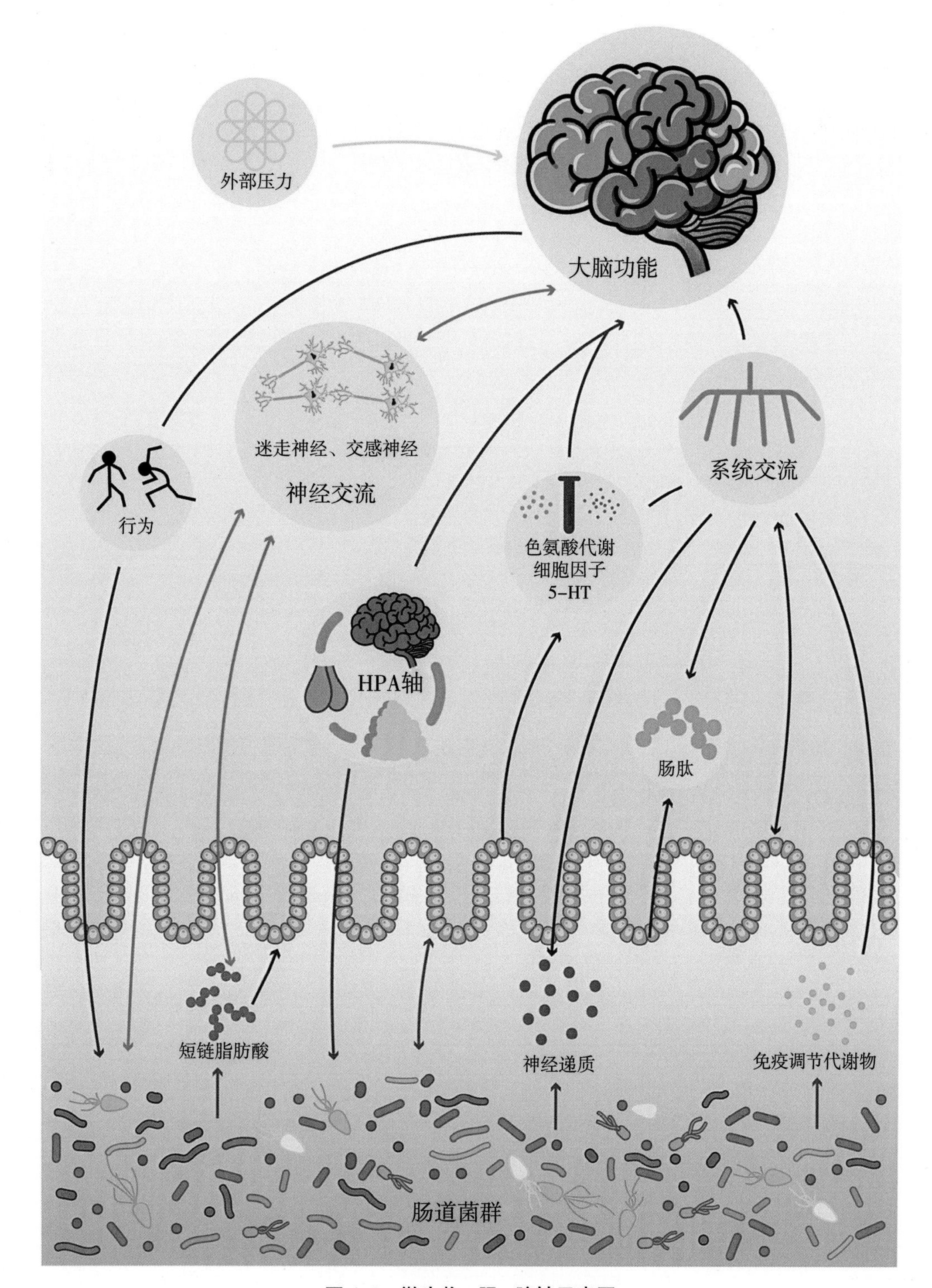

图 1-4 **微生物 - 肠 - 脑轴示意图**

HPA 轴 . 即 the hypothalamic–pituitary–adrenal axis，下丘脑 - 垂体 - 肾上腺轴，是神经内分泌系统的重要部分，参与控制应激反应，并调节许多躯体活动；5-HT.5- 羟色胺

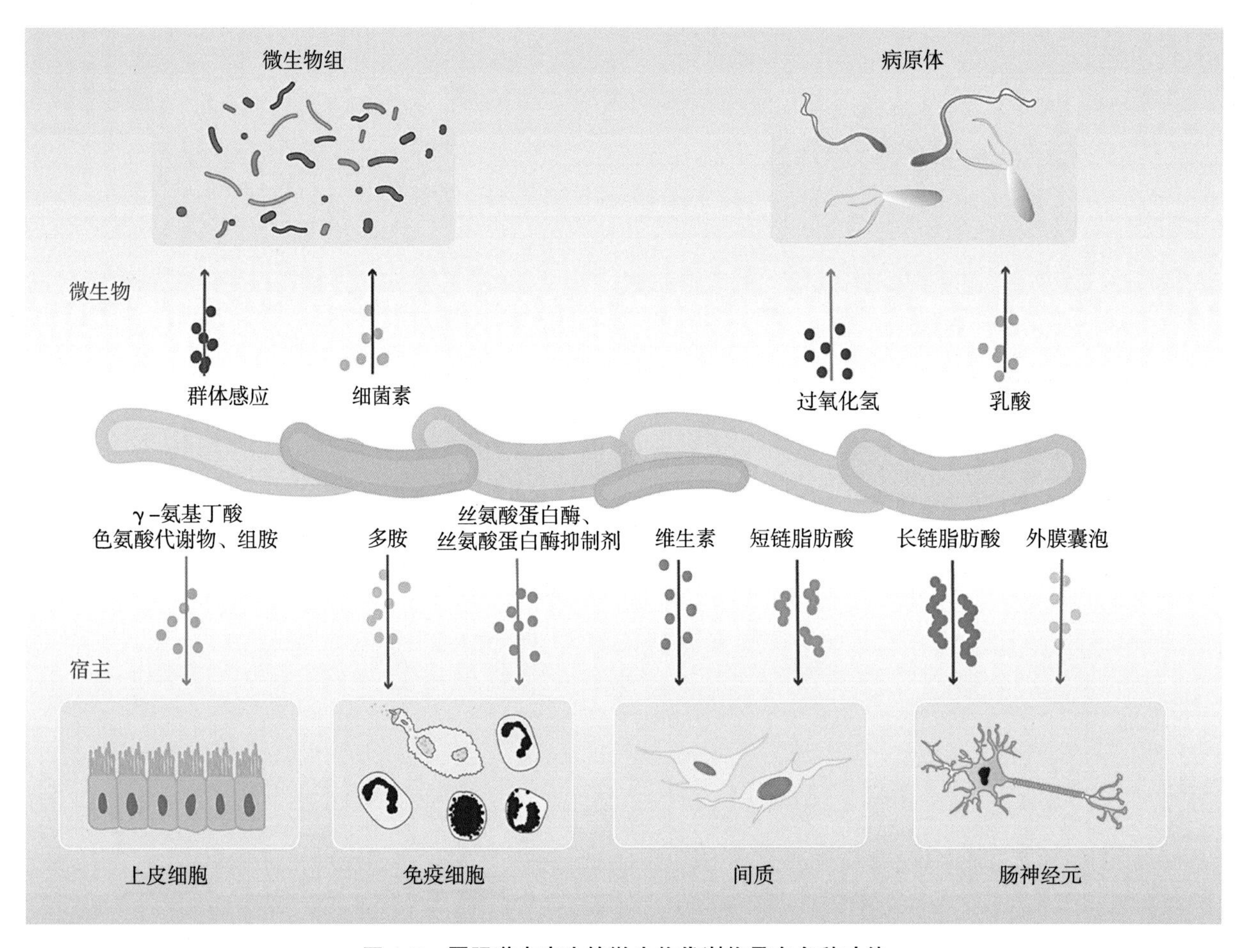

图 1-5　胃肠道中产生的微生物代谢物具有多种功能

微生物代谢产物可以调节微生物内部（微生物 - 微生物）和微生物 - 宿主之间的关系，从而影响人类健康。细菌参与群体感应，并能释放细菌素、过氧化氢和乳酸，从而对肠道微生物群和病原体产生影响。同时，细菌还可以分泌 γ- 氨基丁酸、色氨酸代谢物、组胺、多胺、丝氨酸蛋白酶、丝氨酸蛋白酶抑制剂、维生素、短链脂肪酸、长链脂肪酸和外膜囊泡等，这些代谢产物对宿主上皮细胞、免疫细胞、间质和肠神经元产生影响

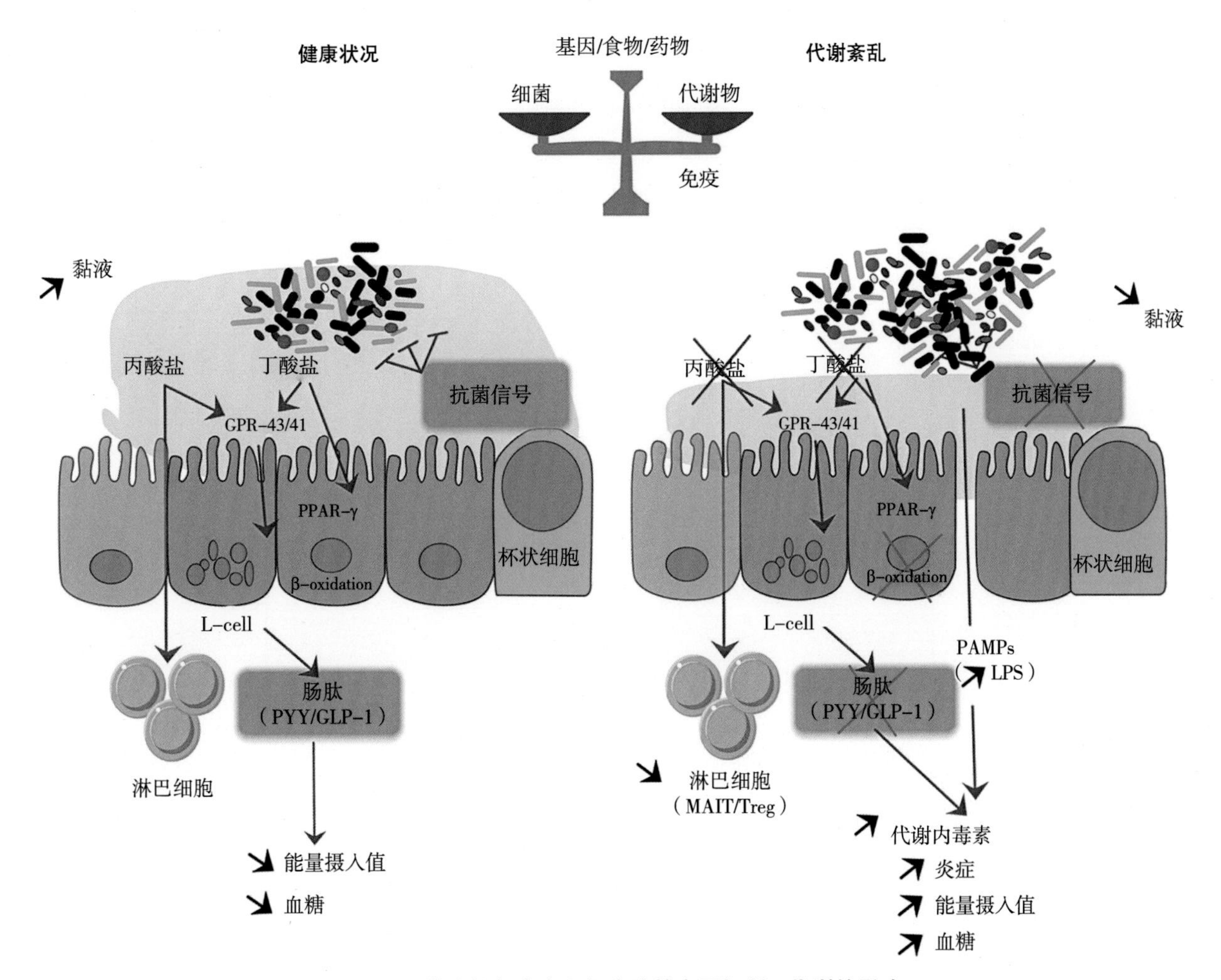

图 1-6 微生物与宿主之间交流的主要机制：代谢的影响

图的左边部分显示，在健康情况下，肠道微生物群的组成与较高的黏液层厚度、抗菌信号的产生以及丁酸和丙酸等不同短链脂肪酸的产生有关。丁酸和丙酸均与肠内分泌 L 细胞上表达的 GPR-43 和 GPR-41 结合，从而刺激胰高血糖素样肽 -1（GLP-1）或多肽 YY（PYY）等肠道肽的分泌。这种作用有助于减少食物摄入和改善葡萄糖代谢。丙酸还可以与淋巴细胞上表达的 GPR-43 结合，以维持宿主的免疫防御。丁酸激活 PPA R-γ 促进 β 氧化和肠道细胞耗氧的增加，该作用有助于维持肠腔厌氧状态。图的右边部分显示，在代谢紊乱过程中，肠道微生物群的变化与较低的黏液层厚度、抗菌防御能力的降低以及丁酸和丙酸的产生减少有关。在这种情况下，L 细胞分泌的肠肽减少。缺乏 PPA R-γ 激活将导致肠道细胞耗氧的减少，肠腔内可供微生物生存的氧气增加，进而促进肠杆菌科的增殖。丙酸的减少也导致包括 MAIT 及 Treg 细胞等在内的特定 T 细胞的丰度降低。总之，微生物环境和代谢物的这种变化将导致 PAMPs 的泄漏，如脂多糖在血液中增加，以及轻度炎症的发生。PPAR-γ. 氧化物酶体增殖物激活受体 -γ；GPR. G 蛋白偶联受体；MAIT. 黏膜相关不变 T 细胞；Treg 细胞 . 调节性 T 细胞；PAMPs. 病原体相关分子模式；LPS. 脂多糖；β-oxidation. β 氧化

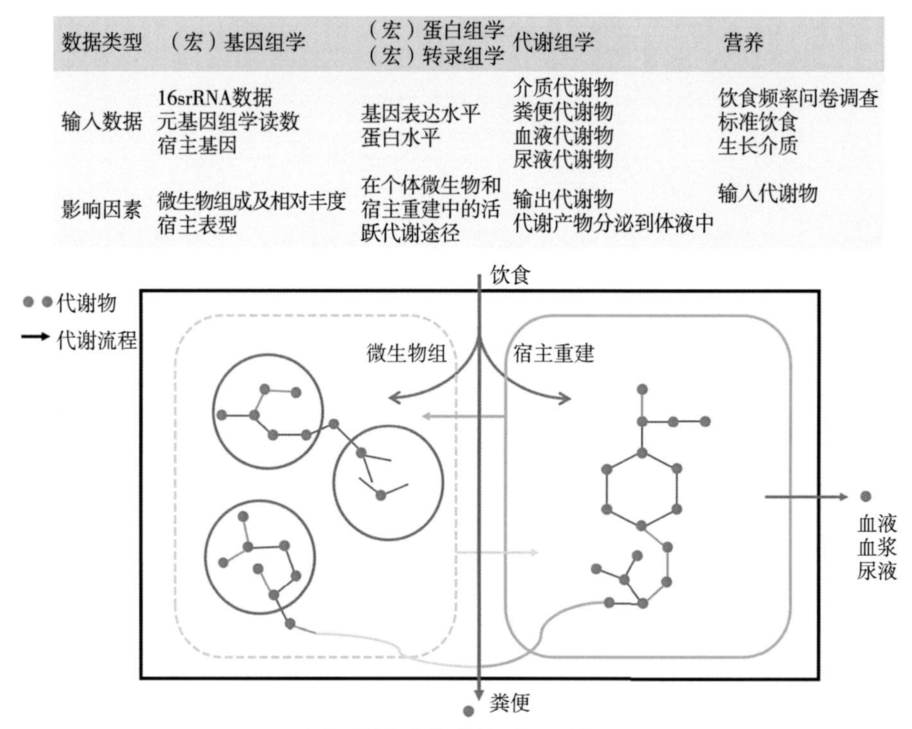

数据类型	（宏）基因组学	（宏）蛋白组学 （宏）转录组学	代谢组学	营养
输入数据	16srRNA数据 元基因组学读数 宿主基因	基因表达水平 蛋白水平	介质代谢物 粪便代谢物 血液代谢物 尿液代谢物	饮食频率问卷调查 标准饮食 生长介质
影响因素	微生物组成及相对丰度 宿主表型	在个体微生物和 宿主重建中的活 跃代谢途径	输出代谢物 代谢产物分泌到体液中	输入代谢物

图 1-7　宿主 - 微生物代谢模型设置的原理图概述

各种数据类型均可以与模型集成，以模拟个性化微生物、营养、体液中检测到的代谢物或基因表达数据。图中不同的颜色代表了模型中可以与不同数据类型关联的部分

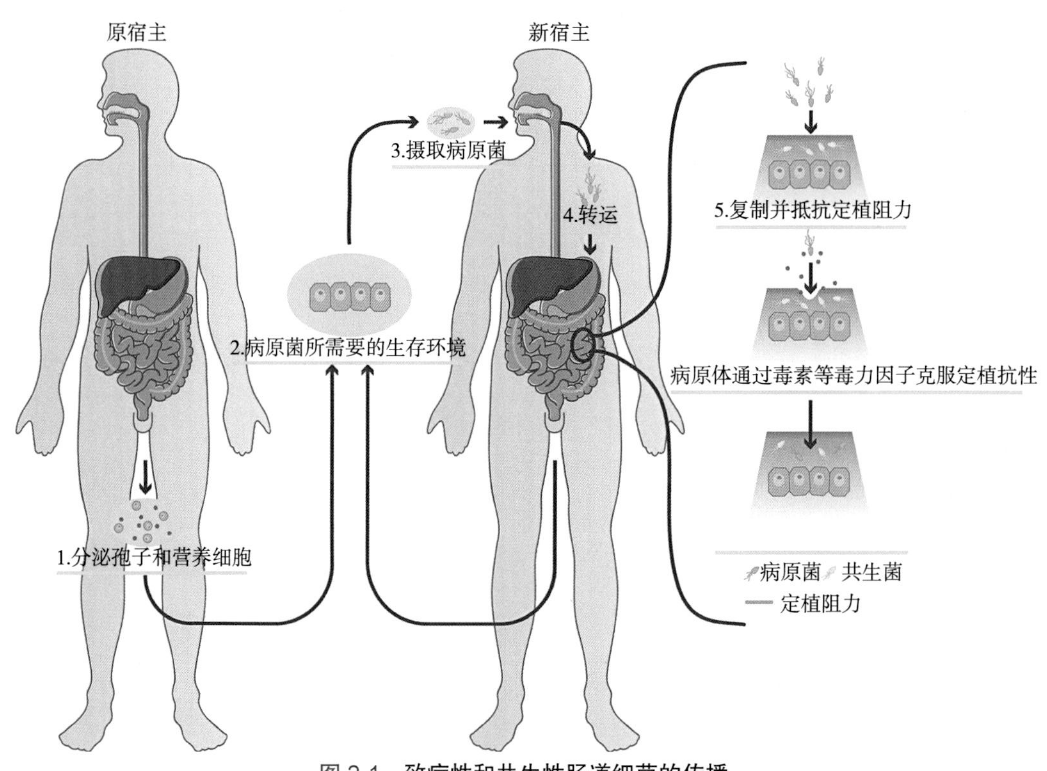

图 2-1　致病性和共生性肠道细菌的传播

肠道病原体和共生菌使用相似的机制在宿主之间传播。细菌从宿主的粪便中排出是传播的第一阶段（步骤 1）。为了促进新宿主的传播和随后的摄取，病原体可能会导致原宿主腹泻。一旦进入外部环境，这些以厌氧为主的细菌就会利用耐氧性、休眠和产生孢子等生存机制来生存和传播。如人、食物、动物和建筑环境，将作为传播的源头发挥作用（步骤 2）。一旦被新宿主摄取（步骤 3），细菌就转移到肠道（步骤 4）。来自新宿主原有菌群的竞争可以阻止定植（步骤 5）；但是，如阻力不足，病原菌就可以定植（步骤 5）。细菌的恢复性具有维持定植抗性和维持肠道健康相关的菌群多样性的功能。病原体可以通过诱导毒力因子（如毒素）的表达来克服定植抗力，这种毒力因子可能导致炎症并扰乱原有的菌群（步骤 5）

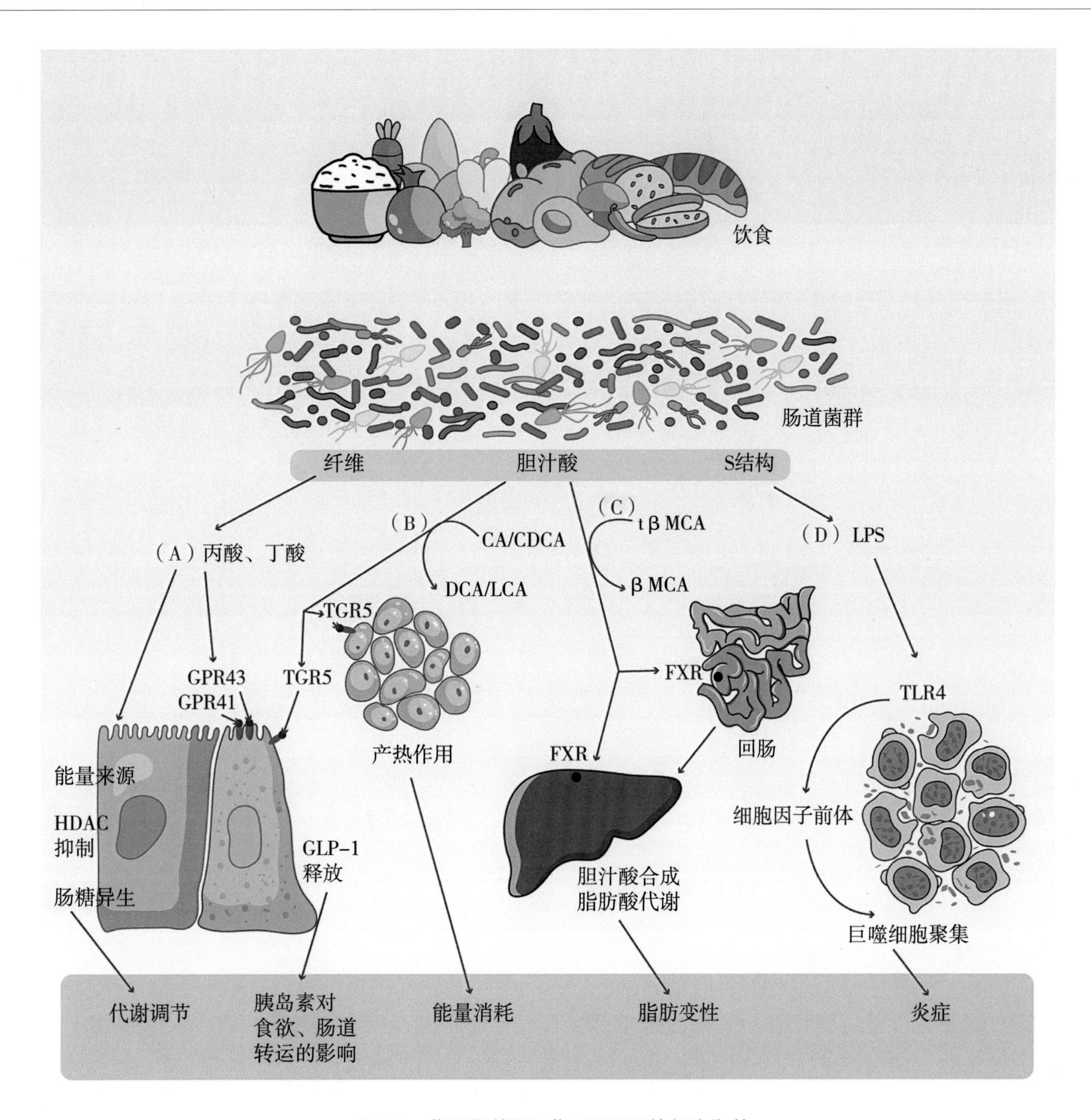

图 2-2　菌群代谢物调节不同组织的新陈代谢

A. 肠道菌群发酵膳食纤维产生短链脂肪酸。短链脂肪酸中的丁酸可作为能量来源并促进肠细胞中 HDAC 的抑制；丙酸和丁酸可刺激肠道糖异生，促进代谢调节。短链脂肪酸还能通过 G 蛋白偶联受体 43 和 41，促进胰高血糖素样肽 -1 的释放，菌群衍生的次级胆汁酸胆石酸和脱氧胆酸也可以通过胆汁酸受体 5 同样促进胰高血糖素样肽 -1 的释放，达到抑制食欲、减少肠道转运的效果。B. 胆石酸和脱氧胆酸可以促进棕色脂肪产热，增加能量消耗。C. 肠道菌群减轻肠道和肝脏的法尼醇 X 受体抑制，从而减少胆汁酸的合成并改变脂肪酸的代谢。D. 脂多糖是一种来源于革兰阴性细菌膜的促炎分子，通过 Toll 样受体 4 促进白色脂肪组织中巨噬细胞的募集和极化，诱导炎症。HDAC. 组蛋白去乙酰化酶；GPR. G 蛋白偶联受体；TGR. 胆汁酸受体；GLP-1. 胰高血糖素样肽 -1；CA. 胆酸；CDCA. 鹅脱氧胆酸；DCA. 脱氧胆酸；LCA. 石胆酸；MCA. 鼠胆酸；FXR. 法尼醇 X 受体；LPS. 脂多糖；TLR4. Toll 样受体 4

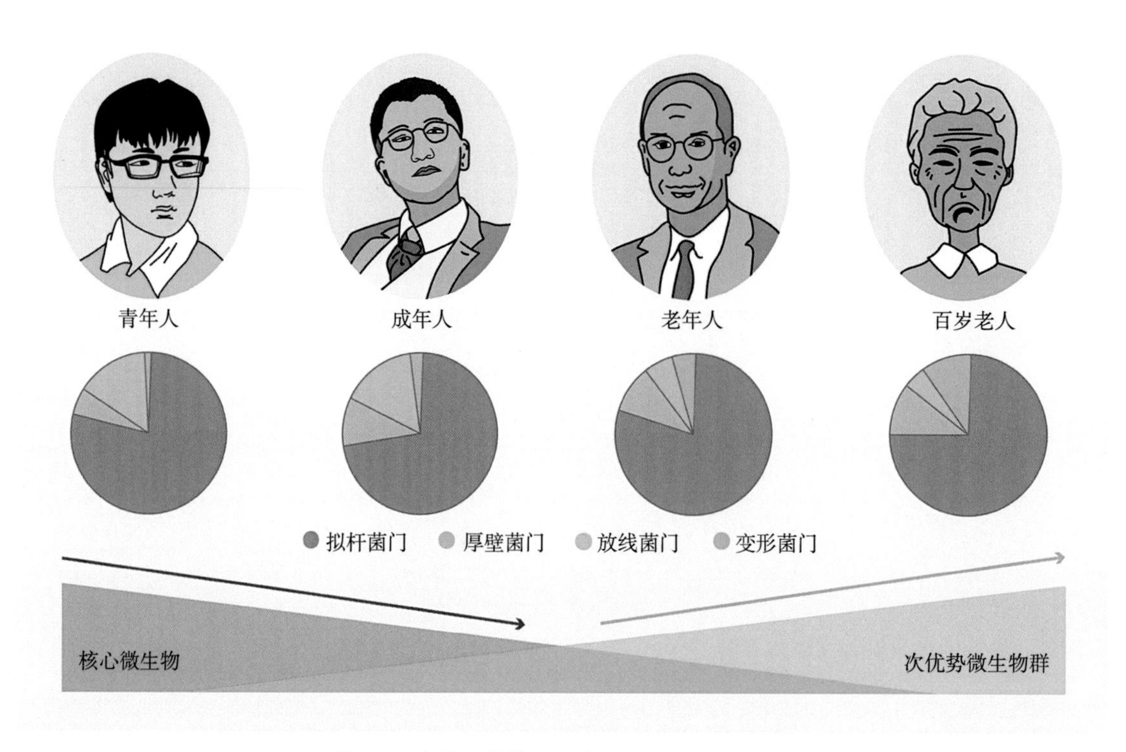

图 2-3　**人类肠道菌群组成随年龄增长而变化**

①青少年核心肠道菌群组成与成年人相似，但青少年的梭菌和双歧杆菌属的水平显著高于成年人；②成人肠道菌群个体差异极大，但以厚壁菌门、拟杆菌门、变形菌门和放线菌门为核心的肠道菌群是健康所必需的；③老年人肠道菌群会出现稳态失衡，多样性降低，肠道菌群的主要构成部分也由厚壁菌门向拟杆菌门转变，拟杆菌种类多样性增加，双歧杆菌种类多样性减少，而梭状芽孢杆菌较丰富；④百岁老人肠道菌群的特点是厚壁菌门的重排和变形杆菌的富集，包括机会性促炎细菌，并出现黏液真杆菌的明显增加

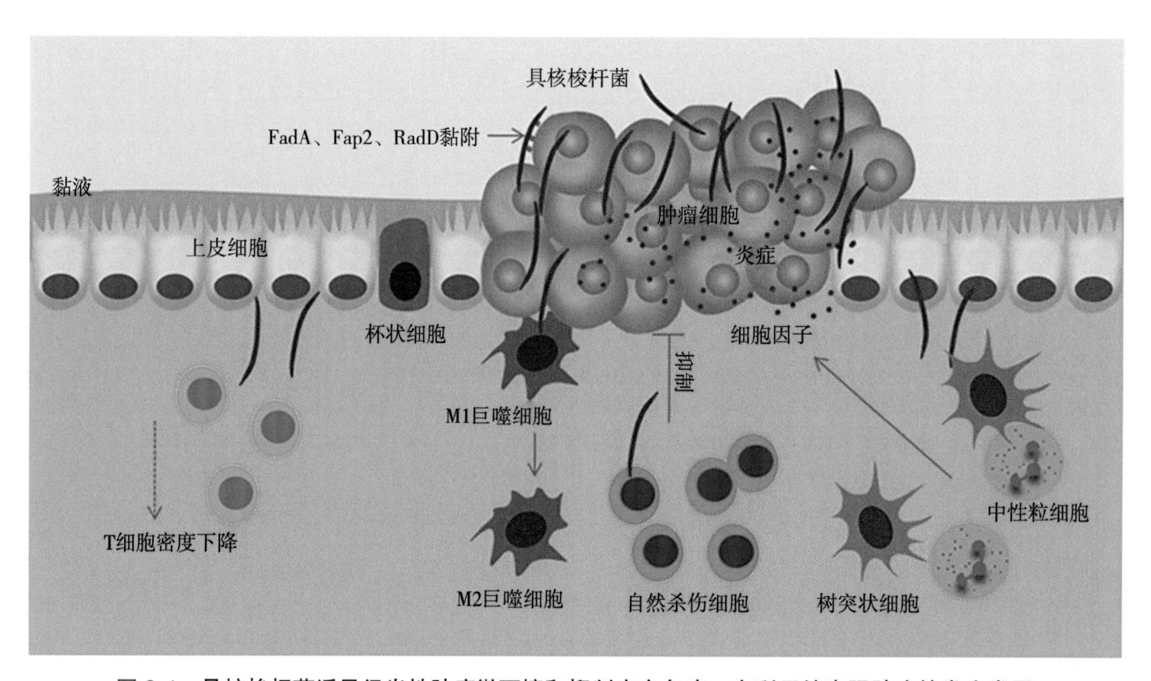

图 2-4　**具核梭杆菌诱导促炎性肿瘤微环境和抑制宿主免疫，有利于结直肠肿瘤的发生发展**

具核梭杆菌通过黏附分子 FadA、Fap2 和 RadD 与结肠上皮结合，侵入黏膜。具核梭杆菌的入侵增加了炎症细胞的浸润，刺激细胞因子的释放，从而促进细胞增殖。此外，具核梭杆菌还与多种免疫细胞相互作用，导致 T 细胞密度降低，巨噬细胞 M2 极化增强，自然杀伤细胞活性抑制，树突状细胞和肿瘤相关中性粒细胞增多，从而抑制抗肿瘤免疫反应，促进肿瘤进展。

［Wu J，Li Q，Fu XS. Fusobacterium nucleatum Contributes to the Carcinogenesis of Colorectal Cancer by Inducing Inflammation and Suppressing Host Immunity. Translational oncology，2019，12（6）:846-851.］

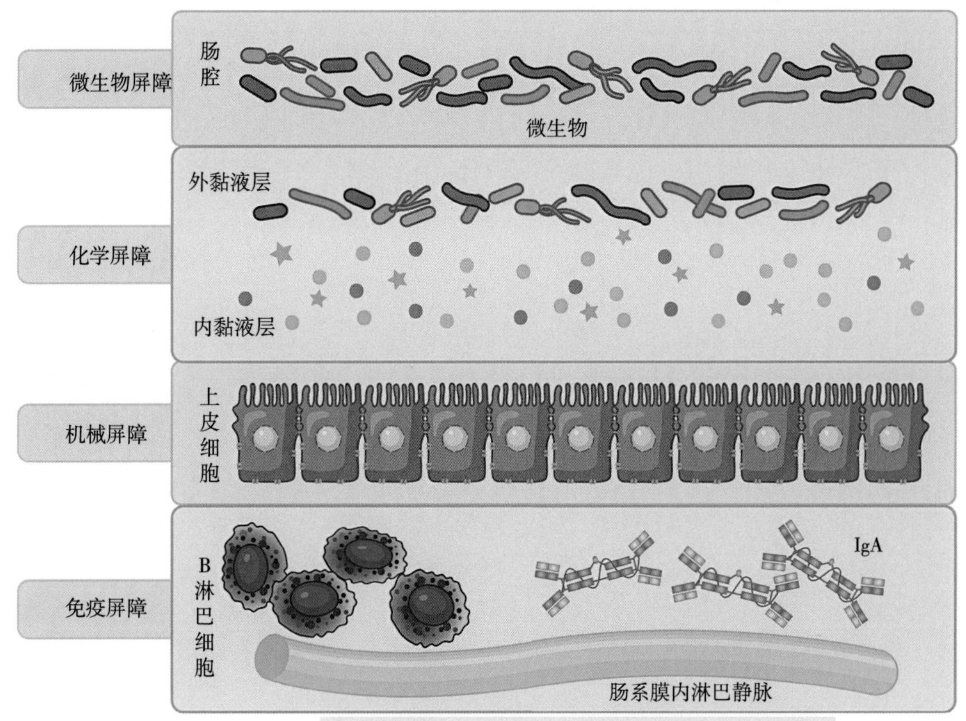

图 3-1　肠道屏障的 4 个组成部分

微生物屏障是指肠道正常菌群，各种菌株相互依赖，有益细菌通过生物拮抗和免疫功能形成宿主生物防御。化学屏障是指消化道分泌胃酸、黏液、黏蛋白、胆汁、糖蛋白、黏多糖、各种消化酶、溶菌酶等化学物质。机械屏障由肠腔和内环境之间的一层柱状上皮细胞组成。上皮细胞之间是细胞间连接复合体，包括紧密连接、黏附连接、桥粒和缝隙连接。免疫屏障由肠道上皮细胞、肠上皮内淋巴组织、固有层淋巴细胞、派尔集合淋巴结、肠系膜淋巴结、其他肠组织和 sIgA 组成

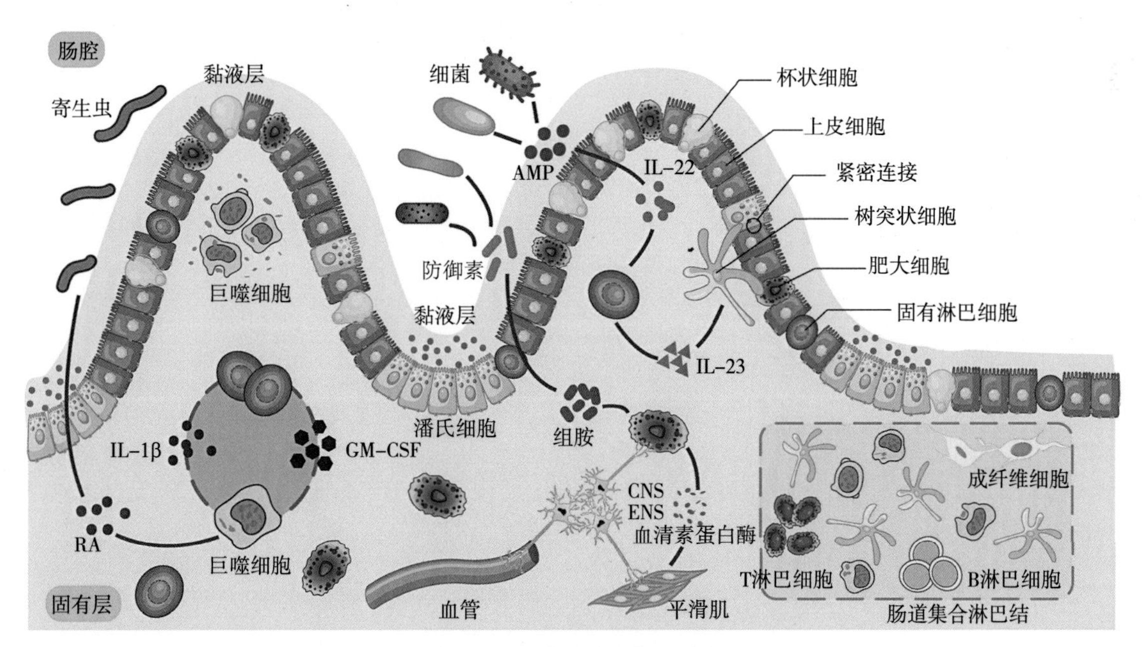

图 3-2　肠道屏障结构和功能

化学屏障中的抗菌肽，如两性调节素，可攻击包括细菌和蠕虫等病原体。机械屏障就像一堵墙，在空间上将入侵的微生物和宿主隔开。肠上皮中有许多类型的细胞可调节肠黏膜免疫功能。肠道屏障被破坏后，肠道细菌从肠腔渗漏到固有层，诱导宿主免疫细胞产生过度免疫反应。巨噬细胞或树突状细胞释放的维 A 酸有助于抵抗蠕虫感染。树突状细胞产生的 IL-23 可调控固有淋巴细胞释放 IL-22，后者可促进上皮细胞分泌 AMP 以应对细菌感染。此外，巨噬细胞来源的 IL-1β 可促进固有淋巴细胞产生粒细胞 - 巨噬细胞集落刺激因子，进一步刺激单核细胞向巨噬细胞分化。RA. 维甲酸；GM-CSF. 粒细胞 - 巨噬细胞集落刺激因子；AMP. 抗菌肽；CNS. 中枢神经系统；ENS. 肠道神经系统

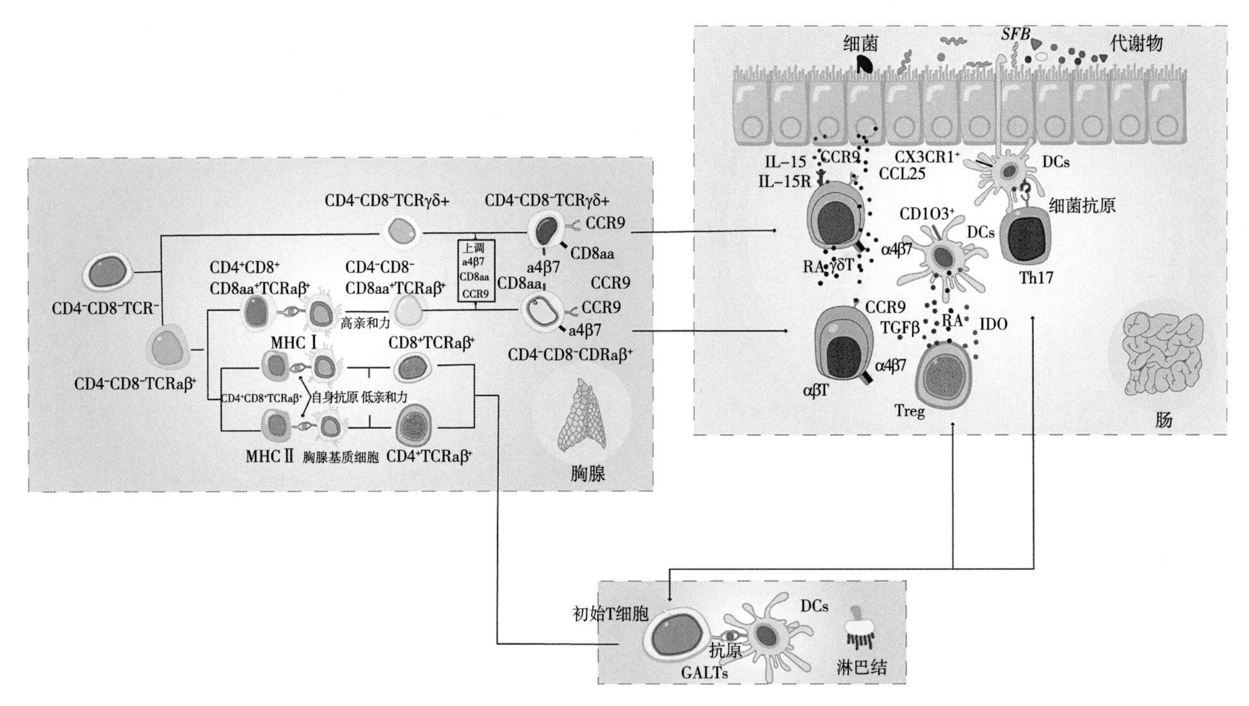

图 3-3 肠道 T 淋巴细胞的发育与成熟

肠道 T 细胞可分为诱导的“传统”（或“a 型”）肠道 T 细胞或“非传统”（或“b 型”）肠道 T 细胞。传统的肠道 T 细胞表达 TCRαβ 和 CD4 或 CD8αβ，并作为 TCR2 的辅助受体。非传统肠道 T 细胞表达 TcRαβ 或 TcRγδ，通常也表达 CD8αα 同源二聚体。传统的 T 细胞来源于胸腺中的 $CD4^-CD8^-$（DN）祖细胞，发展为 SP $CD4^+T$（MHC Ⅰ）细胞或 $CD8^+T$ 细胞（MHC Ⅱ），随后迁移到外周淋巴器官，如淋巴结，在那里它们遇到抗原并获得激活的效应器表型，从而驱动它们迁移到肠道。或者，胸腺中未成熟的三阴性胸腺细胞（$CD4^-CD8^-TCR^-$）分化为双阴性（$CD4^-CD8^-$）、TCRγδ 阳性或 TCRαβ 阳性的肠道 T 细胞前体。CCR9. 趋化因子受体 9；MHC Ⅰ. 主要组织相容性复合体 - Ⅰ；MHC Ⅱ. 主要组织相容性复合体 - Ⅱ；*SFB*. 肠道分节丝状菌；RA. 维甲酸；CCL25. 趋化因子 25；IDO. 双加氧酶；DCS. 树突状细胞；Treg. 调节性 T 细胞；TCR.T 细胞受体；TGF. 转化生长因子；GALT. 肠道相关淋巴组织

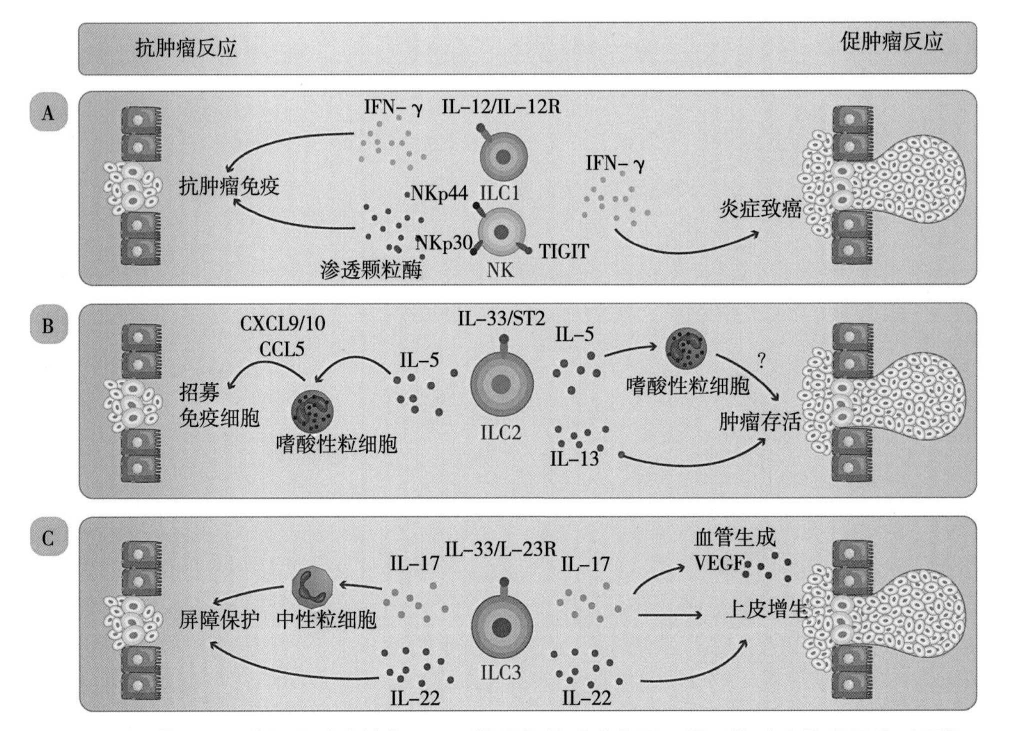

图 3-4 肠道 ILC 既能促进肿瘤的发展，又能发挥抗肿瘤作用，从而使肿瘤的发展达到平衡

NK 细胞和上皮内 ILC1（A）、ILC2（B）和 ILC3（C）通过其激活或抑制受体信号调节这些免疫细胞的功能。ILC 通过分泌细胞因子和细胞毒分子，调节肿瘤微环境，抑制或促进大肠癌的发生和发展。ILC. 固有淋巴细胞；CXCL. 趋化因子配体；TIGIT.T 细胞免疫球蛋白和 ITIM 结构域蛋白；VEGF. 血管内皮生长因子；CCL5. 趋化因子 5

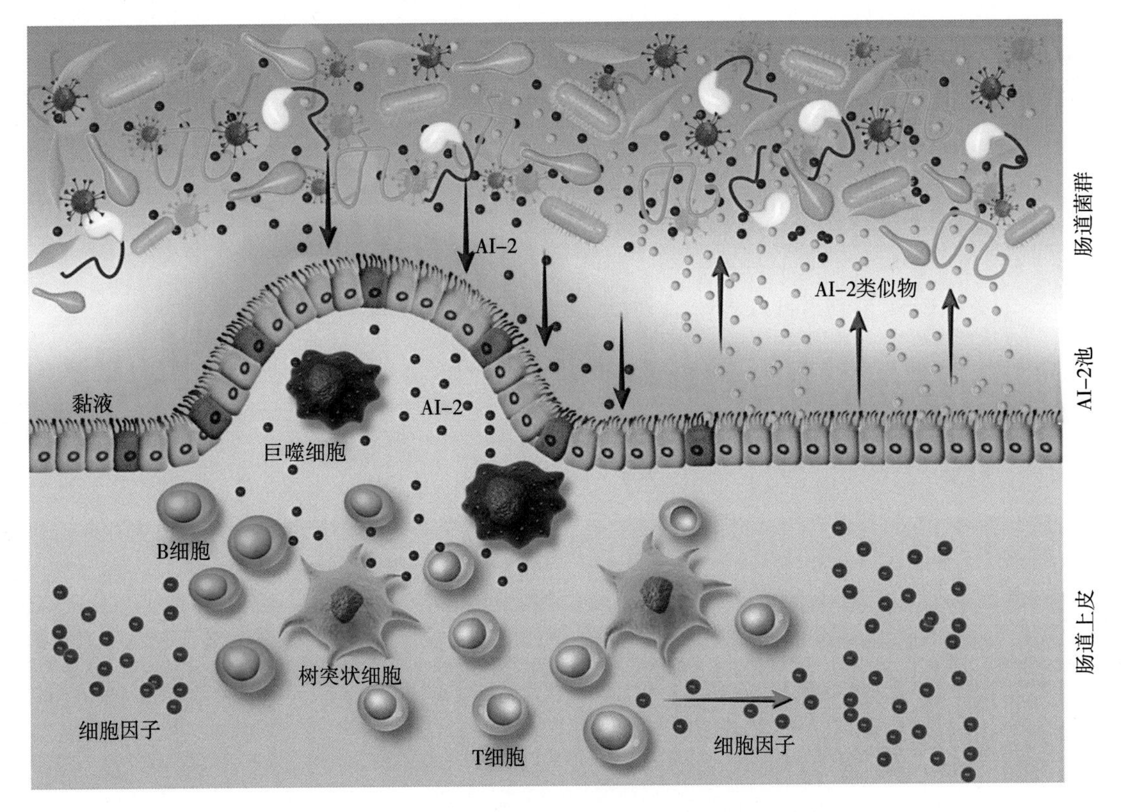

图 4-1　肠道微生物群与宿主之间的界间信号传递

肠道微生物群的群体感应分子（AI-2）通过与 T 淋巴细胞、巨噬细胞、树突状细胞和中性粒细胞等免疫细胞作用，从而与宿主肠道对话。宿主肠道分泌的激素可以作用于肠道细菌，并通过 AI-2 类似物干扰和淬灭细菌的群体感应，从而影响肠道细菌的生物学行为。引自：Li Q，Y Ren，X Fu，Inter-kingdom signaling between gut microbiota and their host. Cell Mol Life Sci，2019, 76（12）: 2383-2389.

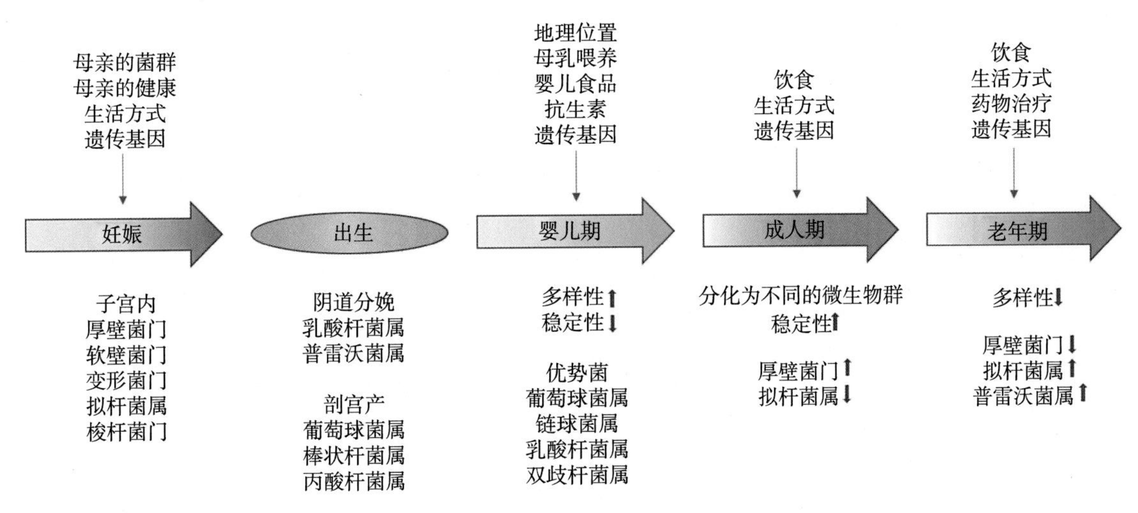

图 5-1　人类生命周期中微生物群的变化

子宫不是无菌环境，在胎盘、胎膜、脐带血和胎粪中都存在细菌。婴儿肠道细菌的定植取决于分娩方式，并从中获得独特的细菌群落。幼儿肠道菌群的组成经常变化，非常多样化且不稳定，随着年龄的增长，肠道菌群变得更稳定。与年轻人相比，老年人肠道菌群多样性降低，并且产生短链脂肪酸的细菌减少，分泌内毒素的革兰阴性细菌增加

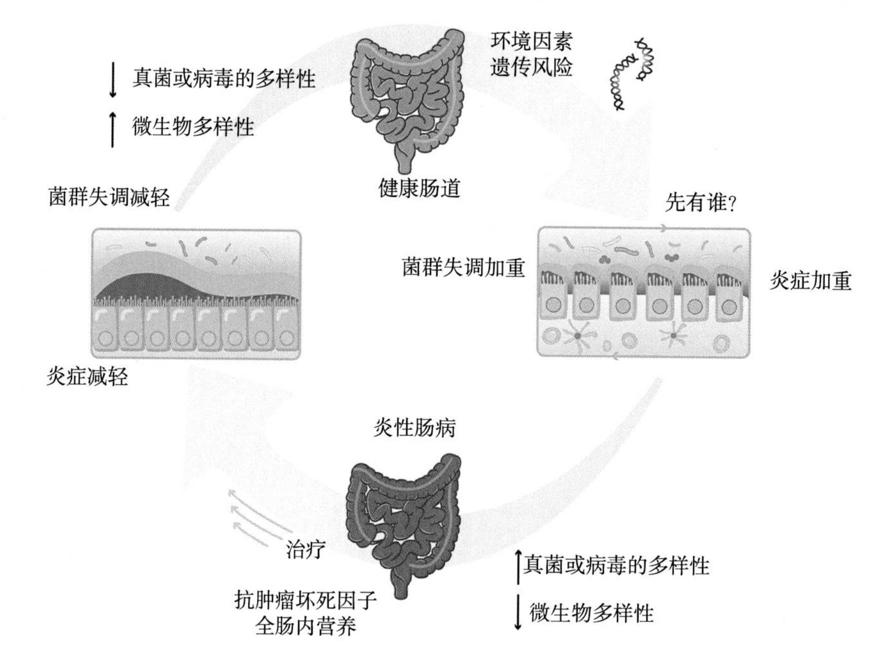

图 5-2 **肠道菌群失调和免疫反应**

在环境因素影响下，具有遗传倾向（例如 NOD2、IL-10R 或 ATG16L1 突变）的健康个体可能发生肠道菌群失调和炎症，但两者的先后顺序尚不清楚。一旦发生微生态失调和炎症反应的正反馈循环，就会导致炎性肠病。用抗肿瘤坏死因子或全肠内营养治疗后，炎症减轻，随后菌群失调好转，再次回到肠道稳态

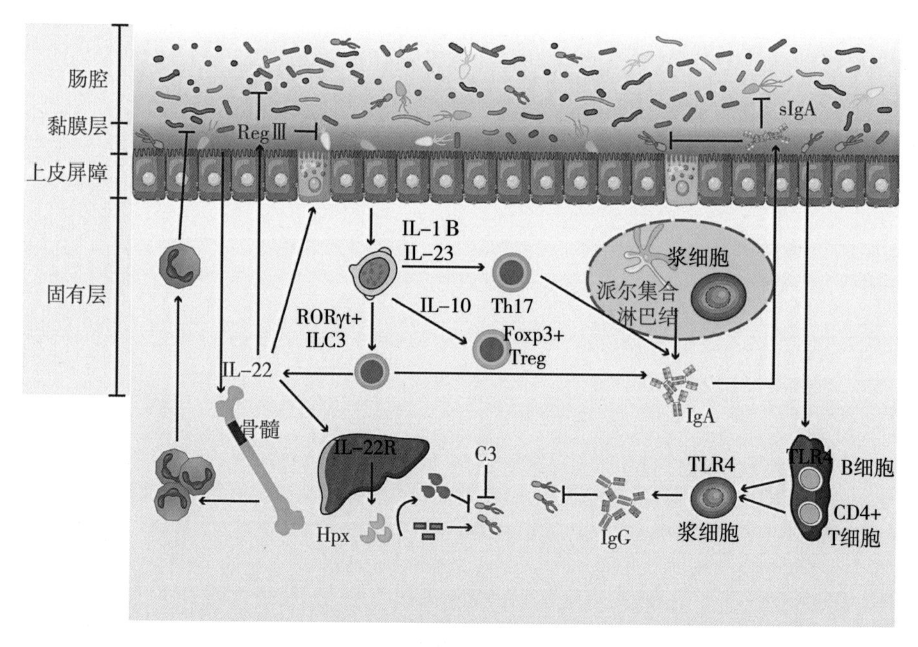

图 5-3 **肠道微生物群对宿主免疫反应的影响**

通过 Toll 样受体，来自肠道共生细菌的分子促进骨髓中嗜中性粒细胞生成和感染后中性粒细胞的动员。肠道共生细菌的存在对于肠道巨噬细胞中 pro-IL-1 的表达很重要。在肠道损伤过程中，肠道共生细菌通过 NLRP3 炎症体将其裂解为成熟的 IL-1，从而促进适当的炎症反应。来自肠道巨噬细胞的 IL-1、IL-23 和 IL-6 对促进黏膜 Th17 细胞应答很重要，而肠道巨噬细胞的 IL-10 和微生物短链脂肪酸参与了稳态条件下的肠道 Tregs 的发育。肠道菌群的存在对于诱导 T 细胞依赖性和独立产生 IgA 抗体至关重要。其中大多数 IgA 抗体对肠道共生体具有特异性，并被分泌到肠腔中，以侵袭性细菌为目标，防止它们穿过上皮屏障。在肠道柠檬酸杆菌或艰难梭菌感染期间，由 3 型固有淋巴细胞产生的 IL-22 可以诱导肝细胞产生血红素结合蛋白和补体 C3，分别抑制全身性移位细菌的生长和促进其清除。在稳态条件下，部分革兰阴性肠道共生菌诱导机体产生能够识别细菌表面抗原的 IgG 抗体，如某些革兰阴性病原体上表达的胞壁质脂蛋白，从而有助于宿主抵御肠道共生体或病原体的全身感染。RegⅢ. 胰岛再生源蛋白Ⅲ；TLR4.toll 样受体 4；RORγt. 维甲酸相关孤儿核受体 γt；Hpx. 血红素结合蛋白；Foxp3. 叉头状转录因子 P3；ILC3.3 型固有淋巴细胞；Treg. 调节性 T 细胞

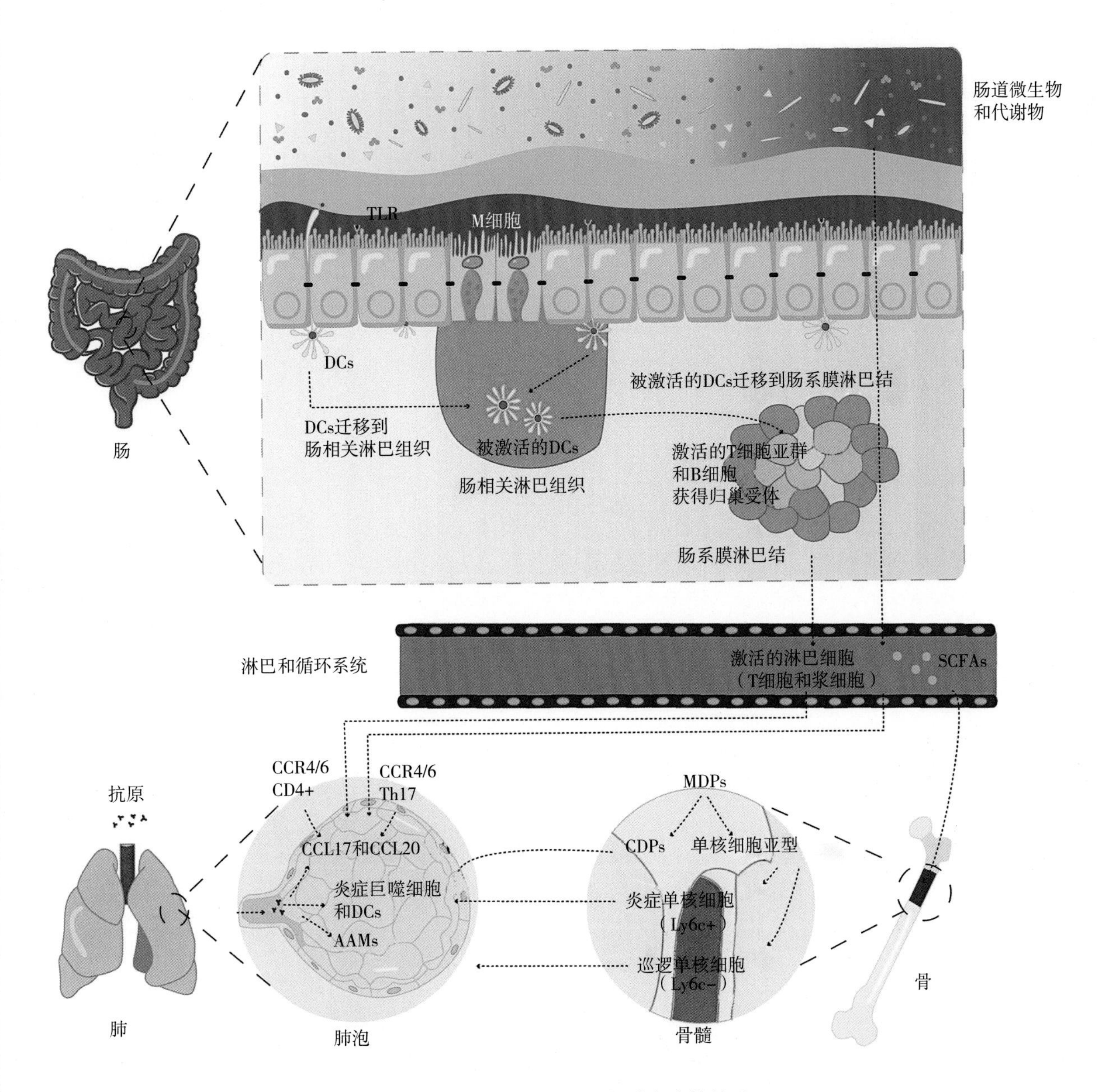

图 5-4　肠道微生物群与肺免疫的关系

肠道细菌与肺免疫之间的相互作用由多种因素介导，包括病原体相关分子模式、模式识别受体、短链脂肪酸、肠道完整性和固有层免疫细胞。在正常状态下，树突状细胞通过微折叠细胞活性、树突延伸和肠道屏障功能从管腔中连续取样，这决定了细菌 /PAMP 易位。在树突状细胞取样后，这些细胞迁移到肠道相关淋巴样组织，然后迁移到肠系膜淋巴结，根据肠道微生物群释放的某些细胞因子来调节淋巴细胞（T 和 B 细胞）的分化和归巢。活化的 T 细胞和 B 细胞通过循环分布在肺部。此外，微生物暴露后肺部产生的 CCL20 和 CCL17 的水平有助于 T 细胞亚群的印记。短链脂肪酸可以渗入骨髓，通过影响树突状细胞祖细胞向炎症或抗炎免疫细胞的分化来影响肺免疫。肺部的炎性巨噬细胞和树突状细胞来源于树突状细胞前体和 Ly6c+ 炎性单核细胞；活化巨噬细胞是肺部的抗炎免疫细胞，来源于 $Ly6c^-$ 单核细胞。DCs. 树突状细胞；TLR.toll 样受体；CCR. 趋化因子受体；CCL. 趋化因子配体；CDPs. 普通树突状细胞前体细胞；MDPs. 巨噬树突细胞祖细胞；AAMs. 选择性激活的巨噬细胞；SCFAs. 短链脂肪酸

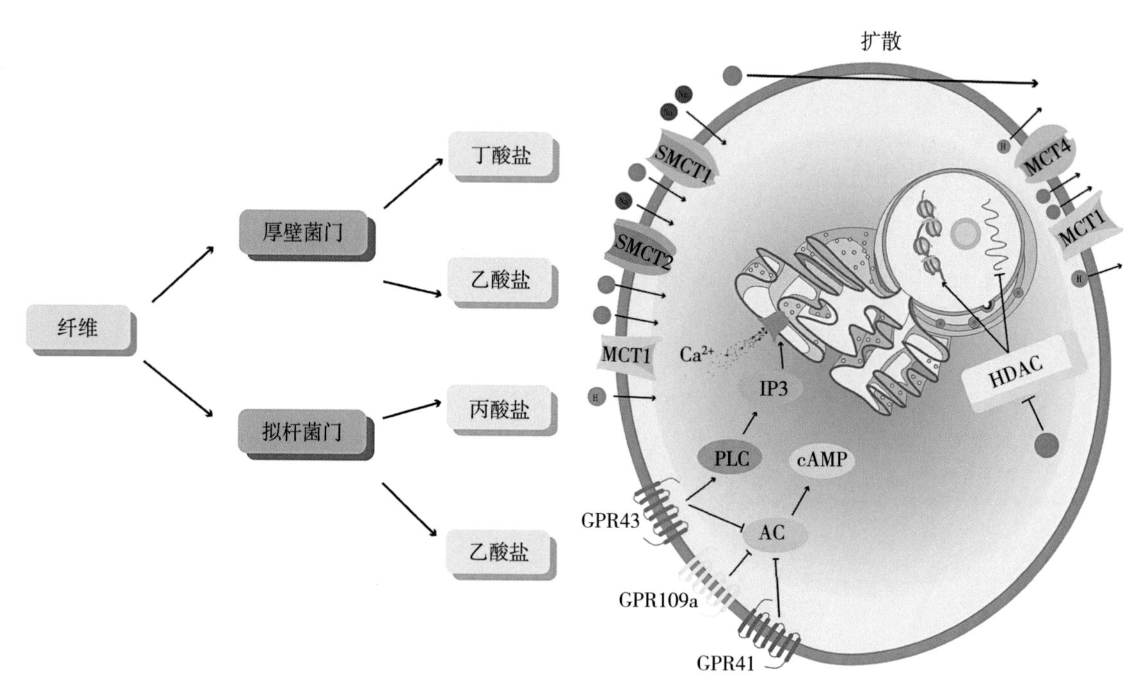

图 6-1　短链脂肪酸的形成、转运和作用机制

SCFAs（包括乙酸盐、丙酸盐和丁酸盐）由细菌对膳食纤维发酵而产生。SCFAs 的吸收通过 3 种机制发生：被动扩散、电中性和生电性摄取。SCFAs 的电荷决定其吸收是通过被动扩散还是通过载体机制进行。例如，当 SCFAs 以质子化形式存在时，SCFAs 的吸收主要为被动扩散；在生理条件下，这是 SCFAs 转运的主要机制。相反，阴离子形式的 SCFAs 的吸收依赖于载体介导，这种吸收通过 4 种转运蛋白进行。MCT1 和 MCT4 是电中性转运蛋白，它们依赖氢，而 SMCT1 和 SMCT2 则依赖钠，分别是生电性转运蛋白和电中性转运蛋白。SCFAs 通过刺激 3 种 G 蛋白偶联受体（GPR41、GPR43 和 GPR109a），以及作为组蛋白去乙酰化酶抑制剂而发挥作用。HDAC. 组蛋白去乙酰化酶；AC. 腺苷酸环化酶；cAMP. 环磷酸腺苷；PLC. 磷脂酶 C；IP3. 三磷酸肌醇；MCT. 单羧酸转运体；SMCT.Na^+ 偶联单羧酸转运蛋白；GPR.G 蛋白偶联受体

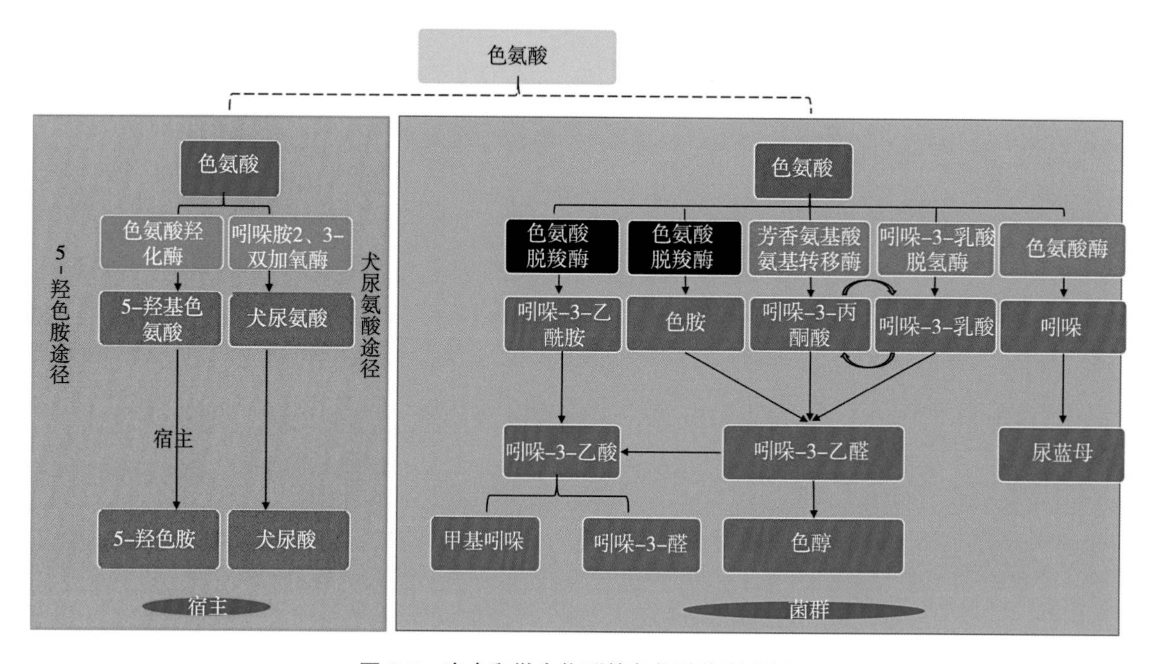

图 6-2　宿主和微生物群的色氨酸代谢途径

在微生物代谢产物中，吲哚和吲哚酸衍生物是肠道内主要的色氨酸微生物代谢产物，肠道微生物产生不同的代谢产物，这些代谢产物是细菌产生催化酶的基础。犬尿氨酸途径和 5- 羟色胺途径是宿主色氨酸代谢的主要途径

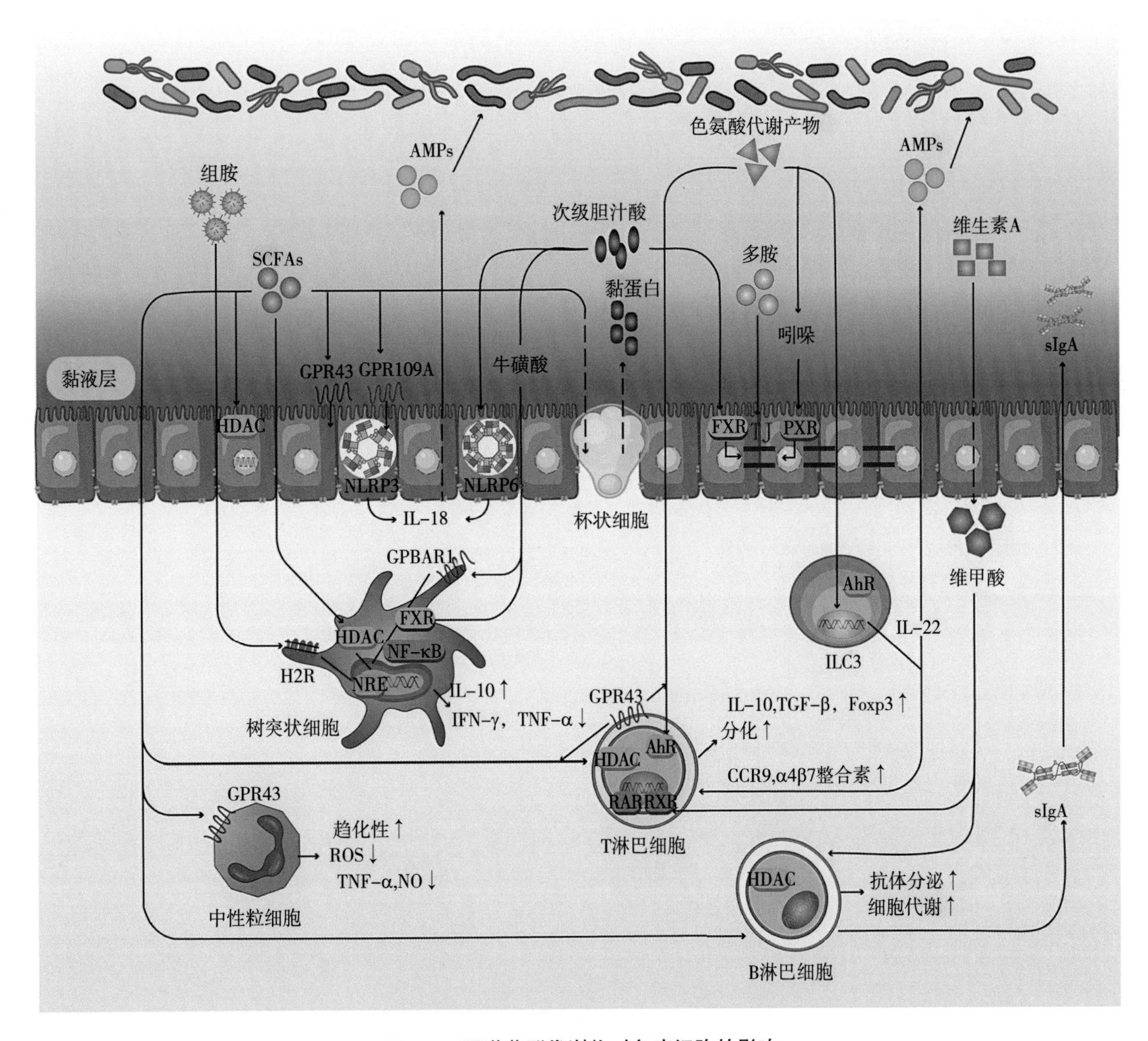

图 6-3　肠道菌群代谢物对免疫细胞的影响

来自微生物或宿主的代谢物参与复杂的宿主 - 微生物相互作用。许多微生物群相关代谢物具有生物活性，如短链脂肪酸、吲哚和次级胆汁酸等，这些代谢物可以通过信号通路与宿主中相应的传感平台反应。生物活性代谢物影响先天性和适应性免疫细胞的成熟、激活、极化和效应功能，从而调节抗炎或促炎反应。AMPs. 抗菌肽；HDAC. 组蛋白去乙酰化酶；NLRP3. 核苷酸结合寡聚化结构域样受体蛋白 3；NLRP6. 核苷酸结合寡聚化结构域样受体蛋白 6；FXR. 法尼醇 X 受体；PXR. 孕烷 X 受体；sIgA. 分泌型免疫球蛋白 A；H2R. 组胺受体 2；NRE. 负调控元件；NF-κB. 核转录因子 -κ B；GPBAR1.G 蛋白偶联胆汁酸受体 1；AhR. 芳香烃受体；RAR. 维甲酸受体；RXR. 维甲酸 X 受体；Foxp3. 叉头转录因子 3；CCR9. 趋化因子受体 9；ILC3.3 型固有淋巴细胞；ROS. 活性氧；SCFAs. 短链脂肪酸；GPR.G 蛋白偶联受体；IL. 白细胞介素；TNF-α. 肿瘤坏死因子 α；TGF. 转化生长因子

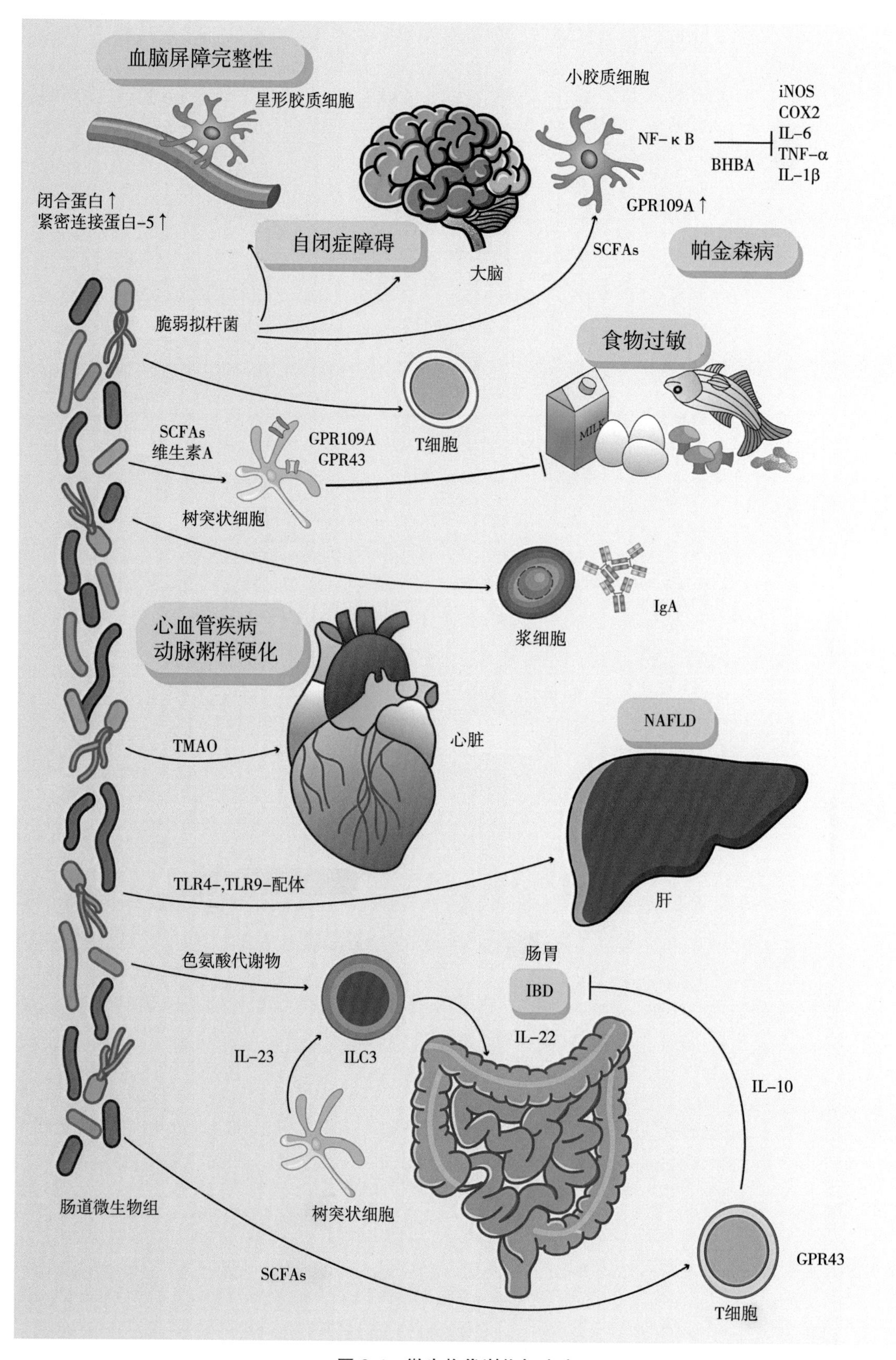

图 6-4　**微生物代谢物与疾病**

微生物代谢物水平改变与免疫介导和免疫相关性疾病有关。代谢物介导的疾病可能局限于胃肠道，如IBD或食物过敏，或影响全身远处器官，如大脑、心脏和肝脏。IBD. 炎性肠病；BHBA. β-羟丁酸；iNOS. 诱导型一氧化氮合成酶；COX2. 环氧化酶2；IgA. 免疫球蛋白A；ILC3. 3型固有淋巴细胞；GPR109A. G蛋白偶联受体109A；GPR43. G蛋白偶联受体43；SCFAs. 短链脂肪酸；TMAO. 三甲胺-N-氧化物；NAFLD. 非酒精性脂肪性肝病；NF-κB. 核因子κB；IL. 白细胞介素；TLR.Toll受体；TNF-α：肿瘤坏死因子α

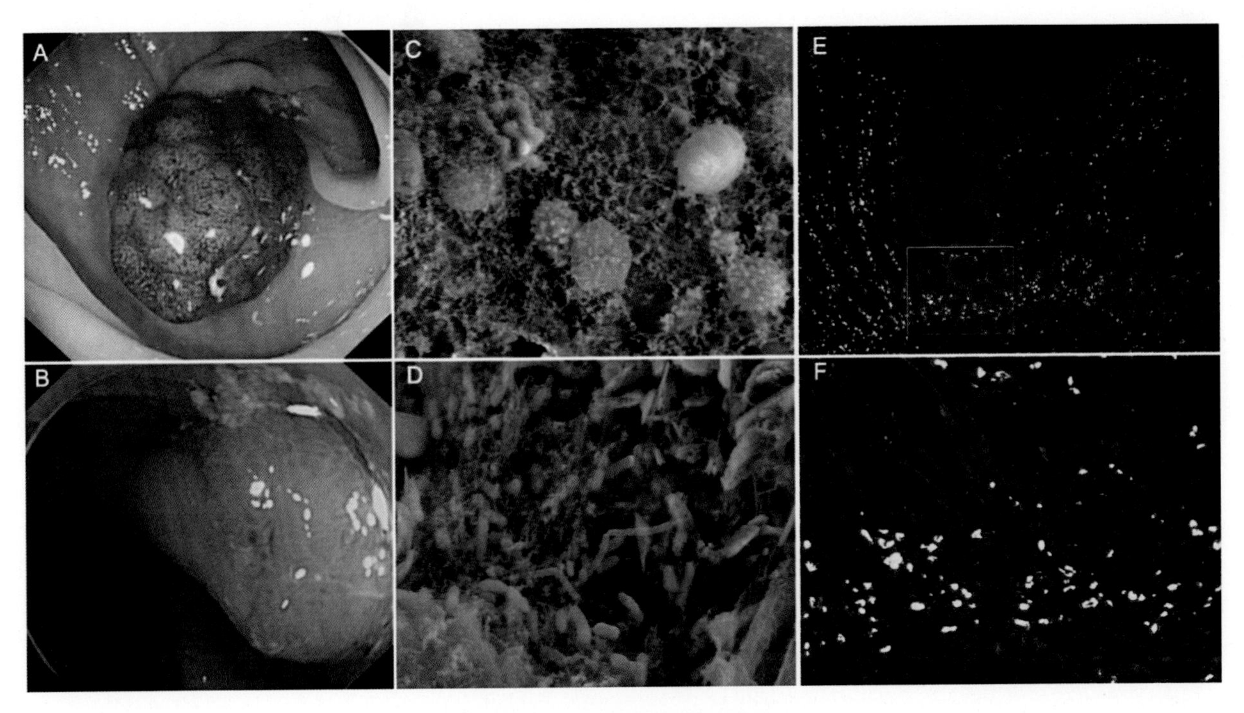

图 7-2 肠道菌群与肠道肿瘤

A、B. 肠镜检查发现人结直肠肿瘤；C、D. 扫描电镜发现肿瘤表面大量球菌及杆菌；E、F. 荧光原位杂交显示大量细菌侵入肠道肿瘤

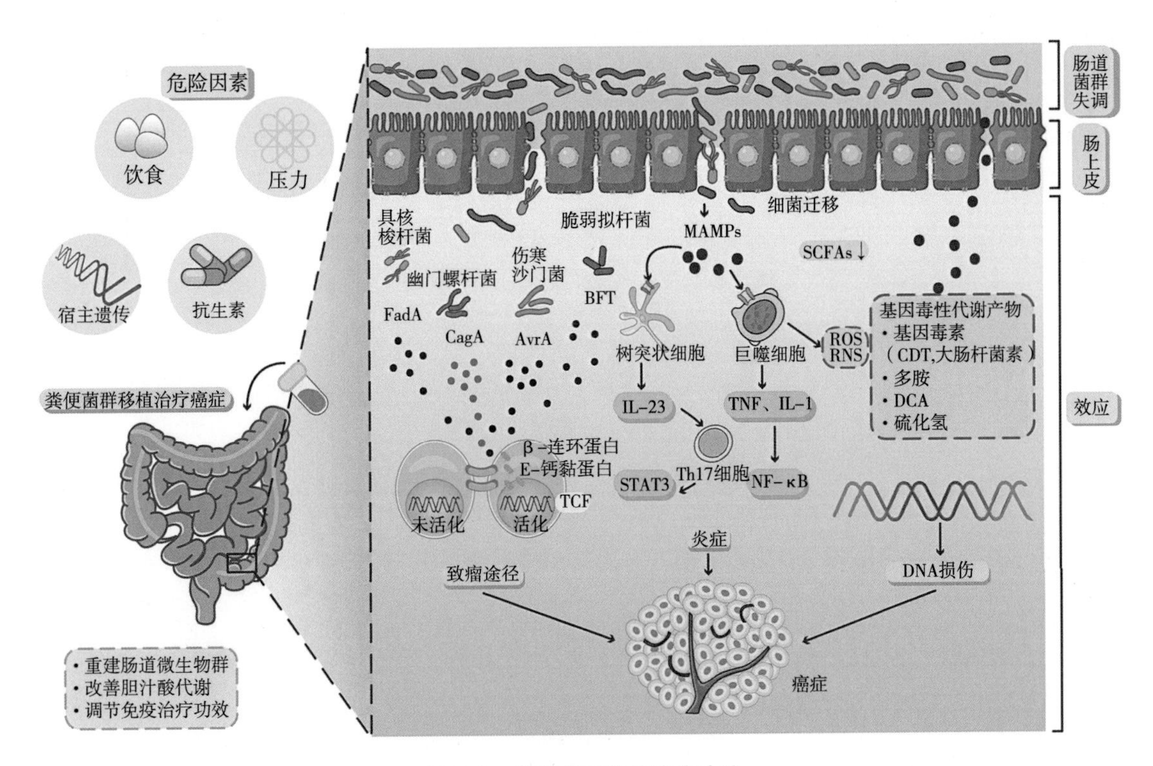

图 7-4 粪便菌群移植治疗癌症

粪菌移植是一种通过重建肠道微生物群、改善胆汁酸代谢和调节免疫治疗功效来治疗癌症的潜在策略。肠道微生物失调和特殊细菌能够通过激活致瘤途径、诱导炎症和破坏宿主 DNA 来影响癌症的发展和进展。特殊的细菌产物，如来自具核梭杆菌的 FadA 毒素、来自幽门螺杆菌的 CagA 蛋白、来自伤寒沙门菌的 AvrA 蛋白和来自产肠毒素脆弱拟杆菌的 BFT，可以促进 β- 连环蛋白与 E- 钙黏蛋白的分离，从而触发 β- 连环蛋白的活化并促进肿瘤的发生。BFT. 脆弱拟杆菌肠毒素；MAMPs. 微生物相关分子模式；IL-23. 白细胞介素 23；TNF. 肿瘤坏死因子；IL-1. 白细胞介素 1；STAT3. 信号转导和转录激活蛋白 3；NF-κB. 核因子 κ B；ROS. 活性氧；RNS. 活性氮；CDT. 细胞致死膨胀毒素；DCA. 脱氧胆酸；CagA. 细胞毒素相关基因 A；SCFAs. 短链脂肪酸；TCF. 转录因子

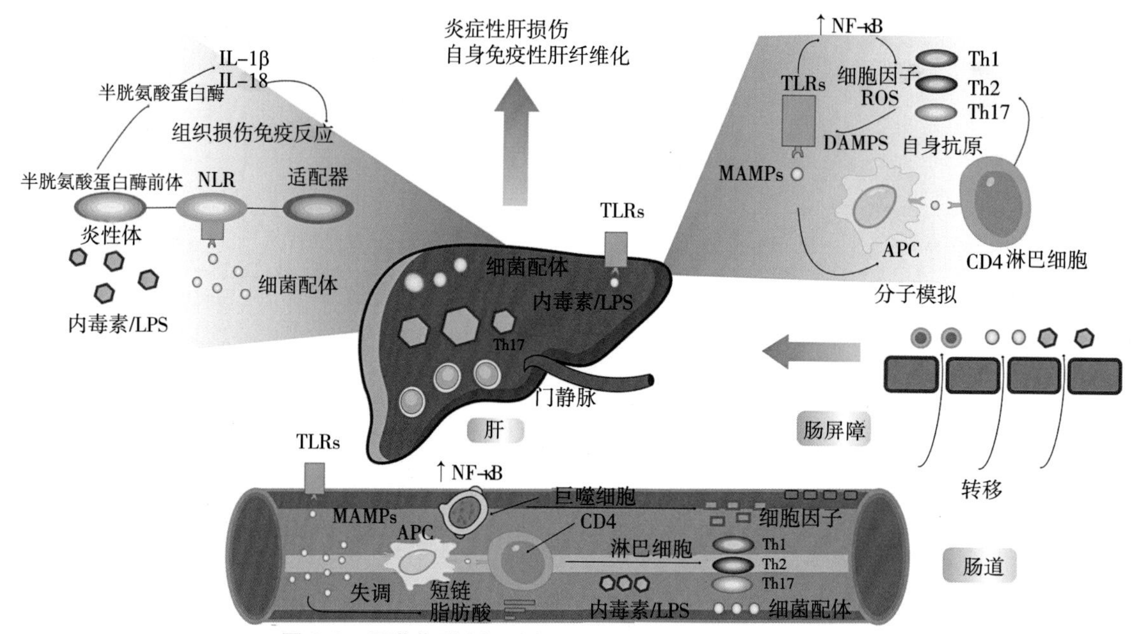

图 8-1 肠道菌群失调对自身免疫性肝炎发生发展的影响

肠道菌群失调产生病原相关分子模式，激活肠道内 Toll 样受体，进而刺激巨噬细胞分泌促炎细胞因子。它们还可以增加抗原呈递细胞上主要组织相容性复合体的表达，使 CD4 淋巴细胞增敏。肠道菌群失调还可产生短链脂肪酸、内毒素、脂多糖和细菌成分等，其可发生易位，进入门静脉并被输送到肝脏，进而激活肝细胞、肝星状细胞等，并产生促炎症细胞因子和活性氧，这些细胞可以产生损伤相关分子模式，在自放大环路中激活 Toll 样受体（右上角放大图）；同时，细菌配体和肠源性内毒素可以激活肝细胞和肝星状细胞内炎症小体的非肥胖糖尿病样受体（左上角放大图）。最终的效应是增加肝脏炎症和肝损伤，易发生自身免疫性炎症和肝纤维化。APC. 抗原呈递细胞；DAMPS. 损伤相关分子模式；ROS. 活性氧；LPS. 脂多糖；MAMPs. 微生物相关分子模式；NF-κ B. 核因子 κ B；NLR. 非肥胖性糖尿病样病变受体；TLRs.Toll 样受体

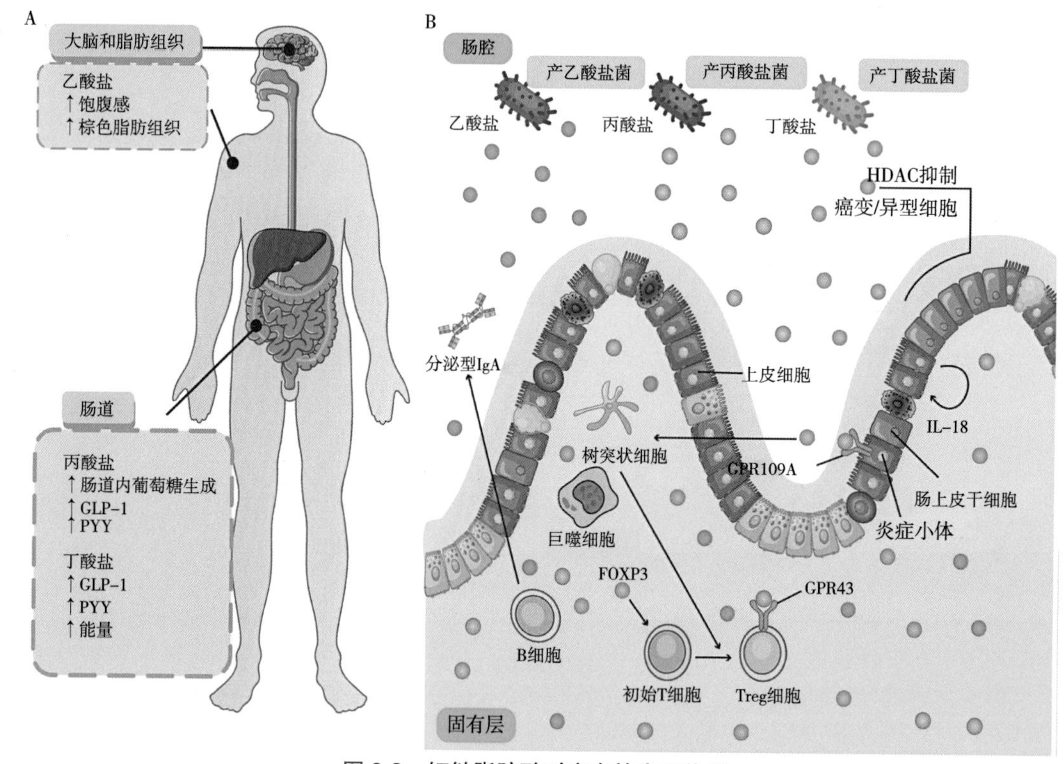

图 8-2 短链脂肪酸对宿主的生理作用

A. 短链脂肪酸（SCFA）介导宿主脂肪代谢，包括增加饱腹感和白色脂肪组织的褐变。丙酸盐是肠道糖异生的前体，丙酸盐和丁酸盐可增加多肽 YY（PYY）和胰高血糖素样肽 1（GLP1）的分泌。B. SCFA 介导黏膜免疫，包括促进 B 细胞发育、调节性 T（Treg）细胞的分化和增殖，通过抑制组蛋白去乙酰化酶（HDAC）的抗增殖作用，通过活化炎性体促进 IL-18 的产生而维持黏膜完整性，以及通过结肠树突状细胞和巨噬细胞促进 Treg 细胞的分化作用。GLP-1. 胰高血糖素样肽 -1；PYY. 多肽 YY；GPR.G 蛋白偶联受体；FOXP3. 叉头状家族转录因子 3；HDAC. 组蛋白去乙酰化酶

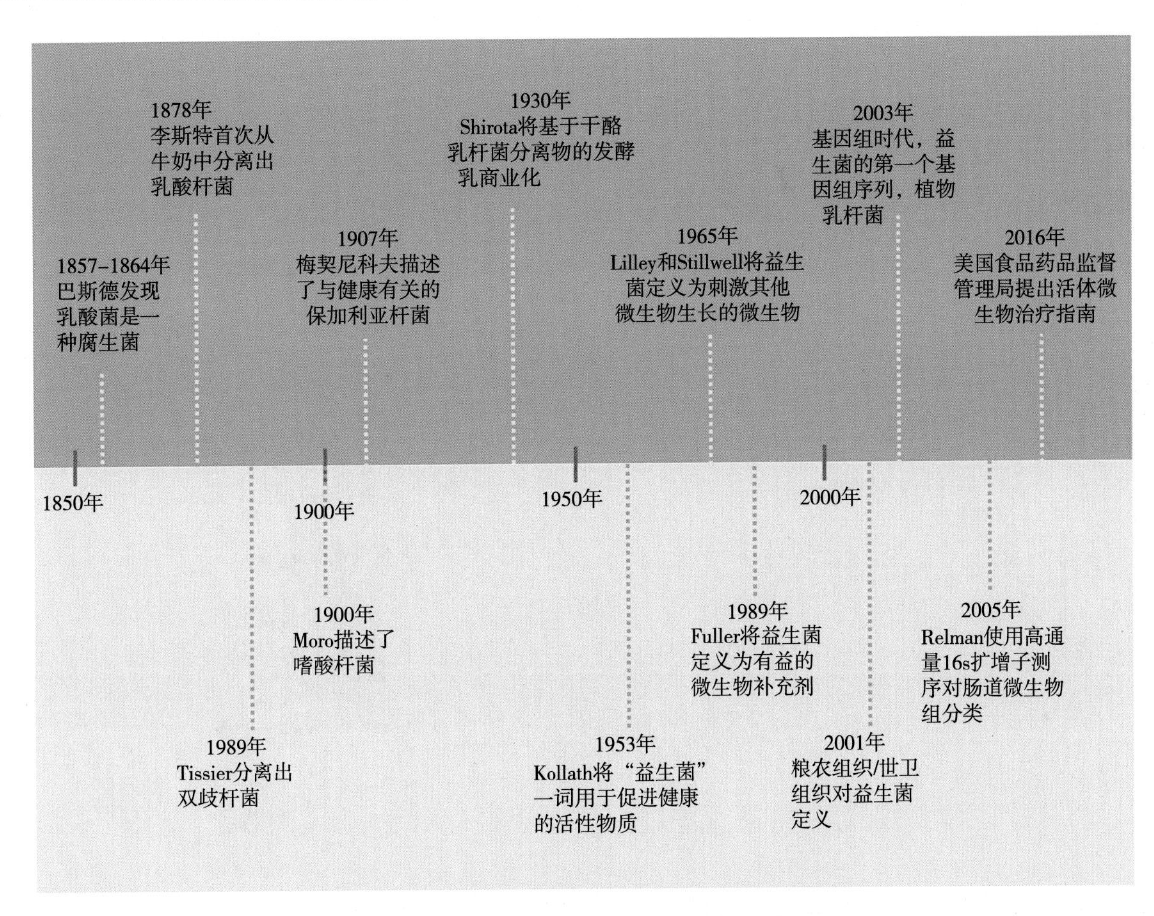

图 9-2 益生菌及二代益生菌历史里程碑时间轴

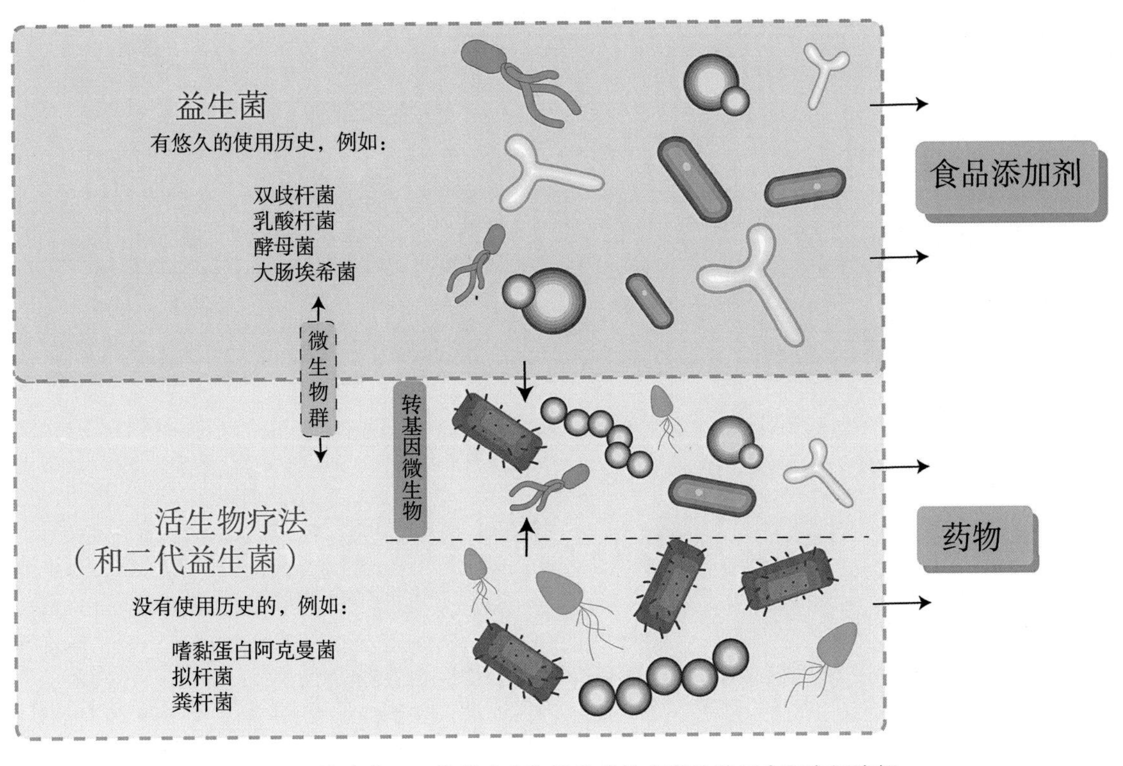

图 9-3 益生菌、二代益生菌和活生物治疗产品的历史和市场途径

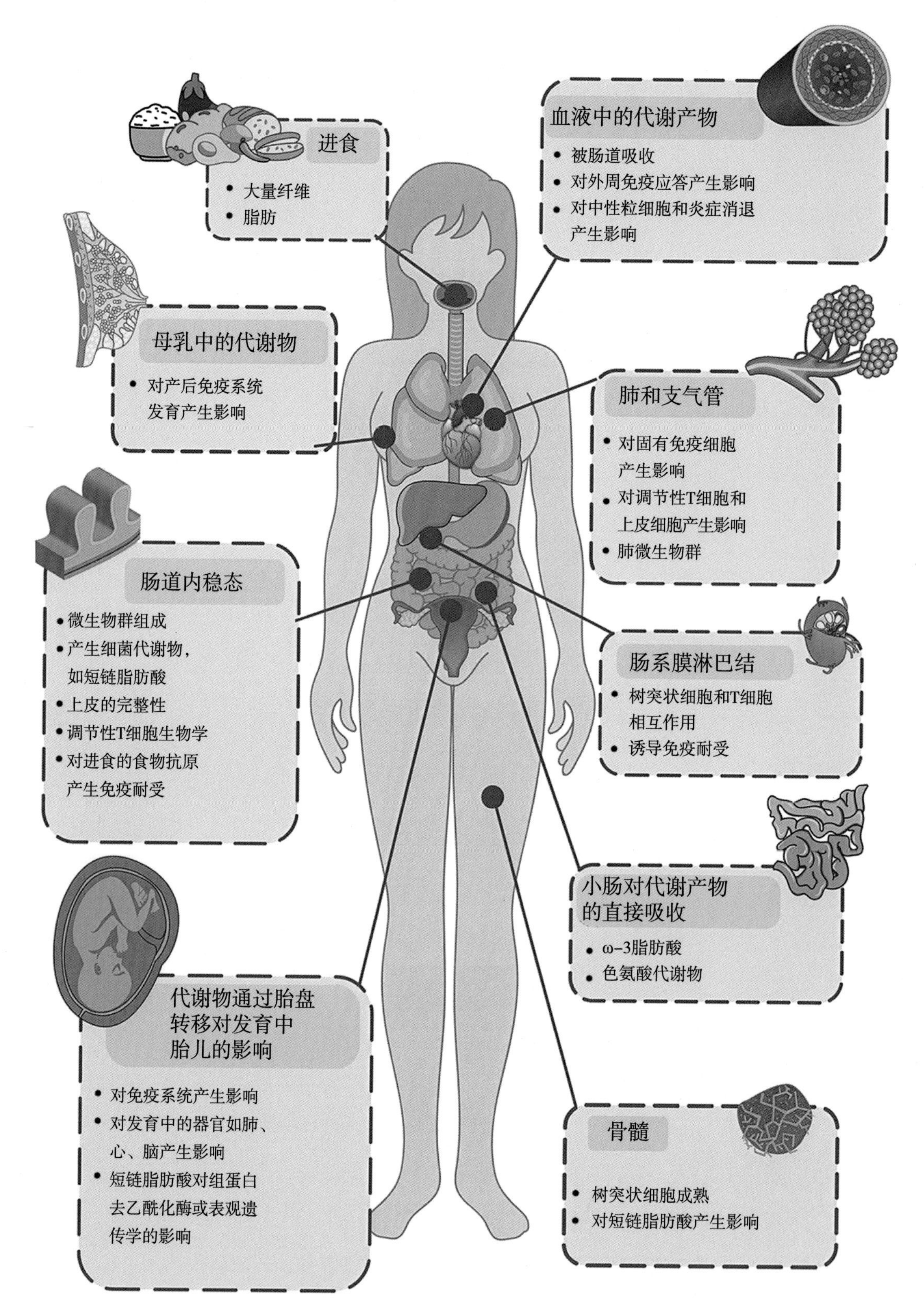

图 10-1　**膳食或细菌代谢产物与人体免疫的相互作用**

在胃肠道中，膳食纤维主要被结肠中的共生细菌消化，这会产生高浓度的短链脂肪酸，例如乙酸盐、丙酸盐和丁酸盐。其他代谢物（例如ω-3 脂肪酸、琥珀酸或犬尿酸）直接在整个胃肠道中消耗和吸收。另外，代谢物可以直接在小肠中吸收。短链脂肪酸（主要是乙酸盐）从肠道转移到血液，借此影响整个人体的骨髓和许多细胞类型。另一个重要的相互作用是代谢产物向发育中的胎儿转移。短链脂肪酸能够穿过胎盘或通过母乳传递，影响基因表达和免疫系统的发育。改自：Thorburn AN，Macia L，Mackay CR. Diet，metabolites，and “western-lifestyle” inflammatory diseases. Immunity, 2014,40（6）:833-842[46]